JACARANDA
MATHS QUEST 9

STAGE 5 NSW SYLLABUS | THIRD EDITION

BEVERLY LANGSFORD WILLING

CATHERINE SMITH

CONTRIBUTING AUTHORS

Vanessa Christman | Gilda De Guzman

Third edition published 2023 by
John Wiley & Sons Australia, Ltd
Level 4, 600 Bourke Street, Melbourne, Vic 3000

First edition published 2011
Second edition published 2014

Typeset in 10.5/13 pt TimesLT Std

ISBN: 978-0-730-38627-8

Front cover images: © alexdndz/Shutterstock
© Marish/Shutterstock; © AlexanderTrou/Shutterstock
© Vector Juice/Shutterstock

Illustrated by various artists, diacriTech and Wiley Composition
Services

Typeset in India by diacriTech

A catalogue record for this
book is available from the
National Library of Australia

NATIONAL LIBRARY OF AUSTRALIA

Printed in Singapore
M WEP220317 210823

The Publishers of this series acknowledge and pay their respects
to Aboriginal Peoples and Torres Strait Islander Peoples as the
traditional custodians of the land on which this resource was
produced.

This suite of resources may include references to (including
names, images, footage or voices of) people of Aboriginal
and/or Torres Strait Islander heritage who are deceased. These
images and references have been included to help Australian
students from all cultural backgrounds develop a better
understanding of Aboriginal and Torres Strait Islander Peoples'
history, culture and lived experience.

It is strongly recommended that teachers examine resources on
topics related to Aboriginal and/or Torres Strait Islander
Cultures and Peoples to assess their suitability for their own
specific class and school context. It is also recommended that
teachers know and follow the guidelines laid down by the
relevant educational authorities and local Elders or community
advisors regarding content about all First Nations Peoples.

All activities in this resource have been written with the safety
of both teacher and student in mind. Some, however, involve
physical activity or the use of equipment or tools. **All due care
should be taken when performing such activities.** To the
maximum extent permitted by law, the author and publisher
disclaim all responsibility and liability for any injury or loss
that may be sustained when completing activities described in
this resource.

The Publisher acknowledges ongoing discussions related to
gender-based population data. At the time of publishing, there
was insufficient data available to allow for the meaningful
analysis of trends and patterns to broaden our discussion of
demographics beyond male and female gender identification.

Contents

NAPLAN practice | online only

Set A Calculator allowed
Set B Non-calculator
Set C Calculator allowed
Set D Non-calculator
Set E Calculator allowed
Set F Non-calculator

About this resource

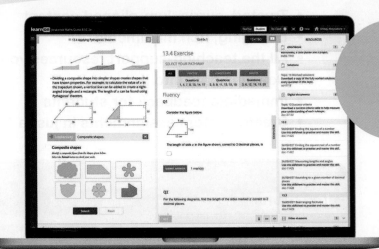

NEW FOR

2024 NSW SYLLABUS

JACARANDA
MATHS QUEST 9

NSW SYLLABUS
THIRD EDITION

Developed by teachers for students

Tried, tested and trusted. The third edition of the *Jacaranda Maths Quest series*, continues to focus on helping teachers achieve learning success for every student — ensuring no student is left behind, and no student is held back.

Because both what and how students learn matter

Learning is personal

Whether students need a challenge or a helping hand, you'll find what you need to create engaging lessons.

Whether in class or at home, students can get unstuck and progress! Scaffolded lessons, with detailed worked examples, are all supported by teacher-led video eLessons. Automatically marked, differentiated question sets are all supported by detailed worked solutions. And Brand-new Quick Quizzes support in-depth skill acquisition.

Learning is effortful

Learning happens when students push themselves. With learnON, Australia's most powerful online learning platform, students can challenge themselves, build confidence and ultimately achieve success.

Learning is rewarding

Through real-time results data, students can track and monitor their own progress and easily identify areas of strength and weakness.

And for teachers, Learning Analytics provide valuable insights to support student growth and drive informed intervention strategies.

Learn online with Australia's most

Everything you need for each of your lessons in one simple view

- Trusted, curriculum-aligned content
- Engaging, rich multimedia
- All the teaching-support resources you need
- Deep insights into progress
- Immediate feedback for students
- Create custom assignments in just a few clicks.

Practical teaching advice and ideas for each lesson provided in teachON

Teaching videos for all lessons

Reading content and rich media including embedded videos and interactivities

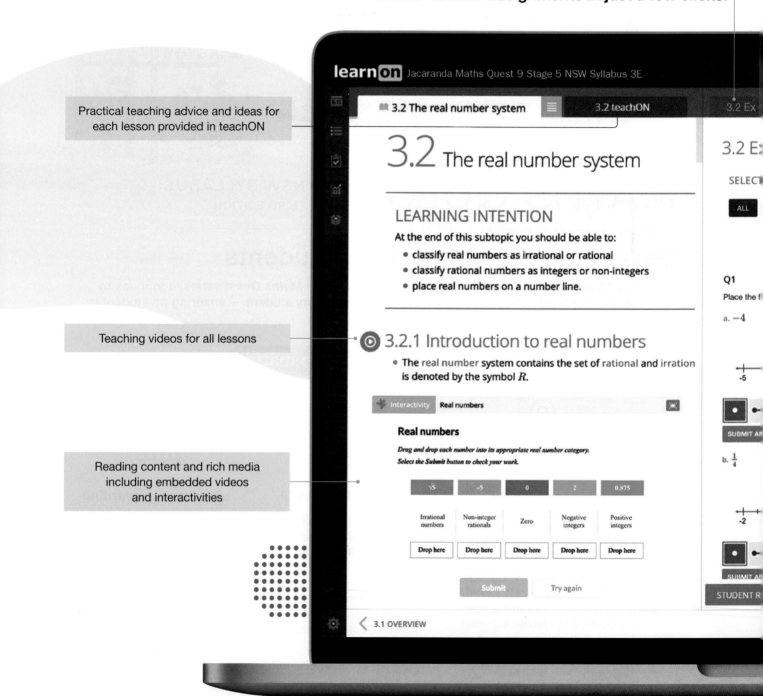

powerful learning tool, learnON

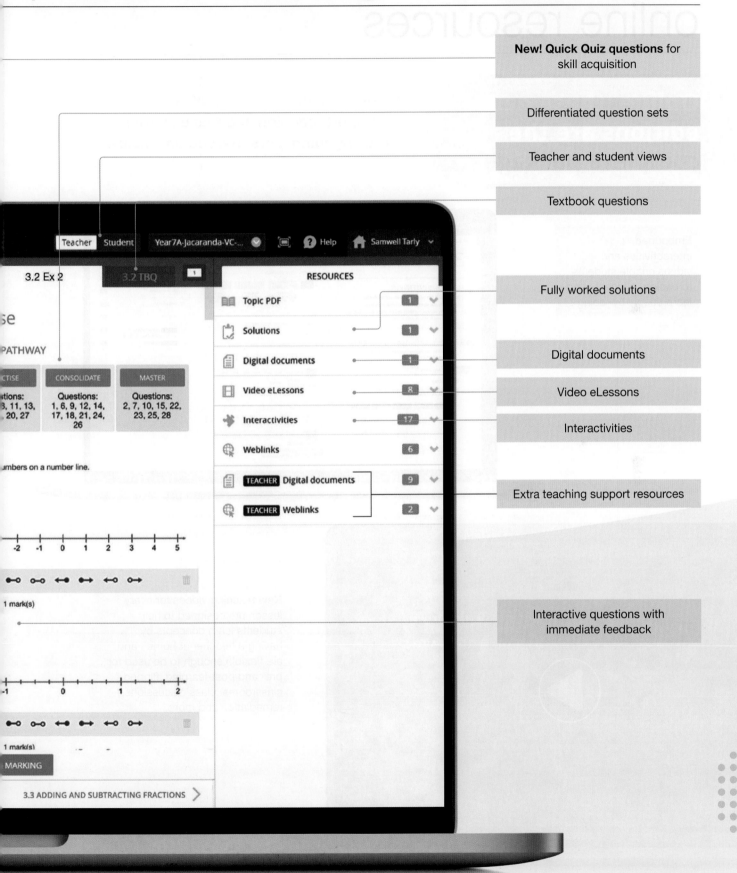

New! Quick Quiz questions for skill acquisition

Differentiated question sets

Teacher and student views

Textbook questions

Fully worked solutions

Digital documents

Video eLessons

Interactivities

Extra teaching support resources

Interactive questions with immediate feedback

Get the most from your online resources

Online, these new editions are the **complete package**

Trusted Jacaranda theory, plus tools to support teaching and make learning more engaging, personalised and visible.

Embedded interactivities and videos enable students to explore concepts and learn deeply by 'doing'.

New teaching videos for every lesson are designed to help students learn concepts by having a 'teacher at home', and are flexible enough to be used for pre- and post-learning, flipped classrooms, class discussions, remediation and more.

Brand new! Quick Quiz questions for skill acquisition in every lesson.

Three differentiated question sets, with immediate feedback in every lesson, enable students to challenge themselves at their own level.

Instant reports give students visibility into progress and performance.

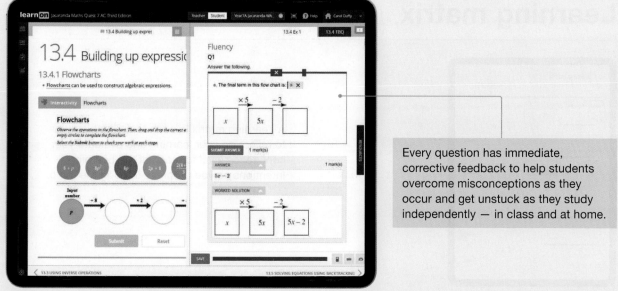

Every question has immediate, corrective feedback to help students overcome misconceptions as they occur and get unstuck as they study independently — in class and at home.

Core–Paths structure made visible

Core–Paths structure visible at the whole-topic or individual-lesson level.

NAPLAN Online Practice

Go online to complete practice NAPLAN tests. There are 6 NAPLAN-style question sets available to help you prepare for this important event. They are also useful for practising your Mathematics skills in general.

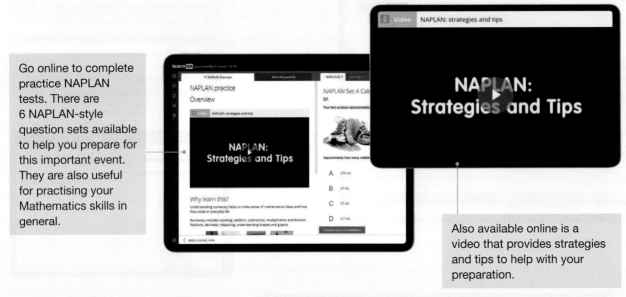

Also available online is a video that provides strategies and tips to help with your preparation.

Learning matrix

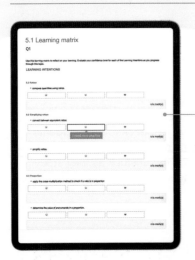

A Learning matrix in each topic enables you to reflect on your learning and evaluate your confidence level for each of the Learning Intentions as you progress through the topic.

A wealth of teacher resources

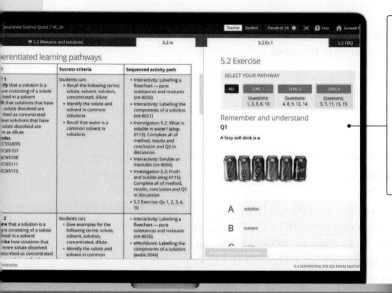

Enhanced teaching-support resources for every lesson, including:

- work programs and curriculum grids
- practical teaching advice
- three levels of differentiated teaching programs
- quarantined topic tests (with solutions)

Customise and assign

An inbuilt testmaker enables you to create custom assignments and tests from the complete bank of thousands of questions for immediate, spaced and mixed practice.

Reports and results

Data analytics and instant reports provide data-driven insights into progress and performance within each lesson and across the entire course.

Show students (and their parents or carers) their own assessment data in fine detail. You can filter their results to identify areas of strength and weakness.

Acknowledgements

The authors and publisher would like to thank the following copyright holders, organisations and individuals for their assistance and for permission to reproduce copyright material in this book.

Subject Outcomes, Objectives and Contents from the NSW Mathematics K–10 Syllabus © Copyright 2019 NSW Education Standards Authority.

Images

• © lobro - stock.adobe.com: **422** • © Shutterstock: **60, 90, 96–98, 138** • © amriphoto.com/Adobe Stock Photos: **21** • © kromkrathog/Adobe Stock Photos: **30** • vicusechka89 - stock.adobe.com: **705** • © Rawpixel.com/Adobe Stock Photos: **3** • hedgehog94 - stock.adobe.com: **314** • © Przemyslaw Iciak - stock.adobe.c: **386** • © Joshua Resnick/Adobe Stock Photos: **12** • Chansak Joe A. - stock.adobe.com: **706** • © BullRun/Adobe Stock Photos: **19** • © master1305/Adobe Stock Photos: **393** • © iMarzi - stock.adobe.com: **385** • © Mary Salen - stock.adobe.com: **421** • © fizkes - stock.adobe.com: **240** • © Microgen/Adobe Stock Photos: **393** • © Christian Blouin/Wirestock Creat: **386** • Solid photos/Adobe Stock Photos: **747** • dikushin - stock.adobe.com: **751** • © weyo - stock.adobe.com: **439** • © Taras Grebinets/Adobe Stock Photos: **15** • © Blue Jean Images/Adobe Stock Photos: **4** • © John Wiley & Sons Australia/ Photo by Renee Bryon: **160** • © John Wiley & Sons Australia/Photo by Renee Bryon: **610** • © John Wiley & Sons: Australia/Photo taken by Renee Bryon, **709** • © Lyn Elms/John Wiley & Sons: **490** • Monkey Business/Adobe Stock Photos: **664** • © NASA/Barney Magrath: **138** • Shutterstock: **671, 672** • Shutterstock/Kuki Ladron de Gu: **131** • Shutterstock/Susan Schmitz: **673** • © 3d_kot/Shutterstock: **633** • © airdynamicShutterstock: **569** • © Alan Poulson Photography/Shutterstock: **759** • © Alex James Bramwell/ Shutterstock: **760** • © Andresr/Shutterstock: **314** • © Andresr/Shutterstock: **632, 743** • © Andrii Yalanskyi/Shutterstock: **428** • © AntonioDiaz/Shutterstock: **643** • © Ariwasabi/Shutterstock: **653** • © Artazum/Shutterstock: **629** • © ARTvektor/Shutterstock: **718** • © BlueBarronPhotoShutterstock: **145** • © Bo Valentino/Shutterstock: **737** • © Bruce Aspley/Shutterstock: **245** • © cassiede alain/Shutterstock: **368** • © CGN089/Shutterstock: **89** • © ChameleonsEye/Shutterstock: **22** • © chrupka/Shutterstock: **715** • © Ciprian Vladut/Shutterstock: **278** • © Dario Sabljak/Shutterstock: **388** • © Daxiao Productions/Shutterstock: **233** • © del-Mar/Shutterstock: **526** • © Diyana DimitrovaShutterstock: **145** • © djgis/Shutterstock: **326** • © DrHitch/Shutterstock: **388** • © Dusan Petkovic/Shutterstock: **654** • © Elena Elisseeva/Shutterstock: **640** • © ESB Professional/Shutterstock: **416** • © Fablok/Shutterstock: **137** • © fizkes/Shutterstock: **46** • © givaga/Shutterstock: **654** • © Ground Picture/Shutterstock: **10, 234** • © hddigital/Shutterstock: **48** • © hidesy/Shutterstock: **2** • © iamlukyeee/Shutterstock: **615** • © imagedb.com/Shutterstock: **667** • © jessicakirshcreative/Shutterstock: **759** • © Jiripravda/Shutterstock: **695** • © JOAT/Shutterstock: **409** • © Kahan Shan/Shutterstock: **413** • © Kaspars Grinvalds/Shutterstock: **415** • © Kitch Bain/Shutterstock: **708** • © Krisana Tongnantree/Shutterstock.com: **585** • © Kummeleon/Shutterstock: **24** • © Liudmila Pleshkun/Shutterstock: **594** • © Madlen/Shutterstock: **658** • © Maxisport/Shutterstock: **594** • © metamorworks/Shutterstock: **686** • © mezzotint/Shutterstock: **646** • © mfauzisaim/Shutterstock: **662** • © Monkey Business Images/Shutterstock: **8, 37** • © Monkey Business Images/Shutterstock: **661** • © Naypong/Shutterstock: **585** • © Neale Cousland/Shutterstock: **660** • © NikoNomadShutterstock: **137, 144** • © Olaf Speier/Shutterstock: **693** • © Omer N Raja/Shutterstock: **628** • © OZBEACHES/Shutterstock: **246** • © Philip Yuan/Shutterstock: **584** • © Piotr Zajc/Shutterstock: **639** • © PlusONE/Shutterstock.com: **602** • © psynovec/Shutterstock: **654** • © quka/Shutterstock: **575** • © Robyn Mackenzie/Shutterstock: **708** • © Sara Julin Ingelmark/Shutterstock: **685** • © Serhiy Stakhnyk/Shutterstock: **210** • © Shutterstock: **209, 499, 521** • © Sinisa Botas/Shutterstock: **703** • © Stefan Schurr/Shutterstock: **443** • © Stuart Monk/Shutterstock: **222** • © Sunflowerey/Shutterstock: **1** • © Suzanne Tucker/Shutterstock: **717** • © szefei/Shutterstock: **45** • © Taki O/Shutterstock: **277, 351** • © Taras Vyshnya/Shutterstock: **422** • © Tetiana Yurchenko/Shutterstock: **616** • © topae/Shutterstock: **650** • © TriffShutterstock: **57** • © UfaBizPhoto/Shutterstock: **38** • © Val Thoermer/Shutterstock: **704** • © wavebreakmedia/Shutterstock: **698** • © weknow/Shutterstock: **198** • © Worytko

Pawel/ Shutterstock: **759** • © XiXinXing/Shutterstock: **13** • © Yellowj/Shutterstock: **663** • © John Wiley & Sons Australia: **696, 697** • ChrisVanLennepPhoto// Adobe Stock Photos: **632** • Olha/Adobe Stock Photos: **646** • RooM The Agency/Adobe Stock Photos: **703** • Vladir09/Shutterstock: **554** • © Adisa/Shutterstock.com: **427** • © Alex Mit/Shutterstock: **234** • © ALPA PROD/Shutterstock: **14** • © Analia Aguilar Camacho/Shutterstock: **539** • © andrea crisante/Shutterstock: **535** • © Andrei Armiagov/Shutterstock: **427** • © Anel Alijagic/Shutterstock: **229** • © antoniodiaz/Shutterstock: **7** • © ARochau/Adobe Stock Photos: **266** • © Ashley Whitworth/Shutterstock: **519** • © Bankoo/Shutterstock: **152** • © CC7/Shutterstock: **424** • © ClaudioValdes/Shutterstock: **219** • © Dario Lo Presti/Shutterstock: **333** • © David Gilder/Shutterstock: **494** • © David H.Seymour/Shutterstock: **552** • © DavideAngelini/Shutterstock: **427** • © Dean Drobot/Shutterstock: **156** • © defpicture/Shutterstock: **520** • © Dimedrol68/Shutterstock: **421** • © Dmitry Natashin/Shutterstock: **89** • © Dragon Images/Shutterstock: **8** • © Drazen Zigic/Shutterstock: **12** • © Edw/Shutterstock: **236** • © Elena Schweitzer/Shutterstock: **423** • © Eric Gevaert/Shutterstock: **443** • © ESB Professional/Shutterstock: **36** • © EyeWire Images/Getty Images: **25** • © f11photo/Adobe Stock Photos: **352** • © freesoulproduction/ Shutterstock: **73** • © Gordon Bell/Shutterstock: **318** • © Gyuszko-Photo/Shutterstock: **17** • © Harvepino/Shutterstock: **32** • © Henri et George/Shutterstock: **236** • © ILYA GENKIN/Shutterstock: **319** • © Image Addict: **560** • © inspiredbyart/Shutterstock: **420** • © ITTIGallery/Shutterstock: **39** • © Janaka Dharmasena/Shutterstock: **410** • © Jennifer Wright/John Wiley & Sons Australia: **541** • © jiris/Shutterstock: **68** • © John Wiley & Sons Australia: **175** • © kavalenkau/Shutterstock: **428** • © Keith Tarrier/Shutterstock: **374** • © Khakimullin Aleksandr/Shutterstock: **236** • © kurhan/Shutterstock: **234** • © Kzenon/Adobe Stock: **6** • © Laitr Keiows/Shutterstock: **509** • © Leah-Anne Thompson/Shutterstock: **332** • © Lerner Vadim/Shutterstock: **354** • © Levente Fazakas/Shutterstock.com: **426** • © lucadp/Shutterstock: **537** • © Luna Vandoorne/Shutterstock: **233** • © Magdalena Kucova/Shutterstock: **500** • © Maria Maarbes/Shutterstock: **28** • © mark higgins/Shutterstock: **355** • © Markus Schröder/Adobe Stock Photos: **50** • © Marques/Shutterstock: **87** • © Michael William/Shutterstock: **228** • © michaeljung/Shutterstock: **14, 16** • © Mila Supinskaya Glashchenko/Shutterstock: **226** • © Nata-Lia/Shutterstock: **552** • © New Africa/Shutterstock: **34** • © Nikodash/Shutterstock: **508** • © No formal credit line required: **358** • © No on-page credit is required.: **388** • © Norman Pogson/Shutterstock: **456** • © only_kim/Adobe Stock Photos: **554** • © Peddalanka Ramesh Babu/Shutterstock: **380** • © PeopleImages/E+/Getty Images/Dean Drobot/Shutterstock: **5** • © Petr Toman/Shutterstock: **430** • © photka/Shutterstock: **412** • © PhotoDisc: **551, 553** • © Photodisc: **470** • © Phovoir/Shutterstock: **486** • © PinkBlue/Shutterstock: **537** • © Pinkyone/Shutterstock: **20** • © Potstock/Shutterstock: **222** • © Pressmaster/Shutterstock: **270** • © Prostock-studio/Adobe Stock Photos: **43** • © pryzmat/Shutterstock: **318** • © puhhha/Shutterstock: **58** • © puhimec/Adobe Stock Photos: **22** • © rachel ko/Shutterstock: **226** • © Rawpixel.com/Shutterstock: **239** • © REDPIXEL.PL/Shutterstock: **38** • © Redshinestudio/Shutterstock: **226** • © Renee Bryon/John Wiley & Sons Australia: **482** • © rh2010/Adobe Stock Photos: **51** • © Robert Kneschke/Shutterstock: **66** • © Robyn Mackenzie/Shutterstock: **419** • © Rosli Othman/Shutterstock.com: **268** • © s74/Shutterstock: **270** • © Samuel Borges Photography/Shutterstock: **437** • © SciePro/Shutterstock: **383** • © Sergei Domashenko/Shutterstock: **30** • © SERGEI PRIMAKOV/Shutterstock: **422** • © Stephen Coburn/Shutterstock: **17** • © StepStock/Shutterstock: **257** • © Steve Salisbury/ABC News: **579** • © StockPhotosLV/Shutterstock: **285** • © Stuart Miles/Shutterstock: **28** • © Studio Mayday/Shutterstock: **26** • © sunsetman/Shutterstock: **430** • © Suzanne Tucker/Shutterstock: **234** • © Taras Vyshnya/Shutterstock: **355** • © Tom Hirtreiter/Shutterstock: **448** • © Torderiul/Shutterstock: **527** • © Tyler Olson/Shutterstock: **294** • © Vitalii Vodolazskyi/Shutterstock/Rtimages/Shutterstock: **10, 49** • © wavebreakmedia/Shutterstock: **240** • © WitR/Shutterstock.com: **570** • © ChiccoDodiFC/Shutterstock: **475** • © Christina Richards/Shutterstock: **482** • © John Wiley & Sons Australia: **175** • © Nadezda Cruzova/Shutterstock: **638** • © Photodisc: Inc, **482** • © Polka Dot Images: **482** • © Renee Bryon/John Wiley & Sons Australia: **475** • © Vereshchagin Dmitry/Shutterstock: **494**

Every effort has been made to trace the ownership of copyright material. Information that will enable the publisher to rectify any error or omission in subsequent reprints will be welcome. In such cases, please contact the Permissions Section of John Wiley & Sons Australia, Ltd.

NAPLAN practice

Go online to complete practice NAPLAN tests. There are 6 NAPLAN-style question sets available to help you prepare for this important event. They are also useful for practising your Mathematics skills in general.

Also available online is a video that provides strategies and tips to help with your preparation.

SET A
Calculator allowed

SET B
Non-calculator

SET C
Calculator allowed

SET D
Non-calculator

SET E
Calculator allowed

SET F
Non-calculator

1 Financial mathematics

LESSON SEQUENCE

LESSON
1.1 Overview

Why learn this?

There are many famous sayings about money. You have probably heard that 'money doesn't grow on trees', 'money can't buy happiness' and 'a fool and his money are soon parted'. Most of us have to be conscious of where our money comes from and where it goes to. Understanding the basic principles of finance is very helpful for managing everyday life. Some of you will already be earning money from a part-time or casual job. Do you know what you spend your money on, or where it goes? Do you have your own savings account? Many of you may already be saving for a new games console, your first car or your dream holiday.

In this topic you will investigate different kinds of employment as well as different investment options for saving your money. Every branch of industry and business, whether large or small, international or domestic, will have to pay their employees either an annual salary or a wage based on an hourly rate. Financial investments are another way for businesses and individuals to make money — it is important to understand how investments work to be able to decide whether an investment is a good idea or not.

A sound knowledge of financial mathematics is essential in a range of careers, including financial consultancy, accountancy, business management and pay administration.

Hey students! Bring these pages to life online

▶ Watch videos

Engage with interactivities

A+ Answer questions and check solutions

Find all this and MORE in jacPLUS ▶

Reading content and rich media, including interactivities and videos for every concept

Extra learning resources

Differentiated question sets

Questions with immediate feedback, and fully worked solutions to help students get unstuck

1. Kira has an annual salary of $85 450.70 (without tax).
   Calculate how much she is paid:
   a. weekly   **b.** fortnightly   **c.** monthly.

2. Ishmael has a casual job at a department store. He is paid $11.50 per hour. He gets double time on a Sunday.
   Calculate his wage for a week in which he worked from 4 pm to 8 pm on Thursday and 12 pm to 6 pm on Sunday.

3. Jose is a tutor. He receives $975 for a week in which he works 30 hours. Calculate his hourly rate of pay.

4. **MC** This table shows a timesheet for Yumi, who works in an electronics store.

Day	Pay rate	Start time	Finish time
Monday	Normal	9 am	1 pm
Tuesday	Normal	9 am	5 pm
Wednesday	Normal	11 am	7 pm
Thursday	Normal	2 pm	7 pm
Friday	Normal	4 pm	8 pm
Saturday	Time-and-a-half	9 am	2 pm
Sunday	Double	12 pm	6 pm

   If Yumi's normal hourly rate is $15.30, the amount she earned for the week is:
   A. $612.00   **B.** $734.00   **C.** $818.55   **D.** $742.05

5. **MC** Marge delivers brochures to letterboxes in her local area. She is paid $47.20 per 1000 brochures delivered. The amount Marge will earn for a delivery of 3570 brochures is:
   A. $132.20   **B.** $165.20   **C.** $168.50   **D.** $177.00

6. **MC** Abena is an author who is paid a royalty of 8.5% on all sales. If a book she wrote sells copies to the value of $12 500 a month, what royalty will she earn in a year?
   A. $1062.50
   B. $12 750.00
   C. $106 250.00
   D. $127 500.00

7. A real estate agent receives 2.5% commission on the first $250 000 of a sale and 4.5% on the rest. What is the commission the real estate agent received on a sale of a $825 000 property?

8. **MC** Charlie is paid a normal rate of $47.30 per hour for a 38-hour working week For his 4 weeks annual leave, Charlie is paid a loading of 17.5%. The total amount that Charlie receives for his 4 weeks annual leave is:
   A. $1258.20   **B.** $1797.40   **C.** $2111.95   **D.** $8447.80

9. The PAYG (Pay As You Go) tax payable on a gross wage of $1850 per week is $491.

   a. What percentage of the gross pay is deducted, correct to 2 decimal places?

   b. Heidi has a gross wage of $1850 per week. She receives $212 in family allowance but has deductions of $375 (superannuation), $100 (private health insurance) and $300 (loan repayment). What is her net pay?

10. The Liang family wants to buy a family car at a cost of $22 700. They pay a deposit of $4500 and borrow the balance at an interest rate of 11.5% p.a. The loan will be paid off with 48 equal monthly payments.

   a. How much interest do they pay?

   b. What will be the total cost of the car, including the interest paid?

   c. How much is each loan repayment?

11. **MC** If $10 000 is invested at a simple interest rate of 6% per annum, the interest earnt for 6 months can be calculated using:

   A. $I = 10\,000 \times \dfrac{6}{100} \times 6$
                B. $I = 10\,000 \times 0.6 \times 6$

   C. $I = 10\,000 \times \dfrac{6}{100} \times \dfrac{1}{2}$
         D. $I = 10\,000 \times \dfrac{1.6}{100} \times \dfrac{1}{2}$

12. If the simple interest charged on a loan of $6400 over 30 months was $1450, what was the percentage rate of interest charged? Give your answer to the nearest whole percentage.

13. **MC** The amount of interest earnt if $112 000 is invested for 4 years at 6.8% p.a. interest compounded annually is:

   A. $23 947.14
      B. $30 464.00
      C. $33 714.59
      D. $142 464.00

14. **MC** The total value of $3500 invested for 18 months at 12% p.a. can be calculated using:

   A. $SI = 3500 \times 12 \times 18$
                B. $SI = 3500 \times 1.12 \times 1.5$

   C. $SI = \dfrac{3500 \times 12 \times 18}{100}$
          D. $SI = \dfrac{3500 \times 12 \times 1.5}{100}$

15. Alecia invests $5000 at 15% p.a. compounded annually and Cooper invests $5000 at 15% p.a. flat rate. After 3 years, how much more will Alecia's investment be than Cooper's investment, to the nearest dollar?

LESSON
1.2 Salaries and wages

▶ 1.2.1 Employees' salaries and wages

eles-4836

- Employees may be paid for their work in a variety of ways — most receive either a **wage** or a **salary**.

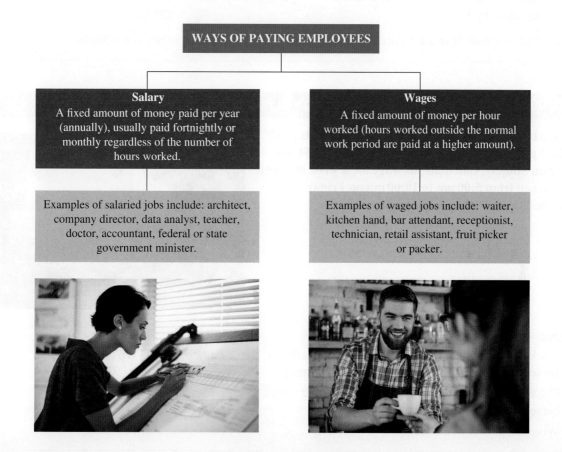

WAYS OF PAYING EMPLOYEES

Salary
A fixed amount of money paid per year (annually), usually paid fortnightly or monthly regardless of the number of hours worked.

Wages
A fixed amount of money per hour worked (hours worked outside the normal work period are paid at a higher amount).

Examples of salaried jobs include: architect, company director, data analyst, teacher, doctor, accountant, federal or state government minister.

Examples of waged jobs include: waiter, kitchen hand, bar attendant, receptionist, technician, retail assistant, fruit picker or packer.

Key points

- **Normal working hours in Australia are 38 hours per week.**
- **There are 52 weeks in a year.**
- **There are 26 fortnights in a year (this value is slightly different for a leap year).**
- **There are 12 months in a year.**
- **A month is not equal to 4 weeks; you must use 1 year divided by 12.**

WORKED EXAMPLE 1 Calculating pay from annual salary

Susan has an annual salary of $63 048.92. Calculate how much she is paid:

a. weekly **b. fortnightly** **c. monthly.**

THINK	WRITE
a. 1. Annual means per year, so divide the salary by 52 because there are 52 weeks in a year.	**a.** Weekly salary $= 63\,048.92 \div 52$ ≈ 1212.48
2. Write the answer in a sentence.	Susan's weekly salary is $1212.48.
b. 1. There are 26 fortnights in a year, so divide the salary by 26.	**b.** Fortnightly salary $= 63\,048.92 \div 26$ ≈ 2424.96
2. Write the answer in a sentence.	Susan's fortnightly salary is $2424.96.
c. 1. There are 12 months in a year, so divide the salary by 12.	**c.** Monthly salary $= 63\,048.92 \div 12$ ≈ 5254.08
2. Write the answer in a sentence.	Susan's monthly salary is $5254.08.

WORKED EXAMPLE 2 Calculating wage given hourly rate of pay

Frisco has casual work at a fast-food store. He is paid $12.27 per hour Monday to Saturday and $24.54 per hour on Sunday. Calculate his wage for a week in which he worked from 5:00 pm to 10:00 pm on Friday and from 6:00 pm to 9:00 pm on Sunday.

THINK	WRITE
1. Work out the number of hours Frisco worked each day. He worked 5 hours on Friday and 3 hours on Sunday.	Friday: $5 \times 12.27 = 61.35$ Sunday: $3 \times 24.54 = 73.62$
2. Calculate the total amount earned by adding the wages earned on Friday and Sunday.	$61.35 + 73.62 = 134.97$
3. Write the answer in a sentence.	Frisco's wage was $134.97.

 Resources

 Interactivity Salaries (int-6067)

Exercise 1.2 Salaries and wages

| 1.2 Quick quiz on | 1.2 Exercise |

Individual pathways

■ PRACTISE	■ CONSOLIDATE	■ MASTER
1, 3, 7, 10, 13, 14, 17	2, 4, 8, 11, 15, 18	5, 6, 9, 12, 16, 19, 20

Fluency

1. **WE1** Johann has an annual salary of $57 482. Calculate how much he is paid:
   a. weekly
   b. fortnightly
   c. monthly.

2. Nigara earns $62 300 per annum. Calculate how much she earns:
   a. weekly
   b. fortnightly
   c. monthly.

3. Calculate the annual salaries of workers with the following weekly incomes.
   a. $368
   b. $892.50
   c. $1320.85

4. Calculate how much is earned per annum by people who are paid fortnightly salaries of:
   a. $995
   b. $1622.46
   c. $3865.31.

5. Compare the following salaries and determine which of each pair is the higher.
   a. $3890 per month or $45 700 per annum
   b. $3200.58 per fortnight or $6700 per month

6. Calculate the hourly rates for these workers.
   a. Rahni earns $98.75 for working 5 hours.
   b. Francisco is paid $54.75 for working $4\frac{1}{2}$ hours.
   c. Nhan earns $977.74 for working a 38-hour week.
   d. Jessica works $7\frac{1}{2}$ hours a day for 5 days to earn $1464.75.

Understanding

7. Henry is a second-year apprentice motor mechanic. He receives an award wage of $12.08 per hour. Jenny, a fourth-year apprentice, earns $17.65 per hour.

   a. Calculate how much Henry earns in a 38-hour week.
   b. Determine how much more Jenny earns in the same period of time.

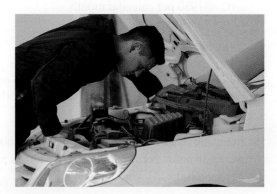

8. **WE2** Juan has casual work for which he is paid $13.17 per hour Monday to Saturday and $26.34 per hour on Sundays. Calculate his total pay for a week in which he worked from 11 am to 5 pm on Thursday and from 2 pm to 7 pm on Sunday.

9. Mimi worked the following hours in one week.

Wednesday	5 pm to 9 pm
Thursday	6 pm to 9 pm
Friday	7 pm to 11 pm

If her pay is $21.79 per hour up to 9 pm and $32.69 per hour after that, calculate her total pay for that week.

10. Decide who earns more money each week: Rhonda, who receives $38.55 an hour for 38 hours of work, or Rob, who receives $41.87 an hour for 36 hours of work.

11. Glenn is a chef. He receives $1076.92 for a week in which he works 35 hours.
Calculate his hourly rate of pay.

12. Zack and Kaylah work in different department stores. Zack is paid $981.77 per week. Kaylah is paid $26.36 per hour. Calculate how many hours Kaylah must work to earn more money than Zack.

13. Calculate what pay each of the following salary earners will receive for each of the periods specified.

   a. Annual salary $83 500, paid each week
   b. Annual salary $72 509, paid each fortnight
   c. Annual salary $57 200, paid each week
   d. Annual salary $105 240, paid each month

Communicating, reasoning and problem solving

14. **MC** When Jack was successful in getting a job as a trainee journalist, he was offered the following choice of four salary packages.

   a. Decide which one Jack should choose

   A. $456 per week
   B. $915 per fortnight
   C. $1980 per calendar month
   D. $23 700 per year

   b. Justify your answer.

15. Julie is considering two job offers for work as a receptionist. Job A pays $878.56 for a 38-hour working week. Job B pays $812.16 for a 36-hour working week. Determine which job has the higher hourly rate of pay. Justify your answer.

16. In his job as a bookkeeper, Minh works 38 hours per week and is paid $32.26 per hour. Michelle, who works 38 hours per week in a similar job, is paid a salary of $55 280 per year. Decide who has the higher-paying job. Justify your answer.

17. A lawyer is offered a choice between two jobs: one with a salary of $74 000 per year and a second that pays $40 per hour. Assuming that the lawyer will work 80 hours every fortnight, decide which job pays the most. Justify your answer.

18. Over the last four weeks Shahni has worked 35, 36, 34 and 41 hours. If she earns $24.45 per hour, calculate how much she earned for each of the two fortnights.

19. Jackson works a 40-hour week (8 hours a day, Monday–Friday) and earns $62 000 per annum.

   a. Calculate Jackson's hourly rate.
   b. If he works on average an extra half an hour per day from Monday to Friday and then another 4 hours over the weekend (for the same annual salary), calculate his actual hourly rate.
   c. If Jackson was earning the hourly rate from part **a.** and was being paid for every hour he actually worked, calculate his potential earnings for the year.

20. Mark saves $10 per week. Phil saves 5 cents in the first week, 10 cents the second week, and doubles the amount each week from that time on. Determine how many weeks it will take for Phil's savings to be worth more than Mark's.

LESSON
1.3 Special rates

LEARNING INTENTION

At the end of this lesson you should be able to:
- calculate total wages when overtime or penalty rates are included
- calculate hours paid when time-and-a-half or double time is involved.

1.3.1 Overtime and penalty rates

eles-4837

- **Overtime** is paid when a wage earner works more than their regular hours each week. These additional payments are often referred to as penalty rates.
- Penalty rates are usually paid for working on weekends, public holidays or at night.
- The extra hours are paid at a higher hourly rate, normally calculated at either time-and-a-half or double time.

Calculating overtime

The overtime hourly rate is usually a multiple of the regular hourly rate. Some examples of overtime rates include:
- **1.5 × regular hourly wage (time-and-a-half)**
- **2 × regular hourly wage (double time)**
- **2.5 × regular hourly wage (double time-and-a-half).**

Regular hourly rate	Overtime hourly rate
$25.00	1.5 × regular hourly rate 1.5 × 25.00 = $37.50

WORKED EXAMPLE 3 Calculating wages with overtime

**Ursula works as a waitress and earns \$23.30 per hour.
Last week she received the normal rate for 30 hours of work
as well as time-and-a-half for 3 hours of overtime
and double time for 5 hours of work on Sunday.
Calculate her total pay.**

THINK	WRITE
1. Calculate Ursula's normal pay for 30 hours.	Normal pay: $30 \times 23.30 = 699.00$
2. Calculate Ursula's pay for 3 hours at time-and-a-half ($1.5 \times$ regular hourly wage).	Overtime: $3 \times 1.5 \times 23.30 = 104.85$
3. Calculate Ursula's pay for 5 hours at double time ($2 \times$ regular hourly wage).	Sunday: $5 \times 2 \times 23.30 = 233.00$
4. Calculate the total amount by adding normal pay and overtime pay.	Total $= 699.00 + 104.85 + 233.00$ $= 1036.85$
5. Write the answer in a sentence.	Ursula's total pay was \$1036.85.

▶ 1.3.2 Time sheets and pay slips

eles-4398

- Employers often use records to monitor the number of working hours of their employees.

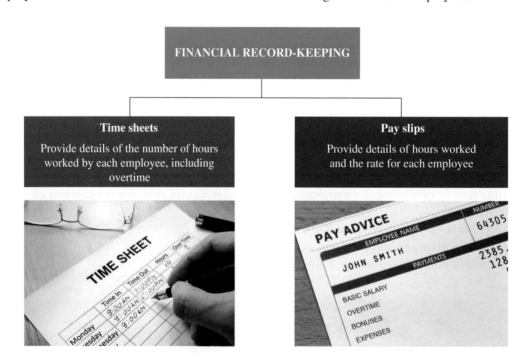

FINANCIAL RECORD-KEEPING

Time sheets
Provide details of the number of hours
worked by each employee, including
overtime

Pay slips
Provide details of hours worked
and the rate for each employee

WORKED EXAMPLE 4 Calculating pay slips showing overtime

Fiona works in a department store, and in the week before Christmas she works overtime. Her time sheet is shown. Fill in the details on her pay slip.

	Start	Finish	Normal hours	Over time 1.5
M	9:00	15:00	6	
T	9:00	17:00	8	
W	9:00	17:00	8	
T	9:00	19:00	8	2
F	9:00	19:00	8	2
S				

Pay slip for: Fiona Lee	Week ending December 21
Normal hours	
Normal rate	$17.95
Overtime hours	
Overtime rate	
Total wage	

THINK

1. Calculate the number of normal hours worked by adding the hours worked on each day of the week.

2. Calculate the number of overtime hours worked by adding the overtime hours worked on Thursday and Friday.

3. Calculate the overtime rate (1.5 × regular hourly wage).

4. Calculate the total pay by multiplying the number of normal hours by the normal rate and adding the overtime amount (calculated by multiplying the number of overtime hours by the overtime rate).

5. Fill in the amounts on the pay slip.

WRITE

Normal hours: $6 + 8 + 8 + 8 + 8 = 38$

Overtime hours: $2 + 2 = 4$

Overtime rate $= 1.5 \times 17.95$
$\qquad\qquad = 26.93$

Total pay $= 38 \times 17.95 + 4 \times 26.93$
$\qquad\qquad = 789.82$

Pay slip for: Fiona Lee	Week ending December 21
Normal hours	38
Normal rate	$17.95
Overtime hours	4
Overtime rate	$26.93
Total wage	$789.82

COMMUNICATING — COLLABORATIVE TASK: Comparing pay rates and conditions

Working in small groups, collect several job advertisements from a mixture of print and digital sources. Compare the pay rates and conditions for the different positions and present a report to the class.

---

on Resources

✦ **Interactivity** Special rates (int-6068)

Individual pathways

■ PRACTISE	■ CONSOLIDATE	■ MASTER
1, 2, 6, 9, 12, 15	3, 4, 7, 10, 13, 16	5, 8, 11, 14, 17

Fluency

1. Calculate the following penalty rates.

   a. Time-and-a-half when the hourly rate is $15.96
   b. Double time when the hourly rate is $23.90
   c. Double time-and-a-half when the hourly rate is $17.40

2. Calculate the following total weekly wages.

   a. 38 hours at $22.10 per hour, plus 2 hours at time-and-a-half
   b. 40 hours at $17.85 per hour, plus 3 hours at time-and-a-half
   c. 37 hours at $18.32 per hour, plus 3 hours at time-and-a-half and 2 hours at double time

3. Julio is paid $956.08 for a regular 38-hour week. Calculate:

   a. his hourly rate of pay
   b. how much he is paid for 3 hours of overtime at time-and-a-half rates
   c. his wage for a week in which he works 41 hours.

4. **WE3** Geoff is a waiter in a cafe and works 8 hours most days. Calculate what he earns for 8 hours of work on the following days.

   a. A Monday, when he receives his standard rate of $21.30 per hour
   b. A Sunday, when he is paid double time
   c. A public holiday, when he is paid double time-and-a-half

5. Albert is paid $870.58 for a 38-hour week. Determine his total wage for a week in which he works 5 extra hours on a public holiday with a double-time-and-a-half penalty rate.

Understanding

6. Jeleesa (aged 16) works at a supermarket on Thursday nights and weekends. The award rate for a 16-year-old is $7.55 per hour. Calculate what she would earn for working:

   a. 4 hours on Thursday night
   b. 6 hours on Saturday
   c. 4 hours on Sunday at double time
   d. the total of the 3 days described in parts a, b and c.

7. Jacob works in a pizza shop and is paid $13.17 per hour.

   a. Jacob is paid double time-and-a-half for public holiday work. Calculate what he earns per hour on public holidays. Give your answer to the nearest cent.

   b. Calculate Jacob's pay for working 6 hours on a public holiday.

8. If Bronte earns $7.80 on normal time, calculate how much she receives per hour:

   a. at time-and-a-half   b. at double time   c. at double time-and-a-half.

9. **WE4** a. Copy and complete the time sheet shown. Calculate the number of hours Susan worked this week.

Day	Pay rate	Start time	Finish time	Hours worked
Monday	Normal	9:00 am	5:00 pm	
Tuesday	Normal	9:00 am	5:00 pm	
Wednesday	Normal	9:00 am	5:00 pm	
Thursday	Normal	9:00 am	5:00 pm	
Friday	Normal	9:00 am	3:00 pm	

   b. Copy and complete Susan's pay slip for this week.

Pay slip for: Susan Jones	Week ending 17 August
Normal hours	
Normal pay rate	$25.60
Overtime hours	0
Overtime pay rate	$38.40
Total pay	

10. a. Manu works in a department store. His time sheet is shown. Copy and complete the table.

Day	Pay rate	Start time	Finish time	Hours worked
Monday	Normal	9:00 am	5:00 pm	
Tuesday	Normal	9:00 am	5:00 pm	
Wednesday	Normal	—		
Thursday	Normal	1:00 pm	9:00 pm	
Friday	Normal	—		
Saturday	Time-and-a-half	8:00 am	12:00 pm	

   b. Copy and complete Manu's pay slip for this week.

Pay slip for: Manu Taumata	Week ending 21 December
Normal hours	
Normal pay rate	$10.90
Overtime hours	
Overtime pay rate (time-and-a-half)	
Total pay	

11. **a.** Eleanor does shift work. Copy and complete their time sheet.

Day	Pay rate	Start time	Finish time	Hours worked
Monday	Normal	7:00 am	3:00 pm	
Tuesday	Normal	7:00 am	3:00 pm	
Wednesday				
Thursday				
Friday	Normal	11:00 pm	7:00 am	
Saturday	Time-and-a-half	11:00 pm	7:00 am	
Sunday	Double time	11:00 pm	7:00 am	

b. Copy and complete Eleanor's pay slip for the week.

Pay slip for: Eleanor Rigby	Week ending 15 September
Normal hours	
Normal pay rate	$16.80
Time-and-a-half hours	
Time-and-a-half pay rate	
Double time hours	
Double time pay rate	
Gross pay	

Communicating, reasoning and problem solving

12. Calculate the following total weekly wages.

    a. 38 hours at $18.40 per hour, plus 2 hours at time-and-a-half
    b. 32 hours at $23.70 per hour plus 6 hours on a Sunday at double time

13. Ruby earns $979.64 for her normal 38-hour week, but last week she also worked 6 hours of overtime at time-and-a-half rates.

    a. Calculate how much extra she earned and give a possible reason for her getting time-and-a-half rates.
    b. Calculate Ruby's total wage.

14. A standard working week is 38 hours and a worker puts in 3 hours overtime at time-and-a-half, plus 2 hours at double time. Calculate the amount of standard work hours equivalent to the total time they have worked.

15. Joshua's basic wage is $22 per hour. His overtime during the week is paid at time-and-a-half. Over the weekend he is paid double time. Calculate Joshua's gross wage in a week when he works his basic 40 hours, together with 1 hour of overtime on Monday, 2 hours of overtime on Wednesday and 4 hours of overtime on Saturday.

16. Lin works 32 hours per week at $22 per hour and is paid overtime for any time worked over the 32 hours per week. In one week Lin worked 42 hours and was paid $814. Overtime is paid at 1.5 times the standard wage. Determine whether Lin was paid the correct amount. If not, then provide the correct amount.

17. The table shows the pay sheet for a small company. If a person works up to 36 hours, the regular pay is $14.50 per hour.

For hours over 36 and up to 40, the overtime is time-and-a-half.

For hours over 40, the overtime is double time.

Copy and complete the table.

	Hours worked	Regular pay	Overtime pay	Total pay
a.	32			
b.	38.5			
c.	40.5			
d.	47.2			

LESSON
1.4 Piecework

LEARNING INTENTION

At the end of this lesson you should be able to:
- calculate the payment of piecework
- calculate the payment of piecework if based on a sliding scale
- calculate the payment of piecework if the rate is based on multiple units.

▶ 1.4.1 Non-wage earnings — piecework

eles-6194

- **Piecework** is a system of payment by which a worker is paid a fixed amount for each job or task they complete.

WORKED EXAMPLE 5 Calculating pay for a task with a fixed rate

Mitchell is washing cars for a fundraiser. He is paid $5.20 per car washed. Calculate the amount Mitchell earns in an afternoon when he washes 24 cars.

THINK	WRITE
1. Multiply the number of cars Mitchell washes by the amount paid for each car.	Amount earned $= 24 \times 5.20$ $= 124.80$
2. Write the answer in a sentence.	Mitchell earns $124.80.

- A person may also be paid on a sliding scale where the pay rate increases as the number of completed tasks increases.

WORKED EXAMPLE 6 Calculating pay for a task with a sliding rate

Angelica is a machinist in a clothing factory. Each week she is paid \$4.28 per garment for the first 180 garments, and \$5.35 per garment thereafter. What will she be paid if she produces 223 garments?

THINK	WRITE
1. Calculate the number of 'extra' garments Angelica makes.	Extra garments $= 223 - 180$ $\qquad\qquad\quad = 43$
2. Calculate her total payment by adding the payment she receives for the first 180 garments to the payment she receives for the extra garments.	Payment $= 180 \times 4.28 + 43 \times 5.35$ $\qquad\quad\; = 1000.45$
3. Write the answer in a sentence.	Angelica earns a total payment of \$1000.45.

- In some cases, piecework is paid for multiple units rather than single units. For example, for letterbox deliveries you may be paid per 1000 deliveries made.

WORKED EXAMPLE 7 Calculating pay given a rate per 1000

Holly is delivering brochures to letterboxes in her local area. She is paid \$43.00 per 1000 brochures delivered. Calculate the amount Holly will earn for a delivery of 3500 brochures.

THINK	WRITE
1. Calculate the number of thousands of brochures Holly will deliver.	$3500 \div 1000 = 3.5$ So Holly will deliver 3.5 thousand brochures.
2. Multiply the number of thousands of brochures delivered by 43 to calculate what Holly will earn.	Holly's pay $= 3.5 \times 43.00$ $\qquad\qquad\; = 150.50$
3. Write the answer in a sentence.	Holly will earn \$150.50.

  Resources

 **Interactivity** Piecework (int-6069)

1.4 Quick quiz on

1.4 Exercise

Individual pathways

■ PRACTISE	■ CONSOLIDATE	■ MASTER
1, 2, 4, 7	3, 5, 8, 9	6, 10, 11, 12

Fluency

1. **WE5** Hitani is paid 65 cents for each teacup she decorates. How much is she paid for decorating 150 teacups?

2. **WE6** Jack makes leather belts. The piece rate is $1.25 each for the first 50 belts and $1.50 thereafter. What is his income for a day in which he produces 68 belts?

3. A production-line worker is paid $1.50 for each of the first 75 toasters assembled, then $1.80 per toaster thereafter. How much does she earn on a day in which she assembles 110 toasters?

4. **WE7** Rudolf earns $42.50 per 1000 leaflets delivered to letterboxes. Calculate what Rudolf will earn for a week in which he delivers 7500 leaflets.

Understanding

5. Dimitri earns $7.20 for each box of fruit picked.
   a. How much does he make for picking 20 boxes?
   b. How many boxes must he pick to earn at least $200?
   c. If he takes 4 hours to pick 12 boxes, what is his hourly rate of pay?

6. Pauline uses her home computer for word processing under contract with an agency. She is paid $3 per page for the first 50 pages, $4 per page from 51 to 100 pages, and $5 per page thereafter. Calculate her total pay for a period in which she prepares:
   a. 48 pages   b. 67 pages   c. 123 pages.

Communicating, reasoning and problem solving

7. Rani delivers bills to letterboxes and is paid $43 per thousand.
   a. How much does she earn for delivering 2500 items?
   b. How many thousands must she deliver to earn at least $1000?
   c. If she takes 6 hours to deliver each thousand on average, what is her hourly rate of pay?

8. Georgio delivers pizzas. He is paid $3 per delivery from 5 pm to 9 pm and $4 per delivery after 9 pm.
   a. How much does he earn on a night in which he makes 12 deliveries by 9 pm and 4 deliveries between 9 pm and 10.30 pm?
   b. What are his average earnings per hour if he has worked from 5 pm to 10.30 pm?

9. A shoemaker is paid $5.95 for each pair of running shoes he can make.

   a. If the shoemaker made 235 pairs of shoes last week, what was the amount paid?
   b. The shoemaker is offered a bonus of 5% if he can make more than 250 pairs of shoes in a week. If he makes 251 pairs, what is the total amount earned, including the bonus?

10. A secretarial assistant gets paid $12 per page that she types. If she manages to type more than 20 pages in a day, she gets a 10% bonus. If she typed 32 pages on Tuesday, how much did she earn?

11. There are both fixed and variable costs associated with some products. Consider the cost of importing a radio from China and selling it in Australia. The costs for a particular company are:
    • import of product $12.50 per unit
    • transportation costs $400 per 1000 units
    • warehouse rental space $1 per unit per month
    • advertising costs $2000 per month (fixed cost).

    a. If this company imports and sells 500 units per month, what is the total cost per month?
    b. At 500 units per month and a selling price of $25.00, what is the total profit per month?

12. What are the advantages and disadvantages of being paid by piecework?

LESSON
1.5 Commissions and royalties

LEARNING INTENTION

At the end of this lesson you should be able to:
• calculate commission on sales
• calculate royalties paid on copyright.

▶ 1.5.1 Non-wage earnings — commissions and royalties

eles-6195
• **Commission** is a method of payment used mainly for salespeople. The commission paid is usually calculated as a percentage of the value of goods sold.
• A **royalty** is a payment made to a person who owns a copyright. For example, a musician who writes a piece of music is paid a royalty on CD and online sales. An author who writes a book is also paid a royalty based on the number of books sold. Royalties are calculated as a percentage of sales.

WORKED EXAMPLE 8 Calculating royalties

Mohamad is a songwriter who is paid a royalty of 12% on all sales of his music. Calculate the royalty that Mohamad earns if a song he writes sells CDs to the value of $150 000.

THINK	WRITE
1. Determine the royalty by calculating 12% of $150 000.	Royalty = 12% of 150 000 = $0.12 \times 150\,000$ = 18 000
2. Write the answer in a sentence.	Mohamad earns $18 000 in royalties.

- Sometimes a salesperson is paid a small wage, called a retainer, plus a percentage of the value of the goods sold.

WORKED EXAMPLE 9 Calculating total wage including commission

Gemma, a car salesperson, is paid a retainer of \$350 per week, plus a commission of 8% of the profits made by the company on cars that she sells.
a. How much does Gemma earn in a week when no sales are made?
b. How much does she earn in a week when \$5000 profit was generated by her sales?

THINK	WRITE
a. If no sales are made, only the retainer is paid.	**a.** Gemma earns \$350.
b. 1. Determine the commission paid by calculating 8% of \$5000.	**b.** Commission $= 8\%$ of 5000 $= 0.08 \times 5000$ $= \$400$
2. Determine the total amount paid by adding the retainer and the commission.	Total earnings $= 350 + 400$ $= \$750$
3. Write the answer in a sentence.	Gemma earns \$750.

- Sometimes the commission is broken into several parts with differing rates.

WORKED EXAMPLE 10 Calculating commission including different rates

A real estate agency receives 2% commission on the first \$300 000 of a sale and 3% on the remainder. How much commission is received on the sale of a \$380 000 property?

THINK	WRITE
1. Calculate the difference between \$380 000 and \$300 000.	$380\,000 - 300\,000 = 80\,000$
2. Calculate 2% of \$300 000.	2% of $300\,000 = 6000$
3. Calculate 3% of \$80 000.	3% of $80\,000 = 2400$
4. Calculate the total commission by adding the commission earned on \$300 000 and the commission earned on \$80 000.	$6000 + 2400 = 8400$
5. Write the answer in a sentence.	The commission received is \$8400.

 Resources

Interactivity Commissions and royalties (int-6070)

1.5 Quick quiz on	1.5 Exercise

Individual pathways

■ PRACTISE	■ CONSOLIDATE	■ MASTER
1, 4, 7, 11, 14	2, 5, 8, 9, 12, 13, 17	3, 6, 10, 15, 16

Fluency

1. **WE8** Danyang is a writer who is paid a royalty of 10% on all sales. Calculate the royalty she earns in a year if a book she writes sells copies to the value of $30 000.

2. A home improvement company pays commission at the rate of 16% on all sales. What would a person earn who had sales to the value of:
   a. $8000
   b. $6972.50?

3. Linda is a car salesperson who is paid a 1.5% commission on her sales. Calculate the amount of money Linda earns in a week where her sales total $95 000.

4. **WE9** Gordon is paid a retainer of $200 per week plus a commission of 6% of the profits made by the company on the goods that he sells.
   a. How much does Gordon earn in a week when no sales are made?
   b. How much does Gordon earn in a week during which a $70 000 profit was generated by his sales?

5. Alfonso gets a retainer of $235 per week plus a commission of $5\frac{1}{2}\%$ on sales. What are his total earnings in a week in which his sales are:
   a. $1000
   b. $4500
   c. $17 384?

6. Bryce is an author. His publisher pays him a fixed allowance of $500 per month, plus $4\frac{1}{2}\%$ royalty on sales. What would be his income for a month in which his book sales totalled:
   a. $0
   b. $2000
   c. $15 000
   d. $23 750?

Understanding

7. **WE10** A real estate agency receives 2% commission on the first $250 000 of a sale and 4% on the rest. How much commission is received on the sale of a $370 000 property?

8. At a second real estate agency, the commission rate is 5% on the first $180 000 of sale price and 2% on the remainder. Calculate the commission on the sale of the $370 000 property.

9. Ingrid's real estate agency pays her 1% commission on the first $500 000 of sale price, then 4% thereafter. How much commission would she receive on the sale of a property worth:
   a. $480 000
   b. $510 000
   c. $735 000?

10. Yanu works for a boat broker who pays him 6% of the first $50 000 of the sale price, then $3\frac{3}{4}$% on the rest. Calculate the commission he receives on the following sales.

    a. $40 000   b. $70 000   c. $395 000

Communicating, reasoning and problem solving

11. Veronica earns $400 per week plus 4% on sales, whereas Francis earns 6% commission only.

    a. How much does each earn on sales of $8400?
    b. What level of sales would yield each the same income?

12. Wolfgang, a car salesman, is paid a weekly retainer of $550, plus 10% of the dealer's profit on each vehicle. Determine find his total income for weeks in which the dealer's profits on vehicles he sold were:

    a. $3500   b. $5980   c. $7036

13. Using the commission table for house sales below, calculate the commission on each of the following sales.

Sale price	Commission	Plus
Between $0 and $80 000	2% of sale price	0
Between $80 001 and $140 000	1.5% of amount over $80 000	$1600 (2% of $80 000)
$140 000 and over	1.1% of amount over $140 000	$2500 (2% of $80 000 + 1.5% of $60 000)

    a. $76 000   b. $122 500   c. $145 000   d. $600 000

14. Mr Hartney is a used car salesman. He receives a basic monthly salary of $2400 together with 5% commission on all sales. Although his sales for the month amounted to $48 300, he also had deductions for insurance ($12.80), association fees ($25.70) and income tax ($1100). Calculate the amount, in dollars, he took home that month.

15. A rock musician makes a royalty on all record sales according to the following formula.

Sales from	Sales to	Royalty rate
0	$100 000	3%
$100 001	$500 000	3.5% on amount over $100 000
$500 001	1 million	4% on amount over $500 000
1 million	and above	5% on amount over 1 million

Calculate the musician's royalties for the following years.

    a. 2007: sales = $456 000   b. 2008: sales = $1 234 500
    c. 2009: sales = $986 400   d. 2010: sales = $2 656 000

16. Four years ago Inka became an employee of TrakRight Tourism, where her starting annual salary was $55 600. After her first year she received a 2% pay rise. The next year she received a 3% pay rise. Last year she received an x% pay rise. If her annual salary is now $61 042, determine the value of x, correct to 1 decimal place.

17. What are the major advantages and disadvantages of getting paid by commission or royalties?

LESSON
1.6 Loadings and bonuses

LEARNING INTENTION

At the end of this lesson you should be able to:
- calculate a wage or salary with an extra loading payment
- calculate annual leave loading
- calculate a bonus based on company performance.

1.6.1 Loadings

eles-6196

- If a wage or salary earner has to work in difficult or hazardous conditions, then the worker may be granted an extra payment or **loading**.
- Most workers are granted a 'holiday loading'. For a 4-week period each year they are paid an extra 17.5% of their usual wage.

WORKED EXAMPLE 11 Calculating pay including a loading payment

Rohan works as an electrician and receives $38.20 per hour for a 36-hour working week. If Rohan works at heights above 3 metres, he receives $2.50 per hour height loading. Calculate Rohan's wage in a week where he works 15 hours at heights above 3 metres.

THINK	WRITE
1. Calculate Rohan's normal weekly wage.	Normal wage $= 36 \times 38.20$ $= \$1375.20$
2. Calculate Rohan's loading for the time he worked at heights above 3 metres.	Loading $= 15 \times 2.5$ $= \$37.50$
3. Calculate Rohan's total wage.	Total wage $= 1375.20 + 37.50$ $= \$1412.70$

WORKED EXAMPLE 12 Calculating annual leave loading

Jelena works as a hairdresser and is paid a normal rate of $19.70 per hour for a 38-hour working week.
a. Calculate Jelena's normal weekly wage.
b. For her 4 weeks annual leave, Jelena is paid a loading of 17.5%. Calculate the amount that Jelena receives in holiday loading.
c. Calculate the total amount that Jelena receives for her 4 weeks annual leave.

THINK	WRITE
a. Calculate Jelena's normal wage by multiplying the hours worked by the hourly rate.	a. Normal wage $= 38 \times 19.70$ $= \$748.60$
b. 1. Calculate 17.5% of Jelena's normal wage.	b. 17.5% of $748.60 = \$131.01$
2. Multiply this amount by 4 to determine the holiday loading.	Holiday loading $= 4 \times 131.01$ $= \$524.04$
c. Calculate the total amount received by multiplying Jelena's normal weekly pay by 4 and adding the holiday loading.	c. Holiday pay $= 4 \times 748.60 + 524.04$ $= \$3518.44$

▶ 1.6.2 Bonuses

eles-6197

- Many people who are employed in managerial positions receive a bonus if the company achieves certain performance targets. The bonus may be a percentage of their annual salary or a percentage of the company's profits.

WORKED EXAMPLE 13 Calculating a performance-based loading

Brooke is the Chief Executive Officer of a fashion company on a salary of $240 000 per year. Brooke will receive a bonus of 1% of her salary for every percentage point that she increases the company profit. If the company profit grows from $3.1 million to $4.4 million in one year, calculate the amount of Brooke's bonus.

THINK	WRITE
1. Calculate the increase in profit.	Increase in profit $= \$4.4m - \$3.1m$ $= \$1.3m$
2. Express the increase in profit as a percentage.	Percentage increase $= \dfrac{1.3}{3.1} \times 100\%$ $= 41.9\%$
3. Calculate this percentage of Brooke's annual salary.	Bonus $= 41.9\%$ of $240 000 $= \$100\,560$
4. Write the answer in a sentence.	Brooke's bonus is $100 560.

 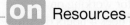 Resources

🧩 **Interactivity** Bonuses (int-6071)

1.6 Quick quiz on	**1.6 Exercise**

Individual pathways

■ PRACTISE	■ CONSOLIDATE	■ MASTER
1, 4, 5, 7, 10	2, 6, 8, 9, 12, 16	3, 11, 13–15

Fluency

1. **WE11** Rashid works as an electrician and receives $35.40 per hour for a 35-hour working week. If Rashid works at heights he receives a height loading of $0.32 per hour. Calculate Rashid's wage in a week where he works 18 hours at heights.

2. Patrick is a railway linesman. If he works in wet weather he is paid a loading of 43 cents per hour. If he normally works a 38-hour working week at $21.02 per hour and 16 hours are spent working in wet weather, determine Patrick's pay for the week.

3. Saci is an industrial cleaner and is paid at the rate of $19.82 per hour. If Saci works in a confined space, she is paid a loading of $0.58 per hour. Calculate Saci's pay for a week in which she worked 38 hours and 19 of those hours were in a confined space.

4. **WE12** Jordan works as the manager of a supermarket and is paid a normal rate of $37.60 per hour for a 38-hour working week.

   a. Calculate Jordan's normal weekly wage.
   b. For her 4 weeks annual leave, Jordan is paid a loading of 17.5%.. Calculate the amount that Jordan receives in holiday loading.
   c. Calculate the total amount that Jordan receives for her 4 weeks annual leave.

Understanding

5. Charlie earns $22.80 per hour for a 38-hour week.

   a. Calculate the amount Charlie will earn in a normal working week.
   b. Calculate the total amount Charlie will receive for his 4 weeks annual leave if he receives a 17.5% holiday loading.

6. Liam is paid $15.95 per hour for a 36-hour working week.

   a. Calculate Liam's weekly wage.
   b. Liam takes one week's holiday, for which he is given a 17.5%loading. Calculate the holiday loading.

7. Karen receives an annual salary of $63 212.

   a. What is her fortnightly pay?
   b. What is she paid for her annual 4-week holiday, for which she receives an extra 17.5% loading?

8. Brian earns $956.46 for a standard 38-hour week and a $27.53 per week allowance for working on scaffolding. Calculate his total pay for a week in which he works on scaffolding and does 4 hours overtime at time and a half.

9. **WE13** Eric is a director of a mining company on a salary of $380 000 per year. Eric is told that at the end of the year he will receive a bonus of 1% of his salary for every percentage point of increase in the company profit. If the company's profit grows from $4.9m to $6.4m in one year, calculate the amount of Eric's bonus.

Communicating, reasoning and problem solving

10. Sally is the manager of a small bakery that employs 12 people. As an incentive to her workers she agrees to pay 15% of the business's profits in Christmas bonuses for her employees. The business makes a profit of $400 000 during the year.

   a. Calculate the total amount that Sally pays in bonuses.
   b. If the bonus is shared equally, what amount does each employee receive as a Christmas bonus?
   c. If one employee earns $42 000 per year, calculate the Christmas bonus as a percentage of annual earnings, correct to 2 decimal places. Explain your answer.

11. Shane, the director of an exercise company, earns a salary of $275 000 a year. Shane gets paid incentives if he is able to increase the company's profit. He gets:
    • 5% if he increases the profit by 0.1–10%
    • 7.5% if he increases the profit by 10.1–20%
    • 10% if he increases the profit by more than 20%.
   If the company's profit grows from $1.2 million to $1.4 million in a year:

   a. explain what percentage incentive Shane will get and why
   b. calculate his salary for the year.

12. Kevin owns a sports store and has 7 staff working for him. He offers each of them a 5.5% end-of-year bonus on any profits over $100 000. This year the store made a profit of $275 000.

   a. Calculate the amount each employee earned in bonuses.
   b. What is the cost to Kevin in total bonuses for the year?
   c. If one employee earned $64 625 including bonuses for the year, what was their base salary?

13. Jimmy is a high-rise window cleaner. He gets paid $15 per window for the first five levels. For the next 15 levels he gets an extra 15% per window, and above this he gets 20% extra as danger money. How much does Jimmy earn for cleaning:

   a. a total of 20 windows on levels 3 to 4
   b. 10 windows on level 4 and a total of 27 windows on levels 10 to 13
   c. a total of 30 windows on levels 11 to 14 and a total of 30 windows on levels 21 to 25?

14. Denise works for a real estate agent. She receives a basic wage of $500 per week plus commission on sales. The rate of commission is variable. For houses up to $600 000, the commission is 0.25%. For houses over $600 000, the commission is an additional 0.10% on the amount over $600 000. How much pay did she receive in the week she sold a house for:

   a. $580 000   b. $830 000

15. When Jack goes on holiday, he is paid $17\frac{1}{2}\%$ holiday loading in addition to his normal pay. When he went on 2 weeks leave, his holiday pay was $1504. What is his normal weekly pay?

16. How are bonuses used to encourage workers?

LESSON
1.7 Taxation and net earnings

▶ 1.7.1 Taxation

eles-6198

- In Australia, people who earn more than $18 200 in a financial year must pay a percentage of their earnings as **income tax**.
- The rates of taxation for Australian residents for 2022–23 are shown in the table below.

Taxable income	Tax on this income
0–$18 200	Nil
$18 201–$45 000	19c for each $1 over $18 200
$45 001–$120 000	$5092 plus 32.5c for each $1 over $45 000
$120 001–$180 000	$29 467 plus 37c for each $1 over $120 000
$180 001 and over	$51 167 plus 45c for each $1 over $180 000

The above rates do not include the Medicare levy of 2.0%.

WORKED EXAMPLE 14 Calculating tax payable

Calculate the amount of tax paid on an annual income of:
a. $22 000
b. $92 000.

THINK	WRITE
a. 1. $22 000 is in the $18 201 to $45 000 bracket.	**a.**
2. The tax payable is 19c (0.19) for every dollar over $18 200.	
3. Calculate the amount over $18 200 by subtracting $18 200 from $22 000.	$22 000 - $18 200 = $3800
4. Apply the rule '19c for every dollar over $18 200'.	Tax payable $= 0.19 \times 3800$ $= 722$
5. Write the answer in a sentence.	The tax payable on $22 000 is $722.
b. 1. $92 000 is in the $45 001 to $120 000 bracket.	**b.**
2. Calculate the amount over $45 000 by subtracting $45 000 from $92 000.	$92 000 - $45 000 = $47 000
3. Apply the rule '$5092 plus 32.5c for each $1 over $45 000'.	Income tax $= 5092 + 0.325 \times 47 000$ $= 20 367$
4. Write the answer in a sentence.	**c.** The tax payable on $92 000 is $20 367.

⊙ 1.7.2 Net earnings
eles-6199
Medicare levy

- Medicare is the scheme that gives Australian residents access to health care.
- Most taxpayers pay 2.0% of their taxable income to pay for this scheme. This is called the Medicare levy.
- People who have private medical insurance can reclaim some of this money.

Pay As You Go (PAYG) taxation

- When you receive your pay, usually as a bank deposit, some of the money has been taken out by the employer to cover your income tax and Medicare levy. This is called 'pay as you go' (PAYG) taxation.
- The initial amount, before tax is taken out, is called your **gross salary** and the amount that you actually receive is called your **net salary**.
- The amount of money to be deducted by the employer each week is published by the Australian Tax Office, as shown in the following table.
- Every taxpayer is required to submit a yearly tax return. Usually the taxpayer's employer will have deducted the correct weekly amount from the employee's gross earnings. If the employee has received income from another source, such as an interest-bearing bank account, this may mean that the employee needs to pay tax on this extra amount. Alternatively, if the employee has some extra tax deductions, this could mean a tax return to the employee.

Calculating net pay

Net pay = gross pay + any allowance − tax withheld − other deductions

This PAYG table shows the tax rates for 2022–23.

PAYG TABLE: Weekly tax withheld ($)					
Gross wage	With tax-free threshold	Gross wage	With tax-free threshold	Gross wage	With tax-free threshold
450	19	950	144	1450	318
500	33	1000	162	1500	335
550	47	1050	179	1550	352
600	58	1100	197	1600	370
650	68	1150	214	1650	381
700	79	1200	231	1700	404
750	90	1250	249	1750	421
800	101	1300	266	1800	439
850	112	1350	283	1850	456
900	127	1400	301	1900	473

Note: Most Australian citizens qualify for the tax-free threshold. For the purposes of this section, apply the tax-free threshold values.

Deductions

- Often other sums of money, such as union fees and private health insurance, are deducted from gross pay.

Family Tax Benefit

- When a family has young or dependent children, the government may pay an allowance called the 'Family Tax Benefit', which is added to a person's gross salary.

WORKED EXAMPLE 15 Calculating net income

Fiona has a gross wage of $900 per week.
a. Use the PAYG table to determine the amount of tax that should be deducted.
b. What percentage of her gross pay is deducted?
c. If Fiona receives $98 in family allowance but has deductions of $71 (superannuation) and $5.50 (union fee), what is her net pay?

THINK	WRITE
a. From the table, PAYG tax payable on a gross wage of $900 per week is $127.	a. $127
b. Calculate 127 as a percentage of 900.	b. $\dfrac{127}{900} \times 100 = 14.11\%$ deducted
c. 1. Fiona receives $98 in family allowance. Add this to her gross weekly wage to determine her total income.	c. Total income $= 900 + 98$ $= \$998$
2. Calculate her total deductions.	Total deductions $= 127 + 71 + 5.50$ $= \$203.50$
3. Calculate her net pay by subtracting her total deductions from her total income.	Net pay $= 998 - 203.50$ $= \$797.50$

1.7 Quick quiz on	**1.7 Exercise**

Individual pathways

■ PRACTISE	■ CONSOLIDATE	■ MASTER
1, 2, 3, 6	4, 5, 7, 9	8, 10–12

Fluency

1. **WE14** Calculate the amount of tax paid to the nearest dollar on an annual income of:

   a. $15 000
   b. $22 000
   c. $44 000
   d. $88 000.

2. **WE15** In the PAYG tax table, look up the amount of tax that must be deducted from each of the following weekly earnings and calculate this as a percentage of the gross pay, correct to 2 decimal places.

   a. $650
   b. $1100
   c. $1550

3. For each of the following weekly pay values, calculate the net pay.

   a. Gross pay $450.00, tax $22.00 and union fees $4.75
   b. Gross pay $550.00, tax $48.00, private health insurance $25.85 and superannuation $53.80
   c. Gross pay $850.00, tax $130.00, loan repayment $160.00 and insurance payment $45.40

Understanding

4. Calculate the net annual salary of a person who has a gross annual salary of $57 200 and a family allowance of $4392.20 with deductions of $17 264 for tax, annual union fees of $262.75 and social club payments of $104.00.

5. Sergio works as a security guard and receives gross pay of $950.00 each week. His tax totals $165 per week. If his other deductions are $60.10 for superannuation and $5.05 for union fees, what is his net pay?

6. Lieng works as an interior decorator and earns $1350 per week.

   a. How much tax should be deducted from her pay each week?
   b. What percentage of her gross pay is her tax?
   c. If Lieng also has deductions of $105 for superannuation, $5.20 for union fees, and $4.00 for a social club, what is her net weekly pay?

Communicating, reasoning and problem solving

7. Yelena works as a chef and is paid $22.86 per hour and works a 35-hour week.

   a. Calculate Yelena's gross weekly earnings.
   b. How much tax should be deducted from Yelena's pay?
   c. What percentage of her gross pay (correct to 2 decimal places) is deducted in tax?
   d. If Yelena also has deductions of $56.20 for superannuation and $22.50 for her health insurance, and she gets $60.00 taken out to pay off her car loan, what is her net pay?
   e. What percentage of her gross pay is her net pay? Give your answer correct to 2 decimal places.

8. Debbie earns $72 000 per year.

   a. Explain why she takes home only $58 133.
   b. Give reasons why this figure could possibly be different again.

9. Jacko works at an IT firm and earns $1725 a week.

   a. How much does he earn a year, gross?
   b. How much tax will he need to pay per year? (Use the yearly table to calculate the tax payable.)
   c. If he has no deductions, how much will he need to pay for the Medicare levy?

10. Tamara works as a swimming instructor and earns $21.50 per hour when working a 38-hour week.

   a. Using the PAYG table, determine the amount of tax that should be deducted from Tamara's salary per week, correct to the nearest dollar.
   b. What percentage of her gross salary is deducted? Give your answer to 1 decimal place.
   c. If Tamara receives $82 per week in family allowance but pays $50 per week towards her superannuation, what is her net weekly pay?

11. Greg started work as an experienced barista in a café and was paid $24 an hour when working a 40-hour week. His weekly tax withheld was $165. After 6 months, he decided to go travelling. If his 6 months of work was all within one financial year, how much money should Greg expect to receive in his tax return? (Assume a compulsory Medicare levy of 2%.)

12. What strategies would you use to remember how to calculate income tax?

LESSON
1.8 Simple interest

LEARNING INTENTION

At the end of this lesson you should be able to:
- calculate the simple interest on a loan or an investment
- apply the simple interest formula to determine the time, the rate or the principal.

▶ 1.8.1 Principal and interest

eles-4399

- When you put money in a financial institution such as a bank or credit union, the amount of money you start with is called the **principal**.
- **Interest** is the fee charged for the use of someone else's money.
- Investors are people who place money in a financial institution — they receive interest from the financial institution in return for the use of their money.
- Borrowers pay interest to financial institutions in return for being given money as a loan.
- The amount of interest is determined by the interest rate.
- Interest rates are quoted as a percentage over a given time period, usually a year.
- The total amount of money, which combines the principal (the initial amount) and the interest earned, is known as the value of the investment.

Value of an investment or loan

$$A = P + I$$

where:
- A = the total amount of money at the end of the investment or loan period
- P = the principal (the initial amount invested or borrowed)
- I = the amount of interest earned.

WORKED EXAMPLE 16 Calculating the total amount

Sam borrowed \$450 000 from a bank for his business. After 30 years Sam had repaid his loan, with a total of \$500 000 paid to the bank. Calculate how much interest Sam paid the bank.

THINK	WRITE
1. Write the known quantities: principal (P) and the total amount (A).	$P = \$450\,000$ $A = \$500\,000$
2. Write the formula to calculate interest paid.	$A = P + I$
3. Substitute the given values into the formula.	$500\,000 = 450\,000 + I$
4. Rearrange the formula to calculate the value of interest.	$I = 500\,000 - 450\,000$ $= 50\,000$
5. Write the answer in a sentence.	Sam paid \$50 000 in interest to the bank.

▶ 1.8.2 Simple interest

eles-4400

- **Simple interest** is the interest paid based on the principal or present value of an investment.
- The principal remains constant and does not change from one period to the next.
- Since the amount of interest paid for each time period is based on the principal, the amount of interest paid is constant.
- Simple interest is usually used for short-term assets or loans.
  - For example, $500 is placed in an account that earns 10% p.a. simple interest every year.
  - This means $50 is paid each year (10% of $500 = $50).
  - After 4 years the total interest paid is $4 \times 50 = \$200$, so the value of the investment after 4 years is $500 + 200 = \$700$.

Time period (years)	Amount of money at the start of the year	Interest for the year
1	$500	$50
2	$550	$50
3	$600	$50
4	$650	$50

- The rate of interest on the investment or loan is $r\%$ p.a. (per annum or per year).
- The duration of the investment or loan is n years.

Formula for simple interest

$$I = Prn$$

where:
- I = the simple interest
- P = the principal
- r = the interest rate per time period
- n = the number of time periods

WORKED EXAMPLE 17 Calculating simple interest

Calculate the amount of simple interest earned over 4 years on an investment of $35 000 that returns 2.1% p.a.

THINK	WRITE
1. Write the simple interest formula.	$I = Prn$
2. State the known values of the variables.	$P = \$35\,000$ $r = \dfrac{2.1}{100} = 0.021$ $n = 4$
3. Substitute the given values into the formula and evaluate the amount of interest.	$I = 35\,000 \times 0.021 \times 4$ $I = 2940$
4. Write the answer in a sentence.	The amount of simple interest over 4 years is $2940.

- Note that the time, n, is in years. Months need to be expressed as a fraction of a year. For example, 27 months = 2 years and 3 months = 2.25 years.

WORKED EXAMPLE 18 Calculating the time of an investment

Karine invests $2000 at a simple interest rate of 4% p.a. Calculate how long she needs to invest this money to earn $120 in interest.

THINK	WRITE
1. Write the simple interest formula.	$I = Prn$
2. State the known values of the variables.	$P = \$2\,000$ $r = \dfrac{4}{100} = 0.04$ $I = \$120$
3. Substitute the given values into the formula and rearrange to calculate the value of the time.	$120 = 2000 \times 0.04 \times n$ $120 = 80n$ $\dfrac{120}{80} = \dfrac{80n}{80}$
4. Change the years into years and months.	$n = 1.5$ $n = 1$ year and 6 months
5. Write the answer in a sentence.	Karine will need to invest the $2000 for 1 year and 6 months.

WORKED EXAMPLE 19 Calculating the total amount of a loan

Zac borrows $3 000 for 2 years at 9% p.a. simple interest.
a. Calculate how much interest Zac is charged over the term of the loan.
b. Calculate the total amount Zac must repay.

THINK	WRITE
a. 1. Write the simple interest formula.	a. $I = Prn$
2. State the known values of the variables.	$P = \$3000$ $r = 0.09$ $n = 2$
3. Substitute the given values into the simple interest formula and evaluate the value of interest.	$I = 3000 \times 0.09 \times 2$ $I = 540$
4. Write the answer in a sentence.	Zac is charged $540 in interest.
b. 1. State the relationship for the value of a loan.	b. $A = P + I$ $A = 3000 + 540$
2. Substitute the values for P and I and calculate the total amount.	$A = 3540$
3. Write the answer in a sentence.	To repay the loan, Zac must pay $3540 in total.

The Carlon-Tozer family needs to buy a new refrigerator at a cost of
$1679. They will pay a deposit of $200 and borrow the balance at a
simple interest rate of 19.5% p.a. The loan will be paid off in 24 equal
monthly payments.

a. Calculate how much money the Carlon-Tozers need to borrow.
b. Calculate how much interest they will pay.
c. Determine the total cost of the refrigerator.
d. Calculate the amount of each payment.

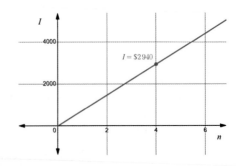

THINK	WRITE
a. 1. Subtract the deposit from the cost to calculate the amount still owing.	a. $1679 - 200 = 1479$
2. Write the answer in a sentence.	They must borrow $1479.
b. 1. State the known values of the variables and write the simple formula.	b. $P = 1479$ $r = 0.195$ $n = 2$ $I = Prn$
2. Substitute the values into the formula to calculate interest (I), by multiplying the values.	$= 1479 \times 0.195 \times 2$ $= 576.81$
3. Write the answer in a sentence.	The interest will be $576.81.
c. 1. Add the interest to the initial cost to calculate the total cost of the refrigerator.	c. $1679.00 + 576.81 = 2255.81$
2. Write the answer in a sentence.	The total cost of the refrigerator will be $2255.81.
d. 1. Subtract the deposit from the total cost to calculate the amount to be repaid.	d. $2255.81 - 200 = 2055.81$.
2. Divide the total payment into 24 equal payments to calculate each repayment.	$2055.81 \div 24 = 85.66$
3. Write the answer in a sentence.	Each payment will be $85.66.

Technology and simple interest

- Digital technologies such as Microsoft Excel or Desmos can be
  helpful in understanding the effects of changing P, r and n. For
  instance, here Desmos has been used to observe how the simple
  interest in Worked example 17 varies with n.

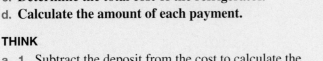

Resources

 Interactivity Simple interest (int-6074)

Exercise 1.8 Simple interest

learnon

1.8 Quick quiz on	1.8 Exercise

Individual pathways

■ PRACTISE	■ CONSOLIDATE	■ MASTER
1, 4, 6, 7, 10, 13, 18, 22	2, 5, 8, 11, 14, 17, 19, 20, 23, 24	3, 9, 12, 15, 16, 21, 25, 26

Fluency

1. **WE16** Alecia invests $400. After 3 years, she withdraws the total value of her investment, which is $496. Calculate the simple interest she earned per year.

2. **WE17** Calculate the amount of simple interest earned on an investment of $4500:

   a. returning 3.5% p.a. for 6 years
   b. returning 2.75% p.a. for 4 years
   c. returning 7.5% p.a. for 2 years
   d. returning 2.5% p.a. for 8 years.

3. Calculate the amount of simple interest earned on an investment of:

   a. $2000, returning 4.5% p.a. for 3 years and 6 months
   b. $18 200, returning 3.6% p.a. for 6 months
   c. $460, returning 2.15% p.a. for 2 years and 3 months
   d. $6700, returning 3.2% p.a. for 7 years.

4. **WE18** Calculate how long it will take an investment of $12 000 to earn the following:

   a. $1260 interest at a simple interest rate of 3.5% p.a.
   b. $2520 interest at a simple interest rate of 4.2% p.a.
   c. $405 interest at a simple interest rate of 2.25% p.a.
   d. $1485 interest at a simple interest rate of 4.5% p.a.

5. Calculate the simple interest rate per annum if:
   a. $3000 earns $270 in 3 years
   b. $480 earns $16.20 in 9 months
   c. $5500 earns $660 in 2.5 years
   d. $2750 earns $748 in 4 years.

6. **MC** If the total interest earned on a $6000 investment is $600 after 4 years, then the annual interest is:

   A. 10%   **B.** 7.5%   **C.** 4%   **D.** 2.5%

7. **WE19** Monique borrows $5000 for 3 years at 8% per annum simple interest.
   a. Calculate how much interest she is charged.   b. Calculate the total amount she must repay.

8. Calculate the simple interest earned on an investment of $15 000 at 5.2% p.a. over 30 months.

9. For each loan in the table, calculate:
   i. the simple interest
   ii. the amount repaid.

	Principal ($)	Interest rate per annum	Time
a.	1000	5%	2 years
b.	4000	16%	3 years
c.	8000	4.5%	48 months
d.	2700	3.9%	2 years 6 months
e.	15 678	9.2%	42 months

10. Calculate the final value of each of the following investments.
    a. $3000 for 2 years at 5% p.a.
    b. $5000 for 3 years at 4.3% p.a.

11. Hasim borrows $14 950 to buy a used car. The bank charges a 9.8% p.a. flat rate of interest over 60 months.
    a. Calculate the total amount he must repay.
    b. Calculate the monthly repayment.

12. Carla borrows $5200 for an overseas trip at 8.9% p.a. simple interest over 30 months. If repayment is made in equal monthly instalments, calculate the amount of each instalment.

Understanding

13. Janan invested $2000 at a simple interest rate of 4% p.a. Calculate how long he needs to invest it in order to earn $200 in interest.

14. If Jodie can invest her money at 8% p.a., calculate how much she needs to invest to earn $2000 in 2 years.

15. If the simple interest charged on a loan of $9800 over 3 years is $2352, determine the percentage rate of interest that was charged.

16. Calculate the missing quantity in each row of the table shown.

	Principal	Rate of interest p.a.	Time	Interest earned
a.	$2000	6%		$240.00
b.	$3760	5.8%		$545.20
c.		7%	3 years	$126.00
d.		4.9%	1 year 9 months	$385.88
e.	$10 000		$1\frac{1}{2}$ years	$1200.00
f.	$8500		42 months	$1041.25

17. **WE20** Mika is buying a used car for $19 998. He has a deposit of $3000 and will pay the balance in equal monthly payments over 4 years. The simple interest rate will be 12.9% p.a.

   a. Calculate how much money he needs to borrow.
   b. Calculate how much interest he will pay.
   c. Determine the total cost of the car.
   d. Determine how many payments he will make.
   e. Calculate the cost of each payment.

Communicating, reasoning and problem solving

18. If a bank offers interest on its savings account of 4.2% p.a. and the investment is invested for 9 months, explain why 4.2 is not substituted into the simple interest formula as the interest rate.

19. Theresa invests $4500 at 5.72% per annum in an investment that attracts simple interest for 6 months. Show that at the end of 6 months she should expect to have $4628.70.

20. A $269 000 business is purchased on an $89 000 deposit with the balance payable over 5 years at an 8.95% p.a. flat rate.

   a. Calculate how much money is borrowed to purchase this business.
   b. Calculate how much interest is charged.
   c. Calculate the total amount that must be repaid.
   d. Determine the size of each of the equal monthly repayments.
   e. Explain two ways in which these payments could be reduced.

21. Nam has $6273 in his bank account at a simple interest rate of 4.86% per annum. After 39 days he calculates that he will have $6305.57 in his account. Did Nam calculate his interest correctly? Justify your answer by showing your calculations.

22. Giang is paid $79.50 in interest for an original investment of $500 for 3 years. Calculate the annual interest rate.

23. A loan is an investment in reverse: you borrow money from a bank and are charged interest. The value of a loan becomes its total cost.
    Jitto wishes to borrow $10 000 from a bank, which charges 11.5% interest per year. If the loan is over 2 years:
    a. calculate the total interest paid
    b. calculate the total cost of the loan.

24. A new sound system costs $3500, but it can be purchased
    for no deposit, followed by 48 equal monthly payments, at a simple interest rate of 16.2% p.a.
    a. Determine the total cost of the sound system.
    b. Under a 'no deposit, no payment for 2 years' scheme,
       48 payments are still required, but the first payment isn't made for 2 years. This will stretch the loan over 6 years. Calculate how much the system will cost using this scheme.
    c. Determine the monthly payments under each of these two schemes.

25. For the following questions, assume that the interest charged on a home loan is simple interest.
    a. Tex and Molly purchase their first home and arrange for a home
       loan of $375 000. Their home loan interest rate rises by 0.25% per annum within the first 6 months of the loan.
       Determine the monthly increase in their repayments. Give your answer to the nearest cent.
    b. Brad and Angel's interest on their home loan is also increased by 0.25% per annum. Their monthly repayments increase by $60.
       Determine the amount of their loan. Give your answer in whole dollars.

26. Juanita sells her car for $10 984. She invests $x\%$ of the money in a bank account at a simple interest rate of 6.68% per annum for 1.5 years. She spends the remainder of the money. At the end of the investment she has exactly enough money to buy a new car for $11 002. Determine the value of x, correct to 2 decimal places.

LESSON
1.9 Introduction to compound interest

LEARNING INTENTION

At the end of this lesson you should be able to:
- calculate the future value of an investment with compound interest
- calculate the interest earned on an investment.

▶ 1.9.1 Compound interest

eles-4842

- Consider $1000 invested for 3 years at 10% p.a. simple interest.
- Each year the value of the investment increases by $100, reaching a total value of $1300.
- The simple interest process can be summarised in the following table.

	Principal	Interest	Total value
Year 1	$1000	$100	$1100
Year 2	$1000	$100	$1200
Year 3	$1000	$100	$1300

Total interest = $300

- For **compound interest**, the future value of the investment is calculated by adding the value of the interest to the present value (principal) at the end of each year.
- The compound interest process can be summarised in the following table.

	Present value	Interest	Future value
Year 1	$1000	$100	$1100
Year 2	$1100	$110	$1210
Year 3	$1210	$121	$1331

Total interest = $331

- The present value or principal increases each year and therefore the amount of interest it earns does so as well.
- The difference between simple interest and compound interest can become enormous over many years.

WORKED EXAMPLE 21 Calculating compound interest

Copy and complete the table to determine the interest paid when $5000 is invested at 11% p.a. compounded annually for 3 years.

	Present value	Interest	Future value
Year 1	$5000		
Year 2			
Year 3			

Total interest =

THINK	WRITE

THINK

1. Interest for year 1 = 11% of $5550

   Calculate the present value for year 2 by adding the interest to the year 1 present value.

2. Interest for year 2 = 11% of $5000
   Calculate the future value at the end of year 2. This is the present value for year 3.

3. Interest for year 3 = 11% of $6160.50.

4. Calculate the interest earned over 3 years by subtracting the year 1 present value from the year 3 future amount.

5. Complete the table.

WRITE

$11\% = \dfrac{11}{100} = 0.11$

$I = 0.11 \times 5000 = 550$
$5000 + 550 = 5550$

$0.11 \times 5550 = 610.50$
$5550 + 610.50 = 6160.50$

$0.11 \times 6160.50 = 677.66$
$6160.50 + 677.66 = 6838.16$

$6838.16 - 5000 = 1838.16$

	Present value	Interest	Future value
Year 1	$5000	$550	$5550
Year 2	$5550	$610.50	$6160.50
Year 3	$6160.50	$677.66	$6838.16

Total interest = $1838.16

- There is a quicker way of calculating the future value of the investment. Look again at Worked example 21. The investment grows by 11% each year, so its value at the end of the year is 111% $\left(\dfrac{111}{100} = 1.11 \right)$ of its value at the start of the year.

$$1.11\% \text{ of } 5000$$
$$= 1.11 \times 5000$$
$$= 5550$$

- This process is repeated each year for 3 years.

$$5000 \xrightarrow{\times 1.11} 5550 \xrightarrow{\times 1.11} 6160.50 \xrightarrow{\times 1.11} 6838.16$$

$$\text{or } 5000 \times 1.11 \times 1.11 \times 1.11 = 6838.16$$

$$\text{or } 5000 \times (1.11)^3 = 6838.16$$

- After 3 years the value of the investment is $6838.16.

WORKED EXAMPLE 22 Calculating the value of an investment

An investment of $2000 receives compounded interest at a rate of 8% p.a. Copy and complete the table to calculate its value after 4 years.

Year	Start of year	End of year
Year 1	$2000	
Year 2		
Year 3		
Year 4		

THINK

1. Interest is compounded at 8%, so at the end of the first year the value is 108% of the initial value. The multiplying factor is 1.08.

2. For the value at the end of year 2, calculate 108% of the amount accumulated in year 1, so calculate 108% of 2160.

3. For the value at the end of year 3, calculate 108% of the amount accumulated in year 2, so calculate 108% of 2332.80.

4. For the value at the end of year 4, calculate 108% of the amount accumulated in year 3, so calculate 108% of 2519.424.

5. Complete the table.

WRITE

$108\% = \dfrac{108}{100} = 1.08$

$1.08 \times 2000 = 2160$
$1.08 \times 2160 = 2332.80$

$1.08 \times 2332.80 = 2519.424$

$1.08 \times 2519.424 = 2720.98$

Year	Start of year	End of year
Year 1	$2000	$2160
Year 2	$2160	$2332.80
Year 3	$2332.80	$2519.42
Year 4	$2519.42	$2720.98

- The repeated multiplication shown in Worked example 22 can be developed into a formula for compound interest.
- In Worked example 22 the principal or present value (*PV*, $2000) was multiplied by 108% four times (because there were 4 years). The future value, *FV*, can be given by the formula:

$$FV = 2000 \times 108\% \times 108\% \times 108\% \times 108\%$$

$$= 2000(100\% + 8\%)^4$$

$$= 2000\left(1 + \frac{8}{100}\right)^4$$

$$= 2000(1.08)^4$$

- With compound interest the future value of an investment is calculated using the following formula.

Formula for compound interest

$$FV = PV(1 + r)^n$$

where:
- *FV* = the future value of the investment
- *PV* = the present value of the investment
- *r* = the interest rate per time period written as a decimal (e.g. 7.5% p.a. is equal to 0.075 p.a.)
- *n* = the number of periods

- In future studies of financial mathematics, the duration of investment under consideration will include monthly, quarterly or weekly investments. At this stage, n will only be considered in terms of the number of years.
- To calculate the amount of compound interest earned or paid, subtract the present value from the future value of the investment.

> ## Calculating compound interest earned
>
> $$I = FV - PV$$
>
> where:
> - I = the amount of interest earned or paid
> - FV = the future value of the investment
> - PV = the present value of the investment

WORKED EXAMPLE 23 Investing an amount earning compound interest

Peter invests $40 000 for 8 years at 7.5% p.a., compounding annually.
a. Calculate how much money Peter has in total at the end of the 8 years.
b. Calculate how much interest Peter earned on the investment.

THINK	WRITE
a. 1. Write the compound interest formula.	a. $FV = PV(1 + r)^n$
2. State the values of the variables PV, r and n.	$PV = 40\,000$ $r = 7.5$ $n = 8$
3. Substitute the values into the formula and calculate the value of the final amount.	$FV = 40\,000(1 + 0.075)^8$ $= 40\,000(1.075)^8$ $= 71\,339.113\,02$
4. Write the answer in a sentence.	The future value of Peter's investment was $71 339.11 (to the nearest cent).
b. 1. Write the relationship between the interest earned, the final amount and the principal.	b. $I = FV - PV$
2. State the values for FV and PV.	$FV = 71\,339.11$ $PV = 40\,000$
3. Substitute the values in the formula and Calculate the value of the interest.	c. $I = 71\,339.11 - 40\,000$ $= 31\,339.11$
4. Write the answer in a sentence.	Peter earned $31 339.11 in interest.

COMMUNICATING — COLLABORATIVE TASK: Compound interest conga

1. Arrange yourself into teams of 3 or more players. The number of players in each team becomes the number of years of a compound interest problem.
2. One person from your class writes a principal value and an interest rate on the board.
3. Using the interest rate and the principal of the investment written on the board, the first member of each group calculates the value of the investment after 1 year.
4. The second member of each group takes the first member's numbers and calculates the value of the investment after 2 years.
5. The rest of the members of the group take it in turns to calculate the interest for the following years.
6. The first team in each round to get the correct answer gets 3 points. The second team to get the correct answer gets 1 point. The team with the most points after the agreed-upon number of rounds is the winner.
7. Repeat the game using different compounding periods (the compounding period is the period of time after which the interest earned is added to the principal) and rotating the roles of the players in each team.
8. As a class, discuss the effect of decreasing compounding periods on the value of the investment.

on Resources

🧩 **Interactivity** Compound interest (int-6075)

Exercise 1.9 Introduction to compound interest learn on

1.9 Quick quiz on	1.9 Exercise

Individual pathways

■ PRACTISE	■ CONSOLIDATE	■ MASTER
1, 3, 5, 7, 11, 12, 15	2, 6, 9, 10, 13, 16	4, 8, 14, 17, 18

Fluency

1. **WE21** Copy and complete the tables shown to determine the interest paid when:

   a. $1000 is invested at 12% p.a. compounded annually for 3 years

	Present value	**Interest**	**Future value**
Year 1	$1000		
Year 2			
Year 3			

   Total interest =

b. $100\,000$ is invested at 9% p.a. compounded annually for 4 years.

	Present value	Interest	Future value
Year 1	$100\,000		
Year 2			
Year 3			
Year 4			

Total interest =

2. **WE22** Copy and complete the tables shown to calculate the final value of each investment.

   a. $5000 invested at 12% p.a. compounded annually for 3 years

	Start of year	End of year
Year 1	$5000	
Year 2		
Year 3		

   b. $200\,000 invested at 7% p.a. compounded annually for 3 years

	Start of year	End of year
Year 1	$200\,000	
Year 2		
Year 3		

3. Calculate the future amount of an investment of $5000 when compound interest is earned at the rate of:

   a. 4% p.a. for 3 years
   b. 3% p.a. for 4 years
   c. 2.75% p.a. for 6 years
   d. 3.5% p.a. for 2 years.

4. Calculate the future value of the following investments when compound interest is earned:

   a. $4650 at 4.5% p.a. for 3 years
   b. $12\,500 at 6.2% p.a. for 5 years
   c. $3560 at 2.4% p.a. for 4 years
   d. $25\,000 at 3.2% p.a. for 10 years.

Understanding

5. **WE23** For each of the following investments, use the compound interest formula to calculate:

   i. the future value
   ii. the amount of interest paid.

   a. $8000 is invested for 8 years at 15% p.a. interest compounding annually
   b. $50\,000 is invested for 4 years at 6% p.a. interest compounding annually
   c. $72\,000 is invested for 3 years at 7.8% p.a. interest compounding annually

6. For each of the following investments, use the compound interest formula to calculate:

   i. the future value
   ii. the amount of interest paid.

   a. $150\,000 is invested for 7 years at 6.3% p.a. interest compounding annually
   b. $3500 is invested for 20 years at 15% p.a. interest compounding annually
   c. $21\,000 is invested for 10 years at 9.2% p.a. interest compounding annually

7. Peter invests $5000 for 3 years at 6% p.a. simple interest. Maria invests the same amount for 3 years at 5.8% p.a. compounding annually.

   a. Calculate the value of Peter's investment on maturity.
   b. Calculate the value of Maria's investment on maturity.
   c. Explain why Maria's investment is worth more, even though she received a lower interest rate.

8. Gianni invests $8000 at 15% p.a. compounded annually. Dylan invests $8000 at a 15% p.a. flat rate. Calculate how much more Gianni's investment is worth than Dylan's after:

   a. 1 year
   b. 2 years
   c. 5 years
   d. 10 years.

9. When Kim's granddaughter was born Kim invested $100 at the rate of 7% p.a. compounding annually. Kim plans to give it to her granddaughter on her eighteenth birthday.
   Calculate how much her granddaughter will receive when she turns 18.

10. Mai's investment account has compounded at a steady 9% for the last 10 years. If it is now worth $68 000, determine how much it was worth:

    a. last year
    b. 10 years ago.

Communicating, reasoning and problem solving

11. Two investment options are available to invest $3000.
    • Invest for 5 years at 5% p.a. compounding monthly.
    • Invest for 5 years at 5% p.a. compounding weekly.
    Explain which option you would choose and why.

12. There are 3 factors that affect the value of a compound interest investment: the initial amount invested the interest rate and the length of the investment.

    a. If the compound interest rate is 10% p.a. and the length of the investment is 2 years, calculate the value of an investment of:

       i. $1000   ii. $2000   iii. $4000.
    b. Comment on the effect that increasing the initial investment amount has on the value of the investment.

13. a. If the principal is $1000 and the compound interest rate is 10% p.a., calculate the value of an investment that lasts for:

       i. 2 years   ii. 4 years   iii. 8 years.
    b. Comment on the effect that increasing the length of the investment has on the value of the investment.

14. **a.** If the principal is $1000 and the length of investment is 5 years, calculate the value of an investment when the compound interest rate is:

   i. 6% p.a.   **ii.** 8% p.a.   **iii.** 10% p.a.

   b. Comment on the effect of increasing the interest rate on the value of the investment.

15. A bank offers a term deposit for 3 years at an interest rate of 8% p.a. with a compounding period of 6 months. Calculate the end value of a $5000 investment under these conditions.

16. A building society offers term deposits at 9% p.a., compounded annually. A credit union offers term deposits at 10% but with simple interest only.

   a. Determine which offer will result in the greatest value after 2 years.
   b. Determine which offer will result in the greatest value after 3 years.
   c. Determine how many years it will take for the compound interest offer to have the greater value.

17. Chris and Jenny each invested $10 000. Chris invested at 6.5% p.a. compounding annually. Jenny took a flat rate of interest. After 5 years, their investments had equal value.

   a. Calculate the value of Chris's investment after 5 years.
   b. Calculate Jenny's interest rate.
   c. Calculate the values of Chris's and Jenny's investments after 6 years.

18. One aspect of compound interest is of great importance to investors: how long does it take for an investment to double in value? Consider a principal of $100 and an annual interest rate of 10% (compounding annually).

   a. Determine how long it would take for this investment to be worth $200.
   b. Evaluate how long it would take for the investment to be worth $400 (to double, and then double again).

LESSON
1.10 Review

1.10.1 Topic summary

Salaries

- A salary is a fixed annual (yearly) payment, regardless of the number of hours worked.
- Salaries are generally paid in weekly, fortnightly or monthly amounts.
  - Normal working hours in Australia are **38 hours** per week.
  - There are **52 weeks** in a year.
  - There are **26 fortnights** in a year (this value is slightly different for a leap year).
  - There are **12 months** in a year.
  e.g. If an annual salary is $52 000:
  - Weekly salary = $52 000 ÷ 52 = $1000
  - Fortnightly salary = $52 000 ÷ 26 = $2000
  - Monthly salary = $52 000 ÷ 12 = $4333.33.

Wages

- A wage is paid at a fixed rate per hour.
- Hours worked outside of the usual 38 or 40 hours per week are usually paid at a higher rate per hour.
  e.g. If the wage rate per hour is $24.50 and the person works 38 hours each week:
  weekly pay = $24.50 × 38 = $931

Special rates

- Special rates are paid when employees work overtime or at weekends, on public holidays or at night.
  e.g. If the hourly pay is $24, then:
  $$\text{time-and-a-half rate} = \$24 \times 1.5$$
  $$= \$36/\text{hour}$$
  $$\text{double time rate} = \$24 \times 2$$
  $$= \$48/\text{hour}$$

FINANCIAL MATHEMATICS

Simple interest

- Interest is the amount earned from investing money with a bank.
- The formula to calculate the final value of a simple interest investment is $I = Prn$, where:
  - P = principal
  - r = the interest rate per time period
  - n = the number of time periods
- The formula to calculate the value of an investment or loan is $A = P + I$, where:
  - A = the total amount of money at the end of the investment or loan period
  - P = the principal
  - I = the amount of interest earned.

Compound interest

- Compound interest is interest that is added to the principal (the amount initially invested) at the end of each year.
- Once the value of the interest is added to the principal, interest is calculated based on the principal's increased value.
- The formula to calculate the final value of a compound interest investment is $FV = PV(1 + r)^n$, where:
  - FV = the future value of the investment
  - PV = the present value of the investment
  - r = the interest rate per time period written as a decimal (e.g. 7.5% p.a. is equal to 0.075 p.a.)
  - n = the number of time periods
- The formula to calculate the compound interest earned is $I = FV - PV$, where:
  - I = the amount of interest earned or paid
  - FV = the future value of the investment
  - PV = the present value of the investment.

Calculating the final amount

- The are different ways to calculate the final amount of an initial investment of $8000 for 10 years, depending on whether it is a simple interest investment or a compound interest investment.

Simple interest investments

- For $8000 over 10 years with a 3% p.a. simple interest rate (also called a flat rate):

  $$I = Prn = 8000 \times 0.03 \times 10 = 2400$$

- Final amount = $8000 + $2400
  $$= \$10\,400$$

Compounding interest investments

- For $8000 over 10 years with a 3% p.a. compound interest rate:

  $$FV = PV(1 + r)^n = 8000(1 + 0.03)^{10}$$
  $$= 8000(1.03)^{10}$$

- Final amount = $10 751.33

1.10.2 Project

Australian currency

Since decimal currency was introduced in Australia in 1966, our notes and coins have undergone many changes. Only our 5c, 10c and 20c coins are still minted as they were back then. The 1c and 2c coins are no longer in circulation, the 50c coin is a different shape, the $1 and $2 notes have been replaced by coins, and our notes have changed from paper to a special type of plastic.

Coins have two sides: an obverse side and a reverse side. The obverse side of all Australian coins depicts our reigning monarch, Queen Elizabeth II, and the year in which the coin was minted. The reverse side depicts a typical Australian feature and sometimes a special commemorative event.

1. Describe what is depicted on the reverse side of each Australian coin.

The following table includes information on Australia's coins currently in circulation. Use the table to answer questions **2** to **4**.

Coin	Diameter (mm)	Mass (g)	Composition
5c	19.41	2.83	75% copper, 25% nickel
10c	23.60	5.65	75% copper, 25% nickel
20c	28.52	11.30	75% copper, 25% nickel
50c	31.51	15.55	75% copper, 25% nickel
$1	25.00	9.00	92% copper, 6% aluminium, 2% nickel
$2	20.50	6.60	92% copper 6% aluminium, 2% nickel

2. Identify the heaviest and lightest type of coin. List every type of coin in order from lightest to heaviest.
3. Identify which coin has the smaller diameter: the 5c coin or the $2 coin. Calculate the difference in size.

The following table displays information on the currency notes currently circulating in Australia. The column on the far right compares the average life of the previously used paper notes with that of the current plastic notes. Use the table to answer questions **4** to **8**.

Note	Date of issue	Size (mm)	Average life of notes (months)	
			Plastic	Paper
$5	07/07/1992 24/04/1995 01/01/2001 01/09/2016	130 × 65	40	6
$10	01/11/1993 20/09/2017	137 × 65	40	8
$20	31/10/1994 09/10/2019	144 × 65	50	10
$50	04/10/1995 18/10/2018	151 × 65	About 100	24
$100	15/05/1996 29/10/2020	158 × 65	About 450	104

4. List the denominations of the notes that are available in Australian currency.
5. State the date on which Australia's first plastic note was issued and the denomination of the note.
6. The $5 note has been issued 4 times, while all other notes have only been issued twice. Suggest a reason for this.
7. Suggest a reason why each note is a different size.
8. The table clearly shows that the plastic notes last about 5 times as long as the paper notes we once used. Suggest a reason why the $50 and $100 notes last longer than the $5 and $10 notes.

 Resources

 Interactivities Crossword (int-2700)
Sudoku puzzle (int-3210)

Exercise 1.10 Review questions

learn on

Fluency

1. Jane earns an annual salary of $45 650. Calculate her fortnightly pay.

2. Express $638.96 per week as an annual salary.

3. Minh works as a casual shop assistant and is paid $8.20 an hour from 3.30 pm to 5.30 pm, Monday to Friday, and $9.50 an hour from 7 am to 12 noon on Saturday. Calculate his total pay for the week.

4. **MC** Below are the pay details for 5 people. Choose who receives the most money.
   A. Billy receives $18.50 per hour for a 40-hour working week.
   B. Jasmine is on an annual salary of $38 400.
   C. Omar receives $1476.90 per fortnight.
   D. Thuy receives $3205 per month.

5. Daniel earns $10 an hour for a regular 38-hour week. If he works overtime, he is paid double time. Calculate how much he would earn if he worked 42 hours in one week.

Understanding

6. Bjorn earns $468.75 per week award wage for a 38-hour week.
   a. Calculate his standard hourly rate.
   b. If he is paid time-and-a-half for normal overtime, calculate his pay for a week in which he works 41 hours.

7. Xana makes gift cards as a hobby and is paid 55 cents for each card.
   a. How much would she earn if she made 50 cards?
   b. How many cards would she need to make to earn $44?

▶

8. Kim sells cakes to the local shop. The shop pays for the ingredients. Kim is paid $3.20 for each cake that she makes and 50 cents for each slice. If Kim makes 5 cakes and 15 slices one week, how much does she earn?

9. A salesperson earns a $250 per week retainer, plus 2% commission on the first $10 000 sale and 1.5% on the remainder. What is their total income for a week in which sales to the value of $18 000 were made?

10. Jennifer is to start a new job selling mobile phones. She is paid commission only at the rate of 17.5% of sales. What value of sales must she make in order to receive commission of $600 in one week?

Communicating, reasoning and problem solving

11. Phillipa is paid an annual salary of $48 800.
    a. Calculate Phillipa's gross weekly salary.
    b. Calculate the total amount Phillipa will receive for her 4 weeks annual leave if she is paid a 17.5% holiday loading.

12. The annual cost of operating the rail system in Sydney is $300 million. The Chief Executive Officer of the rail network is promised a bonus of 0.15% of any saving that she can make to the running of the system. Calculate the bonus paid if the cost is reduced to $275 million.

13. Geoff works as a waiter and is paid $12.50 per hour. If he works 40 hours per week, calculate:
    a. Geoff's gross weekly pay
    b. the tax that Geoff should pay
    c. the percentage of gross pay that Geoff pays in tax
    d. Geoff's net pay if he also has $5.35 deducted for union fees and $100.00 for a loan repayment.

14. a. Calculate the simple interest on an investment of $4000 for 9 months at 4.9% per annum.
    b. To what amount will the investment grow by the end of its term?

15. Calculate the monthly instalment needed to repay a loan of $12 500 over 40 months at 9.75% p.a. simple interest.

16. A simple interest loan of $5000 over 4 years incurred interest of $2300. What interest rate was charged?

17. Daniela is going to invest $16 000 for 2 years at 9% p.a. with interest compounded annually. Calculate:

a. the value of the investment

b. the amount of interest earned.

18. Calculate 3.5% of 900.

19. Rearrange the formula $I = Prn$ to make n the subject.

20. Milos works in a supermarket and earns $11.70 per hour. If Milos works Saturday he is paid at time and a half. State the hourly rate for which Milos works on Saturday.

21. **MC** Frank invests $1000 at 5% p.a. for three years with interest compounded annually. Which of the following calculations will correctly calculate the value of Frank's investment?

A. $1000 \times 5 \times 3$

B. $1000 \times 1.05 \times 3$

C. $1000 \times 1.05 \times 1.05 \times 1.05$

D. $1000 \times 1.05 \times 1.05 \times 1.05 - \$1000$

on To test your understanding and knowledge of this topic, go to your learnON title at www.jacplus.com.au and complete the **post-test**.

Answers

Topic 1 Financial mathematics

Exercise 1.1 Pre-test

1. a. $1643.28   b. $3286.57   c. $7120.89
2. $184
3. $32.50
4. D
5. C
6. B
7. $32 120
8. D
9. a. 26.54%   b. $796
10. a. $8372   b. $31 072   c. $553.58
11. C
12. 9% p.a
13. C
14. D
15. $354

Exercise 1.2 Salaries and wages

1. a. $1105.42   b. $2210.85   c. $4790.17
2. a. $1198.08   b. $2396.15   c. $5191.67
3. a. $19 136   b. $46 410   c. $68 684.20
4. a. $25 870   b. $42 183.96   c. $100 498.06
5. a. $3890 per month   b. $3200.58 per fortnight
6. a. $19.75/h   b. $12.17/h
   c. $25.73/h   d. $39.06/h
7. a. $459.04   b. Jenny earns $211.66 more.
8. $210.72
9. $261.49
10. Rob earns more.
11. $30.77
12. 38 hours
13. a. $1605.77   b. $2788.81
    c. $1100   d. $8770
14. a. B
    b. $915 per fortnight corresponds to the highest annual salary ($23790).
15. Job A has a higher hourly rate.
16. Minh. Minh's annual salary is $63 745.76, which is higher than Michelle's annual salary.
17. $40 per hour. This is the option with the greatest pay.
18. $1735.95, $1833.75
19. a. $29.81   b. $25.64   c. $72 080.58
20. 12 weeks

Exercise 1.3 Special rates

1. a. $23.94   b. $47.80   c. $43.50
2. a. $906.10   b. $794.33   c. $833.56

3. a. $25.16   b. $113.22   c. $1069.30
4. a. $170.40   b. $340.80   c. $426.00
5. $1156.96
6. a. $30.20   b. $45.30
   c. $60.40   d. $135.90
7. a. $32.93   b. $197.55
8. a. $11.70   b. $15.60   c. $19.50
9. a. Mon: 8, Tue: 8, Wed: 8, Thu: 8, Fri: 6
   b.

Pay slip for Susan Jones	Week ending 17 August
Normal hours	38
Normal pay rate	$25.60
Overtime hours	0
Overtime pay rate	$38.40
Total pay	$972.80

10. a. Mon: 8, Tue: 8, Thu: 8, Sat: 4
    b.

Pay slip for Manu Taumata	Week ending 21 December
Normal hours	24
Normal pay rate	$10.90
Overtime hours	4
Overtime pay rate	$16.35
Total pay	$327.00

11. a. Mon: 8, Tue: 8, Fri: 8, Sat: 8, Sun: 8
    b.

Pay slip for Eleanor Rigby	Week ending 17 August
Normal hours	24
Normal pay rate	$16.80
Time-and-a half hours	8
Time-and-a half pay rate	$25.20
Double time hours	8
Double time pay rate	$33.60
Total pay	$873.60

12. a. $754.40   b. $1042.80
13. a. $232.02   b. $1211.66
14. $46\frac{1}{2}$
15. $1155
16. No. Lin should have been paid $1034.
17.

	Hours worked	Regular pay	Overtime pay	Total pay
a.	32	$464	$0	$464
b.	38.5	$522	$54.38	$576.38
c.	40.5	$522	$101.50	$623.50
d.	47.2	$522	$295.80	$817.80

Exercise 1.4 Piecework

1. $97.50
2. $89.50
3. $175.50
4. $318.75
5. a. $144.00   b. 28   c. $21.60
6. a. $144   b. $218   c. $465
7. a. $107.50   b. 24   c. $7.17
8. a. $52   b. $9.45
9. a. $1398.25   b. $1568.12
10. $422.40
11. a. $8950   b. $3550
12. Answers may vary. Advantages could include flexible hours or working from home; disadvantages could be that time per item is often at a very low hourly rate.

Exercise 1.5 Commissions and royalties

1. $3000
2. a. $1280   b. $1115.60
3. $1425
4. a. $200   b. $4400
5. a. $290   b. $482.50   c. $1191.12
6. a. $500   b. $590
   c. $1175   d. $1568.75
7. $9800
8. $12 800
9. a. $4800   b. $5400   c. $14 400
10. a. $2400   b. $3750   c. $15 937.50
11. a. Veronica earns $736; Francis earns $504.
    b. $20 000
12. a. $900   b. $1148   c. $1253.60
13. a. $1520   b. $2237.50
    c. $2555   d. $7560
14. $3676.50
15. a. $15 460   b. $48 725
    c. $36 456   d. $119 800
16. 4.5%
17. Answers may vary. One advantage is that hard work and a successful product or great selling skills will generate more money; one disadvantage is that other factors may influence the success of the product and your income may not reflect your work ethic.

Exercise 1.6 Loadings and bonuses

1. $1244.76
2. $805.64
3. $764.18
4. a. $1428.80   b. $1000.16   c. $6715.36
5. a. $866.40   b. $4072.08
6. a. $574.20   b. $100.49
7. a. $2431.23   b. $5713.39
8. $1135.01

9. $116 326.53
10. a. $60 000   b. $5000   c. 11.90%
11. a. Shane receives a 7.5% incentive as the company's profit has grown by 16.7% (which is between 10.1% and 20%).
    b. $295 625
12. a. $9625   b. $67 375   c. $55 000
13. a. $300   b. $615.75   c. $1057.50
14. a. $1950   b. $2230
15. $640
16. Answers may vary but could include that awarding bonuses encourages employees to work over and above their expected output.

Exercise 1.7 Taxation and net earnings

1. a. 0   b. $722   c. $4902   d. $19 067
2. a. $68, 10.46%   b. $197, 17.91%   c. $352, 22.71%
3. a. $423.25   b. $422.35   c. $514.60
4. $43 961.45
5. $719.85
6. a. $283   b. 20.96%   c. $952.80
7. a. $800.10   b. $101   c. 12.62%
   d. $560.40   e. 70.04%
8. a. Debbie is taxed $13 867 on the $72 000 she earns.
   b. There may be other deductions on Debbie's net pay, such as superannuation, union fees, private health insurance and the Medicare levy.
9. a. $89 700   b. $19 619.50   c. $1794
10. a. $101   b. 12.4%   c. $748
11. $2506.40
12. When calculating income tax, refer to given information such as tables and determine if the tax is to be calculated before or after deductions.

Exercise 1.8 Simple interest

1. $32
2. a. $945   b. $495
   c. $675   d. $900
3. a. $315   b. $327.60
   c. $22.25   d. $1500.80
4. a. 3 years
   b. 5 years
   c. 1.5 years or 1 year and 6 months
   d. 2.75 years or 2 year and 9 months
5. a. 3% p.a.   b. 4.5% p.a.
   c. 4.8% p.a.   d. 6.8% p.a.
6. D
7. a. $1200   b. $6200
8. $1950
9. a. i. $100   ii. $1100
   b. i. $1920   ii. $5920
   c. i. $1440   ii. $9440
   d. i. $263.25   ii. $2963.25
   e. i. $5048.32   ii. $20 726.32

10. a. $3300   **b.** $5645

11. a. $22 275.50   **b.** $371.26

12. $211.90

13. 2.5 years

14. $12 500

15. 8%

16. a. 2 years   **b.** 2.5 years   **c.** $600
     d. $4500.06   **e.** 8%   **f.** 3.5%

17. a. $16 998   **b.** $8770.97   **c.** $28 768.97
     d. 48   **e.** $536.85

18. The interest rate in the simple interest formula needs to be converted from a percentage into a decimal: 4.2% = 0.042.

19. Sample responses can be found in the worked solutions in the online resources.

20. a. $180 000
     b. $80 550
     c. $260 550
     d. $4342.50
     e. Increase the size of the deposit or increase the length of time over which the loan can be repaid.

21. Yes

22. 5.3%

23. a. $2300   **b.** $12 300

24. a. $5768   **b.** $6902   **c.** $120.17, $143.79

25. a. $78.13   **b.** $288 000

26. 91.04

Exercise 1.9 Introduction to compound interest

1. a.

	Present value	Interest	Future value
Year 1	$1000	$120	$1120
Year 2	$1120	$134.40	$1254.40
Year 3	$1254.40	$150.53	$1404.93

Total interest = $404.93

b.

	Present value	Interest	Future value
Year 1	$100 000	$9000	$109 000
Year 2	$109 000	$9810	$118 810
Year 3	$118 810	$10 692.90	$129 502.90
Year 4	$129 502.90	$11 655.26	$141 158.16

Total interest = $41 158.16

2. a.

	Start of year	End of year
Year 1	$5000	$5600
Year 2	$5600	$6272
Year 3	$6272	$7024.64

b.

	Start of year	End of year
Year 1	$200 000	$214 000
Year 2	$214 000	$228 980
Year 3	$228 980	$245 008.60

3. a. $5624.32   **b.** $5627.54
     c. $5883.84   **d.** $5356.13

4. a. $5306.42   **b.** $16 886.23
     c. $3914.26   **d.** $34 256.03

5. a. i. $24 472.18   ii. $16 472.18
     b. i. $63 123.85   ii. $13 123.85
     c. i. $90 196.31   ii. $18 196.31

6. a. i. $230 050.99   ii. $80 050.99
     b. i. $57 282.88   ii. $53 782.88
     c. i. $50 634.40   ii. $29 634.40

7. a. $5900
     b. $5921.44
     c. Maria's principal increases each year.

8. a. 0   **b.** $180
     c. $2090.86   **d.** $12 364.46

9. $337.99

10. a. $62 385.32   **b.** $28 723.93

11. The second option would be the best choice, as the shorter the time between the compounding periods, the greater the interest paid.

12. a. i. $1210   ii. $2420   iii. $4840
     b. Increasing the initial investment will increase the value of the investment because it will have a higher value of interest.

13. a. i. $1210   ii. $1464.10   iii. $2143.59
     b. Increasing the length of the investment will increase the value of the investment because it will have a higher value of interest.

14. a. i. $1338.23   ii. $1469.33   iii. $1610.51
     b. Increasing the interest rate will increase the value of the investment because it will have a higher value of interest.

15. $6326.60

16. a. Simple interest
     b. Simple interest
     c. 4 years

17. a. $13 700.87
     b. 7.4%
     c. Chris, $14 591.42; Jenny, $14 440

18. a. 7.27 years   **b.** 14.55 years

Project

1. 5c coin: echidna
     10c coin: lyrebird
     20c coin: platypus
     50c coin: coat of arms
     $1 coin: five kangaroos
     $2 coin: Walpiri-Anmatyerre Aboriginal Elder Gwoya Jungarai

2. 5c, 10c, $2, $1, 20c, 50c

3. The 5c coin has a smaller diameter. It is smaller by 1.09 mm.

4. $5, $10, $20, $50, $100
5. 7 July 1992, $5
6. Sample responses can be found in the worked solutions in the online resources.
7. The different sizes allow people with vision impairment to tell the difference between each note.
8. The $50 and $100 notes are used less frequently.

Exercise 1.10 Review questions

1. $1755.77
2. $33 225.92
3. $129.50
4. A
5. $460
6. a. $12.34   b. $524.28
7. a. $27.50   b. 80 cards
8. $23.50
9. $570
10. $3428.57
11. a. $938.46   b. $4410.76
12. $37 500
13. a. $500   b. $33
    c. 6.6%   d. $361.65
14. a. $147   b. $4147
15. $414.06
16. 11.5%
17. a. $19 009.60   b. $3009.60
18. 31.5
19. $n = \dfrac{I}{Pr}$
20. $17.55
21. C

2 Indices and surds

LESSON
2.1 Overview

Why learn this?

The earliest mathematics that humans used involved simple counting. Knowing that two goats could be swapped for six chickens was an important part of life twelve thousand years ago, so being able to count was a big deal.

Of course, being able to count is still a big deal today, but over thousands of years we have moved on from counting sheep. Ancient Egypt was the first recorded civilisation to think about fractions. Later on, ancient Greek mathematicians became the first people we know of to think about numbers that can't be expressed as fractions, such as the length of the diagonals of a square of side 1 cm. These thoughts gave rise to the idea of irrational numbers. The value of π (pi) — which is the ratio of the circumference of a circle to its diameter — is one such irrational number. The value of π can only be approximated, to 3.1416 for instance, it has an infinite number of digits in its decimal representation.

As new kinds of numbers have been discovered, new ways of writing them have been developed. These new ways of writing numbers can make it easier to work with numbers like the massively large ones that describe the distances between stars, or the incredibly small ones that describe the size of subatomic particles.

In this topic we will continue our mathematical journey by exploring the ways numbers can be written and represented and the laws of indices that can be applied to calculations. Indices (the plural of index) give us a way of abbreviating multiplication, division and simplify algebraic expressions.

Hey students! Bring these pages to life online

▶ Watch videos

Engage with interactivities

A+ Answer questions and check solutions

Find all this and MORE in jacPLUS

Reading content and rich media, including interactivities and videos for every concept

Extra learning resources

Differentiated question sets

Questions with immediate feedback, and fully worked solutions to help students get unstuck

1. State whether the following expression is True or False.

$$2x^2 \times 3x^3 = 6x^6$$

2. Evaluate $\left(3^2\right)^3$.

3. State whether the following expression is True or False.

$$\left(4x^3y^2\right)^2 = 16x^6y^4$$

4. **PATH** **MC** Select the simplest form of $\sqrt{50}$ from the following list.
   A. $5\sqrt{10}$
   B. $2\sqrt{5}$
   C. $5\sqrt{2}$
   D. $10\sqrt{5}$

5. **PATH** **MC** Select the correct answer for the expression $\sqrt{5} - 2\sqrt{45} + 3\sqrt{20}$.
   A. $7\sqrt{5}$
   B. $\sqrt{5}$
   C. $2\sqrt{70}$
   D. $19\sqrt{5}$

6. **PATH** Evaluate $\dfrac{\sqrt{120} \times \sqrt{40}}{5\sqrt{3}}$.

7. **PATH** Simplify $\dfrac{\sqrt{48a^7}}{\sqrt{27a^3}}$.

8. **PATH** Simplify $2\sqrt[3]{27} + 6\sqrt[4]{81} - \sqrt[5]{243}$.

9. **PATH** **MC** State which expression is equal to $\sqrt[3]{216a^3b^6c^9}$.
   A. $72ab^2c^3$
   B. $72b^3c^6$
   C. $6ab^2c^3$
   D. $6b^3c^6$

10. Simplify $5p^{10}q^3 \times 2p^2q^4$.

11. Evaluate $4\left(a^0 + b^0\right)$.

12. Simplify $\left(3ab^4\right)^2 \times \left(2a^2b\right)^3$.

13. **MC** Select the correct simplification of $(-3)^2 \times \left(3^{-1}\right)^2 \times -(3)^2$.
    A. $-(3)^2$
    B. 3^2
    C. $(-3)^2$
    D. $\left(3^{-1}\right)^2$

14. **PATH** Simplify $\dfrac{\left(2m^2n^{-3}\right)^{-2}}{4\left(mn^{-1}\right)^{-4}}$.

15. **PATH** Simplify $\sqrt[3]{27x^6y^3} \times \sqrt[4]{16x^{12}y^8}$.

LESSON
2.2 Applying the index laws to variables

LEARNING INTENTION

At the end of this lesson you should be able to:
- identify the base and the exponent of a number written in index form
- recall the first three index laws
- use the first three index laws to simplify expressions involving indices.

▶ 2.2.1 Review of index notation

eles-4382

- The product of factors can be written in a shorter form called **index notation** (also known as exponent or power notation).

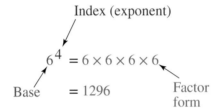

Index (exponent)

$$6^4 = 6 \times 6 \times 6 \times 6$$

Base $= 1296$

Factor form

- 6^4 can be read as '6 to the power of 4'.
- Any composite number can be written as a product of powers of prime factors using a factor tree, or by other methods, such as repeated division.

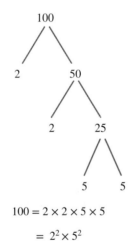

$$100 = 2 \times 2 \times 5 \times 5$$
$$= 2^2 \times 5^2$$

DISCUSSION

Can you use $100 = 2^5 \times 5^2$ and $42 = 2 \times 3 \times 7$ to quickly write 4200 as a product of powers of prime factors?

WORKED EXAMPLE 1 Using index notation to express a product of prime factors

Express 360 as a product of powers of prime factors.

THINK	WRITE
1. Express 360 as a product of a factor pair.	$360 = 6 \times 60$
2. Further factorise 6 and 60.	$= 2 \times 3 \times 4 \times 15$
3. Further factorise 4 and 15 until there are no more composite numbers.	$= 2 \times 3 \times 2 \times 2 \times 3 \times 5$
	$= 2 \times 2 \times 2 \times 3 \times 3 \times 5$
4. Write the answer using index notation. *Note:* The factors are generally expressed with their bases written in ascending order.	$360 = 2^3 \times 3^2 \times 5$

▶ 2.2.2 Multiplication using indices

eles-4383

- Numbers in index form with the same base can be multiplied together by first being written in factor form. For example,

$$5^3 \times 5^2 = (5 \times 5 \times 5) \times (5 \times 5) = 5^5$$
$$\text{or}$$
$$4^4 \times 4^3 = (4 \times 4 \times 4 \times 4) \times (4 \times 4 \times 4) = 4^7$$

- By recognising the pattern, we can multiply terms with the same base by adding the indices. This is the **First Index Law**.
- When more than one base is involved, apply the First Index Law to each base separately.
- The same principle can be applied when multiplying terms with variables.

The First Index Law

$$a^m \times a^n = a^{m+n}$$

For example, $2x^2 \times x^3 = 2x^{2+3} = 2x^5$

WORKED EXAMPLE 2 Simplifying an expression using the First Index Law

Simplify $5e^{10} \times 2e^3$.

THINK	WRITE
1. The order is not important when multiplying, so place the coefficients first.	$5e^{10} \times 2e^3$ $= 5 \times 2 \times e^{10} \times e^3$
2. Simplify by multiplying the coefficients and then applying the First Index Law (add the indices). Write the answer.	$= 10e^{13}$

Simplify $7m^3 \times 3n^5 \times 2m^8n^4$.

THINK	WRITE
1. The order is not important when multiplying, so place the coefficients first and group all similar pronumerals together.	$7m^3 \times 3n^5 \times 2m^8n^4$ $= 7 \times 3 \times 2 \times m^3 \times m^8 \times n^5 \times n^4$
2. Simplify by multiplying the coefficients and applying the First Index Law (add the indices). Write the answer.	$= 42m^{11}n^9$

▶ 2.2.3 Division using indices

eles-4384

- Numbers in index form with the same base can be divided by first being written in factor form. For example,

$$2^6 \div 2^4 = \frac{2 \times 2 \times 2 \times 2 \times 2 \times 2}{2 \times 2 \times 2 \times 2}$$
$$= \frac{\cancel{2} \times \cancel{2} \times \cancel{2} \times \cancel{2} \times 2 \times 2}{\cancel{2} \times \cancel{2} \times \cancel{2} \times \cancel{2}}$$
$$= 2 \times 2$$
$$= 2^2$$

- By recognising the pattern, we can divide terms with the same base by subtracting the indices. This is the **Second Index Law**.
- When the coefficients do not divide evenly, simplify by cancelling.
- The same principle can be applied when dividing terms with variables.

The Second Index Law

$$a^m \div a^n = a^{m-n} \text{ or } \frac{a^m}{a^n} = a^{m-n}$$

For example, $2x^5 \div x^3 = \frac{2x^5}{x^3} = 2x^{5-3} = 2x^2$.

Simplify the following.

a. $\dfrac{15x^6}{5x^2}$

b. $\dfrac{25v^6 \times 8w^9}{10v^4 \times 4w^5}$

THINK	WRITE
a. 1. Simplify by dividing the coefficients and applying the Second Index Law to the indices.	a. $\dfrac{15x^6}{5x^2}$ $= 3x^{(6-2)}$
2. Write the answer	$= 3x^4$

b. 1. Simplify the numerator and the denominator by multiplying the coefficients.

$$\frac{25v^6 \times 8w^9}{10v^4 \times 4w^5}$$
$$= \frac{200v^6w^9}{40v^4w^5}$$

2. Simplify further by dividing the coefficients and applying the Second Index Law (subtract the indices). Write the answer.

$$= \frac{\overset{5}{\cancel{200}}}{\underset{1}{\cancel{40}}} \times \frac{v^6}{v^4} \times \frac{w^9}{w^5}$$
$$= 5v^2w^4$$

WORKED EXAMPLE 5 Simplifying using the Second Index Law with non-cancelling coefficients

Simplify $\dfrac{7t^3 \times 4t^8}{12t^4}$.

THINK

1. Simplify the numerator by multiplying the coefficients and applying the First Index Law (add the indices).

2. Simplify further by dividing the coefficients by the highest common factor. Then apply the Second Index Law (subtract the indices). Write the answer.

WRITE

$$\frac{7t^3 \times 4t^8}{12t^4}$$
$$= \frac{28t^{11}}{12t^4}$$

$$= \frac{28}{12} \times \frac{t^{11}}{t^4}$$
$$= \frac{7t^7}{3}$$

▶ 2.2.4 Zero index

eles-4385

- Any number (except zero) divided by itself is 1.
- This means that, for $a \neq 0$, $\dfrac{a^m}{a^m} = 1$. So, $a^{m-m} = a^0 = 1$.
- In general, any number (except 0) to the power of 0 is equal to 1. This is the **Third Index Law**.

> ### The Third Index Law
>
> $$a^0 = 1, \text{where } a \neq 0$$
>
> For example, $2^0 = 1$.

Evaluate the following.

a. 17^0

b. t^0

c. $(xy)^0$

d. $5x^0$

e. $(5x)^0 + 2$

f. $5^0 + 3^0$

THINK	WRITE
a. Apply the Third Index Law.	a. $17^0 = 1$
b. Apply the Third Index Law.	b. $t^0 = 1$
c. Apply the Third Index Law.	c. $(xy)^0 = 1$
d. Apply the Third Index Law, using the order of operations to evaluate.	d. $5x^0 = 5 \times x^0$ $= 5 \times 1$ $= 5$
e. Apply the Third Index Law, using the order of operations to evaluate.	e. $(5x)^0 + 2 = 1 + 2$ $= 3$
f. Apply the Third Index Law, using the order of operations to evaluate.	f. $5^0 + 3^0 = 1 + 1$ $= 2$

Simplify $\dfrac{9g^7 \times 4g^4}{6g^3 \times 2g^8}$.

THINK	WRITE
1. Simplify the numerator and the denominator by applying the First Index Law.	$\dfrac{9g^7 \times 4g^4}{6g^3 \times 2g^8}$
2. Simplify the fraction further by applying the Second Index Law.	$= \dfrac{36g^{11}}{12g^{11}}$ $= \dfrac{{}^3\cancel{36}g^{11}}{{}^1\cancel{12}g^{11}}$
3. Simplify by applying the Third Index Law.	$= 3g^0$ $= 3 \times 1$
4. Write the answer.	$= 3$

▶ 2.2.5 Cancelling fractions

eles-4386

- Consider the fraction $\dfrac{x^3}{x^7}$. This fraction can be cancelled by dividing the denominator and the numerator by

  the highest common factor (HCF), which is x^3, so $\dfrac{x^3}{x^7} = \dfrac{1}{x^4}$.

  Note: $\dfrac{x^3}{x^7} = x^{-4}$ when we apply the Second Index Law. We will study negative indices in lesson 2.4.

How do the First, Second and Third index laws help with calculations?

WORKED EXAMPLE 8 Using the first three index laws to simplify expressions

Simplify these fractions by cancelling.

a. $\dfrac{x^5}{x^7}$ b. $\dfrac{6x}{12x^8}$ c. $\dfrac{30x^5y^6}{10x^7y^3}$

THINK	WRITE
a. Divide the numerator and denominator by the highest common factor (HCF), which is x^5.	a. $\dfrac{x^5}{x^7} = \dfrac{1}{x^2}$
b. 1. Divide the numerator and denominator by the HCF, which is $6x$.	b. $\dfrac{6x}{12x^8} = \dfrac{6}{12} \times \dfrac{x}{x^8}$ $= \dfrac{1}{2} \times \dfrac{1}{x^7}$
2. Simplify and write the answer.	$= \dfrac{1}{2x^7}$
c. 1. Divide the numerator and denominator by the HCF, which is $10x^5y^3$.	c. $\dfrac{30x^5y^6}{10x^7y^3} = \dfrac{30}{10} \times \dfrac{x^5}{x^7} \times \dfrac{y^6}{y^3}$ $= \dfrac{3}{1} \times \dfrac{1}{x^2} \times \dfrac{y^3}{1}$
2. Simplify and write the answer.	$= \dfrac{3y^3}{x^2}$

on Resources

▶ **Video eLesson**   Index notation (eles-1903)

Interactivities   Review of index form (int-3708)
               First Index Law (int-3709)
               Second Index Law (int-3711)
               Third Index Law (int-3713)

Exercise 2.2 Applying the index laws to variables

2.2 Quick quiz on	2.2 Exercise

Individual pathways

■ PRACTISE	■ CONSOLIDATE	■ MASTER
1, 3, 6, 10, 13, 15, 19, 20, 21, 26	2, 4, 7, 8, 11, 18, 22, 23, 27, 28	5, 9, 12, 14, 16, 17, 24, 25, 29, 30

Fluency

1. **WE1** Use index notation to express each of the following as a product of powers of prime factors.

    a. 12   b. 72   c. 75   d. 240   e. 640   f. 9800

2. **WE2** Simplify each of the following.

    a. $4p^7 \times 5p^4$ b. $2x^2 \times 3x^6$ c. $8y^6 \times 7y^4$ d. $3p \times 7p^7$ e. $12t^3 \times t^2 \times 7t$ f. $6q^2 \times q^5 \times 5q^8$

3. **WE3** Simplify each of the following.

    a. $2a^2 \times 3a^4 \times e^3 \times e^4$ b. $4p^3 \times 2h^7 \times h^5 \times p^3$ c. $3k^2 \times 2k^{11} \times k$ d. $3^4 \times 3^4 \times 3^4$

4. Simplify each of the following.

    a. $2m^3 \times 5m^2 \times 8m^4$ b. $2gh \times 3g^2 h^5$
    c. $5p^4q^2 \times 6p^2q^7$ d. $8u^3w \times 3uw^2 \times 2u^5w^4$

5. Simplify each of the following.

    a. $9y^8 d \times y^5 d^3 \times 3y^4 d^7$
    b. $7b^3c^2 \times 2b^6c^4 \times 3b^5c^3$
    c. $4r^2s^2 \times 3r^6s^{12} \times 2r^8s^4$
    d. $10h^{10}v^2 \times 2h^8v^6 \times 3h^{20}v^{12}$

6. **WE4** Simplify each of the following.

    a. $\dfrac{15p^{12}}{5p^8}$ b. $\dfrac{18r^6}{3r^2}$ c. $\dfrac{45a^5}{5a^2}$ d. $\dfrac{60b^7}{20b}$

7. Simplify each of the following.

    a. $\dfrac{100r^{10}}{5r^6}$ b. $\dfrac{9q^2}{q}$ c. $\dfrac{130d^3}{d^5}$ d. $\dfrac{21t^{11}}{3t^7}$

8. **WE5** Simplify each of the following.

    a. $\dfrac{8p^6 \times 3p^4}{16p^5}$ b. $\dfrac{12b^5 \times 4b^2}{18b^2}$ c. $\dfrac{25m^{12} \times 4n^7}{15m^2 \times 8n}$ d. $\dfrac{27x^9y^3}{12xy^2}$

9. Simplify each of the following.

    a. $\dfrac{12j^8 \times 6f^5}{8j^3 \times 3f^2}$ b. $\dfrac{8p^3 \times 7r^2 \times 2s}{6p \times 14r}$ c. $\dfrac{27a^9 \times 18b^5 \times 4c^2}{18a^4 \times 12b^2 \times 2c}$ d. $\dfrac{81f^{15} \times 25g^{12} \times 16h^{34}}{27f^9 \times 15g^{10} \times 12h^{30}}$

10. **WE6** Evaluate the following.

    a. m^0 b. $6m^0$ c. $(6m)^0$ d. $(ab)^0$ e. $5(ab)^0$

11. Evaluate the following.

    a. w^0x^0 b. 85^0 c. $85^0 + 15^0$ d. $x^0 + 1$ e. $5x^0 - 2$

12. Evaluate the following.

    a. $\dfrac{x^0}{y^0}$ b. $x^0 - y^0$ c. $3x^0 + 11$ d. $3a^0 + 3b^0$ e. $3\left(a^0 + b^0\right)$

13. **WE7** Simplify each of the following.

a. $\dfrac{2a^3 \times 6a^2}{12a^5}$ b. $\dfrac{3c^6 \times 6c^3}{9c^9}$ c. $\dfrac{5b^7 \times 10b^5}{25b^{12}}$ d. $\dfrac{8f^3 \times 3f^7}{4f^5 \times 3f^5}$ e. $\dfrac{9k^{12} \times 4k^{10}}{18k^4 \times k^{18}}$

14. Simplify each of the following.

a. $\dfrac{2h^4 \times 5k^2}{20h^2 \times k^2}$ b. $\dfrac{p^3 \times q^4}{5p^3}$ c. $\dfrac{m^7 \times n^3}{5m^3 \times m^4}$ d. $\dfrac{8u^9 \times v^2}{2u^5 \times 4u^4}$ e. $\dfrac{9x^6 \times 2y^{12}}{3y^{10} \times 3y^2}$

Understanding

15. **WE8** Simplify each of the following by cancelling.

a. $\dfrac{x^7}{x^{10}}$ b. $\dfrac{m}{m^9}$ c. $\dfrac{m^3}{4m^9}$ d. $\dfrac{12x^6}{6x^8}$

16. Simplify each of the following by cancelling.

a. $\dfrac{12x^8}{6x^6}$ b. $\dfrac{24t^{10}}{t^4}$ c. $\dfrac{5y^5}{10y^{10}}$ d. $\dfrac{35x^2y^{10}}{20x^7y^7}$

17. Simplify each of the following by cancelling.

a. $\dfrac{12m^2n^4}{30m^5n^8}$ b. $\dfrac{16m^5n^{10}}{8m^5n^{12}}$ c. $\dfrac{20x^4y^5}{10x^5y^4}$ d. $\dfrac{a^2b^4c^6}{a^6b^4c^2}$

18. Calculate the value of each of the following expressions if $a = 3$.

a. $2a$

b. a^2

c. $2a^2$

d. $a^2 + 2$

e. $a^2 + 2a$

f. $a^2 - a$

Communicating, reasoning and problem solving

19. Explain why x^2 and $2x$ are not the same number. Use an example to illustrate your reasoning.

20. **MC** Answer the following.

a. $12a^8b^2c^4(de)^0f$, when simplified, is equal to:

   A. $12a^8b^2c^4$ **B.** $12a^8b^2c^4f$ **C.** $12a^8b^2f$ **D.** $12a^8b^2$

b. $\left(\dfrac{6}{11}a^2b^7\right)^0 \times - \left(3a^2b^{11}\right)^0 + 7a^0b$, when simplified, is equal to:

   A. $7b$ **B.** $1 + 7b$ **C.** $-1 + 7ab$ **D.** $-1 + 7b$

c. You are told that there is an error in the statement $3p^7q^3r^5s^6 = 3p^7s^6$.
Select the correct expression to replace the left-hand side.

   A. $\left(3p^7q^3r^5s^6\right)^0$ **B.** $\left(3p^7\right)^0q^3r^5s^6$ **C.** $3p^7\left(q^3r^5s^6\right)^0$ **D.** $3p^7\left(q^3r^5\right)^0s^6$

21. **MC** a. You are told that there is an error in the statement $\dfrac{8f^6g^7h^3}{6f^4g^2h} = \dfrac{8f^2}{g^2}$.

Select the correct expression to replace the left hand side.

   A. $\dfrac{8f^6\left(g^7h^3\right)^0}{(6)^0f^4g^2(h)^0}$ **B.** $\dfrac{8\left(f^6g^7h^3\right)^0}{\left(6f^4g^2h\right)^0}$ **C.** $\dfrac{8\left(f^6g^7\right)^0h^3}{\left(6f^4\right)^0g^2h}$ **D.** $\dfrac{8f^6g^7h^3}{\left(6f^4g^2h\right)^0}$

b. Select the option that is equal to $\dfrac{6k^7m^2n^8}{4k^7\left(m^6n\right)^0}$.

   A. $\dfrac{6}{4}$ **B.** $\dfrac{3}{2}$ **C.** $\dfrac{3n^8}{2}$ **D.** $\dfrac{3m^2n^8}{2}$

22. Explain why $5x^5 \times 3x^3$ is not equal to $15x^{15}$.

23. A multiple-choice question requires a student to multiply 5^6 by 5^3. The student is having trouble deciding which of these four answers is correct: 5^{18}, 5^9, 25^{18} or 25^9.

    a. State the correct answer.
    b. Explain your answer by using another example to explain the First Index Law.

24. A multiple-choice question requires a student to divide 5^{24} by 5^8. The student is having trouble deciding which of these four answers is correct: 5^{16}, 5^3, 1^{16} or 1^3.

    a. Determine the correct answer.
    b. Explain your answer by using another example to explain the Second Index Law.

25. a. Calculate the value of $\dfrac{5^7}{5^7}$.
    b. Determine the value of any number divided by itself.
    c. According to the Second Index Law, which deals with exponents and division, $\dfrac{5^7}{5^7}$ should equal 5 raised to what index?
    d. Use an example to explain the Third Index Law.

26. a. For $x^2 x^\triangle = x^{16}$ to be true, determine what number must replace the triangle.
    b. For $x^\triangle x^\bigcirc x^\diamond = x^{12}$ to be true, there are 55 ways of assigning positive whole numbers to the triangle, circle, and diamond. Give at least four of these.

27. a. Determine a pattern in the units digit for powers of 3.
    b. The units digit of 3^6 is 9. Determine the units digit of 3^{2001}.

28. a. Determine a pattern in the units digit for powers of 4.
    b. Determine the units digit of 4^{105}.

29. a. Investigate the patterns in the units digit for powers of 2 to 9.
    b. Predict the units digit for the following.
        i. 2^{35} ii. 3^{16} iii. 8^{51}

30. Write $4^{n+1} + 4^{n+1}$ as a single power of 2.

LESSON
2.3 Simplify algebraic products and quotients using index laws

---

LEARNING INTENTION

At the end of this lesson you should be able to:
- apply the Fourth, Fifth and Sixth index laws to raise a power to another power
- apply the first six index laws to simplify expressions involving indices.

▶ 2.3.1 Raising a power to another power

eles-4387

- $(7^2)^3 = 7^2 \times 7^2 \times 7^2$

    $= 7^{2+2+2}$ (using the First Index Law)

    $= 7^{2 \times 3}$

    $= 7^6$

- When a power is raised to another power the indices are multiplied. This is the **Fourth Index Law**.
- The Fourth Index Law can then be applied to products and quotients of mixed bases to develop the Fifth and Sixth index laws.

The Fourth Index Law

$$(a^m)^n = a^{m \times n}$$

For example, $(x^2)^3 = x^{2 \times 3} = x^6$.

The Fifth Index Law

$$(a \times b)^m = a^m \times b^m$$

For example, $(2x)^3 = 2^3 \times x^3 = 8x^3$.

The Sixth Index Law

$$\left(\frac{a}{b}\right)^m = \frac{a^m}{b^m}$$

For example, $\left(\frac{4x}{3}\right)^2 = \frac{4^2 x^2}{3^2} = \frac{16x^2}{9}$.

WORKED EXAMPLE 9 Simplifying using the Fourth Index Law

Simplify $\left(7^4\right)^8$**, leaving your answer in index form.**

THINK	WRITE
Simplify by applying the Fourth Index Law (multiply the indices).	$\left(7^4\right)^8$ $= 7^{4 \times 8}$ $= 7^{32}$

WORKED EXAMPLE 10 Simplifying using the Fifth Index Law

Simplify the following.

a. $\left(3a^2 b^5\right)^3$

b. $\left(2b^5\right)^2 \times (5b)^3$

THINK	WRITE
a. 1. Write the expression, including all indices.	**a.** $\left(3^1 a^2 b^5\right)^3$
2. Simplify by applying the Fifth Index Law (multiply the indices) for each coefficient and variable inside the brackets.	$= 3^{1 \times 3} a^{2 \times 3} b^{5 \times 3}$ $= 3^3 a^6 b^{15}$
3. Write the answer.	$= 27 a^6 b^{15}$
b. 1. Write the expression, including all indices.	**b.** $\left(2^1 b^5\right)^2 \times \left(5^1 b^1\right)^3$
2. Simplify by applying the Fifth Index Law.	$= 2^{1 \times 2} b^{5 \times 2} \times 5^{1 \times 3} b^{1 \times 3}$ $= 2^2 \times b^{10} \times 5^3 \times b^3$
3. Calculate the number values and bring them to the front. Then use the First Index Law to simplify the powers of b.	$= 4 \times 125 \times b^{10+3}$
4. Write the answer.	$= 500 b^{13}$

WORKED EXAMPLE 11 Simplifying using the Sixth Index Law

Simplify $\left(\dfrac{2a^5}{d^2}\right)^3$**.**

THINK	WRITE
1. Write the expression, including all indices.	$\left(\dfrac{2^1 a^5}{d^2}\right)^3$
2. Simplify by applying the Sixth Index Law for each coefficient and variable inside the brackets.	$= \dfrac{2^{1 \times 3} a^{5 \times 3}}{d^{2 \times 3}}$ $= \dfrac{2^3 a^{15}}{d^6}$
3. Simplify and then write the answer.	$= \dfrac{8 a^{15}}{d^6}$

Resources

Interactivity Fourth Index Law (int-3716)
Fifth and sixth index laws (int-6063)

Exercise 2.3 Simplify algebraic products and quotients using index laws

learn on

2.3 Quick quiz	2.3 Exercise

Individual pathways

■ PRACTISE	■ CONSOLIDATE	■ MASTER
1, 3, 7, 9, 14, 15, 20	2, 5, 8, 10, 12, 16, 17, 21	4, 6, 11, 13, 18, 19, 22, 23

Fluency

1. **WE9** Simplify each of the following.

    a. $(e^2)^3$
    b. $(f^8)^{10}$
    c. $(p^{25})^4$
    d. $(r^{12})^{12}$
    e. $(2a^2)^3$

2. Simplify each of the following.

    a. $(a^2b^3)^4$
    b. $(pq^3)^5$
    c. $(g^3h^2)^{10}$
    d. $(3w^9q^2)^4$
    e. $(7e^5r^2q^4)^2$

3. **WE10** Simplify each of the following.

    a. $(p^4)^2 \times (q^3)^2$
    b. $(r^5)^3 \times (w^3)^3$
    c. $(b^5)^2 \times (n^3)^6$
    d. $(j^6)^3 \times (g^4)^3$
    e. $(2a)^3 \times (b^2)^2$

4. Simplify each of the following.

    a. $(q^2)^2 \times (r^4)^5$
    b. $(h^3)^8 \times (j^2)^8$
    c. $(f^4)^4 \times (a^7)^3$
    d. $(t^5)^2 \times (u^4)^2$
    e. $(i^3)^5 \times (j^2)^6$

5. **WE11** Simplify each of the following.

    a. $\left(\dfrac{3b^4}{d^3}\right)^2$
    b. $\left(\dfrac{5h^{10}}{2j^2}\right)^2$
    c. $\left(\dfrac{2k^5}{3t^8}\right)^3$
    d. $\left(\dfrac{7p^9}{8q^{22}}\right)^2$

6. Simplify each of the following.

    a. $\left(\dfrac{5y^7}{3z^{13}}\right)^3$
    b. $\left(\dfrac{4a^3}{7c^5}\right)^4$
    c. $\left(\dfrac{-4k^2}{7m^6}\right)^3$
    d. $\left(\dfrac{-2g^7}{3h^{11}}\right)^4$

Understanding

7. Simplify each of the following.

    a. $(2^3)^4 \times (2^4)^2$
    b. $(t^7)^3 \times (t^3)^4$
    c. $(a^4)^0 \times (a^3)^7$
    d. $(e^7)^8 \times (e^5)^2$

8. Simplify each of the following.

    a. $\left(g^7\right)^3 \times \left(g^9\right)^2$ b. $\left(3a^2\right) \times \left(2a^6\right)^2$ c. $\left(2d^7\right)^3 \times \left(3d^2\right)^3$ d. $\left(10r^2\right)^4 \times \left(2r^3\right)^2$

9. **MC** $\left(p^7\right)^2 \div p^2$ is equal to:

    A. p^7 B. p^{12} C. p^{16} D. $p^{4.5}$

10. **MC** $\dfrac{\left(w^5\right)^2 \times \left(p^7\right)^3}{\left(w^2\right)^2 \times \left(p^3\right)^5}$ is equal to:

    A. $w^2 p^6$ B. $(wp)^6$ C. $w^{14} p^{36}$ D. $w^2 p^2$

11. **MC** $\left(r^6\right)^3 \div \left(r^4\right)^2$ is equal to:

    A. r^3 B. r^4 C. r^8 D. r^{10}

12. Simplify each of the following.

    a. $\left(a^3\right)^4 \div \left(a^2\right)^3$ b. $\left(m^8\right)^2 \div \left(m^3\right)^4$ c. $\left(n^5\right)^3 \div \left(n^6\right)^2$

    d. $\left(b^4\right)^5 \div \left(b^6\right)^2$ e. $\left(f^7\right)^3 \div \left(f^2\right)^2$ f. $\left(g^8\right)^2 \div \left(g^5\right)^2$

13. Simplify each of the following.

    a. $\left(p^9\right)^3 \div \left(p^6\right)^3$ b. $\left(y^4\right)^4 \div \left(y^7\right)^2$

    c. $\dfrac{\left(c^6\right)^5}{\left(c^5\right)^2}$ d. $\dfrac{\left(f^5\right)^3}{\left(f^2\right)^4}$

    e. $\dfrac{\left(k^3\right)^{10}}{\left(k^2\right)^8}$ f. $\dfrac{\left(p^{12}\right)^3}{\left(p^{10}\right)^2}$

Communicating, reasoning and problem solving

14. a. Replace the triangle with the correct index for the equation $4^7 \times 4^7 \times 4^7 \times 4^7 \times 4^7 = \left(4^7\right)^{\Delta}$.

    b. The expression $\left(p^5\right)^6$ means to write p^5 as a factor how many times?

    c. If you rewrote the expression from part b without any exponents, in the format $p \times p \times p...$, determine how many factors you would need.

    d. Explain the Fourth Index Law.

15. a. Simplify each of the following.

    i. $(-1)^{10}$ ii. $(-1)^7$ iii. $(-1)^{15}$ iv. $(-1)^6$

    b. Write a general rule for the result obtained when -1 is raised to a positive power. Explain your answer.

16. Jo and Danni are having an algebra argument. Jo is sure that $-x^2$ is equivalent to $(-x^2)$, but Danni thinks otherwise. Explain who is correct and justify your answer.

17. A multiple-choice question requires a student to calculate $\left(5^4\right)^3$. The student is having trouble deciding which of these three answers is correct: 5^{64}, 5^{12} or 5^7.

    a. Determine the correct answer.

    b. Explain your answer by using another example to illustrate the Fourth Index Law.

18. a. Without using your calculator, simplify each side of the following equations to the same base and then solve each of them.

    i. $8^x = 32$

    ii. $27^x = 243$

    iii. $1000^x = 100\,000$

    b. Explain why all 3 equations have the same solution.

19. Consider the expression 4^{3^2}. Identify which is the correct answer, 4096 or 262144. Justify your choice.

20. The diameter of a typical atom is so small that it would take about 10^8 atoms arranged in a line to reach a length of just 1 centimetre. Estimate how many atoms are contained in a cubic centimetre. Write this number as a power of 10.

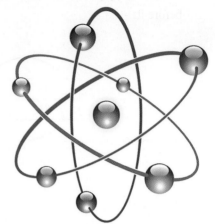

21. Writing a base as a power itself can be used as a way to simplify an expression. Copy and complete the following calculations.

    a. $16^{\frac{1}{2}} = (4^2)^{\frac{1}{2}} = \underline{\hspace{1cm}}$

    b. $343^{\frac{2}{3}} = (7^3)^{\frac{2}{3}} = \underline{\hspace{1cm}}$

22. Simplify the following using index laws.

    a. $8^{\frac{1}{3}}$ b. $27^{\frac{4}{3}}$ c. $125^{-\frac{2}{3}}$ d. $512^{\frac{2}{9}}$

    e. $16^{-\frac{1}{2}}$ f. $4^{-\frac{1}{2}}$ g. $32^{-\frac{1}{5}}$ h. $49^{-\frac{1}{2}}$

23. a. Use the index laws to simplify the following.

    i. $(3^2)^{\frac{1}{2}}$ ii. $(4^2)^{\frac{1}{2}}$ iii. $(8^2)^{\frac{1}{2}}$ iv. $(11^2)^{\frac{1}{2}}$

    b. Use your answers from part a to calculate the value of the following.

    i. $9^{\frac{1}{2}}$ ii. $16^{\frac{1}{2}}$ iii. $64^{\frac{1}{2}}$ iv. $121^{\frac{1}{2}}$

    c. Use your answers to parts a and b to write a sentence describing what happens when you raise a number to a power of one-half.

LESSON
2.4 Applying index laws to numerical and algebraic expressions involving negative-integer indices

LEARNING INTENTION

At the end of this lesson you should be able to:
- apply the Seventh Index Law to evaluate expressions using negative indices
- apply the first seven index laws to simplify expressions involving indices
- apply the index laws to simplify algebraic expressions involving indices (Path).

eles-4388

▶ 2.4.1 Negative indices

- Negative indices occur when the power (or exponent) is a negative number, for example 3^{-2}.
- To explain the meaning or value of negative indices it is useful to consider patterns of numbers written in index form.

For example, in the sequence $3^4 = 81$, $3^3 = 27$, $3^2 = 9$, $3^1 = 3$, $3^0 = 1$ each number is $\dfrac{1}{3}$ of the number before it.

Powers decrease by 1

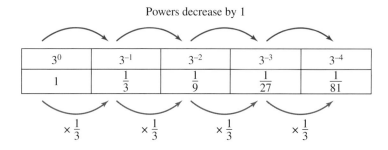

3^4	3^3	3^2	3^1	3^0
81	27	9	3	1

$\times \dfrac{1}{3}$ $\times \dfrac{1}{3}$ $\times \dfrac{1}{3}$ $\times \dfrac{1}{3}$

- It is logical, then, that the next numbers in this sequence are:

$$3^{-1} = \frac{1}{3}, \ 3^{-2} = \frac{1}{9}, \ 3^{-3} = \frac{1}{27}, \ 3^{-4} = \frac{1}{81}$$

Powers decrease by 1

3^0	3^{-1}	3^{-2}	3^{-3}	3^{-4}
1	$\dfrac{1}{3}$	$\dfrac{1}{9}$	$\dfrac{1}{27}$	$\dfrac{1}{81}$

$\times \dfrac{1}{3}$ $\times \dfrac{1}{3}$ $\times \dfrac{1}{3}$ $\times \dfrac{1}{3}$

- Notice that for each negative power, the exponent is applied to the base on the denominator instead of the numerator. For example, $3^{-2} = \dfrac{1}{3^2}$.

The Seventh Index Law

$$a^{-n} = \frac{1}{a^n}, \text{ where } a \neq 0$$

For example, $2^{-3} = \dfrac{1}{2^3} = \dfrac{1}{8}$.

WORKED EXAMPLE 12 Evaluating numerical expressions with negative-integer indices

Evaluate the following.

a. 5^{-2} b. 7^{-1} c. $\left(\dfrac{3}{5}\right)^{-1}$

THINK	WRITE
a. 1. Apply the Seventh Index Law.	a. $5^{-2} = \dfrac{1}{5^2}$
2. Simplify and write the answer.	$= \dfrac{1}{25}$

b. Apply the Seventh Index Law.

b. $7^{-1} = \dfrac{1}{7^1}$

$= \dfrac{1}{7}$

c. 1. Apply the Sixth Index Law, $\left(\dfrac{a}{b}\right)^m = \dfrac{a^m}{b^m}$.

c. $\left(\dfrac{3^1}{5^1}\right) = \dfrac{3^{-1}}{5^{-1}}$

2. Apply the Seventh Index Law, $a^{-n} = \dfrac{1}{a^n}$, to the numerator and denominator.

$= \dfrac{1}{3} \div \dfrac{1}{5}$

3. Simplify and write the answer.

$= \dfrac{1}{3} \times \dfrac{5}{1}$

$= \dfrac{5}{3}$

WORKED EXAMPLE 13 Representing numbers in index form

Write each of the following with negative integers.

a. $\dfrac{1}{125}$

b. $\dfrac{1}{81}$

c. $\dfrac{1}{64}$

THINK

WRITE

a. 1. Change the denominator into index form. 125 is the same as 5^3.

a. $\dfrac{1}{125} = \dfrac{1}{5^3}$

2. Apply the Seventh Index Law, $a^{-n} = \dfrac{1}{a^n}$.

$= 5^{-3}$

b. 1. Change the denominator into index form. 81 is the same as 9^2.

b. $\dfrac{1}{81} = \dfrac{1}{9^2}$

2. Apply the Seventh Index Law, $a^{-n} = \dfrac{1}{a^n}$.

$= 9^{-2}$

c. 1. Change the denominator into index form. 64 is the same as 2^6 4^3 or 8^2.

c. $\dfrac{1}{64} = \dfrac{1}{2^6}$

$= 2^{-6}$

2. Apply the Seventh Index Law, $a^{-n} = \dfrac{1}{a^n}$.

Alternatively,

$\dfrac{1}{64} = 4^{-3}$

$= 8^{-2}$

WORKED EXAMPLE 14 Evaluating algebraic expressions with negative-integer indices

Write the following with positive indices.

a. x^{-3}

b. $5x^{-6}$

c. $\dfrac{x^{-3}}{y^{-2}}$

THINK	WRITE
a. Apply the Seventh Index Law, $a^{-n} = \dfrac{1}{a^n}$.	a. $x^{-3} = \dfrac{1}{x^3}$
b. 1. Write in expanded form. Then apply the Seventh Index Law, $a^{-n} = \dfrac{1}{a^n}$.	b. $5x^{-6} = 5 \times x^{-6}$ $= 5 \times \dfrac{1}{x^6}$
2. Simplify and write the answer.	$= \dfrac{5}{x^6}$
c. 1. Write the fraction using division.	c. $\dfrac{x^{-3}}{y^{-2}}$ $= x^{-3} \div y^{-2}$
2. Apply the Seventh Index Law, $a^{-n} = \dfrac{1}{a^n}$.	$= \dfrac{1}{x^3} \div \dfrac{1}{y^2}$
3. Simplify and write the answer.	$= \dfrac{1}{x^3} \times \dfrac{y^2}{1}$ $= \dfrac{y^2}{x^3}$

WORKED EXAMPLE 15 Simplifying products of powers with positive and negative indices

Simplify the following expressions, writing your answers with positive indices.

a. $x^3 \times x^{-8}$

b. $3x^{-2}y^{-3} \times 5xy^{-4}$

THINK	WRITE
a. 1. Apply the First Index Law, $a^n \times a^m = a^{m+n}$.	a. $x^3 \times x^{-8}$ $= x^{3+-8}$ $= x^{-5}$
2. Write the answer with a positive index.	$= \dfrac{1}{x^5}$

b. 1. Write in expanded form and apply the First Index Law.

b. $3x^{-2}y^{-3} \times 5xy^{-4}$

$= 3 \times 5 \times x^{-2} \times x^1 \times y^{-3} \times y^{-4}$

2. Apply the Seventh Index Law.

$= 15x^{-1}y^{-7}$

$= \dfrac{15}{1} \times \dfrac{1}{x} \times \dfrac{1}{y^7}$

3. Simplify and write the answer.

$= \dfrac{15}{xy^7}$

PATH

WORKED EXAMPLE 16 Simplifying quotients of powers with positive and negative indices

Simplify the following expressions, writing your answers with positive indices.

a. $\dfrac{t^2}{t^{-5}}$

b. $\dfrac{15m^{-5}}{10m^{-2}}$

THINK

a. Apply the Second Index Law, $\dfrac{a^n}{a^m} = a^{n-m}$.

b. 1. Apply the Second Index Law and simplify.

2. Write the answer with positive indices.

WRITE

a. $\dfrac{t^2}{t^{-5}}$

$= t^{2-(-5)}$

$= t^{2+5}$

$= t^7$

b. $\dfrac{15m^{-5}}{10m^{-2}} = \dfrac{15}{10} \times \dfrac{m^{-5}}{m^{-2}}$

$= \dfrac{3}{2} \times m^{-5-(-2)}$

$= \dfrac{3}{2} \times m^{-3}$

$= \dfrac{3}{2} \times \dfrac{1}{m^3}$

$= \dfrac{3}{2m^3}$

DISCUSSION

What kind of strategies can you use to remember the index laws?

---

on Resources

▶ **Video eLesson** Negative indices (eles-1910)

✦ **Interactivity** Negative indices (int-6064)

Exercise 2.4 Applying index laws to numerical and algebraic expressions involving negative-integer indices

learn on

2.4 Quick quiz on	**2.4 Exercise**

Individual pathways

■ PRACTISE	■ CONSOLIDATE	■ MASTER
1, 2, 5, 7, 10, 13, 16, 19	3, 6, 8, 11, 14, 17, 20, 23	4, 9, 12, 15, 18, 21, 22, 24

Fluency

1. Copy and complete the following patterns.

   a. $3^5 = 243$

   $3^4 = 81$

   $3^3 = 27$

   $3^2 =$

   $3^1 =$

   $3^0 =$

   $3^{-1} = \dfrac{1}{3}$

   $3^{-2} = \dfrac{1}{9}$

   $3^{-3} =$

   $3^{-4} =$

   $3^{-5} =$

   b. $5^4 = 625$

   $5^3 =$

   $5^2 =$

   $5^1 =$

   $5^0 =$

   $5^{-1} =$

   $5^{-2} =$

   $5^{-3} =$

   $5^{-4} =$

   c. $10^4 = 10\ 000$

   $10^3 =$

   $10^2 =$

   $10^1 =$

   $10^0 =$

   $10^{-1} =$

   $10^{-2} =$

   $10^{-3} =$

   $10^{-4} =$

2. **WE12a, b** Evaluate each of the following expressions.

   a. 2^{-5} b. 3^{-3} c. 4^{-1} d. 10^{-2}

3. **WE12b, c** Evaluate each of the following expressions.

   a. 5^{-3} b. $\left(\dfrac{1}{7}\right)^{-1}$ c. $\left(\dfrac{3}{4}\right)^{-1}$ d. $\left(\dfrac{3}{4}\right)^{-2}$

4. Evaluate each of the following expressions.

   a. $\left(\dfrac{1}{3}\right)^{-3}$ b. $\left(\dfrac{3}{2}\right)^{-1}$ c. $\left(2\dfrac{1}{4}\right)^{-2}$ d. $\left(\dfrac{2}{7}\right)^{-2}$

5. **WE13** Write each of the following with negative integers.

   a. $\dfrac{1}{4}$ b. $\dfrac{1}{27}$ c. $\dfrac{1}{16}$

6. Write each of the following with negative integers.

   a. $\dfrac{1}{25}$ b. $\dfrac{1}{1000}$ c. $\dfrac{1}{343}$

7. **WE14** **PATH** Write each expression with positive indices.

   a. x^{-4} b. y^{-5} c. z^{-1} d. a^2b^{-3}

8. **PATH** Write each expression with positive indices.

   a. $\left(m^2n^3\right)^{-1}$ b. $\dfrac{x^2}{y^{-2}}$ c. $\dfrac{5}{x^{-3}}$ d. $\dfrac{x^{-2}}{w^{-5}}$ e. $\dfrac{1}{x^{-2}y^{-2}}$

9. PATH Write each expression with positive indices.

a. $a^2b^{-3}cd^{-4}$　　b. $\dfrac{a^2b^{-2}}{c^2d^{-3}}$　　c. $10x^{-2}y$　　d. $3^{-1}x$　　e. $\dfrac{m^{-3}}{x^2}$

Understanding

10. WE15 PATH Simplify the following expressions, writing your answers with positive indices.

a. $a^3 \times a^{-8}$　　b. $m^7 \times m^{-2}$　　c. $m^{-3} \times m^{-4}$
d. $2x^{-2} \times 7x$　　e. $2b^{-2} \times 3b^2$　　f. $ab^{-5} \times 5b^2$

11. PATH Simplify the following expressions, writing your answers with positive indices.

a. $x^5 \times x^{-8}$　　b. $3x^2y^{-4} \times 2x^{-7}y$　　c. $10x^5 \times 5x^{-2}$
d. $x^5 \times x^{-5}$　　e. $10a^2 \times 5a^{-7}$　　f. $10a^{10} \times a^{-6}$

12. PATH Simplify the following expressions, writing your answers with positive indices.

a. $16w^2 \times -2w^{-5}$　　b. $4m^{-2} \times 4m^{-2}$　　c. $\left(3m^2n^{-4}\right)^3$
d. $\left(a^2b^5\right)^{-3}$　　e. $\left(a^{-1}b^{-3}\right)^{-2}$　　f. $\left(5a^{-1}\right)^2$

13. WE16 PATH Simplify the following expressions, writing your answers with positive indices.

a. $\dfrac{x^3}{x^8}$　　b. $\dfrac{x^{-3}}{x^8}$　　c. $\dfrac{x^3}{x^{-8}}$　　d. $\dfrac{x^{-3}}{x^{-8}}$　　e. $\dfrac{6a^2c^5}{a^4c}$

14. PATH Simplify the following expressions, writing your answers with positive indices.

a. $10a^2 \div 5a^8$　　b. $5m^7 \div m^8$　　c. $\dfrac{a^5b^6}{a^5b^7}$　　d. $\dfrac{a^2b^8}{a^5b^{10}}$　　e. $\dfrac{a^{-3}bc^3}{abc}$

15. PATH Simplify the following expressions, writing your answers with positive indices.

a. $\dfrac{4^{-2}ab}{a^2b}$　　b. $\dfrac{m^{-3} \times m^{-5}}{m^{-5}}$　　c. $\dfrac{2t^2 \times 3t^{-5}}{4t^6}$　　d. $\dfrac{t^3 \times t^{-5}}{t^{-2} \times t^{-3}}$　　e. $\dfrac{\left(m^2n^{-3}\right)^{-1}}{\left(m^{-2}n^3\right)^2}$

16. Write the following numbers as powers of 2.

a. 1　　b. 8　　c. 32　　d. 64　　e. $\dfrac{1}{8}$　　f. $\dfrac{1}{32}$

17. Write the following numbers as powers of 4.

a. 1　　b. 4　　c. 64　　d. $\dfrac{1}{4}$　　e. $\dfrac{1}{16}$　　f. $\dfrac{1}{64}$

18. Write the following numbers as powers of 10.

a. 1　　b. 10　　c. 10 000　　d. 0.1　　e. 0.01　　f. 0.000 01

Communicating, reasoning and problem solving

19. a. The result of dividing 3^7 by 3^3 is 3^4. Determine the result of dividing 3^3 by 3^7.
b. Explain what it means to have a negative index.
c. Explain how you write a negative index as a positive index.

20. Indices are encountered in science, where they help to deal with very small and large numbers. The diameter of a proton is 0.000 000 000 000 3 cm.
Explain why it is logical to express this number in scientific notation as 3×10^{-13}.

21. Evaluate the following.

a. Half of 2^{20}　　　　　　　　　　b. One-third of 3^{21}

22. Simplify the following expressions.

a. $\left(2^{-1} + 3^{-1}\right)^{-1}$

b. $\dfrac{3^{400}}{6^{200}}$

23. **PATH** a. When asked to write an expression with positive indices that is equivalent to $x^3 + x^{-3}$, a student gave the answer x^0. Is this answer correct? Explain why or why not.

b. When asked to write an expression with positive indices that is equivalent to $\left(x^{-1} + y^{-1}\right)^{-2}$, a student gave the answer $x^2 + y^2$. Is this answer correct? Explain why or why not.

24. **PATH** a. When asked to write an expression with positive indices that is equivalent to $x^8 - x^{-5}$, a student gave the answer x^3. Is this answer correct? Explain why or why not.

b. Another student said that $\dfrac{x^2}{x^8 - x^5}$ is equivalent to $\dfrac{1}{x^6} - \dfrac{1}{x^3}$. Is this answer correct? Explain why or why not.

LESSON
2.5 Surds (Path)

LEARNING INTENTION

At the end of this lesson you should be able to:
- identify when a square root forms a surd
- place surds in their approximate position on a number line
- simplify expressions containing surds.

▶ 2.5.1 Identifying surds

eles-4393

- In many cases the square root or cube root of a number results in a rational number. For example, $\sqrt{25} = \pm 5$ and $\sqrt[3]{1.728} = 1.2$.
- When the square root of a number is an irrational number, it is called a **surd**.
  For example, $\sqrt{10} \approx 3.162\,277\,660\,17\ldots$ Since $\sqrt{10}$ cannot be written as a fraction, a **recurring decimal** or a **terminating decimal**, it is irrational and therefore it is a surd.
- The value of a surd can be approximated using a number line.
  For example, we know that $\sqrt{21}$ lies between 4 and 5, because it lies between $\sqrt{16}$ (which equals 4) and $\sqrt{25}$ (which equals 5).
- We can show its approximate position on the number line like this:

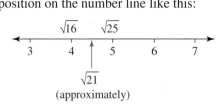

(approximately)

WORKED EXAMPLE 17 Placing surds on a number line

Place $\sqrt{34}$ on a number line.

THINK

1. The next lowest square number that is smaller than 34 is 25. The next highest square number that is larger than 34 is 36.

2. Draw a number line to show the approximate position of $\sqrt{34}$.

WRITE

$\sqrt{34}$ lies between $\sqrt{25}$ and $\sqrt{36}$.

WORKED EXAMPLE 18 Identifying surds

Identify which of the following are surds.

a. $\sqrt{0}$ b. $\sqrt{20}$ c. $-\sqrt{9}$ d. $\sqrt[3]{6}$

THINK

a. $\sqrt{0} = 0$. This is a rational number and therefore not a surd.

b. $\sqrt{20} \approx 4.472\,135\,955\ldots$ This is an irrational number and therefore a surd.

c. $-\sqrt{9} = -3$. This is a rational number and therefore not a surd.

d. $\sqrt[3]{6} \approx 1.817\,120\,592\,83\ldots$ This is an irrational number and therefore a surd.

WRITE

a. $\sqrt{0} = 0$. This is not a surd.

b. $\sqrt{20}$ is a surd.

c. $-\sqrt{9} = -3$. This is not a surd.

d. $\sqrt[3]{6}$ is a surd.

2.5.2 Multiplying and dividing surds

eles-4394

- Multiplication of surds is done by multiplying the numerical parts of each surd under a single radical sign (also known as the radicand).
- Division of surds is done by placing the quotient of the numerical parts of each surd under a single radical sign.

Multiplying surds

$$\sqrt{a} \times \sqrt{b} = \sqrt{ab}$$

where $a \geq 0, b \geq 0$.

Dividing surds

$$\frac{\sqrt{a}}{\sqrt{b}} = \sqrt{\frac{a}{b}}$$

where $a \geq 0, b > 0$.

WORKED EXAMPLE 19 Multiplying surds

Evaluate the following, leaving your answer in surd form.

a. $\sqrt{7} \times \sqrt{2}$
b. $5 \times \sqrt{3}$
c. $\sqrt{5} \times \sqrt{5}$
d. $-2\sqrt{3} \times 4\sqrt{5}$

THINK

a. Apply the rule $\sqrt{a} \times \sqrt{b} = \sqrt{ab}$.

b. Only $\sqrt{3}$ is a surd. It is multiplied by 5, which is not a surd.

c. Apply the rule $\sqrt{a} \times \sqrt{b} = \sqrt{ab}$.

d. Multiply the coefficients. Then multiply the surds.

WRITE

a. $\sqrt{7} \times \sqrt{2} = \sqrt{14}$

b. $5 \times \sqrt{3} = 5\sqrt{3}$

c. $\sqrt{5 \times 5} = \sqrt{25}$
$= 5$

d. $-2\sqrt{3} \times 4\sqrt{5} = -2 \times 4 \times \sqrt{3} \times \sqrt{5}.$
$= -8 \times \sqrt{15}$
$= -8\sqrt{15}$

WORKED EXAMPLE 20 Dividing surds

Evaluate the following, leaving your answer in surd form.

a. $\dfrac{\sqrt{15}}{\sqrt{5}}$
b. $\sqrt{\dfrac{10}{5}}$
c. $\dfrac{\sqrt{20}}{\sqrt{5}}$
d. $\dfrac{-6\sqrt{8}}{4\sqrt{4}}$
e. $\dfrac{5}{\sqrt{5}}$

THINK

a. Apply the rule $\sqrt{a} \div \sqrt{b} = \sqrt{\dfrac{a}{b}}$.

b. Simplify the fraction.

c. Apply the rule $\sqrt{a} \div \sqrt{b} = \sqrt{\dfrac{a}{b}}$.

d. Simplify the whole numbers. Then apply the rule
$\sqrt{a} \div \sqrt{b} = \sqrt{\dfrac{a}{b}}$.

e. Rewrite the numerator as the product of two surds and then simplify.

WRITE

a. $\dfrac{\sqrt{15}}{\sqrt{5}} = \sqrt{\dfrac{15}{5}}$
$= \sqrt{3}$

b. $\sqrt{\dfrac{10}{5}} = \sqrt{2}$

c. $\dfrac{\sqrt{20}}{\sqrt{5}} = \sqrt{\dfrac{20}{5}}$
$= \sqrt{4}$
$= 2$

d. $\dfrac{-6\sqrt{8}}{4\sqrt{4}} = \dfrac{-3\sqrt{2}}{2}$

e. $\dfrac{5}{\sqrt{5}} = \dfrac{\sqrt{5} \times \sqrt{5}}{\sqrt{5}}$
$= \sqrt{5}$

⊙ 2.5.3 Simplifying surds

- Just as a rational number can be written many different ways (e.g. $\dfrac{1}{2} = \dfrac{5}{10} = \dfrac{7}{14}$), so can a surd. It is expected that surds should normally be written in their simplest form.
- A surd is in its simplest form when the number inside the radical sign has the smallest possible value.
- Note that $\sqrt{24}$ can be factorised several ways. For example:

$$\sqrt{24} = \sqrt{2} \times \sqrt{12}$$

$$\sqrt{24} = \sqrt{3} \times \sqrt{8}$$

$$\sqrt{24} = \sqrt{4} \times \sqrt{6}$$

In the last example, $\sqrt{4} = 2$, which means:

$$\sqrt{24} = 2 \times \sqrt{6}$$
$$= 2\sqrt{6}$$

- $2\sqrt{6}$ is $\sqrt{24}$ written in its simplest form.
- To simplify a surd you must determine a factor that is also a perfect square, for example 4, 9, 16, 25, 36 or 49.
- A surd like $\sqrt{22}$, for example, cannot be simplified because 22 has no perfect square factors.
- Surds can be simplified in more than one step.

$$\sqrt{72} = \sqrt{4} \times \sqrt{18}$$
$$= 2\sqrt{18}$$
$$= 2 \times \sqrt{9} \times \sqrt{2}$$
$$= 2 \times 3\sqrt{2}$$
$$= 6\sqrt{2}$$

WORKED EXAMPLE 21 Simplifying surds

Simplify the following surds.

a. $\sqrt{18}$
b. $6\sqrt{20}$

THINK	WRITE
a. 1. Rewrite 18 as the product of two numbers, one of which is a perfect square (9).	a. $\sqrt{18} = \sqrt{9} \times \sqrt{2}$
2. Simplify.	$= 3 \times \sqrt{2}$ $= 3\sqrt{2}$
b. 1. Rewrite 20 as the product of two numbers, one of which is square (4).	b. $6\sqrt{20} = 6 \times \sqrt{4} \times \sqrt{5}$
2. Simplify.	$= 6 \times 2 \times \sqrt{5}$ $= 12\sqrt{5}$

⊳ 2.5.4 Entire surds

- The surd $\sqrt{45}$, when simplified, is written as $3\sqrt{5}$.
  The surd $\sqrt{45}$ is called an entire surd because it is written entirely inside the radical sign. The surd $3\sqrt{5}$, however, is not an entire surd.
- Writing a surd as an entire surd reverses the process of simplification.

WORKED EXAMPLE 22 Writing entire surds

Write $3\sqrt{7}$ as an entire surd.

THINK	WRITE
1. In order to place the 3 inside the radical sign, it has to be written as $\sqrt{9}$.	$3\sqrt{7} = \sqrt{9} \times \sqrt{7}$
2. Apply the rule $\sqrt{a} \times \sqrt{b} = \sqrt{ab}$.	$= \sqrt{63}$

WORKED EXAMPLE 23 Comparing surds

Determine which number is larger, $3\sqrt{5}$ or $5\sqrt{3}$.

THINK	WRITE
1. Write $3\sqrt{5}$ as an entire surd.	$3\sqrt{5} = \sqrt{9} \times \sqrt{5}$ $= \sqrt{45}$
2. Write $5\sqrt{3}$ as an entire surd.	$5\sqrt{3} = \sqrt{25} \times \sqrt{3}$ $= \sqrt{75}$
3. Compare the values of each surd.	$\sqrt{75} > \sqrt{45}$
4. Write your answer.	$5\sqrt{3}$ is the larger number.

⊳ 2.5.5 Addition and subtraction of surds

- Surds can be added or subtracted if they have like terms.
- Surds should be simplified before adding or subtracting like terms.

WORKED EXAMPLE 24 Adding and subtracting surds

Simplify each of the following.

a. $6\sqrt{3} + 2\sqrt{3} + 4\sqrt{5} - 5\sqrt{5}$ **b. $3\sqrt{2} - 5 + 4\sqrt{2} + 9$**

THINK	WRITE
a. Collect the like terms ($\sqrt{3}$ and $\sqrt{5}$).	a. $6\sqrt{3} + 2\sqrt{3} + 4\sqrt{5} - 5\sqrt{5}$ $= 8\sqrt{3} - \sqrt{5}$

84 Jacaranda Maths Quest 9 Stage 5 NSW Syllabus Third Edition

b. Collect the like terms and simplify.

b. $3\sqrt{2} - 5 + 4\sqrt{2} + 9$
$= 3\sqrt{2} + 4\sqrt{2} - 5 + 9$
$= 7\sqrt{2} + 4$

DISCUSSION

Demonstrate why $\sqrt{6} + \sqrt{8}$ and $\sqrt{14}$ are different.

WORKED EXAMPLE 25 Simplifying surd expressions

Simplify $5\sqrt{75} - 6\sqrt{12} + \sqrt{8} - 4\sqrt{3}$.

THINK

1. Simplify $5\sqrt{75}$.

2. Next, simplify $6\sqrt{12}$.

3. Lastly, simplify $\sqrt{8}$.

4. Rewrite the original expression and simplify by adding like terms.

WRITE

$5\sqrt{75} = 5 \times \sqrt{25} \times \sqrt{3}$
$= 5 \times 5 \times \sqrt{3}$
$= 25\sqrt{3}$

$6\sqrt{12} = 6 \times \sqrt{4} \times \sqrt{3}$
$= 6 \times 2 \times \sqrt{3}$
$= 12\sqrt{3}$

$\sqrt{8} = \sqrt{4} \times \sqrt{2}$
$= 2\sqrt{2}$

$5\sqrt{75} - 6\sqrt{12} + \sqrt{8} - 4\sqrt{3}$
$= 25\sqrt{3} - 12\sqrt{3} + 2\sqrt{2} - 4\sqrt{3}$
$= 9\sqrt{3} + 2\sqrt{2}$

DISCUSSION

Are all square root numbers surds?

on Resources

Video eLesson Surds (eles-1906)

Interactivities Simplifying surds (int-6028)
Surds on the number line (int-6029)

2.5 Quick quiz on	2.5 Exercise

Individual pathways

■ PRACTISE	■ CONSOLIDATE	■ MASTER
1, 5, 7, 10, 13, 16, 19, 21, 23, 26, 29, 33, 35, 39	2, 4, 8, 9, 11, 15, 17, 22, 24, 28, 30, 31, 36, 40	3, 6, 12, 14, 18, 20, 25, 27, 32, 34, 37, 38, 41, 42

Fluency

1. Write down the square roots of each of the following.

    a. 1
    b. 4
    c. 0
    d. $\dfrac{1}{9}$
    e. $1\dfrac{9}{16}$

2. Write down the square roots of each of the following.

    a. 0.16
    b. 400
    c. 10 000
    d. $\dfrac{4}{25}$
    e. 1.44

3. Write down the square roots of each of the following.

    a. 20.25
    b. 1 000 000
    c. 0.0009
    d. 256
    e. $2\dfrac{23}{49}$

4. Write down the value of each of the following.

    a. $\sqrt{81}$
    b. $-\sqrt{81}$
    c. $\sqrt{121}$
    d. $-\sqrt{441}$

5. Write down the value of each of the following.

    a. $\sqrt[3]{8}$
    b. $\sqrt[3]{64}$
    c. $\sqrt[3]{343}$
    d. $\sqrt[4]{81}$

6. Write down the value of each of the following.

    a. $\sqrt[5]{1024}$
    b. $\sqrt[3]{125}$
    c. $-\sqrt{49}$
    d. $\sqrt[3]{-27}$

7. **WE17** Place each of the following on a number line.

    a. $\sqrt{3}$
    b. $\sqrt{7}$
    c. $\sqrt{12}$
    d. $\sqrt{32}$

8. **WE18** Identify which of the following are surds.

    a. $-\sqrt[3]{216}$
    b. $\sqrt[3]{-216}$
    c. $-\sqrt{2}$
    d. $1 + \sqrt{2}$

9. Identify which of the following are surds.

    a. 1.32
    b. $1.\overset{..}{3}2$
    c. $\sqrt[4]{64}$
    d. 1.752 16

10. **WE19** Simplify each of the following.

    a. $\sqrt{3} \times \sqrt{7}$
    b. $-\sqrt{3} \times \sqrt{7}$
    c. $2 \times \sqrt{6}$
    d. $2 \times 3\sqrt{7}$
    e. $2\sqrt{7} \times 5\sqrt{2}$

11. Simplify each of the following.

    a. $3\sqrt{7} \times 4$
    b. $\sqrt{7} \times 9$
    c. $-2\sqrt{5} \times -11\sqrt{2}$
    d. $2\sqrt{3} \times 11$
    e. $\sqrt{3} \times \sqrt{3}$

12. Simplify each of the following.

    a. $-3\sqrt{2} \times -5\sqrt{5}$
    b. $2\sqrt{3} \times 4\sqrt{3}$
    c. $\sqrt{6} \times \sqrt{6}$
    d. $\sqrt{11} \times \sqrt{11}$
    e. $\sqrt{15} \times 2\sqrt{15}$

13. **WE20** Simplify each of the following.

   a. $\sqrt{\dfrac{12}{4}}$ b. $-\sqrt{\dfrac{10}{5}}$

   c. $\dfrac{\sqrt{18}}{\sqrt{3}}$ d. $\dfrac{-\sqrt{15}}{-\sqrt{3}}$

14. Simplify each of the following.

   a. $\dfrac{15\sqrt{6}}{5\sqrt{2}}$ b. $\dfrac{15\sqrt{6}}{10}$

   c. $\dfrac{15\sqrt{6}}{\sqrt{3}}$ d. $\dfrac{5}{3}\sqrt{\dfrac{15}{3}}$

15. Simplify each of the following.

   a. $\dfrac{-10\sqrt{10}}{5\sqrt{2}}$ b. $\dfrac{\sqrt{9}}{\sqrt{3}}$ c. $\dfrac{3}{\sqrt{3}}$ d. $\dfrac{7}{\sqrt{7}}$

16. **WE21** Simplify each of the following.

   a. $\sqrt{20}$ b. $\sqrt{8}$ c. $\sqrt{18}$ d. $\sqrt{49}$

17. Simplify each of the following.

   a. $\sqrt{30}$ b. $\sqrt{50}$ c. $\sqrt{28}$ d. $\sqrt{108}$

18. Simplify each of the following

   a. $\sqrt{288}$ b. $\sqrt{48}$ c. $\sqrt{500}$ d. $\sqrt{162}$

19. **WE22** Simplify each of the following.

   a. $2\sqrt{8}$ b. $5\sqrt{27}$ c. $6\sqrt{64}$ d. $7\sqrt{50}$ e. $10\sqrt{24}$

20. Simplify each of the following.

   a. $5\sqrt{12}$ b. $4\sqrt{42}$ c. $12\sqrt{72}$ d. $9\sqrt{45}$ e. $12\sqrt{242}$

21. Write each of the following in the form $\sqrt{a}$ (that is, as an entire surd).

   a. $2\sqrt{3}$ b. $5\sqrt{7}$ c. $6\sqrt{3}$ d. $4\sqrt{5}$ e. $8\sqrt{6}$

22. Write each of the following in the form $\sqrt{a}$ (that is, as an entire surd).

   a. $3\sqrt{10}$ b. $4\sqrt{2}$ c. $12\sqrt{5}$ d. $10\sqrt{6}$ e. $13\sqrt{2}$

23. **WE24** Simplify each of the following.

   a. $6\sqrt{2}+3\sqrt{2}-7\sqrt{2}$ b. $4\sqrt{5}-6\sqrt{5}-2\sqrt{5}$

   c. $-3\sqrt{3}-7\sqrt{3}+4\sqrt{3}$ d. $-9\sqrt{6}+6\sqrt{6}+3\sqrt{6}$

24. Simplify each of the following.

   a. $10\sqrt{11}-6\sqrt{11}+\sqrt{11}$ b. $\sqrt{7}+\sqrt{7}$

   c. $4\sqrt{2}+6\sqrt{2}+5\sqrt{3}+2\sqrt{3}$ d. $10\sqrt{5}-2\sqrt{5}+8\sqrt{6}-7\sqrt{6}$

25. Simplify each of the following.

   a. $5\sqrt{10}+2\sqrt{3}+3\sqrt{10}+5\sqrt{3}$ b. $12\sqrt{2}-3\sqrt{5}+4\sqrt{2}-8\sqrt{5}$

   c. $6\sqrt{6}+\sqrt{2}-4\sqrt{6}-\sqrt{2}$ d. $16\sqrt{5}+8+7-11\sqrt{5}$

26. **WE25** Simplify each of the following.

    a. $\sqrt{8} + \sqrt{18} - \sqrt{32}$

    c. $-\sqrt{12} + \sqrt{75} - \sqrt{192}$

    e. $\sqrt{24} + \sqrt{180} + \sqrt{54}$

    b. $\sqrt{45} - \sqrt{80} + \sqrt{5}$

    d. $\sqrt{7} + \sqrt{28} - \sqrt{343}$

27. Simplify each of the following.

    a. $\sqrt{12} + \sqrt{20} - \sqrt{125}$

    c. $3\sqrt{45} + 2\sqrt{12} + 5\sqrt{80} + 3\sqrt{108}$

    e. $2\sqrt{32} - 5\sqrt{45} - 4\sqrt{180} + 10\sqrt{8}$

    b. $2\sqrt{24} + 3\sqrt{20} - 7\sqrt{8}$

    d. $6\sqrt{44} + 4\sqrt{120} - \sqrt{99} - 3\sqrt{270}$

28. **MC** Choose the correct answer from the given options.

    a. $\sqrt{2} + 6\sqrt{3} - 5\sqrt{2} - 4\sqrt{3}$ is equal to:

       A. $-5\sqrt{2} + 2\sqrt{3}$ **B.** $-3\sqrt{2} + 23$ **C.** $6\sqrt{2} + 2\sqrt{3}$ **D.** $-4\sqrt{2} + 2\sqrt{3}$

    b. $6 - 5\sqrt{6} + 4\sqrt{6} - 8$ is equal to:

       A. $-2 - \sqrt{6}$ **B.** $14 - \sqrt{6}$ **C.** $-2 + \sqrt{6}$ **D.** $-2 - 9\sqrt{6}$

    c. $4\sqrt{8} - 6\sqrt{12} - 7\sqrt{18} + 2\sqrt{27}$ is equal to:

       A. $-7\sqrt{5}$ **B.** $29\sqrt{2} - 18\sqrt{3}$ **C.** $-13\sqrt{2} - 6\sqrt{3}$ **D.** $-13\sqrt{2} + 6\sqrt{3}$

    d. $2\sqrt{20} + 5\sqrt{24} - \sqrt{54} + 5\sqrt{45}$ is equal to:

       A. $19\sqrt{5} + 7\sqrt{6}$ **B.** $9\sqrt{5} - 7\sqrt{6}$ **C.** $-11\sqrt{5} + 7\sqrt{6}$ **D.** $-11\sqrt{5} - 7\sqrt{6}$

Understanding

29. **MC** Choose the correct answer from the given options.

    a. $\sqrt{1000}$ is equal to:

       A. 31.6228 **B.** $50\sqrt{2}$ **C.** $50\sqrt{10}$ **D.** $10\sqrt{10}$

    b. $\sqrt{80}$ in simplest form is equal to:

       A. $4\sqrt{5}$ **B.** $2\sqrt{20}$ **C.** $8\sqrt{10}$ **D.** $5\sqrt{16}$

    c. Choose the surd out of the following that is in simplest form.

       A. $\sqrt{60}$ **B.** $\sqrt{147}$ **C.** $\sqrt{105}$ **D.** $\sqrt{117}$

    d. Choose the surd out of the following that is **not** in simplest form.

       A. $\sqrt{102}$ **B.** $\sqrt{110}$ **C.** $\sqrt{116}$ **D.** $\sqrt{118}$

30. **MC** Choose the correct answer from the given options.

    a. $6\sqrt{5}$ is equal to:

       A. $\sqrt{900}$ **B.** $\sqrt{30}$ **C.** $\sqrt{150}$ **D.** $\sqrt{180}$

    b. Choose the expression out of the following that is **not** equal to the others.

       A. $\sqrt{128}$ **B.** $2\sqrt{32}$ **C.** $8\sqrt{2}$ **D.** $64\sqrt{2}$

    c. Choose the expression out of the following that is **not** equal to the others.

       A. $4\sqrt{4}$ **B.** $2\sqrt{16}$ **C.** 8 **D.** 16

    d. $5\sqrt{48}$ is equal to:

       A. $80\sqrt{3}$ **B.** $20\sqrt{3}$ **C.** $9\sqrt{3}$ **D.** $21\sqrt{3}$

31. Reduce each of the following to simplest form.

    a. $\sqrt{675}$ b. $\sqrt{1805}$ c. $\sqrt{1792}$ d. $\sqrt{578}$

32. Reduce each of the following to its simplest form.

    a. $\sqrt{a^2c}$

    b. $\sqrt{bd^4}$

    c. $\sqrt{h^2jk^2}$

    d. $\sqrt{f^3}$

33. **WE23** Determine which number in each of the following pairs of numbers is larger.

    a. $\sqrt{10}$ or $2\sqrt{3}$

    b. $3\sqrt{5}$ or $5\sqrt{2}$

    c. $10\sqrt{2}$ or $4\sqrt{5}$

    d. $2\sqrt{10}$ or $\sqrt{20}$

34. Write the following sequences of numbers in order from smallest to largest.

    a. $6\sqrt{2}$, 8, $2\sqrt{7}$, $3\sqrt{6}$, $4\sqrt{2}$, $\sqrt{60}$
    b. $\sqrt{6}$, $2\sqrt{2}$, $\sqrt{2}$, 3, $\sqrt{3}$, 2, $2\sqrt{3}$

Communicating, reasoning and problem solving

35. The formula for calculating the speed of a car before it brakes in an emergency is $v = \sqrt{20d}$, where v is the speed in m/s and d is the braking distance in metres.
    Calculate the speed of a car before braking if the braking distance is 32.50 m.
    Write your answer as a surd in its simplest form.

36. A gardener wants to divide their vegetable garden into 10 equal squares, each of area exactly $2\,\text{m}^2$.

    a. Calculate the exact side length of each square. Explain your reasoning.
    b. Determine the side length of each square, correct to 2 decimal places.
    c. The gardener wishes to group the squares in their vegetable garden next to each other so as to have the smallest perimeter possible for the whole garden.

        i. Determine how they should arrange the squares.
        ii. Explain why the exact perimeter of the vegetable patch is $14\sqrt{2}\,\text{m}$.
        iii. Calculate the perimeter, correct to 2 decimal places.

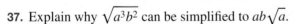

37. Explain why $\sqrt{a^3b^2}$ can be simplified to $ab\sqrt{a}$.

38. Determine the smallest values that a and b can have, given that they are both natural numbers, that $1 < a < b$, and also that $\sqrt{ab}$ is not a surd.

39. Kyle wanted a basketball court in his backyard, but he could not fit a full-sized court in his yard. He was able to get a rectangular court laid with a width of $6\sqrt{2}\,\text{m}$ and a length of $3\sqrt{10}\,\text{m}$. Calculate the area of the basketball court and represent it in its simplest surd form.

40. A netball team went on a pre-season training run. They completed 10 laps of a triangular course that had side lengths of $(200\sqrt{3} + 50)\,\text{m}$, $(50\sqrt{2} + 75\sqrt{3})\,\text{m}$ and $(125\sqrt{2} - 18)\,\text{m}$. Determine the distance they ran, in its simplest surd form.

41. The area of a square is $x\,\text{cm}^2$. Explain whether the side length of the square would be a rational number.

42. To calculate the length of the hypotenuse of a right-angled triangle, use the formula $c^2 = a^2 + b^2$.

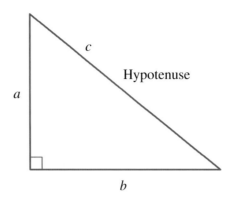

a. Calculate the length of the hypotenuse for triangles with other side lengths as given in the following pairs of values.

    i. $\sqrt{5}, \sqrt{8}$ **ii.** $\sqrt{2}, \sqrt{7}$ **iii.** $\sqrt{15}, \sqrt{23}$

b. Describe any pattern in your answers to part **a**.

c. Calculate the length of the hypotenuse (without calculations) for triangles with the following side lengths.

    i. $\sqrt{1000}, \sqrt{500}$ **ii.** $\sqrt{423}, \sqrt{33}$ **iii.** $\sqrt{124}, \sqrt{63}$

d. Your friend wrote down the following explanation.

    $\sqrt{b} + \sqrt{c} = \sqrt{a}$

Is this answer correct? If not, explain what the correct answer should be.

LESSON
2.6 Introduction to fractional indices (Path)

> **LEARNING INTENTION**
>
> At the end of this lesson you should be able to:
> - identify the symbol $\sqrt{x}$ as the square root of x, the inverse of squaring
> - identify the symbol $\sqrt[3]{x}$ as the cube root of x, the inverse of cubing
> - express a square root of a variable as a power of $\frac{1}{2}$
> - simplify expressions containing square roots and cube roots.

▶ 2.6.1 Square root

eles-6174

- The symbol $\sqrt{}$ means 'square root' — a number that multiplies by itself to give the original number.
- Each number actually has a positive and negative square root. For example, $(2)^2 = 4$ and $(-2)^2 = 4$. Therefore, the square root of 4 is $+2$ or -2.
- $\sqrt{x}$ is defined as the positive square root of x, for $x \geq 0$. $\sqrt{x}$ is undefined for $x < 0$.
- The square root is the inverse of squaring (power 2).
- For this reason, a square root is equivalent to an index of $\frac{1}{2}$.
- In general, $\sqrt{a} = a^{\frac{1}{2}}$.

The square root as an index

$$\sqrt{a} = a^{\frac{1}{2}} \text{ for } a \geq 0$$

WORKED EXAMPLE 26 Simplifying algebraic expressions involving square roots

Evaluate $\sqrt{16p^2}$.

THINK	WRITE
1. We need to obtain the square root of both 16 and p^2.	$\sqrt{16p^2} = \sqrt{16} \times \sqrt{p^2}$
2. Which number is multiplied by itself to give 16? It is 4. Replace the square root sign with a power of $\frac{1}{2}$.	$= 4 \times (p^2)^{\frac{1}{2}}$
3. Use the Fourth Index Law and simplify.	$= 4 \times p^{2 \times \frac{1}{2}}$
	$= 4 \times p^1$
	$= 4p$

⏵ 2.6.2 Cube root

eles-6175

- The symbol $\sqrt[3]{}$ means 'cube root' — a number that multiplies by itself three times to give the original number.
- The cube root is the inverse of cubing (power 3).
- For this reason, a cube root is equivalent to an index of $\frac{1}{3}$.
- In general, $\sqrt[3]{a} = a^{\frac{1}{3}}$.

The cube root as an index

$$\sqrt[3]{a} = a^{\frac{1}{3}}$$

DISCUSSION

Why does a cube root of a number always have the same sign (positive or negative) as the number?

WORKED EXAMPLE 27 Simplifying algebraic expressions involving cube roots

Evaluate $\sqrt[3]{8j^6}$.

THINK	WRITE
1. We need to obtain the cube root of both 8 and j^6.	$\sqrt[3]{8j^6} = \sqrt[3]{8} \times \sqrt[3]{j^6}$
2. Which number, written 3 times and multiplied gives 8? It is 2. Replace the cube root sign with a power of $\frac{1}{3}$.	$= 2 \times (j^6)^{\frac{1}{3}}$
3. Use the Fourth Index Law.	$= 2 \times j^{6 \times \frac{1}{3}}$
4. Simplify.	$= 2 \times j^2$
	$= 2j^2$

- In general terms, $a^{\frac{n}{m}} = \sqrt[m]{a^n}$.

General form

$$\sqrt[m]{a^n} = a^{\frac{n}{m}}$$

 Resources

Interactivities Square roots (int-6066)
Cube roots (int-6065)

Exercise 2.6 Introduction to fractional indices (Path)   learn on

2.6 Quick quiz on	2.6 Exercise

Individual pathways

■ PRACTISE	■ CONSOLIDATE	■ MASTER
1, 2, 4, 6, 7, 10	3, 5, 8, 9, 13, 14	11, 12, 15, 16

Fluency

1. Write the following in surd form.

   a. $x^{\frac{1}{2}}$ b. $y^{\frac{1}{5}}$ c. $z^{\frac{1}{4}}$ d. $(2w)^{\frac{1}{3}}$ e. $7^{\frac{1}{2}}$ f. $(3a)^{\frac{1}{2}}$

2. Write the following in index form.

   a. $\sqrt{15}$ b. $\sqrt{m}$ c. $\sqrt[3]{t}$ d. $\sqrt[3]{w^2}$ e. $\sqrt[5]{n}$ f. $\sqrt[4]{2b}$

3. **WE25** Evaluate the following.

   a. $49^{\frac{1}{2}}$ b. $4^{\frac{1}{2}}$ c. $27^{\frac{1}{3}}$ d. $125^{\frac{1}{3}}$ e. $64^{\frac{1}{3}}$ f. $243^{\frac{1}{5}}$

4. Writing a base as a power itself can be used to simplify an expression. Complete the following calculations.

   a. $16^{\frac{1}{2}} = (4^2)^{\frac{1}{2}} = $ _____

   b. $343^{\frac{2}{3}} = (7^3)^{\frac{2}{3}} = $ _____

5. Simplify the following using index laws.

   a. $8^{\frac{1}{3}}$

   b. $27^{\frac{4}{3}}$

   c. $512^{\frac{2}{9}}$

   d. $16^{-\frac{1}{2}}$

   e. $4^{-\frac{1}{2}}$

   f. $32^{-\frac{1}{5}}$

Understanding

6. a. Use the index laws to simplify the following.

   i. $(3^2)^{\frac{1}{2}}$

   ii. $(4^2)^{\frac{1}{2}}$

   iii. $(8^2)^{\frac{1}{2}}$

   iv. $(11^2)^{\frac{1}{2}}$

   b. Use your answers from part **a** to calculate the values of the following.

   i. $9^{\frac{1}{2}}$

   ii. $16^{\frac{1}{2}}$

   iii. $64^{\frac{1}{2}}$

   iv. $121^{\frac{1}{2}}$

   c. Use your answers to parts **a** and **b** to write a sentence describing what raising a number to a power of one-half does.

7. **WE26** Simplify the following expressions.

   a. $\sqrt{m^2}$

   b. $\sqrt[3]{b^3}$

   c. $\sqrt{36t^4}$

   d. $\sqrt[3]{m^3n^6}$

   e. $\sqrt[3]{125t^6}$

   f. $\sqrt[5]{x^5y^{10}}$

8. Simplify the following expressions.

   a. $\sqrt[4]{a^8m^{40}}$

   b. $\sqrt[3]{216y^6}$

   c. $\sqrt[3]{64x^6y^6}$

   d. $\sqrt{25a^2b^4c^6}$

   e. $\sqrt[7]{b^{49}}$

   f. $\sqrt[3]{b^3} \times \sqrt{b^4}$

9. **MC** a. What does $\sqrt[3]{8000m^6n^3p^3q^6}$ equal?

   A. $2666.6m^2npq^2$ **B.** $20m^2npq^2$ **C.** $20m^3n^0p^0q^3$ **D.** $7997m^2npq^2$

   b. What does $\sqrt[3]{3375a^9b^6c^3}$ equal?

   A. $1125a^3b^2c$ **B.** $1125a^6b3c0$ **C.** $1123a^6b3$ **D.** $15a^3b^2c$

   c. What does $\sqrt[3]{15\,625f^3g^6h^9}$ equal?

   A. $25fg^2h^3$ **B.** $25f^0g^3h^6$ **C.** $25g^3h^6$ **D.** $5208.3fg^2h^3$

Communicating, reasoning and problem solving

10. How would $\sqrt[n]{a^b}$ be written in index form?

11. a. Using the First Index Law, explain how $3^{\frac{1}{2}} \times 3^{\frac{1}{2}} = 3$.

    b. What is another way that $3^{\frac{1}{2}}$ can be written?

    c. Evaluate $\sqrt{3} \times \sqrt{3}$.

    d. How can $\sqrt[n]{a}$ be written in index form?

    e. Without a calculator, evaluate:

    i. $8^{\frac{1}{3}}$

    ii. $32^{\frac{2}{5}}$

12. **a.** Explain why calculating $z^{2.5}$ is a square root problem.
    b. Is $z^{0.3}$ a cube root problem? Justify your reasoning.

13. Mark and Christina are having an algebra argument. Mark is sure that $\sqrt{x^2}$ is equivalent to x, but Christina thinks otherwise. Who is correct?
    Explain how you would resolve this disagreement.

14. Verify that $(-88)^{\frac{1}{3}}$ can be evaluated and explain why $(-88)^{\frac{1}{4}}$ cannot be evaluated.

15. If $n^{\frac{3}{4}} = \dfrac{8}{27}$, what is the value of n?

16. The mathematician Augustus de Morgan enjoyed telling his friends that he was x years old in the year x^2.
    Determine the year of Augustus de Morgan's birth, given that he died in 1871.

LESSON
2.7 Review

2.7.1 Topic summary

INDICES AND SURDS

Surds

- Surds result when the root of a number is an irrational number. e.g. $\sqrt{10} \approx 3.162\ 277\ 660\ 17 ...$
- They are written with a root (radical) sign in them.
- $\sqrt{a} \times \sqrt{b} = \sqrt{ab}$
- $\dfrac{\sqrt{a}}{\sqrt{b}} = \sqrt{\dfrac{a}{b}}$

Index laws

- There are a number of index laws:
- $a^m \times a^n = a^{m+n}$
- $a^m \div a^n = a^{m-n}$
- $a^0 = 1$, where $a \neq 0$
- $(a^m)^n = a^{m \times n}$
- $(a \times b)^m = a^m \times b^m$
- $\left(\dfrac{a}{b}\right)^m = \dfrac{a^m}{b^m}$
- $a^{-n} = \dfrac{1}{a^n}$, where $a \neq 0$
- $a^{\frac{1}{n}} = \sqrt[n]{a}.$
- $a^{\frac{m}{n}} = \sqrt[n]{a^m}$

Square and cube roots

- A square root is the inverse of squaring (raising to the power of 2).
- $\sqrt{4} = \pm 2$ since $2^2 = 4$ and $-(2)^2 = 4$.
- The cube root is the inverse of cubing (raising to the power of 3).
- $\sqrt[3]{8} = 2$ since $2^3 = 8$.

2.7.2 Project

Concentric squares

Consider a set of squares drawn around a central point on a grid, as shown.

These squares can be regarded as concentric squares, because the central point of each square is the point labelled X. The diagram shows four labelled squares drawn on one-centimetre grid paper.

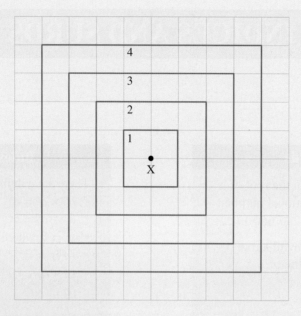

1. Use the diagram to complete the following table, leaving your answers in simplest surd form (if necessary). The first square has been completed for you.

Square	1	2	3	4
Side length (cm)	2			
Diagonal length (cm)	$2\sqrt{2}$			
Perimeter (cm)	8			
Area (cm²)	4			

Observe the patterns in the table and answer the following questions.

2. What would be the side length of the tenth square of this pattern?
3. What would be the length of the diagonal of this tenth square?

Consider a different arrangement of these squares on one-centimetre grid paper, as shown.

These squares all still have a central point, labelled Y.

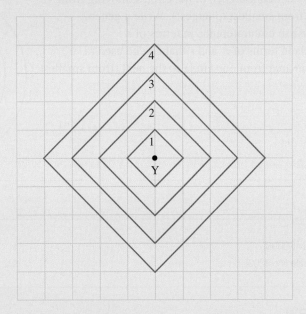

4. Use the diagram to complete the following table for these squares, leaving your answers in simplest surd form if necessary. The first square has been done for you.

Square	1	2	3	4
Side length (cm)	$\sqrt{2}$			
Diagonal length (cm)	2			
Perimeter (cm)	$4\sqrt{2}$			
Area (cm^2)	2			

5. What would be the side length of the tenth square in this pattern of squares?
6. What would be the length of the diagonal of the tenth square?

Use the two diagrams above to answer the following questions.

7. In which of the two arrangements does the square have the greater side length?
8. Which of the two arrangements shows the squares with the greater diagonal length?
9. Compare the area of a square (your choice) in the first diagram with the area of the corresponding square in the second diagram.
10. Compare the perimeter of the same two corresponding squares in each of the two diagrams.
11. Examine the increase in area from one square to the next in these two diagrams. What increase in area would you expect from square 7 to square 8 in each of the two diagrams?

Consider a set of concentric circles around a centre labelled Z, drawn on one-centimetre grid paper, as shown.

12. Investigate the change in circumference of the circles, moving from one circle to the next (from smallest to largest). Write a general formula for this increase in circumference in terms of π.

13. Write a general formula in terms of π and r that can be used to calculate the increase in area from one circle to the next (from smallest to largest).

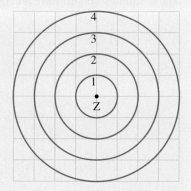

 Resources

 Interactivities Crossword (int-0698)

Sudoku puzzle (int-3202)

Exercise 2.7 Review

learnon

Fluency

1. **PATH** **MC** $\{3\sqrt{2}, 5\sqrt{7}, 9\sqrt{4}, 6\sqrt{10}, 7\sqrt{12}, 12\sqrt{64}\}$ are surds.

   A. $9\sqrt{4}, 12\sqrt{64}$
   B. $3\sqrt{2}$ and $7\sqrt{12}$ only
   C. $3\sqrt{2}, 5\sqrt{7}$ and $6\sqrt{10}$ only
   D. $3\sqrt{2}, 5\sqrt{7}, 6\sqrt{10}$ and $7\sqrt{12}$

2. **PATH** State which of the following are not irrational numbers.

   a. $\sqrt{7}$
   b. $\sqrt{8}$
   c. $\sqrt{81}$
   d. $\sqrt{361}$

3. **PATH** **MC** Choose which of the following is a simplification of the expression $\sqrt{250}$.

   A. $25\sqrt{10}$
   B. $5\sqrt{10}$
   C. $10\sqrt{5}$
   D. $5\sqrt{50}$

4. Simplify each of the following.

   a. $b^7 \times b^3$
   b. $m^9 \times m^2$
   c. $k^3 \times k^5$
   d. $f^2 \times f^8 \times f^4$
   e. $h^4 \times h^5 \times h$

5. Simplify each of the following.

   a. $2q^5 \times 3q^2 \times q^{10}$
   b. $5w^3 \times 7w^{12} \times w^{14}$
   c. $2e^2p^3 \times 6e^3p^5$
   d. $5a^2b^4 \times 3a^8b^5 \times 7a^6b^8$
   e. $2ab^2c^3 \times 3a^2bc^4 \times 5a$

6. Simplify each of the following.

   a. $t^5 \div t$
   b. $r^{19} \div r^{12}$
   c. $p^8 \div p^5$
   d. $\dfrac{f^{17}}{f^{12}}$
   e. $\dfrac{y^{100}}{y^{10}}$

7. Simplify each of the following.

a. $\dfrac{m^{24}}{m^{14}}$
b. $\dfrac{g^4 \times g^5}{g^2}$
c. $\dfrac{x^6 \times x^2 \times x}{x^8}$
d. $\dfrac{d^6 \times d^7 \times d^2}{d^8}$
e. $\dfrac{t^7 \times t \times t^3}{t^2 \times t^4}$

8. Simplify each of the following.

a. $\dfrac{p^5 \times p^3 \times p \times p^4}{p^2 \times p^4 \times p^2}$
b. $\dfrac{16k^{13}}{21} \div \dfrac{8k^9}{42}$
c. $\dfrac{22b^{15}}{c} \div \dfrac{2b^8}{c^6}$

d. $\dfrac{9d^8}{16e^{10}} \div \dfrac{2d^{10}}{e^{16}}$
e. $\dfrac{2a^3 \times ab}{4 \times a \times b}$

9. Simplify each of the following.

a. 5^0
b. 12^0
c. 345^0
d. q^0
e. r^0

10. Simplify each of the following.

a. ab^0
b. $3w^0$
c. $5q^0 - 2q^0$
d. $100s^0 + 99t^0$
e. a^7b^0

11. Simplify each of the following.

a. $v^{10}w^0$
b. prt^0
c. $a^9b^4c^0$

d. $j^8k^0m^3$
e. $4e^2f^0 - 36(a^2b^3)^0$

Understanding

12. Simplify each of the following.

a. $2x^0 \times 3xy^2$
b. $-8(18x^2y^4z^6)^0$

c. $15 - 12x\left(\dfrac{3x}{8}\right)^0$
d. $-4p^0 \times 6(q^2r^3)^0 \div 8(-12q^2)^0$

e. $3(6w^0)^2 \div 2(5w^5)^0$

13. Simplify each of the following.

a. $\left(b^4\right)^2$
b. $\left(a^8\right)^3$
c. $\left(k^7\right)^{10}$
d. $(j^{100})^2$

14. **MC** $\left(\dfrac{4b^4}{d^2}\right)^3$ is equal to:

A. $\dfrac{4b^3}{d^3}$
B. $\dfrac{12b^{12}}{d^6}$
C. $\dfrac{64b^{12}}{d^6}$
D. $\dfrac{64b^7}{d^5}$

15. Simplify each of the following.

a. $(a^5b^2)^3$
b. $(m^7n^{12})^2$
c. $(st^6)^3$
d. $(qp^{30})^{10}$

16. **PATH** Express each of the following in simplest form.

a. 3^{-1}
b. 2^{-2}
c. $(2a)^{-2}$
d. $(4x)^{-3}$

17. **PATH** Write each of the following with positive indices.

a. a^{-1}
b. k^{-4}
c. $4m^3 \div 2m^7$
d. $7x^3y^{-4} \times 6x^{-3}y^{-1}$

18. **PATH** Write each of the following using a negative index.

a. $\dfrac{1}{x}$
b. $\dfrac{2}{y^4}$
c. $z \div z^4$
d. $45p^2q^{-4} \times 3p^{-5}q$

19. PATH Simplify each of the following.

    a. $\sqrt{100}$ **b.** $\sqrt{36}$ **c.** $\sqrt{a^2}$ **d.** $\sqrt{b^2}$ **e.** $\sqrt{49f^4}$

20. PATH Simplify each of the following.

    a. $\sqrt[3]{27}$ **b.** $\sqrt[3]{1000}$ **c.** $\sqrt[3]{x^3}$ **d.** $\sqrt[3]{8d^3}$ **e.** $\sqrt[3]{64f^6g^3}$

Communicating, reasoning and problem solving

21. Write $8^{a+b} + 8^{a+b}$ as a single power of 2.

22. Two students were discussing how to simplify the expression $\left(2^{-2} + 3^{-2}\right)^{-2}$.

One student thought the answer was 97 since $2^4 + 3^4 = 97$.

The other student thought that the expression could not be simplified at all.

Discuss if either student was correct.

23. The radius of a cylinder depends on the volume and height of the cylinder. For a cylinder of a fixed volume, the radius, r, can be found using the relationship

$$r = \frac{k}{\sqrt{h}}$$

where k is a constant and h is the height of the cylinder.

Students are discussing which cylinder, where k was given as 1000, would have the largest radius.
The heights were 121 cm, 144 cm and 169 cm.
What answer should the students give?

24. The formula for calculating the speed of a car before it brakes in an emergency is

$$v = \sqrt{20d}$$

where v is the speed in m/s and d is the braking distance in metres.

    a. Calculate the speed of the car before braking if the braking distance is 35 metres.
        Write your answer as a surd in its simplest form.

    b. Calculate the breaking distance when the speed of the car is 60 km/h.
        Give your answer to 1 decimal place.

on   To test your understanding and knowledge of this topic, go to your learnON title at www.jacplus.com.au and complete the **post-test**.

Answers

Topic 2 Indices and surds

2.1 Pre-test

1. False
2. 729
3. True
4. C
5. B
6. 8
7. $\dfrac{4}{3}a^2$
8. 21
9. C
10. $10p^{12}q^7$
11. 8
12. $72a^8b^{11}$
13. C
14. $\dfrac{n^2}{16}$
15. $6x^5y^3$

2.2 Applying the index laws to variables

1. a. $2^2 \times 3$ b. $2^3 \times 3^2$ c. 3×5^2
   d. $2^4 \times 3 \times 5$ e. $2^7 \times 5$ f. $2^3 \times 5^2 \times 7^2$

2. a. $20p^{11}$ b. $6x^8$ c. $56y^{10}$
   d. $21p^8$ e. $84t^6$ f. $30q^{15}$

3. a. $6a^6e^7$ b. $8p^6h^{12}$ c. $6k^{14}$ d. 3^{12}

4. a. $80m^9$ b. $6g^3h^6$ c. $30p^6q^9$ d. $48u^9w^7$

5. a. $27d^{11}y^{17}$ b. $42b^{14}c^9$ c. $24r^{16}s^{18}$ d. $60h^{38}v^{20}$

6. a. $3p^4$ b. $6r^4$ c. $9a^3$ d. $3b^6$

7. a. $20r^4$ b. $9q$ c. $\dfrac{130}{d^2}$ d. $7t^4$

8. a. $\dfrac{3p^5}{2}$ b. $\dfrac{8b^5}{3}$ c. $\dfrac{5m^{10}n^6}{6}$ d. $\dfrac{9x^8y}{4}$

9. a. $3j^5f^3$ b. $\dfrac{4p^2rs}{3}$ c. $\dfrac{9a^5b^3c}{2}$ d. $\dfrac{20f^6g^2h^4}{3}$

10. a. 1   b. 6   c. 1   d. 1   e. 5

11. a. 1   b. 1   c. 2   d. 2   e. 3

12. a. 1   b. 0   c. 14   d. 6   e. 6

13. a. 1   b. 2   c. 2   d. 2   e. 2

14. a. $\dfrac{h^2}{2}$ b. $\dfrac{q^4}{5}$ c. $\dfrac{n^3}{5}$ d. v^2 e. $2x^6$

15. a. $\dfrac{1}{x^3}$ b. $\dfrac{1}{m^8}$ c. $\dfrac{1}{4m^6}$ d. $\dfrac{2}{x^2}$

16. a. $2x^2$ b. $24t^6$ c. $\dfrac{1}{2y^5}$ d. $\dfrac{7y^3}{4x^5}$

17. a. $\dfrac{2}{5m^3n^4}$ b. $\dfrac{2}{n^2}$ c. $\dfrac{2y}{x}$ d. $\dfrac{c^4}{a^4}$

18. a. 6   b. 9   c. 18   d. 11   e. 15   f. 6

19. Sample responses can be found in the worked solutions in the online resources.

20. a. B   b. D   c. D

21. a. A   b. D

22. Sample responses can be found in the worked solutions in the online resources.

23. a. 5^9

    b. Sample responses can be found in the worked solutions in the online resources.

24. a. 5^{16}

    b. Sample responses can be found in the worked solutions in the online resources.

25. a. 1

    b. 1

    c. 0

    d. Sample responses can be found in the worked solutions in the online resources.

26. a. $\Delta = 14$

    b. Answers will vary, but $\Delta + O + \Diamond$ must sum to 12. Possible answers include: $\Delta = 3, O = 2, \Diamond = 7$; $\Delta = 1, O = 3, \Diamond = 8$; $\Delta = 4, O = 4, \Diamond = 4$; $\Delta = 5, O = 1, \Diamond = 6$.

27. a. The repeating pattern is 1, 3, 9, 7.

    b. 3

28. a. The repeating pattern is 4, 6.

    b. 4

29. a. Sample responses can be found in the worked solutions in the online resources.

    b. i. 8   ii. 1   iii. 2

30. 2^{2n+3}

2.3 Simplify algebraic products and quotients using index laws

1. a. e^6 b. f^{80} c. p^{100} d. r^{144} e. $8a^6$

2. a. a^8b^{12} b. p^5q^{15} c. $g^{30}h^{20}$
   d. $81w^{36}q^8$ e. $49e^{10}r^4q^8$

3. a. p^8q^6 b. $r^{15}w^9$ c. $b^{10}n^{18}$
   d. $j^{18}g^{12}$ e. $8a^3b^4$

4. a. q^4r^{20} b. $h^{24}j^{16}$ c. $a^{21}f^{16}$
   d. $t^{10}u^8$ e. $i^{15}j^{12}$

5. a. $\dfrac{9b^8}{d^6}$ b. $\dfrac{25h^{20}}{4j^4}$ c. $\dfrac{8k^{15}}{27t^{24}}$ d. $\dfrac{49p^{18}}{64q^{44}}$

6. a. $\dfrac{125y^{21}}{27z^{39}}$ b. $\dfrac{256a^{12}}{2401c^{20}}$ c. $\dfrac{-64k^6}{343m^{18}}$ d. $\dfrac{16g^{28}}{81h^{44}}$

7. a. 2^{20} b. t^{33} c. a^{21} d. e^{66}

8. a. g^{39} b. $12a^{14}$ c. $216d^{27}$ d. $40\,000r^{14}$

9. B

10. B

11. D

12. a. a^6 b. m^4 c. n^3
d. b^8 e. f^{17} f. g^6

13. a. p^9 b. y^2 c. c^{20}
d. f^7 e. k^{14} f. p^{16}

14. a. 5
b. 6
c. 30
d. Sample responses can be found in the worked solutions in the online resources.

15. a. i. 1 ii. -1 iii. -1 iv. 1
b. $(-1)^{\text{even}} = 1 (-1)^{\text{odd}} = -1$. Sample responses can be found in the worked solutions in the online resources.

16. Danni is correct. Explanations will vary but should involve $(-x)(-x) = (-x)^2 = x^2$ and $-x^2 = -1 \times x^2 = -x^2$.

17. a. 5^{12}
b. Sample responses can be found in the worked solutions in the online resources.

18. a. i. $x = \dfrac{5}{3}$ ii. $x = \dfrac{5}{3}$ iii. $x = \dfrac{5}{3}$
b. When equating the powers, $3x = 5$.

19. Answers will vary. Possible answers are 4096 and 262144.

20. $10^8 \times 10^8 \times 10^8 = (10^8)^3$ atoms

21. a. 4^1 b. 7^2

22. a. 2^1 b. 3^4 c. $\dfrac{1}{5^2}$ d. 2^2
e. $\dfrac{1}{2^2}$ f. $\dfrac{1}{2}$ g. $\dfrac{1}{2}$ h. $\dfrac{1}{7}$

23. a. i. 3 ii. 4 iii. 8 iv. 11
b. i. 3 ii. 4 iii. 8 iv. 11
c. Raising a number to a power of one-half is the same as calculating the square root of that number.

2.4 Applying index laws to numerical and algebraic expressions involving negative-integer indices

1. a. $3^5 = 243$, $3^4 = 81$, $3^3 = 27$, $3^2 = 9$, $3^1 = 3$, $3^0 = 1$,
$3^{-1} = \dfrac{1}{3}$, $3^{-2} = \dfrac{1}{9}$, $3^{-3} = \dfrac{1}{27}$, $3^{-4} = \dfrac{1}{81}$, $3^{-5} = \dfrac{1}{243}$

b. $5^4 = 625$, $5^3 = 125$, $5^2 = 25$, $5^1 = 5$, $5^0 = 1$, $5^{-1} = \dfrac{1}{5}$,
$5^{-2} = \dfrac{1}{25}$, $5^{-3} = \dfrac{1}{125}$, $5^{-4} = \dfrac{1}{625}$

c. $10^4 = 10\,000$, $10^3 = 1000$, $10^2 = 100$, $10^1 = 10$, $10^0 = 1$,
$10^{-1} = \dfrac{1}{10}$, $10^{-2} = \dfrac{1}{100}$, $10^{-3} = \dfrac{1}{1000}$, $10^{-4} = \dfrac{1}{10\,000}$

2. a. $\dfrac{1}{32}$ b. $\dfrac{1}{27}$ c. $\dfrac{1}{4}$ d. $\dfrac{1}{100}$

3. a. $\dfrac{1}{125}$ b. 7 c. $\dfrac{4}{3}$ d. $\dfrac{16}{9}$

4. a. 27 b. $\dfrac{2}{3}$ c. $\dfrac{16}{81}$ d. $\dfrac{49}{4}$

5. a. 2^{-2} b. 3^{-3} c. 2^{-4} or 4^{-2}

6. a. 5^{-2} b. 10^{-3} c. 7^{-3}

7. a. $\dfrac{1}{x^4}$ b. $\dfrac{1}{y^5}$ c. $\dfrac{1}{z}$
d. $\dfrac{a^2}{b^3}$ e. $\dfrac{1}{m^2 n^3}$

8. a. $\dfrac{1}{m^2 n^3}$ b. $x^2 y^2$ c. $5x^3$
d. $\dfrac{w^5}{x^2}$ e. $x^2 y^2$

9. a. $\dfrac{a^2 c}{b^3 d^4}$ b. $\dfrac{a^2 d^3}{b^2 c^2}$ c. $\dfrac{10y}{x^2}$
d. $\dfrac{x}{3}$ e. $\dfrac{1}{m^3 x^2}$

10. a. $\dfrac{1}{a^5}$ b. m^5 c. $\dfrac{1}{m^7}$
d. $\dfrac{14}{x}$ e. 6 f. $\dfrac{5a}{b^3}$

11. a. $\dfrac{1}{x^3}$ b. $\dfrac{6}{x^5 y^3}$ c. $50x^3$
d. 1 e. $\dfrac{50}{a^5}$ f. $10a^4$

12. a. $\dfrac{-32}{w^3}$ b. $\dfrac{16}{m^4}$ c. $\dfrac{27m^6}{n^{12}}$
d. $\dfrac{1}{a^6 b^{15}}$ e. $a^2 b^6$ f. $\dfrac{25}{a^2}$

13. a. $\dfrac{1}{x^5}$ b. $\dfrac{1}{x^{11}}$ c. x^{11}
d. x^5 e. $\dfrac{6c^4}{a^2}$

14. a. $\dfrac{2}{a^6}$ b. $\dfrac{5}{m}$ c. $\dfrac{1}{b}$
d. $\dfrac{1}{a^3 b^2}$ e. $\dfrac{c^2}{a^4}$

15. a. $\dfrac{1}{16a}$ b. $\dfrac{1}{m^3}$ c. $\dfrac{3}{2t^9}$
d. t^3 e. $\dfrac{m^2}{n^3}$

16. a. 2^0 b. 2^3 c. 2^5 d. 2^6 e. 2^{-3} f. 2^{-5}

17. a. 4^0 b. 4^1 c. 4^3 d. 4^{-1} e. 4^{-2} f. 4^{-3}

18. a. 10^0 b. 10^1 c. 10^4 d. 10^{-1} e. 10^{-2} f. 10^{-5}

19. a. $3^{-4} = \dfrac{1}{3^4}$
b. Answers will vary but should mention that if you are dividing, the power in the numerator is lower than that in the denominator. Sample responses can be found in the worked solutions in the online resources.
c. Sample responses can be found in the worked solutions in the online resources.

20. Answers will vary but should mention that the negative 13 means the decimal point is moved 13 places to the left of 3. Using scientific notation allows the number to be expressed more concisely. Sample responses can be found in the worked solutions in the online resources.

21. a. 2^{19} b. 3^{20}

22. a. $\dfrac{6}{5}$ b. $\left(\dfrac{3}{2}\right)^{200}$

23. a. No. The equivalent expression with positive indices is $\dfrac{x^6+1}{x^3}$.

    b. No. The equivalent expression with positive indices is $\dfrac{(xy)^2}{(x+y)^2}$.

24. a. No. The equivalent expression with positive indices is $\dfrac{x^{13}-1}{x^5}$.

    b. No. The correct equivalent expression is $\dfrac{1}{x^6-x^3}$.

2.5 Surds (Path)

1. a. 1   b. 2   c. 0   d. $\dfrac{1}{3}$ e. $\dfrac{5}{4}$

2. a. 0.4   b. 20   c. 100   d. $\dfrac{2}{5}$ e. 1.2

3. a. 4.5   b. 1000   c. 0.03
   d. 16   e. $\dfrac{11}{7}$

4. a. 9   b. −9   c. 11   d. −21

5. a. 2   b. 4   c. 7   d. 3

6. a. 4   b. 5   c. −7   d. −3

7. a.
      $\sqrt{1}$ $\sqrt{4}$ on a number line from 0 to 2, with $\sqrt{3}$ marked.

  b.
      $\sqrt{1}$ $\sqrt{4}$ $\sqrt{9}$ on a number line from 1 to 3, with $\sqrt{7}$ marked.

  c.
      $\sqrt{4}$ $\sqrt{9}$ $\sqrt{16}$ on a number line from 2 to 4, with $\sqrt{12}$ marked.

  d.
      $\sqrt{25}$ $\sqrt{36}$ on a number line from 4 to 7, with $\sqrt{32}$ marked.

8. c and d

9. c

10. a. $\sqrt{21}$ b. $-\sqrt{21}$ c. $2\sqrt{6}$
   d. $6\sqrt{7}$ e. $10\sqrt{14}$

11. a. $12\sqrt{7}$ b. $9\sqrt{7}$ c. $22\sqrt{10}$
   d. $22\sqrt{3}$ e. 3

12. a. $15\sqrt{10}$ b. 24   c. 6
   d. 11   e. 30

13. a. $\sqrt{3}$ b. $-\sqrt{2}$ c. $\sqrt{6}$ d. $\sqrt{5}$

14. a. $3\sqrt{3}$ b. $\dfrac{3\sqrt{6}}{2}$ c. $15\sqrt{2}$ d. $\dfrac{5\sqrt{5}}{3}$

15. a. $-2\sqrt{5}$ b. $\sqrt{3}$ c. $\sqrt{3}$ d. $\sqrt{7}$

16. a. $2\sqrt{5}$ b. $2\sqrt{2}$ c. $3\sqrt{2}$ d. 7

17. a. $\sqrt{30}$ b. $5\sqrt{2}$ c. $2\sqrt{7}$ d. $6\sqrt{3}$

18. a. $12\sqrt{2}$ b. $4\sqrt{3}$ c. $10\sqrt{5}$ d. $9\sqrt{2}$

19. a. $4\sqrt{2}$ b. $15\sqrt{3}$ c. 48   d. $35\sqrt{2}$ e. $20\sqrt{6}$

20. a. $10\sqrt{3}$ b. $4\sqrt{42}$ c. $72\sqrt{2}$ d. $27\sqrt{5}$ e. $132\sqrt{2}$

21. a. $\sqrt{12}$ b. $\sqrt{175}$ c. $\sqrt{108}$ d. $\sqrt{80}$ e. $\sqrt{384}$

22. a. $\sqrt{90}$ b. $\sqrt{32}$ c. $\sqrt{720}$ d. $\sqrt{600}$ e. $\sqrt{338}$

23. a. $2\sqrt{2}$ b. $-4\sqrt{5}$ c. $-6\sqrt{3}$ d. 0

24. a. $5\sqrt{11}$ b. $2\sqrt{7}$
   c. $10\sqrt{2}+7\sqrt{3}$ d. $8\sqrt{5}+\sqrt{6}$

25. a. $8\sqrt{10}+7\sqrt{3}$ b. $16\sqrt{2}-11\sqrt{5}$
   c. $2\sqrt{6}$ d. $5\sqrt{5}+15$

26. a. $\sqrt{2}$ b. 0   c. $-5\sqrt{3}$
   d. $-4\sqrt{7}$ e. $5\sqrt{6}+6\sqrt{5}$

27. a. $2\sqrt{3}-3\sqrt{5}$ b. $4\sqrt{6}+6\sqrt{5}-14\sqrt{2}$
   c. $29\sqrt{5}+22\sqrt{3}$ d. $9\sqrt{11}-\sqrt{30}$
   e. $28\sqrt{2}-39\sqrt{5}$

28. a. D   b. A   c. D   d. A

29. a. D   b. A   c. C   d. C

30. a. D   b. D   c. D   d. B

31. a. $15\sqrt{3}$ b. $19\sqrt{5}$ c. $16\sqrt{7}$ d. $17\sqrt{2}$

32. a. $a\sqrt{c}$ b. $d^2\sqrt{b}$ c. $hk\sqrt{j}$ d. $f\sqrt{f}$

33. a. $2\sqrt{3}$ b. $5\sqrt{2}$ c. $10\sqrt{2}$ d. $2\sqrt{10}$

34. a. $2\sqrt{7},\ 4\sqrt{2},\ 3\sqrt{6},\ \sqrt{60},\ 8,\ 6\sqrt{2}$
   b. $\sqrt{2},\ \sqrt{3},\ 2,\ \sqrt{6},\ 2\sqrt{2},\ 3,\ 2\sqrt{3}$

35. $5\sqrt{26}$

36. a. $\sqrt{2}$ m
   b. 1.41 m
   c.   i. Any of the arrangements shown would work.

      ii. The length of each square is $\sqrt{2}$ m and there are 14 lengths around each shape.
      iii. 19.80 m

37. $\sqrt{a^3b^2} = \sqrt{a^3} \times \sqrt{b^2}$
             $= \sqrt{a^2} \times \sqrt{a} \times \sqrt{b^2}$
             $= a \times \sqrt{a} \times b$
             $= ab\sqrt{a}$

38. $a = 2$ and $b = 8$.

39. $36\sqrt{5}$ m^2

40. $\left(1750\sqrt{2}+2750\sqrt{3}+320\right)$ m

41. The side length will be rational if x is a perfect square. If x is not a perfect square, the side length will be a surd.

42. a. i. $\sqrt{13}$ **ii.** 3 **iii.** $\sqrt{38}$

b. For side lengths $\sqrt{a}$ and $\sqrt{b}$, the length of the hypotenuse is $\sqrt{a+b}$.

c. i. $\sqrt{1500} = 10\sqrt{15}$

  ii. $2\sqrt{114}$

  iii. $\sqrt{187}$

d. No. The correct answer is $\sqrt{b+a} = \sqrt{c}$.

2.6 Introduction to fractional indices (Path)

1. a. $\sqrt{x}$ **b.** $\sqrt[5]{y}$ **c.** $\sqrt[4]{z}$

  d. $\sqrt[3]{2w}$ **e.** $\sqrt{7}$ **f.** $\sqrt{3a}$

2. a. $15^{\frac{1}{2}}$ **b.** $m^{\frac{1}{2}}$ **c.** $t^{\frac{1}{3}}$

  d. $(w^2)^{\frac{1}{3}}$ **e.** $n^{\frac{1}{5}}$ **f.** $(2b)^{\frac{1}{4}}$

3. a. 7 **b.** 2 **c.** 3
  d. 5 **e.** 4 **f.** 3

4. a. 4^1 **b.** 7^2

5. a. $2^1 = 2$ **b.** $3^4 = 81$ **c.** $2^2 = 4$

  d. $\dfrac{1}{2^2} = \dfrac{1}{4}$ **e.** $\dfrac{1}{2}$ **f.** $\dfrac{1}{2}$

6. a. i. 3 **ii.** 4 **iii.** 8 **iv.** 11

  b. i. 3 **ii.** 4 **iii.** 8 **iv.** 11

  c. Raising a number to a power of one-half is the same as calculating the square root of that number.

7. a. m **b.** b **c.** $6t^2$
  d. mn^2 **e.** $5t^2$ **f.** xy^2

8. a. a^2m^{10} **b.** $6y^2$ **c.** $4x^2y^2$
  d. $5ab^2c^3$ **e.** b^7 **f.** b^3

9. a. B **b.** D **c.** A

10. $a^{\frac{b}{n}}$

11. a. $3^{\frac{1}{2}+\frac{1}{2}} = 3^1$ **b.** $\sqrt{3}$ **c.** 3 **d.** $a^{\frac{1}{n}}$

  e. i. 2 **ii.** 4

12. a. $z^{2.5} = z^{\frac{5}{2}} = (z5)^{\frac{1}{2}} = \sqrt{z^5}$

  b. No, it is the tenth root: $z^{0.3} = z^{\frac{3}{10}} = (z^3)^{\frac{1}{10}} = \left(\sqrt{z^3}\right)^{10}$.

13. Mark is correct: $\sqrt{x^2} = x^{\frac{1}{2}\times 2} = x^1 = x$; x can be a positive or negative number.

14. $\left(-2^3\right)^{\frac{1}{3}} = -2$; answers will vary but should include that we cannot take the fourth root of a negative number.

15. $\dfrac{16}{81}$

16. $\sqrt{1871} \approx 43.25$

  $42^2 = 1764$

  $43^2 = 1849$

  He was 43 years old in 1849. Therefore, he was born in $1849 - 43 = 1806$.

Project

1.

Square	1	2	3	4
Side length (cm)	2	4	6	8
Diagonal length (cm)	$2\sqrt{2}$	$4\sqrt{2}$	$6\sqrt{2}$	$8\sqrt{2}$
Perimeter (cm)	8	16	24	32
Area (cm²)	4	16	36	64

2. 20 cm

3. $20\sqrt{2}$ cm

4.

Square	1	2	3	4
Side length (cm)	$\sqrt{2}$	$2\sqrt{2}$	$3\sqrt{2}$	$4\sqrt{2}$
Diagonal length (cm)	2	4	6	8
Perimeter (cm)	$4\sqrt{2}$	$8\sqrt{2}$	$12\sqrt{2}$	$16\sqrt{2}$
Area (cm²)	2	8	18	32

5. $10\sqrt{2}$ cm

6. 20 cm

7. First arrangement

8. First arrangement

9. The squares in the first diagram are twice the area of each corresponding square in the second diagram.

10. The squares in the first diagram have perimeters that are $\sqrt{2}$ times the perimeter of each corresponding square in the second diagram.

11. An increase of 60 cm² in the first diagram and an increase of 30 cm² for the second diagram.

12. 2π

13. $\pi(2r+1)$

2.7 Review

1. D

2. c and d

3. B

4. a. b^{10} **b.** m^{11} **c.** k^8 **d.** f^{14} **e.** h^{10}

5. a. $6q^{17}$ **b.** $35w^{29}$ **c.** $12e^5p^8$

  d. $105a^{16}b^{17}$ **e.** $30a^4b^3c^7$

6. a. t^4 **b.** r^7 **c.** p^3 **d.** f^5 **e.** y^{90}

7. a. m^{10} **b.** g^7 **c.** x **d.** d^7 **e.** t^5

8. a. p^5 **b.** $4k^4$ **c.** $11b^7c^5$ **d.** $\dfrac{9e^6}{32d^2}$ **e.** $\dfrac{a^3}{2}$

9. a. 1 **b.** 1 **c.** 1 **d.** 1 **e.** 1

10. a. a **b.** 3 **c.** 3 **d.** 199 **e.** a^7

11. a. v^{10} **b.** pr **c.** a^9b^4 **d.** j^8m^3 **e.** $4e^2 - 36$

12. a. $6xy^2$ **b.** -8 **c.** $15 - 12x$
  d. -3 **e.** 54

13. a. b^8 **b.** a^{24} **c.** k^{70} **d.** j^{200}

14. C
15. a. $a^{15}b^6$　　　　　b. $m^{14}n^{24}$

　　c. s^3t^{18}　　　　　d. $q^{10}p^{300}$

16. a. $\dfrac{1}{3}$　　　　　b. $\dfrac{1}{4}$

　　c. $\dfrac{1}{4a^2}$　　　　d. $\dfrac{1}{64x^3}$

17. a. $\dfrac{1}{a}$　　b. $\dfrac{1}{k^4}$　　c. $\dfrac{2}{m^4}$　　d. $\dfrac{42}{y^5}$

18. a. x^{-1}　　　　　b. $2y^{-4}$

　　c. z^{-3}　　　　　d. $135p^{-3}q^{-3}$

19. a. 10　　b. 6　　c. a　　d. b　　e. $7f^2$
20. a. 3　　b. 10　　c. x　　d. $2d$　　e. $4f^2g$
21. $2^{3a+3b+1}$
22. Neither is correct, as $\left(2^{-2}+3^{-2}\right)^{-2}=\dfrac{1296}{169}$.
23. The cylinder with a height of 121 cm has the largest radius (approximately 90.91 cm).
24. a. $10\sqrt{7}$ metres　b. 13.9 metres

3 Numbers of any magnitude

LESSON SEQUENCE

LESSON
3.1 Overview

Why learn this?

Abbreviating numbers allows us to read and write very large and very small numbers a lot easier! Most often than not, we simply give an estimate of a number to convey our information much better.

Ed Sheeran's Divide Tour in 2019 broke all the records, with estimated sales of over 8.9 million tickets. The tour was confirmed as the highest grossing tour of all time, generating $775.6 million.

When we shorten and provide an estimate of a very large number, we emphasise the most important parts of the information. Giving the exact number is irrelevant in this case.

This is also true with very small numbers. The average radius of an atom is 0.000 000 1 mm. This means that if you divide a millimetre into 10 million pieces, that is the size of the radius of an atom! Measuring very small things (like an atom) and very large things (like the solar system) require extra special care, so mathematicians have provided us with ways to do it with more efficiency and accuracy!

1. What is the length of this paperclip?

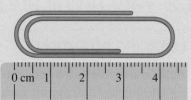

2. Convert 0.344 kilometres to metres.

3. Humans need about 2.6 L of fluids each day. How much fluids are this in mL?

4. Keith's students gave him 3 ribbons with the following measurements. 15 cm, 86 mm and 0.8 m. What is the total length of the ribbons he received?

5. State the limit of accuracy for this measuring cup.

6. Think about some ways an error could be made when measuring the length of an object. Write at least two reasons.

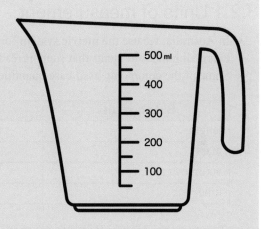

7. The thickness of a paper was measured to be 0.240 mm when the actual thickness was 0.254 mm. What is the absolute percentage error, correct to 2 decimal places?

8. State whether the following statement is True or False. Decimals that do not terminate and also do not recur are irrational numbers.

9. State whether the following statements are True or False.
   a. π is an irrational number.
   b. Every surd is an irrational number.

10. **MC** Select how many significant figures there are in 0.0305.
    A. 1   **B.** 2   **C.** 3   **D.** 4

11. Round 7.43861 to 3 decimal places.

12. Round 7.43861 to 3 significant figures.

13. **MC** Select the correct scientific notation to 3 significant figures for 0.000 000 021 37.
    A. 2.137×10^8 **B.** 2.14×10^8 **C.** 2.14×10^{-8} **D.** 2.137×10^{-8}

14. **MC** Select the series of numbers that is correctly ordered from smallest to largest.
    A. $-\sqrt[3]{8}, -\sqrt[3]{27}, \sqrt[4]{81}$ **B.** $\sqrt{3}, e, 5 \times 10^2$
    C. $2.1 \times 10^{-5}, 2.13 \times 10^2, 2.75 \times 10^{-1}$ **D.** $\sqrt{3}, e, 5 \times 10^2, 3.\dot{1}\dot{4}, \pi, \dfrac{22}{7}$

15. An electron has a mass of $9.109\,381\,9 \times 10^{-31}$ kg, correct to 8 significant figures. A neutron is 1836 times the size of an electron. Evaluate the mass of a neutron correct to 4 significant figures.

LESSON
3.2 Very small and very large measurements

LEARNING INTENTION

At the end of this lesson you should be able to:
- identify and describe the meaning of common prefixes
- establish the meaning of prefixes for very small or very large measurement units
- record measurements using appropriate units
- convert between units of measurements.

▶ 3.2.1 Units of measurement

eles-6223

- In Australia, we use the **metric system** for most quantities. This allows us to have consistency in expressing measurements that support readability and accuracy.
- Some of the commonly used base quantities are listed below.

Base quantity	Base unit	Symbol
length	metres	m
mass	grams	g
volume / capacity	litres	l or L
time	seconds	s

- To write units of measurements, we use:
  - letters (e.g., 'mL' for millilitres)
  - symbols (e.g., 'μm' for micrometres)
  - superscripts (e.g., 'km^2' read as square kilometres)
- To express very large and very small measurements, we use prefixes.
- **Prefixes** make writing multiples and subdivisions (larger and smaller) of any defined unit more convenient.
  - 'kilo' means 'one thousand of', which tells us that kilogram means one thousand grams.
  - 'milli' means 'one thousandth of', which tells us that a millimetre is a thousandth of a metre $\left(\dfrac{1}{1000} \text{ of a metre} \right)$.

Litres can be represented with a small letter, 'l' or a capital letter, 'L'. Most formal content use 'L' as sometimes, 'l' can be confused with the number '1'.

- The commonly used prefixes and their examples are shown in the following table:

Prefix	Symbol	Conversion Factor	Examples
tera	T	10^{12} (1 trillion)	terabyte (TB): a terabyte is 1 trillion bytes. (However, since computers work with binary system using base 2, a terabyte is 2^{40} bytes.)
giga	G	10^9 (1 billion)	gigawatt (GW) power station output gigahertz (GHz) computer processor speed
mega	M	10^6 (1 million)	megawatt (MW) radio transmitter output, megahertz (MHz) radio frequencies
kilo	k	10^3 (1 thousand)	kilometre (km) kilogram (kg) kilowatt (kW)
hecto	h	10^2 (1 hundred)	hectolitre (hl or hL) used to measure wine production
deca	da	10 (ten)	decametre is used less frequently than other units of length
deci	d	10^{-1} (1 tenth)	decilitre (dl or dL) sometimes used for blood sugar levels
centi	c	10^{-2} (1 hundredth)	centimetre (cm) centilitre (cl or cL) wine bottles
milli	m	10^{-3} (1 thousandth)	millimetre (mm) millilitre (ml or mL) milligram (mg) drug doses
micro	μ	10^{-6} (1 millionth)	micrometre (μm) precision in engineering microgram (μg) drug doses microsecond (μs) used in astronomy and in high-tech laboratories that measure data transfer
nano	n	10^{-9} (1 billionth)	nanometre (nm) used to measure wavelengths of light and distances between atoms in molecules nanosecond (ns) used in optics and in determining computer memory speeds.

- Other commonly used units are:
  - hectare (ha) is used to measure bigger land areas:
    $1\ ha = 10\,000\ m^2$
  - tonne (t) is used to measure heavier objects:
    $1\ t = 1000\ kg$
  - nautical mile (M, NM or nmi) is based on the earth's circumference and is equal to one minute of latitude: $1\ NM = 1.852\ km$.
  - light-year (ly) is used to measure the distances in space

- A light year is a unit of astronomical distance equivalent to the distance that light travels in one year, which is $9.4607 \times 10^{12}\ km$.
- It is important to use the appropriate unit to express the distances or sizes of objects or things. The distance between Manila to Sydney is approximately 6 264 000 metres. But it is much more convenient to read and write 6 264 kilometres.
- The consistent structure of the prefixes is understood internationally; thus, it is easy to use in any global industry.

▶ 3.2.2 Taking measurements

eles-6224

- When measuring lengths, it is important to choose units that suit the situation. A suitable unit may be determined by:
  - the visual length of the measurement
  - the context in which the measurement is to be used; for example, builders, carpenters and plumbers work in millimetres and metres.
- A variety of tools can be used to help measure length.
  - A ruler can be used to measure short objects.
  - A tape measure can be used to measure longer objects or distances.
  - A car's odometer can record long measurements, such as the distance between two towns.
  - A picture with a scale, such as a map or a microscope drawing, can be used to measure very large or very small lengths.
- When using a ruler or a tape measure to measure length it is important to ensure that the zero line of the scale markings is always placed at the start of the length being measured.
- When measuring a curved line some useful tools are a piece of string, a trundle wheel or the edge of a piece of paper, slowly rotated around the line.

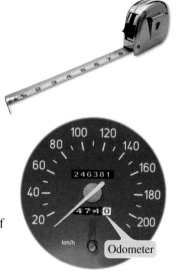

WORKED EXAMPLE 1 Recording measurements using appropriate units

State the measurement marked by the arrow in each of the following. Record your answer in the unit indicated in the brackets.

a.

(mm)

b.

(cm)

THINK	WRITE
a. 1. From left to right count how many centimetres the arrow has passed.	a. 1 cm
2. Multiply that number by 10 to account for the number of mm in each cm (1 cm = 10 mm).	$1 \times 10 = 10$ mm
3. Add the extra millimetres (mm).	$10 + 5 = 15$ mm
b. 1. Count the number of centimetres the arrow has passed. The whole numbers represent the number of cm.	b. 2 cm
2. The extra 4 mm represent $\frac{4}{10}$ cm or 0.4 cm.	0.4 cm
3. Add the whole and part cm.	$2 + 0.4 = 2.4$ cm

- The measurement of volume, weight and time has become extremely accurate due to advances in technology.
- Volume is now measured using advanced laser systems, whereas in the past, some of the volumes that occurred in nature could not be accurately measured.
- Digital scales are used to accurately measure very small or large weights, and atomic clocks can be accurate to the nearest second for millions of years.

COMMUNICATING — COLLABORATIVE TASK: In the olden times

Form groups of 3. Assign a number for each member of the group. Each member will need to determine the answer to the question number corresponding to their assigned number.
1. In the early times, how did the people weigh objects?
2. What were the standard measuring systems used in the early civilisations to measure volume of grains and liquids?
3. How did people keep time before clocks were invented?

Note: You may work with another student from another group assigned with the same topic as you so you can help each other build more information about the topic assigned to you.

Come back to your home group and share what you found out.

▶ 3.2.3 Converting between units of measurements

eles-6225

- The relationships (or ratios) between the metric units of length can be used to **convert** a measurement from one unit to another.

Unit conversion

Units can be converted as shown in the following diagram. The number next to each arrow is called the conversion factor.

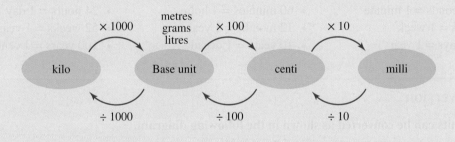

- The following table shows how to convert a **larger unit to a smaller unit**:

	kilo- to base unit (multiply by 1000)	**base unit to centi- (multiply by 100)**	**centi- to milli- (multiply by 10)**
Length	25 km = 25 000 m	5 m = 500 cm	6 cm = 60 mm
Mass	0.75 kg = 750 g	0.5 g = 50 cg	0.3 cg = 3 g
Capacity	3 kL = 3000 L	25 L = 2500 cL	12 cL = 120 mL

- When converting from kilo- to centi-, use the conversion factors in succession:
  × 1000 × 100
- When converting from kilo- to milli-, use the conversion factors in succession:
  × 1000 × 100 × 10

- The following table shows how to convert a **smaller unit to a larger unit:**.

	milli- to centi- (divide by 10)	centi- to base unit (divide by 100)	base unit to kilo- (divide by 1000)
Length	400 mm = 40 cm	750 cm = 7.5 m	25 000 m = 25 km
Mass	52 mg = 5.2 cg	32 cg = 0.32 g	750 g = 0.75 kg
Capacity	3 mL = 0.3 cL	3000 cL = 30 L	25 L = 0.025 kL

- When converting from milli- to the base unit, use the conversion factors in succession: $\div 10 \div 100$.
- When converting from kilo- to milli-, use the conversion factors in succession: $\div 10 \div 100 \div 1000$

WORKED EXAMPLE 2 Converting between units of length

Convert the following lengths to the units shown.

a. 0.234 km = ___ m

b. 24 000 mm = ___ m

THINK

WRITE

a. To convert from kilometres to metres, multiply by 1000, because there are 1000 m for each 1 km.

$$0.234 \text{ km} = (0.234 \times 1000) \text{ m}$$
$$= 234 \text{ m}$$

b. To convert from millimetres to metres, first divide by 10 to convert from millimetres to centimetres, then divide by 100 to convert from centimetres to metres.

$$24\,000 \text{ mm} = (24\,000 \div 10 \div 100) \text{ m}$$
$$= (2400 \div 100) \text{ m}$$
$$= 24 \text{ m}$$

Time

- Time has a standard system that is the same everywhere around the world.

- 60 seconds = 1 minute
- 7 days = 1 week
- 365 days = 1 year
- 60 minutes = 1 hour
- 12 months = 1 year
- 10 years = 1 decade
- 24 hours = 1 day
- 52 weeks = 1 year
- 100 years = 1 century

Time conversion

- **Time units can be converted as shown in the following diagram.**

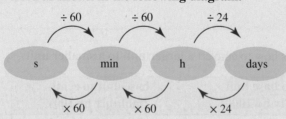

Note: There are 366 days in a leap year. A leap year is divisible by 4 unless it is a century year, when it needs to be divisible by 400. For example, 2004 is a leap year because $\frac{2004}{4} = 501$; 2000 is a leap year because $\frac{2000}{400} = 5$; 3000 is not a leap year because $\frac{3000}{400} = 7.5$.

WORKED EXAMPLE 3 Converting between units of capacity, mass and time

Convert the following volume, weight and time measurements.
a. $400 \, \text{mL} = \underline{\quad} \, \text{L}$ **b.** $1500 \, \text{g} = \underline{\quad} \, \text{kg}$ **c.** $5 \, \text{h} = \underline{\quad} \, \text{s}$

THINK

a. To convert from millilitres (mL) to litres (L), divide by 1000.

b. To convert from grams (g) to kilograms (kg), divide by 1000.

c. To convert from hours to seconds, first multiply by 60 to convert from hours to minutes, then multiply by 60 again to convert from minutes to seconds.

WRITE

a. $400 \, \text{mL} = (400 \div 1000) \, \text{L}$
$= 0.4 \, \text{L}$

b. $1500 \, \text{g} = (1500 \div 1000) \, \text{kg}$
$= 1.5 \, \text{kg}$

c. $5 \, \text{h} = (5 \times 60) \, \text{min}$
$= 300 \, \text{min}$

$300 \, \text{min} = (300 \times 60) \, \text{s}$
$= 18\,000 \, \text{s}$

DISCUSSION

The average thickness of an A4 sheet of paper is between 0.05 mm and 0.10 mm. How would you determine the thickness of a page from your maths book?

 Resources

 Interactivity Converting units of length (int-4011)

Exercise 3.2 Very small and very large measurements   learn

3.2 Quick quiz on	3.2 Exercise

Individual pathways

■ Practice	■ Consolidate	■ Master
1, 4, 7, 9, 14, 17, 19, 22, 24, 25	2, 5, 8, 10, 12, 13, 15, 16, 18, 21, 23	3, 6, 11, 20, 26, 27

Fluency

1. **WE1** State the measurement marked by the arrow in each of the following. Record your answer centimetres (cm).

   a.

   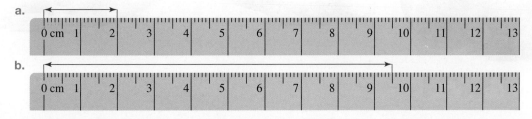

   b.

2. State the measurement marked by the arrow in each of the following. Record your answer in millimetres (mm).

a.
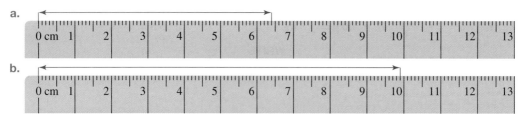

b.
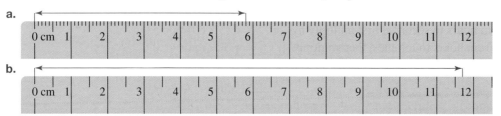

3. Write the measurements of the following lines as accurately as possible.

a.

b.

4. **WE2** Complete each of the following metric conversions.

a. 0.165 km = _____ m
b. 780 cm = _____ m
c. 0.158 m = _____ cm
d. 13 m = _____ km
e. 2500 mL = _____ L
f. 3 L = _____ mL

5. **WE3** Complete each of the following metric conversions.

a. 204 310 cm = ____ km
b. $1\frac{1}{4}$ km = ____ cm
c. 2.2 m = ____ mm

d. 1.35 km = ____ mm
e. 7.5 L = ____ mL
f. $3\frac{3}{4}$ mL = ___ L

6. Complete each of the following metric conversions.

a. 36 000 mm = ____ m
b. 3000 cm = ____ km
c. 35 000 cm = ____ km
d. 5 000 000 g = _____ t
e. 2.4 kL = _____ mL
f. 0.4 cL = ____ mL

7. Convert the following time measurements:

a. 4.5 decades = _____ years
b. 825 years = _____ decades
c. 3.4 centuries = _____ years
d. 485 years = _____ centuries
e. 5 centuries = _____ decades
f. 120 decades = _____ centuries

8. Convert the following time measurements. Give answers to 2 decimal places where necessary.

a. 240 s = _____ min
b. 5.2 min = _____ s
c. 4.6 h = _____ min
d. 320 min = _____ h
e. 30 000 s = _____ h
f. 20.3 h = _____ s

Understanding

9. Which unit of weight — mg, gm or kg — is the most appropriate for each of the following objects?

a.

b.

c.

10. Which unit of capacity — ml, L, kL or ML — is the most appropriate for each of the following objects?

a.

b.

11. The diagram shows sizes in metres, from very small to very large. Think about the average sizes of the following and plot where they go on the number line.

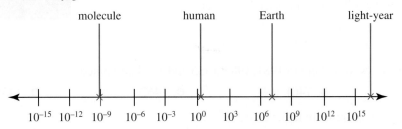

a. length of a didgeridoo
b. length of a midge
c. size of ground wattleseed
d. height of Uluru
e. length of a Barangaroo walkway
f. diameter of Bila (the Sun)

12. Arrange the following lengths in descending order.
a. 89 000 cm, 7.825 m, 98 760 mm, 0.3217 km
b. 0.786 mm, 0.786 km, 0.000 786 m, 0.0786 cm

13. Arrange each of the following in ascending order.
a. 0.15 km, 135 m, 2400 cm
b. 25 cm, 120 mm, 0.5 m
c. 9 m, 10 000 mm, 0.45 km
d. 32 000 cm, 1200 m, 1 km
e. 1.5 m, 150 cm, 0.0015 km
f. 8.25 km, 825 m, 90 000 cm

14. **MC** Which is the larger volume?
A. 1500 mL   B. 3 L   C. 0.465 L   D. 858 mL

15. Arrange the following weights in ascending order.
a. 2500 g, 2.4 kg, 400 000 mg, 0.000 024 t
b. 0.05 kg, 54 g, 52 000 mg, 0.0005 t

16. True or False?
a. 1555 mg > 2 g
b. 3.2 kg = 320 g
c. 45 000 mg < 0.145 kg
d. 500 000 mg > 500 g

17. Fill in the blanks with the appropriate time measurement from the list below. (Use each term only once.)

centuries   decades   hours

weeks   minutes   day

a. The Roman Empire lasted about 5 _____.
b. The internet was invented about 3 _____ ago.
c. A full-time job is considered to be 35–40 _____ per week.
d. There is always at least one low and one high tide every _____.
e. It takes about 2 _____ to cook instant popcorn.
f. A normal pregnancy takes between 37 and 42 _____.

18. Arrange each of the following in descending order.

a. 2 decades, 1.5 centuries, 250 years, 500 months
b. 750 minutes, 12.6 hours, 4250 seconds, 0.56 days, 0.1 weeks

19. **MC** How many days were there in 1981, 1982, 1983 and 1984 combined?

A. 1460   B. 1461   C. 1462   D. 1459

20. **MC** How many hours are there in 3.6 days?

A. 86   B. 86.6   C. 864   D. 86.4

21. **MC** How many days are there in 396 hours?

A. 16.5   B. 16   C. 15.5   D. 15

Communicating, reasoning and problem solving

22. Gemma recorded the length of the measurement marked by the arrow in this ruler as 7.5 cm.

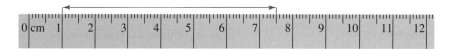

a. Explain the mistake Gemma made in taking the measurement.
b. What is the correct measurement?

23. Your friend has invented a measurement system called Measuro. In the Measuro measurement system, the smallest measurement unit is a *small*. There are 12 *smalls* in a *medium* and 8 *mediums* in a *large*.

a. How many *smalls* are there in 4 *mediums*?
b. How many *smalls* are there in 5 *larges*?
c. If you are 75 *larges* tall, how tall are you in *mediums*?
d. Explain how you convert a length measured in *larges* into *smalls*.
e. Convert 125 *larges* into *smalls*.

24. Explain why it would not be wise to use millimetres to measure the distance between Melbourne and Perth.

25. Wayne's Concreting Crew needs to pour concrete for a driveway. They have calculated that the driveway needs 20 cubic metres $(20\,m^3)$ of concrete.

   a. If 1 cubic metre equals 1000 litres, how many litres of concrete are needed?
   b. How many cubic metres can you cover if you have 2584 litres of concrete?

26. Determine the metric equivalent of each of the following measurements from the imperial measurement system.
   Hint: Convert to centimetres first and then to either metres or kilometres, as appropriate.

1 inch = 2.54 cm	1 mile = 1760 yards
1 foot = 12 inches	1 chain = 22 yards
1 yard = 3 feet	1 chain = 4 rods
1 furlong = 220 yards	1 rod = 25 links

27. a. Determine the metric length of each of the following units and explain what they are used to measure.

   i. A nautical mile
   ii. A light year
   b. Create a poster to display in your classroom that shows the non-metric lengths from part **a** and their equivalent metric lengths.

LESSON
3.3 Absolute and percentage error

LEARNING INTENTION

At the end of this lesson you should be able to:
- determine the precision of a measuring instrument
- determine the absolute error of measuring instruments
- calculate the percentage error of a given measurement.

3.3.1 The precision of a measuring instrument

eles-6226

- The **precision** of a measuring instrument depends on the smallest unit displayed on the measuring device.

If it is a digital device, the smallest unit is the last decimal place, if it is not a decimal device, for example a ruler or a clock face, the smallest unit is the smallest division on the instrument.

In the diagram shown, the paper clip on the left can be measured to the nearest centimetre, a precision of 1 cm, whereas the paperclip on the right, can be measured to the nearest millimetre, a precision of 0.1 cm.

Therefore, the ruler on the right is more precise.

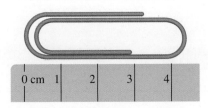

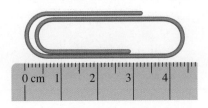

⏵ 3.3.2 Absolute error of measuring instruments

- The following ruler has only centimetre markings, so measurements are approximated to the nearest centimetre. Lengths between 4.5 cm and 5.5 cm are all measured as 5 cm with this ruler.

- The **absolute error** of a measuring instrument is *half of the precision* (the smallest unit marked on the instrument). For the ruler shown, the precision is 1 cm

- The absolute error of the ruler shown is 0.5 cm, which is half of the unit shown on the scale. If an object is measured with this ruler as being 5 cm long, the actual measurement will be between 4.5 cm and 5.5 cm; that is, half a unit above and below the nearest centimetre.

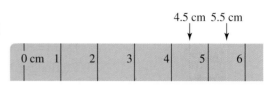

- This can be expressed as 5.0 ± 0.5 cm.

Absolute error of measuring instruments

The absolute error of a measuring instrument is half of the precision or the smallest unit on the instrument.

$$\text{Absolute error} = \frac{1}{2} \times \text{precision}$$

- It is always possible to divide a measurement unit into smaller units. For example, metres can be divided into centimetres, centimetres can be divided into millimetres, millimetres can be divided into micrometres and so on. So the measurements that we take are never **exact**. They are **approximations** only.

WORKED EXAMPLE 4 Calculating the absolute error of measuring instruments

The speedometer shown measures the speed at which a car is travelling.
a. State the absolute error for this speedometer.
b. Provide a range within which the car's actual speed (as indicated by the arrow) lies.

THINK

a. Each marking on the scale represents 5 km/h, so the absolute error is half of this measurement.

b. The arrow is pointing to 80 km/h on the speedometer. The actual speed lies within the range of 80 km/h − 2.5 km/h to 80 km/h + 2.5 km/h.

WRITE

a. Absolute error $= \dfrac{1}{2} \times 5$

$= 2.5$

b. $80 - 2.5 = 77.5$ and
$80 + 2.5 = 82.5$
The actual speed lies within the range of 77.5 km/h to 82.5 km/h.

ⓘ 3.3.3 Types of error

eles-6228

- Errors in measurement can be made due to a number of factors:
  - An inappropriate instrument being used. For example, it would be inappropriate to use a centimetre ruler to measure the distance from Melbourne to Sydney.
  - Human error when reading the measurement. For example, when measuring the length of a school football oval, it is possible for three different students to report three different results.
  - Inaccuracy of the instrument being used. For example, it would be more accurate to measure 20 mL with a measuring cylinder instead of a beaker.

Accuracy of measurement

- The number of decimal places a measurement contains can give an indication of the accuracy of the measurement. For example:
  - A measurement of 5 m indicates that the length is closer to 5 m than it is to either 4 m or 6 m. It has been measured to the nearest metre.
  - A measurement of 5.00 m indicates that the length is closer to 5.00 m than it is to 4.99 m or 5.01 m. It has been measured to the nearest hundredth of a metre or the nearest centimetre.
  - A measurement of 5.000 m has been measured to the nearest thousandth of a metre or the nearest millimetre.
- The number of decimal places a measurement can have is determined by the scale on the measuring device and the units the device has.
- The more decimal places a measurement has, the more accurately it was measured.

WORKED EXAMPLE 5 Calculating absolute error

a. **Richie measured the circumference of the grip of his tennis racquet to be 107 mm. The smallest division on the measuring tape he used was 1 mm. Calculate the absolute error of Richie's measurement.**

b. **Geoff ordered a surfboard that was 1.83 m long. When it arrived, he measured it and found it to be 1.8 m long. If he used a measuring tape on which the smallest division was 1 cm, should he accept that the surfboard was made to his specifications?**

THINK

a. Absolute error $= \dfrac{1}{2} \times$ precision

b. 1. Absolute error $= \dfrac{1}{2} \times$ precision

WRITE

a. Absolute error $= \dfrac{1}{2} \times$ precision

$= \dfrac{1}{2} \times 1$

$= 0.1 \text{ mm}$

b. Absolute error $= \dfrac{1}{2} \times$ precision

$= \dfrac{1}{2} \times 1$

$= 0.5 \text{ cm}$

THINK	WRITE
2. The length recorded was 180 cm and the absolute error was 0.5 cm. Subtract 0.5 from 180 and add 0.5 to 180.	$180 - 0.5 = 179.5$ $180 + 0.5 = 180.5$
3. State the answer.	As the true length could lie between 179.5 and 180.5, Geoff should not accept that the surfboard was made to his specifications.

Percentage error

To calculate the percentage error, divide the absolute error by the measurement and multiply by 100.

Absolute percentage error

$$\text{Absolute percentage error} = \frac{\text{absolute error}}{\text{measurement}} \times 100\%$$

WORKED EXAMPLE 6 Calculating percentage error

A student measured the width of a page of her workbook. Her measurement was 19.5 cm. If she used a ruler on which the smallest interval was 1 cm, calculate:
a. the absolute relative error
b. the absolute percentage error (correct to 2 decimal places).

THINK	WRITE
a. The absolute error is one-half of the precision.	a. $\text{Absolute error} = \frac{1}{2} \times 1$ $= 0.5 \text{ cm}$
b. The absolute percentage error is the absolute error divided by the measurement × 100	b. $\text{Absolute percentage error} = \text{absolute error/measurement} \times 100\%$ $= 0.5/19.5 \times 100\%$ $= 2.56\%$

Exercise 3.3 Absolute and percentage error

3.3 Quick quiz on	3.3 Exercise

Individual pathways

■ Practice	■ Consolidate	■ Master
1, 4, 5, 9, 10, 13	2, 6, 7, 12, 14	3, 8, 11, 15, 16

Fluency

1. **WE4** For each of the scales shown:

   i. state the the absolute error
   ii. write a range to show the values between which the measurement shown on the scale lies.

   a. 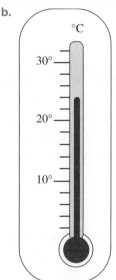   b.   c.

2. For each of the scales shown:

   i. state the absolute error
   ii. write a range to show the values between which the measurement shown on the scale lies.

   a.    b.

   c.

3. For each of the scales shown:
   i. state the absolute error
   ii. write a range to show the values between which the measurement shown on the scale lies.

   a. 2.47 Volts

   b.

   c.

4. Each of the following lengths has been recorded to the nearest centimetre. Provide a range within which each measurement will lie.

   a. 5 cm
   b. 12 cm
   c. 100 cm
   d. 850 cm
   e. 5 m
   f. 2.5 m

5. For each of the following measurements, provide the two limits between which the actual measurement will lie.

   a. The length of a book is 29 cm, correct to the nearest centimetre.
   b. The mass of meat purchased is 250 grams, correct to the nearest 10 grams.
   c. The volume of cough medicine is 10 mL, correct to the nearest 5 mL.
   d. Half an hour, to the nearest minute.

Understanding

6. **MC** Which of the following gives the most accurate measurement?
   A. 7.0
   B. 0.7
   C. 7.7
   D. 7.07

7. **MC** Which of the following gives the least accurate measurement?
   A. 8.000
   B. 8.0
   C. 8
   D. 8.008

8. A measurement of 2.500 m indicates that the measurement is accurate *to the nearest millimetre* (2.500 m = 2500 mm). Determine the unit to which each of the following measurements is accurate.

   a. 35.20 m
   b. 1.30 m
   c. 1.000 km
   d. 8.0 cm
   e. 5.000 m
   f. 2.520 km

9. **WE5** Answer the following.

   a. Jordan measured the length of his bedroom with a measuring tape marked in cm. He found his bedroom to be 395 cm long. What is the absolute error of Jordan's measurement?

   b. Jamie ordered a 30 cm pizza for a party with her friends. However, when it arrived, she found that the pizza was only 29.7 cm wide. Do you think this is a significant error? Why or why not?

10. Calculate the absolute error for each of the following.

    a. The distance from home to school is measured in km
    b. The weight of a packet of lollies is measured in grams.
    c. The scheduled arrival time of the plane from Kansas City is measured in minutes.
    d. The estimated volume of water in a fish tank is measured in litres.

11. You measured your friend's height to be 165 cm. If the absolute error of your measurement is 4 cm, how tall is your friend?

12. `WE6` Complete the table by calculating the absolute error and absolute percentage error. Give your answers correct to 2 decimal places where necessary

	Measurement	Absolute error	Absolute percentage error
a.	245 m		
b.	1.02 m		
c.	18.5 mm		
d.	1 minute		
e.	1 kg 300 g		
f.	0.9 L		

Communicating, reasoning and problem solving

13. a. Explain how you would calculate the absolute error of the ruler shown.

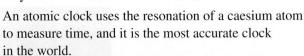

b. A line measured with this ruler is recorded as 23 mm. Between which two measurements will the actual length lie?

14. Explain why any measurement is an approximation and can never be exact.

15. What is the difference between a measurement of 4 km and a measurement of 4.000 km?

16. The accuracy of clocks can vary significantly. Check the clock on your wall, your wristwatch, the clock on your mobile phone and your computer at home. These clocks could be set simultaneously, yet after weeks or months might differ by seconds or even minutes. How do you know which time is the most accurate?

An atomic clock uses the resonation of a caesium atom to measure time, and it is the most accurate clock in the world.

If you go to **time.is** on your phone or computer, you have access to an atomic clock that can compare your computer or mobile-phone time to the atomic clock. There is a problem with this method because of the inaccuracy of synchronisation of your computer or phone with the atomic clock.

All computer or phone clocks are estimates compared to the atomic clock (actual time).

a. Is the accuracy of synchronisation more or less significant if the estimated time is behind or ahead? Explain your answer.

b. If your computer clock's time was one minute ahead of the atomic clock, would the accuracy of synchronisation be significant? Explain your answer.

c. If your computer's clock time was one second behind the atomic clock, would the accuracy of synchronisation be significant? Explain your answer.

LESSON
3.4 Estimating and rounding numbers to a specified degree of accuracy

LEARNING INTENTION

At the end of this lesson you should be able to:
- round numbers to a given number of decimal places
- determine the number of significant figures in a number
- round numbers to a given number of significant figures.

3.4.1 Real numbers

eles-6231

- A **real number** is any number that lies on the number line. Every point on the number line represents a number that is rational or irrational.
- A **rational number** is a number that you can write as a ratio of two integers. This means if you can write the number as a fraction, it is a rational number. Recurring decimals can be written as a fraction. They are therefore rational numbers.
- An **irrational number** is any real number that is NOT a rational number. They are decimals that are non-recurring and non-terminating.

DISCUSSION

$\sqrt{2}$, the diagonal length of a square of side length 1, is an irrational number. The aesthetically pleasing golden ratio, $\phi = \dfrac{1\sqrt{5}}{2}$ is another example of an irrational number. Can you think of other instances in which irrational numbers pop up in your everyday life?

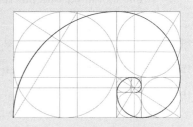

3.4.2 The irrational number π

eles-6232

- Surds are not the only irrational numbers. Most irrational numbers cannot be written as surds. A few of these numbers are so important in mathematics and science that they are given special names.
- The most well known of these numbers is π **(pi)**. It is also called a transcendental number.
- π is irrational; therefore, it cannot be written as a fraction. If you tried to write π as a decimal, you would be writing forever, as the digits never recur and the decimal does not terminate. In the 20th century, computers were used to determine the value of π to 1 trillion decimal places.
- The value of π is very close to (but not equal to) 3.14 or $\dfrac{22}{7}$. Most calculators store an approximate value for π.

⏵ 3.4.3 Rounding to the specified number of decimal places

eles-6233

- When the exact answer in not necessary, rounding numbers makes the numbers shorter and simpler while keeping its value as close as possible to what it was.
- Taking measurements of objects is never exact.
  For example, when we take the length of the string below, we determine where the string is closest to the ruler's marking.

Because the string is closer to 7.5 cm than to 7 cm, the length of this sting is approximated to 7.5 cm.

- Similarly, when we round a number, we determine to which the number is closer to on the number line. For example, think about the number, $\frac{1}{3} = 0.\dot{3} = 0.333\,33\,...$, when it is rounded to the nearest tenths, it is equal to 0.3, because on the number line, it is closer to 0.3 than to 0.4.

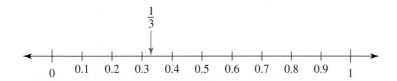

- When we round numbers, we usually write them to either a certain number of decimal places or a certain number of significant figures.

Rounding to a given number of decimal places

1. **Decide how many decimal places you need to round the number to.**
2. **The digit to the right of the last digit that you need to keep is called the critical digit. The critical digit is the deciding digit that tells you whether to round down or to round up.**
   - **If the critical digit is less than 5, leave the last digit as it is ('round down').**
   - **If the critical digit is 5 or more, add one to the value of the last digit ('round up').**
3. **Remove all the remaining digits after the digit you need to keep.**

WORKED EXAMPLE 7 Rounding a number to a given number of decimal places

Round 3.456 734 correct to three decimal places (3 d.p.).

THINK	WRITE
1. Identify the last digit to keep. In this case it is 6.	3.45⑥734
2. Look at the critical digit.	↓
	3.45⑥734
	3.457
3. The critical digit is a 7, so increase the value of the last digit to keep by adding 1. Next, remove the digits after it.	3.457

Write these numbers correct to three decimal places (3 d.p.).

a. $\sqrt{3}$ b. π c. $5.1\dot{9}$ d. $\dfrac{2}{3}$ e. $7.123\,456$

THINK	WRITE
a. $\sqrt{3} = 1.7320\ldots$ The last digit to keep is 2. The critical digit is 0, so, leave 2 as it is.	a. $\sqrt{3} \approx 1.732$
b. $\pi = 3.1415\ldots$ The last digit to keep is 1. The critical digit is 5, so add 1 to the last digit to keep.	b. $\pi \approx 3.142$
c. $5.1\dot{9} = 5.1999\ldots$ The last digit to keep is the third decimal place. The critical digit is 9 and therefore we round up.	c. $5.1\dot{9} \approx 5.200$
d. $\dfrac{2}{3} = 0.6666\ldots$ The last digit to keep is the third decimal place. The critical digit is 6 and therefore we round up.	d. $\dfrac{2}{3} \approx 0.667$
e. $7.123\,456$. The last digit to keep is 3. The critical digit is 4 and therefore we 'round down'.	e. $7.1234 \approx 7.123$

DISCUSSION

What is the effect of rounding during calculations rather than at the end? Does it affect the final result?

3.4.4 Rounding to the specified number of significant figures

eles-6234

- **Significant figures** tell us the number of digits that are important (those that are significant!) for us to determine the accuracy and precision of a measurement we are given. They are used in various scientific and technical measurements.

Determining the number of significant figures

- **All non-zero digits $(1, 2, 3, 4, 5, 6, 7, 8, 9)$ are significant.**
- **Zeros are NOT significant except when:**
  - **they are between two non-zero digits**
  - **they are trailing zeros after the decimal point.**

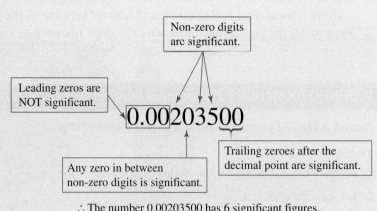

Non-zero digits are significant.

Leading zeros are NOT significant.

Any zero in between non-zero digits is significant.

Trailing zeroes after the decimal point are significant.

$\therefore$ The number 0.00203500 has 6 significant figures.

WORKED EXAMPLE 9 Counting significant figures

State the number of significant figures in each of the following numbers.
a. 25   b. 0.04   c. 3.02   d. 0.100

THINK   **WRITE**

a. Both the digits 2 and 5 are non-zero digits. Therefore, 25 has 2 significant figures.

a. 2 significant figures

b. The leading zeros are not significant. This means, the number has 1 significant figure.

b. 1 significant figure

c. The zero is between two non-zero digits. The number has 3 significant figures.

c. 3 significant figures

d. The leading zero is not significant but the trailing zeros are, because they are after the decimal point. This number has 3 significant figures.

d. 3 significant figures

- Another method of rounding decimals is to write them so that they are correct to a certain number of significant figures.

WORKED EXAMPLE 10 Rounding numbers to five significant figures

Round these numbers so that they are correct to five significant figures.
a. π b. $\sqrt{200}$ c. $0.0\dot{3}$ d. 2530.166

THINK   **WRITE**

a. $\pi \approx 3.14159\ldots$ The first significant figure is 3. Starting with 3, write down the first 5 digits. The critical digit is the 6th digit, which is 9, so add 1 to the value of 5, the 5th digit.

a. 3.1416

b. $\sqrt{200} \approx 14.1421\ldots$ The first significant figure is 1. Starting with 1, write down the first 5 digits. The critical digit is the 6th digit, which is 1, so leave 2, the 5th digit, as it is.

b. 14.142

c. $0.0\dot{3} = 0.030\,303\,0\ldots$ The first significant figure is 3. Starting with 3, write down the first 5 digits. The critical digit is the 6th digit, which is 0, so leave 3, the 5th digit, as it is.

c. 0.030 303

d. $2530.16\ldots$ The first significant figure is 2. Starting with 2, write the first 5 digits. The critical digit is the 6th digit, which is 6, so add 1 to the value of 1, the 5th digit.

d. 2530.2

DISCUSSION

Peta and Shannon are adventurers and have decided to circumnavigate the earth with a combination of trekking and sailing.

They have done some preliminary research and know that the diameter of the earth is 12 742 km. Peta estimates that the circumference of the earth is 40 009 km but Shannon estimates that the circumference is 40 030 km.

Are these two values different enough that the two adventurers need to investigate these values further? Why or why not? Can you suggest a reason why their calculations are different? Use mathematics to explain your answer.

Resources

Interactivities Rounding (int-3730)

Rounding to significant figures (int-6030)

Exercise 3.4 Estimating and rounding numbers to a specified degree of accuracy

learnon

| **3.4 Quick quiz** on | **3.4 Exercise** |

Individual pathways

■ PRACTISE	■ CONSOLIDATE	■ MASTER
1, 4, 7, 10, 12, 15, 18	2, 5, 8, 11, 13, 16, 19	3, 6, 9, 14, 17, 20

Fluency

1. **WE7&8** Write each of the following correct to 3 decimal places.

   a. $\dfrac{\pi}{2}$ b. $\sqrt{5}$ c. $\sqrt{15}$ d. 5.12×3.21 e. $5.\dot{1}$ f. $5.1\dot{5}$

2. Write each of the following correct to 3 decimal places.

   a. $5.\dot{1}\dot{5}$ b. 11.72^2 c. $\dfrac{3}{7}$ d. $\dfrac{1}{13}$ e. $2\dfrac{3}{7}$ f. $0.999\,999$

3. Write each of the following correct to 3 decimal places.

   a. $4.000\,01$ b. $2.79 \div 11$ c. 0.0254 d. $0.000\,913\,6$ e. $5.000\,01$ f. 2342.156

4. **WE9** State how many significant figures there are in each of the following numbers.

   a. 36   b. 207   c. 1631   d. 5.04

5. State how many significant figures there are in each of the following numbers.

   a. 176.2   b. 95.00   c. 0.21   d. 0.01

6. State how many significant figures there are in each of the following numbers.

   a. 0.000 316   b. 0.1007   c. 0.010   d. 0.0512

7. **WE10** Write each number so that it is correct to five significant figures.

   a. $\dfrac{\pi}{2}$ b. $\sqrt{5}$ c. $\sqrt{15}$ d. 5.12×3.21 e. $5.\dot{1}$ f. $5.1\dot{5}$

8. Write each number so that it is correct to five significant figures.

   a. $5.\dot{1}\dot{5}$ b. 11.72^2 c. $\dfrac{3}{7}$ d. $\dfrac{1}{13}$ e. $2\dfrac{3}{7}$ f. $0.999\,999$

9. Write each number so that it is correct to 5 significant figures.

   a. $6.581\,29$ b. $4.000\,01$ c. $2.79 \div 11$ d. 0.0254 e. $0.000\,913\,6$ f. $5.000\,01$

Understanding

10. Write the value of π correct to 4, 5, 6 and 7 decimal places.

11. State whether each of the following numbers has more significant figures than decimal places.

    a. 17.26   b. 0.0032   c. 1.06   d. 0.010 005

12. State whether each statement is True or False.

   a. Every surd is a rational number.
   b. Every surd is an irrational number.
   c. Every irrational number is a surd.
   d. Every surd is a real number.

13. State whether each statement is True or False.

   a. π is a rational number.
   b. π is an irrational number.
   c. π is a surd.
   d. π is a real number.

14. State whether each statement is True or False.

   a. $1.\dot{3}\dot{1}$ is a rational number.
   b. $1.\dot{3}\dot{1}$ is an irrational number.
   c. $1.\dot{3}\dot{1}$ is a surd.
   d. $1.\dot{3}\dot{1}$ is a real number.

Communicating, reasoning and problem solving

15. Write a decimal number that has 3 decimal places but 4 significant figures. Show your working.

16. Explain why the number 12.995 412 3 becomes 13.00 when rounded to 2 decimal places.

17. The rational numbers 3.1416 and $\dfrac{22}{7}$ are both used as approximations to π.

   a. Determine the largest number of decimal places these numbers can be rounded to in order to give the same value.
   b. Explain which of these two numbers gives the best approximation to π.

18. The area of a circle is calculated using the formula $A = \pi \times r^2$, where r is the radius of the circle. Pi (π) is sometimes rounded to 2 decimal places to become 3.14.
    A particular circle has a radius of 7 cm.

   a. Use $\pi = 3.14$ to calculate the area of the circle to 2 decimal places.
   b. Use the π key on your calculator to calculate the area of the circle to 4 decimal places.
   c. Round your answer for part b to 2 decimal places.
   d. Are your answers for parts a and c different? Discuss why or why not.

19. The volume of a sphere (a ball shape) is calculated using the formula $v = \dfrac{4}{3} \times \pi \times r^3$, where r is the radius of the sphere. A beach ball with a radius of 25 cm is bouncing around the crowd at the MCG during the Boxing Day Test.

   a. Calculate the volume of the beach ball to 4 decimal places.
   b. Calculate the volume to 4 decimal places and determine how many significant figures the result has.
   c. Explain whether the calculated volume is a rational number.

20. In a large sample of written English there are about 7 vowels for every 11 consonants. The letter e accounts for about one-third of the occurrence of vowels.
    Explain how many times you would expect the letter e to occur in a passage of 100 000 letters. Round your answer to the nearest 100.

LESSON
3.5 Scientific notation

LEARNING INTENTION

At the end of this lesson you should be able to:
- recognise the need for notation to express very large or very small numbers
- express numbers in scientific notation
- order numbers expressed in scientific notation
- write numbers in decimal notation.

▶ 3.5.1 Using scientific notation

eles-6235

- Scientists often work with extremely large and extremely small numbers. These can range from numbers as large as 1 000 000 000 000 000 000 km (the approximate diameter of our galaxy) down to numbers as small as 0.000 000 000 06 mm (the diameter of an atom).
- Doing calculations with numbers written this way can be challenging — it can be easy to lose track of all of those zeros.
- It is more useful in these situations to use a notation system based on powers of 10. This system is called **scientific notation** (or **standard form**).
- A number is written in scientific notation if it is in this form:

$$m \times 10^n$$

where m is called the **magnitude** and is between 1 and 10; that is, $1 \leq m < 10$; and n is an integer (positive or negative whole number)

- An example looks like this:

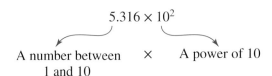

$$5.316 \times 10^2$$

A number between 1 and 10   ×   A power of 10

DIGITAL TECHNOLOGY: Using a calculator for scientific notation

- Numbers written in scientific notation can be entered into a calculator using its scientific notation buttons.
- Some calculators have different ways of displaying numbers written in scientific notation. For example, some calculators display 5.71×10^4 as 5.71E4. When you are writing down scientific notation yourself, you should always show this number as 5.71×10^4.

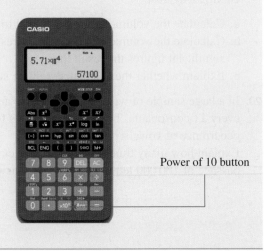

Power of 10 button

Writing numbers in scientific notation

Writing numbers in scientific notation is renaming the number in the form $m \times 10^n$ without changing the value of the number.

Very large numbers: 5 123 000 000
- Think of a way you could write the significant figures of the number so that the magnitude is between 1 and 10. In this case, it will be 5.123.
- Multiply the magnitude by a power of 10 so that the value of the number is retained. The decimal point has moved 9 places to the left. This means, you need to multiply the magnitude by 10^9.

$$5\,123\,000\,000 = 5.123 \times 10^9$$

Very small numbers: 0.007 278
- Think of a way you could write the significant figures of the number so that the magnitude is between 1 and 10. In this case, it will be 7.278.
- Multiply the magnitude by a power of 10 so that the value of the number is retained. The decimal point has moved 3 places to the right. This means, you need to multiply the magnitude by 10^{-3}.

$$0.007\,278 = 7.278 \times 10^{-3}$$

WORKED EXAMPLE 11 Writing numbers in scientific notation

Write each of these numbers in scientific notation then round the answer to 3 significant figures.
a. 827.2   **b.** 51 920 000 000   **c.** 0.0051   **d.** 0.000 000 007 648

THINK	WRITE
a. 1. The magnitude must be written so it is a number between 1 and 10. This means we must move the decimal point 2 places to the left.	8.272 Decimal moves 2 steps left 8.272×10^2
2. Moving the decimal point 2 places left corresponds to the power 10^2. Note that the number of steps moved is equal to the index.	
3. $8.272 \approx 8.27$ (3 significant figures)	8.27×10^2
b. 1. Even though the decimal point is not written, we know that it lies after the final zero.	5.192×10^{10}
2. To get the magnitude lie between 1 and 10 we move the decimal point 10 places to the left. The corresponding power will be 10^{10}.	
3. $5.192 \approx 5.19$ (3 significant figures)	5.19×10^{10}
c. To get the magnitude lie between 1 and 10, we must move the decimal point 3 places to the right. This corresponds to the power 10^{-3}. Note that moving to the right gives a negative index.	5.1×10^{-3}
$5.1 \approx 5.10$ (3 significant figures)	5.10×10^{-3}
d. To get the magnitude lie between 1 and 10 we must move the decimal point 9 places to the right. This corresponds to the power 10^{-9}.	7.648×10^{-9}
$7.648 \approx 7.65$ (3 significant figures)	7.65×10^{-9}

WORKED EXAMPLE 12 Comparing numbers in scientific notation

Arrange the sizes of these planets according to their radius from smallest to largest. The measurements are in km:

Mercury: 2.44×10^3 Venus: 6.052×10^3
Earth: 6.371×10^3 Mars: 3.39×10^3
Jupiter: 6.9911×10^4 Saturn: 5.8232×10^4
Uranus: 2.5362×10^4 Neptune: 2.4622×10^4

THINK	WRITE
a. Look at the powers of 10. The number with the lower power of 10 will be the lower number.	a. Mercury, Venus, Earth and Mars all have 10^3. Jupiter, Saturn, Uranus and Neptune all have 10^4.
b. If they have the same power of 10, then compare the magnitudes to determine the lower number	b. Arranging the decimals from least to greatest is: 2.44 (Mercury) 3.39 (Mars) 6.052 (Venus) 6.371 (Earth)
c. Compare the magnitudes of the numbers with the power of 10^4.	c. Arranging the numbers with the power of 10^4 from least to greatest: 2.4622 (Neptune) 2.5362 (Uranus) 5.8232 (Saturn) 6.9911 (Jupiter)
d. Write the names of the planets according to the size of their radius from smallest to largest.	d. Mercury, Mars, Venus, Earth, Neptune, Uranus, Saturn, Jupiter.

3.5.2 Converting scientific notation to decimal notation

eles-6236

- Converting scientific notation to decimal notation is simply the reverse of the process outlined in section 3.5.1.

WORKED EXAMPLE 13 Writing very large numbers in decimal notation

The following numbers are written in scientific notation. Write them in decimal notation.

a. 7.136×10^2 b. 5.017×10^5 c. 8×10^6

THINK	WRITE
a. The index on the power of 10 is positive 2; therefore, we move the decimal point 2 places to the right.	a. $7.136 \times 10^2 = 713.6$
b. The index on the power of 10 is positive 5; therefore, we move the decimal point 5 places to the right. To do this we will need to add extra zeros.	b. $5.017 \times 10^5 = 501\,700$
c. The index on the power of 10 is positive 6; therefore, we move the decimal point 6 places to the right. Although the decimal point has not been written as part of the scientific notation, we know it lies after the 8. We will need to add extra zeros.	c. $8 \times 10^6 = 8\,000\,000$

WORKED EXAMPLE 14 Writing very small numbers in decimal notation

Write these numbers in decimal notation.

a. 9.12×10^{-1} b. 7.385×10^{-2} c. 6.32×10^{-7}

THINK	WRITE
a. The index on the power of 10 is negative 1; therefore, we move the decimal point 1 place to the left.	$9.12 \times 10^{-1} = 0.912$
b. The index on the power of 10 is negative 2; therefore, we move the decimal point 2 places to the left. To do this we will need to add extra zeros.	$7.3857 \times 10^{-2} = 0.073\,857$
c. The index on the power of 10 is negative 7; therefore, we move the decimal point 7 places to the left. To do this we will need to add extra zeros.	$6.32 \times 10^{-7} = 0.000\,000\,632$

3.5.3 Estimating values by applying index laws to scientific notation

eles-6237

- The index laws that were explored in Topic 2 can also be applied to scientific notation.
- Consider this example.

$$(3.12 \times 10^4) \times (4.2 \times 10^6)$$

We can apply the First Index Law to group the terms as shown and combine terms to create an estimate.

$$\approx 12 \times 10^{10} = 1.2 \times 10^{11}$$

WORKED EXAMPLE 15 Estimating the value of calculations involving scientific notation

Estimate the value of each of the following expressions.

a. $\left(2.9 \times 10^3\right) \times \left(3.1 \times 10^2\right)$ b. $\dfrac{\left(8.2 \times 10^5\right)}{\left(2.1 \times 10^3\right)}$

THINK	WRITE
a. 1. Round the decimals to the nearest whole number.	$\left(3 \times 10^3\right) \times \left(3 \times 10^2\right)$
2. Multiply whole numbers and multiply powers of 10. *Note:* When multiplying indices, the powers are added.	$\left(3 \times 10^3\right) \times \left(3 \times 10^2\right) \approx 9 \times 10^5$
b. 1. Round the decimals to the nearest whole number.	$\dfrac{\left(8 \times 10^5\right)}{\left(2 \times 10^3\right)}$
2. Divide whole numbers and divide powers of 10. *Note:* When dividing indices, the powers are subtracted.	$\dfrac{\left(8 \times 10^5\right)}{\left(2 \times 10^3\right)} \approx 4 \times 10^2$

DISCUSSION

What is the advantage of writing numbers in scientific notation?

Exercise 3.5 Scientific notation

learnon

| **3.5 Quick quiz** | **3.5 Exercise** |

Individual pathways

■ PRACTISE	■ CONSOLIDATE	■ MASTER
1, 4, 6, 7, 10, 13, 14, 15, 20, 24	2, 5, 8, 11, 16, 17, 21, 25	3, 9, 12, 18, 19, 22, 23, 26, 27

Fluency

1. **WE11** Write these numbers in scientific notation.
   a. 5000
   b. 431
   c. 38
   d. 350 000

2. Write these numbers in decimal notation.
   a. 1×10^9
   b. $3.926\,73 \times 10^2$
   c. 5.911×10^2
   d. 5.1×10^3

3. Write these numbers in decimal notation.
   a. 7.34×10^5
   b. 7.1414×10^6
   c. 3.51×10
   d. 8.05×10^4

4. **WE13** Write these numbers in decimal notation.
   a. 6.14×10^2
   b. 6.14×10^3
   c. 6.14×10^4
   d. 3.518×10^2

5. Write these numbers in scientific notation.
   a. 72.5
   b. 725
   c. 7250
   d. 725 000 000

6. Write these numbers in scientific notation, correct to 4 significant figures.
   a. 43.792
   b. 5317
   c. 258.95
   d. 110.11
   e. 1 632 000
   f. 1 million

7. **WE14** Write these numbers in decimal notation.
   a. 2×10^{-1}
   b. 4×10^{-3}
   c. 7×10^{-4}
   d. 3×10^{-2}

8. Write these numbers in decimal notation.
   a. 8.273×10^{-2}
   b. 7.295×10^{-2}
   c. 2.9142×10^{-3}
   d. 3.753×10^{-5}

9. Write these numbers in decimal notation.
   a. 5.29×10^{-4}
   b. 3.3333×10^{-5}
   c. 2.625×10^{-9}
   d. 1.273×10^{-15}

10. Write these numbers in scientific notation.
    a. 0.7
    b. 0.005
    c. 0.000 000 3
    d. 0.000 000 000 01

11. Write these numbers in scientific notation.
    a. 0.231
    b. 0.003 62
    c. 0.000 731
    d. 0.063

12. Write these numbers in scientific notation, correct to 3 significant figures.
    a. 0.006 731
    b. 0.142 57
    c. 0.000 068 3
    d. 0.000 000 005 12
    e. 0.0509
    f. 0.012 46

Understanding

13. **WE12** Write each of the following sets of numbers in ascending order.
    a. 8.31×10^2, 3.27×10^3, 9.718×10^2, 5.27×10^2
    b. 7.95×10^2, 4.09×10^2, 7.943×10^2, 4.37×10^2
    c. 5.31×10^{-2}, 9.29×10^{-3}, 5.251×10^{-2}, 2.7×10^{-3}
    d. 8.31×10^2, 3.27×10^3, 7.13×10^{-2}, 2.7×10^{-3}

14. **WE15** Estimate the value of each of the following expressions.

    a. $(2.8 \times 10^5) \times (1.8 \times 10^4)$

    b. $\dfrac{(6.1 \times 10^7)}{(2.2 \times 10^2)}$

15. One carbon atom weighs 1.994×10^{-23} g.
    a. Write this weight as a decimal.
    b. Calculate how much one million carbon atoms will weigh.
    c. Calculate how many carbon atoms there are in 10 g of carbon. Give your answer correct to 4 significant figures.

16. The distance from Earth to the Moon is approximately 3.844×10^5 km. If you could drive there at a constant speed of 100 km/h, calculate how long it would take. Determine how long that is in days, correct to 2 decimal places.

17. Earth weighs 5.97×10^{24} kg and the Sun weighs 1.99×10^{30} kg. Calculate how many Earths it would take to balance the Sun's weight. Give your answer correct to 2 decimal places.

18. Inside the nucleus of an atom, a proton weighs 1.6703×10^{-28} kg and a neutron weighs 1.6726×10^{-27} kg. Determine which one is heavier and by how much.

19. Earth's orbit has a radius of 7.5×10^7 km and the orbit of Venus has a radius of 5.4×10^7 km. Calculate how far apart the planets are when:

    a. they are closest to each other

    b. they are furthest apart from each other.

Communicating, reasoning and problem solving

20. A USB stick has 8 MB (megabytes) of storage. 1 MB = 1 048 576 bytes.
    a. Determine the number of bytes in 8 MB of storage correct to the nearest 1000.
    b. Write the answer to part a in scientific notation.

21. The basic unit of electric current is the ampere. It is defined as the constant current flowing in 2 parallel conductors 1 metre apart in a vacuum, which produces a force between the conductors of 2×10^7 newtons (N) per metre.
    Complete the following statement. Write the answer as a decimal.
    1 ampere $= 2 \times 10^7$ N/m = _____ N/m

22. **a.** Without performing the calculation, state the power(s) of 10 that you believe the following equations will have when solved.

    i. $5.36 \times 10^7 + 2.95 \times 10^3$ **ii.** $5.36 \times 10^7 - 2.95 \times 10^3$

   b. Evaluate equations **i** and **ii**, correct to 3 significant figures.

   c. Were your answers to part **a** correct? Why or why not?

23. Explain why $2.39 \times 10^{-3} + 8.75 \times 10^{-7} = 2.39 \times 10^{-3}$, correct to 3 significant figures.

24. Distance is equal to speed multiplied by time. If we travelled at 100 km/h it would take us approximately 0.44 years to reach the Moon, 89.6 years to reach Mars, 1460 years to reach Saturn and 6590 years to reach Pluto.

   a. Assuming that there are 365 days in a year, calculate the distance (as a basic numeral) between Earth and:

     i. the Moon   **ii.** Mars   **iii.** Saturn   **iv.** Pluto.

   b. Write your answers to part **a**, correct to 3 significant figures.

   c. Write your answers to part **a** using scientific notation, correct to 3 significant figures.

25. A light-year is the distance that light travels in one year. Light travels at approximately 300 000 km/s.

   a. **i.** Calculate the number of seconds in a year (assuming 1 year = 365 days).

     ii. Write your answer to part **i** using scientific notation.

   b. Calculate the distance travelled by light in one year. Express your answer:

     i. as a basic numeral

     ii. using scientific notation.

   c. The closest star to Earth (other than the Sun) is in the star system Alpha Centauri, which is 4.3 light-years away.

     i. Determine how far this is in kilometres, correct to 4 significant figures.

     ii. If you were travelling at 100 km/h, determine how many years it would take to reach Alpha Centauri.

26. Scientists used Earth's gravitational pull on nearby celestial bodies (for example the Moon) to calculate the mass of Earth. Their answer was that Earth weighs approximately 5.972 sextillion metric tonnes.

   a. Write 5.972 sextillion using scientific notation.

   b. Determine how many significant figures this number has.

27. Atoms are made up of smaller particles called protons, neutrons and electrons. Electrons have a mass of $9.109\,381\,88 \times 10^{-31}$ kilograms, correct to 9 significant figures.

   a. Write the mass of an electron correct to 5 significant figures.

   b. Protons and neutrons are the same size. They are both 1836 times the size of an electron. Use the mass of an electron (correct to 9 significant figures) and your calculator to evaluate the mass of a proton correct to 5 significant figures.

   c. Use the mass of an electron (correct to 3 significant figures) to calculate the mass of a proton, correct to 5 significant figures.

   d. Explain why it is important to work with the original amounts and then round to the specified number of significant figures at the end of a calculation.

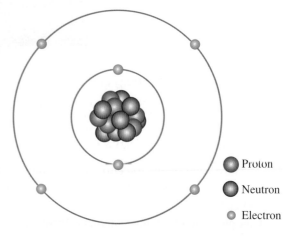

LESSON
3.6 Review

3.6.1 Topic summary

Very small and very large measurements

- The commonly used base units are metres (length), grams (mass), litres (capacity) and seconds (time).
- We use prefixes to express measurements that are very small or very large.
  - 'kilo' means 'one thousand of', which tells us that kilogram means one thousand grams.
  - 'milli' means 'one thousandth of', which tells us that a millimetre is a thousandth of a metre (of a metre).
- Taking measurements require careful consideration of the suitable units and the appropriate measuring equipment or tool used.
- We can convert between units by multiplying or dividing by the appropriate conversion factor.

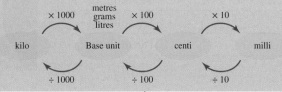

NUMBERS OF ANY MAGNITUDE

Scientific notation

- Scientific notation is used to write very large and very small numbers.
  A number written in scientific notation has the form:
  (number between 1 and 10) × (a power of 10).
- Large numbers have powers of 10 with positive indices, while small numbers have powers of 10 with negative indices.
  For example:
  - 8079 can be written as 8.079×10^3
  - 0.000 15 can be written as 1.5×10^{-4}

Absolute and percentage error

- Because measurements are approximations, we need to keep in mind the absolute error of any measurement taken.
  We use the following formulas:
  - Absolute error = 1/2 × precision

  Absolute percentage error = absolute error/measurement × 100%

Estimating and rounding numbers to a specified degree of accuracy

- Real numbers consist of rational and irrational numbers. If a number can be written as a fraction, it is a rational number. Otherwise, it is an irrational number.
- Rounding numbers makes the number shorter and simpler while keeping it as close to its value as possible.
- When we round numbers, we usually write them to either a certain number of decimal places or a certain number of significant figures.
  - If the critical digit is **less than 5**, leave the last digit as it is ('round down')
  - If the critical digit is **5 or more**, add one to the value of the last digit ('round up')
- All non-zero digits (1, 2, 3, 4, 5, 6, 7, 8, 9) are significant figures. Zeros are NOT significant except when they are between two non-zero digits or that they are trailing zeros after the decimal point.

3.6.2 Project

World top 10 highest paid entertainers in 2022

1.	Genesis	$230 million
2.	Sting	$210 million
3.	Tyler Perry	$175 million
4.	Trey Parker and Matt Stone	$160 million
5.	James L Brooks and Matt Groening	$105 million
6.	Brad Pitt	$100 million
7.	Rolling Stones	$98 million
8.	James Cameron	$95 million
9.	Taylor Swift	$92 million
10.	Bad Bunny	$88 million

Source: forbes.com

1. Explain why James Cameron's earnings written in scientific notation are not 95×10^6.
2. Write the amounts in scientific notation.
3. Sting earned (approximately) $210 million. Write 5 different exact amounts that Sting may have earned. Try to make them as different as possible.
4. What happens to the ranking when the amounts are rounded to the nearest 1 significant figure? To what number of significant figures should the amounts be given so the rankings remain the same?
5. What amount of money could an artist have earned to be the next on the list?
6. Research whether these celebrities have Instagram accounts and determine the number of followers they have. How does this ranking compare to the earnings ranking? What is the same and what is different?
7. If the last person in the number of followers ranking increases his or her followers by 300 people each day, how many followers will he or she have in a month's time?
8. Come up with a list of the world's highest paid sports stars. How do their earnings compare to the ones on the list above?
9. Research the latest top university courses and investigate how the salaries of their possible careers compare to the amount the top artists are paid.

 Resources

 Interactivities Crossword (int-2715)
Sudoku puzzle (int-3213)

Fluency

1. State the measurement marked by the arrow in each of the following. Record your answer using the unit indicated in the brackets.

   a.
   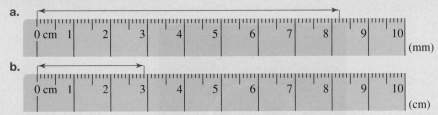
   (mm)

   b.
   (cm)

2. Match the appropriate length, weight or volume measurement with the following pictures.

   a.

   (mm, cm, m, km)

   b.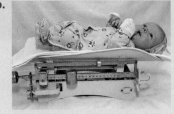

   (g, kg)

   c.

   (mL, L)

3. Convert the following measurements into the units shown in brackets.
   a. 25 cm (mm)   b. 47 mm (cm)   c. 86 m (km)
   d. 1835 cm (km)   e. 1.28 km (cm)   f. 0.0845 km (mm)

4. Convert the following weight and volume measurements into the units shown in brackets.
   a. 30 mL (L)   b. 4.5 L (mL)   c. 0.00089 L (mL)
   d. 1289 mL (L)   e. 1400 g (kg)   f. 5.3 kg (g)

5. Convert the following weight and volume measurements into the units shown in brackets.
   a. 0.385 g (kg)   b. 0.043 kg (g)   c. 52 t (kg)   d. 2150 kg (t)

6. Convert to the measurement indicated in brackets.
   a. 30 cm 22 mm + 42 cm 92 mm (cm)   b. 3.5 cm 20 mm + 42.3 cm 53 mm (mm)
   c. 4 km 2 m + 5.2 km 4 m (km)   d. 80 km 980 m + 5.6 km 225 m (m)
   e. 9.5 kg 354 g + 1520 kg 1555 g (kg)   f. 850.6 kg 25 g + 376.4 kg 254 g (g)

7. Convert the following time measurements as indicated in the brackets. Give your answers to 2 decimal places where necessary.
   a. 30 min (s)   b. 350 s (min)   c. 600 h (days)
   d. 29 days (weeks) e. 7 weeks (days) f. 3 weeks (h)

8. Convert the following time measurements as indicated in the brackets.
   a. 42 months (years)   b. 6.5 years (months)   c. 2 centuries (decades)
   d. 400 years (centuries) e. 3 decades (years)   f. 45 years (decades)

9. For each of the following scales, how much is each interval worth?

a. 20 |_|_|_|_|_|_|_|_|_|_| 30

b. 5 |_|_|_|_|_|_|_|_|_|_| 6

10. A speedometer in a car has markings every 5 km/h.
a. State the absolute error.
b. If the speedometer indicated 55 km/h, within what range could the speed actually lie?

11. The following measurements have been taken with the number of decimal places given according to the smallest markings on the instrument. Calculate the absolute error for each measurement.
a. Actual value = 16.5 m
   Absolute error = _____
b. Actual value = 3.45 litres
   Absolute error = _____
c. Actual value = 256.0 seconds
   Absolute error = _____
d. Actual value = 4.15 cm
   Absolute error = _____

Understanding

12. The distance across a lake is measured by a surveyor to be 3.45 km. When a student measures the distance, she comes up with 3.36 km.
a. What is the absolute error in the surveyor's measurement?
b. What is the percentage error?
c. Between what values would you expect the distance across the lake to lie, based on the surveyor's measurement?
d. Do you think the student should measure again?

13. If the absolute error in measuring the width of a book is 0.2 cm, and you measure the book to be 25.5 cm, what is the actual width of the book?

14. Express the following numbers in scientific notation.
a. 389 670 000
b. 0.004 860 3
c. 406 009 437
d. 0.000 000 100
e. 0.3750
f. 100 300 000 000

15. Express the following numbers in decimal notation.
a. 3.56×10^6
b. 1.245×10^{-9}
c. 1.006×10^{12}
d. 2.16×10^3
e. 9.483×10^{-9}
f. 2.048×10^{-15}

16. Write the following numbers according to the indicated number of significant figures specified in brackets.
    a. 6500.78 (2 significant figures)
    b. 73.000 080 (4 significant figures)
    c. 7398.01 (2 significant figures)
    d. 7000.01 (1 significant figure)
    e. 8.037 (2 significant figures)
    f. 0.000 600 5 (3 significant figures)

17. Write the following numbers in scientific notation and then indicate the number of significant figures.
    a. 35 000   b. 0.000 058 0   c. 0.071   d. 160.7154   e. 5000.1

18. Convert the following measurements as specified and then write the measurements correct to the number of significant figures indicated in brackets.
    a. 1400 cm to m (3 significant figures)
    b. 8.5815 t to kg (4 significant figures)
    c. 76 kg to g (1 significant figure)
    d. 3000.100 L to mL (5 significant figures)
    e. 0.00009 h to min (2 significant figures)
    f. 854 h to days (4 significant figures)

Communicating, reasoning and problem solving

19. Distance = rate × time is an equation used in both mathematics and science. Use this equation to solve the following questions.

    a. A car is traveling at an average speed of 95 km/h for 5.5 hours. How many metres did the car travel?
    b. A family drove 255 km without stopping, and it took them 0.125 days to complete their journey. What was their average speed in km/h?
    c. A sprint car was driving at an average speed of 225 km/h for the entire race. If the track distance was 420 m, how long in seconds did it take the driver to complete one lap?

20. If your body produces about 2.0×10^{11} red blood cells each day, determine how many red blood cells your body has produced after one week.

▶

21. If your body produces about 1.0×10^{10} white blood cells each day, determine how many white blood cells your body produces each hour.

22. If Earth is approximately 1.496×10^8 km from the Sun, and Mercury is about 5.8×10^7 km from the Sun, evaluate the distance from Mercury to Earth.

23. Calculate the closest distance between Mars and Saturn if Mars is 2.279×10^8 km from the Sun and Saturn is 1.472×10^9 km from the Sun.

24. Consider the numbers 6×10^9, 3.4×10^7 and 1.5×10^9.
    a. Place these numbers in order from smallest to largest.
    b. If the population of Oceania is 3.4×10^7 and the world population is 6×10^9, determine the percentage of the world population that is represented by the population of Oceania. Give your answer to 2 decimal places.
    c. Explain what you need to multiply 1.5×10^9 by to get 6×10^9.
    d. Assume the surface area of Earth is approximately $1.5 \times 10^9 \text{km}^2$, and that the world's population is about 6×10^9. If we divided Earth up into equal portions, determine how much surface area each person in the world would get.

25. The total area of the state of New South Wales is $8.01 \times 10^5 \text{ km}^2$. If the total area of Australia is $7.69 \times 10^6 \text{ km}^2$, calculate the percentage of the total area of Australia that is occupied by the state of New South Wales. Write your answer correct to 1 decimal place.

AUSTRALIA MAP

26. A newly discovered colony of bees contains 2.05×10^8 bees. If 0.4% of these bees are estimated to be queen bees, calculate how many queen bees live in the colony.

27. Third-generation fibre optics that can carry up to 10 trillion (1×10^{12}) bits of data per second along a single strand of fibre have recently been tested. Based on the current rate of improvement in this technology, the amount of data that one strand of fibre can carry is tripling every 6 months. This trend is predicted to continue for the next 20 years.
Calculate how many years it will take for the transmission speed along a single strand of fibre to exceed 6×10^{16} bits per second.

28. Mach speed refers to the speed of sound, where Mach 1 is the speed of sound (about 343 metres per second (m/s)). Mach 2 is double the speed of sound, Mach 3 is triple the speed of sound, and so on. Earth has a circumference of about 4.0×10^4 km.

  a. The fastest recorded aircraft speed is 11 270 kilometres per hour (km/h). Express this as a Mach speed, correct to 1 decimal place.

  b. Create and complete a table for three different speeds (Mach 1, 2 and 3) with columns labelled 'Mach number', 'Speed in m/s' and 'Speed in km/h'.

  c. Calculate how long it would take something travelling at Mach 1, 2 and 3 to circle Earth. Give your answer in hours, correct to 2 decimal places.

  d. Evaluate how long it would take for the world's fastest aircraft to circle Earth. Give your answer in hours, correct to 2 decimal places.

  e. Evaluate how many times the world's fastest aircraft could circle Earth in the time it takes an aircraft travelling at Mach 1, 2 and 3 to circle Earth once. Give your answers correct to 2 decimal places.

  f. Explain the relationship between your answers in part e and their corresponding Mach value.

on To test your understanding and knowledge of this topic, go to your learnON title at www.jacplus.com.au and complete the **post-test**.

Answers

Topic 3 Numbers of any magnitude

3.1 Pre-test

1. 4.5 cm
2. 344 metres
3. 2 600 mL
4. 103.6 cm (1 036 mm or 1.036 m)
5. 25 mL
6. Sample answers: inappropriate instrument used, human error in reading the measurement, inaccuracy of the instrument being used.
7. 5.51%
8. True
9. a. True   b. True
10. C
11. 7.439
12. 7.44
13. C
14. B
15. 1.672×10^{-27}

Exercise 3.2 Very small and very large measurements

1. a. 2 cm   b. 9.5 cm
2. a. 64 mm   b. 99 mm
3. a. 5.8 cm   b. 11.7 cm
4. a. 165 m   b. 7.8 m   c. 15.8 cm
   d. 0.013 km   e. 2.5 L   f. 3 000 mL
5. a. 2.0431 km   b. 125 000 cm   c. 2 200 mm
   d. 1 350 000 mm   e. 7500 mL   f. 0.003 75 L
6. a. 36 m   b. 0.03 km   c. 0.35 km
   d. 5 t   e. 2 400 000 mL   f. 4 mL

7. a. 45 years   b. 82.5 decades   c. 340 years
   d. 4.85 centuries   e. 50 decades   f. 12 centuries
8. a. 4 min   b. 312 s   c. 276 min
   d. 5.33 h   e. 8.33 h   f. 73 080 s
9. a. grams   b. kilograms   c. milligrams
10. a. litres   b. millimetres
11 See the image at the bottom of page.*
12. a. 89 000 cm, 0.3217 km, 98 760 mm, 7.825 m
    b. 0.786 km, 0.786 mm = 0.000 786 m = 0.0786 cm
13. a. 2400 cm, 135 m, 0.15 km
    b. 120 mm, 25 cm, 0.5 m
    c. 9 m, 10 000 mm, 0.45 km
    d. 32 000 cm, 1 km, 1200 m
    e. All values are equal.
    f. 825 m, 90 000 cm, 8.25 km
14. B
15. a. 0.000 024 t, 400 000 mg, 2.4 kg, 2500 g
    b. 0.05 kg, 52 000 mg, 54 g, 0.0005 t
16. a. False   b. False   c. True
    d. False
17. a. centuries   b. decades   c. hours
    d. day   e. minutes   f. weeks
18. a. 250 years, 1.5 centuries, 500 months, 2 decades
    b. 0.1 weeks, 0.56 days, 12.6 hours, 750 minutes, 4250 seconds
19. B
20. D
21. A
22. a. Student to notice that the marking did not start at zero but instead at 1.
    b. Start marking is 1 cm. End marking is 7.5 cm. 7.5 minus 1 is equal to 6.5 cm.
23. a. 48 smalls
    b. 480 smalls
    c. 9.375 mediums

*11.

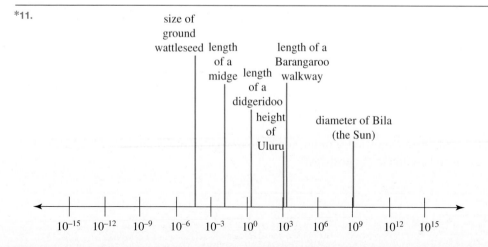

d. This is a small-to-big conversion, which means multiply. First convert large to medium by multiplying by 8, and then to convert from medium to small multiply by 12.

   e. 12 000 smalls

24. The distance from Melbourne to Perth is approximately 2721 km. Large distances are not easy to comprehend in mm, so we use the largest length measurement (km) to express long distances.

25. a. 20 000 L   b. 2.584 m^3

26. 1 foot = 30.48 cm
    1 yard = 91.44 cm
    1 furlong = 201.168 m
    1 mile = 1.609 344 km
    1 chain = 20.1168 m
    1 rod = 5.0292 m
    1 link = 20.1168 cm

27. a. i. 1 nautical mile = 1.852 km; is used for charting and navigating.
      ii. 1 light-year ≈ 9.5 trillion km; is used to measure distance in space.
    b. Answers will vary.

Exercise 3.3 Absolute and percentage error

1. a.i. 0.5   ii. 10.5 mL to 11.5 mL
   b.i. 1   ii. 23 °C to 25 °C
   c.i. 1   ii. 53 °C to 55 °C

2. a.i. 2.5   ii. 82.5 g to 87.5 g
   b.i. 1.25   ii. 62.5 km/h to 67.5 km/h
   c.i. 0.25   ii. 37.25 °C to 37.75 °C

3. a.i. 0.005   ii. 2.465 volts to 2.475 volts
   b.i. 25   ii. 125 g to 175 g
   c.i. 0.2   ii. 5 kg to 5.4 kg

4. a. Between 4.5 cm and 5.5 cm
   b. Between 11.5 cm and 12.5 cm
   c. Between 99.5 cm and 100.5 cm
   d. Between 849.5 cm and 850.5 cm
   e. Between 4.995 m and 5.005 m
   f. Between 2.495 m and 2.505 m

5. a. Between 28.5 cm and 29.5 cm
   b. Between 245 g and 255 g
   c. Between 7.5 mL and 12.5 mL
   d. Between 29.5 min and 30.5 min

6. E

7. C

8. a. 3520 cm (accurate to cm)
   b. 130 cm (accurate to cm)
   c. 1000 m (accurate to m)
   d. 80 mm (accurate to mm)
   e. 5000 mm (accurate to mm)
   f. 2520 m (accurate to m)

9. a. 0.5 cm
   b. No, the true length could lie between 29.5 cm and 30.5 cm.

10. a. 0.5 km   b. 0.5 g   c. 0.5 min   d. 0.5 L

11. The height could lie between 161 and 169 cm.

12. a. Absolute error = 0.5 m
       Absolute percentage error = 0.2%
    b. Absolute error = 0.005 cm
       Absolute percentage error = 0.49%
    c. Absolute error = 0.05 mm
       Absolute percentage error = 0.27%
    d. Absolute error = 0.5 s
       Absolute percentage error = 50%
    e. Absolute error = 0.1 g
       Absolute percentage error = 0.04%
    f. Absolute error = 0.05 L
       Absolute percentage error = 5.56%

13. a. 0.5 mm (half of the smallest unit)
    b. 22.5 mm and 23.5 mm

14. The approximations can always be more exact by measuring to smaller and smaller increments toward infinity. There is no smallest increment.

15. The 4 km measurement is accurate to the km, while the 4.000 km measurement is accurate to the thousandth of a km.

16. a. The time, whether it is behind or ahead, does not affect the significance of the accuracy of synchronisation, because the absolute error calculation forces them to be equal.
    b. The absolute percentage error would be very small and therefore insignificant.
    c. The absolute percentage error would be large and therefore significant.

Exercise 3.4 Estimating and rounding numbers to a specified degree of accuracy

1. a. 1.571   b. 2.236   c. 3.873
   d. 16.435   e. 5.111   f. 5.156

2. a. 5.152   b. 137.358   c. 0.429
   d. 0.077   e. 2.429   f. 1.000

3. a. 4.000   b. 0.254   c. 0.025
   d. 0.001   e. 5.000   f. 2342.156

4. a. 2   b. 3   c. 4   d. 3

5. a. 4   b. 4   c. 2   d. 1

6. a. 3   b. 4   c. 2   d. 3

7. a. 1.5708   b. 2.2361   c. 3.8730
   d. 16.435   e. 5.1111   f. 5.1556

8. a. 5.1515   b. 137.36   c. 0.428 57
   d. 0.076 923 e. 2.4286   f. 1.0000

9. a. 6.5813   b. 4.0000   c. 0.253 64
   d. 0.025 400 e. 0.000 913 60   f. 5.0000

10. 3.1416, 3.14159, 3.141593, 3.1415927

11. a. Yes   b. No   c. Yes   d. No

12. a. False   b. True   c. False   d. True

13. a. False   b. True   c. False   d. True

14. a. True   b. False   c. False   d. True

15. A number of answers are possible. Example answer: 2.004.

16. The 5th digit (5) causes the 4th digit to round up from 9 to 10, which causes the 3rd digit to round up from 9 to 10, which causes the 2nd digit to round up from 2 to 3.

17. a. 2 decimal places   b. 3.1416

18. a. $153.86 \, \text{cm}^2$
    b. $153.9380 \, \text{cm}^2$
    c. $153.94 \, \text{cm}^2$
    d. Yes, because 3.14 is used as an estimate and is not the value of π as used on the calculator.

19. a. $65\,449.8470 \, \text{cm}^3$
    b. 9
    c. Yes, any number with a finite number of decimal places is a rational number.

20. 13 000

Exercise 3.5 Scientific notation

1. a. 5.00×10^3 b. 4.31×10^2
   c. 3.8×10^1 d. 3.5×10^5

2. a. 1 000 000 000   b. 392.673
   c. 591.1   d. 5100

3. a. 734 000   b. 7 141 400
   c. 35.1   d. 80 500

4. a. 614   b. 6140
   c. 61 400   d. 351.8

5. a. 7.25×10^1 b. 7.25×10^2
   c. 7.25×10^3 d. 7.25×10^8

6. a. 4.379×10^1 b. 5.317×10^3 c. 2.590×10^2
   d. 1.101×10^2 e. 1.632×10^6 f. 1.000×10^6

7. a. 0.2   b. 0.004   c. 0.0007   d. 0.03

8. a. 0.082 73   b. 0.072 95
   c. 0.002 914 2   d. 0.000 037 53

9. a. 0.000 529   b. 0.000 033 333
   c. 0.000 000 002 625   d. 0.000 000 000 000 001 273

10. a. 7×10^{-1} b. 5×10^{-3}
    c. 3×10^{-7} d. 1×10^{-11}

11. a. 2.31×10^{-1} b. 3.62×10^{-3}
    c. 7.31×10^{-4} d. 6.3×10^{-2}

12. a. 6.73×10^{-3} b. 1.43×10^{-1} c. 6.83×10^{-5}
    d. 5.12×10^{-9} e. 5.09×10^{-2} f. 1.25×10^{-2}

13. a. $5.27 \times 10^2, 8.31 \times 10^2, 9.718 \times 10^2, 3.27 \times 10^3$
    b. $4.09 \times 10^2, 4.37 \times 10^2, 7.943 \times 10^2, 7.95 \times 10^2$
    c. $2.7 \times 10^{-3}, 9.29 \times 10^{-3}, 5.251 \times 10^{-2}, 5.31 \times 10^{-2}$
    d. $2.7 \times 10^{-3}, 7.13 \times 10^{-2}, 8.31 \times 10^2, 3.27 \times 10^3$

14. a. $\approx 6 \times 10^9$ b. $\approx 3 \times 10^5$

15. a. 0.000 000 000 000 000 000 000 019 94 g
    b. 1.994×10^{-17} g
    c. 5.015×10^{23} atoms

16. 3844 hours $\approx$ 160.17 days

17. 333 333.33

18. The neutron is heavier by 2.3×10^{-30} kg.

19. a. 2.1×10^7 b. 1.29×10^8

20. a. 8 389 000 bytes   b. 8.389×10^6 bytes

21. 0.000 000 2 N/m

22. a. i. Sample responses can be found in the worked solutions in the online resources.
       ii. Sample responses can be found in the worked solutions in the online resources.
    b. i. 5.36×10^7 ii. 5.36×10^7
    c. The answers are the same because 2.95×10^3 is very small compared with 5.36×10^7.

23. 8.75×10^{-7} is such a small amount (0.000 000 875) that, when it is added to 2.39×10^{-3}, it doesn't affect the value when given to 3 significant figures.

24. a. i. 385 440 km
       ii. 78 489 600 km
       iii. 1 278 960 000 km
       iv. 5 772 840 000 km
    b. i. 385 000 km
       ii. 78 500 000 km
       iii. 1 280 000 000 km
       iv. 5 770 000 000 km
    c. i. 3.85×10^5 km
       ii. 7.85×10^7 km
       iii. 1.28×10^9 km
       iv. 5.77×10^9 km

25. a. i. 31 536 000 s   ii. 3.1536×10^7 s
    b. i. 9 460 800 000 000 km   ii. 9.4608×10^{12} km
    c. i. 4.068×10^{13} km   ii. 46 440 000 years

26. a. 5.972×10^{21} b. 4

27. a. 9.1094×10^{-31}
    b. 1.6725×10^{-27}
    c. 1.6726×10^{-27}
    d. It is important to work with the original amounts and leave rounding until the end of a calculation so that the answer is accurate.

Project

1. Because 95 is not between 1 and 10. It should be 9.5×10^7.

2. Genesis: 2.3×10^8, Sting 2.1×10^8, Tyler Perry 1.75×10^8, Trey Parker and Matt Stone 1.6×10^8 (combined), James L Brooks and Matt Groening 1.05×10^8 (combined), Brad Pitt 1×10^8, Rolling Stones 9.8×10^7, James Cameron 9.5×10^7, Taylor Swift 9.2×10^7, Bad Bunny 8.8×10^7

3. Answers may vary. Any amount between 209 500 000 to 210 499 999
   Sample answers are: 209 600 000, 209 900 123, 210 400 000, 210 222 456, 210 499 999

4. The top four all have the same amount ($200 million). The other 6 will be at $100 million.

5. Answers may vary. Any amount less than $88 million when their earnings are rounded to one significant figure only. The ranking would stay the same if the earnings are rounded to 2 significant figures or more.
   Sample answers are: 87 000 000

6. Student answers may vary.
   Sample answer is:

Taylor Swift	258 million
Bad Bunny	45.5 million
Sting	1.4 million
James Cameron	607 thousand
Genesis	132 thousand
–	
James L Brooks	no Instagram
Matt Groening	no Instagram

7. Adding $300 \times 30 = 9000$ to the number of followers of the last person in the ranking.

8. Student answers may vary. Sample answer is:

Lionel Messi:	$130 million
Lebron James:	$121.2 million
Cristiano Ronaldo	$115 million
Neymar	$95 million
Stephen Curry	$92.8 million
Kevin Durrant	$92.1 million
Roger Federer	$90.7 million

9. Sample answer:
   Nursing ($60 000 − $65 000), Business and Management ($106 884), Law and Paralegal Studies ($60 040), Veterinary Studies ($95 000), Dentistry ($145 000), Teacher Education ($73 737), Engineering ($100 000), Medicine ($270 000), Rehabilitation ($100 000 to $120 000), Pharmacy ($99 064).

Exercise 3.6 Review questions

1. a. 82 mm   b. 2.9 cm

2. a. m   b. kg   c. l

3. a. 250 mm   b. 4.7 cm   c. 2.86 m
   d. 360 cm   e. 0.086 km   f. 0.018 35 km
   g. 128 000 cm   h. 84 500 mm

4. a. 0.03 L   b. 4500 mL   c. 0.89 mL
   d. 1.289 L   e. 1.4 kg   f. 5300 g

5. a. 0.000 385 kg   b. 43 g
   c. 52 000 kg   d. 2.15 t

6. a. 83.4 cm   b. 531 mm   c. 9.206 km
   d. 86 805 m   e. 1531.409 kg   f. 1227 279 g

7. a. 1800 s   b. 5.83 min   c. 162 000 s
   d. 0.97 h   e. 25 d   f. 1824 h
   g. 4.14 weeks   h. 49 days   i. 0.21 weeks
   j. 504 h

8. a. 3.5 years   b. 78 months   c. 20 decades
   d. 4 centuries   e. 30 years   f. 4.5 decades

9. a. 0.5   b. 0.05

10. a. 2.5 km/h
    b. Between 52.5 km/h and 57.5 km/h

11. a. 0.05 m
    b. 0.005 L
    c. 0.5 s
    d. 0.005 cm

12. a. 0.005 km
    b. 0.15%
    c. 3.445 and 3.455
    d. Yes, because it falls outside the range in part **c**.

13. 25.3 or 25.7 cm

14. a. 3.8967×10^8 b. 4.8603×10^{-3}
    c. $4.060\,094\,37 \times 10^8$ d. 1.00×10^{-7}
    e. 3.750×10^{-1} f. 1.003×10^{11}

15. a. 3 560 000   b. 0.000 000 001 245
    c. 1 006 000 000 000   d. 2160
    e. 0.000 000 009 483   f. 0.000 000 000 000 002 048

16. a. 6500   b. 73.00   c. 7400
    d. 7000   e. 8.0   f. 0.000 601

17. a. 3.5×10^4 (2 significant figures)
    b. 5.80×10^{-5} (3 significant figures)
    c. 7.1×10^{-2} (2 significant figures)
    d. $1.607\,154 \times 10^2$ (7 significant figures)
    e. 5.0001×10^3 (5 significant figures)

18. a. 14.0 m   b. 8582 kg   c. 80 000 g
    d. 3 000 100   e. 0.0054 min   f. 35.58 days

19. a. 522 500 m   b. 85 km/h   c. 6.72 s

20. 1.4×10^{12}

21. 4.167×10^8

22. 9.16×10^7 km

23. 1.2441×10^9 km

24. a. $3.4 \times 10^7, 1.5 \times 10^9, 6 \times 10^9$
    b. 0.57%
    c. 4
    d. 0.25 km^2

25. 10.4%

26. 820 000 queen bees

27. At the beginning of the 5th year, or in the 5.5 years.

28. a. Mach 9.1
    b.

Mach number	Speed in m/s	Speed in km/h
1	343	1234.8
2	686	2469.6
3	1029	3704.4

    c. At Mach 1 it would take 32.39 hours
    At Mach 2 it would take 16.20 hours.
    At Mach 3 it would take 10.79 hours.
    d. 3.55 hours
    e. The fastest aircraft could circle Earth 9.12 times in the time an aircraft travelling at Mach 1 could circle Earth once.
    The fastest aircraft could circle Earth 4.56 times in the time an aircraft travelling at Mach 2 could circle Earth once.
    The fastest aircraft could circle Earth 3.04 times in the time an aircraft travelling at Mach 3 could circle Earth once.
    f. (Number of times to circle Earth) $\times$ (Mach value) ≈ 9.1

4 Algebraic techniques

LESSON
4.1 Overview

Why learn this?

Most people will tell you that if you want a career in fields like engineering, science, finance and software development, you need to study and do well in algebra at high school and university. Without algebra we wouldn't have landed on the Moon or be able to enjoy the technological marvels we do today, like our smartphones.

Learning algebra in school helps you to develop critical thinking skills. These skills can help with things like problem solving, logic, pattern recognition and reasoning.

Consider these situations:
- You want to work out the cost of a holiday to Japan at the current exchange rate.
- You're driving to a petrol station to refuel your car, but you only have $30 in your pocket. You want to know how much petrol you can afford.
- You're trying to work out whether a new phone plan that charges $5 per day over 24 months for the latest smartphone is worth it.

Working out the answers to these kinds of real-life questions is a direct application of algebraic thinking.

Understanding algebra can help you to make reasoned and well-considered financial and life decisions. It can help you make independent choices and assist you with many everyday tasks. Algebra is an important tool that helps us connect with — and make sense of — the world around us.

Exercise 4.1 Pre-test

1. Simplify the expression $8x - 6 + 3x - 10$.

2. Simplify $\dfrac{9ab^2}{12ab}$.

3. Simplify $8x^2y \div 12xy^2$.

4. **MC** Select the expression that is not equivalent to the other four expressions.

   A. $x - y + z$ **B.** $x - y + z$ **C.** $x - (y - z)$ **D.** $z - (x + y)$

5. Simplify $\dfrac{3}{4x} + \dfrac{5}{2x} - \dfrac{1}{x}$.

6. Expand and simplify the following expressions.

   a. $4(m + 3n) - (2m - n)$ **b.** $-x(y + 3) + y(5 - x)$

7. Expand and simplify the following expression.
   $(3q - r)(2q + r)$.

8. **MC** Students were asked to expand $-2(3x - 4)$, and they gave the following results.
   Student 1: $-6x - 8$
   Student 2: $-6x + 6$
   Student 3: $6x - 8$
   Student 4: $-6x + 8$
   Which student has the correct answer?

   A. Student 1   **B.** Student 2   **C.** Student 3   **D.** Student 4

9. Simplify the following expression.
   $\dfrac{2a}{3} + \dfrac{a}{4}$

10. Simplify the following expression.
    $\dfrac{4y}{15} \times \dfrac{3}{2y^2}$

11. **MC** The equivalent of $(7 - 5x)(2 - 3x)$ is:

    A. $14 - 15x^2$ **B.** $14 - 31x + 15x^2$ **C.** $14 + 15x^2$ **D.** $14 - 31x - 15x^2$

12. **PATH** Expand and simplify the following expression.
    $(2b - 5)^2 - (b + 4)(b - 3)$

13. Factorise the following expressions by taking out a common factor.

    a. $8p^2 + 20p$ **b.** $25x^2 - 10y$ **c.** $6xy + 8x^2y - 3xy^2$

14. **PATH** Factorise the following expression.
    $x^2 + 3x - 4$

15. **PATH** Factorise the following trinomial
    $x^2 - 5x - 14$

LESSON
4.2 Simplifying algebraic expressions

LEARNING INTENTION

At the end of this lesson you should be able to:
- identify whether two terms are like terms
- simplify algebraic expressions by adding and subtracting like terms
- simplify algebraic expressions that involve multiplication and division of algebraic terms.

▶ 4.2.1 Addition and subtraction of like terms

eles-4592

- Simplifying an expression involves writing it in a form with the least number of terms, without $\times$ or $\div$ signs, and with any fractions expressed in their simplest form.
- **Like terms** have identical pronumeral parts (including the power), but may have different coefficients. For example:
  - $5y$ and $10y$ are like terms, but $5y^2$ and $10y$ are not like terms.
  - $3mn^2$ and $-4mn^2$ and like terms, but $3m^2n$ and $-2mn^2$ are not like terms.
- If a term contains more than one pronumeral, the convention is to write the pronumerals in alphabetical order. This makes it easier to identify like terms. For example:
  - $3cb^2a$ and $-7b^2ca$ are like terms and should be written as $3ab^2c$ and $-7ab^2c$ to make this easier to identify.
- We can simplify like terms by adding (or subtracting) the coefficients to form a single term. We cannot simplify any terms that are not like terms. For example:
  - $10xy + 3xy - 7xy = 6xy$.

WORKED EXAMPLE 1 Simplifying algebraic expressions

Simplify the following expressions.

a. $x + 2x + 3x - 5$ **b.** $9a^2b + 5ba^2 + ab^2$ **c.** $-12 - 4c^2 + 10 + 5c^2$

THINK	WRITE
a. 1. Write the expression.	**a.** $x + 2x + 3x - 5$
2. Collect the like terms x, $2x$ and $3x$.	$= 6x - 5$
b. 1. Write the expression.	**b.** $9a^2b + 5ba^2 + ab^2$
2. Collect the like terms $9a^2b$ and $5ba^2$.	$= 14a^2b + ab^2$
c. 1. Write the expression.	**c.** $-12 - 4c^2 + 10 + 5c^2$
2. Collect the like terms $4c^2$ and $5c^2$. The constants -12 and 10 are also like terms.	$= -12 + 10 - 4c^2 + 5c^2$
3. Simplify the expression.	$= -2 + c^2$ or $c^2 - 2$

▶ 4.2.2 Multiplication of algebraic terms

eles-4593

- Any number of algebraic terms can be multiplied together to produce a single term.
- Terms can be multiplied in any order, so it is easiest to multiply the coefficients first, then multiply the pronumerals in alphabetical order.
- Remember that, when multiplying pronumerals, we also add their indices together.

Note: $x = x^1$

$$
\begin{aligned}
4ab \times 3a^2b \times 2a^3 &= 4 \times a \times b \times 3 \times a^2 \times b \times 2 \times a^3 \\
&= 4 \times 3 \times 2 \times a \times a^2 \times a^3 \times b \times b \\
&= 24 \times a^6 \times b^2 \\
&= 24a^6b^2
\end{aligned}
$$

WORKED EXAMPLE 2 Multiplication of algebraic terms

Simplify the following expressions.

a. $4a \times 2b \times a$

b. $7ax \times -6bx \times -2abx$

THINK

a. 1. Write the expression.

    2. Rearrange the expression, writing the coefficients first.

    3. Multiply the coefficients and pronumerals separately.

b. 1. Write the expression.

    2. Rearrange the expression, writing the coefficients first.

    3. Multiply the coefficients and pronumerals separately.

WRITE

a. $4a \times 2b \times a$

    $= 4 \times 2 \times a \times a \times b$

    $= 8a^2b$

b. $7ax \times -6bx \times -2abx$

    $= 7 \times -6 \times -2 \times a \times a \times b \times b \times x \times x \times x$

    $= 84a^2b^2x^3$

▶ 4.2.3 Division of algebraic terms

eles-4594

- When dividing two algebraic terms, first rewrite the terms in fraction form.
- Simplify the fraction by dividing through by the highest common factor of the coefficients and cancelling down the pronumerals.
- Remember that when dividing pronumerals we subtract their indices.

$$
\begin{aligned}
36x^3yz^2 \div 40x^2y^2 &= \frac{36x^3yz^2}{40x^2y^2} \\
&= \frac{\overset{9}{\cancel{36}} \times x \times x \times \cancel{x} \times \cancel{x} \times \cancel{y} \times z \times z}{\underset{10}{\cancel{40}} \times \cancel{x} \times \cancel{x} \times \cancel{x} \times y \times \cancel{y}} \\
&= \frac{9xz^2}{10y}
\end{aligned}
$$

Simplify the following terms.

a. $\dfrac{12xy}{4xz}$

b. $-8ab^2 \div (-16a^2b)$

THINK	WRITE
a. 1. Write the term.	a. $\dfrac{12xy}{4xz}$
2. Cancel the common factors.	$= \dfrac{^3\cancel{12} \times \cancel{x} \times y}{_1\cancel{4} \times \cancel{x} \times z}$
3. Write the answer.	$= \dfrac{3y}{z}$
b. 1. Write the term as a fraction.	b. $\dfrac{-8ab^2}{-16a^2b}$
2. Cancel the common factors.	$= \dfrac{^1\cancel{-8} \times \cancel{a} \times \cancel{b^2}\,b}{_2\cancel{-16} \times_a \cancel{a^2} \times \cancel{b}}$
3. Simplify and write the answer.	$= \dfrac{1}{2} \times \dfrac{1}{a} \times \dfrac{b}{1}$
	$= \dfrac{b}{2a}$

DISCUSSION

Is the expression ab the same as ba? Explain your answer.

 Resources

Exercise 4.2 Simplifying algebraic expressions

4.2 Quick quiz on   **4.2 Exercise**

Individual pathways

■ PRACTISE	■ CONSOLIDATE	■ MASTER
1, 3, 9, 10, 11, 14, 19, 22, 25	2, 4, 5, 7, 12, 15, 17, 20, 23, 26, 27	6, 8, 13, 16, 18, 21, 24, 28, 29

Fluency

1. For each of the following terms, select the terms listed in brackets that are like terms.

   a. $6ab$ $\quad (7a, 8b, 9ab, -ab, 4a^2b^2)$

   b. $-x$ $\quad (3xy, -xy, 4x, 4y, -yx)$

   c. $3az$ $\quad (3ay, -3za, -az, 3z^2a, 3a^2z)$

   d. x^2 $\quad (2x, 2x^2, 2x^3, -2x, -x^2)$

2. For each of the following terms, select the terms listed in brackets that are like terms.

   a. $2x^2y$ $\quad (xy, -2xy, -2xy^2, -2x^2y, -2x^2y^2)$

   b. $3x^2y^5$ $\quad (3xy, 3x^5y^2, 3x^4y^3, -x^2y^5, -3x^2y^5)$

   c. $5x^2\,p^3w^5$ $\quad (-5x^3w^5\,p^3, p^3\,x^2\,w^5, 5xp^3w^5, -5x^2\,p^3w^5, w^5\,p^2\,x^3)$

   d. $-x^2y^5z^4$ $\quad (-xy^5, -y^2z^5x^4, -x+y+z, 4y^5z^4x^2, -2x^2z^4y^5)$

3. **WE1** Simplify the following expressions.

   a. $5x + 2x$

   b. $3y + 8y$

   c. $7m + 12m$

   d. $13q - 2q$

   e. $17r - 9r$

   f. $-x + 4x$

4. Simplify the following expressions.

   a. $5a + 2a + a$

   b. $9y + 2y - 3y$

   c. $7x - 2x + 8x$

   d. $14p - 3p + 5p$

   e. $2q^2 + 7q^2$

   f. $5x^2 - 2x^2$

5. Simplify the following expressions.

   a. $6x^2 + 2x^2 - 3y$

   b. $3m^2 + 2n - m^2$

   c. $-2g^2 - 4g + 5g - 12$

   d. $-5m^2 + 5m - 4m + 15$

   e. $12a^2 + 3b + 4b^2 - 2b$

   f. $6m + 2n^2 - 3m + 5n^2$

6. Simplify the following expressions.

   a. $3xy + 2y^2 + 9yx$

   b. $3ab + 3a^2b + 2a^2b - ab$

   c. $9x^2y - 3xy + 7yx^2$

   d. $4m^2n + 3n - 3m^2n + 8n$

   e. $-3x^2 - 4yx^2 - 4x^2 + 6x^2y$

   f. $4 - 2a^2b - ba^2 + 5b - 9a^2$

7. **MC** Choose which of the following is a simplification of the expression $18p - 19p$.

   A. p $\qquad$ **B.** $-p$ $\qquad$ **C.** p^2 $\qquad$ **D.** -1

8. **MC** Choose which of the following is a simplification of the expression $5x^2 - 8x + 6x - 9$.

   A. $3x - 9$ $\qquad$ **B.** $3x^2 - 9$

   C. $5x^2 + 2x - 9$ $\qquad$ **D.** $5x^2 - 2x - 9$

9. **MC** Choose which of the following is a simplification of the expression $12a - a + 15b - 14b$.

   A. $11a + b$ $\qquad$ **B.** 12 $\qquad$ **C.** $11a - b$ $\qquad$ **D.** $13a + b$

10. **MC** Choose which of the following is a simplification of the expression $-7m^2n + 5m^2 + 3 - m^2 + 2m^2n$.

    A. $-9m^2n + 4m^2 + 3$ $\qquad$ **B.** $-9m^2n + 8$

    C. $-5m^2n - 4m^2 + 3$ $\qquad$ **D.** $-5m^2n + 4m^2 + 3$

11. **WE2** Simplify the following.

    a. $3m \times 2n$ $\quad$ b. $4x \times 5y$ $\quad$ c. $2p \times 4q$ $\quad$ d. $5x \times -2y$ $\quad$ e. $3y \times -4x$

12. Simplify the following.

 a. $-3m \times -5n$ b. $5a \times 2a$ c. $3mn \times 2p$ d. $-6ab \times b$ e. $-5m \times -2mn$

13. Simplify the following.

 a. $-6a \times 3ab$ b. $-3xy \times -5xy \times 2x$ c. $4pq \times -p \times 3q^2$

 d. $4c \times -7cd \times 2c$ e. $-3a^2 \times -5ab^3 \times 2ab^4$

14. **WE3** Simplify the following.

 a. $\dfrac{6x}{2}$ b. $\dfrac{9m}{3}$ c. $\dfrac{12y}{6}$ d. $\dfrac{8m}{2}$ e. $12m \div 3$

15. Simplify the following.

 a. $14x \div 7$ b. $-21x \div 3$ c. $-32m \div 8$ d. $\dfrac{4m}{8}$ e. $\dfrac{6x}{18}$

16. Simplify the following.

 a. $\dfrac{8mn}{18n}$ b. $\dfrac{6ab}{12a^2b}$ c. $\dfrac{28xyz}{14x}$

 d. $\dfrac{2x^2yz}{8xz}$ e. $-7xy^2z^2 \div 11xyz$

Understanding

17. Simplify the following.

 a. $5x \times 4y \times 2xy$ b. $7xy \times 4ax \times 2y$ c. $\dfrac{6x^2y}{12y^2}$ d. $\dfrac{-15x^2ab}{12b^2x^2}$

18. Simplify the following.

 a. $\dfrac{2p^3q^2}{p^3q^2}$ b. $-4a \times -5ab^2 \times 2a$ c. $-a \times 4ab \times 2ba \times b$ d. $2a \times 2a \times 2a \times 2a$

19. Jim buys m pens at p cents each and n books at q dollars each.

 a. Calculate how much Jim spends in:

  i. dollars   ii. cents.

 b. Calculate how much change Jim will have if he starts with $20.

20. At a local discount clothing store 4 shirts and 3 pairs of shorts cost $138 in total. If a pair of shorts costs 2.5 times as much as a shirt, calculate the cost of each kind of clothing.

21. Anthony and Jamila are taking their 3 children to see a movie at the cinema. The total cost for the 2 adults and 3 children is $108. If an adult's ticket is 1.5 times the cost of a child's, calculate the cost of 1 child's ticket plus 1 adult's ticket.

Communicating, reasoning and problem solving

22. Class 9A were given an algebra test. One of the questions is shown below.

 Simplify the following expression: $\dfrac{3ab}{2} \times \dfrac{4ac}{6b} \times 7c$.

 Sean, who is a student in class 9A, wrote his answer as $\dfrac{12aabc}{12b} \times 7c$. Explain why Sean's answer is incorrect, and write the correct answer.

23. Using the appropriate method to divide fractions, simplify the following expression.

$$\frac{15a^2b^2}{16c^4} \div \left(\frac{5ac^2}{4b^3} \times \frac{a}{2b^4} \right)$$

24. Answer the following questions.

   a. Simplify the expression $\dfrac{n}{(n+1)} \times \dfrac{(n+1)}{(n+2)}$.

   b. Simplify the expression $\dfrac{n}{(n+1)} \times \dfrac{(n+1)}{(n+2)} \times \dfrac{(n+2)}{(n+3)}$.

   c. Use the results from parts **a** and **b** to evaluate $\dfrac{1}{2} \times \dfrac{2}{3} \times \dfrac{3}{4} \times \ ... \ \times \dfrac{99}{100}$.

25. a. Fill in the empty bricks in the following pyramids. The expression in each brick is obtained by adding the expressions in the two bricks below it.

   i.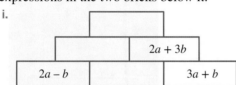

   ii.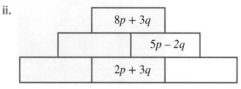

   b. Fill in the empty bricks in the following pyramids. The expression in each brick is obtained by multiplying the expressions in the two bricks below it.

   i.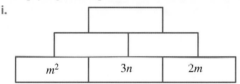

   ii.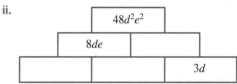

26. For the rectangle shown:

   a. write an expression for the perimeter of the rectangle

   b. show that the expression for the perimeter can be simplified to $6\dfrac{4}{7}w$

   c. calculate the cost (to the nearest cent) to create a wire frame for the rectangle using wire that costs $1.57 per metre, if w equals seven metres.

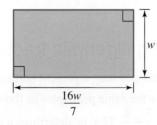

27. A rectangular chocolate block has dimensions x and $(4x - 7)$.

   a. Write an expression for the perimeter (P).
   b. Expand and simplify this expression.
   c. Calculate the perimeter of the chocolate block when $x = 3$.
   d. Explain why x cannot equal 1.

28. A doghouse in the shape of a rectangular prism is to be reinforced with steel edging along its outer edges including the base. The cost of the steel edging is $1.50 per metre. All measurements are in centimetres.

   a. Write an expression for the total length of the straight edges of the frame.
   b. Expand the expression.
   c. Evaluate the cost of steel needed when $x = 80$ cm.

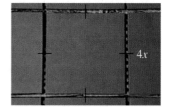

29. A tile manufacturer produces tiles that have the side lengths shown. All measurements are in centimetres.

   a. Write an expression for the perimeter of each shape.
   b. Evaluate the value of x for which the perimeter of the triangular tile is the same as the perimeter of the square tile.

LESSON
4.3 Simplifying algebraic fractions

LEARNING INTENTION

At the end of this lesson you should be able to:
- add and subtract algebraic fractions
- multiply and divide algebraic fractions

4.3.1 Adding and subtracting algebraic fractions

eles-6203

- Algebraic fractions contain pronumerals that may represent particular numbers or changing values.
- Simplifying algebraic fractions follows the same principles as those used for simplifying numerical fraction expressions such as $\frac{2}{5} + \frac{5}{7}$ or $\frac{3}{4} \div \frac{4}{5}$. That is, determine a common denominator for addition and subtraction and use reciprocals for division.
- To add or subtract algebraic fractions we perform the following steps.
   Step 1 Determine the lowest common denominator (LCD) by determining the lowest common multiple (LCM) of the denominators.
   Step 2 Rewrite each fraction as an equivalent fraction with this common denominator.
   Step 3 Add (or subtract) the new numerators.
   Step 4 Simplify.

Worked Example 4 Adding and subtracting simple fractions

Simplify each of the following expressions.

a. $\dfrac{x}{2} + \dfrac{x}{5}$

b. $\dfrac{y}{3} - \dfrac{2y}{7}$

c. $\dfrac{y+1}{5} + \dfrac{x+y}{6}$

THINK

WRITE

a. 1. Write the expression.

2. Determine the lowest common denominator (LCD).
   The lowest common multiple (LCM) of 2 and 5 is 10.

3. Add the numerators.

b. 1. Write the expression.

2. Determine the LCD. The LCM of 3 and 7 is 21.

3. Subtract the numerators.

c. 1. Write the expression.

2. Determine the LCD. The LCM of 5 and 6 is 30.

3. Add the numerators.

4. Expand the grouping symbols in the numerator.

5. Simplify the numerator by collecting like terms.

a. $\dfrac{x}{2} + \dfrac{x}{5}$

$= \dfrac{5x}{10} + \dfrac{2x}{10}$

$= \dfrac{7x}{10}$

b. $\dfrac{y}{3} - \dfrac{2y}{7}$

$= \dfrac{7y}{21} - \dfrac{6y}{21}$

$= \dfrac{y}{21}$

c. $\dfrac{y+1}{5} + \dfrac{x+y}{6}$

$= \dfrac{6(y+1)}{30} + \dfrac{5(x+y)}{30}$

$= \dfrac{6(y+1) + 5(x+y)}{30}$

$= \dfrac{6y + 6 + 5x + 5y}{30}$

$= \dfrac{5x + 11y + 6}{30}$

- If pronumerals appear in the denominator we can treat these separately to their coefficients (numbers).
- In such a case the lowest common denominator (LCD) is found by determining the lowest common multiple (LCM) of the coefficients, then including in the LCD every pronumeral used.

WORKED EXAMPLE 5 Adding and subtracting fractions with pronumeral in denominator

Simplify each of the following expressions.

a. $\dfrac{1}{x} - \dfrac{3}{2x}$

b. $\dfrac{2x}{3y} + \dfrac{3}{5y}$

c. $\dfrac{6}{5z} - \dfrac{3x}{10}$

THINK

WRITE

a. 1. Write the expression.

2. Determine the LCD. The LCM of 1 and 2 is 2. The only pronumeral is x, so include in the LCD. The LCD is $2x$.

3. Subtract the numerators.

a. $\dfrac{1}{x} - \dfrac{3}{2x}$

$= \dfrac{2}{2x} - \dfrac{3}{2x}$

$= -\dfrac{1}{2x}$

b. 1. Write the expression.

b. $\dfrac{2x}{3y} + \dfrac{3}{5y}$

2. Determine the LCD. The LCM of 3 and 5 is 15. The only pronumeral is y so include it in the LCD. The LCD is $15y$.

$= \dfrac{10x}{15y} + \dfrac{9}{15y}$

3. Add the numerators.

$= \dfrac{10x + 9}{15y}$

c. 1. Write the expression.

c. $\dfrac{6}{5z} - \dfrac{3x}{10}$

2. Determine the LCD. The LCM of 5 and 10 is 10. The only pronumeral is z, so include it in the LCD. The LCD is $10z$.

$= \dfrac{12}{10z} - \dfrac{3xz}{10z}$

3. Subtract the numerators.

$= \dfrac{12 - 3xz}{10z}$

▶ 4.3.2 Multiplying and dividing algebraic fractions

eles-6204

- The rules for multiplication and division are the same as for numerical fractions.
- When multiplying algebraic fractions, multiply the numerators and multiply the denominators, then cancel any common factors in the numerator and denominator.

WORKED EXAMPLE 6 Simplifying simple algebraic fractions

Simplify each of the following.

a. $\dfrac{x}{2} \times \dfrac{6}{y}$

b. $\dfrac{y}{2x} \times \dfrac{2z}{3y}$

THINK

WRITE

a. 1. Write the algebraic fractions.

a. $\dfrac{x}{2} \times \dfrac{6}{y}$

2. Multiply the numerators and multiply the denominators.

$= \dfrac{6x}{2y}$

3. Check for common factors in the numerator and denominator and simplify. The numbers 6 and 2 have a common factor of 2.

$= \dfrac{3x}{y}$

b. 1. Write the algebraic fractions.

b. $\dfrac{y}{2x} \times \dfrac{2z}{3y}$

2. Multiply the numerators and multiply the denominators.

$= \dfrac{2yz}{6xy}$

3. Check for common factors in the numerator and denominator and simplify. The numbers 2 and 6 have a common factor of 2. Cancel y.

$= \dfrac{z}{3x}$

- When dividing algebraic fractions, change the division sign to a multiplication sign and write the following fraction as its reciprocal.

Worked Example 7 Dividing simple algebraic fractions

Simplify each of the following.

a. $\dfrac{1}{x} \div \dfrac{4}{x}$

b. $\dfrac{3xy}{2} \div \dfrac{4x}{9y}$

THINK	WRITE
a. 1. Write the algebraic fractions.	a. $\dfrac{1}{x} \div \dfrac{4}{x}$
2. Change the division sign to a multiplication sign and write the second fraction as its reciprocal.	$= \dfrac{1}{x} \times \dfrac{x}{4}$
3. Multiply the numerators and multiply the denominators.	$= \dfrac{x}{4x}$
4. Check for common factors in the numerator and denominator. Cancel x.	$= \dfrac{1}{4}$
b. 1. Write the algebraic fractions.	b. $\dfrac{3xy}{2} \div \dfrac{4x}{9y}$
2. Change the division sign to a multiplication sign and write the second fraction as its reciprocal.	$= \dfrac{3xy}{2} \times \dfrac{9y}{4x}$
3. Multiply the numerators and multiply the denominators.	$= \dfrac{27xy^2}{8x}$
4. Check for common factors in the numerator and denominator. Cancel x.	$= \dfrac{27y^2}{8}$

DISCUSSION

Using numerical values for a, explain the difference between $\dfrac{3a}{9}$ and $\dfrac{9}{3a}$.

Exercise 4.3 Simplifying algebraic fractions

learn

4.3 Quick quiz on

4.3 Exercise

Individual pathways

■ PRACTISE	■ CONSOLIDATE	■ MASTER
1, 4, 7, 8, 12	2, 5, 9, 11, 13, 15	3, 6, 10, 14

Fluency

1. **WE4** Simplify each of the following expressions. Give your answers in simplified fraction form.

a. $\dfrac{x}{3} + \dfrac{x}{4}$

b. $\dfrac{y}{2} - \dfrac{y}{3}$

c. $\dfrac{m}{8} - \dfrac{m}{4}$

d. $\dfrac{x}{6} + \dfrac{x}{12}$

e. $\dfrac{m}{2} + \dfrac{m}{7}$

2. Simplify each of the following expressions. Give your answer in simplified fraction form.

   a. $\dfrac{t}{3} - \dfrac{t}{5}$

   b. $\dfrac{3a}{2} - \dfrac{a}{5}$

   c. $\dfrac{2p}{3} - \dfrac{5p}{6}$

   d. $\dfrac{4p}{5} + \dfrac{p}{3}$

   e. $\dfrac{5x}{6} - \dfrac{2x}{3}$

3. **WE5** Simplify each of the following expressions. Give your answer in simplified fraction form.

   a. $\dfrac{3}{2p} - \dfrac{2}{p}$

   b. $\dfrac{1}{2x} + \dfrac{3}{5x}$

   c. $\dfrac{3}{4m} - \dfrac{7}{2m}$

4. Simplify each of the following expressions. Give your answer in simplified fraction form.

   a. $\dfrac{8}{5b} - \dfrac{5}{4b}$

   b. $\dfrac{11}{6c} - \dfrac{4}{9c}$

   c. $\dfrac{2}{3y} + \dfrac{4}{5y}$

5. **WE6** Simplify each of the following.

   a. $\dfrac{2}{3} \times \dfrac{9}{2}$

   b. $\dfrac{3}{5} \times \dfrac{10}{3}$

   c. $\dfrac{5}{12} \times \dfrac{3}{5}$

   d. $\dfrac{4}{15} \times \dfrac{3}{4}$

   e. $\dfrac{x}{3} \times \dfrac{9}{x}$

6. Simplify each of the following. Give your answer in simplified fraction from.

   a. $\dfrac{4}{y} \times \dfrac{y}{12}$

   b. $\dfrac{4}{3} \times \dfrac{m}{16}$

   c. $\dfrac{n}{9} \times \dfrac{3}{2}$

   d. $\dfrac{7m}{5} \times \dfrac{10}{m}$

   e. $\dfrac{5}{3x} \times \dfrac{x}{15}$

7. Simplify each of the following. Give your answer in simplified fraction form.

   a. $\dfrac{20}{3y} \times \dfrac{6}{5}$

   b. $\dfrac{2x}{3} \times \dfrac{15}{6}$

   c. $\dfrac{4m}{27} \times \dfrac{9}{7m}$

   d. $\dfrac{2}{15} \times \dfrac{7}{p} \times \dfrac{p}{21}$

   e. $\dfrac{x}{22} \times \dfrac{11}{12} \times \dfrac{6}{x}$

Understanding

8. Simplify each of the following.

   a. $\dfrac{10a^4}{5a^2}$

   b. $\dfrac{9a^2 b}{3ab}$

   c. $\dfrac{3ab}{9a^2 b}$

9. **WE7** Simplify each of the following.

   a. $\dfrac{2}{5} \div \dfrac{2}{15}$

   b. $\dfrac{3}{4} \div \dfrac{3}{8}$

   c. $\dfrac{5}{6} \div \dfrac{15}{6}$

   d. $\dfrac{9}{10} \div \dfrac{36}{10}$

   e. $\dfrac{x}{3} \div \dfrac{x}{9}$

10. Simplify each of the following. Give your answer in simplified fraction form where appropriate.

   a. $\dfrac{4}{m} \div \dfrac{12}{m}$

   b. $\dfrac{a}{5} \div \dfrac{a}{20}$

   c. $\dfrac{6}{b} \div \dfrac{20}{b}$

   d. $\dfrac{3a}{14} \div \dfrac{a}{7}$

   e. $\dfrac{21}{4} \div \dfrac{3}{b}$

11. Simplify each of the following. Give your answer in simplified fraction form.

a. $\dfrac{6m}{15} \div \dfrac{2}{3}$

b. $\dfrac{ab}{9} \div \dfrac{a}{24}$

c. $\dfrac{2m}{3p} \div \dfrac{10m}{9pq}$

d. $\dfrac{3}{5} \div \left(\dfrac{10}{m} \div \dfrac{12}{m} \right)$

e. $\dfrac{3x}{8} \div \left(\dfrac{2y}{15} \div \dfrac{y}{4} \right)$

Communicating, reasoning and problem solving

12. Simplify each of the following. Give your answer in simplified fraction form.

a. $\dfrac{3x^2}{8y^5} \div \dfrac{15x^3}{4y}$

b. $\dfrac{a^2b^4}{6} \times \dfrac{9}{a^2b^2}$

c. $\dfrac{3}{x} - \dfrac{1}{2x}$

13. Simplify each of the following. Give your answer in simplified fraction form.

a. $\dfrac{21b^2}{4} \div \dfrac{3}{b^3}$

b. $\dfrac{x}{22y} \times \dfrac{11y^2}{12z} \times \dfrac{6z}{xy}$

c. $\dfrac{3x^3}{8y} \times \dfrac{2y^2}{15} \div \dfrac{y^2}{4}$

14. Consider the triangle shown.

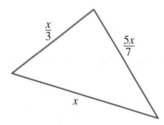

a. Write an expression for the perimeter of the triangle.

b. Show that the perimeter can be simplified to $\dfrac{43}{21}x$

c. If x is 2 metres, calculate the cost (to the nearest dollar) to frame the triangle using timber which costs $4.50 per metre.

15. The following is a student's working.

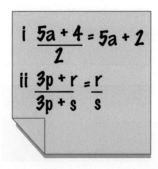

a. What is the student doing wrong?

b. What should the solutions be?

LESSON
4.4 Expanding algebraic expressions and collecting like terms

LEARNING INTENTION

At the end of this lesson you should be able to:
- apply the distributive law to expand and simplify an expression with a single set of brackets
- apply the distributive law to expand and simplify expressions containing two or more sets of brackets.

▶ 4.4.1 The Distributive Law

eles-4595

- There are two ways of calculating the area of the rectangle shown.

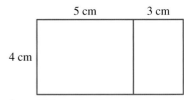

1. The rectangle can be treated as a single shape with a length of 4 cm and a width of $5 + 3$ cm.

$$A = l \times w$$
$$= 4(5 + 3)$$
$$= 4 \times 8$$
$$= 32 \, \text{cm}^2$$

2. The areas of the two smaller rectangles can be added together.

$$A = 4 \times 5 + 4 \times 3$$
$$= 20 + 12$$
$$= 32 \, \text{cm}^2$$

- The length of the rectangle shown is a and its width is $b + c$.

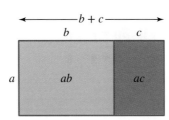

- This rectangle's area can be found in two different ways.
  1. $A = l \times w$
     $$= a(b + c)$$

This expression is described as **factorised** because it shows one number (or factor) multiplied by another. The two factors are a and $(b + c)$.

  2. The areas of the two small rectangles can be added together.

$$A = ab + ac$$

This expression is described as **expanded**, which means that it is written without brackets.
- $a(b + c) = ab + ac$ is called the **Distributive Law**.

The Distributive Law

In order to expand a single set of brackets we apply the distributive law, which states that:

$$a(b + c) = a \times b + a \times c = ab + ac$$

- A helpful way to expand brackets is to draw arrows between factors as you work out each multiplication.

$$x(x - y) = x^2 \dots$$
$$x(x - y) = x^2 - xy$$

WORKED EXAMPLE 8 Expanding a single set of brackets

Use two different methods to calculate the value of the following.

a. $7(5 + 15)$ b. $10(9 - 1)$

THINK

a. 1. Method 1: Work out the brackets first, then evaluate.

2. Method 2: Expand the brackets first, then evaluate.

b. 1. Method 1: Work out the brackets first, then evaluate.

2. Method 2: Expand the brackets first, then evaluate

WRITE

a. $7(5 + 15)$
$= 7 \times 20$
$= 140$

$7(5 + 15)$
$= 7 \times 5 + 7 \times 15$
$= 35 + 105$
$= 140$

b. $10(9 - 1)$
$= 10 \times 8$
$= 80$

$10(9 - 1)$
$= 10 \times 9 + 10 \times (-1)$
$= 90 - 10$
$= 80$

WORKED EXAMPLE 9 Expanding using arrows

Expand the following expressions.

a. $5(x + 3)$ b. $-4y(2x - w)$ c. $3x(5 - 6y + 2y)$

THINK

a. 1. Draw arrows to help with the expansion.

2. Simplify and write the answer.

WRITE

a. $5(x + 3)$

$= 5 \times x + 5 \times 3$
$= 5x + 15$

b. 1. Draw arrows to help with the expansion.

b.

$$-4y(2x - w)$$

2. Simplify and write the answer.

$$= -4y \times 2x - 4y \times (-w)$$
$$= -8xy + 4wy$$

c. 1. Draw arrows to help with the expansion.

c.

$$3x(5 - 6x + 2y)$$

2. Simplify and write the answer.

$$= 3x \times 5 + 3x \times (-6x) + 3x \times 2y$$
$$= 15x - 18x^2 + 6xy$$

▶ 4.4.2 Expanding and simplifying

eles-4596

- When a problem involves expanding that is more complicated, it is likely that the like terms will need to be collected and simplified after you have expanded the brackets.
- When solving these more complicated problems, expand all sets of brackets first, then simplify any like terms that result from the expansion.

WORKED EXAMPLE 10 Expanding and simplifying

Expand and simplify the following expressions by collecting like terms.
a. $4(x - 4) + 5$ **b.** $x(y - 2) + 5x$ **c.** $-x(y - z) + 5x$ **d.** $7x - 6(y - 2x)$

THINK

a. 1. Expand the brackets.

2. Simplify and write the answer.

b. 1. Expand the brackets.

2. Simplify and write the answer.

c. 1. Expand the brackets.

2. Simplify and write the answer. (*Note:* There are no like terms.)

d. 1. Expand the brackets.

2. Simplify and write the answer.

WRITE

a.
$$4(x - 4) + 5$$

$$= 4x - 16 + 5$$
$$= 4x - 11$$

b.
$$x(y - 2) + 5x$$

$$= xy - 2x + 5x$$
$$= xy + 3x$$

c.
$$-x(y - z) + 5x$$

$$= -xy + xz + 5x$$

d.
$$7x - 6(y - 2x)$$

$$= 7x - 6y + 12x$$
$$= 19x - 6y$$

Expand and simplify the following expressions.

a. $5(x + 2y) + 6(x - 3y)$

b. $5x(y - 2) - y(x + 3)$

THINK	WRITE
a. 1. Expand each set of brackets.	a. $5(x + 2y) + 6(x - 3y)$
2. Simplify and write the answer.	$= 5x + 10y + 6x - 18y$ $= 11x - 8y$
b. 1. Expand each set of brackets.	b. $5x(y - 2) - y(x + 3)$
2. Simplify and write the answer.	$= 5xy - 10x - xy - 3y$ $= 4xy - 10x - 3y$

DISCUSSION

Use mathematical reasoning to explain why $(a + b)(c + d) = (c + d)(a + b)$.

 Resources

▶ **Video eLesson** Expanding brackets (eles-1888)

✦ **Interactivities** Expanding brackets (int-6034)

    Like terms (int-6035)

Exercise 4.4 Expanding algebraic expressions and collecting like terms

learn on

4.4 Quick quiz on	4.4 Exercise

Individual pathways

■ PRACTISE	■ CONSOLIDATE	■ MASTER
1, 2, 7, 9, 12, 15, 18	3, 4, 6, 10, 13, 17, 19	5, 8, 11, 14, 16, 20

Fluency

1. **WE8** Use two different methods to determine the value of the following expressions.

   a. $8(10 - 2)$ b. $11(99 + 1)$ c. $-5(3 + 1)$ d. $7(100 - 1)$

2. **WE9** Expand the following expressions. For the first two examples, draw a diagram to represent the expression.

   a. $3(x + 2)$ b. $4(x + 3)$ c. $4(x + 1)$ d. $7(x - 1)$

3. Expand the following expressions.

   a. $-3(p-2)$ **b.** $-(x-1)$ **c.** $3(2b-4)$ **d.** $8(3m-2)$

4. Expand each of the following. For the first two examples, draw a diagram to represent the expression.

   a. $x(x+2)$ **b.** $a(a+5)$ **c.** $x(4+x)$ **d.** $m(7-m)$

5. Expand each of the following.

   a. $2x(y+2)$ **b.** $-3y(x+4)$ **c.** $-b(3-a)$ **d.** $-6a(5-3a)$

6. **WE10** Expand and simplify the following expressions by collecting like terms.

   a. $2(p-3)+4$ **b.** $5(x-5)+8$ **c.** $-7(p+2)-3$
   d. $-4(3p-1)-1$ **e.** $6x(x-3)-2x$

7. Expand and simplify the following expressions by collecting like terms.

   a. $2m(m+5)-3m$ **b.** $3x(p+2)-5$ **c.** $4y(y-1)+7$
   d. $-4p(p-2)+5p$ **e.** $5(x-2y)-3y-x$

8. Expand and simplify the following expressions by collecting like terms.

   a. $2m(m-5)+2m-4$ **b.** $-3p(p-2q)+4pq-1$ **c.** $-7a(5-2b)+5a-4ab$
   d. $4c(2d-3c)-cd-5c$ **e.** $6p+3-4(2p+5)$

Understanding

9. **WE11** Expand and simplify the following expressions.

   a. $2(x+2y)+3(2x-y)$ **b.** $4(2p+3q)+2(p-2q)$ **c.** $7(2a+3b)+4(a+2b)$
   d. $5(3c+4d)+2(2c+d)$ **e.** $-4(m+2n)+3(2m-n)$

10. Expand and simplify the following expressions.

    a. $-3(2x+y)+4(3x-2y)$ **b.** $-2(3x+2y)+3(5x+3y)$ **c.** $-5(4p+2q)+2(3p+q)$
    d. $6(a-2b)-5(2a-3b)$ **e.** $5(2x-y)-2(3x-2y)$

11. Expand and simplify the following expressions.

    a. $4(2p-4q)-3(p-2q)$ **b.** $2(c-3d)-5(2c-3d)$ **c.** $7(2x-3y)-(x-2y)$
    d. $-5(p-2q)-(2p-q)$ **e.** $-3(a-2b)-(2a+3b)$

12. Expand and simplify the following expressions.

    a. $a(b+2)+b(a-3)$ **b.** $x(y+4)+y(x-2)$ **c.** $c(d-2)+c(d+5)$
    d. $p(q-5)+p(q+3)$ **e.** $3c(d-2)+c(2d-5)$

13. Expand and simplify the following expressions.

    a. $7a(b-3)-b(2a+3)$ **b.** $2m(n+3)-m(2n+1)$ **c.** $4c(d-5)+2c(d-8)$
    d. $3m(2m+4)-2(3m+5)$ **e.** $5c(2d-1)-(3c+cd)$

14. Expand and simplify the following expressions.

    a. $-3a(5a+b)+2b(b-3a)$
    b. $-4c(2c-6d)+d(3d-2c)$
    c. $6m(2m-3)-(2m+4)$
    d. $7x(5-x)+6(x-1)$
    e. $-2y(5y-1)-4(2y+3)$

15. **MC** Choose the expression that is the equivalent of $3(a+2b)+2(2a-b)$.

    A. $5a+6b$ **B.** $7a+4b$ **C.** $5(3a+b)$ **D.** $7a+8b$

16. **MC** Choose the expression that is the equivalent of $-3(x-2y)-(x-5y)$.

    A. $-4x+11y$ **B.** $-4x-11y$ **C.** $4x+11y$ **D.** $4x+7y$

17. **MC** Choose the expression that is the equivalent of $2m(n+4) + m(3n-2)$.

    A. $3m + 4n - 8$ **B.** $5mn + 4m$ **C.** $5mn + 10m$ **D.** $5mn + 6m$

Communicating, reasoning and problem solving

18. Three students gave the following incorrect answers when expanding $-5(3x - 20)$.

    i. $-5x - 20$ **ii.** $-8x + 25$ **iii.** $15x - 100$

      a. Explain the errors made by each student.   **b.** Determine the correct answer.

19. In a test, a student expanded brackets and obtained the following answers. Identify and correct the student's errors and write the correct expansions.

    a. $-2(a-5) = -2a - 10$ **b.** $2b(3b-1) = 6b^2 - 1$ **c.** $-2(c-4) = 2c + 8$

20. A student's solution for expanding and simplifying an algebraic expression is shown. Is the student correct? If not, identify where there is an error and correct the solution.

$$4a(5a - 3b) - 2b(a - 8b)$$
$$= 20a - 12ab - 2ba - 16b$$
$$= 20a - 10ab - 16b$$

LESSON
4.5 Expanding binomial products

LEARNING INTENTION

At the end of this lesson you should be able to:
- apply the distributive law to expand binomial products
- simplify the expressions by collecting like terms.

▶ 4.5.1 Expanding binomial factors

eles-4597

- Remember that a **binomial** is an expression containing two terms, for example $x + 3$ or $2y - z^2$. In this section we will look at how two binomials can be multiplied together.
- The rectangle shown has length $a + b$ and width $c + d$.

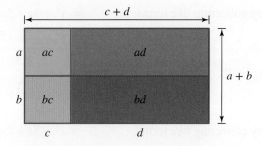

- There are two ways of calculating the area of the large rectangle.
    1. $A = l \times w$
       $$= (a+b) \times (c+d)$$
       $$= (a+b)(c+d)$$
    This is a factorised expression in which the two factors are $(a+b)$ and $(c+d)$.

2. The areas of the four small rectangles can be added together.
$$A = ac + ad + bc + bd$$
So $(a+b)(c+d) = ac + ad + bc + bd$

- There are several methods that can be helpful when remembering how to expand binomial factors. One commonly used method is **FOIL**.
- Each of the letters in FOIL stand for:
  First — multiply the first term in each bracket.
  Outer — multiply the two outer terms of each bracket.
  Inner — multiply the two inner terms of each bracket.
  Last — multiply the last term of each bracket.

FOIL – First Outer Inner Last

$$(a+b)(c+d) = ac + ad + bc + bd$$

Expanding binomial factors

Using FOIL, the expanded product of two binomial factors is given by:

$$(a+b)(c+d) = a \times c + a \times d + b \times c + b \times d = ac + ad + bc + bd$$

It is expected that the result will have four terms.

Note: **It may be possible to simplify like terms after expanding.**

- *Note:* FOIL only works for expanding binomial products. When expanding brackets with more than two terms, each term in the first bracket must be multiplied by each term in the second bracket.

WORKED EXAMPLE 12 Expanding and simplifying two binomial factors

Expand and simplify each of the following expressions.
a. $(x-5)(x+3)$ b. $(x+2)(x+3)$ c. $(2x+2)(2x+3)$

THINK	WRITE
a. 1. Expand the brackets using FOIL.	a. $(x-5)(x+3)$
2. Simplify the expression by collecting like terms.	$= x^2 + 3x - 5x - 15$ $= x^2 - 2x - 15$
b. 1. Expand the brackets using FOIL.	b. $(x+2)(x+3)$
2. Simplify the expression by collecting like terms.	$= x^2 + 3x + 2x + 6$ $= x^2 + 5x + 6$

c. 1. Expand the brackets using FOIL.

2. Simplify the expression by collecting like terms.

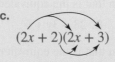

c.

$(2x + 2)(2x + 3)$

$= 4x^2 + 6x + 4x + 6$

$= 4x^2 + 10x + 6$

DISCUSSION

Explain why, when expanded, $(x + y)(2x + y)$ gives the same result as $(2x + y)(x + y)$.

---

on Resources

▶ **Video eLesson**   Expansion of binomial expressions (eles-1908)

🧩 **Interactivity**   Expanding binomial factors (int-6033)

Exercise 4.5 Expanding binomial products

learn on

| 4.5 Quick quiz on | 4.5 Exercise |

Individual pathways

■ PRACTISE	■ CONSOLIDATE	■ MASTER
1, 2, 5, 7, 9, 13	3, 4, 8, 10, 16	6, 11, 12, 14, 15, 17

Fluency

1. Expand and simplify each of the following expressions.
   a. $(a + 2)(a + 3)$ b. $(x + 4)(x + 3)$ c. $(y + 3)(y + 2)$ d. $(m + 4)(m + 5)$ e. $(b + 2)(b + 1)$

2. Expand and simplify each of the following expressions.
   a. $(p + 1)(p + 4)$ b. $(a - 2)(a + 3)$ c. $(x - 4)(x + 5)$ d. $(m + 3)(m - 4)$ e. $(y + 5)(y - 3)$

3. Expand and simplify each of the following expressions.
   a. $(y - 6)(y + 2)$ b. $(x - 3)(x + 1)$ c. $(x - 3)(x - 4)$ d. $(p - 2)(p - 3)$ e. $(x - 3)(x - 1)$

4. Expand and simplify each of the following expressions.
   a. $(2a + 3)(a + 2)$ b. $(c - 6)(4c - 7)$ c. $(7 - 2t)(5 - t)$ d. $(2 + 3t)(5 - 2t)$
   e. $(7 - 5x)(2 - 3x)$ f. $(5x - 2)(5x - 2)$

5. Expand and simplify each of the following expressions.
   a. $(x + y)(z + 1)$ b. $(2x + y)(z + 4)$ c. $(3p + q)(r + 1)$
   d. $(a + 2b)(a + b)$ e. $(2c + d)(c - 3d)$ f. $(x + y)(2x - 3y)$

6. Expand and simplify each of the following expressions.
   a. $(4p - 3q)(p + q)$ b. $(a + 2b)(b + c)$ c. $(3p - 2q)(1 - 3r)$
   d. $(4x - y)(3x - y)$ e. $(p - q)(2p - r)$ f. $(5 - 2j)(3k - 1)$

7. **MC** Choose the expression that is the equivalent of $(4-y)(7+y)$.

A. $28-y^2$　　　　B. $28-3y+y^2$　　　　C. $28-3y-y^2$　　　　D. $11-2y$

8. **MC** Choose the expression that is the equivalent of $(2p+1)(p-5)$.

A. $2p^2-5$　　　　B. $2p^2-11p-5$　　　　C. $2p^2-9p-5$　　　　D. $2p^2-6p-5$

Understanding

9. Expand the following expressions using FOIL, then simplify.

a. $(x+3)(x-3)$　　　　b. $(x+5)(x-5)$　　　　c. $(x+7)(x-7)$
d. $(x-1)(x+1)$　　　　e. $(x-2)(x+2)$　　　　f. $(2x-1)(2x+1)$

10. Expand the following expressions using FOIL, then simplify.

a. $(x+1)(x+1)$　　　　b. $(x+2)(x+2)$　　　　c. $(x+8)(x+8)$
d. $(x-3)(x-3)$　　　　e. $(x-5)(x-5)$　　　　f. $(x-9)(x-9)$

11. Simplify the following expressions.

a. $2.1x(3x+4.7y)-3.1y(1.4x+y)$　　　　b. $(2.1x-3.2y)(2.1x+3.2y)$
c. $(3.4x+5.1y)^2$

12. For the box shown, calculate the following in expanded form:

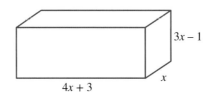

a. the total surface area　　　　b. the volume.

Communicating, reasoning and problem solving

13. For each of the following shapes:

i. write down the area in factor form
ii. expand and simplify the expression
iii. discuss any limitations on the value of x.

a.

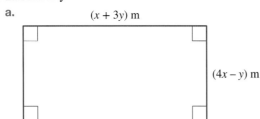

b.

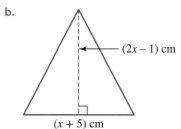

14. Show that the following is true.

$$(a-x)(a+x)-2(a-x)(a-x)-2x(a-x)$$
$$=-(a-x)^2$$

15. A series of incorrect expansions of $(x+8)(x-3)$ are shown below. For each of these incorrect expansions, explain the mistake that has been made.

a. $x^2+11x+24$　　　　b. $x^2+5x+24$　　　　c. x^2-3x

16. Three students' attempts at expanding $(3x+4)(2x+5)$ are shown.

Student A

$$(3x+4)(2x+5)$$
$$= 3x \times 2x + 3x \times 5 + 4 \times 2x + 4 \times 5$$
$$= 6x + 15x + 8x + 20$$
$$= 29x + 20$$
$$= 49x$$

Student B

$$(3x+4)(2x+5)$$
$$= 3x \times 2x + 4 \times 2x + 4 \times 5$$
$$= 6x^2 + 8x + 20$$

Student C

$$(3x+4)(2x+5)$$
$$= 3x \times 2x + 3x \times 5 + 4 \times 2x + 4 \times 5$$
$$= 6x^2 + 15x + 8x + 20$$
$$= 6x^2 + 23x + 20$$

   a. State which student's work is correct.

   b. Copy each of the incorrect answers into your workbook and correct the mistakes in each one as though you were their teacher.

17. Rectangular floor mats have an area of $(x^2 + 2x - 15)$ cm^2.

   a. The length of a mat is $(x + 5)$ cm. Determine an expression for the width of this mat.

   b. If the length of a mat is 70 cm, evaluate its width.

   c. If the width of a mat is 1 m, evaluate its length.

LESSON
4.6 Difference of two squares and perfect squares (Path)

4.6.1 Difference of two squares

eles-4598

- Consider the expansion of $(x+4)(x-4)$:

$$(x+4)(x-4) = x^2 - 4x + 4x - 16 = x^2 - 16$$

- Now consider the expansion of $(x-6)(x+6)$:

$$(x-6)(x+6) = x^2 + 6x - 6x - 36 = x^2 - 36$$

- In both cases the middle two terms cancel each other out, leaving two terms, both of which are perfect squares.
- The two terms that are left are the first value squared minus the second value squared. This is where the phrase **difference of two squares** originates.
- In both cases the binomial terms can be written in the form $(x+a)$ and $(x-a)$. If we can recognise expressions that have this form, we can use the pattern above to quickly expand those expressions. For example: $(x+12)(x-12) = x^2 - 12^2 = x^2 - 144$

Difference of two squares

The difference of two squares rule is used to quickly expand certain binomial products, as long as they are in the forms shown below:

$$(a+b)(a-b) = a^2 - b^2$$

$$(a-b)(a+b) = a^2 - b^2$$

Because the two binomial brackets are being multiplied, the order of the brackets does not affect the final result.

WORKED EXAMPLE 13 Expanding using difference of two squares

Use the difference of two squares rule to expand and simplify each of the following expressions.
a. $(x+8)(x-8)$ b. $(6-3)(6+x)$ c. $(2x-3)(2x+3)$ d. $(3x+5)(5-3x)$

THINK	WRITE
a. 1. Write the expression.	a. $(x+8)(x-8)$
2. This expression is in the form $(a+b)(a-b)$, so the difference of two squares rule can be used. Expand using the formula.	$= x^2 - 8^2$ $= x^2 - 64$
b. 1. Write the expression.	b. $(6-x)(6+x)$
2. This expression is in the form $(a+b)(a-b)$, so the difference of two squares rule can be used. Expand using the formula. *Note:* $36 - x^2$ is not the same as $x^2 - 36$.	$= 6^2 - x^2$ $= 36 - x^2$
c. 1. Write the expression.	c. $(2x-3)(2x+3)$
2. This expression is in the form $(a-b)(a+b)$, so the difference of two squares rule can be used. Expand using the formula. *Note:* $(2x)^2$ and $2x^2$ are not the same. In this case $a = 2x$, so $a^2 = (2x)^2$.	$= (2x)^2 - 3^2$ $= 4x^2 - 9$
d. 1. Write the expression.	d. $(3x+5)(5-3x)$
2. The difference of two squares rule can be used if we rearrange the terms, since $3x + 5 = 5 + 3x$. Expand using the formula.	$(5+3x)(5-3x)$ $= 5^2 - (3x)^2$ $= 25 - 9x^2$

4.6.2 Perfect squares

eles-4599

- A **perfect square** is the result of the square of a whole number. $1 \times 1 = 1, 2 \times 2 = 4$ and $3 \times 3 = 9$, showing that $1, 4$ and 9 are all perfect squares.
- Similarly, $(x+3)(x+3) = (x+3)^2$ is a perfect square because it is the result of a binomial factor multiplied by itself.
- Consider the diagram illustrating $(x+3)^2$. What shape is it?
- The area is given by $x^2 + 3x + 3x + 9 = x^2 + 6x + 9$.
- We can see from the diagram that there are two squares produced $(x^2$ and $3^2 = 9)$ and two rectangles that are identical to each other $(2 \times 3x)$.
- Compare this with the expansion of $(x+6)^2$:

$$(x+6)(x+6) = x^2 + 6x + 6x + 36 = x^2 + 12x + 36$$

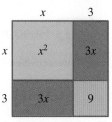

- A pattern begins to emerge after comparing these two expansions. The square of a binomial equals the square of the first term, plus double the product of the two terms plus the square of the second term. For example: $(x+10)(x+10) = x^2 + 2 \times 10 \times x + 10^2 = x^2 + 20x + 100$

Perfect squares

The rule for the expansion of the square of a binomial is given by:

$$(a+b)^2 = (a+b)(a+b) = a^2 + 2ab + b^2$$

$$(a-b)^2 = (a-b)(a-b) = a^2 - 2ab + b^2$$

We can use the above rules to quickly expand binomial products that are presented as squares.

WORKED EXAMPLE 14 Expanding perfect squares

Use the rules for expanding perfect binomial squares to expand and simplify the following.
a. $(x+1)(x+1)$
b. $(x-2)^2$
c. $(2x+5)^2$
d. $(4x-5y)^2$

THINK	WRITE
a. 1. This expression is the square of a binomial.	**a.** $(x+1)(x+1)$
2. Apply the formula for perfect squares: $(a+b)^2 = a^2 + 2ab + b^2$	$= x^2 + 2 \times x \times 1 + 1^2$ $= x^2 + 2x + 1$
b. 1. This expression is the square of a binomial.	**b.** $(x-2)^2$
2. Apply the formula for perfect squares: $(a-b)^2 = a^2 - 2ab + b^2$.	$= (x-2)(x-2)$ $= x^2 - 2 \times x \times 2 \times 2^2$ $= x^2 - 4x + 4$
c. 1. This expression is the square of a binomial.	**c.** $(2x+5)^2$
2. Apply the formula for perfect squares: $(a+b)^2 = a^2 + 2ab + b^2$.	$= (2x)^2 + 2 \times 2x \times 5 + 5^2$ $= 4x^2 + 20x + 25$
d. 1. This expression is the square of a binomial.	**d.** $(4x-5y)^2$
2. Apply the formula for perfect squares: $(a-b)^2 = a^2 - 2ab + b^2$.	$= (4x)^2 - 2 \times 4x \times 5y + (5y)^2$ $= 16x^2 - 40xy + 25y^2$

Resources

 Interactivities Difference of two squares (int-6036)

Exercise 4.6 Difference of two squares and perfect squares (Path)

learn

4.6 Quick quiz on

4.6 Exercise

Individual pathways

■ PRACTISE	■ CONSOLIDATE	■ MASTER
1, 3, 5, 8, 13, 16, 20, 21	2, 6, 9, 11, 14, 17, 22, 23	4, 7, 10, 12, 15, 18, 19, 24, 25

Fluency

1. **WE13** Use the difference of two squares rule to expand and simplify each of the following.
   a. $(x+2)(x-2)$ b. $(y+3)(y-3)$ c. $(m+5)(m-5)$ d. $(a+7)(a-7)$

2. Use the difference of two squares rule to expand and simplify each of the following.
   a. $(x+6)(x-6)$ b. $(p-12)(p+12)$ c. $(a+10)(a-10)$ d. $(m-11)(m+11)$

3. Use the difference of two squares rule to expand and simplify each of the following.
   a. $(2x+3)(2x-3)$ b. $(3y-1)(3y+1)$ c. $(5d-2)(5d+2)$
   d. $(7c+3)(7c-3)$ e. $(2+3p)(2-3p)$

4. Use the difference of two squares rule to expand and simplify each of the following.
   a. $(d-9x)(d+9x)$ b. $(5-12a)(5+12a)$ c. $(3x+10y)(3x-10y)$
   d. $(2b-5c)(2b+5c)$ e. $(10-2x)(2x+10)$

5. **WE14** Use the rule for the expansion of the square of a binomial to expand and simplify each of the following.
   a. $(x+2)(x+2)$ b. $(a+3)(a+3)$ c. $(b+7)(b+7)$ d. $(c+9)(c+9)$

6. Use the rule for the expansion of the square of a binomial to expand and simplify each of the following.
   a. $(m+12)^2$ b. $(n+10)^2$ c. $(x-6)^2$ d. $(y-5)^2$

7. Use the rule for the expansion of the square of a binomial to expand and simplify each of the following.
   a. $(9-c)^2$ b. $(8+e)^2$ c. $2(x+y)^2$ d. $(u-v)^2$

8. Use the rule for expanding perfect binomial squares rule to expand and simplify each of the following.
   a. $(2a+3)^2$ b. $(3x+1)^2$ c. $(2m-5)^2$ d. $(4x-3)^2$

9. Use the rule for expanding perfect binomial squares to expand and simplify each of the following.

   a. $(5a - 1)^2$ **b.** $(7p + 4)^2$ **c.** $(9x + 2)^2$ **d.** $(4c - 6)^2$

10. Use the rule for expanding perfect binomial squares to expand and simplify each of the following.

   a. $(5 + 3p)^2$ **b.** $(2 - 5x)^2$ **c.** $(9x - 4y)^2$ **d.** $(8x - 3y)^2$

Understanding

11. Use the difference of two squares rule to expand and simplify each of the following.

   a. $(x + 3)(x - 3)$ **b.** $(2x + 3)(2x - 3)$ **c.** $(7x - 4)(7x + 4)$

   d. $(2x + 7y)(2x - 7y)$ **e.** $(x^2 + y^2)(x^2 - y^2)$

12. Expand and simplify the following perfect squares.

   a. $(4x + 5)^2$ **b.** $(7x - 3y)^2$ **c.** $(5x^2 - 2y)^2$

   d. $2(x - y)^2$ **e.** $\left(\dfrac{2}{x} + 4x\right)^2$

13. A square has a perimeter of $4x + 12$. Calculate its area.

14. Francis has fenced off a square in her paddock for spring lambs. The area of the paddock is $(9x^2 + 6x + 1)\,\text{m}^2$.

    Using pattern recognition, determine the side length of the paddock in terms of x.

15. A square has an area of $x^2 + 18x + 81$. Determine an expression for the perimeter of this square.

Communicating, reasoning and problem solving

16. Show that $a^2 - b^2 = (a + b)(a - b)$ is true for each of the following.

   a. $a = 5, b = 4$ **b.** $a = 9, b = 1$ **c.** $a = 2,\ b = 7$ **d.** $a = -10, b = -3$

17. Lin has a square bedroom. Her sister Tasneem has a room that is 1 m shorter in length than Lin's room, but 1 m wider.

   a. Show that Lin has the larger bedroom.
   b. Determine how much bigger Lin's bedroom is than Tasneem's bedroom.

18. Expand each of the following pairs of expressions.

   a. **i.** $(x - 4)(x + 4)$ and $(4 - x)(4 + x)$

       ii. $(x - 11)(x + 11)$ and $(11 - x)(11 + x)$

       iii. $(2x - 9)(2x + 9)$ and $(9 - 2x)(9 + 2x)$

   b. State what you notice about the answers to the pairs of expansions above.
   c. Explain how this is possible.

19. Answer the following questions.

   a. Expand $(10k + 5)^2$.
   b. Show that $(10k + 5)^2 = 100k(k + 1) + 25$.
   c. Using part **b**, evaluate 25^2 and 85^2.

20. A large square has been subdivided into two squares and two rectangles.

   a. Write formulas for the areas of these four pieces, using the dimensions a and b marked on the diagram.
   b. Write an equation that states that the area of the large square is equal to the combined area of its four pieces. Do you recognise this equation?

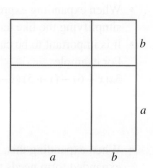

21. Expand each of the following pairs of expressions.

   a. i. $(x-3)^2$ and $(3-x)^2$
      ii. $(x-15)^2$ and $(15-x)^2$
      iii. $(3x-7)^2$ and $(7-3x)^2$

   b. State what you notice about the answers to the pairs of expansions above.
   c. Explain how this is possible.

22. Use the perfect squares rule to quickly evaluate the following.

   a. 27^2 b. 33^2 c. 39^2 d. 47^2

23. Allen is creating a square deck with a square pool installed in the middle of it. The side length of the deck is $(2x+3)$ m and the side length of the pool is $(x-2)$ m. Evaluate the area of the decking around the pool.

24. Ram wants to create a rectangular garden that is 10 m longer than it is wide. Write an expression for the area of the garden in terms of x, where x is the average length of the two sides.

25. The expansion of perfect squares $(a+b)^2 = a^2 + 2ab + b^2$ and $(a-b)^2 = a^2 - 2ab + b^2$ can be used to simplify some arithmetic calculations. For example:

$$97^2 = (100-3)^2$$
$$= 100^2 - 2 \times 100 \times 3 + 3^2$$
$$= 9409$$

Use this method to evaluate the following.

   a. 103^2 b. 62^2 c. 997^2 d. 1012^2 e. 53^2 f. 98^2

LESSON
4.7 Further expansions (Path)

LEARNING INTENTION

At the end of this lesson you should be able to:
- expand multiple sets of brackets and simplify the result.

▶ 4.7.1 Expanding multiple sets of brackets

eles-4600

- When expanding expressions with more than two sets of brackets, expand all brackets first before simplifying the like terms.
- It is important to be careful with signs, particularly when subtracting one expression from another. For example:

$$3x(x+6) - (x+5)(x-3) = 3x^2 + 18x - (x^2 + 5x - 3x - 15)$$
$$= 3x^2 + 18x - (x^2 + 2x - 15)$$
$$= 3x^2 + 18x - x^2 - 2x + 15$$
$$= 2x^2 + 16x + 15$$

- When expanding this, since the entire expression $(x+5)(x-3)$ is being subtracted from $3x(x+6)$, its expanded form needs to be kept inside its own set of brackets. This helps us to remember that, as each term is being subtracted, the sign of each term will switch when opening up the brackets.
- When expanding, it can help to treat the expression as though it has a -1 at the front, as shown.

$$-(x^2 + 2x - 15) = -1(x^2 + 2x - 15) = -x^2 - 2x + 15$$

WORKED EXAMPLE 15 Expanding and simplifying multiple sets of brackets

Expand and simplify each of the following expressions.
a. $(x + 3)(x + 4) + 4(x - 2)$
b. $(x - 2)(x + 3) - (x - 1)(x + 2)$

THINK

a. 1. Expand each set of brackets.

2. Simplify by collecting like terms.

b. 1. Expand and simplify each pair of brackets. Because the second expression is being subtracted, keep it in a separate set of brackets.

2. Subtract all of the second result from the first result. Remember that $-\left(x^2 + x - 2\right) = -1\left(x^2 + x - 2\right)$. Simplify by collecting like terms.

WRITE

a. $(x + 3)(x + 4) + 4(x - 2)$

$= x^2 + 4x + 3x + 12 + 4x - 8$
$= x^2 + 11x + 4$

b. $(x - 2)(x + 3) - (x - 1)(x + 2)$

$= x^2 + 3x - 2x - 6 - \left(x^2 + 2x - x - 2\right)$

$= x^2 + x - 6 - \left(x^2 + x - 2\right)$
$= x^2 + x - 6 - x^2 - x + 2$
$= -4$

WORKED EXAMPLE 16 Further expanding with multiple sets of brackets

Consider the expression $(4x + 3)(4x + 3) - (2x - 5)(2x - 5)$.
a. **Expand and simplify this expression.**
b. **Apply the difference of perfect squares rule to verify your answer.**

THINK

a. 1. Expand each set of brackets.

2. Simplify by collecting like terms.

b. 1. Rewrite as the difference of two squares.

2. Apply the difference of two squares rule, where $a = (4x + 3)$ and $b = (2x - 5)$.

3 Simplify, then expand the new brackets.

4. Simplify by collecting like terms.

WRITE

a. $(4x + 3)(4x + 3) - (2x - 5)(2x - 5)$

$= 16x^2 + 12x + 12x + 9 - \left(4x^2 - 10x - 10x + 25\right)$
$= 16x^2 + 24x + 9 - \left(4x^2 - 20x + 25\right)$
$= 16x^2 + 24x + 9 - 4x^2 + 20x - 25$
$= 12x^2 + 44x - 16$

b. $(4x + 3)(4x + 3) - (2x - 5)(2x - 5)$

$= (4x + 3)^2 - (2x - 5)^2$
$= ((4x + 3) + (2x - 5))((4x + 3) - (2x - 5))$

$= (6x - 2)(2x + 8)$
$= 12x^2 + 48x - 4x - 16$

$= 12x^2 + 44x - 16$

Exercise 4.7 Further expansions (Path)

learn On

4.7 Quick quiz on	4.7 Exercise

Individual pathways

■ PRACTISE	■ CONSOLIDATE	■ MASTER
1, 4, 8, 11, 14	2, 5, 7, 9, 12, 15	3, 6, 10, 13, 16

Fluency

1. **WE15** Expand and simplify each of the following expressions.

    a. $(x+3)(x+5) + (x+2)(x+3)$
    b. $(x+4)(x+2) + (x+3)(x+4)$
    c. $(x+5)(x+4) + (x+3)(x+2)$
    d. $(x+1)(x+3) + (x+2)(x+4)$

2. Expand and simplify each of the following expressions.

    a. $(p-3)(p+5) + (p+1)(p-6)$
    b. $(a+4)(a-2) + (a-3)(a-4)$
    c. $(p-2)(p+2) + (p+4)(p-5)$
    d. $(x-4)(x+4) + (x-1)(x+20)$

3. Expand and simplify each of the following expressions.

    a. $(y-1)(y+3) + (y-2)(y+2)$
    b. $(d+7)(d+1) + (d+3)(d-3)$
    c. $(x+2)(x+3) + (x-4)(x-1)$
    d. $(y+6)(y-1) + (y-2)(y-3)$

4. Expand and simplify each of the following expressions.

    a. $(x+2)^2 + (x-5)(x-3)$
    b. $(y-1)^2 + (y+2)(y-4)$
    c. $(p+2)(p+7) + (p-3)^2$
    d. $(m-6)(m-1) + (m+5)^2$

5. Expand and simplify each of the following expressions.

    a. $(x+3)(x+5) - (x+2)(x+5)$
    b. $(a+5)(a+2) - (a+1)(a+2)$
    c. $(x+3)(x+2) - (x+4)(x+3)$
    d. $(m-2)(m+3) - (m+2)(m-4)$

6. Expand and simplify each of the following expressions.

    a. $(b+4)(b-6) - (b-1)(b+2)$
    b. $(y-2)(y-5) - (y+2)(y+6)$
    c. $(p-1)(p+4) - (p-2)(p-3)$
    d. $(x+7)(x+2) - (x-3)(x-4)$

7. Expand and simplify each of the following expressions.

    a. $(m+3)^2 - (m+4)(m-2)$
    b. $(a-6)^2 - (a-2)(a-3)$
    c. $(p-3)(p+1) - (p+2)^2$
    d. $(x+5)(x-4) - (x-1)^2$

Understanding

8. **WE16** Consider the expression $(x+3)(x+3) - (x-5)(x-5)$.

    a. Expand and simplify this expression.
    b. Apply the difference of perfect squares rule to verify your answer.

9. Consider the expression $(4x-5)(4x-5) - (x+2)(x+2)$.

    a. Expand and simplify this expression.
    b. Apply the difference of perfect squares rule to verify your answer.

10. Consider the expression $(3x - 2y)(3x - 2y) - (y - x)(y - x)$.
    a. Expand and simplify this expression.
    b. Apply the difference of perfect squares rule to verify your answer.

Communicating, reasoning and problem solving

11. Determine the value of x for which $(x + 3) + (x + 4)^2 = (x + 5)^2$ is true.

12. Show that $(p - 1)(p + 2) + (p - 3)(p + 1) = 2p^2 - p - 5$.

13. Show that $(x + 2)(x - 3) - (x + 1)2 = -3x - 7$.

14. Answer the following questions.
    a. Show that $(a^2 + b^2)(c^2 + d^2) = (ac - bd)^2 + (ad + bc)^2$.
    b. Using part a, write $(2^2 + 1^2)(3^2 + 4^2)$ as the sum of two squares and evaluate.

15. Answer the following questions.
    a. Expand $\left(x^2 + x - 1\right)^2$.
    b. Show that $\left(x^2 + x - 1\right)^2 = (x - 1)x(x + 1)(x + 2) + 1$.
    c. i. Evaluate $4 \times 3 \times 2 \times 1 + 1$.
       ii. Determine the value of x if $4 \times 3 \times 2 \times 1 + 1 = (x - 1)x(x + 1)(x + 2) + 1$.

16. Answer the following questions.
    a. Expand $(a + b)(d + e)$.
    b. Expand $(a + b + c)(d + e + f)$. Draw a diagram to illustrate your answer.

LESSON
4.8 The highest common factor

4.8.1 Factorising expressions

eles-4601

- **Factorising** a number or term means writing it as the product of a pair of its factors.
  For example, $18ab$ could be factorised as $3a \times 6b$.
- Factorising an algebraic expression is the opposite process to expanding.
- $3(x + 6)$ is the product of 3 and $(x + 6)$ and is the factorised form of $3x + 18$.
- In this section we will start with the expanded form of an expression, for example $6x + 12$, and work to re-write it in factorised form, in this case $6(x + 2)$.

WORKED EXAMPLE 17 Factorising terms

Complete the following factorisations.

a. $15x = 5x \times$ _____   b. $15x^2 = 3x \times$ _____   c. $-18ab^2 = 6a \times$ _____

THINK	WRITE
a. Divide $15x$ by $5x$.	a. $\dfrac{15x}{5x} = 3$ So, $15x = 5x \times 3$.
b. Divide $15x^2$ by $3x$.	b. $\dfrac{15x^2}{3x} = 5x$ So, $15x^2 = 3x \times 5x$.
c. Divide $-18ab^2$ by $6a$.	c. $\dfrac{-18ab^2}{6a} = -3b^2$ So, $-18ab^2 = 6a \times (-3b^2)$.

- The **highest common factor (HCF)** of two or more numbers is the largest factor that divides into all of them. The highest common factor of 18 and 27 is 9, since it is the biggest number that divides evenly into both 18 and 27.
- We can determine the HCF of two or more algebraic terms by determining the highest common factor of the coefficients, as well as any pronumerals that are common to all terms. For example, the HCF of $30xyz$, $25x^2z$ and $15xz^2$ is $5xz$. It can help to write each term in expanded form so that you can determine all of the common pronumerals.

WORKED EXAMPLE 18 Determining the highest common factor of two terms

Determine the highest common factor for each of the following pairs of terms.

a. $25a^2b$ and $10ab$ b. $3xy$ and $-3xz$

THINK	WRITE
a. 1. Write $25a^2b$ in expanded form.	a. $25a^2b = 5 \times 5 \times a \times a \times b$
2. Write $10ab$ in expanded form.	$10ab = 2 \times 5 \times a \times b$
3. Write the HCF.	HCF $= 5ab$
b. 1. Write $3xy$ in expanded form.	b. $3xy = 3 \times x \times y$
2. Write $-3xz$ in expanded form.	$-3xz = -1 \times 3 \times x \times z$
3. Write the HCF.	HCF $= 3x$

▶ 4.8.2 Factorising expressions by determining the highest common factor

eles-4602

- To factorise an expression such as $15x^2yz + 12xy^2$, determine the highest common factor of both terms. Writing each term in expanded form can help identify what is common to both terms.

$$15x^2yz + 12xy^2 = 3 \times 5 \times x \times x \times y \times z + 3 \times 4 \times x \times x \times y \times y$$
$$= 3 \times x \times x \times y \times 5 \times x \times x \times z + 3 \times x \times x \times y \times 4 \times y$$

- We can see that $3xy$ is common to both terms. To complete the factorisation, place $3xy$ on the outside of a set of brackets and simplify what is left inside the brackets.

$$= 3 \times x \times y \,(5 \times x \times z + 4 \times y)$$
$$= 3xy\,(5xz + 4y)$$

- By removing the HCF from both terms, we have factorised the expression.
- The answer can always be checked by expanding out the brackets and making sure it produces the original expression.

WORKED EXAMPLE 19 Factorising by first determining the HCF

Factorise each expression by first determining the HCF.
a. $5x + 15y$
b. $-14xy - 7y$
c. $6x^2y + 9xy^2$

THINK	WRITE
a. 1. The HCF is 5.	a. $5x + 15y = 5(\qquad)$
2. Divide each term by 5 to determine the binomial.	$= 5(x + 3y)$
3. Check the answer by expanding.	$5(x + 3y) = 5x + 15y$ (correct)
b. 1. The HCF is $7y$ or $-7y$, but $-7y$ makes things a little simpler.	b. $-14xy - 7y = -7y(\qquad)$
2. Divide each term by $-7y$ to determine the binomial.	$= -7y(2x + 1)$
3. Check the answer by expanding.	$-7y(2x + 1) = -14xy - 7y$ (correct)
c. 1. The HCF is $3xy$.	c. $6x^2y + 9xy^2 = 3xy(\qquad)$
2. Divide each term by $3xy$ to determine the binomial.	$= 3xy(2x + 3y)$
3. Check the answer by expanding.	$3xy(2x + 3y) = 6x^2y + 9xy^2$ (correct)

DISCUSSION

How do you determine the factors of terms within algebraic expressions?

---

on Resources

▶ **Video eLesson**   Factorisation (eles-1887)

🧩 **Interactivity**   Highest common factor (int-6037)

Exercise 4.8 The highest common factor

| 4.8 Quick quiz   on | 4.8 Exercise |

Individual pathways

■ PRACTISE	■ CONSOLIDATE	■ MASTER
1, 3, 7, 11, 13, 16, 21, 23, 26	2, 4, 6, 8, 10, 14, 17, 18, 22, 24, 27	5, 9, 12, 15, 19, 20, 25, 28

Fluency

1. **WE17** Complete each of the following factorisations by writing in the missing factor.

   a. $8a = 4 \times \underline{\hspace{1cm}}$
   b. $8a = 2a \times \underline{\hspace{1cm}}$
   c. $12x^2 = 4x \times \underline{\hspace{1cm}}$
   d. $-12x^2 = 3x^2 \times \underline{\hspace{1cm}}$
   e. $3x^2 = x \times \underline{\hspace{1cm}}$
   f. $15a^2b = ab \times \underline{\hspace{1cm}}$

2. Complete each of the following factorisations by writing in the missing factor.

   a. $12x = -4 \times \underline{\hspace{1cm}}$
   b. $10\,mn = 10\,n \times \underline{\hspace{1cm}}$
   c. $10\,mn = -10 \times \underline{\hspace{1cm}}$
   d. $a^2b^2 = ab \times \underline{\hspace{1cm}}$
   e. $30x^2 = 10x \times \underline{\hspace{1cm}}$
   f. $-15\,mn^2 = -3\,m \times \underline{\hspace{1cm}}$

3. **WE18** Determine the highest common factor (HCF) of each of the following.

   a. 4 and 12
   b. 6 and 15
   c. 10 and 25
   d. 24 and 32
   e. 12, 15 and 21

4. Determine the highest common factor (HCF) of each of the following.

   a. 25, 50 and 200
   b. 17 and 23
   c. $6a$ and $12ab$
   d. $14xy$ and $21xz$
   e. $60pq$ and $30q$

5. Determine the highest common factor (HCF) of each of the following.

   a. $50cde$ and $70fgh$
   b. $6x^2$ and $15x$
   c. $6a$ and $9c$
   d. $5ab$ and 25
   e. $3x^2y$ and $4x^2z$

6. **MC** Choose which of the following pairs has a highest common factor of $5m$.

   A. $2m$ and $5m$
   B. $5m$ and m
   C. $25mn$ and $15lm$
   D. $20m$ and $40m$

7. **WE19** Factorise each of the following expressions.

   a. $4x + 12y$
   b. $5m + 15n$
   c. $7a + 14b$
   d. $7m - 21n$
   e. $-8a - 24b$
   f. $8x - 4y$

8. Factorise each of the following expressions.

   a. $-12p - 2q$
   b. $6p + 12pq + 18q$
   c. $32x + 8y + 16z$
   d. $16m - 4n + 24p$
   e. $72x - 8y + 64\,pq$
   f. $15x^2 - 3y$

9. Factorise each of the following expressions.

   a. $5p^2 - 20q$
   b. $5x + 5$
   c. $56q + 8p^2$
   d. $7p - 42x^2y$
   e. $16p^2 + 20q + 4$
   f. $12 + 36a^2b - 24b^2$

10. Factorise each of the following expressions.

    a. $9a + 21b$
    b. $4c + 18d^2$
    c. $12p^2 + 20q^2$
    d. $35 - 14m^2n$
    e. $25y^2 - 15x$

11. Factorise each of the following expressions.

    a. $16a^2 + 20b$
    b. $42m^2 + 12n$
    c. $63p^2 + 81 - 27y$
    d. $121a^2 - 55b + 110c$
    e. $10 - 22x^2y^3 + 14xy$

12. Factorise each of the following expressions.

    a. $18a^2bc - 27ab - 90c$
    b. $144p + 36q^2 - 84pq$
    c. $63a^2b^2 - 49 + 56ab^2$
    d. $22 + 99p^3q^2 - 44p^2r$
    e. $36 - 24ab^2 + 18b^2c$

13. Factorise each of the the following expressions.

    a. $-x + 5$ b. $-a + 7$ c. $-b + 9$ d. $-2m - 6$ e. $-6p - 12$ f. $-4a - 8$

14. Factorise each of the following expressions.

    a. $-3n^2 + 15m$ b. $-7x^2y^2 + 21$ c. $-7y^2 - 49z$
    d. $-12p^2 - 18q$ e. $-63m + 56$ f. $-12m^3 - 50x^3$

15. Factorise each of the following expressions.

    a. $-9a^2b + 30$ b. $-15p - 12q$ c. $-18x^2 + 4y^2$
    d. $-3ab + 18m - 21$ e. $-10 - 25p^2 - 45q$ f. $-90m^2 + 27n + 54p^3$

16. Factorise each of the following expressions.

    a. $a^2 + 5a$ b. $14q - q^2$ c. $18m + 5m^2$
    d. $6p + 7p^2$ e. $7n^2 - 2n$ f. $7p - p^2q + pq$

17. Factorise each of the following expressions.

    a. $xy + 9y - 3y^2$ b. $5c + 3c^2d - cd$ c. $3ab + a^2b + 4ab^2$
    d. $2x^2y + xy + 5xy^2$ e. $5p^2q^2 - 4pq + 3p^2q$ f. $6x^2y^2 - 5xy + x^2y$

18. Factorise each of the following expressions.

    a. $5x^2 + 15x$ b. $24m^2 - 6m$ c. $32a^2 - 4a$
    d. $-2m^2 + 8m$ e. $-5x^2 + 25x$ f. $-7y^2 + 14y$

19. Factorise each of the following expressions.

    a. $-3a^2 + 9a$ b. $-12p^2 - 2p$ c. $-26y^2 - 13y$
    d. $4m - 18m^2$ e. $-6t + 36t^2$ f. $-8p - 24p^2$

Understanding

20. A large billboard display is in the shape of a rectangle as shown. The billboard has 3 regions (A, B, C) with dimensions in terms of x, as shown.

    a. Calculate the total area of the billboard. Give your answer in factorised form.
    b. Determine an expression for the area of each region of the billboard. Write the expression in its simplest form.

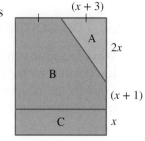

21. **PATH** Consider the expression $3ab^2 + 18ab + 27a$.

    a. Factorise this expression by taking out the highest common factor from each term.
    b. Use the rule for perfect squares to fully factorise this expression.

22. **PATH** Consider the expression $x^2(x + 4) + 8x(x + 4) + 16(x + 4)$.

    a. Factorise this expression by taking out the highest common factor from each term.
    b. Use the rule for perfect squares to fully factorise this expression.

Communicating, reasoning and problem solving

23. **PATH** A question on Marcia's recent Algebra test was, 'Using factorisation, simplify the following expression: $a^2(a - b) - b^2(a - b)$'. Marcia's answer was $(a + b)(a - b)^2$. If Marcia used the difference of two squares rule to get her solution, explain the steps she took to get that answer.

24. **PATH** Prove that, as long as $a \neq 0$, then $(x + a)^2 \neq x^2 + a^2$.

25. **PATH** Using the fact that any positive number, n, can be written as $\left(\sqrt{n}\right)^2$, factorise the following expressions using the difference of two squares rule.

    a. $x^2 - 13$ b. $4x^2 - 17$ c. $(x + 3)^2 - 10$

26. **PATH** Evaluate $\dfrac{x^2 + 2xy + y^2}{4x^2 - 4y^2} \div \dfrac{x^2 + xy}{8xy - 8x^2}$.

27. **PATH** It has been said that, for any two numbers, the product of their LCM and HCF is the same as the product of the two numbers themselves. Show whether this is true.

28. **PATH** a. Factorise $36x^2 - 100y^2$ by first taking out the common factor and then using the difference of two squares rule.

   b. Factorise $36x^2 - 100y^2$ by first using the difference of two squares rule and then taking out the common factor.

   c. Comment on whether you got the same answer for parts **a** and **b**.

LESSON
4.9 The highest common binomial factor (Path)

LEARNING INTENTION

At the end of this lesson you should be able to:
- factorise expressions by taking out a common binomial factor
- factorise expressions by grouping terms.

---

eles-4603

▶ 4.9.1 The common binomial factor

- When factorising an expression, we look for the highest common factor(s) first.
- It is possible for the HCF to be a binomial expression.
- Consider the expression $7(a - b) + 8x(a - b)$. The binomial expression $(a-b)$ is a common factor to both terms. Factorising this expression looks like this:

$$7(a - b) + 8x(a - b) = 7 \times (a-b) + 8x \times (a-b)$$
$$= (a-b)(7 + 8x)$$

WORKED EXAMPLE 20 Factorising by determining a binomial common factor

Factorise each of the following expressions.
a. $5(x + y) + 6b(x + y)$ b. $2b(a - 3b) - (a - 3b)$

Note: In both of these expressions the HCF is a binomial factor.

THINK	WRITE
a. 1. The HCF is $(x + y)$.	a. $\dfrac{5(x+y)}{x+y} = 5, \dfrac{6b(x+y)}{x+y} = 6b$
2. Divide each term by $(x + y)$ to determine the binomial.	Therefore, $5(x+y) + 6b(x+y)$ $= (x+y)(5 + 6b)$

b. **1.** The HCF is $(a - 3b)$.

2. Divide each term by $(a - 3b)$ to determine the binomial.

b. $\dfrac{2b(a-3b)}{a-3b} = 2b, \quad \dfrac{-1(a-3b)}{a-3b} = -1$

Therefore,

$2b(a-3b) - (a-3b)$
$= 2b(a-3b) - 1(a-3b)$
$= (a-3b)(2b-1)$

4.9.2 Factorising by grouping in pairs

eles-4604

- If an algebraic expression has four terms and no common factors in any of its terms, it may be possible to group the terms in pairs and determine a common factor in each pair.
- Consider the expression $10x + 15 - 6ax - 9a$.
- We can attempt to factorise by grouping the first two terms and the last two terms:

$$10x + 15 - 6ax - 9a = 5 \times 2x + 5 \times 3 - 3a \times 2x - 3a \times 3$$
$$= 5(2x+3) - 3a(2x+3)$$

- Once a common factor has been taken out from each pair of terms, a common binomial factor will appear. This common binomial factor can also be factorised out.

$$5(2x+3) - 3a(2x+3) = (2x+3)(5-3a)$$

- Thus the expression $10x + 15 - 6ax - 9a$ can be factorised to become $(2x+3)(5-3a)$.
- It is worth noting that it doesn't matter which terms are paired up first — the final result will still be the same.

$$10x + 15 - 6ax - 9a = 10x - 6ax + 15 - 9a$$
$$= 2x \times 5 + 2x \times -3a + 3 \times 5 + 3 \times -3a$$
$$= 2x(5-3a) + 3(5-3a)$$
$$= (5-3a)(2x+3)$$
$$= (2x+3)(5-3a)$$

WORKED EXAMPLE 21 Factorising by grouping in pairs

Factorise each of the following expressions by grouping the terms in pairs.

a. $5a + 10b + ac + 2bc$ **b.** $x - 3y + ax - 3ay$ **c.** $5p + 6q + 15pq + 2$

THINK

a. **1.** Write the expression.

2. Take out the common factor $a + 2b$.

b. **1.** Write the expression.

WRITE

a. $5a + 10b + ac + 2bc$
$5a + 10b = 5(a+2b)$
$ac + 2bc = c(a+2b)$

$= 5(a+2b) + c(a+2b)$
$= (a+2b)(5+c)$

b. $x - 3y + ax - 3ay$
$x - 3y = 1(x-3y)$
$ax - 3ay = a(x-3y)$

2. Take out the common factor $x - 3y$.

$$= 1(x - 3y) + a(x - 3y)$$
$$= (x - 3y)(1 + a)$$

c. 1. Write the expression.

c. $5p + 6q + 15pq + 2$

2. There are no simple common factors. Write the terms in a different order.

$$= 5p + 15pq + 6q + 2$$
$$5p + 15pq = 5p(1 + 3q)$$
$$6q + 2 = 2(3q + 1)$$

3. Take out the common factor $1 + 3q$.
Note: $1 + 3q = 3q + 1$.

$$= 5p(1 + 3q) + 2(3q + 1)$$
$$= 5p(1 + 3q) + 2(1 + 3q)$$

$$= (1 + 3q)(5p + 2)$$

 Resources

Interactivity Common binomial factor (int-6038)

Exercise 4.9 The highest common binomial factor (Path)   learn

| 4.9 Quick quiz on | 4.9 Exercise |

Individual pathways

■ PRACTISE	■ CONSOLIDATE	■ MASTER
1, 3, 6, 9, 12	2, 4, 7, 10, 13	5, 8, 11, 14

Fluency

1. **WE20** Factorise each of the following expressions.

a. $2(a + b) + 3c(a + b)$ **b.** $4(m + n) + p(m + n)$ **c.** $7x(2m + 1) - y(2m + 1)$
d. $4a(3b + 2) - b(3b + 2)$ **e.** $z(x + 2y) - 3(x + 2y)$

2. Factorise each of the following expressions.

a. $12p(6 - q) - 5(6 - q)$ **b.** $3p^2(x - y) + 2q(x - y)$ **c.** $4a^2(b - 3) + 3b(b - 3)$
d. $p^2(q + 2p) - 5(q + 2p)$ **e.** $6(5m + 1) + n^2(5m + 1)$

3. **WE21** Factorise each of the following expressions by grouping the terms in pairs.

a. $xy + 2x + 2y + 4$ **b.** $ab + 3a + 3b + 9$ **c.** $xy - 4y + 3x - 12$
d. $2xy + x + 6y + 3$ **e.** $3ab + a + 12b + 4$ **f.** $ab - 2a + 5b - 10$

4. Factorise each of the following expressions by grouping the terms in pairs.

a. $m - 2n + am - 2an$ **b.** $5 + 3p + 15a + 9ap$ **c.** $15mn - 5n - 6m + 2$
d. $10pq - q - 20p + 2$ **e.** $6x - 2 - 3xy + y$ **f.** $16p - 4 - 12pq + 3q$

5. Factorise each of the following expressions by grouping the terms in pairs.

a. $10xy + 5x - 4y - 2$ **b.** $6ab + 9b - 4a - 6$ **c.** $5ab - 10ac - 3b + 6c$
d. $4x + 12y - xz - 3yz$ **e.** $5pr + 10qr - 3p - 6q$ **f.** $ac - 5bc - 2a + 10b$

Understanding

6. Simplify the following expressions using factorising.

   a. $\dfrac{ax + 2ay + 3az}{bx + 2by + 3bz}$

   b. $\dfrac{10(3x-4) + 2y(3x-4)}{7a(10+2y) - 5(10+2y)}$

7. Use factorising by grouping in pairs to simplify the following expressions.

   a. $\dfrac{3x + 6 + xy + 2y}{6 + 2y + 18x + 6xy}$

   b. $\dfrac{5xy + 10x + 3ay + 6a}{15bx - 10x + 9ab - 6a}$

8. Use factorising by grouping in pairs to simplify the following expressions.

   a. $\dfrac{6x^2 + 15xy - 4x - 10y}{6xy + 4x + 15y^2 + 10y}$

   b. $\dfrac{mp + 4mq - 4np - 16nq}{mp + 4mq + 4np + 16nq}$

Communicating, reasoning and problem solving

9. Using the method of rectangles to expand, show how $a(m+n) + 3(m+n)$ equals $(a+3)(m+n)$.

10. Fully factorise $6x + 4x^2 + 6x + 9$ by grouping in pairs. Discuss what you noticed about this factorisation.

11. a. Write out the product $5(x+2)(x+3)$ and show that it also corresponds to the diagram shown.

   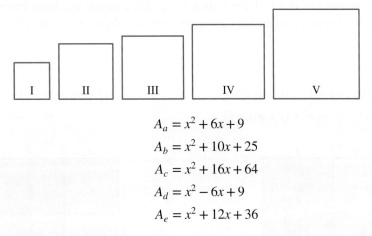

   b. Explain why $5(x+2)(x+3)$ is equivalent to $(5x+10)(x+3)$. Use bracket expansion and a labelled diagram to support your answer.

   c. Explain why $5(x+2)(x+3)$ is equivalent to $(x+2)(5x+15)$. Use bracket expansion and a labelled diagram to support your answer.

12. A series of five squares of increasing size and a list of five area formulas are shown.

   a. Use factorisation to calculate the side length that correlates to each area formula.

   b. Using the area given and the side lengths found, match the squares below with the appropriate algebraic expression of their area.

   $$A_a = x^2 + 6x + 9$$
   $$A_b = x^2 + 10x + 25$$
   $$A_c = x^2 + 16x + 64$$
   $$A_d = x^2 - 6x + 9$$
   $$A_e = x^2 + 12x + 36$$

   c. If $x = 5\,\text{cm}$, use the formula given to calculate the area of each square.

13. Fully factorise both sets and brackets in the expression $(-12xy + 27x + 8y - 18) - (-8xy + 18x + 12y - 27)$ using grouping in pairs, then factorise the result.

14. The area formulas shown relate to either squares or rectangles.

   i. $9s^2 + 48s + 64$
   ii. $25s^2 - 4$
   iii. $s^2 + 4s + 3$
   iv. $4s^2 - 28s - 32$

   a. Without completing any algebraic operations, examine these formulas and work out which ones belong to squares and which ones belong to rectangles. Explain your answer.
   b. Factorise each formula and classify it as a square or rectangle. Check your classifications against your answer to part a.

LESSON
4.10 Factorising monic quadratic expressions (Path)

LEARNING INTENTION

At the end of this lesson you should be able to:
- recognising monic quadratic expressions
- factorising monic quadratic expressions by determining two numbers that add to the middle term and multiply to the end term (the constant term).

▶ 4.10.1 Factorising monic quadratics

eles-6205

- A **quadratic trinomial** is an expression of the form $ax^2 + bx + c$, where a, b and c are constants (except zero).
- If the expression contains $1x^2$, that is if $a = 1$, then it is called a **monic** quadratic trinomial.
- The area model of the binomial expansion can be reversed to determine a pattern for factorising a general quadratic expression.

   For example:

$$(x+f)(x+h) = x^2 + fx + hx + fh$$
$$= x^2 + (f+h)x + fh$$

$$(x+4)(x+3) = x^2 + 4x + 3x + 12$$
$$= x^2 + 7x + 12$$

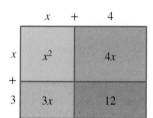

- To factorise a general quadratic, look for factors of c that sum to b.

$$x^2 + bx + c = (x+f)(x+h)$$

Factors of c that sum to b

For example, $x^2 + 7x + 12 = (x+3)(x+4)$
$$3 + 4 = 7$$
$$3 \times 4 = 12$$

WORKED EXAMPLE 22 Factorising monic quadratic expressions

Factorise the following quadratic expressions.
a. $x^2 + 5x + 6$
b. $x^2 + 10x + 24$

THINK	WRITE
a. 1. The general quadratic expression has the pattern $x^2 + 5x + 6 = (x+f)(x+h)$, where f and h are a factor pair of 6 that add to 5.	**a.**

Factors of 6	Sum of factors
1 and 6	7
2 and 3	5

Calculate the sums of factor pairs of 6. The factors of 6 that add to 5 are 2 and 3, as shown in blue.

a. 2. Substitute the values of f and h into the expression in its factorised form.

$$x^2 + 5x + 6 = (x+2)(x+3)$$

b. 1. The general quadratic expression has the pattern $x^2 + 10x + 24 = (x+f)(x+h)$, where f and h are a factor pair of 24 that add to 10.

b.

Factors of 24	Sum of factors
1 and 24	25
2 and 12	14
3 and 8	11
4 and 6	10

Calculate the sums of factor pairs of 24. The factors of 24 that add to 10 are 4 and 6, as shown in blue.

b. 2. Substitute the values of f and h into the expression in its factorised form.

$$x^2 + 10x + 24 = (x+4)(x+6)$$

WORKED EXAMPLE 23 Factorising monic quadratic expressions

Factorise the following expressions.
a. $x^2 - 9x + 18$
b. $x^2 + 6x - 16$

THINK	WRITE
a. 1. The general quadratic expression has the pattern $x^2 - 9x + 18 = (x+f)(x+h)$, where f and h are a factor pair of 18 that add to -9.	**a.**

Calculate the sums of factor pairs of 18. As shown in blue, -3 and -6 are factors of 18 that add to -9.

Factors of 18	Sum of factors
1 and 18	19
-1 and -18	-19
2 and 9	11
-2 and -9	-11
3 and 6	9
-3 and -6	-9

2. Substitute the values of f and h into the expression in its factorised form.

$$x^2 - 9x + 18 = (x - 3)(x - 6)$$

b. 1. The general quadratic expression has the pattern $x^2 + 6x - 16 = (x + f)(x + h)$, where f and h are a factor pair of 24 that add to 10.

Calculate the sums of factor pairs of -16. As shown in blue, -2 and 8 are factors of -16 that add to 6.

b.

Factors of -16	Sum of factors
1 and -16	-15
-1 and 16	15
2 and -8	-6
-2 and 8	6

2. Substitute the values of f and h into the expression in its factorised form.

$$x^2 - 6x + 16 = (x - 2)(x + 8)$$

 Resources

Interactivites Factorising monic quadratics (int-6092)

Factorise trinomials (int-0748)

Exercise 4.10 Factorising monic quadratic expressions (Path)

learn on

4.10 Quick quiz	4.10 Exercise

Individual pathways

■ PRACTISE	■ CONSOLIDATE	■ MASTER
1, 5, 8, 11, 15, 18	2, 6, 10, 12, 14, 16, 19	3, 4, 7, 9, 13, 17

Fluency

1. Expand the following expressions, giving your answer in the form $ax^2 + bx + c$.

   a. $(x + 4)(x + 2)$
   b. $(x - 2)(x - 4)$
   c. $(x - 4)(x - 5)$

2. Expand the following expressions, putting the powers in descending order.

   a. $(m - 1)(m - 5)$
   b. $(t + 8)(t + 11)$
   c. $(t - 10)(t - 20)$

3. Expand the following expressions, putting powers in descending order.

   a. $(v + 5)(v - 8)$
   b. $(v - 5)(v + 8)$
   c. $(x + 7)(x - 2)$

4. Match the following factorised expressions with their expanded forms.

Factorised expression	Expanded forms
i. $(z + 3)(z - 6)$	A. $z^2 - 10z + 21$
ii. $(z + 4)(z + 4)$	B. $z^2 + 3z - 88$
iii. $(z + 11)(z - 8)$	C. $z^2 - 3z - 18$
iv. $(z - 3)(z - 7)$	D. $z^2 + 8z + 16$

5. **WE22** Factorise the following expressions.

   a. $x^2 + 4x + 3$
   b. $x^2 - 4x + 3$
   c. $x^2 + 12x + 11$

6. Factorise the following expressions.
   a. $a^2 - 6a + 5$
   b. $n^2 + 8n + 16$
   c. $n^2 + 10n + 16$

7. Select the correct factorisations of the following expressions.

   a. $y^2 - 12y + 27$
   b. $x^2 - 13x + 42$
   c. $t^2 - 8t + 12$

8. **WE23** Factorise each of the following.
   a. $x^2 + 3x - 18$
   b. $x^2 - 3x - 18$
   c. $x^2 - 2x - 15$

9. Factorise each of the following.
   a. $x^2 + 2x - 15$
   b. $n^2 - 13n - 14$
   c. $n^2 + 2n - 35$

10. Factorise each of the following.
    a. $v^2 + 5v - 6$
    b. $v^2 - 5v - 6$
    c. $t^2 + 4t - 12$
    d. $t^2 - 5t - 14$

Understanding

11. Consider the quadratic trinomial $x^2 + 7x + c$ where c is a positive integer.
    a. Factorise the expression if $c = 6$.
    b. What other positive whole number values can c take if the expression is to be factorised? Factorise the expression for each of these values.

12. Consider the quadratic trinomial $x^2 + 3x + c$ where c is a negative integer.
    a. Factorise the expression if $c = -4$.
    b. Determine three more values of c for which the expression can be factorised, and factorise each one.

13. In the following, take out a common factor and then factorise the trinomials.
    a. $2x^2 - 10x - 28$
    b. $4x^2 + 28x + 40$

14. In the following, take out a common factor and then factorise the trinomials.
    a. $-2x^2 - 2x + 24$
    b. $5x^2 - 40x + 75$

15. **MC** When factorised, $x^2 - 3x - 18$ is equal to:
    A. $(x - 3)(x - 6)$
    B. $(x - 3)(x + 6)$
    C. $(x + 3)(x + 6)$
    D. $(x + 3)(x - 6)$

Communicating, reasoning and problem solving

16. Show that $x^2 + 8x + 10$ has no factors if only whole numbers can be used.

17. To factorise the quadratic trinomial $6x^2 + 9x + 3$, the expression is rewritten as an equivalent form consisting of four terms, $6x^2 + \square x + \square x + 3$, that are then grouped using the 'two and two' method, $6x^2 + \square x + \square x + 3$.

    Investigate all the possible combinations for $\square x + \square x = 9x$, and show that there is only one combination that will factorise the trinomial appropriately.

18. Rectangular floor mats have an area of $x^2 + 2x - 15$.

    a. If the length of the mat is $(x + 5)\,\text{cm}$, determine an expression for the width.
    b. If the length of the mat is 70 cm, what is the width?
    c. If the width of the mat is 1 m, what is the length?

19. A particular rectangle has an area of $120\,\text{m}^2$.

    a. Generate a list of possible whole number dimensions for the rectangle.
    b. The area of the rectangle can also be expressed as $A = x^2 + 2x - 48$. Determine algebraic expressions for the length and the width of the rectangle in terms of x.
    The length is the larger dimension.
    c. Use your list of possible dimensions and your algebraic expressions for width and length to determine the dimensions of the rectangle.
    Hint: The value of x in the length and width must be the same.

LESSON
4.11 Review

4.11.1 Topic summary

Algebraic techniques

Simplifying algebraic expressions

- An algebraic expression can be simplified by adding or subtracting like terms.
- Two terms are considered like terms if they have exactly the same pronumeral component.
  - $3x$ and $-10x$ are **like** terms.
  - $5x^2$ and $11x$ are **not like** terms.
  - $4abc$ and $-3cab$ are **like** terms.
  - e.g. The expression $3xy + 10xy - 5xy$ has 3 like terms and can be simplified to $8xy$.

  The expression $4a \times 2b \times a$ can be simplified to $8a^2b$.
  The expression $12xy \div 4xz$ can be simplified to $\dfrac{3y}{z}$.

Expanding brackets

- We can expand a single set of brackets using the **Distributive Law**:
  $$a(b + c) = ab + ac$$
- The product of binomial factors can be expanded using **FOIL**:
  $$(a + b)(c + d) = ac + ad + bc + bd$$
  e.g. $(x + 3)(y - 4) = xy - 4x + 3y - 12$

Factorising

- Factorising is the opposite process to expanding.
- An expression is factorised by determining the highest common factor of each term.
  e.g. $9xy + 15xz - 21x^2$
  $= 3x(3y + 5z - 7x)$
  The highest common factor of each term is $3x$.

Special cases (Path)

- We can also expand if we recognise the following special cases.
  - Difference of two squares:
    $$(a + b)(a - b) = a^2 - b^2$$
  - Perfect squares:
    $$(a + b)^2 = a^2 + 2ab + b^2$$
    $$(a - b)^2 = a^2 - 2ab + b^2$$
  e.g. $(x - 6)(x + 6) = x^2 - 36$
  $(2x + 5)^2 = 4x^2 + 20x + 25$
  $(x - 2)^2 = x^2 - 4x + 4$

Grouping in pairs

- When presented with an expression that has four terms with no common factor, we can factorise by grouping the terms in pairs.
  e.g. $6xy + 8y - 12xz - 16z$
  $= 2y(3x + 4) - 4z(3x + 4)$
  $= (3x + 4)(2y - 4z)$

Factorising monic quadratic expressions (Path)

- When factorising quadratics, look for factors of c that sum to b.
  $$x^2 + bx + c = (x + f)(x + h)$$

  Factors of c that sum to b

  e.g. $x^2 + 7x + 12 = (x + 3)(x + 4)$
  $3 + 4 = 7$
  $3 \times 4 = 12$

4.11.2 Project

Quilt squares

People all over the world are interested in quilt making, which involves stitching pieces of fabric together and inserting stuffing between layers of stitched-together fabric. When making a quilt, the fabric can be arranged and sewn in a variety of ways to create attractive geometric designs. Because of the potential for interesting designs in quilt-making, quilts are often used as decorative objects.

Medini is designing a quilt. She is sewing pieces of differently coloured fabric together to make a block, then copying the block and sewing the blocks together in a repeated pattern.

Making your own quilt

A scaled diagram of the basic block that Medini is using to make her quilt is shown. The letters indicate the colours of the fabric that make up the block: yellow, black and white. The yellow and white pieces are square, while the black pieces are rectangular. The finished blocks are sewn together in rows and columns.

y	b	y
b	w	b
y	b	y

Trace or copy the basic block shown onto a sheet of paper. Repeat this process until you have nine blocks. Colour in each section and cut out all nine blocks.

1. Place all nine blocks together to form a 3×3 square. Draw and colour a scaled diagram of your result.
2. Describe the feature created by arranging the blocks in the manner described in question 1. Observe the shapes created by the different colours.

Medini sold her design to a company that now manufactures quilts made from 100 of these blocks. Each quilt covers an area of $1.44\,\text{m}^2$. Each row and column has the same number of blocks. Answer the following, ignoring any seam allowances.

3. Calculate the side length of each square block.
4. If the entire quilt has an area of $1.44\,\text{m}^2$, what is the area of each block?
5. Determine the dimensions of the yellow, black and white pieces of fabric of each block.
6. Calculate the area of the yellow, black and white pieces of fabric of each block.
7. Determine the total area of each of the three different colours required to construct this quilt.
8. Due to popular demand, the company that manufactures these quilts now makes them in different sizes. A customer can specify the approximate quilt area, the three colours they want and the number of blocks in the quilt, but the quilt must be either square or rectangular. Come up with a general formula that would let the company quickly work out the areas of the three coloured fabrics in each block. Give an example of this formula. Draw a diagram on a separate sheet of paper to illustrate your formula.

 Resources

 Interactivities Crossword (int-0699)
  Sudoku puzzle (int-3203)

Exercise 4.11 Review questions

Fluency

1. **MC** What is $\dfrac{3x}{5} - \dfrac{x}{4}$ simplified?

   A. $\dfrac{7x}{20}$ B. $\dfrac{2x}{20}$ C. $\dfrac{2x}{1}$ D. $\dfrac{4x}{20}$

2. **MC** What does $\dfrac{3}{p} \div \dfrac{6}{p}$ equal?

   A. 2 B. $\dfrac{1}{2}$ C. $12p$ D. $12p^2$

3. **MC** What is the equivalent of $6 - 4x(x+2) + 3x$?

   A. $6 - 4x^2 - 5x$ B. $6 + 4x^2 + 5x$
   C. $6 - 4x^2 + 5x$ D. $6 + 4x^2 - 5x$

4. Simplify the following expressions by collecting like terms.

   a. $8p + 9p$ b. $5y^2 + 2y - 4y^2$
   c. $9s^2t - 12s^2t$ d. $11c^2d - 2cd + 5dc^2$
   e. $n^2 - p^2q - 3p^2q + 6$ f. $8ab + 2a^2b^2 - 5a^2b^2 + 7ab$

5. Simplify the following expressions.

   a. $6a \times 2b$ b. $2ab \times b$ c. $2xy \times 4yx$
   d. $\dfrac{4x}{12}$ e. $18 \div 4b$

6. Expand these expressions.

   a. $5(x+3)$ b. $-(y+5)$ c. $-x(3-2x)$ d. $-4m(2m+1)$

7. **MC** **PATH** Choose the equivalent of $(3-a)(3+a)$.

   A. $9 + a^2$ B. $9 - a^2$ C. $3 + a^2$ D. $3 - a^2$

8. **MC** **PATH** Choose the equivalent of $(2y+5)^2$.

   A. $4y^2 + 20y + 25$ B. $4y^2 + 10y + 25$ C. $2y^2 + 20y + 25$ D. $2y^2 + 20y + 5$

9. **MC** **PATH** Select what $6(a+2b) - x(a+2b)$ equals when it is factorised.

   A. $6 - x(a+2b)$ B. $(6-x)(a+2b)$ C. $6(a+2b-x)$ D. $(6+x)(a-2b)$

Understanding

10. Expand and simplify these expressions by collecting like terms.

    a. $3(x-2) + 9$ b. $-2(5m-1) - 3$
    c. $4m(m-3) + 3m - 5$ d. $7p - 2 - (3p+4)$

11. Expand and simplify the following expressions.

    a. $3(a+2b) + 2(3a+b)$ b. $-4(2x+3y) + 3(x-2y)$
    c. $2m(n+6) - m(3n+1)$ d. $-2x(3-2x) - (4x-3)$

12. Expand and simplify these expressions.
   a. $(x+4)(x+5)$
   b. $(m-2)(m+1)$
   c. $(3m-2)(m-5)$
   d. $(2a+b)(a-3b)$

13. **PATH** Expand and simplify the following expressions.
   a. $(x+4)(x-4)$
   b. $(9-m)(9+m)$
   c. $(x+y)(x-y)$
   d. $(1-2a)(1+2a)$

14. **PATH** Expand and simplify the following expressions.
   a. $(x+5)^2$
   b. $(m-3)^2$
   c. $(4x+1)^2$
   d. $(2-3y)^2$

15. Expand and simplify these expressions.
   a. $(x+2)(x+1)+(x+3)(x+2)$
   b. $(m+7)(m-2)+(m+3)^2$
   c. $(x+6)(x+2)-(x+3)(x-1)$
   d. $(b-7)^2-(b-3)(b-4)$

16. **PATH** Expand and simplify the following expressions.
   a. $(x+2)^2+(x+3)^2$
   b. $(x-2)^2-(x-3)^2$
   c. $(x+4)^2-(x-4)^2$

17. Factorise each of the following by determining common factors.
   a. $6x+12$
   b. $6x^2+12x^3y$
   c. $8a^2-4b$
   d. $16x^2-24xy$
   e. $-2x-4$
   f. $b^2-3b+4bc$

18. **PATH** Simplify each of the following expressions using common factor techniques.
   a. $5(x+y)-4a(x+y)$
   b. $7a(b+5c)-6c(b+5c)$
   c. $15x(d+2e)+25xy(d+2e)$
   d. $2x+2y+ax+ay$
   e. $6xy+4x-6y-4$
   f. $pq-r+p-rq$

19. Simplify the following expressions.
   a. $\dfrac{x}{2}+\dfrac{x}{5}$
   b. $\dfrac{x}{6}-\dfrac{x}{12}$
   c. $\dfrac{3x}{4}-\dfrac{5x}{8}$
   d. $\dfrac{5}{12m}+\dfrac{7}{8m}$
   e. $\dfrac{a}{6}\times\dfrac{12}{a}$
   f. $\dfrac{2}{y}\times\dfrac{y}{8}$

20. Simplify the following expressions.
   a. $\dfrac{7}{2x}\times\dfrac{x}{14}$
   b. $\dfrac{1}{3p}\times\dfrac{6p}{5}$
   c. $\dfrac{a}{12}\times\dfrac{4}{5}\times\dfrac{10}{a}$
   d. $\dfrac{a}{2}\div\dfrac{a}{4}$
   e. $\dfrac{5}{m}\div\dfrac{30}{m}$
   f. $\dfrac{5m}{6y}\div\dfrac{15m}{3xy}$

21. Simplify the following expressions.
   a. c^2+5c+4
   b. x^2-6x-7
   c. x^2-4x+4
   d. $p^2+10p-24$
   e. q^2+q-42
   f. $x^2-2x-24$

22. **PATH** Factorise the following expressions.
   a. $y^2-10y+24$
   b. x^2+3x+2
   c. $c^2-11c-26$
   d. $m^2-7m+10$
   e. $x^2+6x-27$
   f. $m^2+24m+44$

Communicating, reasoning and problem solving

23. A rectangular rug has a length of $3x$ cm and a width of x cm.
    a. Write an expression for its perimeter.
    b. Write an expression for its area.
    c. **i.** If its side length is increased by y cm, write an expression for its new side length.
       ii. Write an expression for its new perimeter and expand.
       iii. Calculate the perimeter when $x = 90$ cm and $y = 30$ cm.
       iv. Write an expression for its new area and expand.
       v. Calculate the area when $x = 90$ cm and $= 30$ cm.

24. A rectangular garden bed has a length of 15 m and a width of 8 m. It is surrounded by a path of width p m.
    a. Write the total area of the garden bed and path in factorised form.
    b. Expand the expression you found in part **a.**
    c. Determine the area of the path in terms of p.
    d. Write an equation that could be solved to Determine the width of the path if the area of the path is 200 m^2.

25. This large sign appears in a parking lot at the entrance to car park 5. It has a uniform width, with the dimensions as shown.
    Write an algebraic expression for the area of the front of the sign.

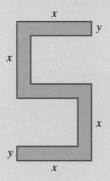

Answers

Topic 4 Algebraic techniques

4.1 Pre-test

1. $11x - 16$
2. $\dfrac{3b}{4}$
3. $\dfrac{2x}{3y}$
4. D
5. $\dfrac{9}{4x}$
6. a. $4(m + 3n) - (2m - n) = 2m + 13n$
   b. $-x(y + 3) + y(5 - x) = -2xy - 3x + 5y$
7. $6q^2 + qr - r^2$
8. D
9. $\dfrac{11a}{12}$
10. $\dfrac{2}{5y}$
11. B
12. $3b^2 - 21b + 37$
13. a. $4p(2p + 5)$ b. $5(5x^2 - 2y)$ c. $xy(6 + 8x - 3y)$
14. $(x + 4)(x - 1)$
15. $x^2 - 5x - 14$

4.2 Simplification of algebraic expressions

1. a. $9ab, -ab$ b. $4x$
   c. $-3za, -az$ d. $2x^2, -x^2$

2. a. $-2x^2y$ b. $-x^2y^5, -3x^2y^5$
   c. $p^3x^2w^5, -5x^2p^3w^5$ d. $4y^5z^4x^2, -2x^2z^4y^5$

3. a. $7x$ b. $11y$ c. $19m$
   d. $11q$ e. $8r$ f. $3x$

4. a. $8a$ b. $8y$ c. $13x$
   d. $16p$ e. $9q^2$ f. $3x^2$

5. a. $8x^2 - 3y$ b. $2m^2 + 2n$
   c. $-2g^2 + g - 12$ d. $-5m^2 + m + 15$
   e. $12a^2 + b + 4b^2$ f. $3m + 7n^2$

6. a. $12xy + 2y^2$ b. $2ab + 5a^2b$
   c. $16x^2y - 3xy$ d. $m^2n + 11n$
   e. $-7x^2 + 2x^2y$ f. $-3a^2b - 9a^2 + 5b + 4$

7. B
8. D
9. A
10. D
11. a. $6mn$ b. $20xy$ c. $8pq$ d. $-10xy$ e. $-12xy$
12. a. $15mn$ b. $10a^2$ c. $6mnp$ d. $-6ab^2$ e. $10\,m^2n$
13. a. $-18a^2b$ b. $30x^3y^2$ c. $-12p^2q^3$
    d. $-56c^3d$ e. $30a^4b^7$
14. a. $3x$ b. $3m$ c. $2y$ d. $4m$ e. $4m$

---

15. a. $2x$ b. $-7x$ c. $-4m$ d. $\dfrac{m}{2}$ e. $\dfrac{x}{3}$

16. a. $\dfrac{4m}{9}$ b. $\dfrac{1}{2a}$ c. $2yz$ d. $\dfrac{xy}{4}$ e. $\dfrac{-7yz}{11}$

17. a. $40x^2y^2$ b. $56ax^2y^2$ c. $\dfrac{x^2}{2y}$ d. $\dfrac{-5a}{4b}$

18. a. 2 b. $40a^3b^2$ c. $-8a^3b^3$ d. $16a^4$
19. a. i. $(0.01\,mp + nq)$ dollars   ii. $(mp + 100\,nq)$ cents
    b. $20 - (0.01mp + nq)$ dollars
20. Shirt $= \$12$ each
    Shorts $= \$30$ each
21. $\$45$
22. The correct answer is $7a^2c^2$. Sample responses can be found in the worked solutions in the online resources.
23. $\dfrac{3b^9}{2c^6}$

24. a. $\dfrac{n}{(n + 2)}$ b. $\dfrac{n}{(n + 3)}$ c. $\dfrac{1}{100}$

25. a. i.
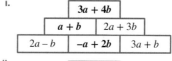

   ii.
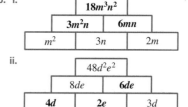

   b. i.

		$18m^3n^2$		
	$3m^2n$		$6mn$	
m^2		$3n$		$2m$

   ii.

		$48d^2e^2$		
	$8de$		$6de$	
$4d$		$2e$		$3d$

26. a. $P = \dfrac{x}{3} + x + \dfrac{5x}{7}$

   b. Sample responses can be found in the worked solutions in the online resources.
   c. $\$18$

27. a. $P = 2w + \dfrac{32w}{7}$

   b. Sample responses can be found in the worked solutions in the online resources.
   c. $\$72.22$

28. a. $P = 2x + 2(4x - 7)$
   b. $P = 10x - 14$
   c. 16 units
   d. If $x = 1$ the rectangle will have a negative value perimeter, which is not possible.

29. a. $l = 4(x + 40) + 8(x + 20)$
   b. $l = 12x + 320$
   c. $\$19.20$

30. a. $P_1 = (10x + 12)$ cm   b. $x = 2$
    $P_2 = 16x$ cm

4.3 Simplifying algebraic fractions

1. a. $\dfrac{7x}{12}$ b. $\dfrac{y}{6}$ c. $-\dfrac{m}{8}$ d. $\dfrac{x}{4}$ e. $\dfrac{9m}{14}$

2. a. $\dfrac{2t}{15}$ b. $\dfrac{13a}{10}$ c. $-\dfrac{p}{6}$ d. $\dfrac{17p}{15}$ e. $\dfrac{x}{6}$

3. a. $-\dfrac{1}{2p}$ b. $\dfrac{11}{10x}$ c. $-\dfrac{11}{4m}$

4. a. $\dfrac{7}{20b}$ b. $\dfrac{25}{18c}$ c. $\dfrac{22}{15y}$

5. a. 3 b. 2 c. $\dfrac{1}{4}$ d. $\dfrac{1}{5}$ e. 3

6. a. $\dfrac{1}{3}$ b. $\dfrac{m}{12}$ c. $\dfrac{n}{6}$ d. 14 e. $\dfrac{1}{9}$

7. a. $\dfrac{8}{y}$ b. $\dfrac{5x}{3}$ c. $\dfrac{4}{21}$ d. $\dfrac{2}{45}$ e. $\dfrac{1}{4}$

8. a. $2a^2$ b. $3a$ c. $\dfrac{1}{3a}$

9. a. 3 b. 2 c. $\dfrac{1}{3}$ d. $\dfrac{1}{4}$ e. 3

10. a. $\dfrac{1}{3}$ b. 4 c. $\dfrac{3}{10}$ d. $\dfrac{3}{2}$ e. $\dfrac{7b}{4}$

11. a. $\dfrac{3m}{5}$ b. $\dfrac{8b}{3}$ c. $\dfrac{3q}{5}$ d. $\dfrac{18}{25}$ e. $\dfrac{45x}{64}$

12. a. $\dfrac{1}{10xy^4}$ b. $\dfrac{3b^2}{2}$ c. $\dfrac{5}{2x}$

13. a. $\dfrac{7b^5}{4}$ b. $\dfrac{1}{4}$ c. $\dfrac{x^3}{5y}$

14. a. $P = \dfrac{x}{3} + x + \dfrac{5x}{7}$

    b. $\dfrac{43}{21}x$

    c. $18

15. a. The student is not factorising the expression before dividing.

    b. The student should have left the answers as they are because there are no common factors to factorise the expressions.

4.4 Expanding algebraic expressions and collecting like terms

1. a. 64 b. 1100 c. -20 d. 693

2. a. $3x + 6$ b. $4x + 12$

	3x	6			4x	12
3				4		
	x	2			x	3

   c. $4x + 4$ d. $7x - 7$

3. a. $-3p + 6$ b. $-x + 1$ c. $6b - 12$ d. $24m - 16$

4. a. $x^2 + 2x$

	x^2	$2x$
x		
	x	2

   b. $a^2 + 5a$

	a^2	$5a$
a		
	a	5

   c. $4x + x^2$ d. $7m - m^2$

5. a. $2xy + 4x$ b. $-3xy - 12y$ c. $-3b + ab$ d. $-30a + 18a^2$

6. a. $2p - 2$ b. $5x - 17$ c. $-7p - 17$ d. $-12p + 3$ e. $6x^2 - 20x$

7. a. $2m^2 + 7m$ b. $3px + 6x - 5$ c. $4y^2 - 4y + 7$ d. $-4p^2 + 13p$ e. $4x - 13y$

8. a. $2m^2 - 8m - 4$ b. $-3p^2 + 10pq - 1$ c. $-30a + 10ab$ d. $7cd - 12c^2 - 5c$ e. $-2p - 17$

9. a. $8x + y$ b. $10p + 8q$ c. $18a + 29b$ d. $19c + 22d$ e. $2m - 11n$

10. a. $6x - 11y$ b. $9x + 5y$ c. $-14p - 8q$ d. $-4a + 3b$ e. $4x - y$

11. a. $5p - 10q$ b. $-8c + 9d$ c. $13x - 19y$ d. $-7p + 11q$ e. $-5a + 3b$

12. a. $2ab + 2a - 3b$ b. $2xy + 4x - 2y$ c. $2cd + 3c$ d. $2pq - 2p$ e. $5cd - 11c$

13. a. $5ab - 21a - 3b$ b. $5m$ c. $6cd - 36c$ d. $6m^2 + 6m - 10$ e. $9cd - 8c$

14. a. $-15a^2 + 2b^2 - 9ab$ b. $-8c^2 + 3d^2 + 22cd$ c. $12m^2 - 20m - 4$ d. $-7x^2 + 41x - 6$ e. $-10y^2 - 6y - 12$

15. B

16. A

17. D

18. a. i. The student did not multiply both terms.

       ii. The student used addition instead of multiplication.

       iii. The student did not change negative and positive signs.

    b. $100 - 15x$

19. a. $-2(a - 5) = 2a - 10$ is incorrect because the student did not change the multiplied negative signs for -2×-5 to a positive sign. The correct answer is $-2a + 10$.

    b. $2b(3b - 1) = 6b^2 - 1$ is incorrect because the student did not multiply -1 and $2b$ together. The correct answer is $6b^2 - 2b$.

    c. $-2(c - 4) = 2c + 8$ is incorrect because the student left out the negative sign when multiplying -2 and c. The correct answer is $-2c + 8$.

20. No, the answer is $20a^2 - 14ab + 16b^2$

4.5 Expanding binomial products

1. a. $a^2 + 5a + 6$
   b. $x^2 + 7x + 12$
   c. $y^2 + 5y + 6$
   d. $m^2 + 9m + 20$
   e. $b^2 + 3b + 2$

2. a. $p^2 + 5p + 4$
   b. $a^2 + a - 6$
   c. $x^2 + x - 20$
   d. $m^2 - m - 12$
   e. $y^2 + 2y - 15$

3. a. $y^2 - 4y - 12$
   b. $x^2 - 2x - 3$
   c. $x^2 - 7x + 12$
   d. $p^2 - 5p + 6$
   e. $x^2 - 4x + 3$

4. a. $2a^2 + 7a + 6$
   b. $4c^2 - 31c + 42$
   c. $2t^2 - 17t + 35$
   d. $-6t^2 + 11t + 10$
   e. $15x^2 - 31x + 14$
   f. $25x^2 - 20x + 4$

5. a. $xz + x + yz + y$
   b. $2xz + 8x + yz + 4y$
   c. $3pr + 3p + qr + q$
   d. $a^2 + 3ab + 2b^2$
   e. $2c^2 - 5cd - 3d^2$
   f. $2x^2 - xy - 3y^2$

6. a. $4p^2 + pq - 3q^2$
   b. $ab + ac + 2b^2 + 2bc$
   c. $3p - 9pr - 2q + 6qr$
   d. $12x^2 - 7xy + y^2$
   e. $2p^2 - pr - 2pq + qr$
   f. $15k - 5 - 6jk + 2j$

7. C

8. C

9. a. $x^2 - 9$
   b. $x^2 - 25$
   c. $x^2 - 49$
   d. $x^2 - 1$
   e. $x^2 - 4$
   f. $4x^2 - 1$

10. a. $x^2 + 2x + 1$
    b. $x^2 + 4x + 4$
    c. $x^2 + 16x + 64$
    d. $x^2 - 6x + 9$
    e. $x^2 - 10x + 25$
    f. $x^2 - 18x + 81$

11. a. $6.3x^2 + 5.53xy - 3.1y^2$
    b. $4.41x^2 - 10.24y^2$
    c. $11.56x^2 + 34.68xy + 26.01y^2$

12. a. Surface area $= 38x^2 + 14x - 6$
    b. Volume $= 12x^3 + 5x^2 - 3x$

13. a. i. $((x + 3y)(4x - y))\,\text{m}^2$
       ii. $(4x^2 + 11xy - 3y^2)\,\text{cm}^2$
       iii. Both brackets must be positive; therefore,
       $x > -3y,\ x > \dfrac{y}{4}$
    b. i. $\dfrac{(2x - 1)(x + 5)}{2}$
       ii. $\dfrac{2x^2 + 9x - 5}{2}$
       iii. Sample responses can be found in the worked solutions in the online resources.

14. Sample responses can be found in the worked solutions in the online resources.

15. a. Negative sign ignored
    b. Negative sign ignored
    c. Distributive law not used

16. a. Student C
    b. Corrections to the students' answers are shown in bold.
       Student A:
       $(3x + 4)(2x + 5)$
       $= 3x \times 2x + 3x \times 5 + 4 \times 2x + 4 \times 5$
       $= \mathbf{6x^2} + \mathbf{23x} + 20$
       (Also, $29x + 20$ does not equal $49x$.)

Student B:
$(3x + 4)(2x + 5)$
$= 3x \times 2x + \mathbf{3x \times 5} + 4 \times 2x + 4 \times 5$
$= 6x^2 + \mathbf{15x} + 8x + 20$
$= 6x^2 + \mathbf{23x} + 20$

17. a. $w = (x - 3)\,\text{cm}$
    b. $w = 62\,\text{cm}$
    c. $l = 108\,\text{cm}$

4.6 Difference of two squares and perfect squares (Path)

1. a. $x^2 - 4$
   b. $y^2 - 9$
   c. $m^2 - 25$
   d. $a^2 - 49$

2. a. $x^2 - 36$
   b. $p^2 - 144$
   c. $a^2 - 100$
   d. $m^2 - 121$

3. a. $4x^2 - 9$
   b. $9y^2 - 1$
   c. $25d^2 - 4$
   d. $49c^2 - 9$
   e. $4 - 9p^2$

4. a. $d^2 - 81x^2$
   b. $25 - 144a^2$
   c. $9x^2 - 100y^2$
   d. $4b^2 - 25c^2$
   e. $100 - 4x^2$

5. a. $x^2 + 4x + 4$
   b. $a^2 + 6a + 9$
   c. $b^2 + 14b + 49$
   d. $c^2 + 18c + 81$

6. a. $m^2 + 24m + 144$
   b. $n^2 + 20n + 100$
   c. $x^2 - 12x + 36$
   d. $y^2 - 10y + 25$

7. a. $81 - 18c + c^2$
   b. $64 + 16e + e^2$
   c. $2x^2 + 4xy + 2y^2$
   d. $u^2 - 2uv + v^2$

8. a. $4a^2 + 12a + 9$
   b. $9x^2 + 6x + 1$
   c. $4m^2 - 20m + 25$
   d. $16x^2 - 24x + 9$

9. a. $25a^2 - 10a + 1$
   b. $49p^2 + 56p + 16$
   c. $81x^2 + 36x + 4$
   d. $16c^2 - 48c + 36$

10. a. $25 + 30p + 9p^2$
    b. $4 - 20x + 25x^2$
    c. $81x^2 - 72xy + 16y^2$
    d. $64x^2 - 48xy + 9y^2$

11. a. $x^2 - 9$
    b. $4x^2 - 9$
    c. $49x^2 - 16$
    d. $4x^2 - 49y^2$
    e. $x^4 - y^4$

12. a. $16x^2 + 40x + 25$
    b. $49x^2 - 42xy + 9y^2$
    c. $25x^4 - 20x^2y + 4y^2$
    d. $2x^2 - 4xy + 2y^2$
    e. $\dfrac{4}{x^2} + 16 + 16x^2$

13. $(x^2 + 6x + 9)\,\text{units}^2$

14. $(3x + 1)\,\text{m}$

15. Perimeter $= 4x + 36$

16. Sample responses can be found in the worked solutions in the online resources.

17. a. Sample responses can be found in the worked solutions in the online resources.
    b. Lin's bedroom is larger by $1\,\text{m}^2$.

18. a. i. $x^2 - 16$ and $16 - x^2$
       ii. $x^2 - 121$ and $121 - x^2$
       iii. $4x^2 - 81$ and $81 - 4x^2$
    b. The answers to the pairs of expansions are the same, except that the negative and positive signs are reversed.
    c. This is possible because when a negative number is multiplied by a positive number, it becomes negative. When expanding a DOTS in which the expressions have different signs, the signs will be reversed.

19. a. $100k^2 + 100k + 25$
    b. $(10k + 5)^2 = 100 \times k \times k + 100 \times k + 25$
       $= 100k(k + 1) + 25$

c. $25^2 = (10 \times 2 + 5)^2$
Let $k = 2$.
$$25^2 = 100k(k + 1) + 25$$
$$= 100 \times 2 \times (2 + 1) + 25$$
$$= 625$$
$$85^2 = (10 \times 8 + 5)^2$$
Let $k = 8$.
$$85^2 = 100k(k + 1) + 25$$
$$= 100 \times 8 \times (8 + 1) + 25$$
$$= 7225$$

20. a. $A_1 = a^2$ units^2

$A_2 = ab$ units^2

$A_3 = ab$ units^2

$A_4 = b^2$ units^2

b. $A = a^2 + ab + ab + b^2$

$= a^2 + 2ab + b^2$

This is the equation for perfect squares.

21. a. **i.** $x^2 - 6x + 9$ and $9 - 6x + x^2$

ii. $x^2 - 30x + 225$ and $225 - 30x + x^2$

iii. $9x^2 - 42x + 49$ and $49 - 42x + 9x^2$

b. The answers to the pairs of expansions are the same.

c. This is possible because when a negative number is multiplied by itself, it becomes positive. When expanding a perfect square in which the two expressions are the same, the negative signs cancel out and result in the same answer.

22. a. 729   **b.** 1089   **c.** 1521   **d.** 2209

23. $(3x^2 + 16x + 5)$ m^2

24. $(x^2 - 25)$ m^2

25. a. 10 609   **b.** 3844   **c.** 994 009

     d. 1 024 144   **e.** 2809   **f.** 9604

4.7 Further expansions (Path)

1. a. $2x^2 + 13x + 21$ **b.** $2x^2 + 13x + 20$
   c. $2x^2 + 14x + 26$ **d.** $2x^2 + 10x + 11$

2. a. $2p^2 - 3p - 21$ **b.** $2a^2 - 5a + 4$
   c. $2p^2 - p - 24$ **d.** $2x^2 + 19x - 36$

3. a. $2y^2 + 2y - 7$ **b.** $2d^2 + 8d - 2$
   c. $2x^2 + 10$ **d.** $2y^2$

4. a. $2x^2 - 4x + 19$ **b.** $2y^2 - 4y - 7$
   c. $2p^2 + 3p + 23$ **d.** $2m^2 + 3m + 31$

5. a. $x + 5$ **b.** $4x + 8$
   c. $-2x - 6$ **d.** $3m + 2$

6. a. $-3b - 22$ **b.** $-15y - 2$
   c. $8p - 10$ **d.** $16x + 2$

7. a. $4m + 17$ **b.** $-7a + 30$
   c. $-6p - 7$ **d.** $3x - 21$

8. a. $16x - 16$

b. Sample responses can be found in the worked solutions in the online resources.

9. a. $15x^2 - 44x + 21$

b. Sample responses can be found in the worked solutions in the online resources.

10. a. $8x^2 - 10xy + 3y^2$

b. Sample responses can be found in the worked solutions in the online resources.

11. $x = -4$

12. $(p^2 + p - 2) + (p^2 - 2p - 3)$

$= p^2 + p^2 + p - 2p - 2 - 2$

$= 2p^2 - p - 5$

13. $(x + 2)(x - 3) - (x + 1)^2 = (x^2 - x - 6) - (x^2 + 2x + 1)$

$= -3x - 7$

14. a. Sample responses can be found in the worked solutions in the online resources.

b. $a = 2$, $b = 1$, $c = 3$, $d = 4$
$$(2^2 + 1^2)(3^2 + 4^2) = (2 \times 3 - 1 \times 4)^2 + (2 \times 4 - 1 \times 3)^2$$
$$= 4 + 121$$
$$= 125$$

15. a. $x^4 + 2x^3 - x^2 - 2x + 1$

b. Sample responses can be found in the worked solutions in the online resources.

c. **i.** 25

ii. $x = 2$

16. a. $ad + ae + bd + be$

b. $(a + b + c)(d + e + f) = ad + ae + bd + be + cd + ce + af + bf + cf$

	a	b	c
d	ad	bd	cd
e	ae	be	ce
f	af	bf	cf

4.8 The highest common factor

1. a. $2a$ **b.** 4   **c.** $3x$
   d. -4 **e.** $3x$ **f.** $15a$

2. a. $-3x$ **b.** m **c.** $-mn$
   d. ab **e.** $3x$ **f.** $5n^2$

3. a. 4   **b.** 3   **c.** 5   **d.** 8   **e.** 3

4. a. 25   **b.** 1   **c.** $6a$ **d.** $7x$ **e.** $30q$

5. a. 10   **b.** $3x$ **c.** 3   **d.** 5   **e.** x^2

6. C

7. a. $4(x + 3y)$ **b.** $5(m + 3n)$
   c. $7(a + 2b)$ **d.** $7(m - 3n)$
   e. $-8(a + 3b)$ **f.** $4(2x - y)$

8. a. $-2(6p + q)$ **b.** $6(p + 2pq + 3q)$
   c. $8(4x + y + 2z)$ **d.** $4(4m - n + 6p)$
   e. $8(9x - y + 8pq)$ **f.** $3(5x^2 - y)$

9. a. $5(p^2 - 4q)$ **b.** $5(x + 1)$
   c. $8(7q + p^2)$ **d.** $7(p - 6x^2y)$
   e. $4(4p^2 + 5q + 1)$ **f.** $12(1 + 3a^2b - 2b^2)$

10. a. $3(3a + 7b)$ **b.** $2(2c + 9d^2)$
   c. $4(3p^2 + 5q^2)$ **d.** $7(5 - 2m^2n)$
   e. $5(5y^2 - 3x)$

11. a. $4(4a^2 + 5b)$ **b.** $6(7m^2 + 2n)$
   c. $9(7p^2 + 9 - 3y)$ **d.** $11(11a^2 - 5b + 10c)$
   e. $2(5 - 11x^2y^3 + 7xy)$

12. a. $9(2a^2bc - 3ab - 10c)$ **b.** $12(12p + 3q^2 - 7pq)$
  c. $7(9a^2b^2 - 7 + 8ab^2)$ **d.** $11(2 + 9p^3q^2 - 4p^2r)$
  e. $6(6 - 4ab^2 + 3b^2c)$

13. a. $-(x - 5)$ **b.** $-(a - 7)$ **c.** $-(b - 9)$
  d. $-2(m + 3)$ **e.** $-6(p + 2)$ **f.** $-4(a + 2)$

14. a. $-3(n^2 - 5m)$ **b.** $-7(x^2y^2 - 3)$ **c.** $-7(y^2 + 7z)$
  d. $-6(2p^2 + 3q)$ **e.** $-7(9m - 8)$ **f.** $-2(6m^3 + 25x^3)$

15. a. $-3(3a^2b - 10)$ **b.** $-3(5p + 4q)$
  c. $-2(9x^2 - 2y^2)$ **d.** $-3(ab - 6m + 7)$
  e. $-5(2 + 5p^2 + 9q)$ **f.** $-9(10m^2 - 3n - 6p^3)$

16. a. $a(a + 5)$ **b.** $q(14 - q)$ **c.** $m(18 + 5m)$
  d. $p(6 + 7p)$ **e.** $n(7n - 2)$ **f.** $p(7 - pq + q)$

17. a. $y(x + 9 - 3y)$ **b.** $c(5 + 3cd - d)$
  c. $ab(3 + a + 4b)$ **d.** $xy(2x + 1 + 5y)$
  e. $pq(5pq - 4 + 3p)$ **f.** $xy(6xy - 5 + x)$

18. a. $5x(x + 3)$ **b.** $6m(4m - 1)$ **c.** $4a(8a - 1)$
  d. $-2m(m - 4)$ **e.** $-5x(x - 5)$ **f.** $-7y(y - 2)$

19. a. $-3a(a - 3)$ **b.** $-2p(6p + 1)$ **c.** $-13y(2y + 1)$
  d. $2m(2 - 9m)$ **e.** $-6t(1 - 6t)$ **f.** $-8p(1 + 3p)$

20. a. $2(x + 3)(4x + 1)$
  b. $A = x(x + 3)$
   $B = (5x + 2)(x + 3)$
   $C = 2x(x + 3)$

21. a. $3a(b^2 + 6b + 9)$
  b. $3a(b + 3)^2$

22. a. $(x + 4)(x^2 + 8x + 16)$
  b. $(x + 4)^3$

23. Sample responses can be found in the worked solutions in the online resources.

24. Sample responses can be found in the worked solutions in the online resources.

25. a. $\left(x + \sqrt{13}\right)\left(x - \sqrt{13}\right)$
  b. $\left(2x - \sqrt{17}\right)\left(2x + \sqrt{17}\right)$
  c. $\left(x + 3 + \sqrt{10}\right)\left(x + 3 - \sqrt{10}\right)$

26. -2

27. True. Sample responses can be found in the worked solutions in the online resources.

28. a. $4(3x - 5y)(3x + 5y)$
  b. $4(3x - 5y)(3x + 5y)$
  c. Yes, the answers are the same.

4.9 The highest common binomial factor (Path)

1. a. $(a + b)(2 + 3c)$ **b.** $(m + n)(4 + p)$
  c. $(2m + 1)(7x - y)$ **d.** $(3b + 2)(4a - b)$
  e. $(x + 2y)(z - 3)$

2. a. $(6 - q)(12p - 5)$ **b.** $(x - y)(3p^2 + 2q)$
  c. $(b - 3)(4a^2 + 3b)$ **d.** $(q + 2p)(p^2 - 5)$
  e. $(5m + 1)(6 + n^2)$

3. a. $(y + 2)(x + 2)$ **b.** $(b + 3)(a + 3)$
  c. $(x - 4)(y + 3)$ **d.** $(2y + 1)(x + 3)$
  e. $(3b + 1)(a + 4)$ **f.** $(b - 2)(a + 5)$

4. a. $(m - 2n)(1 + a)$ **b.** $(5 + 3p)(1 + 3a)$
  c. $(3m - 1)(5n - 2)$ **d.** $(10p - 1)(q - 2)$
  e. $(3x - 1)(2 - y)$ **f.** $(4p - 1)(4 - 3q)$

5. a. $(2y + 1)(5x - 2)$ **b.** $(2a + 3)(3b - 2)$
  c. $(b - 2c)(5a - 3)$ **d.** $(x + 3y)(4 - z)$
  e. $(p + 2q)(5r - 3)$ **f.** $(a + 5b)(c - 2)$

6. a. $\dfrac{a}{b}$ **b.** $\dfrac{3x - 4}{7a - 5}$

7. a. $\dfrac{x + 2}{2 + 6x}$ **b.** $\dfrac{y + 2}{3b - 2}$

8. a. $\dfrac{3x - 2}{3y + 2}$ **b.** $\dfrac{m - 4n}{m + 4n}$

9. Sample responses can be found in the worked solutions in the online resources.

10. $(2x + 3)^2$ This is a perfect square.

11. a. $5(x + 2)(x + 3) = 5(x^2 + 5x + 6) = 5x^2 + 25x + 30$

	x	3	x	3	x	3	x	3	x	3
x	x^2	$3x$	x^2	$3x$	x^2	$3x$	x^2	$3x$	x^2	$3x$
2	$2x$	6	$2x$	6	$2x$	6	$2x$	6	$2x$	6

  b. $5(x + 2)(x + 3) = (5 \times x + 2 \times 5)(x + 3)$
   $= (5x + 10)(x + 3)$

	x	2	x	2	x	2	x	2	x	2
x	x^2	$2x$	x^2	$2x$	x^2	$2x$	x^2	$2x$	x^2	$2x$
3	$3x$	6	$3x$	6	$3x$	6	$3x$	6	$3x$	6

  c. $5(x + 2)(x + 3) = (5 \times x + 3 \times 5)(x + 2)$
   $= (5x + 15)(x + 2)$

	$5x$	15
x	$5x^2$	$15x$
2	$10x$	30

12. a. Side length$_a = x + 3$
   Side length$_b = x + 5$
   Side length$_c = x + 8$
   Side length$_d = x - 3$
   Side length$_e = x + 6$

  b. $A_a = \text{II} = (x + 3)^2$
   $A_b = \text{III} = (x + 5)^2$
   $A_c = \text{V} = (x + 8)^2$
   $A_d = \text{I} = (x - 3)^2$
   $A_e = \text{IV} = (x + 6)^2$

  c. $A_a = \text{II} = 64 \text{ cm}^2$
   $A_b = \text{III} = 100 \text{ cm}^2$
   $A_c = \text{V} = 169 \text{ cm}^2$
   $A_d = \text{I} = 4 \text{ cm}^2$
   $A_e = \text{IV} = 121 \text{ cm}^2$

13. $-(x + 1)\left(4y - 9\right)$

14. a. i. Square, because it is a perfect square.
  ii. Rectangle, because it is a DOTS.
  iii. Rectangle, because it is a trinomial.
  iv. Rectangle, because it is a trinomial.

b. i. $(3s + 8)^2$ ii. $(5s + 2)(5s - 2)$
  iii. $(s + 1)(s + 3)$ iv. $4(s - 8)(s + 1)$

4.10 Factorising monic quadratic expressions (Path)

1. a. $x^2 + 6x + 8$ **b.** $x^2 - 6x + 8$ **c.** $x^2 - 9x + 20$

2. a. $m^2 - 6m + 5$ **b.** $t^2 + 19t + 88$ **c.** $t^2 - 30t + 200$

3. a. $v^2 - 3v - 40$ **b.** $v^2 + 3v - 40$ **c.** $x^2 + 5x - 14$

4. i. C **ii.** D
  iii. B **iv.** A

5. a. $(x + 1)(x + 3)$ **b.** $(x - 1)(x - 3)$ **c.** $(x + 1)(x + 11)$

6. a. $(a - 5)(a - 1)$ **b.** $(n + 4)^2$ **c.** $(n + 8)(n + 2)$

7. a. $(y - 9)(y - 3)$ **b.** $(x - 7)(x - 6)$ **c.** $(t - 6)(t - 2)$

8. a. $(x + 6)(x - 3)$ **b.** $(x - 6)(x + 3)$ **c.** $(x - 5)(x + 3)$

9. a. $(x + 5)(x - 3)$ **b.** $(n - 14)(n + 1)$
  c. $(n + 7)(n - 5)$

10. a. $(v + 6)(v - 1)$ **b.** $(v - 6)(v + 1)$
  c. $(t + 6)(t - 2)$ **d.** $(t - 7)(t + 2)$

11. a. $(x + 1)(x + 6)$
  b. c can also take the values of 10 and 12.
   $x^2 + 7x + 10 = (x + 5)(x + 2)$
   $x^2 + 7x + 12 = (x + 3)(x + 4)$

12. a. $(x - 1)(x + 4)$
  b. Various answers are possible. These are a selection:
   $x^2 + 3x - 10 = (x + 5)(x - 2)$
   $x^2 + 3x - 18 = (x + 6)(x - 3)$
   $x^2 + 3x - 28 = (x + 7)(x - 4)$

13. a. $2(x + 2)(x - 7)$
  b. $4(x + 2)(x + 5)$

14. a. $-2(x + 4)(x - 3)$
  b. $5(x - 5)(x - 3)$

15. D

16. $x^2 + 8x + 10$
  The factor pairs of 10 are: 1 and 10; 2 and 5.
  Neither of these factor pairs have a sum of 8.

17. $3x + 6x = 9x$
  $6x^2 + 3x + 6x + 3 = 3x(2x + 1) + 3(2x + 1)$
  $= (3x + 3)(2x + 1)$

18. a. $(x - 3)$ cm **b.** 62 cm **c.** 108 cm

19. a. Possible dimensions for a rectangle with an area of
   120 m² are:
   1×120 m
   2×60 m
   3×40 m
   4×30 m
   5×24 m
   6×20 m
   8×15 m
   10×12 m
  b. $l = x + 8$
   $w = x - 6$
  c. 6 m $\times 20$ m

Project

1. Solution not required.

2. Students need to describe the feature created by arranging the blocks.

3. 0.12 m

4. 0.0144 m²

5. Yellow: 0.03 m × 0.03 m
  Black: 0.03 m × 0.06 m
  White: 0.06 m × 0.06 m

6. Yellow: 0.0009 m²
  Black: 0.0018 m²
  White: 0.0036 m²

7. Yellow: 0.0036 m²
  Black: 0.0072 m²
  White: 0.0036 m²

8. Students are required to derive a general formula that would allow the company to determine the area of the three coloured fabrics in each block. They should give an example and illustrate it with the help of a diagram.

4.11 Review questions

1. A

2. B

3. A

4. a. $17p$ **b.** $y^2 + 2y$ **c.** $-3s^2t$
  d. $16c^2d - 2cd$ **e.** $n^2 - 4p^2q + 6$ **f.** $15ab - 3a^2b^2$

5. a. $12ab$ **b.** $2ab^2$ **c.** $8x^2y^2$
  d. $\dfrac{x}{3}$ **e.** $\dfrac{9}{2b}$

6. a. $5x + 15$ **b.** $-y - 5$
  c. $-3x + 2x^2$ **d.** $-8m^2 - 4m$

7. B

8. A

9. B

10. a. $3x + 3$ **b.** $-10m - 1$
  c. $4m^2 - 9m - 5$ **d.** $4p - 6$

11. a. $9a + 8b$ **b.** $-5x - 18y$
  c. $-mn + 11m$ **d.** $4x^2 - 10x + 3$

12. a. $x^2 + 9x + 20$ **b.** $m^2 - m - 2$
  c. $3m^2 - 17m + 10$ **d.** $2a^2 - 5ab - 3b^2$

13. a. $x^2 - 16$ **b.** $81 - m^2$
  c. $x^2 - y^2$ **d.** $1 - 4a^2$

14. a. $x^2 + 10x + 25$ **b.** $m^2 - 6m + 9$
  c. $16x^2 + 8x + 1$ **d.** $4 - 12y + 9y^2$

15. a. $2x^2 + 8x + 8$ **b.** $2m^2 + 11m - 5$
  c. $6x + 15$ **d.** $-7b + 37$

16. a. $2x^2 + 10x + 13$ **b.** $2x - 5$
  c. $16x$

17. a. $6(x + 2)$ **b.** $6x^2(1 + 2xy)$
  c. $4\left(2a^2 - b\right)$ **d.** $8x(2x - 3y)$
  e. $-2(x + 2)$ **f.** $b(b - 3 + 4c)$

18. a. $(5 - 4a)(x + y)$ **b.** $(7a - 6c)(b + 5c)$
  c. $5x(3 + 5y)(d + 2e)$ **d.** $(2 + a)(x + y)$
  e. $2(x - 1)(3y + 2)$ **f.** $(p - r)(q + 1)$

19. a. $\dfrac{7x}{10}$　　b. $\dfrac{x}{12}$　　c. $\dfrac{x}{8}$

　　d. $\dfrac{31}{24m}$　　e. 2　　f. $\dfrac{1}{4}$

20. a. $\dfrac{1}{4}$　　b. $\dfrac{2}{5}$　　c. $\dfrac{2}{3}$

　　d. 2　　e. $\dfrac{1}{6}$　　f. $\dfrac{x}{6}$

21. a. $(c+1)(c+4)$

　　b. $(x-7)(x+1)$

　　c. $(x-2)(x-2) = (x-2)^2$

　　d. $(p-2)(p+12)$

　　e. $(q+7)(q-6)$

　　f. $(x-6)(x+4)$

22. a. $(y-6)(y-4)$　　b. $(x+1)(x+2)$

　　c. $(c-13)(c+2)$　　d. $(m-2)(m-5)$

　　e. $(x+9)(x-3)$　　f. $(m+22)(m+2)$

23. a. $8x$ cm

　　b. $3x^2$ cm^2

　　c. i. $(3x+y)$ cm　　ii. $(8x+2y)$ cm

　　　iii. 780 cm　　iv. $(3x^2+xy)$ cm^2

　　　v. 27000 cm^2

24. a. $(8+2p)(15+2p)$ m^2　　b. $\left(4p^2+46p+120\right)$ m^2

　　c. $\left(4p^2+46p\right)$ m^2　　d. $4p^2+46p = 200$

25. $5xy - 4y^2$

5 Linear equations

LESSON
5.1 Overview

Why learn this?

Mathematical equations are all around us. Learning how to solve equations lets us work out unknown values. Being able to manipulate and solve equations has many uses in many fields, including chemistry, medicine, economics and commerce, to name just a few.

Being able to solve linear equations is a useful skill for many everyday tasks. Linear equations can be used when converting temperature from Celsius to Fahrenheit, when working out how to balance chemical equations, or when converting between different currencies. Linear equations can also be used when working out fuel consumption for a road trip, deciding on the right dosage of medicine, or budgeting for a holiday.

Any calculation of rates will use some kind of linear equation. Linear equations can also be used to help predict future trends involving growth or decay. An understanding of the underlying principles for solving linear equations can also be applied to the solution of other mathematical models.

Hey students! Bring these pages to life online

- Watch videos
- Engage with interactivities
- Answer questions and check solutions

Find all this and MORE in jacPLUS

Reading content and rich media, including interactivities and videos for every concept

Extra learning resources

Differentiated question sets

Questions with immediate feedback, and fully worked solutions to help students get unstuck

1. Solve the linear equation $\dfrac{x}{3} - 2 = -1$.

2. Solve the linear equation $0.4x - 2.6 = 6.2$.

3. **MC** Select which of the following equations has the solution $x = 2$.
   A. $x + 3 = -5$
   B. $3 - x = 5$
   C. $3(x + 1) = 2x + 5$
   D. $\dfrac{3}{x} = 6$

4. Solve the following linear equations.
   a. $\dfrac{3x + 1}{4} = -5$
   b. $\dfrac{3 - z}{2} = -4$

5. Solve the equation $-3(2a - 4) = -12$.

6. Dylan is solving the equation $4(y + 3) = -10$. State whether it is true or false that Dylan will calculate the correct value of y if he subtracts 3 from both sides of the equation then divides both sides by 4.

7. **MC** Select the correct value for m in the equation $0.4(m - 8) = 0.6$.
   A. $m = -6.5$
   B. $m = 6.5$
   C. $m = 7.8$
   D. $m = 9.5$

8. **MC** Select the first step to solve the equation $2x + 1 = 11 - 3x$.
   A. Add $2x$ to both sides.
   B. Add 1 to both sides.
   C. Add $3x$ to both sides.
   D. Add 11 to both sides.

9. Solve the following equations.
   a. $5(a + 2) = 4a + 12$
   b. $0.2(b - 6) = 5.2 + b$

10. **MC** Choose the equation that matches the statement 'subtracting 6 from 5 multiplied by a certain number gives a result of -8'.
    A. $6 - 5x = -8$
    B. $5x - 6 = -8$
    C. $5 - 6x = -8$
    D. $6x - 5 = -8$

11. The cost of renting a car is given by $c = 75d + 0.4k$, where $d =$ number of days rented and $k =$ number of kilometres driven. Toni has $340 to spend on a car for three days. Calculate the total distance she can travel.

12. Solve the value of x in the statement 'x is multiplied by 5 and then 4 is subtracted. The result is the same as three times x minus 10'.

13. If $A = \dfrac{1}{2}bh$, determine the value of A if $b = 12$ and $h = 5$.

14. The formula for the perimeter, P, of a rectangle is $P = 2(l + w)$ where l is the length and w the width of the rectangle. Find the length of a rectangle 8 cm wide with a perimeter of 48 cm.

15. For the formula, $v = u + at$, find the value of t if $v = 30$ and $a = 10$ when $u = 8$.

LESSON
5.2 Solving linear equations

LEARNING INTENTION

At the end of this lesson you should be able to:
- recognise inverse operations
- solve multi-step equations with pronumerals on one side of the equation
- solve equations with algebraic fractions.

▶ 5.2.1 What is a linear equation?

eles-4689

- An **equation** is a mathematical statement that contains an equals sign (=).
- The **expression** on the left-hand side of the equals sign has the same value as the expression on the right-hand side of the equals sign.
- A **linear equation** is an equation for a straight or 'linear' line.
- Solving a linear equation means determining a value for the pronumeral that makes the statement true.
- 'Doing the same thing' to both sides of an equation (also known as applying the **inverse operation**) ensures that the expressions on either side of the equals sign remain equal, or **balanced**.

WORKED EXAMPLE 1 Using substitution to check equations

For each of the following equations, determine whether $x = 10$ is a solution.

a. $\dfrac{x+2}{3} = 6$ b. $2x + 3 = 3x - 7$ c. $x^2 - 2x = 9x - 10$

THINK	WRITE
a. 1. Substitute 10 for x in the left-hand side of the equation.	a. $\text{LHS} = \dfrac{x+2}{3}$ $= \dfrac{10+2}{3}$ $= \dfrac{12}{3}$ $= 4$
2. Write the right-hand side.	$\text{RHS} = 6$
3. Is the equation true? Does the left-hand side equal the right-hand side?	$\text{LHS} \neq \text{RHS}$
4. State whether $x = 10$ is a solution.	$x = 10$ is not a solution.
b. 1. Substitute 10 for x in the left-hand side of the equation.	b. $\text{LHS} = 2x + 3$ $= 2(10) + 3$ $= 23$
2. Substitute 10 for x in the right-hand side of the equation.	$\text{RHS} = 3x - 7$ $= 3(10) - 7$ $= 23$
3. Is the equation true? Does the left-hand side equal the right-hand side?	$\text{LHS} = \text{RHS}$
4. State whether $x = 10$ is a solution.	$x = 10$ is a solution.

---

c. 1. Substitute 10 for x in the left-hand side of the equation.

c. $LHS = x^2 - 2x$
$= 10^2 - 2(10)$
$= 100 - 20$
$= 80$

2. Substitute 10 for x in the right-hand side of the equation.

$RHS = 9x - 10$
$= 9(10) - 10$
$= 90 - 10$
$= 80$

3. Is the equation true? Does the left-hand side equal the right-hand side?

$LHS = RHS$

4. State whether $x = 10$ is a solution.

$x = 10$ is a solution.

▶ 5.2.2 Solving multi-step equations

eles-4690

- If an equation performs two operations on a pronumeral, it is known as a **two-step equation**.
- If an equation performs more than two operations on a pronumeral, it is known as a **multi-step equation**.
- To solve multi-step equations, first determine the order in which the operations were performed.
- Once the order of operations is determined, perform the inverse of those operations, in the reverse order, to both sides of the equation. This will keep the equation balanced.
- Each inverse operation must be performed one step at a time.
- These steps can be applied to any equation with two or more steps, as shown in the worked examples that follow.

Inverse operations

The inverse operation of an operation has the effect of undoing the original operation.

Operation	Inverse operation
+	−
−	+
×	÷
÷	×

WORKED EXAMPLE 2 Solving equations using inverse operations

Solve the following linear equations.

a. $3x + 1 = 11$

b. $1 - \dfrac{x}{3} = 11$

c. $\dfrac{2x}{3} + \dfrac{x}{4} = -11$

d. $\dfrac{-x - 0.3}{2} = -1.1$

THINK

a. 1. Subtract 1 from both sides of the equation.

2. Divide both sides by 3.

WRITE

a. $3x + 1 = 11$
$3x + 1 - 1 = 11 - 1$

$\dfrac{3x}{3} = \dfrac{10}{3}$

3. Write the value for x.

Note: It is preferable to write fractions in improper form rather than as mixed numbers.

$$x = \frac{10}{3}$$

b. 1. Subtract 1 from both sides of the equation.

b. $1 - \dfrac{x}{3} = 11$

$$1 - \frac{x}{3} - 1 = 11 - 1$$

2. Multiply both sides by -3.

$$\frac{-x}{3} \times (-3) = 10 \times (-3)$$

3. Write the value for x.

$$x = -30$$

Alternative method:

b. 1. Subtract 1 from both sides.

b. $1 - \dfrac{x}{3} = 11$

$$1 - \frac{x}{3} - 1 = 11 - 1$$

2. Multiply both sides by 3.

$$\frac{-x}{3} \times 3 = 10 \times 3$$

3. Divide both sides by -1.

$$\frac{-x}{-1} = \frac{30}{-1}$$

4. Write the value for x.

$$x = -30$$

c. 1. Add like terms to simplify the left-hand side of the equation. To do this, you will need to find the lowest common denominator.

c. $\dfrac{2x}{3} + \dfrac{x}{4} = -11$

$$\frac{2x}{3} \times \frac{4}{4} + \frac{x}{4} \times \frac{3}{3} = -11$$

$$\frac{8x + 3x}{12} = -11$$

2. Multiply both sides by 12.

$$\frac{11x}{12} \times 12 = -11 \times 12$$

3. Divide both sides by 11.

$$\frac{11x}{11} = \frac{-132}{11}$$

4. Write the value for x.

$$x = -12$$

d. 1. Multiply both sides of the equation by 2.

d. $\dfrac{-x - 0.3}{2} = -1.1$

$$\frac{-x - 0.3}{2} \times 2 = -1.1 \times 2$$

2. Add 0.3 to both sides.

$$-x - 0.3 + 0.3 = -2.2 + 0.3$$

3. Divide both sides by -1.

$$\frac{-x}{-1} = \frac{-1.9}{-1}$$

4. Write the value for x.

$$x = 1.9$$

▶ 5.2.3 Algebraic fractions with the pronumeral in the denominator

eles-4691

- If an equation uses fractions and the pronumeral is in the denominator, there is an extra step involved in determining the solution.
- Consider the following equation.

$$\frac{4}{x} = \frac{3}{2}$$

- In order to solve this equation, first multiply both sides by x.

$$\frac{4}{x} \times x = \frac{3}{2} \times x$$

$$4 = \frac{3x}{2}$$

$$\text{or} \quad \frac{3x}{2} = 4$$

- The pronumeral is now in the numerator, and the equation is easy to solve.

$$\frac{3x}{2} = 4$$

$$3x = 8$$

$$x = \frac{8}{3}$$

- Alternatively, the equation can be solved using the equivalent ratios method.

$$\frac{4}{x} = \frac{3}{2}$$

- This equation can be written as the equivalent ratio.

$$\frac{x}{4} = \frac{2}{3}$$

- Now the equation can be solved by multiplying both sides by 4.

$$\frac{x}{4} \times 4 = \frac{2}{3} \times 4$$

$$x = \frac{8}{3}$$

WORKED EXAMPLE 3 Solving equations with pronumerals in the denominator

Solve each of the following linear equations.

a. $\dfrac{3}{a} = \dfrac{4}{5}$　　　　b. $\dfrac{5}{b} = -2$

THINK	WRITE
a. 1. Write the equation	a. $\dfrac{3}{a} = \dfrac{4}{5}$
2. Multiply both sides by a.	$3 = \dfrac{4a}{5}$
3. Multiply both sides by 5.	$15 = 4a$
4. Divide both sides by 4.	$a = \dfrac{15}{4}$

b. 1. Write the equation.

b. $\dfrac{5}{b} = -2$

2. Multiply both sides by b.

$5 = -2b$

3. Divide both sides by -2.

$\dfrac{5}{-2} = b$ or $\dfrac{-5}{2} = b$

 Resources

▶ **Video eLesson** Solving linear equations (eles-1895)

🧩 **Interactivity** Using algebra to solve problems (int-3805)

Exercise 5.2 Solving linear equations

learnon

5.2 Quick quiz on	**5.2 Exercise**

Individual pathways

■ PRACTISE	■ CONSOLIDATE	■ MASTER
1, 3, 6, 8, 12, 16, 21, 23, 26, 29	2, 4, 9, 10, 13, 14, 17, 19, 20, 24, 27, 30	5, 7, 11, 15, 18, 22, 25, 28, 31, 32

Fluency

1. **WE1** For each of the following equations, determine whether $x = 6$ is a solution.

   a. $x + 3 = 7$

   b. $2x - 5 = 7$

   c. $x^2 - 2 = 38$

   d. $\dfrac{6}{x} + x = 7$

   e. $\dfrac{2(x+1)}{7} = 2$

   f. $3 - x = 9$

2. For each of the following equations, determine whether $x = 6$ is a solution.

   a. $x^2 + 3x = 39$

   b. $3(x+2) = 5(x-4)$

   c. $x^2 + 2x = 9x - 6$

   d. $x^2 = (x+1)^2 - 14$

   e. $(x-1)^2 = 4x + 1$

   f. $5x + 2 = x^2 + 4$

3. **WE2a** Solve each of the following linear equations. Check your answers by substitution.

   a. $x - 43 = 167$

   b. $x + 286 = 516$

   c. $58 + x = 81$

   d. $209 - x = 305$

   e. $5x = 185$

   f. $60x = 1200$

4. **WE2b** Solve each of the following linear equations. Check your answers by substitution.

   a. $5x = 250$

   b. $\dfrac{x}{23} = 6$

   c. $\dfrac{x}{17} = 26$

   d. $\dfrac{x}{9} = 27$

   e. $y - 16 = -31$

   f. $5.5 + y = 7.3$

5. **WE2c** Solve each of the following linear equations. Check your answers by substitution.

   a. $y - 7.3 = 5.5$

   b. $6y = 14$

   c. $0.9y = -0.05$

   d. $\dfrac{y}{5} = 4.3$

   e. $\dfrac{y}{7.5} = 23$

   f. $\dfrac{y}{8} = -1.04$

6. Solve each of the following linear equations.

   a. $2y - 3 = 7$
   b. $2y + 7 = 3$
   c. $5y - 1 = 0$
   d. $6y + 2 = 8$
   e. $7 + 3y = 10$
   f. $8 + 2y = 12$

7. Solve each of the following linear equations.

   a. $15 = 3y - 1$
   b. $-6 = 3y - 1$
   c. $6y - 7 = 140$
   d. $4.5y + 2.3 = 7.7$
   e. $0.4y - 2.7 = 6.2$
   f. $600y - 240 = 143$

8. Solve each of the following linear equations.

   a. $3 - 2x = 1$
   b. $-3x - 1 = 5$
   c. $-4x - 7 = -19$
   d. $1 - 3x = 19$
   e. $-5 - 7x = 2$
   f. $-8 - 2x = -9$

9. Solve each of the following linear equations.

   a. $9 - 6x = -1$
   b. $-5x - 4.2 = 7.4$
   c. $2 = 11 - 3x$
   d. $-3 = -6x - 8$
   e. $-1 = 4 - 4x$
   f. $35 - 13x = -5$

10. Solve each of the following linear equations.

   a. $7 - x = 8$
   b. $8 - x = 7$
   c. $5 - x = 5$
   d. $5 - x = 0$
   e. $15.3 = 6.7 - x$
   f. $5.1 = 4.2 - x$

11. Solve each of the following linear equations.

   a. $9 - x = 0.1$
   b. $140 - x = 121$
   c. $-30 - x = -4$
   d. $-5 = -6 - x$
   e. $-x + 1 = 2$
   f. $-2x - 1 = 0$

12. Solve each of the following linear equations.

   a. $\dfrac{x}{4} + 1 = 3$
   b. $\dfrac{x}{3} - 2 = -1$
   c. $\dfrac{x}{8} = \dfrac{1}{2}$

   d. $-\dfrac{x}{3} = 5$
   e. $5 - \dfrac{x}{2} = -8$
   f. $4 - \dfrac{x}{6} = 11$

13. Solve each of the following linear equations.

   a. $\dfrac{2x}{3} = 6$
   b. $\dfrac{5x}{2} = -3$
   c. $-\dfrac{3x}{4} = -7$
   d. $-\dfrac{8x}{3} = 6$
   e. $\dfrac{2x}{7} = -2$
   f. $-\dfrac{3x}{10} = -\dfrac{1}{5}$

14. Solve each of the following linear equations.

   a. $\dfrac{z - 1}{3} = 5$
   b. $\dfrac{z + 1}{4} = 8$
   c. $\dfrac{z - 4}{2} = -4$

   d. $\dfrac{6 - z}{7} = 0$
   e. $\dfrac{3 - z}{2} = 6$
   f. $\dfrac{-z - 50}{22} = -2$

15. Solve each of the following linear equations.

   a. $\dfrac{z - 4.4}{2.1} = -3$
   b. $\dfrac{z + 2}{7.4} = 1.2$
   c. $\dfrac{140 - z}{150} = 0$

   d. $\dfrac{-z - 0.4}{2} = -0.5$
   e. $\dfrac{z - 6}{9} = -4.6$
   f. $\dfrac{z + 65}{73} = 1$

16. Solve each of the following linear equations.

   a. $\dfrac{5x + 1}{3} = 2$
   b. $\dfrac{2x - 5}{7} = 3$
   c. $\dfrac{3x + 4}{2} = -1$

   d. $\dfrac{4x - 13}{9} = -5$
   e. $\dfrac{4 - 3x}{2} = 8$
   f. $\dfrac{1 - 2x}{6} = -10$

17. Solve each of the following linear equations.

a. $\dfrac{-5x-3}{9}=3$

b. $\dfrac{-10x-4}{3}=1$

c. $\dfrac{4x+2.6}{5}=8.8$

d. $\dfrac{5x-0.7}{-0.3}=-3.1$

e. $\dfrac{1-0.5x}{4}=-2.5$

f. $\dfrac{-3x-8}{14}=\dfrac{1}{2}$

18. **WE3** Solve each of the following linear equations.

a. $\dfrac{2}{x}=\dfrac{1}{2}$

b. $\dfrac{3}{x}=7$

c. $\dfrac{-4}{x}=\dfrac{7}{2}$

d. $\dfrac{5}{x}=\dfrac{-3}{4}$

e. $\dfrac{0.4}{x}=\dfrac{9}{2}$

f. $\dfrac{8}{x}=1$

19. Solve each of the following linear equations.

a. $\dfrac{-4}{x}=\dfrac{2}{3}$

b. $\dfrac{-6}{x}=\dfrac{-4}{5}$

c. $\dfrac{1.7}{x}=\dfrac{1}{3}$

d. $\dfrac{6}{x}=-1$

e. $\dfrac{4}{x}=\dfrac{-15}{22}$

f. $\dfrac{50}{x}=\dfrac{-35}{43}$

20. **MC** Answer the following questions.

a. Determine the solution to the equation $82-x=44$.

A. $x=126$ B. $x=-126$ C. $x=122$ D. $x=38$

b. Determine the solution to the equation $5x-12=-62$.

A. $x=-14.8$ B. $x=14.8$ C. $x=10$ D. $x=-10$

c. Determine the solution to the equation $\dfrac{x-1}{2}=5.3$.

A. $x=9.6$ B. $x=10.6$ C. $x=11.6$ D. $x=2$

21. Solve each of the following linear equations.

a. $3a+7=4$

b. $5-b=-5$

c. $4c-4.4=44$

d. $\dfrac{d-4}{67}=0$

e. $5-3e=-10$

f. $\dfrac{2f}{3}=8$

22. Solve each of the following linear equations.

a. $100=6g+4.2$

b. $\dfrac{h+2}{6}=5.5$

c. $452i-124=-98$

d. $\dfrac{6j-1}{17}=0$

e. $\dfrac{12-k}{5}=4$

f. $\dfrac{l-5.2}{3.4}=1.5$

Understanding

23. Write each of the following worded statements as a mathematical sentence and then solve for the unknown value.

a. Seven is added to the product of x and 3, which gives the result of 4.

b. Four is divided by x and this result is equivalent to $\dfrac{2}{3}$.

c. Three is subtracted from x and this result is divided by 12 to give 25.

24. Driving lessons are usually quite expensive, but a discount of $15 per lesson is given if a family member belongs to the automobile club.
If 10 lessons cost $760 after the discount, calculate the cost of each lesson before the discount.

25. Anton lives in Australia. His friend Utan lives in the USA. Anton's home town of Horsham experienced one of the hottest days on record with a temperature of 46.7 °C. Utan said that his home town had experienced a day hotter than that, with the temperature reaching 113 °F.

The formula for converting Celsius to Fahrenheit is $F = \dfrac{9}{5}C + 32$. State whether Utan was correct.

Show full working.

Communicating, reasoning and problem solving

26. If the expression $\dfrac{12}{x-4}$ always results in a positive integer value, explain how you would determine the possible values for x.

27. Santo solved the linear equation $9 = 5 - x$. His second step was to divide both sides by -1. Trudy, his mathematics buddy, said she multiplied both sides by -1. Explain why they are both correct.

28. Determine the mistake in the following working and explain what is wrong.

$$\frac{x}{5} - 1 = 2$$
$$x - 1 = 10$$
$$x = 11$$

29. Sweet-tooth Sami goes to the corner store and buys an equal number of 25-cent and 30-cent lollies for a total of $16.50. Determine the amount of lollies he bought.

30. In a cannery, cans are filled by two machines that together produce 16 000 cans during an 8-hour shift. If the newer machine of the two produces 340 more cans per hour than the older machine, evaluate the number of cans produced by each machine in an eight-hour shift.

31. General admission to a music festival is $55 for an adult ticket, $27 for a child and $130 for a family of two adults and two children.

a. Evaluate how much you would save by buying a family ticket instead of two adult and two child tickets.
b. Determine if it is worthwhile buying a family ticket if a family has only one child.

32. A teacher asks her students to determine the value of n in the diagram shown. Use your knowledge of linear equations to solve this problem.

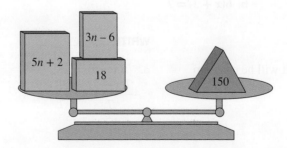

LESSON
5.3 Solving linear equations with brackets

LEARNING INTENTION

At the end of this lesson you should be able to:
- solve equations involving brackets.

▶ 5.3.1 Linear equations with brackets

eles-4692

- Consider the equation $3(x + 5) = 18$. There are two methods for solving this equation.

Method 1:

Start by dividing both sides by 3.

$$\frac{3(x+5)}{3} = \frac{18}{3}$$
$$x + 5 = 6$$
$$x = 1$$

Method 2:

Start by expanding the brackets.

$$3(x + 5) = 18$$
$$3x + 15 = 18$$
$$3x = 3$$
$$x = 1$$

- In this case, method 1 works well because 3 divides exactly into 18.
- Now try the equation $7(x + 2) = 10$.

Method 1:

Start by dividing both sides by 7.

$$\frac{7(x+2)}{7} = \frac{10}{7}$$
$$x + 2 = \frac{10}{7}$$
$$x = -\frac{4}{7}$$

Method 2:

Start by expanding the brackets.

$$7(x + 2) = 10$$
$$7x + 14 = 10$$
$$7x = -4$$
$$x = \frac{-4}{7}$$

- In this case, method 2 works well because it avoids the addition or subtraction of fractions.
- For each equation you need to solve, try both methods and choose the method that works best.

WORKED EXAMPLE 4 Solving equations with brackets

Solve each of the following linear equations.
a. $7(x - 5) = 28$ b. $6(x + 3) = 7$

THINK	WRITE
a. 1. 7 is a factor of 28, so it will be easier to divide both sides by 7.	a. $7(x - 5) = 28$ $\frac{7(x-5)}{7} = \frac{28}{7}$
2. Add 5 to both sides.	$x - 5 = 4$
3. Write the value of x.	$x = 9$

---

b. 1. 6 is not a factor of 7, so it will be easier to expand the brackets first.

b.
$$6(x + 3) = 7$$
$$6x + 18 = 7$$

2. Subtract 18 from both sides.

$$6x + 18 = 7 - 18$$
$$6x = -11$$

3. Divide both sides by 6.

$$x = \frac{11}{6}$$

 Resources

 Interactivity Linear equations with brackets (int-6039)

Exercise 5.3 Solving linear equations with brackets

learn on

5.3 Quick quiz on

5.3 Exercise

Individual pathways

■ PRACTISE	■ CONSOLIDATE	■ MASTER
1, 3, 5, 13, 16	2, 7, 9, 11, 12, 14, 17	4, 6, 8, 10, 15, 18

Fluency

1. Solve each of the following linear equations.
   a. $5(x - 2) = 20$
   b. $4(x + 5) = 8$
   c. $6(x + 3) = 18$
   d. $5(x - 41) = 75$
   e. $8(x + 2) = 24$
   f. $3(x + 5) = 15$

2. Solve each of the following linear equations.
   a. $5(x + 4) = 15$
   b. $3(x - 2) = -12$
   c. $7(x - 6) = 0$
   d. $-6(x - 2) = 12$
   e. $4(x + 2) = 4.8$
   f. $16(x - 3) = 48$

3. Solve each of the following linear equations.
   a. $6(b - 1) = 1$
   b. $2(m - 3) = 3$
   c. $2(a + 5) = 7$
   d. $3(m + 2) = 2$
   e. $5(p - 2) = -7$
   f. $6(m - 4) = -8$

4. Solve each of the following linear equations.
   a. $-10(a + 1) = 5$
   b. $-12(p - 2) = 6$
   c. $-9(a - 3) = -3$
   d. $-2(m + 3) = -1$
   e. $3(2a + 1) = 2$
   f. $4(3m + 2) = 5$

5. Solve each of the following linear equations.
   a. $9(x - 7) = 82$
   b. $2(x + 5) = 14$
   c. $7(a - 1) = 28$
   d. $4(b - 6) = 4$
   e. $3(y - 7) = 0$
   f. $-3(x + 1) = 7$

6. Solve each of the following linear equations.
   a. $-6(m + 1) = -30$
   b. $-4(y + 2) = -12$
   c. $-3(a - 6) = 3$
   d. $-2(p + 9) = -14$
   e. $3(2m - 7) = -3$
   f. $2(4p + 5) = 18$

Understanding

7. Solve each of the following linear equations. Round the answers to 3 decimal places where appropriate.

   a. $2(y+4)=-7$ b. $0.3(y+8)=1$ c. $4(y+19)=-29$
   d. $7(y-5)=25$ e. $6(y+3.4)=3$ f. $7(y-2)=8.7$

8. Solve each of the following linear equations. Round the answers to 3 decimal places where appropriate.

   a. $1.5(y+3)=10$ b. $2.4(y-2)=1.8$ c. $1.7(y+2.2)=7.1$
   d. $-7(y+2)=0$ e. $-6(y+5)=-11$ f. $-5(y-2.3)=1.6$

9. **MC** Select the best first step for solving the equation $7(x-6)=23$.

   A. Add 6 to both sides.   **B.** Subtract 7 from both sides.
   C. Divide both sides by 23.   **D.** Expand the brackets.

10. **MC** Select which one of the following is closest to the solution for the equation $84(x-21)=782$.

    A. $x=9.31$ **B.** $x=9.56$
    C. $x=30.31$ **D.** $x=-11.69$

11. In 1974 a mother was 6 times as old as her daughter. If the mother turned 50 in the year 2000, calculate the year in which the mother's age was double her daughter's age.

12. New edging is to be placed around a rectangular children's playground. The width of the playground is x metres and the length is 7 metres longer than the width.

    a. Write down an expression for the perimeter of the playground. Write your answer in factorised form.
    b. If the amount of edging required is 54 m, determine the dimensions of the playground.

Communicating, reasoning and problem solving

13. Explain the two possible methods for solving equations in factorised form.

14. Juanita is solving the equation $2(x-8)=10$. She performs the following operations to both sides of the equation in order: $+8, \div 2$. Explain why Juanita will not find the correct value of x using her order of inverse operations, then solve the equation correctly.

15. As your first step to solve the equation $3(2x-7)=18$, you are given three options:
    - Expand the brackets on the left-hand side.
    - Add 7 to both sides.
    - Divide both sides by 3.

    Explain which of these options is your least preferred.

16. Five times the sum of four and an unknown number is equal to 35. Evaluate the value of the unknown number.

17. Oscar earns $55 more than Josue each week, but Hector earns three times as much as Oscar. If Hector earns $270 a week, determine how much Oscar and Josue earn each week.

18. A school wants to hire a bus to travel to a football game. The bus will take 28 passengers, and the school will contribute $48 towards the cost of the trip. The price of each ticket is $10. If the hiring of the bus is $300 + 10% of the total cost of all the tickets, evaluate the cost per person.

LESSON
5.4 Solving linear equations with pronumerals on both sides

LEARNING INTENTION

At the end of this lesson you should be able to:
- solve equations by collecting pronumerals on one side of the equation before solving.

▶ 5.4.1 Linear equations with pronumerals on both sides

eles-4693

- If pronumerals occur on both sides of an equation, the first step in solving the equation is to move all of the pronumerals so that they appear on only one side of the equation.
- When solving equations, it is important to remember that whatever you do to one side of an equation you must also do to the other side.

WORKED EXAMPLE 5 Solving equations with pronumerals on both sides

Solve each of the following linear equations.

a. $5y = 3y + 3$

b. $7x + 5 = 2 - 4x$

c. $3(x + 1) = 14 - 2x$

d. $2(x + 3) = 3(x + 7)$

THINK	WRITE
a. 1. $3y$ is smaller than $5y$. This means it is easiest to subtract $3y$ from both sides of the equation.	**a.** $\begin{aligned} 5y &= 3y + 3 \\ 5y - 3y &= 3y + 3 - 3y \\ 2y &= 3 \end{aligned}$
2. Divide both sides by 2.	$y = \dfrac{3}{2}$
b. 1. $-4x$ is smaller than $7x$. This means it is easiest to add $4x$ to both sides of the equation.	**b.** $\begin{aligned} 7x + 5 &= 2 - 4x \\ 7x + 5 + 4x &= 2 - 4x + 4x \\ 11x + 5 &= 2 \end{aligned}$
2. Subtract 5 from both sides.	$\begin{aligned} 11x + 5 - 5 &= 2 - 5 \\ 11x &= -3 \end{aligned}$
3. Divide both sides by 11.	$x = \dfrac{-3}{11}$
c. 1. Expand the brackets. **2.** $-2x$ is smaller than $3x$. This means it is easiest to add $2x$ to both sides of the equation. **3.** Subtract 3 from both sides. **4.** Divide both sides by 5.	**c.** $3(x + 1) = 14 - 2x$ $3x + 3 = 14 - 2x$ $3x + 3 + 2x = 14 - 2x + 2x$ $5x + 3 = 14$ $\begin{aligned} 5x + 3 - 3 &= 14 - 3 \\ 5x &= 11 \end{aligned}$ $x = \dfrac{11}{5}$

d. 1. Expand the brackets.

d.

$$2(x + 3) = 3(x + 7)$$
$$2x + 6 = 3x + 21$$

2. $2x$ is smaller than $3x$. This means it is easiest to subtract $2x$ from both sides of the equation.

$$2x + 6 - 2x = 3x + 21 - 2x$$
$$6 = x + 21$$

3. Subtract 21 from both sides.

$$6 - 21 = x + 21 - 21$$
$$-15 = x$$

4. Write the answer with the pronumeral written on the left-hand side.

$$x = -15$$

---

on Resources

▶ **Video eLesson** Solving linear equations with pronumerals on both sides (eles-1901)

Exercise 5.4 Solving linear equations with pronumerals on both sides

learn on

5.4 Quick quiz on	**5.4 Exercise**

Individual pathways

■ PRACTISE	■ CONSOLIDATE	■ MASTER
1, 3, 5, 8, 14, 17	2, 6, 10, 12, 15, 18	4, 7, 9, 11, 13, 16, 19, 20

Fluency

1. WE5a Solve each of the following linear equations.

   a. $5y = 3y - 2$ **b.** $6y = -y + 7$ **c.** $10y = 5y - 15$

   d. $25 + 2y = -3y$ **e.** $8y = 7y - 45$ **f.** $15y - 8 = -12y$

2. Solve each of the following linear equations.

   a. $7y = -3y - 20$ **b.** $23y = 13y + 200$ **c.** $5y - 3 = 2y$

   d. $6 - 2y = -7y$ **e.** $24 - y = 5y$ **f.** $6y = 5y - 2$

3. MC Select the first step for solving the equation $3x + 5 = -4 - 2x$.

   A. Add $3x$ to both sides.   **B.** Add 5 to both sides.

   C. Add $2x$ to both sides.   **D.** Subtract $2x$ from both sides.

4. MC Select the first step for solving the equation $6x - 4 = 4x + 5$.

   A. Subtract $4x$ from both sides.   **B.** Add $4x$ to both sides.

   C. Subtract 4 from both sides.   **D.** Add 5 to both sides.

5. WE5b Solve each of the following linear equations.

   a. $2x + 3 = 8 - 3x$ **b.** $4x + 11 = 1 - x$ **c.** $x - 3 = 6 - 2x$

   d. $4x - 5 = 2x + 3$ **e.** $3x - 2 = 2x + 7$ **f.** $7x + 1 = 4x + 10$

---

6. Solve each of the following linear equations.

    a. $5x + 3 = x - 5$ b. $6x + 2 = 3x + 14$ c. $2x - 5 = x - 9$

    d. $10x - 1 = -2x + 5$ e. $7x + 2 = -5x + 2$ f. $15x + 3 = 7x - 3$

7. Solve each of the following linear equations.

    a. $x - 4 = 3x + 8$ b. $2x + 9 = 7x - 1$ c. $-2x + 7 = 4x + 19$

    d. $-3x + 2 = -2x - 11$ e. $11 - 6x = 18 - 5x$ f. $6 - 9x = 4 + 3x$

8. **MC** Determine the solution to $5x + 2 = 2x + 23$.

    A. $x = 3$ **B.** $x = -3$ **C.** $x = 5$ **D.** $x = 7$

9. **MC** Determine the solution to $3x - 4 = 11 - 2x$.

    A. $x = 15$ **B.** $x = 7$ **C.** $x = 3$ **D.** $x = 5$

10. **WE5c,d** Solve each of the following.

    a. $5(x - 2) = 2x + 5$ b. $7(x + 1) = x - 11$ c. $2(x - 8) = 4x$

    d. $3(x + 5) = x$ e. $6(x - 3) = 14 - 2x$ f. $9x - 4 = 2(3 - x)$

11. Solve each of the following.

    a. $4(x + 3) = 3(x - 2)$ b. $5(x - 1) = 2(x + 3)$ c. $8(x - 4) = 5(x - 6)$

    d. $3(x + 6) = 4(2 - x)$ e. $2(x - 12) = 3(x - 8)$ f. $4(x + 11) = 2(x + 7)$

Understanding

12. Aamir's teacher gave him the following algebra problem and told him to solve it.

$$3x + 7 = x^2 + k = 7x + 15$$

Suggest how you can you help Aamir calculate the value of k.

13. A classroom contained an equal number of boys and girls. Six girls left to play hockey, leaving twice as many boys as girls in the classroom. Determine the original number of students present.

Communicating, reasoning and problem solving

14. Express the information in the diagram shown as an equation, then show that $n = 29$ is the solution.

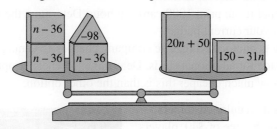

15. The block shown has a width of 1 unit and length of $(x + 1)$ units.

    a. Draw two rectangles with different sizes, each using 3 of the blocks shown.

    b. Show that the areas of both of the rectangles you have drawn using the blocks shown are the same. Explain how the areas of these rectangles relate to expanding brackets.

16. Explain what the difficulty is when trying to solve the equation $4(3x - 5) = 6(4x + 2)$ without expanding the brackets first.

17. This year Tom is 4 times as old as his daughter. In 5 years' time he will be only 3 times as old as his daughter. Determine the age of Tom and his daughter now.

18. If you multiply an unknown number by 6 and then add 5, the result is 7 less than the unknown number plus 1, multiplied by 3. Evaluate the unknown number.

19. You are looking into getting a business card printed for your new store. A local printing company charges $250 for the materials and an hourly rate for labour of $40.

a. If h is the number of hours of labour required to print the cards, construct an equation for the cost of the cards, C.
b. You have budgeted $1000 for this printing job. Determine the number of hours of labour that you can afford. Give your answer to the nearest minute.
c. The company estimates that it can print 1000 cards an hour. Determine the number of cards that you can afford to get printed with your current budget.
d. An alternative to printing is photocopying. The company charges 15 cents per side for the first 10 000 cards, then 10 cents per side for additional cards. Determine the cheaper option to get 18 750 single-sided cards made. Determine how much cheaper this is than the other option.

20. A local games arcade offers its regular customers the following deal. For a monthly fee of $40, players can play 25 $2 games of their choice. Extra games cost $2 each. After a player has played 50 games in a month, all further games are $1.

a. If Iman has $105 to spend in a month, determine the number of games she can play if she takes up the special deal.
b. Determine the amount of money Iman will save by taking up the special deal compared to playing the same number of games at $2 a game.

LESSON
5.5 Solving worded problems and substituting into formulas

⊙ 5.5.1 Solving problems with linear equations

eles-4694

- When pronumerals are not provided in a problem, you need to introduce a pronumeral to the problem.
- When translating a worded problem into an equation, make sure that the order of the operations is presented correctly.

WORKED EXAMPLE 6 Translating worded expressions into linear equations

Write linear equations for each of the following statements, using x to represent the unknown value. Solve the equations.

a. Three more than seven times a certain number is zero. Determine the number.

b. The sum of three consecutive integers is 102. Determine the three numbers.

c. The difference between two numbers is 11. Determine the numbers if, when eight is added to each of the two numbers, the larger result will be double the value of the smaller result.

THINK	WRITE
a. 1. Let x be the number.	a. $x =$ unknown number
2. 7 times the number is $7x$. Increasing the value of $7x$ by 3 gives $7x + 3$. This expression equals 0. Write the equation.	$7x + 3 = 0$
3. Solve the equation using inverse operations.	$7x + 3 - 3 = 0 - 3$ $$\frac{7x}{7} = \frac{-3}{7}$$
4. Write the value of x.	$$x = \frac{-3}{7}$$
b. 1. Let x be the first number.	b. $x =$ smallest number
2. The next two consecutive integers are $(x + 1)$ and $(x + 2)$. The sum of all three integers equals 102. Write the equation.	$x + (x + 1) + (x + 2) = 102$

3. Solve the equation using inverse operations.

$$3x + 3 - 3 = 102 - 3$$
$$\frac{3x}{3} = \frac{99}{3}$$
$$x = 33$$

4. Write the value of x and the next 2 integers.

The three numbers are 33, 34 and 35.

c. 1. Let x be the larger number.

c. x = larger number

2. The difference between the 2 numbers is 11. This means the smaller number is 11 less than x. Write this as $(x - 11)$.

$(x - 11)$ = smaller number

3. Increasing both numbers by 8 makes the larger number $(x + 8)$ and the smaller number $(x - 11 + 8)$.

The larger number becomes double the smaller number.

$$(x + 8) = 2(x - 11 + 8)$$
$$x + 8 = 2(x - 3)$$

4. Solve the equation using inverse operations.

$$x + 8 - 8 = 2x - 6 - 8$$
$$x - x = 2x - 14 - x$$
$$0 + 14 = x - 14 + 14$$
$$x = 14$$

5. Write the value of x and the other number.

The two numbers are 14 and 3.

A taxi charges \$3.60 plus \$1.38 kilometre for any trip in Sydney. If Elena's taxi ride cost \$38.10, calculate the distance she travelled.

THINK	WRITE
1. The distance travelled by Elena has to be found. Define the pronumeral.	Let x = the distance travelled (in kilometres).
2. It costs 1.38 to travel 1 kilometre, so the cost to travel x kilometres = $1.38x$. The fixed cost is \$3.60. Write an expression for the total cost.	Total cost = $3.60 + 1.38x$
3. Let the total cost = 38.10.	$3.60 + 1.38x = 38.10$
4. Solve the equation.	$3.60 + 1.38x = 38.10$ $1.38x = 34.50$ $x = \dfrac{34.50}{1.38}$ $= 25$
5. Write the solution in words.	Elena travelled 25 kilometres.

In a basketball game, Hao scored 5 more points than Seve. If they scored a total of 27 points between them, calculate how many points each of them scored.

THINK	WRITE
1. Define the pronumeral.	Let Seve's score be x.
2. Hao scored 5 more than Seve. Write an expression for Hao's score.	Hao's score is $x + 5$.
3. Hao and Seve scored a total of 27 points. Write this as an equation.	$x + (x + 5) = 27$
4. Solve the equation.	$2x + 5 = 27$ $2x = 22$ $x = 11$
5. Since $x = 11$, Seve's score is 11. Substitute this into the expression to work out Hao's score.	Hao's score $= x + 5$ $= 11 + 5$ $= 16$
6. Write the answer in words.	Seve scored 11 points and Hao scored 16 points.

A collection of 182 marbles is owned by four friends. Pat has twice the number of marbles that Quentin has, and Rachel has 20 fewer marbles than Pat. If Sam has two-thirds the number of marbles that Rachel has, determine who has the second-largest number of marbles.

THINK	WRITE
a. 1. Let p represent the number of marbles Pat has.	$p =$ number of Pat's marbles
Pat has twice the number of marbles that Quentin has.	$\left(\dfrac{p}{2}\right) =$ number of Quentin's marbles
Rachel has 20 fewer marbles than Pat.	$p - 20 =$ number of Rachael's marbles
Sam has $\dfrac{2}{3}$ of the number of marbles that Rachel has.	$\dfrac{2}{3}(p - 20) =$ number of Sam's marbles
2. Let the sum of all marbles equal 182.	$p + \left(\dfrac{p}{2}\right) + (p - 20) + \dfrac{2}{3}(p - 20) = 182$
3. Simplify the left-hand side of the equation by adding like terms.	$p + \dfrac{p}{2} + p + \dfrac{2p}{3} - 20 - \dfrac{40}{3} = 182$ $\dfrac{6p}{6} + \dfrac{3p}{6} + \dfrac{6p}{6} + \dfrac{4p}{6} - \dfrac{120}{6} - \dfrac{80}{6} = 182$

▶

4. Solve using inverse operations.

$$\frac{19p - 200}{6} \times 6 = 182 \times 6$$

$$19p - 200 + 200 = 1092 + 200$$

$$\frac{19p}{19} = \frac{1292}{19}$$

$$p = 68$$

5. Calculate the number of marbles that each friend has.

Pat has 68 marbles.

Quentin has $\dfrac{68}{2} = 34$ marbles.

Rachel has $68 - 20 = 48$ marbles.

Sam has $\dfrac{2}{3} \times 48 = 32$ marbles.

6. Write down the second-highest number of marbles and the name of the person with that number of marbles.

Rachel has the second-highest number of marbles. She has 48 marbles.

⏵ 5.5.2 Substituting into formulas

eles-6200

- Formulas are equations that are used for specific purposes. They are generally written in terms of two or more pronumerals, also known as 'variables'.
- To find the value of a pronumeral in a formula, we need the values of every other pronumeral in the formula.
- Substituting these values into the formula gives a linear equation, which we can solve.
- Care must be taken with the units of the variables and, if given, included in the answer.

WORKED EXAMPLE 10 Solving linear equations from substitution

The formula for the perimeter of a rectangle is $P = 2(l + h)$ where l is the length and h the height of the rectangle.
a. Find the perimeter of a rectangle with length 24 cm and height 20 cm.
b. A rectangle, with a perimeter of 120 cm, has a height of 15 cm. What is the length of the rectangle?
c. Find the height of a rectangle if the perimeter is 2.5 metres and it is 90 cm in length.

THINK

a. 1. Write the given formula.
2. State the value of the known variables.
3. Substitute into the formula.
4. Simplify and evaluate.
5. Write the answer, including units.

WRITE

a. $P = 2(l + h)$
$l = 24$, $h = 20$
$P = 2(24 + 20)$
$P = 2 \times 44$
$P = 88$
The perimeter of the rectangle is 88 cm.

b. **1.** Write the given formula. $P = 2(l + h)$
2. State the value of the known variables. $P = 120, h = 15$
3. Substitute into the formula. $120 = 2(l + 15)$
4. Solve the linear equation. $60 = l + 15$
5. Write the answer, including units. $l = 45$
The length of the rectangle is 45 cm.

c. **1.** Write the given formula. $P = 2(l + h)$
2. State the value of the known variables. $P = 2.5$ metres, $l = 90$ cm
Change the units to cm. $P = 250$ cm, $l = 90$ cm
3. Substitute into the formula. $250 = 2(90 + h)$
4. Solve the linear equation. $125 = 90 + h$
5. Write the answer, including units. $h = 35$
The height of the rectangle is 35 cm.

DISCUSSION

Why is it important to define the pronumeral when using a linear equation to solve a problem?

Exercise 5.5 Solving worded problems and substituting into formulas

learnon

5.5 Quick quiz on	5.5 Exercise

Individual pathways

■ PRACTISE	■ CONSOLIDATE	■ MASTER
1, 3, 7, 9, 13, 14, 18, 20, 23, 25	2, 4, 8, 10, 15, 19, 21, 24	5, 6, 11, 12, 16, 17, 22

Fluency

1. **WE6** Write linear equations for each of the following statements, using x to represent the unknown, without solving the equations.

   a. When 3 is added to a certain number, the answer is 5.
   b. Subtracting 9 from a certain number gives a result of 7.
   c. Multiplying a certain number by 7 gives 24.
   d. A certain number divided by 5 gives a result of 11.
   e. Dividing a certain number by 2 equals −9.

2. Write linear equations for each of the following statements, using x to represent the unknown, without solving the equations.

   a. Subtracting 3 from 5 times a certain number gives a result of −7.
   b. When a certain number is subtracted from 14 and the result is then multiplied by 2, the final result is −3.
   c. When 5 is added to 3 times a certain number, the answer is 8.
   d. When 12 is subtracted from 2 times a certain number, the result is 15.
   e. The sum of 3 times a certain number and 4 is divided by 2. This gives a result of 5.

3. **MC** Select the equation that matches the following statement.
   A certain number, when divided by 2, gives a result of -12.

   A. $x = \dfrac{-12}{2}$ **B.** $2x = -12$ **C.** $\dfrac{x}{2} = -12$ **D.** $\dfrac{x}{12} = -2$

4. **MC** Select the equation that matches the following statement.
   Dividing 7 times a certain number by -4 equals 9.

   A. $\dfrac{x}{-4} = 9$ **B.** $\dfrac{-4x}{7} = 9$ **C.** $\dfrac{7+x}{-4} = 9$ **D.** $\dfrac{7x}{-4} = 9$

5. **MC** Select the equation that matches the following statement.
   Subtracting twice a certain number from 8 gives 12.

   A. $2x - 8 = 12$ **B.** $8 - 2x = 12$ **C.** $2 - 8x = 12$ **D.** $8 - (x + 2) = 12$

6. **MC** Select the equation that matches the following statement.
   When 15 is added to a quarter of a number, the answer is 10.

   A. $15 + 4x = 10$ **B.** $10 = \dfrac{x}{4} + 15$ **C.** $\dfrac{x + 15}{4} = 10$ **D.** $15 + \dfrac{4}{x} = 10$

7. If $v = u + at$, calculate the value of:
   a. v when $u = 6$, $a = 4$ and $t = 5$
   b. u when $v = 25$, $a = 3$ and $t = 4$
   c. a when $v = 42$, $u = 15$ and $t = 3$
   d. t when $v = 16$, $u = 4$ and $a = 5$.

8. If $T = a + (n - 1)d$, calculate the value of:
   a. T when $a = 10$, $n = 12$, $d = 2$
   b. a when $T = 38$, $n = 13$, $d = 3$
   c. n when $T = 30$, $a = -2$, $d = 4$
   d. d when $T = 12$, $a = 32$, $n = 5$.

Understanding

9. When a certain number is added to 3 and the result is multiplied by 4, the answer is the same as when that same number is added to 4 and the result is multiplied by 3. Determine the number.

10. **WE7** John is three times as old as his son Jack. The sum of their ages is 48. Calculate how old John is.

11. In one afternoon's shopping Seedevi spent half as much money as Georgia, but $6 more than Amy. If the three of them spent a total of $258, calculate how much Seedevi spent.

12. The rectangular blocks of land shown have the same area. Determine the dimensions of each block and use these dimensions to calculate the area.

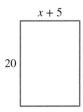

Communicating, reasoning and problem solving

13. **WE 8** A square pool is surrounded by a paved area that is 2 metres wide. If the area of the paving is $72\,\text{m}^2$, determine the length of the pool. Show all working.

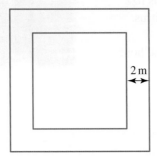

14. Maria is paid \$11.50 per hour plus \$7 for each jacket that she sews. If she earned \$176 for a particular eight-hour shift, determine the number of jackets she sewed during that shift.

15. Mai hired a car for a fee of \$120 plus \$30 per day. Casey's rate for his car hire was \$180 plus \$26 per day. If their final cost and rental period is the same for both Mai and Casey, determine the rental period.

16. **WE9** The cost of producing an album on CD is quoted as \$1200 plus \$0.95 per CD. If Maya has a budget of \$2100 for her debut album, determine the number of CDs she can make.

17. Joseph wants to have some flyers for his grocery business delivered. Post Quick quotes a price of \$200 plus 50 cents per flyer, and Fast Box quotes \$100 plus 80 cents per flyer.
    a. If Joseph needs to order 1000 flyers, determine the distributor that would be cheaper to use.
    b. Determine the number of flyers that Joseph need to get delivered for the cost to be the same for either distributor.

18. **WE10** The formula for the perimeter of a trapezium is $P = \dfrac{1}{2}(a+b)h$ where a and b are the lengths of the parallel sides and h the perpendicular height.
    a. Determine the perimeter of a trapezium with parallel sides of 16 cm and 10 cm and the perpendicular height of 5 cm.
    b. A trapezium, with parallel sides of 9 cm and 5 cm, has a perimeter of 42 cm. What is the height of the trapezium?
    c. Calculate the length of one of the parallel sides of a trapezium if the other parallel side is 52 cm, the perpendicular height is 12 cm and the perimeter of the trapezium is 8 metres.

19. The formula to convert temperatures measured in Fahrenheit to Celsius is $C = \dfrac{5}{9}(F-32)$.
    a. Convert 86 °F to degrees Celsius.
    b. If the temperature is 20 °C, what would be the equivalent in Fahrenheit?
    c. Jemima lives in Sydney. Fred lives in Los Angeles where temperatures are measured in Fahrenheit. Jemima says the temperature that day was 25 °C.
    Fred thinks that, at 25 °C, it is below freezing point. Is Fred correct?

20. A certain number is multiplied by 8 and then 16 is subtracted. The result is the same as 4 times the original number minus 8. Evaluate the number.

21. Carmel sells three different types of healthy drinks: herbal, vegetable and citrus fizz. In an hour she sells 4 herbal, 3 vegetable and 6 citrus fizz drinks for $60.50.

In the following hour she sells 2 herbal, 4 vegetable and 3 citrus fizz drinks. In the third hour she sells 1 herbal, 2 vegetable and 4 citrus. The total amount in cash sales for the three hours is $136.50.

Carmel made $7 less in the third hour than she did in the second hour of sales.

Evaluate Carmel's sales in the fourth hour if she sells 2 herbal, 3 vegetable and 4 citrus fizz drinks.

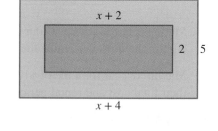

22. A rectangular swimming pool is surrounded by a path that is enclosed by a pool fence. All measurements given are in metres and are not to scale in the diagram shown.

   a. Write an expression for the area of the entire fenced-off section.
   b. Write an expression for the area of the path surrounding the pool.
   c. If the area of the path surrounding the pool is $34\,\text{m}^2$, evaluate the dimensions of the swimming pool.
   d. Determine the fraction of the fenced-off area taken up by the pool.

23. The cost to rent a car is given by the formula $C = 50d + 0.2k$, where $d =$ the number of days rented and $k =$ the number of kilometres driven. Lin has $300 to spend on car rental for her four-day holiday.

Calculate the distance that she can drive on her holiday.

24. A cyclist pumps up a bike tyre that has a slow leak. The volume of air (in cm^3) after t minutes is given by the formula $V = 24\,000 - 300t$.

   a. Determine the volume of air in the tyre when it is first filled.
   b. Write out the equation for the tyre's volume and solve it to work out how long it would take the tyre to go completely flat.

25. The total surface area of a cylinder is given by the formula $T = 2\pi r^2 + 2\pi rh$, where $r =$ radius and $h =$ height. A car manufacturer wants its engines' cylinders to have a radius of 4 cm and a total surface area of $400\,\text{cm}^2$. Show that the height of the cylinder is approximately 11.92 cm, correct to 2 decimal places.

LESSON
5.6 Review

5.6.1 Topic summary

LINEAR EQUATIONS

Inverse operations

- Solving a linear equation means determining a value for the pronumeral that makes the statement true.
- Solving equations requires inverse operations.
- The inverse operation has the effect of undoing the original operation.

Operation	Inverse operation
+	−
−	+
×	÷
÷	×

- To keep an equation balanced, inverse operations are done on both sides of the equation.

e.g.
$$3x + 1 = 11$$
$$3x + 1 - 1 = 11 - 1$$
$$\frac{3x}{3} = \frac{10}{3}$$
$$x = \frac{10}{3}$$

Equations with brackets

- The number in front of the brackets indicates the multiple of the expression inside the brackets.
- Brackets can be removed by dividing both sides by the multiple.

e.g.
$$3(x + 5) = 18$$
$$\frac{3(x + 5)}{3} = \frac{18}{3}$$
$$(x + 5) = 6$$

- Alternatively, brackets can be removed by expanding (multiplying) out the expression.

e.g.
$$3(x + 5) = 18$$
$$3x + 15 = 18$$

Equations with pronumerals on both sides

- Pronumerals must be moved so that they appear only on one side using inverse operations.
- Once this is done, you can solve the equation.

e.g.
$$5y = 3y + 3$$
$$5y - 3y = 3y + 3 - 3y$$
$$2y = 3$$
$$y = \frac{3}{2}$$

Pronumeral in the denominator

- The pronumeral needs to be expressed as a numerator.
- This can be done by multiplying both sides of the equation by the pronumeral.

e.g.
$$\frac{4}{x} \times x = \frac{3}{2} \times x$$

- Alternatively, you can flip both sides of the equation.

e.g.
$$\frac{4}{x} = \frac{3}{2}$$
becomes $\frac{x}{4} = \frac{2}{3}$.

Worded problems

- Worded problems should be translated into equations.
- Pronumerals must be defined in the answer if this is not already done in the question.

e.g. The equation for the worded problem 'When 6 is subtracted from a certain number, the result is 15' is 'x − 6 = 15'.

Substituting into formulas and rearranging

- To determine the value of a pronumeral in a formula, we need the values of every other pronumeral in the formula.
- Sometimes we need to rearrange the formula to solve for a particular pronumeral.

e.g.
$$P = 2(l + h)$$
When $P = 16$ and $h = 3$:
$$16 = 2(l + 3)$$
$$8 = l + 3$$
$$l = 5$$

5.6.2 Project

Forensic science

Scientific studies have been conducted on the relationship between a person's height and the measurements of a variety of body parts. One study has suggested that there is a general relationship between a person's height and the humerus bone in their upper arm, and that this relationship is slightly different for men and women.

According to this study, there is a general trend indicating that $h = 3.08l + 70.45$ for men, and $h = 3.36l + 57.97$ for women, where h represents body height in centimetres and l the length of the humerus in centimetres.

Imagine the following situation.

A decomposed body has been in the bushland outside your town. A team of forensic scientists suspects that the body could be the remains of either Alice Brown or James King, both of whom have been missing for several years. From the descriptions provided by their Missing Persons files, Alice is 162 cm tall and James is 172 cm tall. The forensic scientists hope to identify the body based on the length of the body's humerus.

1. Based on the relationship suggested by the study previously mentioned, complete both of the following tables using the equations provided. Calculate body height to the nearest centimetre.

Body height (men)

Length of humerus, l (cm)	20	25	30	35	40
Body height, h (cm)					

Body height (women)

Length of humerus, l (cm)	20	25	30	35	40
Body height, h (cm)					

2. On a piece of graph paper, draw the first quadrant of a Cartesian plane. Since the length of the humerus is the independent variable, place it on the x-axis. Place the dependent variable, body height, on the y-axis.

3. Plot the points from the two tables onto the set of axes drawn in question 2. Join the points with straight lines, using different colours to represent men and women.

4. Describe the shape of the two graphs.

5. Measure the length of your humerus. Use your graph to predict your height. How accurate is the measurement?

6. The two lines of your graph will intersect if extended. At what point does this occur? Comment on this value.

The forensic scientists measured the length of the humerus of the decomposed body and found it to be 33 cm.

7. Using methods covered in this activity, is it more probable that the body is that of Alice or of James? Justify your decision with mathematical evidence.

Exercise 5.6 Review questions

learnon

Fluency

1. **MC** Select the linear equation represented by the sentence 'When a certain number is multiplied by 3, the result is 5 times that certain number plus 7'.
   A. $3x + 7 = 5x$
   B. $5(x + 7) = 3x$
   C. $5x + 7 = 3x$
   D. $5x = 3x + 7$

2. **MC** Calculate the solution to the equation $\dfrac{x}{3} = 5$.

   A. $x = -15$
   B. $x = 15$
   C. $x = 1\dfrac{2}{3}$
   D. $x = 3$

3. **MC** Determine the solution to the equation $7 = 21 + x$.
   A. $x = 28$
   B. $x = -28$
   C. $x = -14$
   D. $x = 14$

4. **MC** Calculate the solution to the equation $5x + 3 = 37$.
   A. $x = 8$
   B. $x = -8$
   C. $x = 6.8$
   D. $x = 106$

5. **MC** Determine the solution to the equation $8 - 2x = 22$.
   A. $x = 11$
   B. $x = 15$
   C. $x = -15$
   D. $x = -7$

6. **MC** Determine the solution to the equation $4x + 3 = 7x - 33$.

   A. $x = -12$
   B. $x = 12$
   C. $x = \dfrac{36}{11}$
   D. $x = \dfrac{30}{11}$

7. **MC** Calculate the solution to the equation $7(x - 15) = 28$.
   A. $x = 11$
   B. $x = 19$
   C. $x = 20$
   D. $x = 6.14$

8. Identify which of the following are linear equations.
   a. $5x + y^2 = 0$
   b. $2x + 3 = x - 2$
   c. $\dfrac{x}{2} = 3$
   d. $x^2 = 1$
   e. $\dfrac{1}{x} + 1 = 3x$
   f. $8 = 5x - 2$
   g. $5(x + 2) = 0$
   h. $x^2 + y = -9$
   i. $r = 7 - 5(4 - r)$

9. Solve each of the following linear equations.
   a. $3a = 8.4$
   b. $a + 2.3 = 1.7$
   c. $\dfrac{b}{21} = -0.12$
   d. $b - 1.45 = 1.65$
   e. $b + 3.45 = 0$
   f. $7.53b = 5.64$

10. Solve each of the following linear equations.

   a. $\dfrac{2x-3}{7} = 5$

   b. $\dfrac{5-x}{2} = -4$

   c. $\dfrac{-3x-4}{5} = 3$

   d. $\dfrac{6}{x} = 5$

   e. $\dfrac{4}{x} = \dfrac{3}{5}$

   f. $\dfrac{x+1.7}{2.3} = -4.1$

11. Solve each of the following linear equations.

   a. $5(x-2) = 6$

   b. $7(x+3) = 40$

   c. $4(5-x) = 15$

   d. $6(2x+3) = 1$

   e. $4(x+5) = 2x - 5$

   f. $3(x-2) = 7(x+4)$

Understanding

12. Liz has a packet of 45 lolly snakes. She saves 21 to eat tomorrow, but she rations the remainder so that she can eat 8 snakes every hour until today's share of snakes is gone.
   a. Write a linear equation in terms of the number of hours, h, to represent this situation.
   b. Determine the number of hours it will take Liz to eat today's share.

13. Solve each of the following linear equations.
   a. $11x = 15x - 2$
   b. $3x + 4 = 16 - x$
   c. $5x + 2 = 3x + 8$
   d. $8x - 9 = 7x - 4$
   e. $2x + 5 = 8x - 7$
   f. $3 - 4x = 6 - x$

14. Translate the following sentences into algebraic equations. Use x to represent the number in question.
   a. Twice a certain number is equal to 3 minus that certain number.
   b. When 8 is added to 3 times a certain number, the result is 19.
   c. Multiplying a certain number by 6 equals 4.
   d. Dividing 10 by a certain number equals one more than dividing that number by 6.
   e. Multiply a certain number by 2, then add 5. Multiply this result by 7. This expression equals 0
   f. Twice a certain distance travelled is 100 metres more than that certain distance travelled plus 50 metres.

15. Takanori decides to go on a holiday. He travels a certain distance on the first day, twice that distance on the second day, three times that distance on the third day and four times that distance on the fourth day. If Takanori's total journey is 2000 km, calculate the distance he travelled on the third day.

Communicating, reasoning and problem solving

16. The formula for displacement is $s = ut + \dfrac{1}{2}at^2$ where t is the time in seconds, s is the displacement in metres, u is the initial velocity and a is the acceleration due to gravity.

   a. Calculate s when $u = 16.5\,\text{m/s}$, $t = 2.5$ seconds and $a = 9.8\,\text{m/s}^2$.

   b. A body has a displacement of 137.2 metres after 4 seconds.
   Determine the initial velocity if $a = 9.8\,\text{m/s}^2$.

17. The formula for the period (T) of a pendulum in seconds is $T = 2\pi\sqrt{\dfrac{L}{g}}$,

where L is the length in metres of the pendulum and $g = 9.81\ \text{m/s}^2$ is the acceleration due to gravity.

Determine the period of a pendulum, to 1 decimal place, in a grandfather clock with a pendulum length of 154 cm.

18. Saeed is comparing two car rental companies, Golden Ace Rental Company and Silver Diamond Rental Company. Golden Ace Rental Company charges a flat rate of $38 plus $0.20 per kilometre. The Silver Diamond Rental Company charges a flat rate of $30 plus $0.32 per kilometre. Saeed plans to rent a car for three days.

   a. Write an equation for the cost of renting a car for three days from the Golden Ace Rental Company, in terms of the number of kilometres travelled, k.
   b. Write an algebraic equation for the cost of renting a car for three days from the Silver Diamond Rental Company, in terms of the number of kilometres travelled, k.
   c. Evaluate the number of kilometres Saeed would have to travel so that the cost of hiring from each company is the same.

19. Frederika has $24 000 saved to pay for a holiday. Her travel expenses are $5400 and her daily expenses are $260.
   a. Write down an equation for the cost of her holiday if she stays for d days.
   b. Determine the number of days Frederika can spend on holiday if she wants to be able to buy a new laptop for $ 2500 when she gets back from her holidays.

20. A company that makes bottled orange juice buys their raw materials from two sources. The first source provides liquid with 6% orange juice, whereas the second source provides liquid with 3% orange juice. The company wants to make 1-litre bottles that have 5% orange juice. Let $x =$ the amount of liquid (in litres) that the company buys from the first source.
   a. Write an expression for the amount of orange juice from the first supplier, given that x is the amount of liquid.
   b. Write an expression for the amount of liquid from the second supplier, given that x is the amount of liquid used from the first supplier.
   c. Write an expression for the amount of orange juice from the second supplier.
   d. Write an equation for the total amount of orange juice in a mixture of liquids from the two suppliers, given that 1 litre of bottled orange juice will be mixed to contain 5% orange juice.
   e. Determine the quantity of the first supplier's liquid that the company uses.

21. Jayani goes on a four-day bushwalk. She travels a certain distance on the first day, half of that distance on the second day, a third of that distance on the third day and a quarter of that distance on the fourth day.

   If Jayani's total journey is 50 km, evaluate the distance that she walked on the first day.

22. Svetlana goes on a five-day bushwalk, travelling the same relative distances as Jayani travelled in question **22** (a certain amount, then half that amount, then one third of that, then one quarter of that, and finally one fifth of that).

   If Svetlana's journey is also 50 km, determine the distance that she travelled on the first day.

23. An online bookstore advertises its shipping cost to Australia as a flat rate of $20 for up to 10 books. Their major competitor offers a flat rate of $12 plus $1.60 per book.

   Determine the number of books you would have to buy (6, 7, 8, 9 or 10) for the first bookstore's shipping cost to be a better deal.

on To test your understanding and knowledge of this topic, go to your learnON title at www.jacplus.com.au and complete the **post-test**.

Answers

Topic 5 Linear equations

Exercise 5.1 Pre-test

1. 3
2. 22
3. C
4. a. $x = -7$ b. $z = 11$
5. 4
6. False
7. D
8. C
9. a. $a = 2$ b. $b = -8$
10. B
11. 287.5 km
12. -3
13. $A = 30$
14. 16 cm
15. $t = 2.2$

Exercise 5.2 Solving linear equations

1. a. No   b. Yes   c. No
   d. Yes   e. Yes   f. No

2. a. No   b. No   c. Yes
   d. No   e. Yes   f. No

3. a. $x = 210$ b. $x = 230$ c. $x = 23$
   d. $x = -96$ e. $x = 37$ f. $x = 20$

4. a. $x = 50$ b. $x = 138$ c. $x = 442$
   d. $x = 243$ e. $y = -15$ f. $y = 1.8$

5. a. $y = 12.8$ b. $y = 2\dfrac{1}{3}$ c. $y = -\dfrac{1}{18}$
   d. $y = 21.5$ e. $y = 172.5$ f. $y = -8.32$

6. a. $y = 5$ b. $y = -2$ c. $y = 0.2$
   d. $y = 1$ e. $y = 1$ f. $y = 2$

7. a. $y = 5\dfrac{1}{3}$ b. $y = -1\dfrac{2}{3}$ c. $y = 24.5$
   d. $y = 1.2$ e. $y = 22.25$ f. $y = \dfrac{383}{600}$

8. a. $x = 1$ b. $x = -2$ c. $x = 3$
   d. $x = -6$ e. $x = -1$ f. $x = \dfrac{1}{2}$

9. a. $x = 1\dfrac{2}{3}$ b. $x = -2.32$ c. $x = 3$
   d. $x = -\dfrac{5}{6}$ e. $x = 1\dfrac{1}{4}$ f. $x = 3\dfrac{1}{13}$

10. a. $x = -1$ b. $x = 1$ c. $x = 0$
    d. $x = 5$ e. $x = -8.6$ f. $x = -0.9$

11. a. $x = 8.9$ b. $x = 19$ c. $x = -26$
    d. $x = -1$ e. $x = -1$ f. $x = -\dfrac{1}{2}$

12. a. $x = 8$ b. $x = 3$ c. $x = 4$
    d. $x = -15$ e. $x = 26$ f. $x = -42$

13. a. $x = 9$ b. $x = -1\dfrac{1}{5}$ c. $x = 9\dfrac{1}{3}$
    d. $x = -2\dfrac{1}{4}$ e. $x = -7$ f. $x = \dfrac{2}{3}$

14. a. $z = 16$ b. $z = 31$ c. $z = -4$
    d. $z = 6$ e. $z = -9$ f. $z = -6$

15. a. $z = -1.9$ b. $z = 6.88$ c. $z = 140$
    d. $z = 0.6$ e. $z = -35.4$ f. $z = 8$

16. a. $x = 1$ b. $x = 13$ c. $x = -2$
    d. $x = -8$ e. $x = -4$ f. $x = 30\dfrac{1}{2}$

17. a. $x = -6$ b. $x = -\dfrac{7}{10}$ c. $x = 10.35$
    d. $x = 0.326$ e. $x = 22$ f. $x = -5$

18. a. $x = 4$ b. $x = \dfrac{3}{7}$ c. $x = -1\dfrac{1}{7}$
    d. $x = -6\dfrac{2}{3}$ e. $x = \dfrac{4}{45}$ f. $x = 8$

19. a. $x = -6$ b. $x = 7.5$ c. $x = 5.1$
    d. $x = -6$ e. $x = -5\dfrac{13}{15}$ f. $x = -61\dfrac{3}{7}$

20. a. D   b. D   c. C

21. a. $a = -1$ b. $b = 10$ c. $c = 12.1$
    d. $d = 4$ e. $e = 5$ f. $f = 12$

22. a. $g = 15\dfrac{29}{30}$ b. $h = 31$ c. $i = \dfrac{13}{226}$
    d. $j = \dfrac{1}{6}$ e. $k = -8$ f. $l = 10.3$

23. a. -1 b. 6   c. 303

24. $91

25. No. 46.7 °C ≈ 116.1 °F

26. For a positive integer, $(x-4)$ must be a factor of 12.
    $x = 5, 6, 7, 8, 10, 16$

27. Sample responses can be found in the worked solutions in the online resources.

28. The mistake is in the second line: the -1 should have been multiplied by 5.

29. 60 lollies

30. Old machine: 6640 cans; new machine: 9360 cans

31. a. $34
    b. Yes, a saving of $7

32. 17

Exercise 5.3 Solving linear equations with brackets

1. a. $x = 6$ b. $x = -3$ c. $x = 0$
   d. $x = 56$ e. $x = 1$ f. $x = 0$

2. a. $x = -1$ b. $x = -2$ c. $x = 6$
   d. $x = 0$ e. $x = -0.8$ f. $x = 6$

3. a. $b = 1\dfrac{1}{6}$ b. $m = 4\dfrac{1}{2}$ c. $a = -1\dfrac{1}{2}$
   d. $m = -1\dfrac{1}{3}$ e. $p = \dfrac{3}{5}$ f. $m = 2\dfrac{2}{3}$

4. a. $a = -1\dfrac{1}{2}$ b. $p = 1\dfrac{1}{2}$ c. $a = 3\dfrac{1}{3}$

   d. $m = -2\dfrac{1}{2}$ e. $a = -\dfrac{1}{6}$ f. $m = -\dfrac{1}{4}$

5. a. $x = 16\dfrac{1}{9}$ b. $x = 2$ c. $a = 5$

   d. $b = 7$ e. $y = 7$ f. $x = -3\dfrac{1}{3}$

6. a. $m = 4$ b. $y = 1$ c. $a = 5$
   d. $p = -2$ e. $m = 3$ f. $p = 1$

7. a. $y = -7.5$ b. $y = -4.667$ c. $y = -26.25$
   d. $y = 8.571$ e. $y = -2.9$ f. $y = 3.243$

8. a. $y = 3.667$ b. $y = 2.75$ c. $y = 1.976$
   d. $y = -2$ e. $y = -3.167$ f. $y = 1.98$

9. D

10. C

11. 1990

12. a. $[2(2x + 7)]$ m

   b. Width 10 m, length 17 m

13. Expand out the brackets or divide by the factor.

14. $x = 13$. Sample responses can be found in the worked solutions in the online resources.

15. Adding 7 to both sides is the least preferred option, as it does not resolve the subtraction of 7 within the brackets.

16. 3

17. Oscar: $90, Josue: $35

18. $20

Exercise 5.4 Solving linear equations with pronumerals on both sides

1. a. $y = -1$ b. $y = 1$ c. $y = -3$

   d. $y = -5$ e. $y = -45$ f. $y = \dfrac{8}{27}$

2. a. $y = -2$ b. $y = 20$ c. $y = 1$

   d. $y = -1\dfrac{1}{5}$ e. $y = 4$ f. $y = -2$

0. C

4. A

5. a. $x = 1$ b. $x = -2$ c. $x = 3$
   d. $x = 4$ e. $x = 9$ f. $x = 3$

6. a. $x = -2$ b. $x = 4$ c. $x = -4$

   d. $x = \dfrac{1}{2}$ e. $x = 0$ f. $x = -\dfrac{3}{4}$

7. a. $x = -6$ b. $x = 2$ c. $x = -2$

   d. $x = 13$ e. $x = -7$ f. $x = \dfrac{1}{6}$

8. D

9. C

10. a. $x = 5$ b. $x = -3$ c. $x = -8$

   d. $x = -7\dfrac{1}{2}$ e. $x = 4$ f. $x = \dfrac{10}{11}$

11. a. $x = -18$ b. $x = 3\dfrac{2}{3}$ c. $x = \dfrac{2}{3}$

   d. $x = -1\dfrac{3}{7}$ e. $x = 0$ f. $x = -15$

12. $k = -3$

13. 24

14. $3(n - 36) - 98 = -11n + 200$. Sample responses can be found in the worked solutions in the online resources.

15. a.

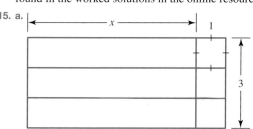

   b. Area of first rectangle $= (x + 1) \times 3$
   $$= 3x + 3$$
   Area of second rectangle $= (x + 1 + x + 1 + x + 1) \times 1$
   $$= 3x + 3$$
   Both rectangles have an area of $3x + 3$.

16. You cannot easily divide the left-hand side by 6 or the right-hand side by 4.

17. Daughter = 10 years, Tom = 40 years

18. The unknown number is -3.

19. a. $C = 40h + 250$

   b. 18 hours, 45 minutes

   c. 18 750

   d. The printing is cheaper by $1375.

20. a. 65 games b. $25

Exercise 5.5 Solving worded problems and substituting into formulas

1. a. $x + 3 = 5$ b. $x - 9 = 7$ c. $7x = 24$

   d. $\dfrac{x}{5} = 11$ e. $\dfrac{x}{2} = -9$

2. a. $5x - 3 = -7$ b. $2(14 - x) = -3$
   c. $3x + 5 = 8$ d. $2x - 12 = 15$

   e. $\dfrac{3x + 4}{2} = 5$

3. C

4. D

5. B

6. B

7. a. $v = 26$ b. $u = 13$ c. $a = 9$ d. $t = 2.4$

8. a. $T = 32$ b. $a = 2$ c. $n = 9$ d. $d = -5$

9. 0

10. 36 years old

11. $66

12. 20×15; 30×10; area = 300 square units

13. 7 m

14. 12 jackets

15. 15 days

16. 947 CDs

17. a. Post Quick (cost = $700)

   b. The cost is nearly the same for 333 flyers ($366.50 and $366.40).

18. a. 65 cm   b. 6 cm   c. $81\frac{1}{3}$ cm

19. a. 30 °C

   b. 68 °F

   c. 25 °C is equivalent to 77 °F.
   Fred is not correct as 25 °C = 77 °F, which is well above the freezing point of 32 °F.

20. 2

21. $42.50

22. a. $A_{\text{fenced}} = (5x + 20)$ m^2 b. $A_{\text{path}} = (3x + 16)$ m^2

   c. $l = 8$ m, $w = 2$ m   d. $\frac{8}{25}$ m

23. 500 km

24. a. 24 000 cm^3 b. 80 minutes

25. Sample responses can be found in the worked solutions in the online resources.

Project

1. Body height (men)

Length of humerus, l (cm)	20	25	30	35	40
Body height, h (cm)	132	147	163	178	194

Body height (women)

Length of humerus, l (cm)	20	25	30	35	40
Body height, h (cm)	125	142	159	176	192

2. and 3.

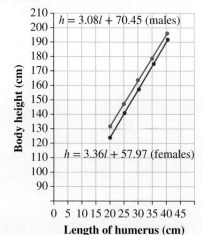

3. Linear

4. Results will vary for each student.

5. (44.6, 207.8)

6. The height of the male body with $l = 33$ cm:
   $h = 3.08 \times 33 + 70.45 = 172.09$ cm
   The height of the female body with $l = 33$ cm:
   $h = 3.36 \times 33 + 57.97 = 168.85$ cm
   The estimated height for a male body is very close to the known height of James King; the estimated height for a female body is more different from the known height of Alice Brown. Therefore, the body is more likely to be that of James King.

Exercise 5.6 Review questions

1. C

2. B

3. C

4. C

5. D

6. B

7. B

8. b, c, f, g, i

9. a. $a = 2.8$ b. $a = -0.6$ c. $b = -2.52$
   d. $b = 3.1$ e. $b = -3.45$ f. $b = 0.749$

10. a. $x = 19$ b. $x = 13$ c. $x = -6\frac{1}{3}$
    d. $x = 1\frac{1}{5}$ e. $x = 6\frac{2}{3}$ f. $x = -11.13$

11. a. $x = 3\frac{1}{5}$ b. $x = 2\frac{5}{7}$ c. $x = 1\frac{1}{4}$
    d. $x = -1\frac{5}{12}$ e. $x = -12\frac{1}{2}$ f. $x = -8\frac{1}{2}$

12. a. $8h + 21 = 45$ b. 3 hours

13. a. $x = \frac{1}{2}$ b. $x = 3$ c. $x = 3$
    d. $x = 5$ e. $x = 2$ f. $x = -1$

14. a. $2x = 3 - x$ b. $3x + 8 = 19$ c. $6x = 4$
    d. $\frac{10}{x} - 1 = \frac{x}{6}$ e. $7(2x + 5) = 0$ f. $2x - 100 = x + 50$

15. 600 km

16. a. $s = 71.875$ m   b. $u = 14.7$ m/s

17. 2.5 seconds

18. a. $C_G = 114 + 0.20k$
    b. $C_S = 90 + 0.32k$
    c. 200 km

19. a. $5400 + 260d = C_H$ b. 61 days

20. a. $0.06x$ b. $(1 - x)$
    c. $0.03(1 - x)$ d. $0.06x + 0.03(1 - x) = 0.05$
    e. 0.667 or 66.7%

21. 24 km

22. $21\frac{123}{137} \approx 21.9$ km

23. The online bookstore's fee is a better price than their major competitor's fee for all of the options given: 6, 7, 8, 9 and 10 books.

6 Quadratic equations (Path)

LESSON SEQUENCE

LESSON
6.1 Overview

Why learn this?

Galileo did a lot of important things in his lifetime. One of these was showing that objects launched into the air at an angle follow a parabolic path as they fall back to the ground. You can see this for yourself every time you throw a basketball, kick a football or hit a golf ball. As a ball moves through the air, gravity acts on it, causing it to follow a parabolic curved path. The parabolic curve that it follows can be described mathematically by quadratic equations. Every ball that is thrown, every arrow shot from a bow and every diver launched from a springboard has a quadratic equation that describes that object's position in space and also in time.

Quadratic equations also describe the parabolic shape of radio telescope dishes such as the one at Parkes in New South Wales. The curved shape of these telescope dishes allows radio waves from space objects like quasars, galaxies and nebulae to be focused at a receiver to get a stronger signal. It was even used to relay transmissions between NASA and Apollo 11 during the first moon landing. On a much smaller scale, parabolic satellite dishes allow us to get clear internet and television. The bend in a banana, the loop of a rollercoaster, the curve of some suspension bridges or the path of a droplet of water from a fountain — these are all examples of the parabolas and quadratic equations that are all around us.

Hey students! Bring these pages to life online

Watch videos

Engage with interactivities

Answer questions and check solutions

Find all this and MORE in jacPLUS

Reading content and rich media, including interactivities and videos for every concept

Extra learning resources

Differentiated question sets

Questions with immediate feedback, and fully worked solutions to help students get unstuck

Exercise 6.1 Pre-test

1. **MC** The expression $(a-4)^2 - 25$ factorises to:
   A. $(a-9)(a+1)$
   B. $(a+9)(a-1)$
   C. $(a-7)(a+3)$
   D. $(a-4)(a+4)-25$

2. State whether the following is True or False.
   A monic quadratic trinomial is an expression in the form $ax^2 + bx + c$, where a, b and c are constants.

3. Factorise $m^2 + 2m - 8$.

4. Factorise the equation $2w^2 - 10w + 12$ in its simplest form.

5. **MC** Select the solutions to the quadratic equation $-(x-3)^2 + 4 = 0$.
   A. $x = 3$ and $x = -4$
   B. $x = -3$ and $x = 4$
   C. $x = 3$ and $x = 4$
   D. $x = 5$ and $x = 1$

6. Factorise $x(x+5) - 6$.

7. Express the given quadratic equation in general form.

$$2x(x-3) + 4x^2 = 7(2-3x)$$

8. Determine whether $x = 4$ is a solution to the equation $x^2 = 4x + 4$.

9. Determine whether $x = -3$ is a solution to the equation $x^2 - 45 = 12x$.

10. Solve the equation $x^3 = 125$.

11. Solve the equation $x^3 + 8 = 0$.

12. The solution to the equation $4x^2 + 1 = 0$ is $x = -\dfrac{1}{2}$. Is the statement true or false?

13. Solve the quadratic equation $(x-3)(x+2) = 0$.

14. Solve the equation $3x(4-3x)(2x+5) = 0$.

15. Solve the quadratic equation $x^2 = 27x - 180$.

LESSON
6.2 Quadratic equations and cubic equations of the form $ax^3 = k$

LEARNING INTENTION

At the end of this lesson you should be able to:
- identify equations that are quadratic
- rearrange quadratic equations into the general form $ax^2 + bx + c = 0$
- solve quadratic equations of the form $ax^2 + c = 0$
- solve cubic equations of the form $ax^3 = k$.

▶ 6.2.1 Quadratic equations

eles-4929

- **Quadratic equations** are equations that contain an x^2 term. They may also contain the variable x and constants, but cannot contain any higher powers of x such as x^3.
- Examples of quadratic equations include $x^2 + 2x - 7 = 0$, $2x^2 = 18$ and $x^2 = 5x$.
- The general form of a quadratic equation is $ax^2 + bx + c = 0$.
- Any quadratic equation can be expressed in general form by rearranging and combining like terms.

WORKED EXAMPLE 1 Rearranging equations into general quadratic form

Rearrange the following quadratic equations so that they are in general form and state the values of a, b and c.

a. $5x^2 - 2x + 3 = 2x^2 + 4x - 12$

b. $\dfrac{x^2}{2} - \dfrac{1}{6} = x\left(\dfrac{x}{3}\right) - 4$

c. $x(3 - 2x) = 4(x - 6)$

THINK		WRITE	
a. 1. Write the equation.		a.	$5x^2 - 2x + 3 = 2x^2 + 4x - 12$
2. Subtract $2x^2$ from both sides of the equation.			$3x^2 - 2x + 3 = 4x - 12$
3. Subtract $4x$ from both sides of the equation.			$3x^2 - 6x + 3 = -12$
4. Add 12 to both sides of the equation. The equation is now in the general form $ax^2 + bx + c = 0$.			$3x^2 - 6x + 15 = 0$
5. Write the values of a, b and c.			$a = 3, b = -6, c = 15$
b. 1. Write the equation.		b.	$\dfrac{x^2}{2} - \dfrac{1}{6} = x\left(\dfrac{x}{3}\right) - 4$
2. Expand the bracket. Lowest common denominator (LCD) $= 6$.			$\dfrac{x^2}{2} - \dfrac{1}{6} = \dfrac{x^2}{3} - 4$
3. Multiply through by 6.			$\dfrac{x^2}{2} \times \dfrac{6}{1} - \dfrac{1}{6} \times \dfrac{6}{1} = \dfrac{x^2}{3} \times \dfrac{6}{1} - 4 \times 6$

4. Simplify each fraction.	$3x^2 - 1 = 2x^2 - 24$
5. Collect like terms on the left-hand side. The equation is now in the general form $ax^2 + bx + c = 0$.	$x^2 - 1 = -24$ $x^2 + 23 = 0$
6. Write the values of a, b and c.	$a = 1$, $b = 0$, $c = 23$

c. 1. Write the equation.	c.	$x(3 - 2x) = 4(x - 6)$
2. Expand the brackets.		$3x - 2x^2 = 4x - 24$
3. To collect like terms on the left-hand side of the equation, subtract $4x$ from both sides.		$-2x^2 - x = -24$
4. Add 24 to both sides.		$-2x^2 - x + 24 = 0$
5. Multiply all terms by -1 to make the x^2 term positive. The equation is now in the general form $ax^2 + bx + c = 0$.		$2x^2 + x - 24 = 0$
6. Write the values of a, b and c.		$a = 2$, $b = 1$, $c = -24$

▶ 6.2.2 Solving equations of the form $ax^2 + c = 0$

eles-4930

- Solving a quadratic equation means determining the values of x that satisfy the equation.
- Some quadratic equations have two solutions, some have only one solution, and some have no solutions.
- Quadratic equations of the form $ax^2 + c = 0$ can be solved using square roots. To solve such an equation, rearrange it so that the x^2 term is isolated on one side of the equation. Then take the square root of both sides.

WORKED EXAMPLE 2 Solving equations of the form $ax^2 + c = 0$

Solve the following equations.
a. $2x^2 - 18 = 0$ b. $x^2 + 9 = 0$ c. $3x^2 + 4 = 4$

THINK	WRITE
a. 1. Write the equation. The aim is to make x the subject.	a. $2x^2 - 18 = 0$
2. Add 18 to both sides.	$2x^2 = 18$
3. Divide both sides by 2.	$x^2 = 9$
4. Take the square root of both sides.	$x = 3, -3$
5. Write the solutions.	Two solutions: $x = 3$, $x = -3$
b. 1. Write the equation. The aim is to make x the subject.	b. $x^2 + 9 = 0$
2. Take 9 from both sides.	$x^2 = -9$
3. Take the square root of both sides.	-9 has no square root.
4. Write the solutions.	The equation has no solution.
c. 1. Write the equation. The aim is to make x the subject.	c. $3x^2 + 4 = 4$
2. Take 4 from both sides.	$3x^2 = 0$

▶

3. Divide both sides by 3.

$$x^2 = 0$$

4. Take the square root of both sides. Zero has only one square root.

$$x = 0$$

One solution: $x = 0$

5. Write the solutions.

6.2.3 Confirming the solutions to quadratic equations

eles-6206

- Solutions to quadratic equations can be checked by substitution.
- Substitute the solution(s) into the equation, and if the RHS of the equation is equal to the LHS, the solutions are correct.

WORKED EXAMPLE 3 Confirming solutions to quadratic equations

Determine whether any of the following are solutions to the quadratic equation $x^2 = 4x - 3$.

a. $x = 1$ b. $x = 2$ c. $x = 3$

THINK	WRITE
a. 1. Substitute $x = 1$ into both sides of the equation.	a. $x^2 = 4x - 3$
2. Evaluate the left-hand side.	$\begin{aligned} \text{LHS} &= (1)^2 \\ &= 1 \end{aligned}$
3. Evaluate the right-hand side.	$\begin{aligned} \text{RHS} &= 4(1) - 3 \\ &= 4 - 3 \\ &= 1 \end{aligned}$
4. Since the left-hand side equals the right-hand side, $x = 1$ is a solution. Write the answer.	$x = 1$ is a solution to the equation.
b. 1. Substitute $x = 2$ into both sides of the equation.	b. $x^2 = 4x - 3$
2. Evaluate the left-hand side.	$\begin{aligned} \text{LHS} &= (2)^2 \\ &= 4 \end{aligned}$
3. Evaluate the right-hand side.	$\begin{aligned} \text{RHS} &= 4(2) - 3 \\ &= 8 - 3 \\ &= 5 \end{aligned}$
4. The left-hand side does not equal the right-hand side, so $x = 2$ is not a solution. Write the answer.	$x = 2$ is not a solution to the equation.
c. 1. Substitute $x = 3$ into both sides of the equation.	c. $x^2 = 4x - 3$
2. Evaluate the left-hand side.	$\begin{aligned} \text{LHS} &= (3)^2 \\ &= 9 \end{aligned}$
3. Evaluate the right-hand side.	$\begin{aligned} \text{RHS} &= 4(3) - 3 \\ &= 12 - 3 \\ &= 9 \end{aligned}$
4. The left-hand side equals the right-hand side, so $x = 3$ is a solution. Write the answer.	$x = 3$ is a solution to the equation.

▶ 6.2.4 Solving equations of the form $ax^3 = k$

eles-6207

- Cubic equations of the form $ax^3 = k$ can be solved using cube roots.
- To solve, rearrange the equation so the x^3 term is by itself. Then take the cube root of both sides.

WORKED EXAMPLE 4 Solving equations of the form $ax^3 = k$

Solve the following equations.

a. $2x^3 = 16$
b. $x^3 + 27 = 0$
c. $8x^3 = 125$
d. $4x^3 = 25$, giving your answer to 2 decimal places.

THINK	WRITE
a. 1. Write the equation.	$2x^3 = 16$
2. Divide both sides by 2.	$x^3 = 8$
3. Take the cube root of both sides.	$x = \sqrt[3]{8}$
4. Write the solution.	$x = 2$ The solution is $x = 2$.
b. 1. Write the equation.	$x^3 + 27 = 0$
2. Subtract 27 from both sides.	$x^3 = -27$
3. Take the cube root of both sides.	$x = \sqrt[3]{-27}$
4. Write the solution.	$x = -3$ The solution is $x = -3$.
c. 1. Write the equation.	$8x^3 = 125$
2. Divide both sides by 8.	$x^3 = \dfrac{125}{8}$
3. Take the cube root of both sides.	$x = \sqrt[3]{\dfrac{125}{8}}$
4. Write the solution.	$x = \dfrac{5}{2}$ The solution is $x = 2.5$.
d. 1. Write the equation.	$4x^3 = 25$
2. Divide both sides by 4.	$x^3 = \dfrac{25}{4}$
3. Take the cube root of both sides. Neither of these numbers are cubic numbers. Use a calculator to approximate the cube root.	$x = \sqrt[3]{\dfrac{25}{4}}$ $x = 1.84202$
4. Write the solution, correct to 2 decimal places.	The solution is $x \approx 1.84$.

Exercise 6.2 Quadratic equations and cubic equations of the form $ax^3 = k$

6.2 Quick quiz on	6.2 Exercise

Individual pathways

■ PRACTISE	■ CONSOLIDATE	■ MASTER
1, 5, 7, 11, 13, 16	2, 3, 8, 10, 12, 14, 17	4, 6, 9, 15, 18

Fluency

1. **WE1** Rearrange each of the following quadratic equations so that they are in the form $ax^2 + bx + c = 0$.
   a. $3x - x^2 + 1 = 5x$ b. $5(x-2) = x(4-x)$ c. $x(5-2x) = 6\left(5-x^2\right)$

2. Rearrange each of the following quadratic equations so that they are in the form $ax^2 + bx + c = 0$.
   a. $4x^2 - 5x + 2 = 6x - 4x^2$ b. $5(x-12) = x(4-2x)$ c. $x^2 - 4x - 16 = 1 - 4x$

3. **MC** Select which of the following is a quadratic equation.
   A. $2x - 1 = 0$ **B.** $2^x - 1 = 0$
   C. $x^2 - x = 1 + x^2$ **D.** $x^2 - x = 1 - x^2$

4. **MC** Select which of the following is not a quadratic equation.
   A. $x(x-1) = 2x - 1$ **B.** $-3x^2 + 2x = 1$
   C. $3(x+2) + 5(x+3) = 2(x+1)$ **D.** $2(x-1) + 3x = x(2x-3)$

5. **WE2** Solve the following quadratic equations.
   a. $x^2 - 16 = 0$ b. $2x^2 + 18 = 0$ c. $3x^2 + 2x = x(x+2)$

6. Solve the following quadratic equations.
   a. $2(x^2 + 7) = 16$ b. $x^2 + 17 = 13$ c. $-3x^2 + 17 = 5$

Understanding

7. **WE3** Determine whether $x = 4$ is a solution to the following equations.
   a. $x^2 = x + 12$ b. $x^2 = 3x + 1$ c. $x^2 = 4x$

8. Determine whether $x = -3$ is a solution to the following equations.
   a. $x^2 = x + 12$ b. $x^2 = 3x + 1$ c. $x^2 = 4x$

9. Determine whether $x = 0$ is a solution to the following equations.
   a. $x^2 = x + 12$ b. $x^2 = 3x + 1$ c. $x^2 = 4x$

10. Determine whether the following are solutions to the equation $5(x-1)^2 + 7 = 27$.
    a. $x = -1$ b. $x = 1$ c. $x = 0$

11. **WE4** Solve the following cubic equations, giving your answers correct to 2 decimal places if necessary.
    a. $2x^3 = 54$ b. $x^3 + 8 = 0$ c. $9x^3 = 25$

12. Solve the following cubic equations, giving your answers correct to 2 decimal places if necessary.
    a. $5x^3 = 625$ b. $27x^3 + 1 = 0$ c. $3x^3 = 9$

Communicating, reasoning and problem solving

13. Is $(x+4)^2$ equal to $x^2 + 16$? Explain using a numerical example.

14. Consider the following diagram.

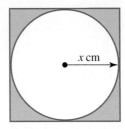

a. Show that the shaded area can be represented by $A = x^2(4 - \pi)$.
b. If the shaded area $= 10 \text{ cm}^2$, calculate the value of x correct to 3 significant figures.
c. Why can't x be equal to -3.41?

15. Explain why $x^2 + 7x + 4 = 7x$ has no solutions.

16. Five squares of increasing size and five area formulas are given below.

a. Use factorisation to determine the side length that correlates to each area formula.
b. Using the area given and side lengths found, match the squares below with the appropriate algebraic expression for their area.

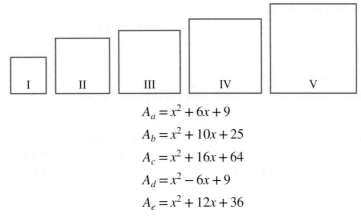

$$A_a = x^2 + 6x + 9$$
$$A_b = x^2 + 10x + 25$$
$$A_c = x^2 + 16x + 64$$
$$A_d = x^2 - 6x + 9$$
$$A_e = x^2 + 12x + 36$$

c. If $x = 5$ cm, use the formula given to calculate the area of each square.

17. The sum of 5 and the square of a number is 41.

a. Determine the number.
b. Is there more than one possible number? If so, determine the other number and explain why there is more than one.

18. Two less than a number is squared and its result is tripled. The difference between this and 7 is 41.

a. Write the algebraic equation.
b. Determine both possible answers that satisfy the criteria.

LESSON
6.3 The Null Factor Law

LEARNING INTENTION

At the end of this lesson you should be able to:
- recall the Null Factor Law
- solve quadratic equations using the Null Factor Law.

▶ 6.3.1 Using the Null Factor Law to solve equations

eles-4931

- The **Null Factor Law** states that if the product of two or more factors is zero, then at least one of the factors must be zero.

The Null Factor Law

If $a \times b = 0$, then:

$a = 0$ or $b = 0$ or both a and b equal 0.

- For example, if $(x - 5)(x - 2) = 0$, then

$$(x - 5) = 0 \text{ or } (x - 2) = 0 \text{ or both } x - 5 = x - 2 = 0.$$

$$\text{If } x - 5 = 0 \quad \text{If } x - 2 = 0$$
$$x = 5 \qquad x = 2$$

Both $x = 5$ and $x = 2$ make the equation true.
- This can be checked by substituting the values into the original equation.
For example:

$$\text{If } x = 5, \text{ then:} \qquad\qquad \text{If } x = 2, \text{ then:}$$
$$(x - 5)(x - 2) = (5 - 5)(x - 2) \qquad (x - 5)(x - 2) = (x - 5)(2 - 2)$$
$$= 0(x - 2) \qquad\qquad\qquad = (x - 5)0$$
$$= 0 \qquad\qquad\qquad\qquad = 0$$

- *Note:* Quadratic equations can have a maximum of two solutions. This is known as the **Fundamental Theorem of Algebra**.

WORKED EXAMPLE 5 Using the Null Factor Law to solve quadratics

Solve each of the following quadratic equations.
a. $(x - 2)(2x + 1) = 0$ b. $(4 - 3x)(6 + 11x) = 0$

c. $x(x - 3) = 0$ d. $(x - 1)^2 = 0$

THINK

a. 1. The product of 2 factors is 0, so apply the Null Factor Law.

   2. One of the factors must equal zero.

   3. Solve the equations.

WRITE

a. $(x - 2)(2x + 1) = 0$

Either $x - 2 = 0$ or $2x + 1 = 0$

$$x = 2 \qquad\qquad 2x = -1$$
$$= -\frac{1}{2}$$

4. Write the solutions.

$x = -\dfrac{1}{2}, x = 2$

b. 1. The product of two factors is 0, so apply the Null Factor Law.

b. $(4 - 3x)(6 + 11x) = 0$

2. One of the factors must equal zero.

Either $4 - 3x = 0$ or $6 + 11x = 0$

3. Solve the equations.

$$-3x = -4 \qquad\qquad 11x = -6$$
$$x = \dfrac{4}{3} \qquad\qquad x = -\dfrac{6}{11}$$

4. Write the solutions.

$x = \dfrac{4}{3}, -\dfrac{6}{11}$

c. 1. The product of two factors is 0, so apply the Null Factor Law.

c. $x(x - 3) = 0$

2. One of the factors must equal zero.

Either $x = 0$ or $x - 3 = 0$

3. Solve the equations.

$x = 3$

4. Write the solutions.

$x = 0, 3$

d. 1. Write the equation.

d. $(x - 1)^2 = 0$

2. Rewrite the squared factors as the product of two factors.

$(x - 1)(x - 1) = 0$

3. Apply the Null Factor Law.

Either $x - 1 = 0$ or $x - 1 = 0$

4. Solve the equations. The two solutions are the same.

$x = 1 \qquad\qquad x = 1$

5. Write the solution.

$x = 1$

---

on Resources

▶ **Video eLesson** The Null Factor Law (eles-2312)

✦ **Interactivity** The Null Factor Law (int-6095)

Exercise 6.3 The Null Factor Law

learn on

| **6.3 Quick quiz** on | **6.3 Exercise** |

Individual pathways

■ PRACTISE	■ CONSOLIDATE	■ MASTER
1, 6, 9, 12	2, 4, 7, 10, 13	3, 5, 8, 11, 14

Fluency

1. WE5 Solve each of the following quadratic equations.

a. $(x - 2)(x + 3) = 0$
b. $(2x + 4)(x - 3) = 0$
c. $(x + 2)(x - 3) = 0$
d. $(2x + 5)(4x + 3) = 0$
e. $(x + 4)(2x + 1) = 0$

2. Solve each of the following quadratic equations.

    a. $(2x-1)(x+30)=0$ b. $(2x+1)(3-x)=0$ c. $(1-x)(3x-1)=0$

    d. $x(x-2)=0$ e. $\left(x+\dfrac{1}{3}\right)\left(2x-\dfrac{1}{2}\right)=0$

3. Solve each of the following quadratic equations.

    a. $(5x-1.5)(x+2.3)=0$ b. $\left(2x+\dfrac{1}{3}\right)\left(2x-\dfrac{1}{3}\right)=0$ c. $(x-2)^2=0$

    d. $x(4x-15)=0$ e. $(x+4)^2=0$

4. The Null Factor Law can be extended to products of more than two factors. Use this to determine all the solutions to the following equations.

    a. $(x-2)(x+2)(x+3)=0$ b. $(x+2)(x+2)(2x-5)=0$
    c. $(x+2)(x+2)(x+4)=0$ d. $x(x+2)(3x+12)=0$

5. The Null Factor Law can be extended to products of more than two factors. Use this to determine all the solutions to the following equations.

    a. $(2x-2.2)(x+2.4)(x+2.6)=0$ b. $(2x+6)\left(x+\dfrac{1}{2}\right)(9x-15)=0$

    c. $3(x-3)^2=0$ d. $(x+1)(x-2)^2=0$

Understanding

6. **MC** Select the solutions to $(2x-4)(x+7)=0$.

    A. $x=4,\ x=7$ B. $x=4,x=-7$
    C. $x=2,\ x=7$ D. $x=2,x=-7$

7. **MC** The Null Factor Law cannot be applied to the equation $x\left(x+\dfrac{1}{2}\right)\left(\dfrac{x}{2}-1\right)=1$ because:

    A. there are more than two factors.
    B. the right-hand side equals 1.
    C. the first factor is a simple x-term.
    D. the third term has x in a fraction.

8. Rewrite the following equations so that the Null Factor Law can be used. Then solve the resulting equation.
    a. $x^2+10x=0$
    b. $2x^2-14x=0$
    c. $25x^2-40x=0$

Communicating, reasoning and problem solving

9. Explain why $x^2-x-6=0$ has the same solution as $-2x^2+2x+12=0$.

10. a. Explain why $x=2$ is not a solution of the equation $(2x-2)(x+2)=0$.
     b. Determine the solutions to the equation.

11. Explain what is the maximum number of solutions a quadratic such as $x^2-x-56=0$ can have.

12. A bridge is constructed with a supporting structure in the shape of a parabola, as shown in the diagram. The origin $(0, 0)$ is at the left-hand edge of the bridge, which is 100 m long.

    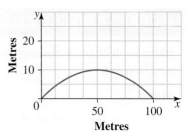

    a. Identify the maximum height of the bridge support.
    b. If the equation of the support is $y = ax(b - x)$, determine the values of a and b. (*Hint:* Let $y = 0$.)
    c. Calculate the height of the support when $x = 62$.

13. Consider a ball thrown upwards so that it reaches a height h of metres after t seconds. The expression $20t - 4t^2$ represents the height of the ball, in metres, after t seconds.

    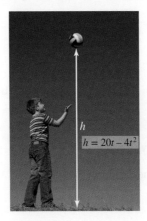

    h

    $h = 20t - 4t^2$

    a. Factorise the expression for height.
    b. Calculate the height of the ball after:
       i. 1 second   ii. 5 seconds.
    c. Explain if factorising the expression made it easier to evaluate.

14. The product of 7 more than a number and 14 more than that same number is 330. Determine the possible values of the number.

LESSON
6.4 Solving quadratic equations with two terms

LEARNING INTENTION

At the end of this lesson you should be able to:
* solve quadratic equations with two terms using the Null Factor Law.

▶ 6.4.1 Solving quadratic equations of the form $ax^2 + c = 0$

eles-4932

* In section 6.2.2, we saw how to solve equations in the form $ax^2 + c = 0$ using algebra. In most cases, we can use the Null Factor Law to solve them instead.
* To use the Null Factor Law the quadratic needs to be factorised and the RHS of the equation must be 0.

WORKED EXAMPLE 6 Using the Null Factor Law to solve $ax^2 + c = 0$

Solve each of the following quadratic equations.

a. $x^2 - 1 = 0$

b. $2x^2 - 18 = 0$

THINK	WRITE
a. 1. Write the equation: $x^2 - 1$ is the difference of two squares.	**a.** $x^2 - 1 = 0$
2. Factorise the left-hand side using the difference of two squares rule.	$(x+1)(x-1) = 0$
3. Apply the Null Factor Law.	$x + 1 = 0 \quad x - 1 = 0$ $x = -1 \quad\quad x = 1$
4. State the solutions.	$x = 1, x = -1$ (This can be abbreviated to $x = \pm 1$.)
b. 1. Write the equation.	**b.** $\quad 2x^2 - 18 = 0$
2. Take out the common factor.	$2(x^2 - 9) = 0$
3. Divide both sides of the equation by 2.	$x^2 - 9 = 0$
4. Factorise the left-hand side using the difference of two squares rule.	$(x+3)(x-3) = 0$
5. Apply the Null Factor Law.	$x + 3 = 0 \quad x - 3 = 0$ $x = -3 \quad\quad x = 3$
6. State the solutions.	$x = \pm 3$

▶ 6.4.2 Solving quadratic equations of the form $ax^2 + bx = 0$

eles-4933

- The equation $ax^2 + bx = 0$ can be solved easily using the Null Factor Law.
- To solve an equation of this form, factorise the LHS to $x(ax + b) = 0$ and then apply the Null Factor Law.
- Since both terms on the left-hand side involve the variable x, one solution will always be $x = 0$.

WORKED EXAMPLE 7 Solving equations in the form $ax^2 + bx = 0$

Solve each of the following equations.

a. $x^2 + 4x = 0$

b. $-2x^2 - 4x = 0$

THINK	WRITE
a. 1. Write the equation.	**a.** $x^2 + 4x = 0$
2. Factorise by taking out a common factor of x.	$x(x+4) = 0$
3. Apply the Null Factor Law.	$x = 0 \quad\quad x + 4 = 0$ $x = 0 \quad\quad\quad x = -4$
4. Write the solutions.	$x = -4, \quad\quad x = 0$
b. 1. Write the equation.	**b.** $-2x^2 - 4x = 0$
2. Factorise by taking out the common factor of $-2x$.	$-2x(x+2) = 0$

3. Apply the Null Factor Law.	$-2x = 0$ $x + 2 = 0$
	$x = 0$ $x = -2$
4. Write the solutions.	$x = -2,$ $x = 0$

WORKED EXAMPLE 8 Solving word problems using Null Factor Law

If the square of a number is multiplied by 5, the answer is 45. Calculate the number.

THINK	WRITE
1. Define the number.	Let x be the number.
2. Write an equation that can be used to determine the number.	$5x^2 = 45$
3. Transpose to make the right-hand side equal to zero. Solve the equation.	$5x^2 - 45 = 0$
	$5(x^2 - 9) = 0$
	$x^2 - 9 = 0$
	$x^2 - 3^2 = 0$
	$(x + 3)(x - 3) = 0$
	$x + 3 = 0$ or $x - 3 = 0$
	$x = -3$ or $x = 3$
4. Write the answer in a sentence.	The number is either 3 or -3.

Exercise 6.4 Solving quadratic equations with two terms learn

6.4 Quick quiz on	6.4 Exercise

Individual pathways

■ PRACTISE	■ CONSOLIDATE	■ MASTER
1, 4, 7, 9, 12, 16	2, 5, 8, 10, 13, 17	3, 6, 11, 14, 15, 18

Fluency

1. Solve each of the following quadratic equations using the Null Factor Law.

   a. $x^2 - 9 = 0$ b. $x^2 - 16 = 0$ c. $2x^2 - 18 = 0$
   d. $2x^2 - 50 = 0$ e. $100 - x^2 = 0$

2. Solve each of the following quadratic equations using the Null Factor Law.

   a. $49 - x^2 = 0$ b. $3x^2 - 27 = 0$ c. $5x^2 - 20 = 0$
   d. $x^2 + 6 = 0$ e. $2x^2 + 18 = 0$

3. Solve each of the following quadratic equations using the Null Factor Law.

   a. $-x^2 + 9 = 0$ b. $-3x^2 + 48 = 0$ c. $-4x^2 + 100 = 0$
   d. $x^2 = 0$ e. $-x^2 = 0$

4. **WE7** Solve each of the following equations.

   a. $x^2 + 6x = 0$

   b. $x^2 - 8x = 0$

   c. $x^2 + 9x = 0$

   d. $x^2 - 11x = 0$

   e. $2x^2 - 12x = 0$

5. Solve each of the following equations.

   a. $2x^2 - 15x = 0$

   b. $3x^2 - 2x = 0$

   c. $4x^2 + 7x = 0$

   d. $2x^2 - 5x = 0$

   e. $x^2 + x = 0$

6. Solve each of the following equations.

   a. $4x^2 - x = 0$

   b. $-x^2 - 5x = 0$

   c. $-2x^2 - 24x = 0$

   d. $-x^2 + 18x = 0$

   e. $x^2 - 2.5x = 0$

7. **MC** Select the solutions to $4x^2 - 36 = 0$.

   A. $x = 3$ and $x = -3$

   B. $x = 9$ and $x = -9$

   C. $x = 1$ and $x = -1$

   D. $x = 2$ and $x = -2$

8. **MC** Select the solutions to $x^2 - 5x = 0$.

   A. $x = 1$ and $x = 5$

   B. $x = 0$ and $x = -5$

   C. $x = 0$ and $x = 5$

   D. $x = -1$ and $x = 5$

Understanding

9. **WE8** If the square of a number is multiplied by 2, the answer is 32. Calculate the number.

10. A garden has two vegetable plots. One plot is a square; the other plot is a rectangle with one side 3 m shorter than the side of the square and the other side 4 m longer than the side of the square. Both plots have the same area.
    Sketch a diagram and determine the dimensions of each plot.

11. The square of a number is equal to 10 times the same number. Calculate the number.

Communicating, reasoning and problem solving

12. Explain why the equation $x^2 + 9 = 0$ cannot be solved using the Null Factor Law.

13. Solve the equation $m^2x^2 - n^2 = 0$ using the Null Factor Law.

14. Explain why the equation $x^2 + bx = 0$ can be solved using the Null Factor Law.

15. A quadratic expression may be written as $ax^2 + bx + c$. Using examples, explain why b and c may take any values, but a cannot equal zero.

16. The square of a number that is then tripled is equal to 24 times the same number. Calculate the number.

17. The sum of 6 times the square of a number and 72 times the same number is equal to zero. Calculate the number.

18. A rectangular horse paddock has a width of x and a length that is 30 metres longer than its width. The area of the paddock is $50x$ square metres.

    a. Write the algebraic equation for the area described.

    b. Factorise the equation.

    c. Solve the equation.

    d. Can you use all your solutions? Explain why or why not.

LESSON
6.5 Solving monic quadratic equations

LEARNING INTENTION

At the end of this lesson you should be able to:
- solve monic quadratic equations using factorisation and the Null Factor Law.

▶ 6.5.1 Solving quadratic equations of the form $x^2 + bx + c = 0$

eles-6208

- In lesson 4.10 we saw how to factorise monic quadratic expressions in the form $x^2 + bx + c$.
- Factorising monic quadratic expressions is demonstrated in Worked examples 9 and 10.
- If the expression can be factorised, then the quadratic equation $x^2 + bx + c = 0$ can be solved using the Null Factor Law.
- The quadratic equation may need to be transposed so that the right-hand side equals zero.

WORKED EXAMPLE 9 Factorising and solving quadratic equations

a. Factorise the expression $x^2 + 9x + 8$.
b. Hence, solve the quadratic equation $x^2 + 9x + 8 = 0$.

THINK

a. 1. Determine two numbers that multiply to 8 and add to 9.

  2. $1 \times 8 = 8$ and $1 + 8 = 9$

  3. Write the expression in factorised form.

b. 1. Write the equation.

  2. Substitute the factors from part **a**.

  3. Apply the Null Factor Law.

  4. Solve the equation.

  5. Write the solutions.

WRITE

a.

Factors of 8	Sum of factors
1 and 8	9
2 and 4	6

$x^2 + 9x + 8 = (x + 1)(x + 8)$

b. $x^2 + 9x + 8 = 0$

$(x + 1)(x + 8) = 0$

$x + 1 = 0$ or $x + 8 = 0$

$x = -1 \quad x = -8$

$x = -1, x = -8$

WORKED EXAMPLE 10 Factorising and solving quadratic equations with negative terms

a. Factorise the expression $x^2 - x - 6$.
b. Hence, solve the quadratic equation $x^2 - x - 6 = 0$.

THINK

a. 1. Determine two numbers that multiply to −6 and add to −1.

  2. $2 \times (-3) = -6, 2 + (-3) = -1$

WRITE

a.

Factors of −6	Sum of factors
−2 and 3	1
2 and −3	−1
−1 and 6	5
1 and −6	−5

3. Write the expression in factorised form. $x^2 - x - 6 = (x-3)(x+2)$

b. 1. Write the equation. b. $x^2 - x - 6 = 0$

2. Substitute the factors from part **a**. $(x-3)(x+2) = 0$

3. Apply the Null Factor Law. $x - 3 = 0$ or $x + 2 = 0$

4. Solve the equation. $x = 3$ or $x = -2$

5. Write the solutions. $x = -2, x = 3$

WORKED EXAMPLE 11 Solving monic quadratic equations

Solve the following quadratic equations.

a. $x^2 - 5x - 6 = 0$ **b.** $x^2 + 14x = 15$

THINK	WRITE
a. 1. Write the equation.	a. $x^2 - 5x - 6 = 0$
2. Factorise $x^2 - 5x - 6$.	$(x-6)(x+1) = 0$
3. Apply the Null Factor Law.	$x - 6 = 0 \quad x + 1 = 0$
4. Solve the equations.	$x = 6 \qquad x = -1$
5. Write the solutions.	$x = -1, \quad x = 6$
b. 1. Write the equation and transpose so that the right-hand side equals 0.	b. $\quad x^2 + 14x = 15$ $x^2 + 14x - 15 = 0$
2. Factorise $x^2 - 14x - 15$.	$(x+15)(x-1) = 0$
3. Apply the Null Factor Law.	$x + 15 = 0 \quad x - 1 = 0$ $x = -15 \quad x = 1$
4. Solve the linear equations and write the solutions.	$x = -15, \quad x = 1$

 Resources

 Interactivity Factorising monic quadratics (int-6092)

Exercise 6.5 Solving monic quadratic equations

| 6.5 Quick quiz on | 6.5 Exercise |

Individual pathways

■ PRACTISE	■ CONSOLIDATE	■ MASTER
1, 2, 3, 5, 8, 10, 15	4, 6, 11, 13, 16, 18	7, 9, 12, 14, 17, 19, 20

Fluency

1. **WE9**

   a. Factorise the quadratic expression $x^2 + 4x + 3$.
   b. Hence, solve $x^2 + 4x + 3 = 0$.

2. a. Factorise the quadratic expression $x^2 + 10x + 16$.
   b. Hence, solve $x^2 + 10x + 16 = 0$.

3. **WE10**

   a. Factorise the quadratic expression $x^2 + 3x - 18$.
   b. Hence, solve $x^2 + 3x - 18 = 0$.

4. a. Factorise the quadratic expression $x^2 - 3x - 18$.
   b. Hence, solve $x^2 - 3x - 18 = 0$.

5. **WE11** Solve each of the following quadratic equations.

   a. $n^2 + 2n - 35 = 0$
   b. $v^2 + 5v - 6 = 0$
   c. $v^2 - 5v - 6 = 0$
   d. $t^2 + 4t - 12 = 0$
   e. $t^2 - 5t - 14 = 0$

6. Solve each of the following quadratic equations.

   a. $x^2 - x - 20 = 0$
   b. $x^2 + x - 20 = 0$
   c. $n^2 + n - 90 = 0$
   d. $n^2 - 3n - 70 = 0$
   e. $x^2 - 4x - 5 = 0$

7. Solve each of the following quadratic equations.

   a. $x^2 - 6x + 8 = 0$
   b. $x^2 + 6x + 8 = 0$
   c. $x^2 + 6x + 5 = 0$
   d. $x^2 + x - 6 = 0$
   e. $x^2 + 2x - 15 = 0$

8. Solve each of the following quadratic equations.

   a. $x^2 + 4x + 4 = 0$
   b. $x^2 + 2x - 24 = 0$
   c. $x^2 - 5x - 24 = 0$
   d. $x^2 - x - 12 = 0$
   e. $x^2 + 13x + 12 = 0$

9. Solve each of the following quadratic equations.

   a. $x^2 - 10x = 11$
   b. $x^2 + x = 20$
   c. $x^2 + 29x = -100$
   d. $x^2 - 15x = -50$
   e. $0 = x^2 - 2x - 8$

Understanding

10. **MC** Select the solutions to the quadratic equation $x^2 + 2x - 8 = 0$.

    A. $x = 4$ and $x = -2$
    B. $x = -4$ and $x = 2$
    C. $x = -4$ and $x = 4$
    D. $x = -2$ and $x = 2$

11. **MC** Select the solutions to the quadratic equation $x^2 - 7x - 8 = 0$.

    A. $x = 1$ and $x = 8$
    B. $x = -1$ and $x = -8$
    C. $x = -8$ and $x = 1$
    D. $x = -1$ and $x = 8$

12. **MC** Select the values of x that are the solutions to the quadratic equation $x^2 - 5x - 14 = 0$.

   A $2, 7$ **B** $2, -7$ **C** $-2, 7$ **D** $-2, -7$

13. **a.** Expand and simplify $(x + 3)^2 - 25$.
    b. Hence, solve $(x + 3)^2 - 25 = 0$.

14. Solve the equation $x(x + 3) - 8(x + 4) = 4$.

15. The rectangle shown has an area of $45\,\text{cm}^2$. By solving a quadratic equation, determine the dimensions of the rectangle.

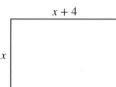

Communicating, reasoning and problem solving

16. Consider the quadratic equation $x^2 + 7x + c = 0$, where c is a positive integer.
    a. Solve the equation if $c = 6$.
    b. Determine what other positive integer values c can take if the equation can be solved.
    Justify your answer.

17. Consider the quadratic equation $x^2 + 3x + c = 0$, where c is a negative integer.
    a. Solve the equation if $c = -4$.
    b. Determine what other negative integer values c can take if the equation can be solved.
    Justify your answer.

18. Show that $x^2 + 8x + 10 = 0$ has no solutions if only integers can be used.

19. Show that $x^2 + 12x + 24 = 0$ has no solutions if only integers can be used.

20. The number of diagonals in a polygon is given by the formula $D = \dfrac{n}{2}(n - 3)$, where n is the number of sides in the polygon.

   a. Determine how many diagonals there are in:
      i. a triangle   **ii.** a square   **iii.** a decagon.
   b. Determine what type of polygon has:
      i. 20 diagonals   **ii.** 170 diagonals.

LESSON
6.6 Applications of quadratic equations

LEARNING INTENTION

At the end of this lesson you should be able to:
- recognise quadratics in real-life contexts
- evaluate solutions to ensure feasible answers.

▶ 6.6.1 Quadratic equations in real life

eles-6209

- There are many situations in science, engineering, economics and other fields where quadratic equations can be applied to determining solutions to problems.
- Solutions to quadratic equations need to be evaluated to ensure feasible answers. For example, time and length need to be positive in real-life situations.

WORKED EXAMPLE 12 Writing and solving worded problems

When 10 is added to the square of a positive number, the result is equal to 3 times the number subtracted from twice its square. Determine the number.

THINK	WRITE
1. Define the unknown quantity.	Let x be the positive number.
2. Translate each part of the sentence into an algebraic expression or term.	The square of a positive number: x^2 Add 10 to the (positive number)^2: $x^2 + 10$ Three times the number: $3x$ Three times the number is subtracted from twice its square: $2x^2 - 3x$
3. Form a quadratic equation and rearrange it so that it is in the form $ax^2 + bx + c = 0$.	$x^2 + 10 = 2x^2 - 3x$ $10 = x^2 - 3x$ $x^2 - 3x - 10 = 0$
4. Solve the equation by factorising.	$(x - 5)(x + 2) = 0$ $x - 5 = 0$ or $x + 2 = 0$ $x = 5 \qquad x = -2$
5. Evaluate the result.	The number is positive, so $x = 5$ is the only valid solution.
6. Answer the question in a sentence.	The number is 5.

WORKED EXAMPLE 13 Determining a feasible answer

The distance travelled by an accelerating skier is given by the formula $d = 3t + t^2$, where t is the time in seconds and d is the distance in metres.

If the distance travelled was 130 m, for how long was the skier travelling?

THINK	WRITE
1. Write the equation.	$d = 3t + t^2$
2. Substitute the given value, $d = 130$, into the equation.	$130 = 3t + t^2$
3. Rearrange the equation so that it is in the form $ax^2 + bx + c = 0$.	$0 = 3t + t^2 - 130$ $t^2 + 3t - 130 = 0$
4. Factorise the left-hand side of the equation, using the factors of -130 that add up to 3.	$-130: -10 + 13 = 3$ $(t - 10)(t + 13) = 0$
5. Solve the equation by using the Null Factor Law.	Either $t - 10 = 0$ or $t + 13 = 0$ $t = 10$ $t = -13$
6. Evaluate the result.	The only feasible answer is $t = 10$, because time is always positive.
7. Answer the question in a sentence.	The skier was travelling for 10 seconds.

Exercise 6.6 Applications of quadratic equations

learn on

6.6 Quick quiz on	6.6 Exercise

Individual pathways

■ PRACTISE	■ CONSOLIDATE	■ MASTER
1, 3, 5, 8	2, 6, 9, 11	4, 7, 10, 12

Fluency

1. **WE12** The product of two consecutive numbers is 42. What are the numbers?

2. The product of two consecutive odd numbers is 143. What are the numbers?

3. If the shaded area is 180 square units, determine the value of x.

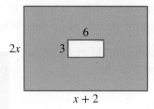

4. If these triangles have the same area, what is the value of x?

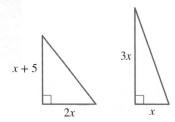

Understanding

5. Determine the side lengths of this right-angled triangle. Show your working.

6. Determine the side lengths of this right-angled triangle. Show your working.

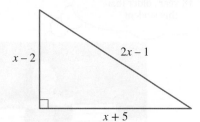

7. A sheet of A4 paper is approximately 30 cm by 21 cm. Four corners are cut from a sheet as shown, and the remaining paper is folded to make a box with a base area of 286 cm². What are the dimensions of the box?

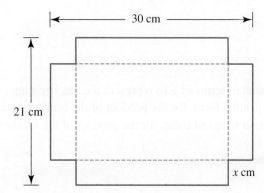

Communicating, reasoning and problem solving

8. A gardener maintains a large rectangular rose garden of length 15 m and width 12 m. He plans to lay a wide pathway around the garden that will require 160 m² of paving. What will be the width of the pathway?

9. **WE12** A cricket ball is struck high into the air. After t seconds its height, h, is given by the formula $h = 20t - 5t^2$.

    a. What is its height after 1 second?
    b. When will the ball strike the ground?

10. Each term in the sequence $-3, -1, 1, 3, 5...$ is 2 more than the previous term. The sum S of the first n terms in the sequence is given by the formula $S = n^2 - 4n$.

    a. What is the sum of the first 4 terms?
    b. How many terms are needed to make a sum of 77?

11. The sum, S, of the first n positive integers is given by the formula

$$S = \frac{n}{2}(n + 1)$$

    How many positive integers would you need to give a total of 91?

12. A girl, her brother and her teacher are shown below. The product of the teacher's and brother's ages is 256.

    a. Write an algebraic expression in terms of g to represent the teacher's age.
    b. Write an expression, in expanded form, for the product of the brother's and teacher's ages.
    c. Write a quadratic equation, in simplest form, for the product of the brother's and teacher's ages
    d. Determine the value of g.
    e. State the ages of the girl, her brother and the teacher.

LESSON
6.7 Review

6.7.1 Topic summary

QUADRATIC EQUATIONS (PATH)

Quadratic equations and cubic equations of the form $ax^3 = k$

- Quadratic equations are equations in which the largest power of x is 2. They contain an x^2 term.
- The general form of a quadratic equation is $ax^2 + bx + c = 0$, where $a \neq 0$.
- Quadratic equations for which $a = 1$ are called monic quadratic equations.
- The largest number of solutions a quadratic equation can have is 2.
- Cubic equations of the form $ax^3 = k$ can be solved using cube roots.

Factorising and solving monic quadratic equations

- To factorise a monic quadratic $x^2 + bx + c$, look for factors of c that sum to b.

- If the expression can be factorised, then the quadratic equation $x^2 + bx + c = 0$ can be solved using the Null Factor Law.

Applications of quadratic equations

- There are many real-life situations where quadratic equations can be applied to determining solutions to problems.

- Solutions to quadratic equations need to be evaluated to ensure feasible answers.

Null Factor Law

- If $a \times b = 0$, then:
  $a = 0$ or $b = 0$ or $a = b = 0$.

Solving quadratic equations with two terms

- The equation $ax^2 + bx = 0$ can be solved easily using the Null Factor Law.

- To solve an equation of this form, factorise to $x(ax + b) = 0$ and then apply the Null Factor Law.

6.7.2 Project

Constructing a parabola

The word *parabola* comes from the Greek language and means 'thrown', because it is the path followed by a projectile in flight. Notice that the water streams shown in the photo are moving in the path of a parabola. This is also the name of the graph of a quadratic equation. This investigation explores the technique of folding paper to display the shape of a parabola. The instructions are given below. Take care with each step to ensure your finished product is a well-constructed parabola that can be used in later parts of this investigation.

Forming a parabola by folding paper

- Take a sheet of A4 paper. Cut it into two pieces by dividing the longer side into two. Only one of the halves is required for this investigation.
- Along one of the longer sides of your piece of paper, mark points that are equally spaced 1 cm apart. Start with the first point being on the very edge of the paper.
- Turn over the piece of paper and mark a point, X, 3 cm above the centre of the edge that has the markings on the reverse side.
- Fold the paper so that the first point you marked on the edge touches point X. Make a sharp crease and open the paper flat.
- Fold the paper again so that the second mark touches the point X. Crease and unfold again.
- Repeat this process until all the marks have been folded to touch point X.
- With the paper flat and the point X facing up, you should notice the shape of a parabola appearing in the creases.

1. Trace the curve with a pencil.
   The point X is called the focus of the parabola. Consider the parabola to represent a mirror. Rays of light from the focus would hit the mirror (parabola) and be reflected. The angle at which each ray hits the mirror is the same size as the angle at which it is reflected.

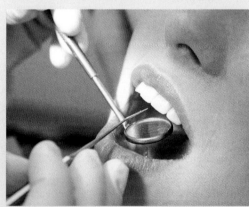

2. Using your curve traced from your folding activity, accurately draw a series of lines to represent rays of light from the point X to the parabola (mirror). Use a protractor to carefully measure the angle each line makes with the mirror and draw the path of these rays after reflection in the mirror.

3. Draw a diagram to describe your finding from question **2**. Provide a brief comment on your description.
4. Retrace your parabola onto another sheet of paper. Take a point other than the focus and repeat the process of reflection of rays of light from this point by the parabolic mirror.
5. Draw a diagram to describe your finding from question **4**. Provide a brief comment on your description.
6. Give examples of where these systems could be used in society.

Exercise 6.7 Review questions

Fluency

1. **MC** Which of the following is *not* a quadratic equation?

   A. $x^2 - 1 = 0$ **B.** $x^2 - 1 + 2x = 0$ **C.** $x^2 - \frac{1}{2} = 0$ **D.** $x^2 - \frac{1}{x} = 0$

2. **MC** Which of the following is in general form?

   A. $x^2 + 1 = 0$ **B.** $x^2 - 1 = 2$ **C.** $2x^2 = x^2 + 3$ **D.** $x^2 = x^2 + 1$

3. **MC** The Null Factor Law cannot be applied to the equation $x(x+3)(x-2) = 1$ because:
   A. the first factor is a simple x term.   **B.** there are more than two factors.
   C. the right hand side equals 1.   **D.** the second term is positive.

4. **MC** If the solutions to the quadratic equation $(x-3)(x-b) = 0$ are 3 and -5, then b is equal to:
   A. 5   **B.** -5 **C.** 3   **D.** -3

5. **MC** The solutions to $3x^2 - 27 = 0$ are:
   A. $x = 3$ and $x = -3$.   **B.** $x = 9$ and $x = -9$.
   C. $x = 1$ and $x = -1$.   **D.** $x = 2$ and $x = -2$.

6. **MC** The quadratic equation $x^2 - 4x + 3 = 0$ has solutions at:
   A. $x = -3$ and $x = -1$.   **B.** $x = -3$ and $x = 1$.
   C. $x = 3$ and $x = -1$.   **D.** $x = 3$ and $x = 1$.

7. **MC** The quadratic equation $9(x+9)^2 - 9 = 0$ has:
   A. the solutions 9 and -9.   **B.** the solutions $x = 1$ and -1.
   C. the solutions 8 and 10.   **D.** the solutions -8 and -10.

Understanding

8. Determine the solutions to the quadratic equation $(x-3)(2x+8) = 0$.

9. Determine the solutions to the equation $(4-x)(2x-7) = 0$.

10. Determine the solutions to the following quadratic equations.
    a. $2x^2 - 98 = 0$ **b.** $5x^2 - 125 = 0$ **c.** $2x^2 + 14x = 0$
    d. $x^2 - 4x - 5 = 0$ **e.** $x^2 - 11x + 10 = 0$ **f.** $x^2 - 9x - 10 = 0$

11. Solve the following equations, identifying those with no real solutions.
    a. $x^2 + 11x + 10 = 0$ **b.** $3x^2 + 6x = 0$ **c.** $-2x^2 - 1 = 0$

▶

12. Determine the solutions to the following equations.

    a. $(x+2)^2 - 16 = 0$ b. $4(x-3)^2 - 36 = 0$ c. $(x+1)^2 = 25$

13. Solve the following equations.

    a. $x^3 = 64$ b. $x^3 = 1$

14. Solve the following equations.

    a. $3x^3 = 3000$ b. $8x^3 = 1$

15. Solve the following equations, giving your answers correct to 2 decimal places.

    a. $5x^3 + 72 = 0$ b. $8x^3 - 25 = 0$

Communicating, reasoning and problem solving

16. The distance travelled by a motorbike is given by the formula $d = 18t + 2t^2$, where t is the time in seconds and d is the distance in metres.
    How long would it take the motorbike to travel a distance of 180 m?

17. When 20 is subtracted from the square of a certain number, the result is 8 times that number. Determine the number, which is negative.

18. A ball is thrown from the balcony of an apartment building. The ball is h metres above the ground when it is a horizontal distance of x metres from the building. The path of the ball follows the rule $h = -x^2 + 3x + 28$.
    How far from the building will the ball land?

19. An astronaut needs to determine a positive number such that twice the number plus its square gives 35.
    a. Write this as an equation.
    b. Solve the equation to determine the answer for him.

20. A window cleaner was 3 metres off the ground, cleaning the windows of a high-rise building, when he dropped his bucket.
    The height, h metres, of the bucket above the ground t seconds after it is dropped is given by the equation $h = 3 - t^2 - 2t$.
    a. What is the value of h when the bucket hits the ground?
    b. Rewrite the equation, replacing h with the value it takes when the bucket hits the ground.
    c. Solve the equation.
    d. Determine the time it takes for the bucket to reach the ground.

on   To test your understanding and knowledge of this topic, go to your learnON title at www.jacplus.com.au and complete the **post-test**.

Answers

Topic 6 Quadratic equations (Path)

Exercise 6.1 Pre-test

1. A
2. False
3. $(m-2)(m+4)$
4. $2(w-3)(w-2)$
5. D
6. $(x+6)(x-1)$
7. $6x^2 + 15x - 14 = 0$
8. Not a solution
9. Yes, solution to equation
10. $x = 5$
11. $x = -2$
12. False
13. $x = -2, 3$
14. $x = 0, \dfrac{4}{3}, -\dfrac{5}{2}$
15. $x = 12, 15$

Exercise 6.2 Quadratic equations and cubic equations of the form $ax^3 = k$

1. a. $a = 1, b = 2, c = -1$
   b. $a = 1, b = 1, c = -10$
   c. $a = 4, b = 5, c = -30$
2. a. $a = 8, b = -11, c = 2$
   b. $a = 2, b = 1, c = -60$
   c. $a = 1, b = 0, c = -17$
3. D
4. C
5. a. $x = 4, x = -4$
   b. There is no solution.
   c. $x = 0$
6. a. $x = 1, x = -1$
   b. There is no solution.
   c. $x = 2, x = -2$

7. a. Y b. N c. Y
8. a. Y b. N c. N
9. a. N b. N c. Y
10. a. Y b. N
11. a. 3 b. -2 c. 1.41
12. a. 5 b. $-\dfrac{1}{3}$ c. 1.44

13. When $x = 2$:
$$(x+4)^2 = x^2 + 8x + 16$$
$$= 2^2 + 8 \times 2 + 16$$
$$= 4 + 16 + 16$$
$$= 36$$
$$x^2 + 16 = 2^2 + 16$$
$$= 4 + 16$$
$$= 20$$
The two expressions are not equal.

14. a. Area of square $= (2x)^2 = 4x^2$
    Area of circle $= \pi x^2$
    $A =$ Area of square $-$ Area of circle
    $A = 4x^2 - \pi x^2$
    $A = x^2(4 - \pi)$

    b. $10 = x^2(4 - \pi)$
    $$\dfrac{10}{4 - \pi} = x^2$$
    $$x = \sqrt{\dfrac{10}{4 - \pi}} \approx 3.41 \text{ cm}$$

    c. x is a length, and lengths do not have negative values.

15. If $x^2 + 7x + 4 = 7x$, then $x^2 + 4 = 0$, which means that $x^2 = -4$ and, finally, $x = \sqrt{-4}$.
    As the square root of a negative number is not a real number, x has no solutions in the real number range.

16. a. Side length_a $= x + 3$
    Side length_b $= x + 5$
    Side length_c $= x + 8$
    Side length_d $= x - 3$
    Side length_e $= x + 6$

    b. I $= (x - 3)^2$
    II $= (x + 3)^2$
    III $= (x + 5)^2$
    IV $= (x + 6)^2$
    V $= (x + 8)^2$

17. a. 6 or -6
    b. There are two possible numbers, as the square root of a positive number, in this case 36, has two possible solutions, one positive and one negative.

18. a. $3(x - 2)^2 - 7 = 41$
    $7 - 3(x - 2)^2 = 41$ has no solution.
    b. -2 and 6

Exercise 6.3 The Null Factor Law

1. a. $x = -3, x = 2$ b. $x = -2, x = 3$
   c. $x = -2, x = 3$ d. $x = -2\dfrac{1}{2}, x = -\dfrac{3}{4}$
   e. $x = -4, x = -\dfrac{1}{2}$
2. a. $x = \dfrac{1}{2}, x = -30$ b. $x = -\dfrac{1}{2}, x = 3$
   c. $x = 1, x = \dfrac{1}{3}$ d. $x = 0, x = 2$
   e. $x = -\dfrac{1}{3}, x = \dfrac{1}{4}$
3. a. $x = -2.3, x = 0.3$ b. $x = -\dfrac{1}{6}, x = \dfrac{1}{6}$
   c. $x = 2$ d. $x = 0, x = \dfrac{15}{4}$
   e. $x = -4$
4. a. $x = 2, x = -2, x = -3$ b. $x = -2, x = 2.5$
   c. $x = -2, x = -4$ d. $x = 0, x = -2, x = -4$

5. a. $x = 1.1, x = -2.4, x = -2.6$

   b. $x = -3, x = -\dfrac{1}{2}, x = 1\dfrac{2}{3}$

   c. $x = 3$

   d. $x = -1, x = 2$

6. D

7. B

8. a. $x(x + 10) = 0$; $x = 0 \text{ or } x = -10$

   b. $2x(x - 7) = 0$; $x = 0 \text{ or } x = 7$

   c. $5x(5x - 8) = 0$; $x = 0 \text{ or } x = \dfrac{8}{5}$

9. By dividing both sides of the equation $-2x^2 + 2x + 12 = 0$ by -2, we get $(x - 3)(x + 2) = 0$. Since the equations are equivalent, they have the same solution(s).

10. a. When $x = 2$, the first bracket equals 2 and the second bracket equals 4; therefore, the product is 8.

    b. -2 and 1

11. A quadratic can have a maximum of two solutions, because a quadratic can at most be factorised into two separate pairs of brackets, each of which represents one solution.

12. a. 10 m

    b. $y = \dfrac{1}{250}x(100 - x)$; $a = \dfrac{1}{250}, b = 100$

    c. 9.424 m

13. a. $h = -4t(t - 5)$

    b. i. 16 m

    ii. 0 m

    c. Sample responses can be found in the worked solutions in the online resources.

14. -29 or 8

Exercise 6.4 Solving quadratic equations with two terms

1. a. $x = -3, x = 3$ b. $x = -4, x = 4$
   c. $x = -3, x = 3$ d. $x = -5, x = 5$
   e. $x = -10, x = 10$

2. a. $x = -7, x = 7$ b. $x = -3, x = 3$
   c. $x = -2, x = 2$ d. No real solutions
   e. No real solutions

3. a. $x = -3, x = 3$ b. $x = -4, x = 4$
   c. $x = -5, x = 5$ d. $x = 0$
   e. $x = 0$

4. a. $x = 0, x = -6$ b. $x = 0, x = 8$
   c. $x = 0, x = -9$ d. $x = 0, x = 11$
   e. $x = 0, x = 6$

5. a. $x = 0, x = 7.5$ b. $x = 0, x = \dfrac{2}{3}$

   c. $x = 0, x = -1\dfrac{3}{4}$ d. $x = 0, x = 2\dfrac{1}{2}$

   e. $x = 0, x = -1$

6. a. $x = 0, x = \dfrac{1}{4}$ b. $x = 0, x = -5$

   c. $x = 0, x = -12$ d. $x = 0, x = 18$
   e. $x = 0, x = 2.5$

7. A

8. C

9. 4 or -4

10. The square plot is 12 m × 12 m; the rectangular plot is 16 m × 9 m.

11. The number is 0 or 10.

12. $x^2 + 9$ cannot be factorised.

13. $x = \pm \dfrac{n}{m}$

14. x cannot be isolated, so the only way to solve the equation will be to factorise it and use the Null Factor Law.

15. If $a = 0$, then the expression is not quadratic.

16. 0 or 8

17. -12 or 0

18. a. $x(x + 30) = 50x$

    b. $x(x - 20) = 0$

    c. $x = 0 \text{ or } x = 20$

    d. No, x cannot be 0, because the width has to have a positive value.

Exercise 6.5 Solving monic quadratics equations

1. a. $(x + 1)(x + 3)$ b. $x = -1 \text{ or } x = -3$

2. a. $(x + 2)(x + 8)$ b. $x = -2 \text{ or } x = -8$

3. a. $(x + 6)(x - 3)$ b. $x = -6 \text{ or } x = 3$

4. a. $(x - 6)(x + 3)$ b. $x = 6 \text{ or } x = -3$

5. a. $n = -7, n = 5$ b. $v = -6, v = 1$
   c. $v = 6, v = -1$ d. $t = -6, t = 2$
   e. $t = -2, t = 7$

6. a. $x = -4, x = 5$ b. $x = -5, x = 4$
   c. $n = -10, n = 9$ d. $n = -7, n = 10$
   e. $x = -1, x = 5$

7. a. $x = 2, x = 4$ b. $x = -2, x = -4$
   c. $x = -1, x = -5$ d. $x = 2, x = -3$
   e. $x = 3, x = -5$

8. a. $x = -2$ b. $x = 4, x = -6$
   c. $x = 8, x = -3$ d. $x = -3, x = 4$
   e. $x = -12, x = -1$

9. a. $x = 11, x = -1$ b. $x = 4, x = -5$
   c. $x = -25, x = -4$ d. $x = 5, x = 10$
   e. $x = 4, x = -2$

10. B

11. D

12. C

13. a. $x^2 + 6x - 16$ b. $x = -8, x = 2$

14. $x = -4, x = 9$

15. 9 m by 5 m

16. a. $x = -1, x = -6$ b. 10 or 12

17. a. $x = -4, x = 1$

    b. $-10, -18, -28$ are some of the possible values of c, since one number will be positive and the other number will be negative.

18. No two integers multiply to 10 and add to 8.

19. No two integers multiply to 24 and add to 12.

20. a i. 0   ii. 2   iii. 35

    b i. Octagon   ii. Icosahedron

Exercise 6.6 Applications of quadratic equations

1. 6 and 7 or −7 and −6
2. 11 and 13 or −13 and −11
3. $x = 9$
4. $x = 10$
5. 3, 4, 5
6. 5, 12, 13
7. Length = 22 cm, width = 13 cm, height = 4 cm
8. 2.5m
9. a. 15m   b. 4 seconds
10. a. 0   b. 11 terms
11. 13
12. a. $g + 18$
    b. $g^2 + 12g - 108$
    c. $g^2 + 12g - 364 = 0$
    d. $g = 14$
    e. The girl is 14 years old, her brother is 8 years old and her teacher is 32 years old.

Project

1-5. Students need to construct a parabola by folding paper with the directions given in this project. Draw incident and reflected rays of light and use a protractor to carefully measure the angle each line makes with the mirror. Also comment on the diagram.
6. Parabolas could be seen in real life in, for example, water shot by the fountain in a parabolic path, a ball thrown into the air, bridges, headlights, satellite dishes or telescopes.

Exercise 6.7 Review questions

1. D
2. A
3. C
4. B
5. A
6. D
7. D
8. $x = -4, x = 3$
9. $x = 3.5, x = 4$
10. a. $x = -7, x = 7$ b. $x = -5, x = 5$
    c. $x = -7, x = 0$ d. $x = -1, x = 5$
    e. $x = 1, x = 10$ f. $x = -1, x = 10$
11. a. $x = -10, x = -1$ b. $x = -2, x = 0$
    c. No real solutions
12. a. $x = -6, x = 2$ b. $x = 0, x = 6$
    c. $x = -6, x = 4$
13. a. 4   b. 1
14. a. 10   b. $\dfrac{1}{2}$
15. a. −2.43   b. 1.46
16. 6 s
17. $x = -2$
18. 7 m
19. a. $x^2 + 2x = 35$ b. 5
20. a. $h = 0$ b. $0 = 3 - t^2 - 2t$
    c. $t = -3$ or $t = 1$ d. 1 s

7 Linear relationships

LESSON SEQUENCE

7.1 Overview

Why learn this?

We live in a world full of patterns, and recognising them help us understand the world around us a lot better. In many aspects of life, one quantity depends on another quantity, and the relationship between them can be described by equations. These equations can then be used in mathematical modelling to gain an understanding of the situation — we can draw graphs and use them to analyse, interpret and explain the relationship between the variables, and to make predictions about the future.

Scientists, engineers, health professionals and financial analysts all rely heavily on mathematical equations to model real-life situations and solve problems. Linear graphs are used extensively to represent trends, for example in the stock market or when considering the population growth of various countries. In day-to-day life, linear relationships help us compare phone rates, convert between currencies, and distinguish between increasing or decreasing consumption or sales.

Because of the many uses of linear graphs, it is important to understand the basic concepts that you will study in this topic, such as gradient, how to calculate the distance between two points, and how to identify linear relationships.

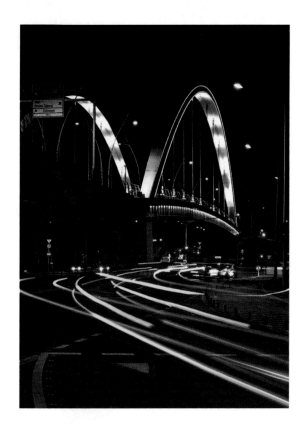

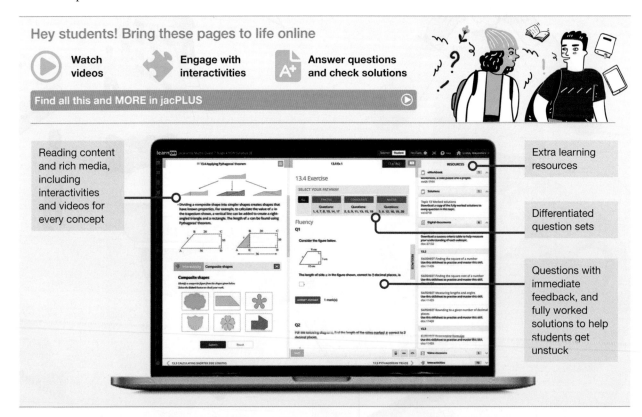

Hey students! Bring these pages to life online

▶ Watch videos

🧩 Engage with interactivities

A⁺ Answer questions and check solutions

Find all this and MORE in jacPLUS ▶

Reading content and rich media, including interactivities and videos for every concept

Extra learning resources

Differentiated question sets

Questions with immediate feedback, and fully worked solutions to help students get unstuck

1. In which quadrant does the point $(-3, 0)$ lie?

2. Using the table shown, match each point in the left-hand column with the line in the right-hand column that passes through that point.

Point	Line
a. $(3, -1)$	**A.** $y = 3 - 2x$
b. $(0, 4)$	**B.** $2y = x$
c. $(2, 1)$	**C.** $x + y = 2$
d. $(-1, 5)$	**D.** $y = x + 4$

3. **MC** Select the rule that corresponds to the table of values shown.

x	-3	-2	-1	-0	1
y	2	1	0	-1	-2

    A. $y = -x + 1$ **B.** $y = x + 1$ **C.** $y = -x - 1$ **D.** $y = x - 1$

4. Calculate the gradient of the line passing through the points $\left(-3, \dfrac{1}{2}\right)$ and $\left(5, -\dfrac{7}{2}\right)$.

5. **MC** Select the gradient of the line $3x + 4y = 12$.

    A. $\dfrac{4}{3}$ **B.** 3 **C.** 4 **D.** $-\dfrac{3}{4}$

6. Match the gradients and y-intercepts to the rules given by $y = mx + c$.

Gradient and y-intercept	Rule
a. $m = \dfrac{1}{2}, c = 3$	**A.** $2y - 6x = 4$
b. $m = \dfrac{1}{4}, c = 0$	**B.** $y = \dfrac{1}{2}x + 3$
c. $m = 4, c = -1$	**C.** $4y = x$
d. $m = 3, c = 2$	**D.** $y = 4x - 1$

7. **a.** Determine the rule for a straight line that passes through the origin and point $(2.4, -0.6)$.
   b. Determine the rule for a straight line that has an x-intercept of -20 and a y-intercept of $= 400$.

8. **MC** Select the equation of the linear graph shown.

   A. $y = 2x - 6$

   B. $y = 2x - 8$

   C. $y = -2x - 6$

   D. $y = -2x - 4$

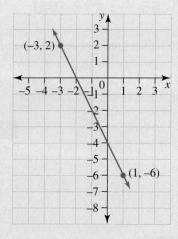

9. A student plays the game Blue Blobs. She has seven blobs on her screen as shown in the diagram.

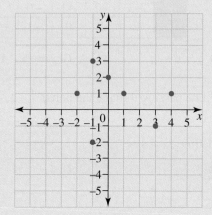

   The student types in an equation for a line that will pass through four blobs.

   a. Determine the equation of the straight line that passes through four blobs.

   b. The student then types in the equation $x = -1$. State the coordinate of the blob that both lines would hit.

10. Determine the value of a so that the point $M\left(-\dfrac{1}{2}, -1\dfrac{1}{2}\right)$ is the midpoint of the segment joining points A $(3, 2)$ and B $(-4, a)$.

11. Calculate the exact distance between the points $(2, 5)$ and $(-3, 7)$.

12. Calculate the distance between the point $(4, -3)$ and the origin.

13. Is the following statement true or false?

    'The straight line $y = 2x + 7$ is parallel to the straight line $2x + y = 9$.'

14. Determine the y-intercept of the linear equation $2x + 3y = 12$.

15. **MC** Select the point that lies on the line $3x - 4y = 18$.

    A. $(4, -3)$ B. $(3, -4)$ C. $(2, 3)$ D. $(2, -3)$

LESSON
7.2 Plotting linear graphs

LEARNING INTENTION

At the end of this lesson you should be able to:
- sketch graphs of lines on the Cartesian plane using a table of values
- determine if a point lies on the graph of an equation.

7.2.1 Plotting linear graphs on the Cartesian plane

eles-4779

- The **Cartesian plane** is divided into four regions (quadrants) by the x- and y-axes, as shown.
- Every point in the plane is described exactly by a pair of coordinates (x, y). The point P $(3, 2)$ is marked on the diagram.

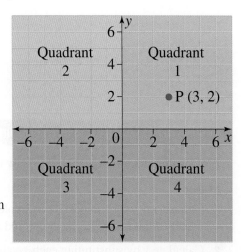

Plotting linear graphs from a rule

- A graph that forms a straight line is called a **linear graph**.
- An equation written in the form $y = mx + c$ is in the gradient–intercept form of a linear equation.
  For example, $y = 4x - 5$ or $y = x + 2$.
- To draw a graph of a linear equation, we need to determine the pairs of x- and y-values that make the equation true and plot them on the Cartesian plane. A minimum of two points are needed to plot a linear graph.
- We can use a table of values to determine these points.

Plotting a linear graph from a rule on a Cartesian plane

1. **Draw a table of values and assign values for x.**
2. **Substitute the x-values into the rule to calculate the y-values.**
3. **Plot the points on the Cartesian plane.**
4. **Draw a line through the points and label the graph.**

WORKED EXAMPLE 1 Graphing lines using a table of values

Draw the graph specified by the rule $y = x + 2$ for the x-values $-3, -2, -1, 0, 1, 2, 3$.

THINK	WRITE
1. Draw a table and write in the required x-values.	

x	−3	−2	−1	0	1	2	3
y							

2. Substitute each x-value into the rule $y = x + 2$ to obtain the corresponding y-value.
When $x = -3, y = -3 + 2 = -1$.
When $x = -2, y = -2 + 2 = 0$ etc.
Write the y-values into the table.

x	−3	−2	−1	0	1	2	3
y	−1	0	1	2	3	4	5

3. Plot the points from the table.

4. Join the points with a straight line and label
   the graph with its equation: $y = x + 2$.

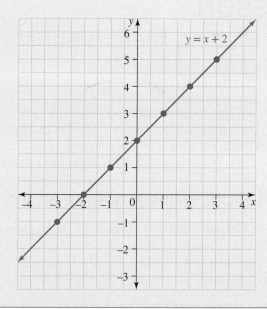

WORKED EXAMPLE 2 Drawing linear graphs by plotting two points

Plot two points and use them to draw the linear graph of $y = 2x - 1$.

THINK

1. Choose any two x-values, for example $x = -2$ and $x = 3$. You
   usually want to assign a negative and a positive value.

2. Calculate y by substituting each x-value into $y = 2x - 1$.
   $x = -2$: $y = 2 \times -2 - 1 = -5$
   $x = 3$: $y = 2 \times 3 - 1 = 5$

3. Plot the points $(-2, -5)$ and $(3, 5)$.

4. Draw a line through the points and label the graph with its
   equation.

WRITE

x	-2	3
y	-5	5

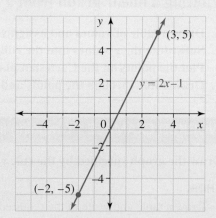

⦿ 7.2.2 Points on a line

eles-4781

- Consider the line that has the rule $y = 2x + 3$ as shown in the graph.
  If $x = 1$, then $y = 2(1) + 3$
  $\qquad\qquad\qquad = 5$
  so the point $(1, 5)$ lies on the line $y = 2x + 3$.
- The points $(1, -3)$ $(1, 9)$, $(1, 12)$ … are not on the line, but lie above or below it.
- The point $(-6, 4)$ is in quadrant 2.
  If $x = -6$: $y = 2(-6) + 3$
  $\qquad\qquad\quad y = -9$
  $\qquad\qquad\quad y \neq 5$
  This shows that the point does not lie on the line.

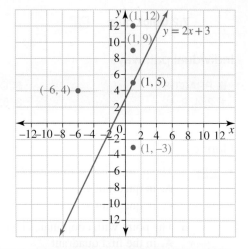

WORKED EXAMPLE 3 Determining whether a point lies on a line

Determine whether the point $(2, 4)$ lies on the line given by:

a. $y = 3x - 2$
b. $x + y = 5$.

THINK	WRITE
a. 1. Substitute $x = 2$ into the equation $y = 3x - 2$ and solve for y.	a. $y = 3x - 2$ $x = 2$: $y = 3(2) - 2$ $\qquad\quad = 6 - 2$ $\qquad\quad = 4$
2. When $x = 2$, $y = 4$, so the point $(2, 4)$ lies on the line. Write the answer.	The point $(2, 4)$ satisfies the equation $y = 3x - 2$. The point lies on the line.
b. 1. Substitute $x = 2$ into the equation $x + y = 5$ and solve for y.	b. $x + y = 5$ $x = 2$: $2 + y = 5$ $\qquad\qquad y = 3$
2. The point $(2, 3)$ lies on the line, but the point $(2, 4)$ does not. Write the answer.	The point $(2, 4)$ does not satisfy the equation $x + y = 5$. The point does not lie on the line.

DISCUSSION

In linear equations, what does the coefficient of x determine?

 Resources

⬥ **Interactivity** Plotting linear graphs (int-3834)

Exercise 7.2 Plotting linear graphs

7.2 Quick quiz on	7.2 Exercise

Individual pathways

■ PRACTISE	■ CONSOLIDATE	■ MASTER
1, 2, 4, 9, 12	3, 5, 8, 11, 13	6, 7, 10, 14

Fluency

1. **MC** a. The point with coordinates $(-2, 3)$ is:
   - **A.** in quadrant 1
   - **C.** in quadrant 3
   - **B.** in quadrant 2
   - **D.** in quadrant 4

   b. The point with coordinates $(-1, -5)$ is:
   - **A.** in the first quadrant
   - **C.** in the third quadrant
   - **B.** in the second quadrant
   - **D.** in the fourth quadrant

   c. The point with coordinates $(0, -2)$ is:
   - **A.** in the third quadrant
   - **C.** on the x-axis
   - **B.** in the fourth quadrant
   - **D.** on the y-axis

2. **WE1** For each of the following rules:

   i. complete the table of values   ii. draw the linear graph.

x	-3	-2	-1	0	1	2	3
y							

   a. $y = x$ b. $y = 2x + 2$ c. $y = 3x - 1$ d. $y = -2x$

Understanding

3. **WE2** By first plotting two points, draw the linear graph given by each of the following.

   a. $y = -x$ b. $y = \dfrac{1}{2}x + 4$ c. $y = -2x + 3$ d. $y = x - 3$

4. **WE3** Determine whether these points lie on the graph of $y = 2x - 5$.

   a. $(3, 1)$ b. $(-1, 3)$ c. $(0, 5)$ d. $(5, 5)$

5. Determine whether the given point lies on the given line.

   a. $y = -x - 7, (1, -8)$ b. $y = 3x + 5, (0, 5)$ c. $y = x + 6, (-1, 5)$
   d. $y = 5 - x, (8, 3)$ e. $y = -2x + 11, (5, -1)$ f. $y = x - 4, (-4, 0)$

6. **MC** Select the line that passes through the point $(2, -1)$.

   A. $y = -2x + 5$ **B.** $y = 2x - 1$ **C.** $y = -2x + 1$ **D.** $x + y = 1$

7. Match each point with the line passing through that point.

Point	Line
a. $(1, 1)$	**A.** $x + y = 4$
b. $(1, 3)$	**B.** $2x - y = 1$
c. $(1, 6)$	**C.** $y = 3x - 7$
d. $(1, -4)$	**D.** $y = 7 - x$

Communicating, reasoning and problem solving

8. The line through $(1, 3)$ and $(0, 4)$ passes through every quadrant except one. State the quadrant through which this line does not pass. Explain your answer.

9. Consider the equation $y = x + 1$.
   a. Determine which quadrant(s) the line passes through.
   b. Show that the point $(1, 3)$ does not lie on the line.

10. Explain the process of how to check whether a point lies on a given line.

11. Using the coordinates $(-1, -3)$, $(0, -1)$ and $(2, 3)$, show that a rule for the linear graph is $y = 2x - 1$.

12. Consider the pattern of squares on the grid shown.
    Determine the coordinates of the centre of the 20th square.

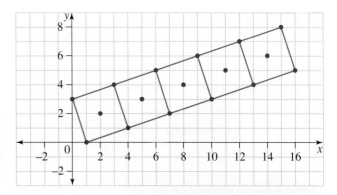

13. It is known that the mass of a certain kind of genetically modified tomato increases linearly over time. The following results were recorded.

Time, t (weeks)	1	4	6	9	16
Mass, m (grams)	6	21	31	46	81

   a. Plot the given points on a Cartesian plane.
   b. Determine the rule relating mass with time.
   c. Show that the mass of one of these tomatoes is 101 grams after 20 weeks.

14. As a particular chemical reaction proceeds, the temperature increases at a constant rate. The graph represents the same chemical reaction with and without stirring. Interpret the graph and explain how stirring affects the reaction.

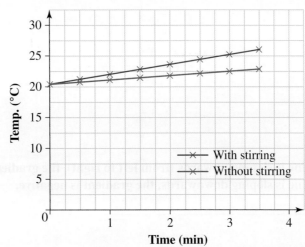

LESSON
7.3 Features of linear graphs

▶ 7.3.1 The gradient (*m*)

eles-4782

- The **gradient** of a line is a measure of the steepness of its slope.
- The symbol for the gradient is *m*.
- The gradient of an interval AB is defined as the rise (distance up) divided by the run (distance across).

Gradient

$$m = \frac{\text{rise}}{\text{run}}$$

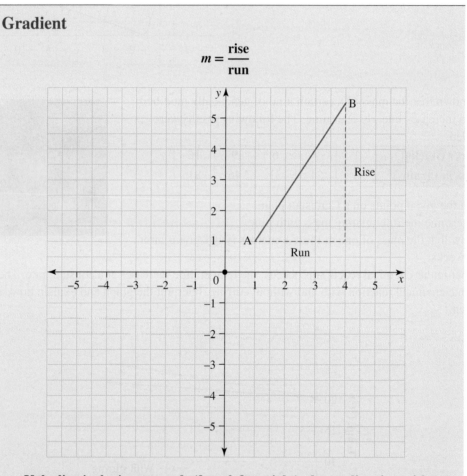

- **If the line is sloping upwards (from left to right), the gradient is positive.**
- **If the line is sloping downwards, the gradient is negative.**

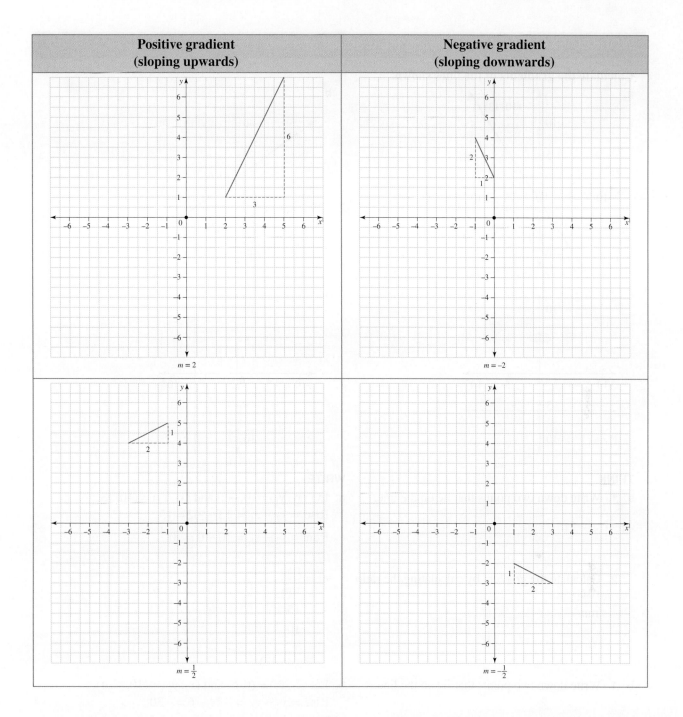

Positive gradient (sloping upwards)	Negative gradient (sloping downwards)
$m = 2$	$m = -2$
$m = \dfrac{1}{2}$	$m = -\dfrac{1}{2}$

The gradients of horizontal and vertical lines

- The gradient of a horizontal line is zero, since it does not have any upwards or downwards slope.

$$m = \frac{\text{rise}}{\text{run}} = \frac{0}{\text{run}} = 0$$

- The gradient of a vertical line is undefined, as the run is zero and you cannot divide by zero.

$$m = \frac{\text{rise}}{\text{run}} = \frac{\text{rise}}{0} = \text{undefined}$$

Calculate the gradients of the lines shown.

a.
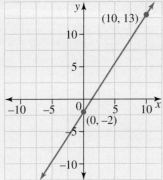

b.
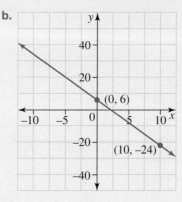

c.
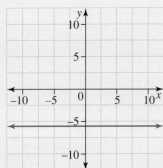

d.
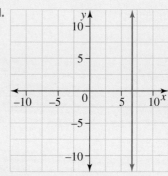

THINK	**WRITE**
a. 1. Write down two points that lie on the line.	a. Let $(x_1, y_1) = (0, -2)$ and $(x_2, y_2) = (10, 13)$. Rise $= y_2 - y_1 = 13 - -2 = 15$ Run $= x_2 - x_1 = 10 - 0 = 10$
2. Calculate the gradient by calculating the ratio $\dfrac{\text{rise}}{\text{run}}$.	$m = \dfrac{\text{rise}}{\text{run}}$ $= \dfrac{15}{10}$ $= \dfrac{3}{2}$ or 1.5
b. 1. Write down two points that lie on the line.	b. Let $(x_1, y_1) = (0, 6)$ and $(x_2, y_2) = (10, -24)$. Rise $= y_2 - y_1 = -24 - 6 = -30$ Run $= x_2 - x_1 = 10 - 0 = 10$
2. Calculate the gradient.	$m = \dfrac{\text{rise}}{\text{run}}$ $= \dfrac{-30}{10}$ $= -3$
c. 1. Write down two points that lie on the line.	c. Let $(x_1, y_1) = (5, -6)$ and $(x_2, y_2) = (10, -6)$.

2. There is no rise between the two points.

$$\text{Rise} = y_2 - y_1$$
$$= -6 - -6 = 0$$
$$\text{Run} = x_2 - x_1$$
$$= 10 - 5 = 5$$

3. Calculate the gradient.
Note: The gradient of a horizontal line is always zero. The line has no slope.

$$m = \frac{\text{rise}}{\text{run}}$$
$$= \frac{0}{5}$$
$$= 0$$

d. 1. Write down two points that lie on the line.

d. Let $(x_1, y_1) = (7, 10)$ and $(x_2, y_2) = (7, -3)$.

2. The vertical distance between the selected points is 13 units. There is no run between the two points.

$$\text{Rise} = y_2 - y_1 = -3 - 10 = 13$$
$$\text{Run} = x_2 - x_1 = 7 - 7 = 0$$

3. Calculate the gradient.
Note: The gradient of a vertical line is always undefined.

$$m = \frac{\text{rise}}{\text{run}}$$
$$= \frac{13}{0}$$
$$= \text{undefined}$$

Determining the gradient of a line passing through two points

- Suppose a line passes through the points $(1, 4)$ and $(3, 8)$, as shown in the graph.
- By completing a right-angled triangle, it can be seen that:
  - the rise (difference in y-values): $8 - 4 = 4$
  - the run (difference in x-values): $3 - 1 = 2$
  - to determine the gradient:

$$m = \frac{8 - 4}{3 - 1}$$
$$= \frac{4}{2}$$
$$= 2$$

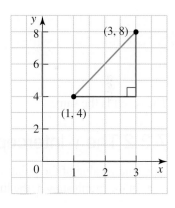

Gradient of a line passing through two points

- **In general, if a line passes through the points (x_1, y_1) and (x_2, y_2), then we can determine the gradient using the formula:**

$$m = \frac{y_2 - y_1}{x_2 - x_1}$$

Calculate the gradient of the line passing through the points $(-2, 5)$ and $(1, 14)$.

THINK	WRITE
1. Let the two points be (x_1, y_1) and (x_2, y_2).	$(-2, 5) = (x_1, y_1), (1, 14) = (x_2, y_2)$
2. Write the formula for gradient.	$m = \dfrac{y_2 - y_1}{x_2 - x_1}$
3. Substitute the coordinates of the given points into the formula and evaluate.	$m = \dfrac{14 - 5}{1 - -2}$ $= \dfrac{9}{1 + 2}$ $= \dfrac{9}{3}$ $= 3$
4. Write the answer.	The gradient of the line passing through $(-2, 5)$ and $(1, 14)$ is 3.

Note: If you were to switch the order of the points and let $(x_1, y_1) = (1, 14)$ and $(x_2, y_2) = (-2, 5)$, then the gradient could be calculated as shown.

$$m = \dfrac{y_2 - y_1}{x_2 - x_1}$$
$$= \dfrac{5 - 14}{-2 - 1}$$
$$= \dfrac{-9}{-3}$$
$$= 3$$

The result is the same.

▶ 7.3.2 The *y*-intercept

eles-4783

- The *y*-value of the point where a line cuts the *y*-axis is called the **y-intercept.**
- In the graph shown, the line cuts the *y*-axis at the point (0, 2), so the *y*-intercept is 2.

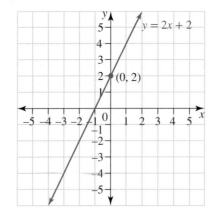

- The *y*-intercept of any line is easily found by substituting $x = 0$ and calculating the *y*-value.

Determine the *y*-intercepts of the lines whose linear rules are given, and state the coordinates of the *y*-intercept.

a. $y = -4x + 7$ b. $5y + 2x = 10$ c. $y = 2x$ d. $y = -8$

THINK

WRITE

a. 1. To calculate the *y*-intercept, substitute $x = 0$ into the equation.

a. $y = -4x + 7$
$y = -4(0) + 7$

2. Solve for *y*.

$y = 7$

3. Write the coordinates of the *y*-intercept.

y-intercept: $(0, 7)$

b. 1. To calculate the *y*-intercept, substitute $x = 0$ into the equation.

b. $5y + 2x = 10$
$5y + 2(0) = 10$

2. Solve for *y*.

$5y = 10$
$y = 2$

3. Write the coordinates of the *y*-intercept.

y-intercept: $(0, 2)$

c. 1. To calculate the *y*-intercept, substitute $x = 0$ into the equation.

c. $y = 2x$
$y = 2(0)$
$y = 0$

2. Write the coordinates of the *y*-intercept.

y-intercept: $(0, 0)$

d. The value of *y* is -8 regardless of the *x*-value.

d. *y*-intercept: $(0, -8)$

DISCUSSION

Why is the *y*-intercept of a graph found by substituting $x = 0$ into the equation?

Can you use this information to easily determine the *y*-intercept of the line $y = \dfrac{1}{7}x - \dfrac{3}{4}$?

 Resources

▶ **Video eLesson**   Gradient (eles-1889)

Interactivities   The gradient (int-3836)
Linear graphs (int-6484)

Individual pathways

■ PRACTISE	■ CONSOLIDATE	■ MASTER
1, 4, 6, 8, 11, 14	2, 5, 7, 9, 12, 15	3, 10, 13, 16, 17

Fluency

1. **WE4** Calculate the gradients of the lines shown.

   a. 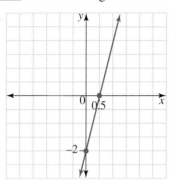   b. 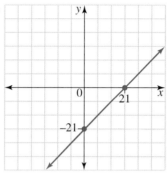   c.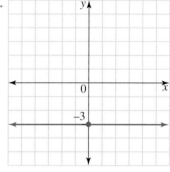

2. Calculate the gradients of the lines shown.

   a. 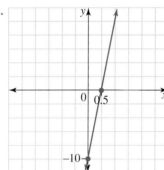   b. 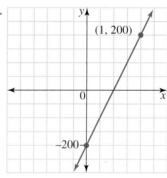   c.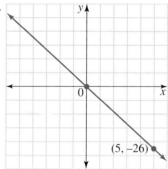

3. Calculate the gradients of the lines shown.

   a. 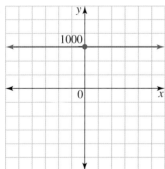   b. 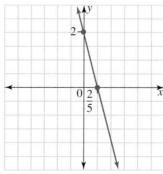   c.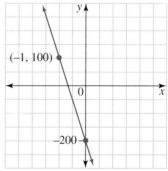

4. **WE5** Calculate the gradients of the lines passing through the following pairs of points.
   a. $(2, 10)$ and $(4, 22)$
   b. $(1, -2)$ and $(3, -10)$
   c. $(-3, 0)$ and $(7, 0)$
   d. $(-4, -7)$ and $(1, -1)$
   e. $(0, 4)$ and $(4, -4.8)$
   f. $(-2, 122)$ and $(1, -13)$

5. Calculate the gradients of the lines passing through the following pairs of points.
   a. $(2, 3)$ and $(17, 3)$
   b. $(-2, 2)$ and $(2, 2.4)$
   c. $(1, -5)$ and $(5, -15.4)$
   d. $(-12, -7)$ and $(8.4, -7)$
   e. $(-2, -17.7)$ and $(0, 0.3)$
   f. $(-3, 3.4)$ and $(5, 2.6)$

6. **WE6** Determine the y-intercepts of the lines whose rules are given below.
   a. $y = 5x + 23$
   b. $y = 54 - 3x$
   c. $y = 3(x - 2)$
   d. $y = 70 - 2x$

7. Determine the y-intercepts of the lines whose rules are given below.
   a. $y = \dfrac{1}{2}(x + 2)$
   b. $y = \dfrac{x}{2} + 5.2$
   c. $y = 100 - x$
   d. $y = 100$

Understanding

8. **MC** Select which of the following statements about linear graphs is false.
   A. A gradient of zero means the graph is a horizontal line.
   B. A gradient can be any real number.
   C. A linear graph can have two y-intercepts.
   D. In the form $y = mx + c$, the y-intercept equals c.

9. Identify the y-intercept of the line $y = mx + c$.

10. Determine the coordinates of the y-intercepts of the lines with the following rules.
    a. $y = -6x - 10$
    b. $3y + 3x = -12$
    c. $7x - 5y + 15 = 0$
    d. $y = 7$
    e. $x = 9$

Communicating, reasoning and problem solving

11. Explain why the gradient of a vertical line is undefined.

12. Explain why the gradient of a horizontal line is zero.

13. Show the gradient of the line passing through the points (a, b) and (c, d) is $\dfrac{d - b}{c - a}$.

14. When using the gradient to draw a line, does it matter if you rise before you run or run before you rise? Explain.

15. Consider the graph shown.
    a. Determine a general formula for the gradient m in terms of x, y and c.
    b. Rearrange your formula to make y the subject. Discuss what you notice about this equation.

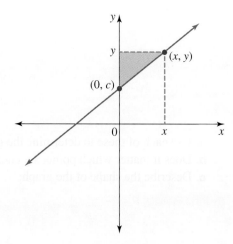

16. The price per kilogram for three different types of meat is illustrated in the graph shown.

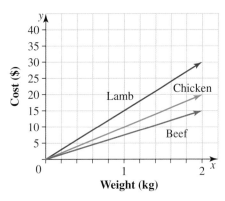

a. Calculate the gradient (using units) for each graph.

b. Determine the cost of 1 kg of each type of meat.

c. Evaluate the cost of purchasing:

   i. 1 kg of lamb   ii. 0.5 kg of chicken   iii. 2 kg of beef.

d. Calculate the total cost of the order in part **c**.

e. Complete the table below to confirm your answer from part **d**.

Meat type	Cost per kilogram ($/kg)	Weight required (kg)	Cost = $/kg × kg
Lamb		1	
Chicken		0.5	
Beef		2	
		Total cost	

17. Three right-angled triangles have been superimposed on the graph shown.

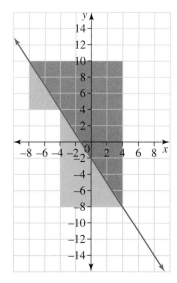

a. Use each of these to determine the gradient of the line.

b. Does it matter which points are chosen to determine the gradient of a line? Explain your answer.

c. Describe the shape of the graph.

LESSON
7.4 The equation of a straight line, $y = mx + c$

LEARNING INTENTION

At the end of this lesson you should be able to:
- determine the equation of a straight line given the gradient and the y-intercept
- determine the equation of a straight line given the gradient and a point
- determine the equation of a straight line given two points
- determine the equation of a straight line from a graph.

⏵ 7.4.1 The equation of a straight line

eles-4784

- The gradient–intercept or slope–intercept form of a straight line is

$$y = mx + c$$

  where m is the gradient and c is the y-intercept of the line.
- It is very easy to state the equation of a straight line if its gradient and y-intercept are known.

WORKED EXAMPLE 7 Determining the equation of a line given the gradient and y-intercept

Determine the equation of the line with a gradient of -2 and y-intercept of 3.

THINK	WRITE
1. Write the equation of a straight line.	$y = mx + c$
2. Substitute the values $m = -2$, $c = 3$ to write the equation.	$m = -2,\ c = 3$ $y = -2x + 3$

⏵ 7.4.2 Determining the equation of a straight line given the gradient and the coordinates of a point

eles-4785

- If the gradient (m) and any single point on a straight line are known, the y-intercept can be calculated algebraically.
- The gradient and y-intercept can then be used to determine the equation of a straight line.

WORKED EXAMPLE 8 Determining the equation of a line given the gradient and a point

Determine the equation of a straight line that goes through the point $(1, -3)$, if its gradient is -2.

THINK	WRITE
1. Write the gradient–intercept form of the equation of a straight line.	$y = mx + c$
2. Substitute the value $m = -2$.	$y = -2x + c$
3. Since the line passes through the point $(1, -3)$, substitute $x = 1$ and $y = -3$ into $y = -2x + c$ to calculate the value of c.	When $x = 1$, $y = -3$. $-3 = -2 \times 1 + c$

▶

4. Solve for c.	$-3 = -2 + c$
	$-3 + 2 = c$
	$c = -1$
5. Use the given m and the calculated c to write the rule.	$y = -2x - 1$

▶ 7.4.3 Determining the equation of a straight line given two points

eles-6201

- If two points on a straight line are known, then the gradient (m) can be calculated using the formula $m = \dfrac{y_2 - y_1}{x_2 - x_1}$, or $m = \dfrac{\text{rise}}{\text{run}}$.

- Using the gradient and one of the points, the equation can be found using the method in Worked example 8.

WORKED EXAMPLE 9 Determining the rule of a line given two points

Determine the equation of the straight line passing through the points $(-1, 6)$ and $(3, -2)$.

THINK	WRITE
1. Write the gradient–intercept form of the equation of a straight line.	$y = mx + c$
2. Write the formula for calculating the gradient, m.	$m = \dfrac{y_2 - y_1}{x_2 - x_1}$
3. Let $(x_1, y_1) = (-1, 6)$ and $(x_2, x_2) = (3, -2)$. Substitute the values into the formula and determine the value of m.	$m = \dfrac{-2 - 6}{3 - -1}$
	$= \dfrac{-2 - 6}{3 + 1}$
	$= -\dfrac{8}{4}$
	$= -2$
4. Substitute the value of m into the equation.	$y = -2x + c$
5. Select either of the two points, say $(3, -2)$, and substitute into $y = -2x + c$.	Point $(3, -2)$:
	$-2 = -2(3) + c$
	$-2 = -6 + c$
6. Solve for c.	$c = 4$
7. Write the rule using the values $c = 4$ and $m = -2$.	$y = -2x + 4$

▶ 7.4.4 Determining the equation of a straight line from a graph

eles-6202

- The equation of a straight line can be determined from a graph by observing the y-intercept and the gradient.
- Any two points on a straight line will determine the gradient.

Determine the equation of the linear graph shown.

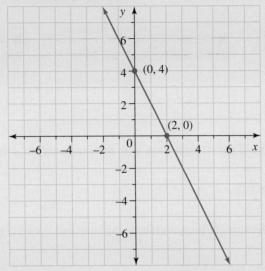

THINK

1. Form a right-angled triangle from the interval on the Cartesian plane.

WRITE

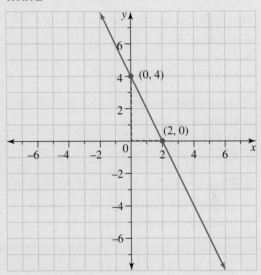

2. Determine the rise from left to right (vertical distance).

The rise is -4.

3. Determine the run from left to right (horizontal distance).

The run is 2.

4. Calculate the gradient using $m = \dfrac{\text{rise}}{\text{run}}$.

$$m = \dfrac{-4}{2}$$
$$= -2$$

5. Determine the y-value of the point where the line crosses the y-axis.

The line cuts the y-axis at $(0, 4)$. The y-intercept is 4.

6. Use the values of m and c to write the equation.

The equation of the line is $y = -2x + 4$.

 Resources

▶ **Video eLesson** The equation of a straight line (eles-2313)

7.4 Quick quiz on	7.4 Exercise

Individual pathways

■ PRACTISE	■ CONSOLIDATE	■ MASTER
1, 3, 5, 10, 15, 16, 19, 22, 25	2, 4, 6, 8, 11, 13, 17, 20, 23, 26, 27	7, 9, 12, 14, 18, 21, 24, 28, 29

Fluency

1. **WE7** Determine the equations of the straight lines with the gradients and y-intercepts given.
   a. Gradient $= 4$, y-intercept $= 2$
   b. Gradient $= -4$, y-intercept $= 1$
   c. Gradient $= 4$, y-intercept $= 8$

2. Determine the equations of the straight lines with the gradients and y-intercepts given.
   a. Gradient $= 6$, y-intercept $= 7$
   b. Gradient $= -2.5$, y-intercept $= 6$
   c. Gradient $= 45$, y-intercept $= 135$

3. Determine the equation for each straight line passing through the origin and with the gradient given.
   a. Gradient $= -2$
   b. Gradient $= 4$
   c. Gradient $= 10.5$

4. Determine the equation for each straight line passing through the origin and with the gradient.
   a. Gradient $= -20$
   b. Gradient $= 1.07$
   c. Gradient $= 32$

5. **WE8** Determine the equation of the straight lines with:
   a. gradient $= 1$, point $= (3, 5)$
   b. gradient $= -1$, point $= (3, 5)$
   c. gradient $= -4$, point $= (-3,\ 4)$
   d. gradient $= 2$, point $= (5, -3)$

6. Determine the equation of the straight lines with:
   a. gradient $= -5$, point $= (13, 5)$
   b. gradient $= 2$, point $= (10, -3)$
   c. gradient $= -6$, point $= (2, -1)$
   d. gradient $= -1$, point $= (-2, 0.5)$

7. Determine the equation of the straight lines with:
   a. gradient $= 6$, point $= (-6, -6)$
   b. gradient $= -3.5$, point $= (3, 5)$
   c. gradient $= 1.2$, point $= (2.4, -1.2)$
   d. gradient $= 0.2$, point $= (1.3, -1.5)$

8. Determine the equations of the straight lines with:
   a. gradient $= -4$, x-intercept $= -6$
   b. gradient $= 2$, x-intercept $= 3$
   c. gradient $= -2$, x-intercept $= 2$

9. Determine the equation of the straight lines with:
   a. gradient $= 5$, x-intercept $= -7$
   b. gradient $= 1.5$, x-intercept $= 2.5$
   c. gradient $= 0.4$, x-intercept $= 2.4$.

10. **WE9** Determine the equation for each straight line passing through the given points.

 a. $(-6, 11)$ and $(6, 23)$
 b. $(1, 2)$ and $(-5, 8)$
 c. $(4, 11)$ and $(6, 11)$
 d. $(3, 6.5)$ and $(6.5, 10)$

11. Determine the equation for each straight line passing through the given points.

 a. $(1.5, 2)$ and $(6, -2.5)$
 b. $(-7, 3)$ and $(2, 4)$
 c. $(25, -60)$ and $(10, 30)$
 d. $(5, 100)$ and $(25, 500)$

12. Determine the equation for each straight lines passing through the given points.

 a. $(1, 3)$ and $(3, 1)$
 b. $(2, 5)$ and $(-2, 6)$
 c. $(9, -2)$ and $(2, -4)$
 d. $(1, 4)$ and $(-0.5, 3)$

13. Determine the rules for the linear graphs that have the following x- and y-intercepts.

 a. x-intercept $= -3$, y-intercept $= 3$ b. x-intercept $= 4$, y-intercept $= 5$
 c. x-intercept $= 1$, y-intercept $= 6$ d. x-intercept $= -40$, y-intercept $= 35$

14. Determine the rules for the linear graphs that have the following x- and y-intercepts.

 a. x-intercept $= -8$, y-intercept $= 8$ b. x-intercept $= 3$, y-intercept $= 6$
 c. x-intercept $= -7$, y-intercept $= -3$ d. x-intercept $= -200$, y-intercept $= 50$

15. Determine the rule for each straight line passing through the origin and:

 a. the point $(4, 7)$ b. the point $(5, 5)$
 c. the point $(-4, 8)$ d. the point $(-1.2, 3.6)$
 e. the point $(-22, 48)$ f. the point $(-105, 35)$.

16. **WE10** Determine the equation of the line shown on each of the following graphs.

 a.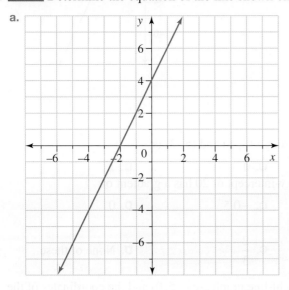

 b.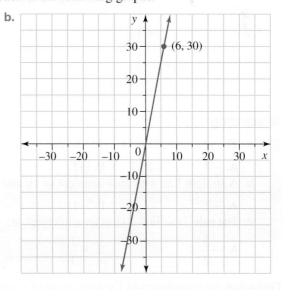

17. Determine the equation of the line shown on each of the following graphs.

a.

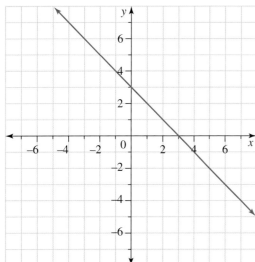

b.

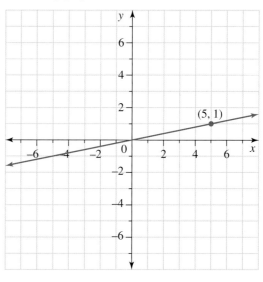

18. Determine the equation of the line shown on each of the following graphs.

a.

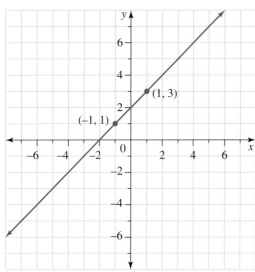

b.

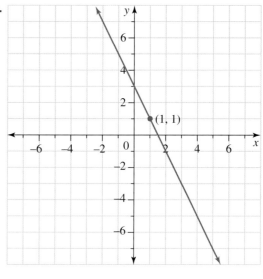

Understanding

19. **MC** a. The gradient of the straight line that passes through (3, 5) and (5, 3) is:

    A. -2 **B.** -1 **C.** 0 **D.** 1

  b. A straight line with an x-intercept of 10 and a y-intercept of 20 has a gradient of:

    A. -2 **B.** -1 **C.** -0.5 **D.** 0

  c. The rule $2y - 3x = 20$ has an x-intercept at:

    A. $-\dfrac{3}{2}$ **B.** $-\dfrac{2}{3}$ **C.** $\dfrac{2}{3}$ **D.** None of these

20. Given that the coordinates of the x-intercept of a straight-line graph are $(-5, 0)$ and the coordinates of the y-intercept are $(0, -12)$:

  a. determine the equation of the straight line   b. calculate the value of y when $x = 19.3$.

21. a. Determine the equation of the straight line shown in the graph. Use the fact that when $x = 5$, $y = 7$.
   b. Determine the x-intercept.

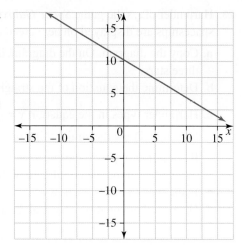

Communicating, reasoning and problem solving

22. The graph shows the carbon dioxide (CO_2) concentration in the atmosphere, measured in parts per million (ppm).

   a. If the trend follows a linear pattern, determine the equation for the line.
   b. Explain why c cannot be read directly from the graph.
   c. Infer the CO_2 concentration predicted for 2020. State the assumption made when determining this value.

23. Show that the equation for the line that passes through the point (3, 6) parallel to the line through the points $(0, -7)$ and $(4, -15)$ is $y = -2x + 12$.

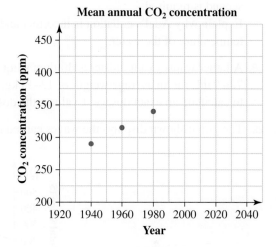

Mean annual CO_2 concentration

24. a. Determine the equations for line A and line B as shown in this graph.
   b. Write the point of intersection between line A and line B and mark it on the Cartesian plane.
   c. Show that the equation of the line that is perpendicular to line B and passes through the point $(-4, 6)$ is $y = x + 10$.

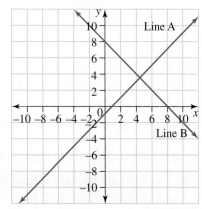

25. The graph shown describes the mass in kilograms of metric cups of water. Write a rule to describe the mass of water (m) relative to the number of cups (c).

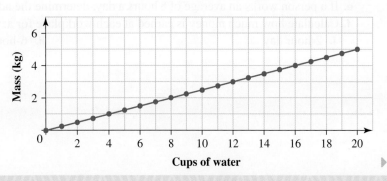

26. Harpinder plays the game *Space Galaxy* on her phone. The stars and spaceships are displayed on her screen as shown.

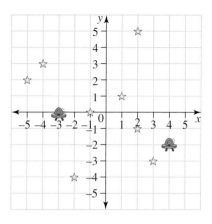

   a. Copy the diagram. On the diagram, draw a straight line that will hit three stars.
   b. Determine the equation of the straight line that will hit three stars.
   c. Harpinder types in the equation $y = \dfrac{1}{2}x + \dfrac{1}{2}$ and manages to hit two stars. Draw the straight line on your diagram.
   d. If Harpinder types in the equation from part **b** and the equation from part **c**, determine the coordinate of the star that both lines will hit.
   e. If she types $y = 2$, state how many stars will she hit.
   f. Give another equation of a straight line that will hit two stars.

27. The temperature of water in a kettle is 15 °C before the temperature increases at a constant rate for 20 seconds to reach boiling point (100 °C). Adel argues that $T = 5t + 15$ describes the water temperature, citing the starting temperature of 15 °C and that to reach 100 °C in 20 seconds an increase of 5 °C for every second is required.
   Explain why Adel's equation is incorrect and devise another equation that correctly describes the temperature of the water.

28. A father wants to administer a children's liquid painkiller to his child. The recommended dosage is a range, 7.5–9 mL for an average weight of 12–14 kg. The child weighs 12.8 kg. The father uses a linear relationship to calculate an exact dosage. Evaluate the dosage that the father calculates.

29. The graph shown displays the wages earned in three different workplaces.

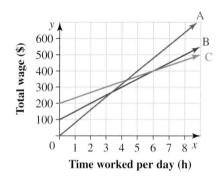

   a. Identify the set allowance for each workplace.
   b. Determine the hourly rates for each workplace.
   c. Using your answers from parts **a** and **b**, determine linear equations that describe the wages at each workplace.
   d. Match each working lifestyle below to the most appropriate workplace.

      i. Working lifestyle 1: Earn the most money possible while working at most 4 hours in a day.
      ii. Working lifestyle 2: Earn the most money possible while working an 8-hour day.

   e. If a person works an average of 8 hours a day, determine the advantage of workplace C.
   f. Calculate how much money is earned at each workplace for a:
      i. 2-hour day
      ii. 6-hour day.

LESSON
7.5 Sketching linear graphs

LEARNING INTENTION

At the end of this lesson you should be able to:
- sketch linear graphs using the x- and y-intercepts
- sketch linear graphs using the gradient and the y-intercept
- sketch vertical and horizontal lines.

- There are different strategies that can help us sketch the graphs of different linear equations.
- In Lesson 7.2, we showed that we can use a table of values to find points of the line and draw the graph of the equation. This lesson will show us two other methods.

▶ 7.5.1 Graphing using the x- and y-intercepts

eles-4786

- To sketch a linear graph all you need is two points on the line. Once you have found two points, the line can be drawn through those points.
- One method to sketch a linear graph is to determine both the **x-intercept** and the **y-intercept**.
- To determine the intercepts of a graph:
  x-intercept: let $y = 0$ and solve for x
  y-intercept: let $x = 0$ and solve for y.
- If the x- and y-intercepts are both zero (at the origin), you would need to choose another value for x and then calculate y to get another point on the graph.

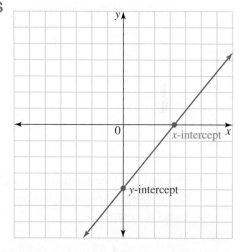

WORKED EXAMPLE 11 Sketching graphs using the x- and y-intercepts

Sketch the graphs of the following equations using the x- and y-intercepts.

a. $2y + 3x = 6$ b. $y = \dfrac{4}{5}x + 5$ c. $y = 2x$.

THINK

a. 1. Write the equation.

   2. To calculate the y-intercept, let $x = 0$.
      Write the coordinates of the y-intercept.

   3. To calculate the x-intercept, let $y = 0$.
      Write the coordinates of the x-intercept.

WRITE

a. $2y + 3x = 6$

$x = 0:\quad 2y + 3 \times 0 = 6$
$\qquad\qquad\qquad 2y = 6$
$\qquad\qquad\qquad\ y = 3$
$y\text{-intercept: } (0, 3)$

$y = 0:\quad 2 \times 0 + 3x = 6$
$\qquad\qquad\qquad 3x = 6$
$\qquad\qquad\qquad\ x = 2$
$x\text{-intercept: } (2, 0)$

4. Plot and label the *x*- and *y*-intercepts on a Cartesian plane and rule a straight line through them.
Label the graph.

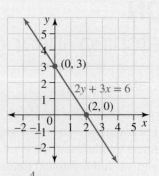

b. 1. Write the equation.

b. $y = \dfrac{4}{5}x + 5$

2. The equation is in the form $y = mx + c$, so the *y*-intercept is the constant term, *c*.

$c = 5$
y-intercept: $(0, 5)$

3. To calculate the *x*-intercept, let $y = 0$. Write the coordinates of the *x*-intercept.

$y = 0$: $\quad y = \dfrac{4}{5}x + 5$

$0 = \dfrac{4}{5}x + 5$

$-5 = \dfrac{4}{5}x$

$x = -\dfrac{25}{4} \left(= -6\dfrac{1}{4} \right)$

x-intercept: $\left(-\dfrac{25}{4}, 0 \right)$

4. Plot and label the intercepts on a Cartesian plane and rule a straight line through them.
Label the graph.

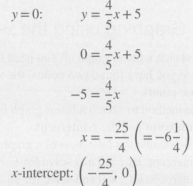

c. 1. Write the equation.

c. $y = 2x$

2. To calculate the *y*-intercept, let $x = 0$.
Write the coordinates of the *y*-intercept.

$x = 0$: $\quad y = 2 \times 0$
$\quad\quad\quad\quad = 0$
y-intercept: $(0, 0)$

3. The *x*- and *y*-intercepts are the same point, $(0, 0)$, so one more point is required.
Choose any value for *x*, such as $x = 3$.
Substitute and write the coordinates of the point.

$x = 3$: $\quad y = 2 \times 3$
$\quad\quad\quad\quad = 6$
Another point: $(3, 6)$

4. Plot the points, then rule and label the graph.

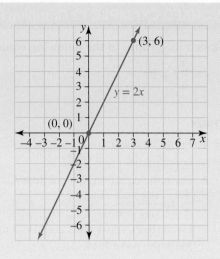

▶ 7.5.2 Graphing using the gradient and the *y*-intercept

eles-4787

- To use this method, the gradient and the *y*-intercept must be known.
- The line is drawn by plotting the *y*-intercept, and using the rise and the run to find other points on the line.

A line interval of gradient $3 \left(= \dfrac{3}{1} \right)$ can be drawn with a rise of 3 and a run of 1.	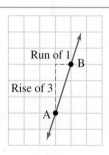
A line interval of gradient $-2 \left(= \dfrac{-2}{1} \right)$ can be drawn with a rise downwards of 2 and a run of 1.	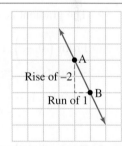
A line interval of gradient $\dfrac{3}{5}$ can be shown with a rise of 3 and a run of 5.	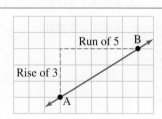

Sketch the graphs of the following equations using the gradient and the *y*-intercept.

a. $y = \dfrac{3}{4}x + 2$

b. $4x + 2y = 3.$

THINK

WRITE

a. 1. From the equation, the *y*-intercept is 2. Plot the point $(0, 2)$.

2. From the equation, the gradient is $\dfrac{3}{4}$, so $\dfrac{\text{rise}}{\text{run}} = \dfrac{3}{4}$.
From $(0, 2)$, run 4 units and rise 3 units. Mark the point P $(4, 5)$.

3. Draw a line through $(0, 2)$ and P $(4, 5)$. Label the graph.

a.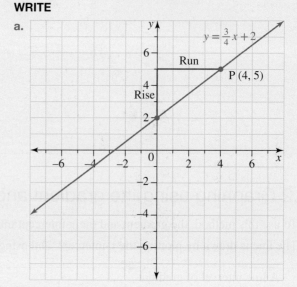

b. 1. Write the equation in gradient–intercept form: $y = mx + c$.
From the equation, $m = -2$ and $c = \dfrac{3}{2}$.
Plot the point $\left(0, \dfrac{3}{2}\right)$.

b. $4x + 2y = 3$
$2y = 3 - 4x$
$y = \dfrac{3}{2} - 2x$
$y = -2x + \dfrac{3}{2}$

2. The gradient is -2, so $\dfrac{\text{rise}}{\text{run}} = \dfrac{-2}{1}$.
From $\left(0, \dfrac{3}{2}\right)$, run 1 units and rise -2 units (i.e. go down 2 units). Mark the point P $\left(1, -\dfrac{1}{2}\right)$.

3. Draw a line through $\left(0, \dfrac{3}{2}\right)$ and P $\left(1, -\dfrac{1}{2}\right)$. Label the graph.

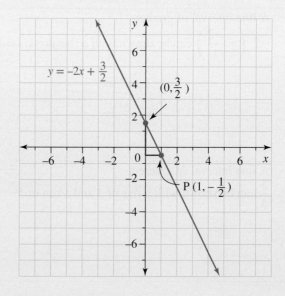

⊙ 7.5.3 Horizontal and vertical lines

eles-4788

Horizontal lines ($y = c$)

- Horizontal lines are expressed in the form $y = c$, where c is the y-intercept.
- In horizontal lines the y-value remains the same regardless of the x-value.
- This can be seen by looking at a table of values, like the one shown.

x	-2	0	2	4
y	c	c	c	c

- Plotting these points gives a horizontal line, as shown in the graph.
- Horizontal lines have a gradient of 0. They do not rise or fall.

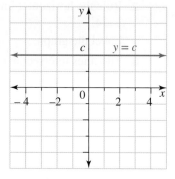

Vertical lines ($x = a$)

- Vertical lines are expressed in the form $x = a$, where a is the x-intercept.
- In vertical lines the x-value remains the same regardless of the y-value.
- This can be seen by looking at a table of values, like the one shown.

x	a	a	a	a
y	-2	0	2	4

- Plotting these points gives a vertical line, as shown in the graph.
- The run of the graph is 0, so using the formula $m = \dfrac{\text{rise}}{\text{run}}$ involves dividing by zero, which cannot be done.
  The gradient is said to be **undefined**.

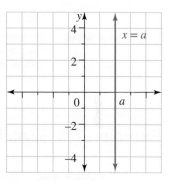

WORKED EXAMPLE 13 Sketching horizontal and vertical lines

On a pair of axes, sketch the graphs of the following equations and label the point of intersection of the two lines.

a. $x = -3$ 　　　　　　　　　　 b. $y = 4$

THINK

a. 1. The line $x = -3$ is in the form $x = a$.
   This is a vertical line.

2. Rule the vertical line where $x = -3$.
   Label the graph.

WRITE

a.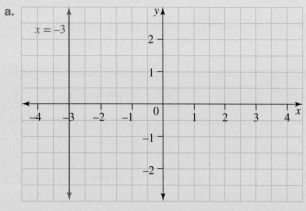

b. 1. The line $y = 4$ is in the form $y = c$.
This is a horizontal line.

2. Rule the horizontal line where $y = 4$.
Label the graph.

3. The lines intersect at $(-3, 4)$.

b.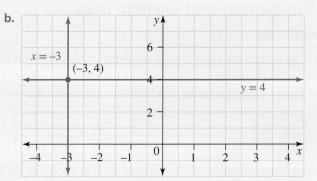

---

on Resources

▶ **Video eLessons** Sketching linear graphs (eles-1919)
Sketching linear graphs using the gradient–intercept method (eles-1920)

✦ **Interactivities** The intercept method (int-3840)
The gradient–intercept method (int-3839)
Vertical and horizontal lines (int-6049)

Exercise 7.5 Sketching linear graphs

learn on

7.5 Quick quiz **on**	7.5 Exercise

Individual pathways

■ PRACTISE	■ CONSOLIDATE	■ MASTER
1, 3, 5, 7, 9, 11, 15, 20	4, 8, 12, 13, 16, 17, 21	2, 6, 10, 14, 18, 19, 22

Fluency

1. **WE11** Using the x- and y-intercept method, sketch the graphs of the following equations.

   a. $5y - 4x = 20$
   b. $y = x + 2$
   c. $y = -3x + 6$

2. Using the x- and y-intercept method, sketch the graphs of the following equations.

   a. $3y + 4x = -12$
   b. $x - y = 5$
   c. $2y + 7x - 8 = 0$

3. **WE12** Using the gradient–intercept method, sketch the graphs of the following equations.

   a. $y = x - 7$
   b. $y = 2x + 2$
   c. $y = -2x + 2$

4. Using the gradient–intercept method, sketch the graphs of the following equations.

   a. $y = \dfrac{1}{2}x - 1$
   b. $y = 4 - x$
   c. $y = -x - 10$

5. **WE13** On a Cartesian plane, sketch the graphs of the following equations.

   a. $y = 4$
   b. $y = -3$

6. On a Cartesian plane, sketch the graphs of the following equations.

   a. $y = -12.5$

   b. $y = \dfrac{4}{5}$

7. Sketch the graphs of the following equations.

   a. $x = 2$

   b. $x = -6$

8. Sketch the graphs of the following equations.

   a. $x = -2.5$

   b. $x = \dfrac{3}{4}$

9. Sketch the graphs of the following equations.

   a. $y = 3x$

   b. $y = -2x$

10. Sketch the graphs of the following equations.

    a. $y = \dfrac{3}{4}x$

    b. $y = -\dfrac{1}{3}x$

Understanding

11. **MC** Select which of the following statements about the rule $y = 4$ is false.

    A. The gradient $m = 0$.
    B. The y-intercept is at $(0,\ 4)$.
    C. The graph is parallel to the x-axis.
    D. The point $(4,\ 2)$ lies on this graph.

12. **MC** Select which of the following statements is not true about the equation $y = -\dfrac{3}{5}x$.

    A. The graph passes through the origin.
    B. The gradient $m = -\dfrac{3}{5}$.
    C. The x-intercept is at $x = 0$.
    D. The graph can be sketched using the x- and the y-intercept method.

13. $2x + 5y = 20$ is a linear equation in the form $ax + by = c$.

    a. Rearrange this equation into the form $y = mx + c$.
    b. State the gradient.
    c. State the x- and y-intercepts.
    d. Sketch this straight line.

14. A straight line has an x-intercept of -3 and a y-intercept of 5.

    a. State the gradient.
    b. Draw the graph.
    c. Write the equation in the form:
       i. $y = mx + c$
       ii. $ax + by = c$.

Communicating, reasoning and problem solving

15. Consider the relationship $4x - 3y = 24$.

    a. Rewrite this relationship, making y the subject.
    b. Show that the x- and y-intercepts are at $(6, 0)$ and $(0, -8)$ respectively.
    c. Sketch a graph of this relationship.

16. Consider the relationship $ax + by = c$.

    a. Rewrite the relationship, making y the subject.

    b. If a, b and c are positive integer values, explain how the gradient is negative.

17. Josie accidently spilled a drink on her work. Part of her calculations were smudged. The line $y = \frac{1}{2}x + \frac{3}{4}$ was written in the form $ax + 4y = 3$. Show that the value of $a = -2$.

18. Explain why the descriptions 'right 3 up 2', 'right 6 up 4', 'left 3 down 2', 'right $\frac{3}{2}$ up 1' and 'left 1 down $\frac{2}{3}$,' all describe the same gradient for a straight line.

19. A straight line passes through the points (3, 5) and (6, 11).

    a. Determine the slope of the line

    b. Determine the equation of the line

    c. State the coordinates of another point that lies on the same line.

20. a. Match the descriptions given below with their corresponding line.

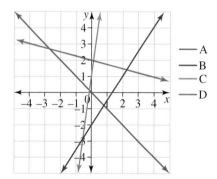

    i. Straight line with a y-intercept of 1 and a positive gradient

    ii. Straight line with a gradient of $1\frac{1}{2}$

    iii. Straight line with a gradient of -1

  b. Write a description for the unmatched graph.

21. a. Sketch the linear equation $y = -\frac{5}{7}x - \frac{3}{4}$:

    i. using the y-intercept and the gradient

    ii. using the x- and y-intercepts

    iii. using two other points.

  b. Compare and contrast the methods and generate a list of advantages and disadvantages for each method used in part **a**. Explain which method you think is best. Give your reasons.

22. Consider these two linear graphs.

$$y - ax = b \text{ and } y - cx = d.$$

Show that if these two graphs intersect where both x and y are positive, then $a > c$ when $d > b$.

LESSON
7.6 Technology and linear graphs

LEARNING INTENTION

At the end of this lesson you should be able to:
- graph lines using digital technologies
- investigate important features of linear graphs
- interpret gradient, including gradients of parallel and perpendicular lines.

7.6.1 Graphing with technology

eles-4789

- There are many digital technologies that can be used to graph linear relationships.
- The Desmos Graphing Calculator is a free graphing tool that can be found online.
- Other commonly used digital technologies include Microsoft Excel and graphing calculators.
- Digital technologies can help identify important features and patterns in graphs.

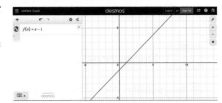

Sketching a linear graph using technology

- Depending on the choice of digital technology used, the steps involved to produce a linear graph may vary slightly.
- Most graphing calculators have an entry (or input) line to type in the equation of the line you wish to sketch.
- When using the Desmos Graphing Calculator you can simply type $y = x + 1$ into the input box to produce its graph.
- When using technologies such as a Texas Instruments or CASIO graphing calculator you will need to open a Graphs page first. The entry line for these types of technologies has a template that needs to be followed carefully. Both of these graphing calculators begin with $f1(x) =$ or $y1 =$, which is effectively saying $y =$.
- Thus, to draw the graph of $y = x + 1$, you would simply enter '$x + 1$'.
- The screen shows the graph of $y = x + 1$ sketched on a TI-Nspire CAS calculator.

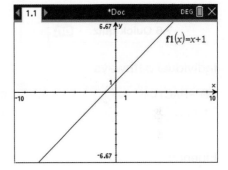

7.6.2 Graphing parallel lines

eles-4790

- Lines with equal gradients form **parallel lines**. For a pair of parallel lines, $m_1 = m_2$.
- Digital technologies can be used to help visualise this concept. For example, $y = 3x + 1$, $y = 3x - 4$ and $y = 3x$ are all parallel lines, because $m = 3$.
- Select the digital technology of your choice and sketch these three parallel lines on the same set of axes.
- If m_1 is not equal to m_2, then the two lines are not parallel and therefore the two lines intersect.

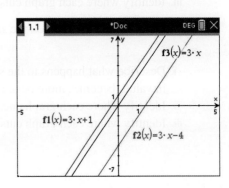

▶ 7.6.3 Graphing perpendicular lines

eles-4791

- A special case of intersecting lines is when they meet at a right angle.
- Lines that meet at right angles are called **perpendicular lines**.
- The product of the gradients of two perpendicular lines is equal to -1.
  For a pair of perpendicular lines, $m_1 \times m_2 = -1$.
- Digital technologies can be used to help visualise this concept.
  For example, $y = 2x + 1$ and $y = -\dfrac{1}{2}x + 6$ are perpendicular,

  because $2 \times -\dfrac{1}{2} = -1$.

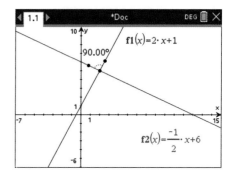

- Select a digital technology of your choice and sketch these two lines on the same set of axes.

---

on Resources

 Interactivity Parallel lines (int-3841)

Exercise 7.6 Technology and linear graphs

learn on

| 7.6 Quick quiz **on** | 7.6 Exercise |

Individual pathways

■ PRACTISE	■ CONSOLIDATE	■ MASTER
1, 4, 9, 13	2, 5, 6, 10, 12, 14	3, 7, 8, 11, 15

Fluency

Use technology wherever possible to answer the following questions.

1. Sketch the following graphs on the same Cartesian plane.

   a. $y = x$ b. $y = 2x$ c. $y = 3x$

   i. Describe what happens to the steepness of the graph (the gradient of the line) as the coefficient of x increases in value.

   ii. Identify where each graph cuts the x-axis (its x-intercept).

   iii. Identify where each graph cuts the y-axis (its y-intercept).

2. Sketch the following graphs on the same Cartesian plane.

   a. $y = -x$ b. $y = -2x$ c. $y = -3x$

   i. Describe what happens to the steepness of the graph as the magnitude of the coefficient of x decreases in value (becomes more negative).

   ii. Identify where each graph cuts the x-axis (its x-intercept).

   iii. Identify where each graph cuts the y-axis (its y-intercept).

3. Identify the correct word or words from the options given and rewrite the following sentences using the correct option.

   a. If the coefficient of x is (**positive/negative**), then the graph will have an upward slope to the right. That is, the gradient of the graph is (**positive/negative**).

   b. If the coefficient of x is negative, then the graph will have a (**downward/upward**) slope to the right. That is, the gradient of the graph is (**positive/negative**).

   c. The bigger the magnitude of the coefficient of x (more positive or more negative), the (**bigger/smaller**) the steepness of the graph.

   d. If there is no constant term in the equation, the graph (**will/will not**) pass through the origin.

Understanding

4. Sketch the graphs shown on the same Cartesian plane and answer the following questions for each graph.

   a. Is the coefficient of x the same for each graph? If so, state the coefficient.
   b. State whether the steepness (gradient) of each graph differs.
   c. Identify where each graph cuts the x-axis (its x-intercept).
   d. Identify where each graph cuts the y-axis (its y-intercept).

      i. $y = x$ ii. $y = x + 2$ iii. $y = x - 2$

5. Sketch the graphs shown on the same Cartesian plane and answer the following questions for each graph.

   a. Is the coefficient of x the same for each graph? If so, state the coefficient.
   b. State whether the steepness (gradient) of each graph differs.
   c. Identify where each graph cuts the x-axis (its x-intercept).
   d. Identify where each graph cuts the y-axis (its y-intercept).

      i. $y = -x$ ii. $y = -x + 2$ iii. $y = -x - 2$

6. Identify the correct word or words from the options given and rewrite the following sentences using the correct option.

   a. For a given set of linear graphs, if the coefficient of x is (**the same/different**), the graphs will be parallel.

   b. The constant term in the equation is the (**y-intercept/x-intercept**) or where the graph cuts the (**y-axis/x-axis**).

   c. The (**y-intercept/x-intercept**) can be found by substituting $x = 0$ into the equation.

   d. The (**y-intercept/x-intercept**) can be found by substituting $y = 0$ into the equation.

7. On the same Cartesian plane, sketch the following graphs.

   a. $y = x + 5$ b. $y = -x + 5$ c. $y = 3x + 5$ d. $y = -\dfrac{2}{5}x + 5$

      i. Is the coefficient of x the same for each graph? If so, state the coefficient.
      ii. State whether the steepness (gradient) of each graph differ.
      iii. Write down the gradient of each linear graph.
      iv. Identify where each graph cuts the x-axis (its x-intercept).
      v. Identify where each graph cuts the y-axis (its y-intercept).

8. Identify the correct word or words from the options given and rewrite the following sentences using the correct option.

   a. One of the general forms of the equation of a linear graph is $y = mx + c$, where m is the (**steepness/x-coordinate**) of the graph. We call the steepness of the graph the *gradient*.

   b. The value of c is the (**x-coordinate/y-coordinate**) where the graph cuts the (**x-axis/y-axis**).

   c. All linear graphs with the (**same/different**) gradient are (**parallel/perpendicular**).

   d. All linear graphs that have the same y-intercept pass through (**the same/different**) point on the y-axis.

9. For each of the following lines, identify the gradient and the y-intercept.

   a. $y = 2x$ b. $y = x + 1$ c. $y = -3x + 5$ d. $y = \dfrac{2}{3}x - 7$

Communicating, reasoning and problem solving

10. Using technology, draw the graphs of the lines $y = 3(x - 1) + 5$, $y = 2(x - 1) + 5$ and $y = -\dfrac{1}{2}(x - 1) + 5$. Describe what they have in common and how they differ from each other.

11. A phone company charges $2.20 for international calls of 1 minute or less and $0.55 for each additional minute. Using technology, draw a graph of the cost of calls that last for whole numbers of minutes. Explain all the important values needed to sketch the graph.

12. Shirly walks dogs after school for extra pocket money. She determines that she can use the equation $P = -15 + 10N$ to calculate her profit (in dollars) each week.

    a. Explain the real-world meaning of the numbers -15 and 10 and the variable N.
    b. What is the minimum number of dogs that Shirly must walk in order to earn a profit?
    c. Using technology, sketch the equation.

13. Plot the points $(6, 3.5)$ and $(-1, -10.5)$ using technology and:

    a. determine the equation of the line
    b. sketch the graph, showing x- and y-intercepts
    c. calculate the value of y when $x = 8$
    d. calculate the value of x when $y = 12$.

14. A school investigating the price of a site licence for their computer network found that it would cost $1750 for 30 computers and $2500 for 60 computers.

    a. Using technology, determine a linear equation that represents the cost of a site licence in terms of the number of computers in the school.
    b. Determine is the y-intercept of the linear equation and explain how it relates to the cost of a site licence.
    c. Calculate the cost for 200 computers.
    d. Evaluate how many computers you could connect for $3000.

15. Dylan starts his exercise routine by jogging to the gym, which burns 325 calories. He then pedals a stationary bike burning 3.8 calories a minute.

    a. Graph this data using technology.
    b. Evaluate how many calories Dylan has burnt after 15 minutes of pedalling.
    c. Evaluate how long it took Dylan to burn a total of 450 calories.

LESSON
7.7 Practical applications of linear graphs

LEARNING INTENTION

At the end of this lesson you should be able to:
- determine the linear rule from a table of values
- model real-life situations using equations.

7.7.1 Determining a linear rule from a table of values

cles-4792

- If two **variables** have a linear relationship, then as one variable increases, the other increases (or decreases) at a steady rate.
- The table shown gives an example of a linear relationship.

x	0	1	2	3
y	5	8	11	14

- Each time x increases by 1, y increases by 3.
- The linear rule relating x and y is $y = 3x + 5$.

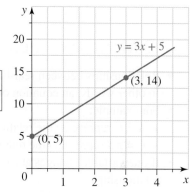

- Consider the relationship depicted in the table shown.

x	0	1	2	3
y	7	5	3	1

- Each time x increases by 1, y decreases by 2.
- The linear rule in this case is $y = -2x + 7$.

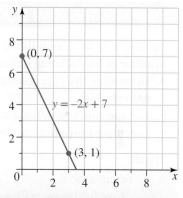

WORKED EXAMPLE 14 Determining a rule from a table of values

Determine the rule connecting x and y in each of the following table of values.

a.
x	0	1	2	3
y	-3	2	7	12

b.
x	3	4	5	6
y	12	11	10	9

THINK

a. 1. y increases at a steady rate, so this is a linear relationship. Write the rule.

2. To calculate m: y increases by 5 each time x increases by 1. Write the value of the gradient.

3. To determine c: From the table, when $x = 0$, $y = -3$. Write the value of the y-intercept.

4. Write the rule.

WRITE

a. $y = mx + c$

$m = 5$

$c = -3$

$y = 5x - 3$

b. 1. y decreases at a steady rate, so this is a linear relationship. Write the rule.

b. $y = mx + c$

2. To calculate m: decreases by 1 each time x increases by 1.
Write the value of the gradient.

$m = -1$

3. To determine c: From the table, when $x = 3$, $y = 12$.
To calculate the y-intercept, substitute the x- and y-values of one of the points, and solve for c.

$(3, 12):$ $y = -x + c$
$12 = -(3) + c$
$12 = -3 + c$
$c = 15$

4. Write the rule.

$y = -x + 15$

7.7.2 Modelling linear relationships

eles-4793

- Relationships between real-life variables are often modelled (described) by a mathematical equation. In other words, an equation or formula is used to link the two variables.
- For example:
  - $A = l^2$ represents the relationship between the area and the side length of a square
  - $C = \pi d$ represents the relationship between the circumference and the diameter of a circle.
- If one variable changes at a constant rate compared to the other, then the two variables have a linear relationship.

WORKED EXAMPLE 15 Modelling linear relationships in real-life problems

An online bookstore sells a certain textbook for \$21 and charges \$10 for delivery, regardless of the number of books being delivered.
a. Determine the rule connecting the cost (\$$C$) with the number of copies of the textbook delivered (n).
b. Use the rule to calculate the cost of delivering 35 copies of the textbook.
c. Calculate how many copies of the textbook can be delivered for \$1000.

THINK

a. 1. Set up a table.
Cost for 1 copy $= 21 + 10$
Cost for 2 copies $= 2(21) + 10$
$= 52$

2. The cost rises steadily, so there is a linear relationship. Write the rule.

3. To calculate the gradient, use the formula $m = \dfrac{y_2 - y_1}{x_2 - x_1}$ with the points $(1, 31)$ and $(2, 52)$.

4. To determine the value of c, substitute $C = 31$ and $n = 1$.

5. Write the rule.

WRITE

a.

n	1	2	3
C	31	52	73

$C = mn + c$

$m = \dfrac{52 - 31}{2 - 1}$
$= 21$
$C = 21n + c$

$(1, 31):$ $31 = 21(1) + c$
$c = 10$

$C = 21n + 10$

b. 1. Substitute $n = 35$ and calculate the value of C.	**b.** $C = 21(35) + 10$ $= 735 + 10$ $= 745$
2. Write the answer.	The cost including delivery for 35 copies is $\$745$.
c. 1. Substitute $C = 1000$ and calculate the value of n.	**c.** $1000 = 21n + 10$ $21n = 990$ $n = 47.14$
2. You cannot buy 47.14 books, so round down. Write the answer.	For $\$1000, 47$ copies of the textbook can be bought and delivered.

- In Worked example 15, compare the rule $C = 21n + 10$ with the original question. It is clear that the $21n$ refers to the cost of the textbooks (a variable amount, depending on the number of copies) and that 10 refers to the fixed (constant) delivery charge.
- In this case C is called the **dependent variable**, because it depends on the number of copies of the textbooks (n).
- The variable n is called the **independent variable** because it is the variable that may explain the changes in the dependent variable.
- When graphing numerical data, the dependent variable is plotted on the vertical axis and the independent variable is plotted on the horizontal axis.

 Resources

 Interactivity Dependent and independent variables (int-6050)

Exercise 7.7 Practical applications of linear graphs   learn on

7.7 Quick quiz on   **7.7 Exercise**

Individual pathways

■ PRACTISE	■ CONSOLIDATE	■ MASTER
1, 3, 5, 6, 11, 16	2, 7, 9, 12, 13, 17	4, 8, 10, 14, 15, 18

Fluency

1. **WE14** Determine the linear rule linking the variables in each of the following tables.

a.
x	0	1	2	3
y	-5	1	7	13

b.
x	0	1	2	3
y	8	5	2	-1

2. Determine the linear rule linking the variables in each of the following tables.

a.
x	0	1	2	3
y	4	6	8	10

b.
x	0	1	2	3
y	1.1	2.0	2.9	3.8

3. Determine the linear rule linking the variables in each of the following tables.

a.

x	2	3	4	5
y	7	10	13	16

b.

x	5	6	7	8
y	12	11	10	9

4. Determine the linear rule linking the variables in each of the following tables.

a.

t	3	4	5	6
v	18	15	12	9

b.

d	1	2	3	4
C	11	14	17	20

5. **MC** Sasha and Fiame hire a car. They are charged a fixed fee of $150 for hiring the car and then $25 per day. They hire the car for d days.

Select which one of the following rules describes the number of days the car is hired and the total cost, C, they would be charged for that number of days charged.

A. $C = 25d$ B. $C = 150d$ C. $C = 175d$ D. $C = 25d + 150$

Understanding

6. **WE15** Fady's bank balance has increased in a linear manner since he started his part-time job. After 20 weeks of work his bank balance was at $560 and after 21 weeks of work it was at $585.

   a. Determine the rule that relates the size of Fady's bank balance, A, and the time (in weeks) worked, t.
   b. Use the rule to calculate the amount in Fady's account after 200 weeks.
   c. Use the rule to identify the initial amount in Fady's account.

7. The cost of making a shoe increases as the size of the shoe increases. It costs $5.30 to make a size 6 shoe, and $6.40 to make a size 8 shoe. Assuming that a linear relationship exists:

   a. determine the rule relating cost (C) to shoe size (s)
   b. calculate much it costs to produce a size 12 shoe.

8. The number of books in a library (N) increases steadily with time (t). After 10 years there are 7200 publications in the library, and after 12 years there are 8000 publications.

   a. Determine the rule predicting the number of books in the library.
   b. Calculate how many books were there after 5.5 years.
   c. Calculate how many books will there be after 25 years.

9. A skyscraper can be built at a rate of 4.5 storeys per month.

   a. Calculate how many storeys will be built after 6 months.
   b. Calculate how many storeys will be built after 24 months.

10. The Nguyens' water tank sprang a leak and has been losing water at a steady rate. Four days after the leak occurred, the tank contained 552 L of water, and ten days later it held only 312 L.

   a. Determine the rule linking the amount of water in the tank (w) and the number of days (t) since the leak occurred.
   b. Calculate the amount of water that was initially in the tank.
   c. If water loss continues at the same rate, determine when the tank will be empty.

Communicating, reasoning and problem solving

11. The pressure inside a boiler increases steadily as the temperature increases. For each 1 °C increase in temperature, the pressure increases by 10 units, and at a temperature of 100 °C the pressure is 1200 units. If the maximum pressure allowed is 2000 units, show that the temperature cannot exceed 180 °C.

12. After 11 pm a taxi company charges a $3.50 flag fall plus $2.57 for each kilometre travelled.
    a. Determine the linear rule connecting the cost, C, and the distance travelled, d.
    b. Calculate how much an 11.5 km trip will cost.
    c. If you have $22 in your pocket, calculate how far you can afford to travel, correct to 1 decimal place.

13. A certain kind of eucalyptus tree grows at a linear rate for its first 2 years of growth. If the growth rate is 5 cm per month, show that the tree will be 1.07 m tall after 21.4 months.

14. A software company claims that its staff can fix 22 bugs each month. They are working on a project to fix a program that started out with 164 bugs.
    a. Determine the linear rule connecting the number of bugs left, N, and the time in months, t, from the beginning of the project.
    b. Calculate how many bugs will be left after 2 months.
    c. Determine how long it will be until there are only 54 bugs left.
    d. Determine how long it will take to eliminate all of the bugs. Justify your answer.

15. Michael produces and sells prints of his art at a local gallery. For each print run his profit (P) is given by the equation $P = 200n - 800$, where n is the number of prints sold.
    a. Sketch the graph of this rule.
    b. Identify the y-intercept. Determine what it represents in this situation.
    c. Identify the x-intercept. Determine what it represents in this example.
    d. Identify the gradient of the graph. Determine what this means in this situation.

16. The cost of a taxi ride is $3.50 flag fall plus $2.14 for each kilometre travelled.
    a. Determine the linear rule connecting the cost, C, and the distance travelled, d.
    b. Calculate how much an 11.5 km trip will cost.
    c. Calculate how much a 23.1 km trip will cost.

17. Theo is going on holiday to Japan. One yen (¥) buys 0.0127 Australian dollars (A$).
    a. Write an equation that converts Australian dollars to Japanese yen (¥), where A represents amount of Australian dollars and Y represents amount of yen.
    b. Using the equation from part a, calculate how many yen Theo will receive in exchange if they have A$2500.
    c. There is a commission to be paid on exchanging currency. Theo needs to pay 2.8% for each Australian dollar they exchange into yen. Write down an equation that calculates the total amount of yen Theo will receive. Write your equation in terms of Y_T (total amount of yen) and $A (Australian dollars).

18. Burchill and Cody need to make a journey to the other branch of their store across town. The traffic is very busy at this time of the day, so Burchill catches a train that travels halfway, then walks the rest of the way. Cody travels by bike the whole way. The bike path travels along the train line and then follows the road to the other branch of their store.
    Cody's bike travels twice as fast as Burchill's walking speed, and the train travels 4 times faster than Cody's bike. Evaluate who arrives at the destination first.

LESSON
7.8 Midpoint of a line segment and distance between two points

7.8.1 Calculating the midpoint of a line segment

eles-4794
- The point halfway between two end points of a line segment is called the **midpoint**.

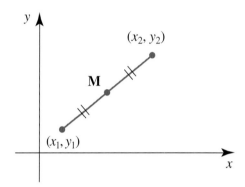

- The coordinates of the midpoint, M, of a line can be found by calculating the average of the x- and y-coordinates of the end points.
- The diagram shows two points, $(4, 3)$ and $(10, 7)$. We want to determine the midpoint, M.
  - Consider the x-coordinates.
    The average of 4 and 10 is $\dfrac{4 + 10}{2} = 7$.
  - Consider the y-coordinates.
    The average of 3 and 7 is $\dfrac{3 + 7}{2} = 5$.
  - The midpoint, M, of the interval is $(7, 5)$.
- Similarly, for two general points, (x_1, y_1) and (x_2, y_2), the midpoint, M (x, y), can be found by averaging the x- and y-coordinates.
  - $x = \dfrac{x_1 + x_2}{2}$
  - $y = \dfrac{y_1 + y_2}{2}$

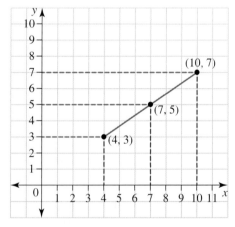

The midpoint of a line segment (Path)

The midpoint, M, of the line segment joining the points (x_1, y_1) and (x_2, y_2) is:

$$M = \left(\frac{x_1 + x_2}{2}, \frac{y_1 + y_2}{2} \right)$$

WORKED EXAMPLE 16 Calculating the midpoint of a line segment

Calculate the midpoint of the line segment joining $(5,\ 9)$ and $(-3, 11)$.

THINK	WRITE
1. Average the x-values: $\dfrac{x_1 + x_2}{2}$.	$\begin{aligned} x &= \frac{5 - 3}{2} \\ &= \frac{2}{2} \\ &= 1 \end{aligned}$
2. Average the y-values: $\dfrac{y_1 + y_2}{2}$.	$\begin{aligned} y &= \frac{9 + 11}{2} \\ &= \frac{20}{2} \\ &= 10 \end{aligned}$
3. Write the answer.	The midpoint is $(1, 10)$.

WORKED EXAMPLE 17 Determining the coordinates of a point given the midpoint and another point of an interval

M $(7, 2)$ is the midpoint of the line segment AB. If the coordinates of A are $(1, -4)$, determine the coordinates of B.

THINK	WRITE
1. Let B have the coordinates (x, y).	A $(1, -4)$, B (x, y), M $(7, 2)$
2. The midpoint is $(7,\ 2)$, so the average of the x-values is 7. Solve for x.	$\begin{aligned} \frac{1 + x}{2} &= 7 \\ 1 + x &= 14 \\ x &= 13 \end{aligned}$
3. The average of the y-values is 2. Solve for y.	$\begin{aligned} \frac{-4 + y}{2} &= 2 \\ -4 + y &= 4 \\ y &= 8 \end{aligned}$
4. Write the answer.	The coordinates of point B are $(13, 8)$.

⏵ 7.8.2 The distance between two points

eles-4795

- The distance between two points on the Cartesian plane is calculated using Pythagoras' theorem applied to a right-angled triangle.
- The diagram shows two points $(4, 3)$ and $(7, 8)$. The length of the interval joining the points can be found by drawing a right-angled triangle with the interval as the hypotenuse.
  - The horizontal distance from 4 to 7 is 3 since $7 - 4 = 3$.
  - The vertical distance from 3 to 8 is 5 since $8 - 3 = 5$.
  - Let the length of the hypotenuse be d.
    Using Pythagoras' theorem:

$$d^2 = 3^2 + 5^2$$
$$d = \sqrt{9 + 25}$$
$$= \sqrt{34}$$

- The distance between the points is exactly $\sqrt{34}$, or approximated to 5.83 (to 2 d.p.).

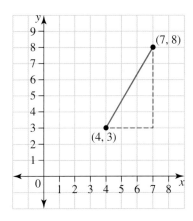

- Similarly, the distance between two general points (x_1, y_1) and (x_2, y_2) can be found using Pythagoras' theorem.

Distance between two points (Path)

The distance, d, of the line segment joining the points (x_1, y_1) and (x_2, y_2) is:

$$d = \sqrt{(x_2 - x_1)^2 + (y_2 - y_1)^2}$$

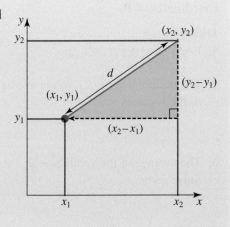

Calculate the distance between the points $(-1, 3)$ and $(4, 5)$:

a. exactly

b. correct to 3 decimal places.

THINK

WRITE

a. 1. Draw a diagram showing the right-angled triangle and determine the horizontal and vertical distances.

a.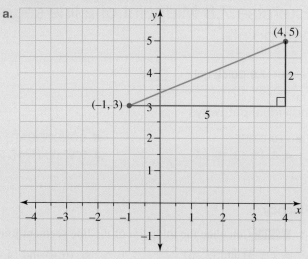

2. Let the distance between the two points be d and apply Pythagoras' theorem.

$$d^2 = 2^2 + 5^2$$
$$= 4 + 25$$
$$= 29$$

3. Write the answer.

$$d = \sqrt{29}$$

Alternatively, using the distance formula (Path):

1. Write the formula for the distance between two points.

$$d = \sqrt{(x_2 - x_1)^2 + (y_2 - y_1)^2}$$

2. Let $(x_1, y_1) = (-1, 3)$ and $(x_2, y_2) = (4, 5)$. Substitute the x- and y-values into the equation.

$$d = \sqrt{(4 - -1)^2 + (5 - 3)^2}$$

3. Simplify.

$$d = \sqrt{5^2 + 2^2}$$
$$d = \sqrt{25 + 4}$$
$$d = \sqrt{29}$$

4. Write the answer.

The exact distance between points $(-1, 3)$ and $(4, 5)$ is $\sqrt{29}$.

b. 1. Write $\sqrt{29}$ as a decimal to 4 decimal places.

b. $\sqrt{29} = 5.3851$

2. Write the answer correct to 3 decimal places.

$d \approx 5.385$

Exercise 7.8 Midpoint of a line segment and distance between two points

learnon

| **7.8 Quick quiz** | **7.8 Exercise** |

Individual pathways

■ PRACTISE	■ CONSOLIDATE	■ MASTER
1, 4, 8, 11, 13, 16, 20	2, 5, 9, 12, 14, 17, 21	3, 6, 7, 10, 15, 18, 19, 22

Fluency

1. **WE16** Calculate the midpoint of the line segment joining each of the following pairs of points.

   a. $(1, 3)$ and $(3, 5)$
   b. $(6, 4)$ and $(4, -2)$
   c. $(2, 3)$ and $(12, 1)$

2. Calculate the midpoint of the line segment joining each of the following pairs of points.

   a. $(6, 3)$ and $(10, 15)$
   b. $(4, 2)$ and $(-4, 8)$
   c. $(0, -5)$ and $(-2, 9)$

3. Calculate the midpoint of the line segment joining each of the following pairs of points.

   a. $(8, 2)$ and $(-18, -6)$
   b. $(-3, -5)$ and $(7, 11)$
   c. $(-8, -3)$ and $(8, 27)$

4. Calculate the midpoint of the segment joining each of the following pairs of points.

   a. $(7, -2)$ and $(-4, 13)$
   b. $(0, 22)$ and $(-6, -29)$
   c. $(-15, 8)$ and $(-4, 11)$
   d. $(-3, 40)$ and $(0, -27)$

5. **WE17** Determine the value of a in each series of points so that the point M is the midpoint of the line segment joining points A and B.

   a. A $(-2, a)$, B $(-6, 5)$, M $(-4, 5)$
   b. A $(a, 0)$, B $(7, 3)$, M $\left(8, \dfrac{3}{2}\right)$

   c. A $(3, 3)$, B $(4, a)$, M $\left(3, \dfrac{1}{2}, -6\dfrac{1}{2}\right)$
   d. A $(-4, 4)$, B $(a, 0)$, M $(-2, 2)$

6. M is the midpoint of the line interval AB. Determine the coordinates of B if:

   a. A $= (0, 0)$ and M $= (2, 3)$
   b. A $= (2, 3)$ and M $= (0, 0)$
   c. A $= (-3, 2)$ and M $= (4, 2)$
   d. A $= (3, -1)$ and M $= (-2, -2)$.

7. Determine the equation of a line that has a gradient of 5 and passes through the midpoint of the line segment joining $(-1, -7)$ and $(3, 3)$.

8. **WE18** Calculate the distance between each of the following pairs of points.

   a. $(4, 5)$ and $(1, 1)$
   b. $(7, 14)$ and $(15, 8)$
   c. $(2, 4)$ and $(2, 3)$

9. Calculate the distance between each of the following pairs of points.

   a. $(12, 8)$ and $(10, 8)$ b. $(14, 9)$ and $(2, 14)$ c. $(5, -13)$ and $(-3, -7)$

10. Calculate the distance between each of the following pairs of points.

    a. $(-14, -9)$ and $(-10, -6)$ b. $(0, 1)$ and $(-15, 9)$ c. $(-4, -8)$ and $(1, 4)$

11. Calculate the distance between the following pairs of points, correct to 3 decimal places.

    a. $(-14, 10)$ and $(-8, 14)$ b. $(6, -7)$ and $(13, 6)$ c. $(-11, 1)$ and $(2, 2)$

12. Calculate the distance between the following pairs of points, correct to 3 decimal places

    a. $(9, 0)$ and $(5, -8)$ b. $(2, -7)$ and $(-2, 12)$ c. $(9, 4)$ and $(-10, 0)$

Understanding

13. Calculate the perimeter of each figure shown, giving your answers correct to 3 decimal places.

a.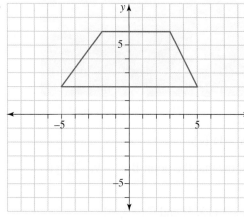

b.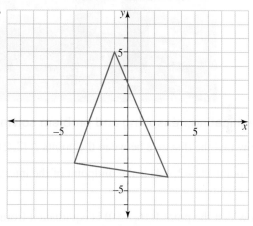

14. Calculate the perimeter of each triangle shown, giving your answers correct to 3 decimal places.

a.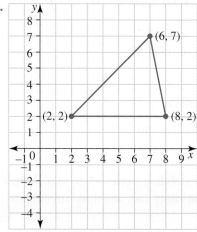

b.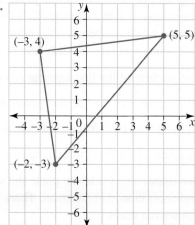

15. Two hikers are about to hike from A to B (shown on the map). Calculate the straight-line distance from A to B.

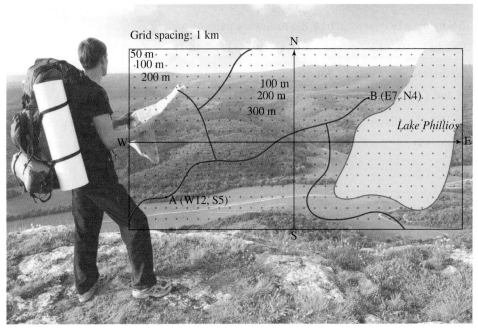

Grid spacing: 1 km

A (W12, S5)

B (E7, N4)

Lake Phillios

Communicating, reasoning and problem solving

16. Show that the distance between the points A $(2, 2)$ and B $(6, -1)$ is 5.

17. The point M $(-2, -4)$ is the midpoint of the interval AB. Show that the point B is $(-9, -2)$, given A is $(5, -6)$.

18. Show that the point B $(6, -10)$ is equidistant from the points A $(15, 3)$ and C $(-7, -1)$.

19. Answer the following questions.
    a. Plot the following points on a Cartesian plane: A $(-1, -4)$, B $(2, 3)$, C $(-3, 8)$ and D $(4, -5)$.
    b. Show that the midpoint of the interval AC is $(-2, 2)$.
    c. Calculate the exact distance between the points A and C.
    d. If B is the midpoint of an interval CM, determine the coordinates of point M.
    e. Show that the gradient of the line segment AB is $\dfrac{7}{3}$.
    f. Determine the equation of the line that passes through points B and D.

20. Explain what type of triangle $\triangle ABC$ is if it has vertices A $(-4, 1)$, B $(2, 3)$ and C $(0, -3)$.

21. Calculate the gradient of the line through the points $(-1, \ 3)$ and $(3 + 4t, 5 + 2t)$.

22. A map of a town drawn on a Cartesian plane shows the main street extending from $(-4, 5)$ to $(0, -7)$. There are five streetlights positioned in the street. There is one streetlight at either end, and three streetlights spaced evenly down the street.
    Give the position of the five lights in the street.

LESSON
7.9 Review

7.9.1 Topic summary

LINEAR RELATIONSHIPS

Gradient and intercepts

- The gradient of a straight line is given the label m. It can be calculated using the following formula:

$$m = \frac{\text{rise}}{\text{run}} = \frac{y_2 - y_1}{x_2 - x_1}$$

- If the line is sloping upwards (from left to right), the gradient is positive.
- If the line is sloping downwards, the gradient is negative.
- The y-intercept of a line can be determined by letting $x = 0$, then solving for y.
- The x-intercept of a line can be determined by letting $y = 0$, then solving for x.

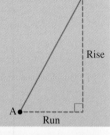

Equation of a straight line

- The equation of a straight line is $y = mx + c$ where:
  - m is the gradient
  - c is the y-intercept.

Sketching linear graphs

- To sketch a linear graph, all you need are two points that the line passes through.
- The x- and y-intercept method involves determining both axis intercepts, then drawing the line through them.
- The gradient and y-intercept method involves plotting the y-intercept and then one other point (usually $x = 1$) using the gradient.

Horizontal and vertical lines

- Horizontal lines are of the form $y = c$. They have a gradient of zero ($m = 0$).
- Vertical lines are of the form $x = a$. They have an undefined gradient.

Midpoint and distance

- The midpoint of two points is found by averaging the x- and y- coordinates.
- The distance between two points is found using Pythagoras' theorem.

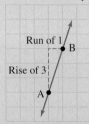 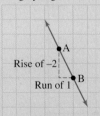 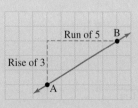

Midpoint and distance (Path)

- The midpoint of two points, (x_1, y_1) and (x_2, y_2), is

$$M = \left(\frac{x_2 + x_1}{2}, \frac{y_2 + y_1}{2} \right).$$

- The distance between two points, (x_1, y_1) and (x_2, y_2), is

$$d = \sqrt{(x_2 - x_1)^2 + (y_2 - y_1)^2}.$$

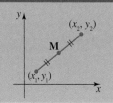

7.9.2 Project

Path of a billiard ball

The path of a billiard ball can be mapped using mathematics. A billiard table can be represented by a rectangle, a ball by a point and its path by line segments. In this investigation, we will look at the trajectory of a single ball, unobstructed by other balls. *Note:* A billiard table has a pocket at each of its corners and one in the middle of each of its long sides.

Consider the path of a ball that is hit on its side, from the lower left-hand corner of the table, so that it travels at a 45° angle from the corner of the table. Assume that the ball continues to move, rebounding from the sides and stopping only when it comes to a pocket. Diagrams of the table drawn on grid paper are shown. Each grid square has a side length of 0.25 metres.

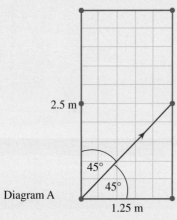

Diagram A

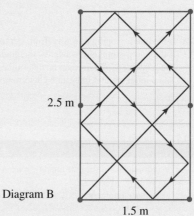

Diagram B

Diagram A shows the trajectory of a ball that has been hit on its side at a 45° angle, from the lower left-hand corner of a table that is 2.5 m long and 1.25 m wide. Note that, because the ball has been hit on its side at a 45° angle, it travels diagonally through each square in its path, from corner to corner. Diagram B shows the trajectory of a ball on a 2.5 m by 1.25 m table.

1. In Diagram B, how many times does the ball rebound off the sides before going into a pocket?
2. A series of eight tables of different sizes are drawn on grid paper in the following diagrams. For each table, determine the trajectory of a ball hit at 45° on its side, from the lower left-hand corner of the table. Draw the path each ball travels until it reaches a pocket.

a.

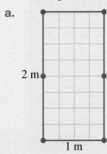

b.

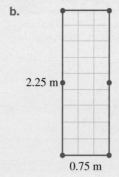

c.

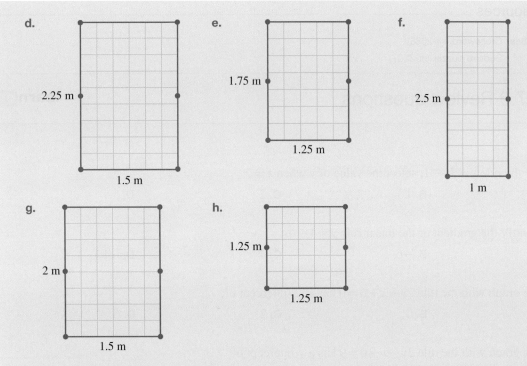

d.   2.25 m   1.5 m

e.   1.75 m   1.25 m

f.   2.5 m   1 m

g.   2 m   1.5 m

h.   1.25 m   1.25 m

3. Determine for which tables the ball will travel through the simplest path. What is special about the shape of these tables?

4. Determine for which table the ball will travel through the most complicated path. What is special about this path? Draw another table (and path of the ball) with the same feature.

5. Determine for which tables the ball will travel through a path that does not cross itself. Draw another table (and path of the ball) with the same feature.

6. Consider the variety of table shapes. Will a ball hit on its side from the lower left-hand corner of a table at 45° always end up in a pocket (assuming it does not run out of energy)?

Simplify matters a little and consider a billiard table with no pockets in the middle of the long sides. Use a systematic way to look for patterns for tables whose dimensions are related in a special way.

7. Draw a series of billiard tables of length 3 m. Increase the width of these tables from an initial value of 0.25 m in increments of 0.25 m. Investigate the final destination (the pocket the ball lands in) of a ball hit from the lower left-hand corner. Complete the table.

	Length of table (m)	Width of table (m)	Destination pocket
a.	3	0.25	
b.	3	0.5	
c.	3	0.75	
d.	3	1	
e.	3	1.25	
f.	3	1.5	
g.	3	1.75	
h.	3	2	

8. How can you predict (without drawing a diagram) the destination pocket of a ball hit from the lower left-hand corner of a table that is 3 m long? Provide an illustration to verify your prediction.

Exercise 7.9 Review questions

Fluency

1. **MC** For the rule $y = 3x - 1$, select the value of y when $x = 2$.

   A. -1 B. 1 C. 2 D. 5

2. **MC** Identify the gradient of the linear rule $y = 4 - 6x$.

   A. 6 B. -6 C. 4 D. -4

3. **MC** The graph with the rule $2y - x + 6 = 0$ has an x-intercept of:

   A. -2 B. 0 C. 2 D. 6

4. **MC** The graph with the rule $2y - x + 6 = 0$ has a y-intercept of:

   A. -6 B. -3 C. 0 D. 3

5. **MC** Consider a linear graph that goes through the points $(6, -1)$ and $(0, 5)$. The gradient of this line is:

   A. 5 B. -5 C. 1 D. -1

6. **MC** A straight line passes through the points $(2, 1)$ and $(5, 4)$. Its rule is:

   A. $y = x - 1$ B. $y = x + 1$ C. $y = 2x$ D. $4y = 5x$

7. **MC** The rule for a line whose gradient is -4 and y-intercept $= 8$ is:

   A. $y = -4x + 32$ B. $y = -4x + 8$ C. $y = 4x - 32$ D. $y = 4x - 8$

8. **MC** Identify which of the following linear rules will *not* intersect with the straight line defined by $y = 3x$.

   A. $y = 3x + 2$ B. $y = -3x + 1$ C. $y = -3x + 2$ D. $y = -2x + 1$

9. **MC** If $y = 2x + 1$, select from the following the point that could not be on the line.

   A. $(3, 7)$ B. $(-3, -5)$ C. $(0, 1)$ D. $(-3, 0)$

10. **MC** The solution to $y = 3x + 1$ and $y = -3x + 1$ is:

    A. $(0, 1)$ B. $(1, 0)$ C. $\left(0, -\dfrac{1}{3}\right)$ D. $\left(-\dfrac{1}{3}, 0\right)$

11. Write down the gradient and the y-intercept of the following linear graphs.

    a. $y = 8x - 3$ b. $y = 5 - 9x$ c. $2x + y - 6 = 0$

    d. $4x - 2y = 0$ e. $y = \dfrac{2x - 1}{3}$

12. **MC** Determine which of the following lines are parallel to the line with the equation $y = 6 - x$. There could be more than one correct answer.

   A. $y + x = 4$ **B.** $y = 13 - x$ **C.** $2y - 2x = 1$ **D.** $x + 2y - 4 = 0$

13. **MC** Determine which of the following rules will yield a linear graph. There could be more than one correct answer.

   A. $3y = -5x - \dfrac{1}{2}$ **B.** $y = 3x^2 - 2\dfrac{1}{2}$ **C.** $3x + 4y + 6 = 0$ **D.** $2y = 4^2x + 9^2$

Understanding

14. Determine the gradients of the lines shown.

   a.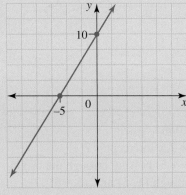

   b.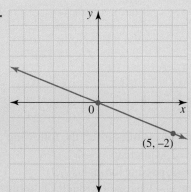

   c.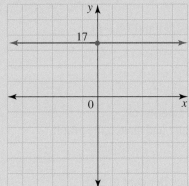

15. Calculate the gradients of the lines passing through the following pairs of points.

   a. $(2, -3)$ and $(4, 1)$ **b.** $(0, -5)$ and $(4, 0)$.

16. For each of the following rules, state the gradient and the y-intercept.

   a. $y = -3x + 7$ **b.** $2y - 3x = 6$ **c.** $y = -\dfrac{2}{5}x$ **d.** $y = 4$

17. For the following rules, use the gradient–intercept method to sketch linear graphs.

   a. $y = -x + 5$ **b.** $y = 4x - 2.5$ **c.** $y = \dfrac{2}{3}x - 1$ **d.** $y = 3 - \dfrac{5}{4}x$

18. For the following rules, use the x- and y-intercept method to sketch linear graphs.

   a. $y = -6x + 25$ **b.** $y = 20x + 45$ **c.** $2y + x = -5$ **d.** $4y + x - 2.5 = 0$

19. For the following rules, use an appropriate method to sketch linear graphs.

    a. $y = -3x$ **b.** $y = \dfrac{1}{4}x$ **c.** $y = -2$ **d.** $x = 3$

20. Determine the ruled of the lines shown.

    a.

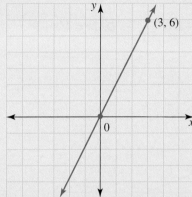

    b.

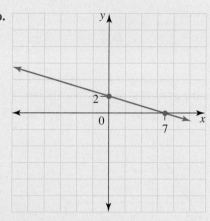

21. Determine the linear rules given the following pieces of information.

    a. Gradient $= 2$, y-intercept $= -7$ **b.** Gradient $= 2$, x-intercept $= 7$

    c. Gradient $= 2$, passing through $(7, 9)$ **d.** Gradient $= -5$, passing through the origin

    e. y-intercept $= -2$, passing through $(1, -3)$ **f.** Passing through $(1, 5)$ and $(5, -6)$

    g. x-intercept $= 3$, y-intercept $= -3$ **h.** y-intercept $= 5$, passing through $(-4, 13)$

22. Determine the midpoint of the line interval joining the points $(-2, 3)$ and $(4, -1)$.

23. Calculate the distance between the points $(1, 1)$ and $(4, 5)$.

Communicating, reasoning and problem solving

24. Louise owes her friend Sula $400 and agrees to pay her back $15 per week.

    a. State a linear rule that demonstrates this debt reduction schedule and sketch the graph.

    b. Calculate how many weeks it takes Louise to repay the debt.

    c. Calculate how much she owes after 15 weeks.

    d. Evaluate how many repayments Louise needs to make to reduce her debt to $85.

25. A bushwalker is 40 km from their base camp when they decide to head back.
    If they are able to walk 3.5 km each hour:

    a. determine the linear rule that describes this situation and sketch its graph

    b. calculate how long, correct to 1 decimal place, it will take them to reach base camp

    c. calculate how far they will have walked in 6.5 hours.

26. Udaya is writing test questions. She has already written 25 questions and can write a further 5 questions per hour.
    a. Represent this information as a linear equation where t hours is the time spent writing test questions and n is the number of questions written.
    b. Predict the total number of questions written after a further 8 hours assuming the same linear rule.
    c. Calculate how long, to the nearest minute, it will take Udaya to have written 53 questions.

27. Catherine earns a daily rate of $200 for working in her mother's store. She receives $5 for each necklace that she sells.
    a. Write an equation to show how much money (m) Catherine earned for the day after selling (n) necklaces.
    b. Draw a graph of the equation that you created in part a, showing the two intercepts.
    c. Explain which part of the line applies to her earnings.
    d. Explain which part of the line does not apply to her earnings.

28. Calculate the gradient of the line passing through the points $(2, 3)$ and $(6 + 4t, 5 + 2t)$. Write your answer in simplest form.

29. Determine the point on the line $y = 2x + 7$ that is also 5 units above the x-axis.

30. An experiment was conducted to collect data for two variables p and t. The data are presented in the table shown.

p	$-\dfrac{1}{2}$		$\dfrac{1}{2}$	3.6
t	$2\dfrac{1}{4}$	1.75		-5.95

It is known that the relationship between p and t is a linear one. Determine the two values missing from the table.

on   To test your understanding and knowledge of this topic go to your learnON title at www.jacplus.com.au and complete the **post-test**.

Answers

Topic 7 Linear relationships

7.1 Pre-test

1. On the x-axis (because $y = 0$).
2. a. C b. D
   c. B d. A
3. C
4. -0.5
5. D
6. a. B b. C
   c. D d. A
7. a. $y = -\dfrac{1}{4}x$ b. $y = 20x + 400$
8. D
9. a. $y = -x + 2$ b. $(-1, 3)$
10. $a = -5$
11. $\sqrt{29}$
12. 5
13. False
14. y-intercept $= 4$, or point $(0, 4)$
15. D

7.2 Plotting linear graphs

1. a. B b. C c. D
2. a. i. $y = x$

x	-3	-2	-1	0	1	2	3
y	-3	-2	-1	0	1	2	3

ii. $y = x$

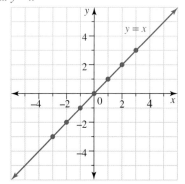

b. i. $y = 2x + 2$

x	-3	-2	-1	0	1	2	3
y	-4	-2	0	2	4	6	8

ii. $y = 2x + 2$

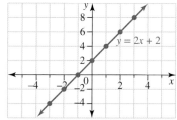

c. i. $y = 3x - 1$

x	-3	-2	-1	0	1	2	3
y	-10	-7	-4	-1	2	5	8

ii. $y = 3x - 1$

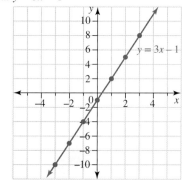

d. i. $y = -2x$

x	-3	-2	-1	0	1	2	3
y	6	4	2	0	-2	-4	-6

ii. $y = -2x$

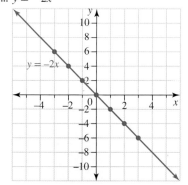

3. a. Sample points shown; $y = -x$

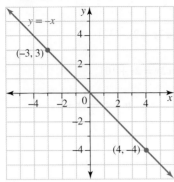

b. Sample points shown; $y = \frac{1}{2}x + 4$

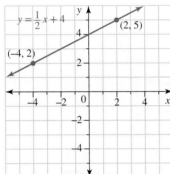

c. Sample points shown; $y = -2x + 3$

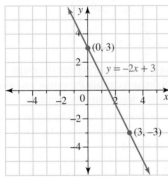

d. Sample points shown; $y = x - 3$

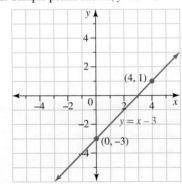

4. a. Yes   **b.** No   **c.** No   **d.** Yes

5. a. Yes   **b.** Yes   **c.** Yes
    d. No   **e.** No   **f.** No

6. D

7. a. B   **b.** A   **c.** D   **d.** C

8. By plotting the points on the Cartesian plane and joining them to make a line it can be seen that this line does not pass through the third quadrant.

9. a. The first, second and third quadrants.

    b. Answers will vary. Sample response: Substitute point $(1, 3)$ into the line equation. $3 \neq 1 + 1$.

10. Answers will vary. Sample response: Substitute the point into the line equation. If the LHS equals the RHS of the equation, then the point lies on the line. Alternatively, draw the graph of the line and check the point on the Cartesian plane.

11. Answers will vary. Sample response: Plot the points on the Cartesian plane and draw a line through the points. Determine the equation of the line. Check all points lie on the line by substituting into the line equation.

12. $(59, 21)$

13. a.

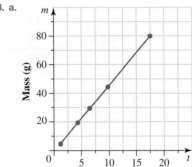

    b. $m = 5t + 1$

    c. Sample responses can be found in the worked solutions in the online resources.

14. Stirring increases the rate of reaction.

7.3 Features of linear graphs

1. a. 4   **b.** 1   **c.** 0

2. a. 20   **b.** 400   **c.** -5.2

3. a. 0   **b.** -5 **c.** -300

4. a. 6   **b.** -4 **c.** 0
    d. $\frac{6}{5}$ **e.** -2.2 **f.** -45

5. a. 0   **b.** 0.1   **c.** -2.6
    d. 0   **e.** 9   **f.** -0.1

6. a. 23   **b.** 54   **c.** -6 **d.** 70

7. a. 1   **b.** 5.2   **c.** 100   **d.** 100

8. C

9. y-intercept $= c$

10. a. $(0, -10)$ **b.** $(0, -4)$ **c.** $(0, 3)$
    d. $(0, 7)$ **e.** No y-intercept

11. There is a rise and no run; $\frac{\text{rise}}{0} = $ undefined.

12. There is no rise and a run; $\frac{0}{\text{run}} = 0$

13. Sample responses can be found in the worked solutions in the online resources.

14. It does not matter if you rise before you run or run before you rise, as long as you take into account whether the rise or run is negative.

15. a. $m = \dfrac{y-c}{x}$ **b.** $y = mx + c$

16. a. Lamb: $m = \$15/\text{kg}$, Chicken: $m = \$10/\text{kg}$,
Beef: $m = \$7.50/\text{kg}$

b. Lamb: $\$15$, Chicken: $\$10$, Beef: $\$7.50$

c. i. $\$15$ **ii.** $\$5$ **iii.** $\$15$

d. $\$35$

e.

Meat type	Cost per kilogram ($/kg)	Weight required (kg)	Cost = $/kg × kg
Lamb	15	1	15
Chicken	10	0.5	5
Beef	7.50	2	15
		Total cost	35

17. a. $m = -\dfrac{3}{2}$

b. It does not matter which points are chosen to determine the gradient of the graph because the gradient will always remain the same.

c. Straight line with a y-intercept of $(0, -2)$ and a slope of $-\dfrac{3}{2}$

7.4 The equation of a straight line, $y = mx + c$

1. a. $y = 4x + 2$ **b.** $y = -4x + 1$ **c.** $y = 4x + 8$

2. a. $y = 6x + 7$
b. $y = -2.5x + 6$
c. $y = 45x + 135$

3. a. $y = -2x$ **b.** $y = 4x$ **c.** $y = 10.5x$

4. a. $y = -20x$ **b.** $y = 1.07x$ **c.** $y = 32x$

5. a. $y = x + 2$ **b.** $y = -x + 8$
c. $y = -4x - 8$ **d.** $y = 2x - 13$

6. a. $y = -5x + 70$ **b.** $y = 2x - 23$
c. $y = -6x + 11$ **d.** $y = -x - 1.5$

7. a. $y = 6x + 30$ **b.** $y = -3.5x + 15.5$
c. $y = 1.2x - 4.08$ **d.** $y = 0.2x - 1.76$

8. a. $y = -4x - 24$ **b.** $y = 2x - 6$
c. $y = -2x + 4$

9. a. $y = 5x + 35$ **b.** $y = 1.5x - 3.75$
c. $y = 0.4x - 0.96$

10. a. $y = x + 17$ **b.** $y = -x + 3$
c. $y = 11$ **d.** $y = x + 3.5$

11. a. $y = -x + 3.5$

b. $y = \dfrac{1}{9}x + \dfrac{34}{9}$ (or $9y = x + 34$)

c. $y = -6x + 90$
d. $y = 20x$

12. a. $y = -x + 4$
b. $y = -0.25x + 5.5$

c. $y = \dfrac{2}{7}x - \dfrac{32}{7}$ (or $7y = 2x - 32$)

d. $y = \dfrac{2}{3}x + \dfrac{10}{3}$ (or $3y = 2x + 10$)

13. a. $y = x + 3$ **b.** $y = \dfrac{-5}{4}x + 5$

c. $y = -6x + 6$ **d.** $y = \dfrac{7}{8}x + 35$

14. a. $y = x + 8$ **b.** $y = -2x + 6$

c. $y = \dfrac{-3}{7}x - 3$ **d.** $y = \dfrac{1}{4}x + 50$

15. a. $y = \dfrac{7}{4}x$ **b.** $y = x$ **c.** $y = -2x$

d. $y = -3x$ **e.** $y = \dfrac{-24}{11}x$ **f.** $y = \dfrac{-x}{3}$

16. a. $y = 2x + 4$ **b.** $y = 5x$

17. a. $y = -x + 3$ **b.** $y = \dfrac{1}{5}x$

18. a. $y = x + 2$ **b.** $y = -2x + 3$

19. a. B   **b.** A   **c.** D

20. a. $y = -2.4x - 12$ **b.** $y = -58.32$

21. a. $y = -0.6x + 10$ **b.** $\left(16\dfrac{2}{3}, 0\right)$

22. a. The equation is $y = \dfrac{5}{4}x - 2135$

b. Values of c cannot be read directly from the graph because the graph doesn't contain the origin, and since c is found at $x = 0$, we need to use the equation.

c. In 2020 the concentration of CO_2 will be 390 ppm. The assumption is that the concentration of CO_2 will continue to follow this linear pattern.

23. $m = \dfrac{-15 + 7}{4 - 0} = -2$

$y = -2x + c$
$6 = -6 + c$
$c = 12$
The equation is $y = -2x + 12$.

24. a. Line A $y = x - 1$, Line B: $y = -x + 8$

b. $(4.5, 3.5)$

c. Sample responses can be found in the worked solutions in the online resources. Possible methods include graphing on a Cartesian plane or using algebra.
Line B : $y = -x + 8, m = -1, m_\perp = 1$
$y = x + c$
$6 = -4 + c$
$c = 10$

25. $m = \dfrac{1}{4}C$

26. a, c

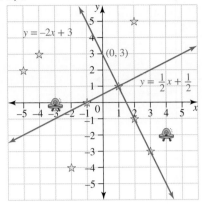

b. $y = -2x + 3$ **c.** $(1, 1)$ **d.** 1 star

e. Sample responses can be found in the worked solutions in the online resources. Possible responses could include the following.
$y = -x - 1$, $x = 2$

27. Sample responses can be found in the worked solutions in the online resources. Responses should include the correct equation: $T = 4.25t + 15$.

28. 8.1 mL

29. a. A: $0, B: $100, C: $200

b. A: $80, B: $50 C: $33.33

c. A: $w = 80h$, B: $w = 50h + 100$, C: $w = 33.3h + 200$

d. i. C   **ii.** A

e. The advantage of location C is that it has the highest minimum pay, so you would be guaranteed at least $200 per day, regardless of how many hours you work.

f. i. A: $160, B: $200, C: $266.66

ii. A: $480, B: $400, C: $400

7.5 Sketching linear graphs

1. a. $5y - 4x = 20$

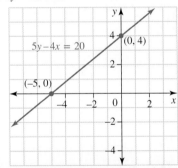

b. $y = x + 2$

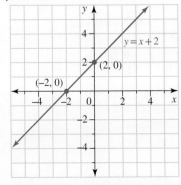

c. $y = -3x + 6$

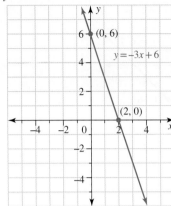

2. a. $3y + 4x = -12$

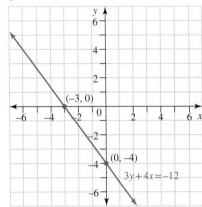

b. $x - y = 5$

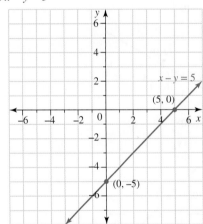

c. $2y + 7x - 8 = 0$

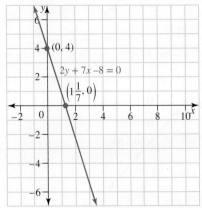

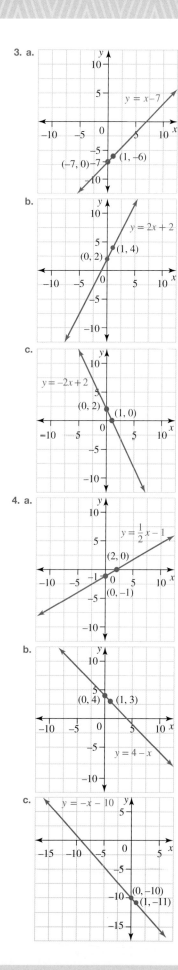

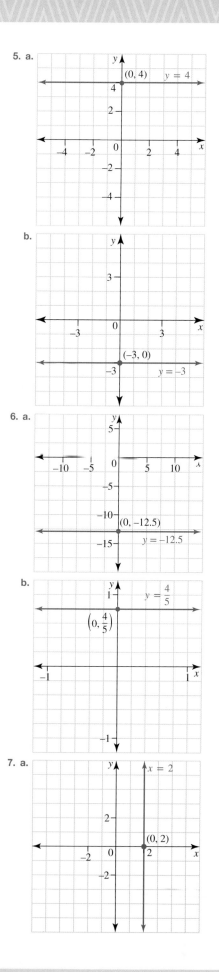

b.

$x = -6$

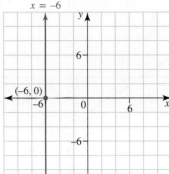

8. a.

$x = -2.5$

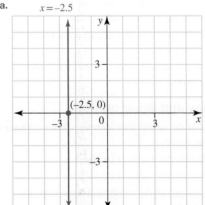

b.

$x = \dfrac{3}{4}$

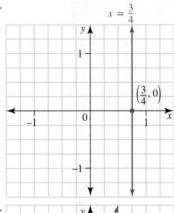

9. a.

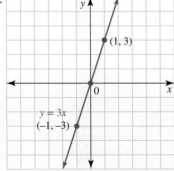

b.

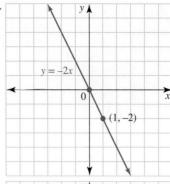

10. a.

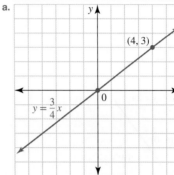

b.

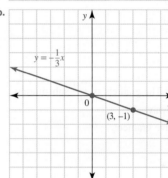

11. D

12. D

13. a. $y = -\dfrac{2x}{5} + 4$

b. $-\dfrac{2}{5}$

c. The x-intercept is 10, the y-intercept is 4.

d.

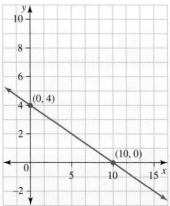

14. a. $m = \dfrac{\text{rise}}{\text{run}} = \dfrac{5}{3}$

b.

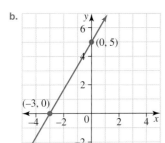

c. i. $y = \dfrac{5x}{3} + 5$ **ii.** $-5x + 3y = 15$

15. a. $y = \dfrac{4}{3}x - 8$

b. $y = 0,\ 4x = 24,\ x = 6$
$x = 0,\ -3y = 24,\ y = -8$

c.

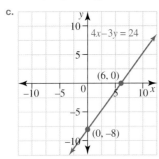

16. a. $y = \dfrac{-ax + c}{b}$

b. Sample response: The gradient is $\dfrac{-a}{b}$ substituting positive values always
results in a negative gradient.

17. Sample response:

$4y = \dfrac{4}{2}x + 3$

$4y = 2x + 3$

$-2x + 4y = 3$

18. Sample response: All descriptions use the idea that a
gradient is equal to $\dfrac{\text{rise}}{\text{sun}}$, which equals $\dfrac{2}{3}$ in all of these
cases.

19. a. The gradient is 2. **b.** The equation of the line is
$y = 2x - 1$.

c. One point is $(1, 1)$.

20. a. i. C **ii.** B **iii.** A

b. Straight line y-intercept of $(0, 2)$ with a slope of $-\dfrac{1}{4}$

21. a. i.

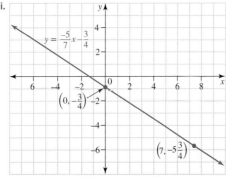

ii.

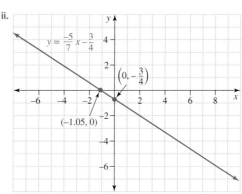

iii. Sample responses can be found in the worked
solutions in the online resources.

b. Sample responses can be found in the worked solutions
in the online resources.

22. Sample responses can be found in the worked solutions in
the online resources.

7.6 Technology and linear graphs

1. i. As the size of the coefficient increases, the steepness of
the graph increases.

ii. Each graph cuts the x-axis at $(0, 0)$.

iii. Each graph cuts the y-axis at $(0, 0)$.

2. i. As the magnitude of the coefficient decreases, the
steepness of the graph increases.

ii. Each graph cuts the x-axis at $(0, 0)$.

iii. Each graph cuts the y-axis at $(0, 0)$.

3. a. positive, positive **b.** downward, negative
c. bigger **d.** will

4. a. Yes, 1

b. No, the lines are parallel.

c. i. $(0, 0),\ x = 0$

ii. $(-2, 0),\ x = -2$

iii. $(2, 0),\ x = 2$

d. i. $(0, 0),\ y = 0$

ii. $(0, 2),\ y = 2$

iii. $(0, -2),\ y = -2$

5. a. Yes, -1

b. No, the lines are parallel.

c. i. $(0, 0),\ x = 0$

ii. $(2, 0),\ x = 2$

iii. $(-2, 0),\ x = -2$

d. i. $(0, 0),\ y = 0$

ii. $(0, 2),\ y = 2$

iii. $(0, -2),\ y = -2$

6. a. same **b.** y-intercept, y-axis
c. y-intercept **d.** x-intercept

7. i. No

ii. Yes

iii. a. 1 **b.** -1

c. 3 **d.** $-\dfrac{2}{5}$

iv. a. $(-5, 0)$, i.e. $x = -5$

b. $(5, 0)$, i.e. $x = 5$

c. $\left(-\dfrac{5}{3}, 0\right)$, i.e. $x = -\dfrac{5}{3}$

d. $\left(\dfrac{25}{2}, 0\right)$, i.e. $x = \dfrac{25}{2}$

v. a. $(0, 5)$, i.e. $y = 5$

b. $(0, 5)$, i.e. $y = 5$

c. $(0, 5)$, i.e. $y = 5$

d. $(0, 5)$, i.e. $y = 5$

8. a. steepness   b. y-coordinate, y-axis

c. same, parallel   d. the same

9. a. gradient = 2, y-intercept = (0, 0), $y = 0$

b. gradient = 1, y-intercept = (0, 1), $y = 1$

c. gradient = -3, y-intercept = (0, 5), $y = 5$

d. gradient = 2/3, y-intercept = (0, -7), $y = -7$

10. Sample response: They all have in common the point $(1, 5)$. Their gradients are all different.

11. Answers should show the y-intercept $(0, 2.2)$, point $(1, 2.75)$ and gradient 0.55.

12. a. N is the number of dogs walked; $-\$15$ is Shirly's starting cost and out-of-pocket expense before she walks a dog, and she earns \$10 for every dog walked.

b. She needs to walk at least two dogs a week before she can make a profit.

c.

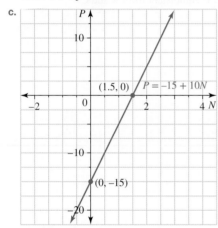

13. a. $y = 2x - 8.5$

b.

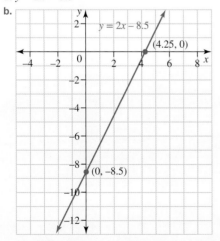

c. $y = 7.5$

d. $x = 10.25$

14. a. $C = 25n + 1000$

b. $(0, 1000)$. This is the initial cost of the site licence.

c. \$6000

d. 80

15. a. Sample response: Possible methods include plotting the points $(0, 325)$ and $(1, 328.8)$ or using the equation $c = 3.8m + 325$.

b. 382 calories

c. 32.9 minutes

7.7 Practical applications of linear graphs

1. a. $y = 6x - 5$ b. $y = -3x + 8$

2. a. $y = 2x + 4$ b. $y = 0.9x + 1.1$

3. a. $y = 3x + 1$ b. $y = -x + 17$

4. a. $v = -3t + 27$ b. $C = 3d + 8$

5. D

6. a. $A = 25t + 60$ b. \$5060

c. \$60

7. a. $C = 0.55s + 2.0$ b. \$8.60

8. a. $N = 400t + 3200$ b. 5400

c. 13 200

9. a. 27   b. 108

10. a. $w = -40t + 712$ b. 712 L

c. 18 days

11. Sample responses can be found in the worked solutions in the online resources.

12. a. $C = 2.57d + 3.50$

b. \$33.06

c. 7.20 km

13. $5 \times 21.4 = 107\,\text{cm} = 1.07\,\text{m}$

14. a. $N = -22t + 164$ b. 120

c. 5   d. 7.5 months

15. a. See the figure at the foot of the page.*

   b. $(0, -800)$, $y = -800$. Fixed costs are $800.

   c. $(4, 0)$, $x = 4$, this is the break-even amount

   d. 200, this is the sale price per print that contributes to profit

16. a. $C = 2.14d + 3.50$

   b. $28.11

   c. $52.93

17. a. $Y = 78.7A$ b. ¥196 750   c. $Y_T = 76.5A$

18. Cody arrives first.

7.8 Midpoint of a line segment and distance between two points

1. a. $(2, 4)$ b. $(5, 1)$ c. $(7, 2)$

2. a. $(8, 9)$ b. $(0, 5)$ c. $(-1, 2)$

3. a. $(-5, -2)$ b. $(2, 3)$ c. $(0, 12)$

4. a. $\left(1\frac{1}{2}, 5\frac{1}{2}\right)$ b. $\left(-3, -3\frac{1}{2}\right)$

   c. $\left(-9\frac{1}{2}, 9\frac{1}{2}\right)$ d. $\left(-1\frac{1}{2}, 6\frac{1}{2}\right)$

5. a. 5   b. 9   c. -16 d. 0

6. a. $(4, 6)$ b. $(-2, -3)$

   c. $(11, 2)$ d. $(-7, -3)$

7. $y = 5x - 7$

8. a. 5   b. 10   c. 1

9. a. 2   b. 13   c. 10

10. a. 5   b. 17   c. 13

11. a. 7.211   b. 14.765   c. 13.038

12. a. 8.944   b. 19.416   c. 19.416

13. a. 24.472   b. 25.464

14. a. 17.788   b. 25.763

15. 21.024 km

16. $d = 5$

17. $-2 = \dfrac{x + 5}{2} \Rightarrow x = -9$

   $-4 = \dfrac{-6 + y}{2} \Rightarrow y = -2$

   The required point is $(-9, -2)$.

18. $d_{AB} = 5\sqrt{10}$

   $d_{BC} = 5\sqrt{10}$

   Therefore, B is equidistant from A and C.

*15. a.

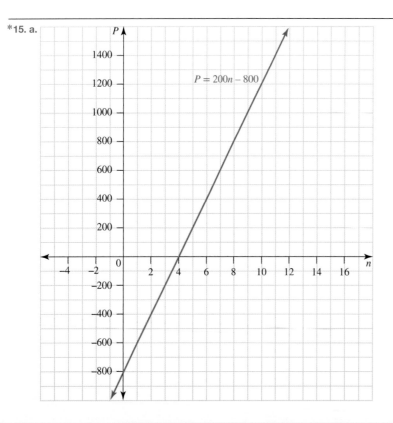

$P = 200n - 800$

19. a.

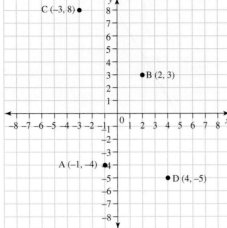

b. $M = (-2, 2)$

c. $2\sqrt{37}$

d. $(7, -2)$

e. $\dfrac{3 - (-4)}{2 - (-1)} = \dfrac{7}{3}$

f. $y = -4x + 11$

20. Sample responses can be found in the worked solutions in the online resources.

$$d_{AB} = 2\sqrt{10}$$
$$d_{BC} = 2\sqrt{10}$$
$$d_{AC} = 4\sqrt{2}$$

Side length AB is equal to side length BC but not equal to side length AC. Therefore, $\triangle ABC$ is an isosceles triangle.

21. $\dfrac{1}{2}$

22. $(-4, 5), (-3, 2), (-2, -1), (-1, -4), (0, -7)$

Project

1. 6

2. Sample responses can be found in the worked solutions in the online resources.

3. **a** and **h**

4. **e**

5. **a, b, c** and **h**

6. Yes

7.

	Length of table (m)	Width of table (m)	Destination pocket
a	3	0.25	Far left
b	3	0.5	Far left
c	3	0.75	Far left
d	3	1	Far right
e	3	1.25	Far left
f	3	1.5	Far left
g	3	1.75	Far right
h	3	2	Close right

8. Sample responses can be found in the worked solutions in the online resources.

7.9 Review questions

1. D

2. B

3. D

4. B

5. D

6. A

7. B

8. A

9. D

10. A

11. a. gradient $= 8$, y-intercept -3

b. gradient $= -9$, y-intercept 5

c. gradient $= -2$, y-intercept 6

d. gradient $= 2$, y-intercept 0

e. gradient $= \dfrac{2}{3}$, y-intercept $-\dfrac{1}{3}$

12. A, B

13. A, C, D

14. a. 2   b. $-\dfrac{2}{5}$ c. 0

15. a. 2   b. $\dfrac{5}{4}$

16. a. $m = -3$, $c = 7$

b. $m = \dfrac{3}{2}$, $c = 3$

c. $m = -\dfrac{2}{5}$, $c = 0$

d. $m = 0$, $c = 4$

17. a.

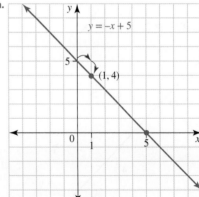

b.

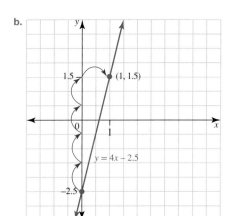

c.

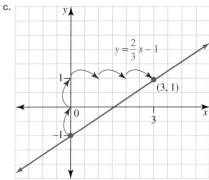

d.

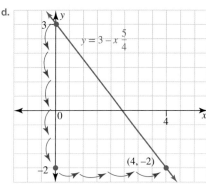

18. a.

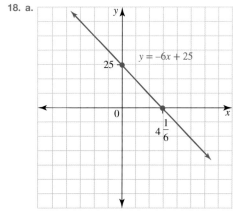

b.

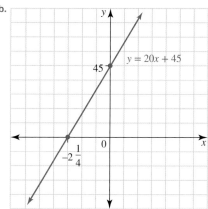

c.

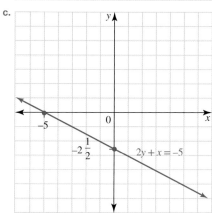

d.

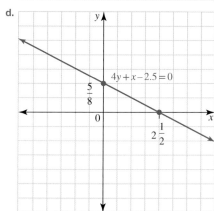

19. a.

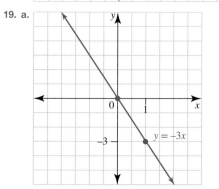

b.

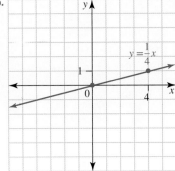

$y = \frac{1}{4}x$

c.

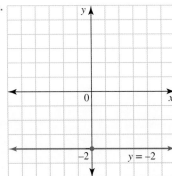

$y = -2$

d.

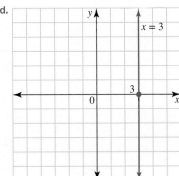

$x = 3$

20. a. $y = 2x$ **b.** $y = 2 - \frac{2}{7}x$

21. a. $y = 2x - 7$ **b.** $y = 2x - 14$
   c. $y = 2x - 5$ **d.** $y = -5x$
   e. $y = -x - 2$ **f.** $y = -2.75x + 7.75$
   g. $y = x - 3$ **h.** $y = -2x + 5$

22. (1.1)

23. 5 units

24. a. $y = 400 - 15x$

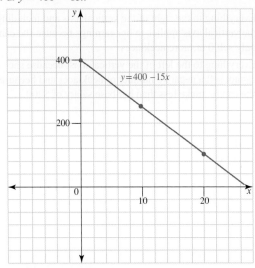

$y = 400 - 15x$

b. 26.7 or 27 weeks
c. \$175
d. 21 repayments

25. a. $y = 40 - 3.5x$

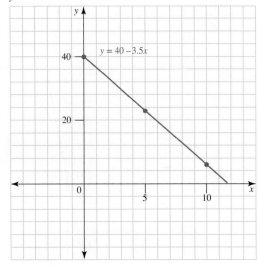

$y = 40 - 3.5x$

b. 11.4 hours
c. 22.75 km

26. a. $n = 5t + 25$ **b.** 65   **c.** 336 minutes

27. a. $m = 5n + 200$

b. See the figure at the foot of the page.*

c. Everything to the right of the vertical axis and including the vertical axis applies to her earnings because she can only sell 0 or more necklaces.

d. Everything to the left of the vertical axis, because she cannot sell a negative number of necklaces and because she cannot make less than her base salary of $200.

28. $\dfrac{1}{2}$

29. $(-1, 5)$

30. $p = -0.25$, $t = 0.25$

*27. b.

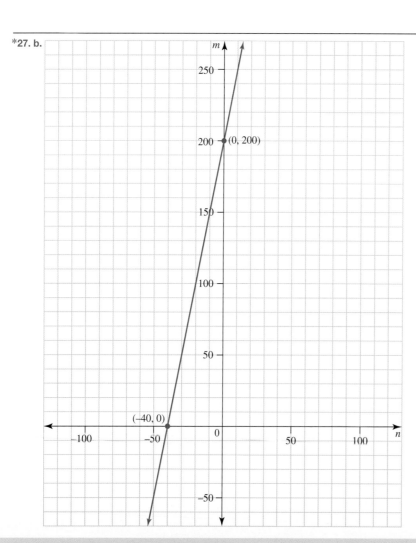

Semester review 1

The learnON platform is a powerful tool that enables students to complete revision independently and allows teachers to set mixed and spaced practice with ease.

Student self-study

Review the **Course Content** to determine which topics and lessons you studied throughout the year. Notice the green bubbles showing which elements were covered.

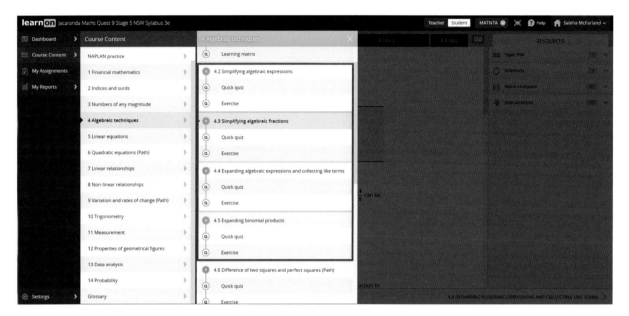

Review your results in **My Reports** and highlight the areas where you may need additional practice.

Use these and other tools to help identify areas of strengths and weakness and target those areas for improvement.

Teachers

It is possible to set questions that span multiple topics. These assignments can be given to individual students, to groups or to the whole class in a few easy steps.

Go to **Menu** and select **Assignments** and then **Create Assignment**. You can select questions from one or many topics simply by ticking the boxes as shown below.

Once your selections are made, you can assign to your whole class or subsets of your class, with individualised start and finish times. You can also share with other teachers.

More instructions and helpful hints are available at www.jacplus.com.au.

8 Non-linear relationships

LESSON
8.1 Overview

Why learn this?

As seen in the previous topic, we live in a world full of shapes. Scientists, engineers, health professionals and financial analysts all rely heavily on mathematical equations to model real-life situations and solve problems.

Most relationships do not produce a constant rate of change. Anything that does not have a constant rate of change has a non-linear relationship and will produce a non-linear graph. Some of the most common curves we see each day are in the arches of bridges. The arch shape is actually a parabolic function. Because of its shape, it is very strong and stable. Architects and structural engineers use arches extensively in buildings and other structures. In space, satellites orbiting Earth follow an elliptical path, while the orbits of planets are almost circular.

Because of the many uses of linear and non-linear graphs, it is important to understand the basic concepts that you will study in this topic, and how to identify linear and non-linear relationships.

1. State the equation of the axis of symmetry for the graph shown.

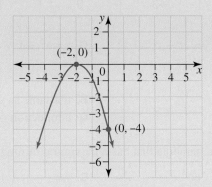

2. **MC** For the graph $y = -x^2 + 4$, select the correct turning point of the parabola.
   A. a maximum at $(0, -4)$ B. a minimum at $(4, 0)$
   C. a maximum at $(-4, 0)$ D. a maximum at $(0, 4)$

3. Sketch the graph of $y = x^2 - 9$.

4. Sketch the graph of $y = 3^x$.

5. **MC** Which of the following equations is a parabola?
   A. $y = 2^x + 7$ B. $y = 7 - 2^x$ C. $y = 7 - 2x^2$ D. $y = 7x + 2$

6. State the y-intercept of $y = 2x^2 - 8$.

7. State the y-intercept of $y = 5^x$.

8. Identify the coordinates of the turning point for the graph shown.

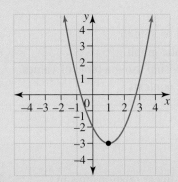

9. **MC** Identify the x-intercepts of the graph shown.
   A. 0 and 1   B. 0 and 2
   C. 0 and 4   D. 1 and 2

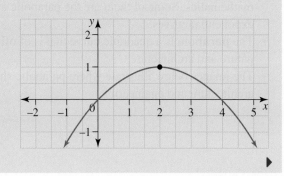

10. **MC** Select the equation that matches this table of values.

x	-3	-2	-1	0	1	2
y	-8	-3	0	1	0	-3

  A. $y = (1 - x)^2$ B. $y = x^2 + x$ C. $y = x^2 - 1$ D. $y = 1 - x^2$

11. **MC** The graph of $y = 2x^2$ is:
  A. wider than $y = x^2$
  C. narrower than $y = x^2$
  B. the same as $y = x^2$
  D. a reflection of $y = x^2$ in the x-axis

12. **MC** Select the equation for a parabola that has a maximum turning point and is narrower than $y = x^2$.

  A. $y = 2x^2$ B. $y = \dfrac{1}{2}x^2$ C. $y = (-x)^2 + 1$ D. $y = -2x^2$

13. **MC** Select the equation of a quadratic relation in the form $y = kx^2$ that passes through $(-2, 1)$.

  A. $y = 4x^2$ B. $y = 2x^2$ C. $y = -2x^2$ D. $y = \dfrac{1}{4}x^2$

14. For the graph of the equation $y = -9 - x^2$, state the coordinates of the turning point.

15. **MC** The graph of $y = x^2$ has been reflected in the x-axis and translated 2 units down.
  The new equation is:

  A. $y = x^2 + 2$ B. $y = x^2 - 2$ C. $y = -x^2 + 2$ D. $y = -x^2 - 2$

LESSON
8.2 Quadratic relationships and graphs

LEARNING INTENTION

At the end of this lesson you should be able to:
- identify the features of a parabolic graph
- graph a simple quadratic relationship using a table of values
- use quadratic graphs in real-life contexts.

⊙ 8.2.1 Parabolic graphs

eles-6242

- There are many examples of non-linear relationships in mathematics. Some of them are the parabolic and the exponential curves.
- The **parabola** is a quadratic curve that is of often found in nature and architecture.
- The images below illustrate where parabolic shapes are used in architecture and engineering.

- When quadratic functions are graphed, they produce curved lines called parabolas.
- The graph shown is a typical parabola with features as listed below.

  - The dotted line is the **axis of symmetry**; the parabola is the same on either side of this line.
  - The **turning point**, or **vertex**, is the point where the graph changes direction. It is located on the axis of symmetry.
  - The turning point is a local minimum if it is the lowest point on the graph, and a local maximum if it is the highest point on the graph.
  - The x-intercept(s) is where the graph crosses (or sometimes just touches) the x-axis. Not all parabolas have x-intercepts.
  - The y-intercept is where the graph crosses the y-axis. All parabolas have one y-intercept.

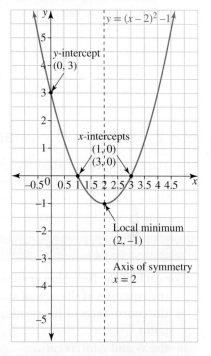

WORKED EXAMPLE 1 Identifying the features of parabolas

For each of the following graphs:
  i. **state the equation of the axis of symmetry**
 ii. **state the coordinates of the turning point or vertex**
iii. **indicate whether it is a maximum or a minimum turning point.**

a.

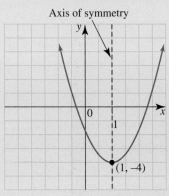

b.

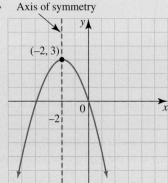

THINK	WRITE
a. i. State the equation of the vertical line that cuts the parabola in half.	**a.** i. The axis of symmetry is $x = 1$.
ii. State the turning point, or vertex.	ii. The turning point is at $(1, -4)$.
iii. Determine the nature of the turning point by observing whether it is the highest or lowest point of the graph.	iii. Minimum turning point.
b. i. State the equation of the vertical line that cuts the parabola in half.	**b.** i. The axis of symmetry is $x = -2$.
ii. State the turning point, or vertex.	ii. The turning point is at $(-2, 3)$.
iii. Determine the nature of the turning point by observing whether it is the highest or lowest point of the graph.	iii. Maximum turning point.

⏵ 8.2.2 The x- and y-intercepts

eles-6243

- The **x-intercept** is where the graph crosses (or just touches) the x-axis.
- The **y-intercept** is where the graph crosses the y-axis. All parabolas have one y-intercept.
- When sketching a parabola, the x-intercepts (if any) and the y-intercept should always be marked on the graph, with their respective coordinates.

WORKED EXAMPLE 2 Determining the features of a parabola

For each of the following graphs, state:
 i. **the equation of the axis of symmetry**
 ii. **the coordinates of the turning point and whether the point is a maximum or a minimum turning point**
 iii. **the x- and y-intercepts.**

a.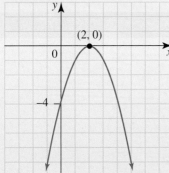

b.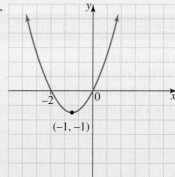

THINK	WRITE
a. i. State the equation of the vertical line that cuts the parabola in half.	**a.** i. The axis of symmetry is $x = 2$.
ii. State the turning point and its nature; that is, determine whether it is the highest or lowest point of the graph.	ii. The maximum turning point is at $(2, 0)$.
iii. Observe where the parabola crosses the x-axis. In this case, the graph touches the x-axis when $x = 2$, so there is only one x-intercept.	iii. The x-intercept is 2. It occurs at the point $(2, 0)$.
Observe where the parabola crosses the y-axis.	The y-intercept is -4. It occurs at the point $(0, -4)$.

b. **i.** State the equation of the vertical line that cuts the parabola in half.

    ii. State the turning point and its nature; that is, determine whether it is the highest or lowest point of the graph.

    iii. Observe where the parabola crosses the x-axis.

        Observe where the parabola crosses the y-axis.

b. **i.** The axis of symmetry is $x = -1$.

    ii. Minimum turning point is at $(-1, -1)$.

    iii. The x-intercepts are -2 and 0. They occur at the points $(-2, 0)$ and $(0, 0)$.

        The y-intercept is 0. It occurs at the point $(0, 0)$.

▶ 8.2.3 Plotting points to graph quadratic functions

eles-4937

- If there is a rule relating x and y, a table of values can be used to determine actual coordinates.
- When drawing straight line graphs, a minimum of two points is required. For parabolas there is *no minimum* number of points, but between 6 and 12 points is a reasonable number.
- The more points used, the 'smoother' the parabola will appear. The points should be joined with a smooth curve, not ruled.
- Ensure that points plotted include (or are near) the main features of the parabola, namely the axis of symmetry, the turning point and the x- and y-intercepts.
- Occasionally a list of x-values will be provided and the corresponding y-values can be calculated. In the following example, the set of x-values is specified as $-4 \leq x \leq 2$.

WORKED EXAMPLE 3 Plotting points to graph quadratic functions

Plot the graph of $y = x^2 + 2x - 3$, $-4 \leq x \leq 2$ and hence state:
a. the equation of the axis of symmetry
b. the coordinates of the turning point and whether it is a maximum or a minimum
c. the x- and y-intercepts.

THINK

1. Write the equation.

2. Complete a table of values by substituting into the equation each integer value of x from -4 to 2. For example, when $x = -4$, $y = (-4)^2 + 2 \times (-4) - 3 = 5$.

3. List the coordinates of the points.

4. Draw and label a set of axes, plot the points listed and join the points to form a smooth curve.

WRITE/DRAW

$y = x^2 + 2x - 3$

x	-4	-3	-2	-1	0	1	2
y	5	0	-3	-4	-3	0	5

$(-4, 5), (-3, 0), (-2, -3), (-1, -4), (0, -3),$
$(1, 0), (2, 5)$

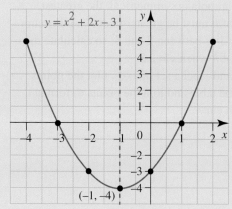

<table>
<tr><td>a.</td><td>State the equation of the line that divides the parabola exactly into two halves.</td><td>a. The axis of symmetry is $x = -1$.</td></tr>
<tr><td>b.</td><td>Identify the point where the graph turns or changes direction, and decide whether it is the highest or lowest point of the graph. State the coordinates of this point.</td><td>b. Minimum turning point is at $(-1, -4)$.</td></tr>
<tr><td>c. 1.</td><td>State the x-coordinates of the points where the graph crosses the x-axis.</td><td>c. The x-intercepts are at -3 and 1. They occur at the points $(-3, 0)$ and $(1, 0)$.</td></tr>
<tr><td>2.</td><td>State the y-coordinate of the point where the graph crosses the y-axis.</td><td>The y-intercept is at -3. It occurs at the point $(0, -3)$.</td></tr>
</table>

- A rule relating x and y will be occasionally provided. From this rule, pairs of x- and y-values can be calculated. In the following example, the rule is given as $h = -5x^2 + 25x$.
- Graphs can be drawn using technology or by hand.

WORKED EXAMPLE 4 Using graphs of quadratic functions

Rudie, the cannonball chicken, was fired out of a cannon. His path could be traced by the equation $h = -5x^2 + 25x$, where h is Rudie's height, in metres, above the ground and x is the horizontal distance, in metres, from the cannon.
Plot the graph for $0 \le x \le 5$ and use it to determine the maximum height of Rudie's path.

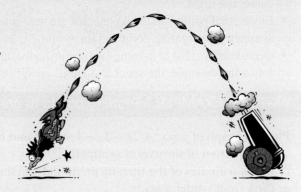

THINK	WRITE
1. Write the equation.	$h = -5x^2 + 25x$
2. Complete a table of values by substituting into the equation each integer value of x from 0 to 5. For example, when $x = 0$, $h = -5 \times 0^2 + 25 \times 0 = 0$.	<table><tr><td>x</td><td>0</td><td>1</td><td>2</td><td>3</td><td>4</td><td>5</td></tr><tr><td>h</td><td>0</td><td>20</td><td>30</td><td>30</td><td>20</td><td>0</td></tr></table>
3. List the coordinates of the points.	$(0, 0), (1, 20), (2, 30), (3, 30), (4, 20), (5, 0)$
4. As a parabola is symmetrical, the greatest value of h must be greater than 30 and occurs when x lies between 2 and 3, so calculate the value of h when $x = 2.5$.	When $x = 2.5$, $h = 5 \times (2.5)^2 + 25 \times 2.5$ $\qquad = 31.25$
5. Draw and label a set of axes, plot the points from the table and join the points to form a smooth curve.	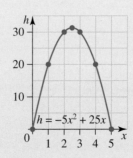

6. The maximum height is the value of h at the highest point of the graph.	$h = 31.25$
7. Answer the question in a sentence.	The maximum height of Rudie's path is 31.25 metres.

 Resources

 Interactivity The y-intercept (int-3837)

Exercise 8.2 Quadratic relationships and graphs   learn

8.2 Quick quiz on	**8.2 Exercise**

Individual pathways

■ PRACTISE	■ CONSOLIDATE	■ MASTER
1, 4, 5, 9, 10, 11, 16, 19	2, 6, 8, 12, 14, 17, 20	3, 7, 13, 15, 18, 21, 22

Fluency

1. **WE1** For each of the graphs below:

   i. state the equation of the axis of symmetry
   ii. give the coordinates of the turning point
   iii. indicate whether it is a minimum or maximum turning point.

   a.
   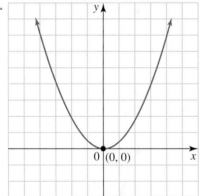

   b.
   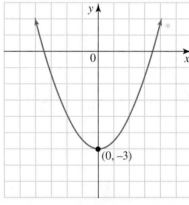

   c.
   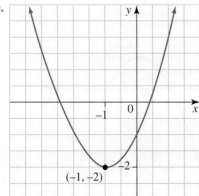

   d.
   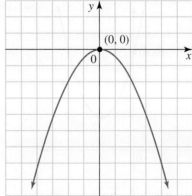

e.

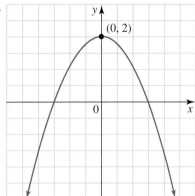

(0, 2)

f.

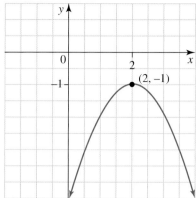

2

(2, −1)

−1

2. For each of the following graphs, state:

   i. the equation of the axis of symmetry

   ii. the coordinates of the turning point

   iii. whether the turning point is a maximum or minimum.

a.

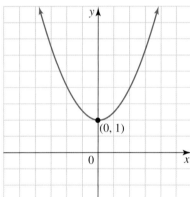

(0, 1)

b.

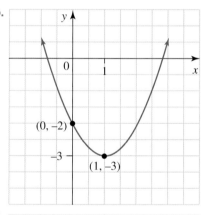

1

(0, −2)

−3

(1, −3)

c.

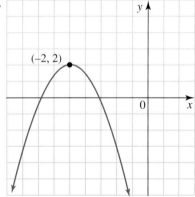

(−2, 2)

d.

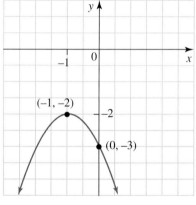

−1

(−1, −2)

−2

(0, −3)

e.

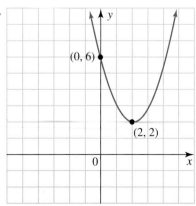

(0, 6)

(2, 2)

f.

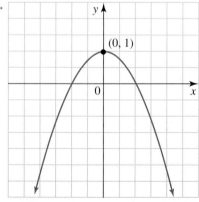

(0, 1)

3. **WE2** For each of the following graphs, state:
   i. the equation of the axis of symmetry
   ii. the coordinates of the turning point and whether the point is a maximum or a minimum
   iii. the *x*- and *y*-intercepts.

a.
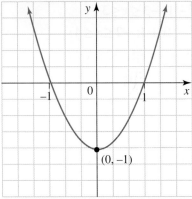

b.
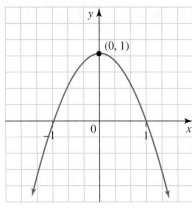

c.
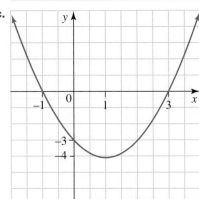

d.
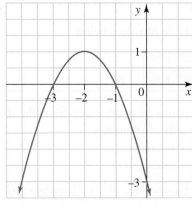

e.
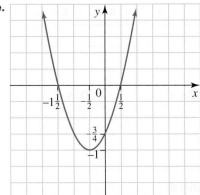

f.
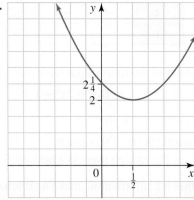

4. **MC** a. Identify the axis of symmetry for the graph shown.

   A. $x = 0$　　　**B.** $x = -2$　　　**C.** $x = -4$　　　**D.** $y = 0$

   b. Identify the coordinates of the turning point for the graph.
   A. $(0, 0)$　　　**B.** $(-2, 4)$　　　**C.** $(-2, -4)$　　　**D.** $(2, -4)$

   c. Identify the *y*-intercept.
   A. 0　　　　　**B.** −2　　　　　**C.** −4　　　　　**D.** 2

   d. Identify the *x*-intercepts.
   A. 0 and 4　　　**B.** 0 and 2　　　**C.** −2 and 0　　　**D.** −4 and 0

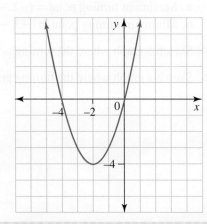

For questions **5** to **7**, plot the graph of the quadratic equation and hence state:

  i. the equation of the axis of symmetry

  ii. the coordinates of the turning point and whether it is a maximum or a minimum

  iii. the x- and y-intercepts.

5. **WE3** **a.** $y = x^2 + 8x + 15,\ -7 \le x \le 0$ **b.** $y = x^2 - 1,\ -3 \le x \le 3$

6. **a.** $y = x^2 - 4x,\ -1 \le x \le 5$ **b.** $y = x^2 - 2x + 3,\ -2 \le x \le 4$

7. **a.** $y = x^2 + 12x + 35,\ -9 \le x \le 0$ **b.** $y = -x^2 + 4x + 5,\ -2 \le x \le 6$

8. Consider the equations for $-3 \le x \le 3$:

  i. $y = x^2 + 2$

  ii. $y = x^2 + 3$

    a. Make a table of values and plot the points on the same set of axes.

    b. State the equation of the axis of symmetry for each equation.

    c. State the x-intercepts for each equation.

Understanding

9. Consider the table of values below.

x	−4	−3	−2	−1	0	1	2	3	4
y	12	5	0	−3	−4	−3	0	5	12

    a. Plot these points on graph paper. State the shape of the graph.

    b. Locate the axis of symmetry.

    c. Locate the y-intercept.

    d. Locate the x-intercept(s).

10. Consider the function $y = x^2 + x$. Complete this table of values for the function.

x	−6	−4	−2	0	2	4	6	8
y				0				

11. **MC** Select which of the following rules is not a parabola.

  A. $y = -2x^2$ **B.** $y = 2x^2 - x$ **C.** $y = -2 \div x^2$ **D.** $y = -2 + x^2$

12. Consider the graph of $y = -x^2$.

    a. State the turning point of this graph.

    b. State whether the turning point is a maximum or a minimum.

13. Given the following information, make a sketch of the graph involved.

    a. Maximum turning point $= (-2, -2)$, y-intercept $= (0, -6)$

    b. Minimum turning point $= (-3, -2)$, x-intercept $(1, 0)$, and $(-7, 0)$

14. Sketch a graph where the turning point and the x-intercept are the same. Suggest a possible equation.

15. Sketch a graph where the turning point and the y-intercept are the same. Suggest a possible equation.

Communicating, reasoning and problem solving

16. On a set of axes, sketch a parabola that has no x-intercepts and has an axis of symmetry $x = -2$. Explain if the parabola has a maximum or minimum turning point, or both.

17. If the axis of symmetry of a parabola is $x = -4$ and one of the x-intercepts is at $(10, 0)$, show that the other x-intercept is at $(-18, 0)$.

18. If the x-intercepts of a parabola are at $(-2, 0)$ and $(5, 0)$, show that the axis of symmetry is $x = 1.5$.

19. Consider the parabola given by the rule $y = x^2$ and the straight line given by $y = x$. Show that the two graphs meet at $(0, 0)$ and $(1, 1)$.

20. **a.** The axis of symmetry of a parabola is $x = -4$. If one x-intercept is -10, determine the other x-intercept.
    b. Suggest a possible equation for the parabola.

21. **WE4** A missile was fired from a boat during a test. The missile's path could be traced by the equation $h = -\dfrac{1}{2}x^2 + x$, where h is the missile's height above the ground, in kilometres, and x is the horizontal distance from the boat, in kilometres.

    Plot the graph for $0 \leq x \leq 2$ and use it to determine the maximum height of the missile's path, in metres.

22. SpaceCorp sent a lander to Mars to measure the temperature change over a period of time. The results were plotted on a set of axes, as shown.

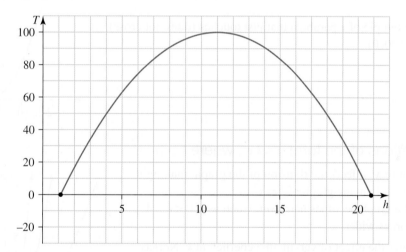

    From the graph it can be seen that the temperature change follows the quadratic rule $T = -h^2 + 22h - 21$, where T is the temperature in degrees Celsius, and the time elapsed, h, is in hours.

    a. State the initial recorded temperature on Mars.
    b. State when temperature measured $0\,°C$.
    c. Identify when the highest temperature was recorded.
    d. State the highest temperature recorded.

LESSON
8.3 Quadratic equations of the form $y = kx^2$

LEARNING INTENTION

At the end of this lesson you should be able to:
- sketch parabolas for quadratic equations of the form $y = kx^2$
- identify and describe the features of a parabola of the form $y = kx^2$ including the vertex, axis of symmetry, x- and y-intercepts, and concavity
- describe the effect of changing the value of k on parabolas of the form $y = kx^2$.

⊳ 8.3.1 Quadratic equations

eles-4938

The graph of the quadratic equation $y = x^2$

- The simplest parabola, $y = x^2$, is shown.
  - Both the x- and y-axes are clearly indicated, along with their scales.
  - The turning point, or vertex, $(0, 0)$ is indicated.
  - The x- and y-intercepts are indicated. For this graph they are all $(0, 0)$.
  - This is an example of a parabola that just touches (does not cross) the x-axis at $(0, 0)$.

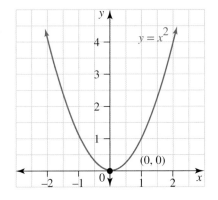

Quadratic equations of the form $y = kx^2$, where $k > 0$

- The coefficient of x^2 affects the **dilation** of the graph, making it wider or narrower than the graph of $y = x^2$.
- If $k > 1$ then the graph becomes narrower, whereas if $0 < k < 1$, the graph becomes wider.
- The graph shows the effect of varying the coefficient k. Notice that $y = 3x^2$ is narrower than $y = x^2$ and $y = \frac{1}{4}x^2$ is wider than $y = x^2$.
- When $k > 0$, the parabola is concave up.

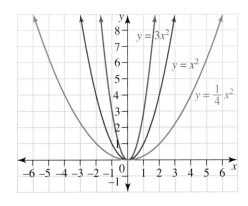

Parabolas of the form $y = kx^2$, where $k > 0$

For all parabolas of the form $y = kx^2$, where $k > 0$:
- the axis of symmetry is $x = 0$
- the turning point, or vertex, is $(0, 0)$
- the x-intercept is $(0, 0)$
- the y-intercept is $(0, 0)$
- the shape of the parabola is concave up, or a U shape (∪).

On the same set of axes, sketch the graph of $y = x^2$ and $y = 3x^2$, marking the coordinates of the turning point and the intercepts. State which graph is narrower.

THINK	WRITE/DRAW
1. Write the equation of the first graph.	$y = x^2$
2. State its axis of symmetry.	The axis of symmetry is $x = 0$.
3. State the coordinates of the turning point.	The turning point is $(0, 0)$.
4. State the intercepts.	The x-intercept is $(0, 0)$ and the y-intercept is also $(0, 0)$.
5. Calculate the coordinates of one other point.	When $x = 1$, $y = 1$; $(1, 1)$
6. Write the equation of the second graph.	$y = 3x^2$
7. State its axis of symmetry.	The axis of symmetry is $x = 0$.
8. State the coordinates of the turning point.	The turning point is $(0, 0)$.
9. State the intercepts.	The x-intercept is 0 and the y-intercept is 0.
10. Calculate the coordinates of one other point.	When $x = 1$, $y = 3$; $(1, 3)$
11. Sketch the graphs, labelling the turning point.	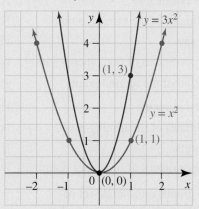
12. State which graph is narrower.	The graph of $y = 3x^2$ is narrower.

8.3.2 Quadratic equations of the form $y = kx^2$ where $k < 0$

eles-4940

- When $k < 0$, the graph is inverted; that is, it is $\cap$ shaped and is concave down.
- The coefficient of x^2 affects the dilation of the graph, making it wider or narrower than the graph of $y = -x^2$.
- If $-1 < k < 0$, the graph is wider than $y = -x^2$.
- If $k < -1$, the graph is narrower than $y = -x^2$.
  - The graph shows the effect of varying the coefficient k.
    Notice that $y = -3x^2$ is narrower than $y = -x^2$ and $y = -\dfrac{1}{4}x^2$ is wider than $y = -x^2$.

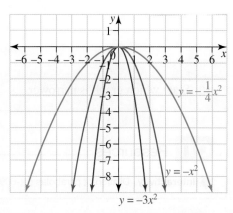

Parabolas of the form $y = kx^2$, where $k < 0$

For all parabolas of the form $y = kx^2$, where $k < 0$:
- the axis of symmetry is $x = 0$
- the turning point, or vertex, is $(0, 0)$
- the x-intercept is $(0, 0)$
- the y-intercept is $(0, 0)$
- the shape of the parabola is concave down or an upside-down U shape ($\cap$).

WORKED EXAMPLE 6 Graphing quadratic equations of the form $y = kx^2$, where $k < 0$

On the same set of axes sketch the graphs of $y = -x^2$ and $y = -2x^2$, marking the coordinates of the turning point and the intercept. State which graph is narrower.

THINK	WRITE/DRAW
1. Write the equation of the first graph.	$y = -x^2$
2. State its axis of symmetry.	The axis of symmetry is $x = 0$.
3. State the coordinates of the turning point.	The turning point is $(0, 0)$.
4. State the intercepts.	The x-intercept is $(0, 0)$ and the y-intercept is also $(0, 0)$.
5. Calculate the coordinates of one other point.	When $x = 1$, $y = -1$; $(1, -1)$
6. Write the equation of the second graph.	$y = -2x^2$
7. State its axis of symmetry.	The axis of symmetry is $x = 0$.
8. State the coordinates of the turning point.	The turning point is $(0, 0)$.
9. State the intercepts.	The x-intercept is $(0, 0)$ and the y-intercept is also $(0, 0)$.
10. Calculate the coordinates of one other point.	When $x = 1$; $y = -2$; $(1, -2)$
11. Sketch the two graphs on a single set of axes, labelling the turning point (as well as intercepts and maximum).	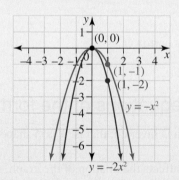
12. State which graph is narrower.	The graph of $y = -2x^2$ is narrower.

 Resources

🧩 **Interactivity** Dilation of parabolas (int-6096)

Exercise 8.3 Quadratic equations of the form $y = kx^2$

8.3 Quick quiz **on**	8.3 Exercise

Individual pathways

■ PRACTISE	■ CONSOLIDATE	■ MASTER
1, 5, 7, 11, 14	2, 4, 8, 12	3, 6, 9, 10, 13

Fluency

1. **WE5** On the same set of axes, sketch the graph of $y = x^2$ and $y = 4x^2$, marking the coordinates of the turning point and the intercepts. State which graph is narrower.

2. On the same set of axes, sketch the graph of $y = x^2$ and $y = \frac{1}{2}x^2$, marking the coordinates of the turning point and the intercepts. State which graph is narrower.

3. Sketch the graph of the following table. State the equation of the graph.

x	−4	−3	−2	−1	0	1	2	3	4
y	4	2.25	1	0.25	0	0.25	1	2.25	4

4. a. **WE6** Using the same set of axes, sketch the graphs of $y = -x^2$ and $y = -0.5x^2$, marking the coordinates of the turning point and the intercepts. State which graph is narrower.
   b. Using the same set of axes, sketch the graph of $y = -5x^2$, marking the coordinates of the turning point and the intercepts. State which graph is the narrowest.

5. **MC** a. The graph of $y = -3x^2$ is:
   A. wider than $y = x^2$.　　　　　　　　　　**B.** narrower than $y = x^2$.
   C. the same width as $y = x^2$.　　　　　　　**D.** a reflection of $y = x^2$ in the x-axis.

   b. The graph of $y = \frac{1}{3}x^2$ is:
   A. wider than $y = x^2$.　　　　　　　　　　**B.** narrower than $y = x^2$.
   C. the same width as $y = x^2$.　　　　　　　**D.** a reflection of $y = x^2$ in the x-axis.

   c. The graph of $y = \frac{1}{2}x^2$ is:
   A. wider than $y = \frac{1}{4}x^2$.　　　　　　　**B.** narrower than $y = \frac{1}{4}x^2$.
   C. the same width as $y = \frac{1}{4}x^2$.　　　　**D.** a reflection of $y = \frac{1}{4}x^2$ in the x-axis.

6. Match each of the following parabolas with the appropriate equation from the list.
   i. $y = 3x^2$
   ii. $y = -x^2$
   iii. $y = 4x^2$
   iv. $y = \frac{1}{2}x^2$
   v. $y = -4x^2$
   vi. $y = -2x^2$

a.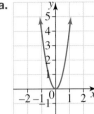

b.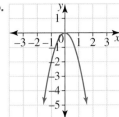

c.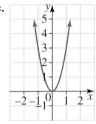

d.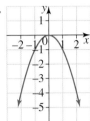

e.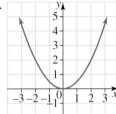

f.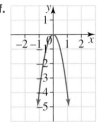

Understanding

7. Write an equation for a parabola that has a minimum turning point and is narrower than $y = x^2$.

8. Write an equation for a parabola that has a maximum turning point and is wider than $y = x^2$.

9. Determine the equation of a quadratic equation if it has an equation of the form $y = kx^2$ and passes through:

   a. $(1, 3)$ **b.** $(-1, -1)$.

10. Consider the equation $y = -3.5x^2$. Calculate the values of y when x is:

   a. 10 **b.** -10 **c.** -3 **d.** 1.5 **e.** -2.2.

Communicating, reasoning and problem solving

11. **a.** Sketch the following graphs on the same axes: $y = x^2$, $y = 2x^2$ and $y = -3x^2$. Shade the area between the two graphs above the x-axis and the area inside the graph below the x-axis. Describe the shape that has been shaded.

   b. Sketch the following graphs on the same axes: $y = x^2$, $y = \dfrac{1}{3}x^2$ and $y = -4x^2$. Shade the area inside the graphs of $y = x^2$ and $y = -4x^2$. Also shade the area between the graph of $y = \dfrac{1}{3}x^2$ and the x-axis. Describe the shape that you have drawn.

12. The photograph shows the parabolic shape of a skate ramp. The rule of the form $y = ax^2$ describes the shape of the ramp. If the top of the ramp has coordinates $(3, 6)$, determine a possible equation that describes the shape. Justify your answer.

13. The total sales of a fast food franchise vary as the square of the number of franchises in a given city. Let S be the total sales (in millions of dollars per month) and f be the number of franchises. If sales are \$25 million when $f = 4$, then:

   a. show that the equation relating S and f is $S = 1\,562\,500f^2$
   b. determine the number of franchises needed to (at least) double the sales from \$25 000 000.

14. The amount of power (watts) in an electric circuit varies as the square of the current (amperes). That is $P = kx^2$, where P is the power in watts, and x is the current in amperes.
   If the power is 100 watts when the current is 2 amperes, calculate:

   a. the power when the current is 4 amperes
   b. the power when the current is 5 amperes.

LESSON
8.4 Quadratic equations of the form $y = kx^2 + c$

LEARNING INTENTION

At the end of this lesson you should be able to:
- identify and describe the key features of a parabola of the form $y = kx^2 + c$ including the vertex, axis of symmetry, x- and y-intercepts, and concavity
- sketch graphs of the form $y = kx^2 + c$
- describe the changes that transform the graph of $y = x^2$ into $y = kx^2 + c$.

▶ 8.4.1 Sketching parabolas of the form $y = kx^2 + c$

eles-4941

- If $k > 0$, the shape of the parabola is concave up.
- If $k < 0$, the shape of the parabola is concave down.
- The coefficient of the x^2 affects the dilation of the graph, making it wider or narrower as seen in lesson 8.3
- A parabola with the equation $y = x^2 + c$ is simply the graph of $y = x^2$ moved vertically c units.
- Adding a constant, c, **translates** the graph vertically.
- If $c > 0$, the y value increases, and the graph is translated vertically upwards by c units.
- If $c < 0$, the y value decreases, and the graph is translated vertically downwards by c units.
- The coordinates of the turning point, or vertex, are $(0, c)$

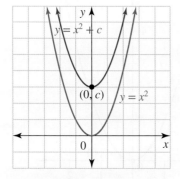

This graph shows the effect of adding c units to the curve $y = x^2$.

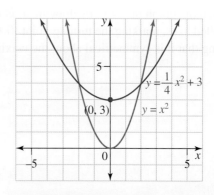

This graph shows the effect of the dilation and translation of the graph. The curve $y = \dfrac{1}{4}x^2 + 3$ is $y = x^2$ made wider since $0 < k < 1$ and translated upwards by 3 units.

For each part of the question, sketch the graph of $y = x^2$, then, on the same axes, sketch the given graph, clearly labelling the turning point or vertex.

a. $y = x^2 + 2$ b. $y = -x^2 - 3$

THINK	WRITE/DRAW
a. 1. Sketch the graph of $y = x^2$ by drawing a set of labelled axes, marking the turning point $(0, 0)$ and noting that it is symmetrical about the y-axis.	**a.** 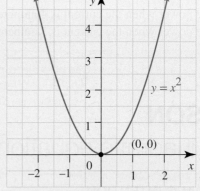
2. Determine the turning point of $y = x^2 + 2$ by adding 2 to the y-coordinate of the turning point of $y = x^2$.	The turning point, or vertex, of $y = x^2 + 2$ is $(0, 2)$.
3. Using the same axes as for the graph of $y = x^2$, sketch the graph of $y = x^2 + 2$, marking the turning point and making sure that it is the same width as the graph of $y = x^2$. (The coefficient of x^2 is the same for both graphs.)	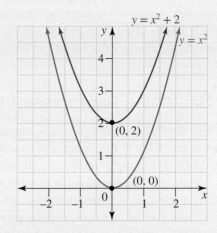
b. 1. Sketch the graph of $y = x^2$ by drawing a set of labelled axes, marking the turning point $(0, 0)$ and noting that it is symmetrical about the y-axis.	**b.** 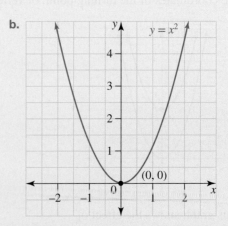

2. Determine the turning point of $y = -x^2 - 3$ by subtracting 3 from the y-coordinate of the turning point of $y = x^2$.

The turning point of $y = -x^2 - 3$ is $(-0, -3)$.

3. Using the same axes as for the graph of $y = x^2$, sketch the graph of $y = -x^2 - 3$, marking the turning point, inverting the graph and making sure that the graph is the same width as the graph of $y = x^2$.

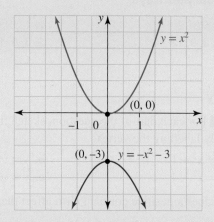

WORKED EXAMPLE 8 Sketching equations of the form $y = kx^2 + c$

Sketch the graph of $y = -x^2 + 4$, drawing clearly labelled axes and marking the axis of symmetry, vertex and y-intercept.
State whether the vertex is a maximum or minimum.

THINK

1. The equation of the axis of symmetry is the same as for $y = x^2$.

2. The vertex has been moved up c units and is $(0, c)$.

3. The coefficient of x^2 is negative so the graph is concave down.

4. Draw clearly labelled axes, mark the vertex and draw the graph.

WRITE/DRAW

The axis of symmetry is $x = 0$.

The vertex is $(0, 4)$.

Maximum point

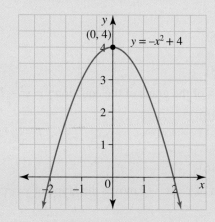

---

on Resources

⬧ **Interactivity** Vertical translation: $y = x^2 + c$ (int-1192)

Exercise 8.4 Quadratic equations of the form $y = kx^2 + c$

learnon

8.4 Quick quiz on	8.4 Exercise

Individual pathways

■ PRACTISE	■ CONSOLIDATE	■ MASTER
1, 4, 7, 8, 9, 12	2, 3, 5, 10, 14, 16, 19	6, 11, 13, 15, 17, 18, 20

Fluency

1. **WE7** For each part of the question, sketch the graph of $y = x^2$, then, on the same axes, sketch the given graph, clearly labelling the vertex.
   a. $y = x^2 + 1$ b. $y = x^2 + 4$ c. $y = x^2 - 1$
   d. $y = x^2 - 4$ e. $y = -x^2 + 1$ f. $y = -x^2 - 1$
2. How does a positive constant term affect the graph?
3. How does a negative constant term affect the graph?
4. **WE8** Sketch each of the following graphs on clearly labelled axes, marking the axis of symmetry, vertex and y-intercept of each one. State whether the vertex is a maximum or a minimum.
   a. $y = x^2 + 2$ b. $y = x^2 - 5$ c. $y = -x^2 + 3$
   d. $y = -x^2 + 4$ e. $y = -x^2 - 3$ f. $y = x^2 - \frac{1}{2}$
5. Sketch the following graphs, indicating the vertex and estimating the x-intercepts for each graph.
   a. $y = 4 - x^2$ b. $y = -4 - x^2$ c. $y = 1 - x^2$
6. a. Does the vertex change if there is a negative number in front of the x^2 term in the equation $y = x^2$?
   b. How does a negative coefficient of x^2 affect the graph?
   c. What is the axis of symmetry for all the graphs in this exercise?
7. **MC** a. The vertex for the graph of the equation $y = -x^2 + 8$ is:
   A. (0, 0)
   B. (−1, 8)
   C. (0, 8)
   D. (0, −8)

   b. The vertex of the graph of the equation $y = x^2 - 16$ is:
   A. (1, −16)
   B. (−16, 1)
   C. (0, 0)
   D. (0, −16)

   c. The graph of $y = x^2 - 7$ moves the graph of $y = x^2$ in the following way:
   A. up 1
   B. down 1
   C. up 7
   D. down 7

   d. The y-intercept of the graph of $y = -x^2 - 6$ is:
   A. 1
   B. −1
   C. 6
   D. −6

Understanding

8. Match each of the following parabolas with the appropriate equation from the list below.

i. $y = x^2 - 3$ **ii.** $y = x^2 + 3$ **iii.** $y = 3 - x^2$

iv. $y = x^2 + 2$ **v.** $y = -x^2 + 2$ **vi.** $y = -x^2 - 2$

a.
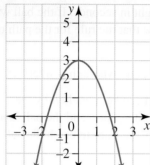

b.
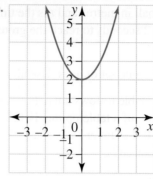

c.
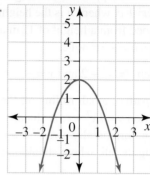

d.
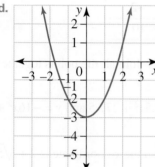

e.
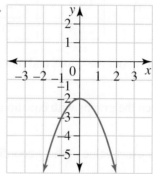

f.
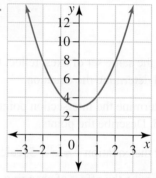

9. The vertical cross-section through the top of the mountain called the Devil's Tower can be approximated by the graph $y = -x^2 + 5$. Sketch the graph. If the x-axis represents sea level, and both x and y are in kilometres, determine the maximum height of the mountain.

10. The cross-section of a large bowl can be given by the rule $y = 3x^2 - 243$, where both x (measured across the bowl) and y (the depth of the bowl) are measured in centimetres.

    a. By factorising the rule, determine the points where $y = 0$. What are these points called?
    b. If the bowl's rim occurs at the point where $y = 0$, determine the greatest depth of the bowl.
    c. What is the width of the bowl at its rim?

11. The photo shows an imaginary line drawn across the surface of a lake. A vertical cross-section of the lake is taken at the line, such that the depth of the lake can be approximated by the graph $y = x^2 - 12$, where x and y are in metres.

    a. What would be a suitable 'domain' (set of possible x-values) for this graph?
    b. Sketch the graph over this domain.
    c. What is the greatest depth of the lake?
    d. What is the width of the lake along the white line shown in the photo?

Communicating, reasoning and problem solving

12. Show that the equation of the parabola that is of the form $y = x^2 + c$ and passes through:

    a. $(2, 1)$ is $y = x^2 - 3$

    b. $(-3, 1)$ is $y = x^2 - 10$.

13. A ball is dropped from a height and its descent follows the downward path of the parabola given by $h = kt^2 + c$, where h is the height of the ball in metres and t is the time of flight in seconds. If the ball was dropped from a height of 12 m and took 2 seconds to reach the ground, show that the rule for the path of the ball is $h = -3t^2 + 12$.

14. Sketch $y = x^2 - 2$ and $y = -x^2 + 2$ on the same set of axes. Use algebra to explain where they intersect.

15. The path of a ball rolling off the end of a table follows a parabolic curve and can be modelled by the equation $y = kx^2 + c$. A student rolls a ball off a tabletop that is 128 cm above the floor, and the ball lands 80 cm horizontally away from the desk. If the student sets a cup 78 cm above the floor to catch the ball during its fall, where should the cup be placed?

16. On a set of axes, plot the following points.

x	0	1	−3	−2
y	2	4	20	10

    On the same set of axes, reflect the points about the y-axis and plot the points. Join all the points together and determine the equation.

17. Determine the intersection points between the graphs with equations $y = -2.5x^2 + 277.5$ and $y = -16x^2 + 555$. Describe the points.

18. For the equation $y = x^2 - 5$, calculate the exact value of the y-coordinate when the x-coordinate is $2 + \sqrt{5}$.

19. Describe the effect of changing k and c, separately, in the rule $y = kx^2 + c$.

20. A rocket is released from a rocket launcher on the ground and travels in an upward parabolic path to a maximum height of 125 metres before falling and landing on the ground 10 metres from where it was launched.
    Determine the equation of the path taken by the rocket.

LESSON
8.5 Exponential relationships and graphs

LEARNING INTENTION

At the end of this lesson you should be able to:
- draw a graph of an exponential curve using a table of values
- recognise and sketch exponential equations
- describe the features of an exponential curve, including the y-intercept and the asymptote.

▶ 8.5.1 Exponential equations

eles-6238

- An exponential graph is a non-linear graph whose equation is $y = a^x$, where $a > 0$ and $a \neq 1$, and x is the power or exponent.
- We can create a table of values to sketch the graph of an exponential equation.
- Worked example 9 demonstrates how to draw a graph of the exponential curve $y = 2^x$ by completing a table of values and plotting points.
- Remember: $2^{-3} = \dfrac{1}{2^3}$

WORKED EXAMPLE 9 Using a table of values to draw the graph of an exponential equation

Complete the table of values below and use it to draw the graph of $y = 2^x$.

x	−3	−2	−1	0	1	2	3
y							

THINK

1. Substitute each x-value into the function $y = 2^x$ to obtain the corresponding y-value.

WRITE/DRAW

x	−3	−2	−1	0	1	2	3
y	$\dfrac{1}{8}$	$\dfrac{1}{4}$	$\dfrac{1}{2}$	1	2	4	8

2. Draw a set of axes and plot the points from the table. Join them with a smooth curve.

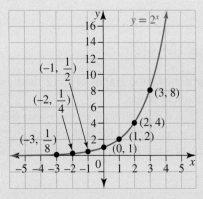

- The graph in Worked example 9 has several important features.
  - The y-intercept is 1.
  - The value of y is always greater than zero.
  - As x increases, y becomes very large.
  - As x decreases, y gets closer to but never reaches zero. So the graph gets closer to but never touches the x-axis. The x-axis (or the line $y = 0$) is called a horizontal **asymptote**.

8.5.2 Comparing exponential curves

eles-6239

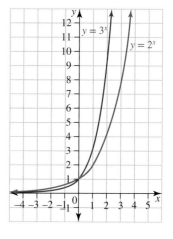

- Drawing the graphs of exponential curves allows the curves to be compared.
- When comparing exponential curves, consider:
  - the shape of the curve
  - the y-intercept
  - the nature of the curve for very large and very small values of x
  - reflections in the x- or y- axis
  - the equation of the asymptote.
- The diagram shows the graphs of $y = 2^x$ and $y = 3^x$.
- Comparing these two graphs:
  - the y-intercept is 1 for both graphs
  - $y = 0$ is an asymptote for both graphs
  - the graph of $y = 3^x$ climbs more steeply than the graph of $y = 2^x$.
- In general, for $a > 1$, as a increases, the curve climbs more steeply or quickly

Reflection in the x-axis

- The diagram at right shows the graphs of $y = 2^x$ and $y = -2^x$.
- The graphs have identical shape.
- The graph of $y = -2^x$ is a reflection about the x-axis of the graph $y = 2^x$.
- The x-axis ($y = 0$) is an asymptote for both graphs.
- In general, the graph of $y = -k^x$ is a reflection in the x-axis of the graph of $y = k^x$.

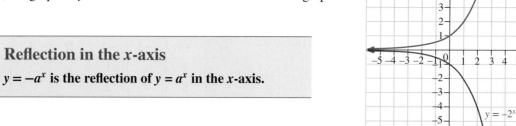

> **Reflection in the x-axis**
>
> $y = -a^x$ **is the reflection of** $y = a^x$ **in the x-axis.**

Reflection in the y-axis

- The diagram at right shows the graphs of $y = 2^x$ and $y = 2^{-x}$.
- The graphs have identical shape.
- The graph of $y = 2^{-x}$ is a reflection about the x-axis of the graph $y = 2^x$.
- The x-axis ($y = 0$) is an asymptote for both graphs.
- In general, the graph of $y = -k^x$ is a reflection in the y-axis of the graph of $y = a^x$.

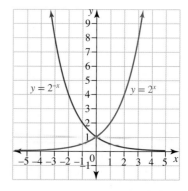

> **Reflection in the y-axis**
>
> $y = a^{-x}$ **is the reflection of** $y = a^x$ **in the y-axis.**

DISCUSSION

Will the graph of an exponential function always have a horizontal asymptote? Why?

WORKED EXAMPLE 10 Comparing graphs of exponential equations

Sketch and describe similarities and differences between the following pairs of graphs.
a. $y = 2^x$ and $y = -2^x$
b. $y = 2^x$ and $y = 2^{-x}$

THINK	WRITE/DRAW
a. 1. Substitute each x-value into the function $y = 2^x$ to obtain the corresponding y-value. Substitute each x-value into the function $y = -2^x$ to obtain the corresponding y-value.	<table><tr><td>x</td><td>−3</td><td>−2</td><td>−1</td><td>0</td><td>1</td><td>2</td><td>3</td></tr><tr><td>$y = 2^x$</td><td>$\frac{1}{8}$</td><td>$\frac{1}{4}$</td><td>$\frac{1}{2}$</td><td>1</td><td>2</td><td>4</td><td>8</td></tr><tr><td>$y = -2^x$</td><td>$-\frac{1}{8}$</td><td>$-\frac{1}{4}$</td><td>$-\frac{1}{2}$</td><td>−1</td><td>−2</td><td>−4</td><td>−8</td></tr></table>
2. Draw a set of axes and plot the points from the table. Join them with a smooth curve.	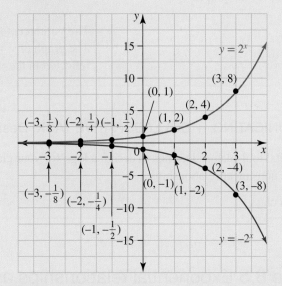
3. Compare the graphs.	The graphs of $y = 2^x$ and $y = -2^x$ are the same shape. The y-intercept of $y = 2^x$ is 1, and the y-intercept of $y = -2^x$ is −1. The graph of $y = -2^x$ is the graph of $y = 2^x$ reflected in the x-axis.
b. 1. Substitute each x-value into the function $y = 2^x$ to obtain the corresponding y-value. Substitute each x-value into the function $y = 2^{-x}$ to obtain the corresponding y-value.	<table><tr><td>x</td><td>−3</td><td>−2</td><td>−1</td><td>0</td><td>1</td><td>2</td><td>3</td></tr><tr><td>$y = 2^x$</td><td>$\frac{1}{8}$</td><td>$\frac{1}{4}$</td><td>$\frac{1}{2}$</td><td>1</td><td>2</td><td>4</td><td>8</td></tr><tr><td>$y = 2^{-x}$</td><td>8</td><td>4</td><td>2</td><td>1</td><td>$\frac{1}{2}$</td><td>$\frac{1}{4}$</td><td>$\frac{1}{8}$</td></tr></table>

2. Draw a set of axes and plot the points from the table. Join them with a smooth curve.

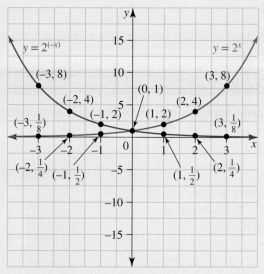

3. Compare the graphs.

The graphs of $y = 2^x$ and $y = 2^{-x}$ are the same shape.
The y-intercept for both graphs is 1.
The graph of $y = 2^{-x}$ is the graph of $y = 2^x$ reflected in the y-axis.

COMMUNICATING — COLLABORATIVE TASK: Designing a logo

Use a digital tool such as Desmos to design a logo for your class or sporting team. Use a combination of linear and non-linear relationships.

Share as a class.

Exercise 8.5 Exponential relationships and graphs

learn on

8.5 Quick quiz on	8.5 Exercise

Individual pathways

■ PRACTISE	■ CONSOLIDATE	■ MASTER
1, 4, 6, 9	2, 3, 5, 7, 10	8, 11, 12, 13

Fluency

1. **WE9** Complete the table below and use the table to plot the graph of $y = 3^x$ for $-3 \leq x \leq +3$.

x	−3	−2	−1	0	1	2	3
y							

2. If $x = 1$, determine the value of y when:
   a. $y = 2^x$
   b. $y = 3^x$
   c. $y = 4^x$
   d. $y = 10^x$
   e. $y = a^x$.

3. Using graphing technology, sketch the graphs of $y = 2^x$, $y = 3^x$ and $y = 4^x$ on the same set of axes.

a. What do the graphs have in common?
b. How does the value of the base (2, 3, 4) affect the graph?
c. Predict where the graph $y = 8^x$ would lie and sketch it in.

4. a. Using graphing technology, sketch the graphs of:

   i. $y = 2^x$ and $y = -2^x$
   ii. $y = 3^x$ and $y = -3^x$
   iii. $y = 6^x$ and $y = -6^x$.

   b. What is the relationship between these pairs of graphs?

5. a. Using graphing technology, sketch the graphs of:

   i. $y = 2^x$ and $y = 2^{-x}$
   ii. $y = 3^x$ and $y = 3^{-x}$
   iii. $y = 6^x$ and $y = 6^{-x}$.

   b. What is the relationship between these pairs of graphs?

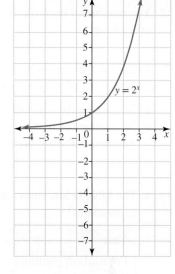

Understanding

6. Given the graph of $y = 2^x$, sketch on the same axes the graphs of:

   a. $y = -2^x$
   b. $y = 2^{-x}$

7. Given the graph of $y = 3^x$, sketch on the same axes the graph of $y = -3^x$.

8. Given the graph of $y = 4^x$, sketch on the same axes the graph of $y = 4^{-x}$.

9. Match each of the following graphs with its correct label.

   a. $y = 2^x$
   b. $y = 3^x$
   c. $y = -4^x$
   d. $y = 5^{-x}$

   i.
   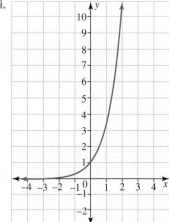

   ii.
   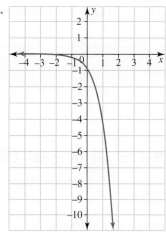

   iii.
   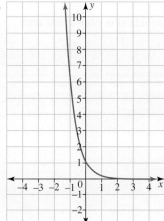

   iv.
   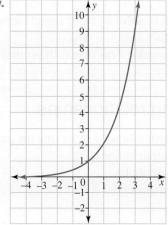

Communicating, reasoning and problem solving

10. By considering the graph of $y = 3^x$, sketch the graph of $y = -3^{-x}$.

11. The graph of $y = 16^x$ can be used to solve for x in the exponential equation $16^x = 32$. Draw a graph of $y = 16^x$ and use it to solve $16^x = 32$.

12. The number of bacteria, N, in a certain culture is reduced by a third every hour, so:

$$N = N_0 \times \left(\tfrac{1}{3}\right)^t$$

where t is the time in hours after 12 noon on a particular day. Initially there are 10 000 bacteria present.

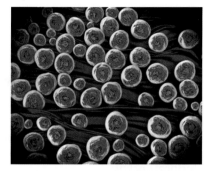

   a. Determine the value of N_0.

   b. Determine the number of bacteria, correct to the nearest whole number, in the culture when:

      i. $t = 2$ **ii.** $t = 5$ **iii.** $t = 10$.

13. **a.** The table below shows the population of a city between 1850 and 1930. Is the population growth exponential?

Year	1850	1860	1870	1880	1890	1900	1910	1920	1930
Population (million)	1.0	1.3	1.69	2.197	2.856	3.713	4.827	6.275	8.157

   b. What is the common ratio in part **a**?
   c. What is the annual percentage increase?
   d. Estimate the population in 1895.
   e. Estimate the population in 1980.

LESSON
8.6 Applications of linear and non-linear relationships

LEARNING INTENTION

At the end of this lesson you should be able to:
- identify and sketch graphs of linear and non-linear relationships
- use graphing tools to solve a pair of simultaneous equations where one equation is non-linear and interpret the results
- apply non-linear relationships in real-life contexts.

8.6.1 Identifying relationships

eles-6240

- We can identify relationships by observing the general graphs:
  - $y = mx + c$ for straight lines
  - $y = kx^2 + c$ for parabolas
  - $y = a^x$ for exponential curves.

Identify and sketch the following equations.

a. $y = 3x$

b. $y = 3x^2$

c. $y = 3^x$

THINK	WRITE

a. 1. Write the curve.

$y = 3x$

2. Choose the general pattern.

$y = mx + c$

3. Identify and state the features of the curve.

$y = 3x$ is a straight line, with a gradient of 3 and passing through the origin $(0, 0)$.

4. Sketch the curve $y = 3x$.

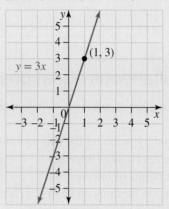

b. 1. Write the curve.

$y = 3x^2$

2. Choose the general pattern.

$y = kx^2 + c$

3. Identify and state the features of the curve.

$y = 3x^2$ is a parabola with vertex $(0, 0)$ and axis of symmetry $x = 0$ passing through the point $(1, 3)$.

4. Sketch the curve $y = 3x^2$.

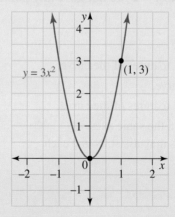

c. 1. Write the curve.

$y = 3^x$

2. Choose the general pattern.

$y = a^x$

3. Identify and state the features of the curve.

$y = 3^x$ is an exponential graph passing through the points $(0, 1)$ and $(1, 3)$, and the x-axis is an asymptote.

4. Sketch the curve $y = 3^x$.

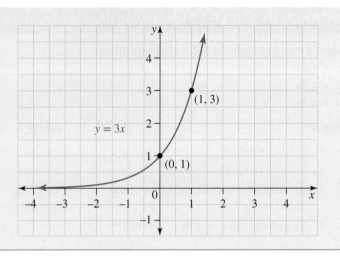

8.6.2 Solving simultaneous equations with one non-linear equation

eles-6241

- Solving a pair of simultaneous equations is to determine the coordinates of any point or points that satisfy both equations.
- The solution can be found by graphing the two equations and identifying the coordinates of the point or points where the two curves cross or intersect.
- Not all graphs will have points of intersection.
- The accuracy of the solution depends on having an accurate graph.
- Digital technology is a useful tool when graphing.

WORKED EXAMPLE 12 Solving simultaneous equations by graphing

a. **Determine the point of intersection of $y = x^2 - 2$ and $y = 2x + 1$.**
b. **Explain why $y = -2$ and $y = 2^x$ do not have any point or points of intersection.**

THINK

a. 1. Write the first curve.

   2. Identify the general features of the curve.

   3. Write the second curve.

   4. Identify the general features of the curve.

   5. Sketch the curves on the same axes. Plotting extra points may assist. Name each curve.

WRITE

a. $y = x^2 - 2$

Parabola, vertex $(0, -2)$, concave up

$y = 2x + 1$

Straight line passing through $(0, 1)$ and $(1, 3)$

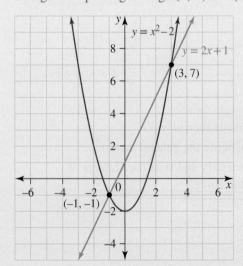

6. On the graph, locate the points of intersection and write the solutions.

The points of intersection are $(-1, -1)$ and $(3, 7)$.

b. 1. Graph the curve $y = 2^x$ and identify its key features: y-intercept at $(0, 1)$; asymptote $y = 0$.

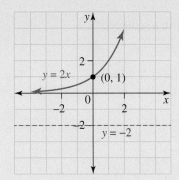

2. Sketch $y = -2$.
3. Comment on the results.

The exponential graph is always positive, so there are no points of intersection with the straight line, $y = -2$.

⊳ 8.6.3 Non-linear relationships in real-life contexts

eles-6244

- Exponential graphs can be used to model many real-life situations involving natural growth.
- The general equation for exponential growth is $y = k \times a^x$ where:
  - $a > 1$, the growth rate per hour
  - k is the initial quantity that is growing
  - x is used to represent time.
- In real-life problems, time cannot be negative.

WORKED EXAMPLE 13 Real-life contexts using exponential graphs

The number of bacteria, N, in a Petri dish after x hours is given by the equation $N = 50 \times 2^x$.
  a. Identify the initial number of bacteria in the Petri dish.
  b. Determine the number of bacteria in the Petri dish after 3 hours.
  c. Draw the graph of the function of N against x.
  d. Use the graph to estimate the length of time it will take for the initial number of bacteria to treble.

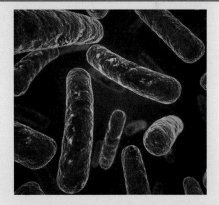

THINK	WRITE/DRAW
a. 1. Write the equation.	a. $N = 50 \times 2^x$
2. Substitute $x = 0$ into the given formula and evaluate. (Notice that this is the value of x for equations of the form $y = k \times a^x$.)	When $x = 0$, $N = 50 \times 2^0$ $= 50 \times 1$ $= 50$
3. Write the answer in a sentence.	The initial number of bacteria in the Petri dish is 50.

▶

b. **1.** Substitute $x = 3$ into the formula and evaluate.

b. When $x = 3$, $N = 50 \times 2^3$
$$= 50 \times 8$$
$$= 400$$

2. Write the answer in a sentence.

After 3 hours there are 400 bacteria in the Petri dish.

c. **1.** Calculate the value of N when $x = 1$ and $x = 2$.

c. At $x = 1$, $N = 50 \times 2^1$
$$= 50 \times 2$$
$$= 100$$

At $x = 2$, $N = 50 \times 2^2$
$$= 50 \times 4$$
$$= 200$$

2. Draw a set of axes, labelling the horizontal axis as x and the vertical axis as N.

3. Plot the points generated by the answers to parts **a**, **b** and **c 1**.

4. Join the points plotted with a smooth curve.

5. Label the graph.

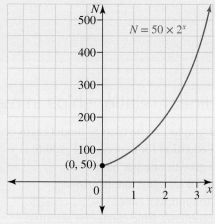

d. **1.** Determine the number of bacteria required.

d. Number of bacteria $= 3 \times 50$
$$= 150$$

2. Draw a horizontal line from $N = 150$ to the curve and from this point draw a vertical line to the x-axis. This will help us to the estimate the time taken for the number of bacteria to treble.

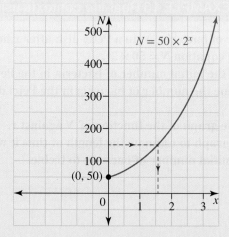

3. Write the answer in a sentence.

The time taken will be approximately 1.6 hours.

Exercise 8.6 Applications of linear and non-linear relationships **learn** on

| 8.6 Quick quiz   on | 8.6 Exercise |

Individual pathways

■ PRACTISE	■ CONSOLIDATE	■ MASTER
1, 4, 7, 10, 13, 16	2, 5, 8, 11, 14, 17	3, 6, 9, 12, 15, 18

Fluency

1. **WE11** Identify and sketch the following curves.

   a. $y = 2x$ b. $y = 2x^2$ c. $y = 2^x$

2. Identify and sketch the following curves.

   a. $y = 4x$ b. $y = 4x^2$ c. $y = 4^x$

3. Identify and sketch the following curves.

   a. $y = -5x$ b. $y = -5x^2$ c. $y = -5^x$

4. **MC** Which of the following equations is a parabola?

   A. $y = 3^x + 7$ B. $y = 7 - 3^x$ C. $y = 7 - 3x^2$ D. $y = 7x + 3$

5. Is the following statement true or false?
   The graph of the curve $y = 2x^2$ is wider than the curve $y = x^2$

6. Is the following statement true or false?
   The graph of the curve $y = 2^x$ is steeper than the curve $y = 3^x$

Understanding

7. **WE12a** Determine the point(s) of intersection of $y = x^2$ and $y = x + 6$.

8. Determine the point(s) of intersection of $y = x^2$ and $y = 4x - 4$.

9. Determine the point(s) of intersection of $y = x^2 - 2$ and $y = x - 10$.

10. **WE12b** Explain how many points the curves $y = 2^x$ and $y = 4$ have in common.

11. Determine the point(s) of intersection of the curves $y = 2^x$ and $y = 1 - x$.

12. Determine the point(s) of intersection the curves $y = -2^x$ and $y = 2x - 1$.

Communicating, reasoning and problem solving

13. **WE13** The number of bacteria, N, in a dish after x hours is given by the equation

$$N = 20 \times 3^x$$

   a. Determine how many bacteria were in the dish initially.
   b. Determine the number of bacteria in the dish after 2 hours.
   c. Sketch the graph of the function of N against x.
   d. Use the graph to estimate the length of time it will take for the initial number of bacteria to treble.

14. A growth of a population, P, after x hours is given by the equation

$$P = 40 \times 2^x$$

a. Determine the initial size of the population.
b. Determine the size of the population after 5 hours.
c. Sketch the graph of the function of P against x.
d. Use the graph to estimate the length of time it will take for the initial population to treble in size.

15. An investment, $\$P$, after x years is given by the equation

$$P = 20\,000 \times 1.08^x$$

a. How much was the original investment?
b. Determine the value of the investment after 2 years.
c. Sketch the graph of the function of P against x.
d. Use the graph to estimate the time it will take for the original investment to increase by 25%. Give your answer correct to the nearest year.

16. A spurt of water emerging from an outlet just below the surface of an ornamental fountain follows a parabolic path described by the equation $h = -x^2 + 8x$, where h is the height of the water and x is the horizontal distance from the outlet in metres.
Plot the graph of $h = -x^2 + 8x$, for $0 \le x \le 8$, and hence state:

a. the coordinates of the vertex of the parabola
b. the maximum height of the water above the surface of the fountain
c. the horizontal distance from the outlet when the water is at a height of 10 metres.
Give your answer to 1 decimal place.

17. A railway bridge has an arch below it that can be modelled by the equation $h = 6x - x^2$ where h is the height of the bridge, in metres, above the land.
Plot the graph of $h = 6x - x^2$, for $0 \le x \le 6$ and hence state:

a. the coordinates of the vertex of the arch
b. the maximum height of the bridge
c. the width of the arch 5 metres above the land, to 1 decimal place.

18. Jack and his dog were playing outside. Jack was throwing a stick in the air for his dog to catch. The height of the stick, in metres, followed the equation $h = 20t - 5t^2$, where t is time in minutes.

a. The graph of h is a parabola. State if the parabola is concave up or concave down.
Explain your answer.
b. Plot the graph of $h = 20t - 5t^2$ for $0 \le t \le 4$ and hence calculate the maximum height reached by the stick.
c. For approximately how long is the stick 10 metres above the ground?

LESSON
8.7 Review

8.7.1 Topic summary

NON-LINEAR RELATIONSHIPS

The parabola

- The graph of a quadratic equation is called a parabola.

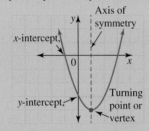

Axis of symmetry

- Every parabola has an axis of symmetry. This is the vertical line that passes through the parabola's turning point.

Transformations of the parabola

- The parabola with the equation $y = x^2$ has a turning point at $(0,0)$.
- A **dilation** from the x-axis stretches the parabola.
- A **dilation** of factor k from the x-axis produces the equation $y = kx^2$ and:
  - will produce a narrow graph for $k > 0$
  - will produce a wider graph for $0 < k < 1$
  - for $k < 0$ (k is negative), the graph is **reflected** in the x-axis.

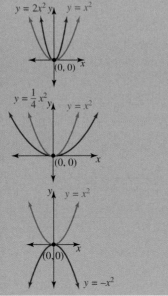

Intercepts

- All parabolas have a single y-intercept. The location of this y-intercept can be found by letting $x = 0$.
- Parabolas can have two, one or no intercepts.
- The x-intercepts can be found by letting $y = 0$ and solving for x. If the equation can't be solved, there is no x-intercept.

Parabolas of the form $y = kx^2$ and $y = kx^2 + c$

- Parabolas of the form $y = kx^2$ have a turning point at the origin.
- If $k > 0$, the graph is ∪-shaped and the turning point is a local minimum.
- If $k < 0$, the graph is ∩-shaped and the turning point is a local maximum.
- Larger values of k create narrower graphs. Smaller values of k create wider graphs.

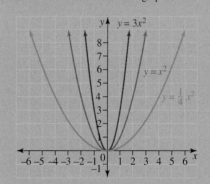

- Parabolas of the form $y = kx^2 + c$ have a vertical translation. A positive c moves the curve up and a negative c moves the curve down.

Exponential relationships

- Equations of the form $y = a^x$ are called exponential functions.
- $y = a^x$ has:
  - a horizontal asymptote at $y = 0$
  - a y-intercept at $(0, 1)$ for $a > 0$ and at $(0, -1)$ for $a < 0$.
- The curve $y = -a^x$ is a reflection of $y = a^x$ in the x-axis ($a > 1$).
- The curve $y = a^{-x}$ is a reflection of $y = a^x$ in the y-axis ($a > 1$).

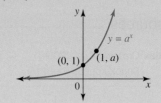

8.7.2 Project

Shaping up!

Many beautiful patterns are created by starting with a single function or relation and transforming and repeating it over and over.

In this task you will apply what you have learned about functions, relations and transformations (dilations, reflections and translations) to explore mathematical patterns.

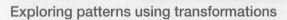

Exploring patterns using transformations

1. **a.** On the same set of axes, draw the graphs of:

   i. $y = x^2 - 4x + 1$ **ii.** $y = x^2 - 3x + 1$ **iii.** $y = x^2 - 2x + 1$

   iv. $y = x^2 + 2x + 1$ **v.** $y = x^2 + 3x + 1$ **vi.** $y = x^2 + 4x + 1$

   b. Describe the pattern formed by your graphs. Use mathematical terms such as intercepts, turning point, shape and transformations.

   What you have drawn is referred to as a family of curves — curves in which the shape of the curve changes if the values of a, b and c in the general equation $y = ax^2 + bx + c$ change.

   c. Explore the family of parabolas formed by changing the values of a and c. Comment on your findings.

   d. Explore exponential functions belonging to the family of curves with equation $y = ka^x$, families of cubic functions with equations $y = ax^3$ or $y = ax^3 + bx^2 + cx + d$, and families of quartic functions with equations $y = ax^4$ or $y = ax^4 + bx^3 + cx^2 + dx + e$. Comment on your findings.

   e. Choose one of the designs shown below and recreate it (or a simplified version of it). Record the mathematical equations used to complete the design.

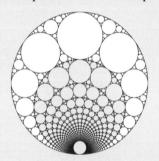
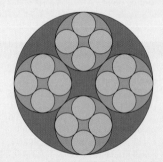

Coming up with your design

2. Use what you know about transformations to functions and relations to create your own design from a basic graph. You could begin with a circle, add some line segments and then repeat the pattern with some change. Record all the equations and restrictions you use.
   It may be helpful to apply your knowledge of inverse functions too.
   A digital technology will be very useful for this task.
   Create a poster of your design to share with the class.

 Resources

 Interactivities Crossword (int-0708)
Sudoku puzzle (int-3215)

Fluency

Questions **1** to **4** refer to the graph shown

1. **MC** The axis of symmetry for the graph shown is:

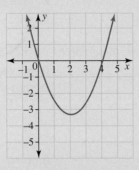

A. $x = 0$ B. $x = 2$ C. $x = 4$ D. $x = -4$

2. **MC** The coordinates of the vertex for the graph are:
   A. $(0, 0)$ B. $(2, 4)$ C. $(2, -4)$ D. $(-2, -4)$

3. **MC** The y-intercept is:
   A. 0 B. -2 C. -4 D. 2

4. **MC** The x-intercepts are:
   A. 2 and -4 B. 0 and 2 C. 0 and -2 D. 0 and 4

5. **MC** The graph of $y = -4x^2$ is:
   A. wider than $y = x^2$ B. narrower than $y = x^2$
   C. the same width as $y = x^2$ D. a reflection of $y = x^2$ in the x-axis

6. **MC** The graph of $y = \frac{1}{2}x^2$ is:

   A. wider than $y = x^2$ B. narrower than $y = x^2$
   C. the same width as $y = x^2$ D. a reflection of $y = x^2$ in the x-axis

7. **MC** Compared to the graph of $y = x^2$, the graph of $y = -2x^2$ would be:
   A. half as wide   B. twice as wide
   C. moved 2 units to the right   D. moved 2 units up

8. **MC** The vertex of the graph of the equation $y = -x^2 + 6$ is:
   A. $(0, 0)$ B. $(-1, 6)$ C. $(0, 6)$ D. $(0, -6)$

9. **MC** The vertex of the graph of the equation $y = x^2 - 12$ is:
   A. $(1, -12)$ B. $(-12, 1)$ C. $(0, 12)$ D. $(0, -12)$

10. **MC** The graph of $y = x^2 - 4$ moves the graph of $y = x^2$ in the following way:

   A. up 4

   B. down 4

   C. no change

   D. up 2

11. **MC** The y-intercept of the graph of $y = -x^2 - 3$ is:

   A. 1

   B. -1

   C. 3

   D. -3

12. **MC** Compared to the graph of $y = x^2$, the graph of $y = x^2 - 2$ would be:

   A. moved 2 units to the left

   B. moved 2 units to the right

   C. moved 2 units up

   D. moved 2 units down

Understanding

13. **MC** The asymptote for $y = 3^{-x}$ is:

   A. $x = -1$

   B. $x = 0$

   C. $y = 3$

   D. $y = 0$

14. **MC** The y-intercept of the curve $y = 5 \times 2^x$ is:

   A. $(0, 5)$

   B. $(5, 0)$

   C. $(0, 2)$

   D. $(0, 7)$

15. **MC** The graph of $y = -3 \times 2^x$ is best represented by:

   A.
   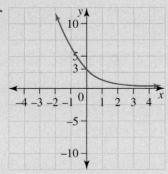

   B.
   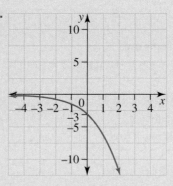

   C.
   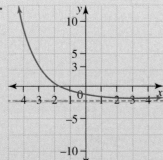

   D.
   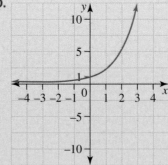

16. Match each equation with its correct graph.

   i. $y = 2^x$ **ii.** $y = 3^x$ **iii.** $y = -4^x$ **iv.** $y = 5^{-x}$

a.
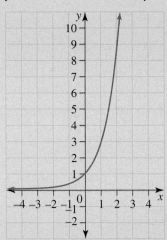

b.
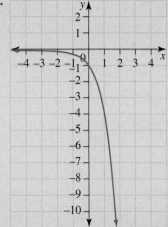

c.
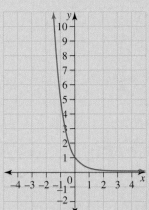

d.
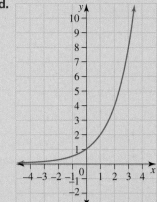

17. Match each equation with its correct graph

   i. $y = x^2 + 3$ **ii.** $y = 2x + 3$ **iii.** $y = 3^x$ **iv.** $y = 3x^2$

a.
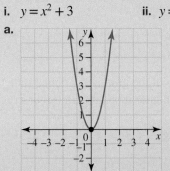

b.
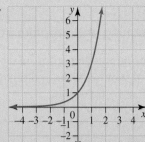

c.
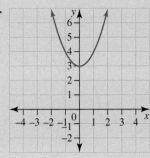

d.
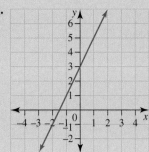

18. For each of the following graphs, state the equation of the axis of symmetry, the coordinates of the vertex, whether the point is a maximum or a minimum, and the x- and y-intercepts.

a.
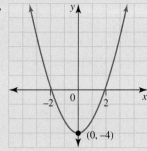

b.
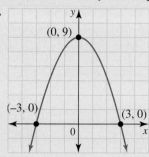

c.
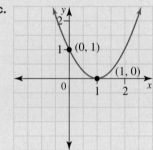

d.
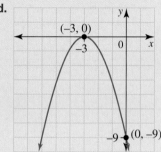

e.
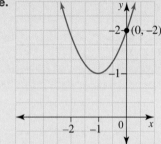

f.
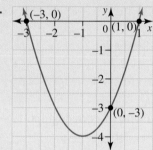

Communicating, reasoning and problem solving

19. For each of the following equations, draw a table of values, use it to plot the graph, and hence state the equation of the axis of symmetry, the coordinates of the vertex and whether it is a maximum or a minimum, and the x- and y-intercepts.

a. $y = x^2 - 4x,\ -2 \le x \le 6$

b. $y = -x^2 - 2x + 8,\ -5 \le x \le 3$

c. $y = 2x^2 - 4x + 4,\ -2 \le x \le 3$

d. $y = -x^2 + 6x - 5,\ 0 \le x \le 6$

20. Sketch each of the following graphs, labelling the axis of symmetry, the vertex and the intercepts, and stating the type of vertex.

a. $y = 2x^2$

b. $y = \frac{1}{2}x^2$

c. $y = -4x^2$

d. $y = -\frac{1}{3}x^2$

21. Sketch each of the following graphs, labelling the axis of symmetry, the vertex and the y-intercept, and stating the nature of the vertex.

a. $y = x^2 + 2$ **b.** $y = x^2 - 4$

c. $y = x^2 + 5$ **d.** $y = x^2 - 3$

22. A missile was fired from a boat during a test. The missile's path could be traced by the equation $h = -\dfrac{1}{4}x^2 + x$, where h is the missile's height above the water in kilometres, and x is the horizontal distance from the boat in kilometres.

Plot the graph for $0 \le x \le 4$ and use it to determine the maximum height of the missile's path, in metres.

23. The height, h metres, of a golf ball t seconds after it is hit is given by the equation $h = 4t - t^2$.

a. Sketch the graph of the path of the ball.
b. Calculate the maximum height the golf ball reaches.
c. Determine how long it takes for the ball to reach the maximum height.
d. Determine how long it is before the ball lands on the ground after it has been hit.

24. A soccer ball is kicked upwards in the air. The height, h metres, t seconds after the kick is modelled by the quadratic equation $h = -5t^2 + 20t$.

a. Sketch the graph of this relationship.
b. Determine how many seconds the ball is in the air.
c. Determine the maximum height reached by the ball.
d. Calculate how many seconds the ball is above a height of 10 metres.

25. A growth of a population, P, after x hours is given by the equation

$$P = 20 \times 1.5^x$$

a. Determine the initial size of the population.
b. Determine the size of the population after 2 hours.
c. Sketch the graph of the function of P against x by plotting points for $0 \le x \le 5$.
d. Use the graph to estimate the length of time it will take for the initial population to increase in size to 80. Give your answer to the nearest half hour.

 To test your understanding and knowledge of this topic, go to your learnON title at www.jacplus.com.au and complete the **post-test**.

Answers

Topic 8 Non-linear relationships

8.1 Pre-test

1. $x = -2$
2. D
3.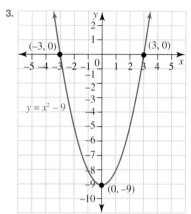
4.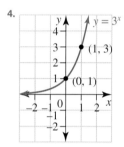
5. C
6. $(0, -8)$
7. $(0, 1)$
8. $(1, -3)$
9. C
10. D
11. C
12. D
13. D
14. $(0, -9)$
15. D

8.2 Quadratic relationships and graphs

1. a. Axis of symmetry: $x = 0$, turning point $(0, 0)$, minimum
   b. Axis of symmetry: $x = 0$, turning point $(0, -3)$, minimum
   c. Axis of symmetry: $x = -1$, turning point $(-1, -2)$, minimum
   d. Axis of symmetry: $x = 0$, turning point $(0, 0)$, maximum
   e. Axis of symmetry: $x = 0$, turning point $(0, 2)$, maximum
   f. Axis of symmetry: $x = 2$, turning point $(2, -1)$, maximum

2. a. $x = 0$, $(0, 1)$, minimum
   b. $x = 1$, $(1, -3)$, minimum
   c. $x = -2$, $(-2, 2)$, maximum
   d. $x = -1$, $(-1, -2)$, maximum
   e. $x = 2$, $(2, 2)$, minimum
   f. $x = 0$, $(0, 1)$, maximum

3. a. $x = 0$; TP $(0, -1)$, minimum; x-intercepts are -1 and 1, y-intercept is -1.
   b. $x = 0$; TP $(0, 1)$, maximum; x-intercepts are -1 and 1, y-intercept is 1.
   c. $x = 1$; TP $(1, -4)$, minimum; x-intercepts are -1 a and 3, y-intercept is -3.
   d. $x = -2$; TP $(-2, 1)$, maximum; x-intercepts are -3 and -1, y-intercept is -3.
   e. $x = -\dfrac{1}{2}$; TP $\left(-\dfrac{1}{2}, -1\right)$, minimum; x-intercepts are $-1\dfrac{1}{2}$ and $\dfrac{1}{2}$, y-intercept is $-\dfrac{3}{4}$.
   f. $x = \dfrac{1}{2}$; TP $\left(\dfrac{1}{2}, 2\right)$, minimum; no x-intercepts, y-intercept is $2\dfrac{1}{4}$.

4. a. B   b. C   c. A   d. D

5. a. $y = x^2 + 8x + 15$, $-7 \le x \le 0$
   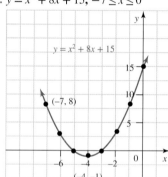

   i. $x = -4$
   ii. $(-4, -1)$, minimum
   iii. The x-intercepts are -5 and -3; the y-intercept is 15.

   b. $y = x^2 - 1$, $-3 \le x \le 3$
   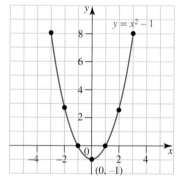

   i. $x = 0$
   ii. $(0, -1)$, minimum
   iii. The x-intercepts are -1 and 1; the y-intercept is -1.

6. a. $y = x^2 - 4x, \; -1 \le x \le 5$

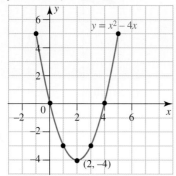

$y = x^2 - 4x$

(2, −4)

  i. $x = 2$

  ii. $(2, -4)$, minimum

  iii. The x-intercepts are 0 and 4; the y-intercept is 0.

b. $y = x^2 - 2x + 3, \; -2 \le x \le 4$

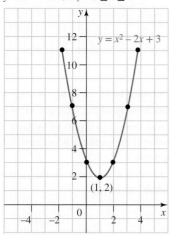

$y = x^2 - 2x + 3$

(1, 2)

  i. $x = 1$

  ii. $(1, 2)$, minimum

  iii. No x-intercepts, y-intercept is 3.

7. a. $y = x^2 + 12x + 35, \; -9 \le x \le 0$

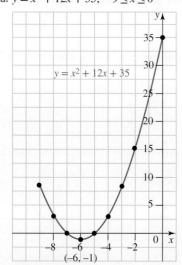

$y = x^2 + 12x + 35$

(−6, −1)

  i. $x = -6$

  ii. $(-6, -1)$, minimum

  iii. The x-intercepts are −7 and −5; the y-intercept is 35.

b. $y = -x^2 + 4x + 5, \; -2 \le x \le 6$

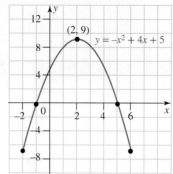

(2, 9)

$y = -x^2 + 4x + 5$

  i. $x = 2$

  ii. $(2, 9)$, maximum

  iii. The x-intercepts are −1 and 5; the y-intercept is 5.

8.

x	−3	−2	−1	0	1	2	3
$y = x^2 + 2$	11	6	3	2	3	6	12
$y = x^2 + 3$	12	7	4	3	4	7	12

a.

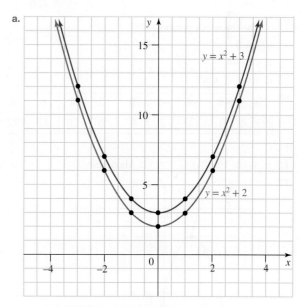

$y = x^2 + 3$

$y = x^2 + 2$

b. Axis of symmetry: $x = 0$ for both equations.

c. No x-intercepts for either equation.

9. a.

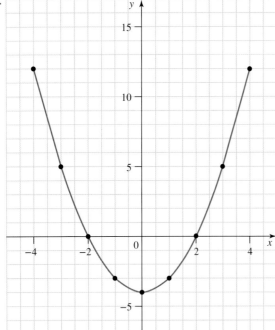

The shape is a parabola.

b. $x = 0$

c. $y = -4$

d. $x = -2$, $x = 2$

10.

x	-6	-4	-2	0	2	4	6	8
y	30	12	2	0	6	20	42	72

11. C

12. a. $(0, 0)$ **b.** Maximum

13. a.

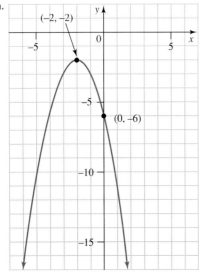

b.

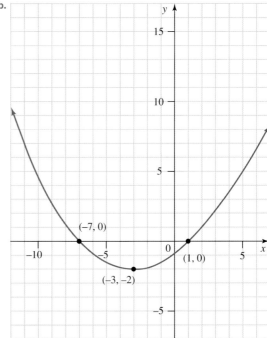

14. Answers will vary but must be of the form $y = a(x - h)^2$.

15. Answers will vary but must be of the form $y = ax^2 + c$.

16. Answers will vary. If the parabola is upright, it has a minimum turning point. If the parabola is inverted, it has a maximum turning point.

17. Axis of symmetry crosses halfway between the x-intercepts. Let the unknown x-intercept be at p.

$$-4 = \frac{10 + p}{2}$$
$$-8 = 10 + p$$
$$p = -18$$

Therefore, the other x-intercept is at $(-18, 0)$.

18. Let $(a, 0)$ be the point at which the axis of symmetry crosses the x-axis.

As a will lie halfway between the x-intercepts $(-2, 0)$ and $(5, 0)$:

$$a = \frac{-2 + 5}{2} = \frac{3}{2} = 1.5$$

Therefore, the axis of symmetry has the equation $x = 1.5$.

19.

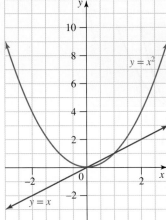

By inspection or algebra, the graphs meet at $(0, 0)$ and again at $(1, 1)$.

20. a. $\dfrac{x_2 + 10}{2} = -4$

$x_2 = 2$

b. Answers will vary but must be of the form $y = a(x - 2)(x + 10)$.

21.

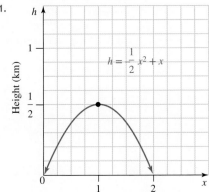

The maximum height is 500 m.

22. a. $-21\,°C$ **b.** 1 hour and at 21 hours

   c. $t = 11$ hours   **d.** $100\,°C$

8.3 Quadratic equations of the form $y = kx^2$

1.

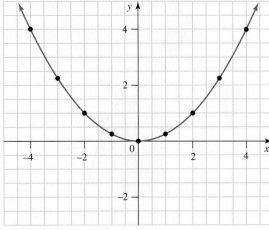

$y = 4x^2$ is narrower.
The turning point for each is at $(0, 0)$; the x-intercept and y-intercept are 0 for both.

2.

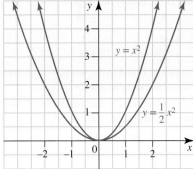

$y = x^2$ is narrower.
The turning point for each is at $(0, 0)$; the x-intercept and y-intercept are 0 for both.

3.

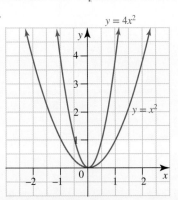

$y = 0.25x^2$

4. a, b. See the figure at the foot of the page.*

5. a. B   **b.** A   **c.** B

6. a. iii   **b.** vi   **c.** i
   d. ii   **e.** iv   **f.** v

7. Sample responses can be found in the worked solutions in the online resources; for example, $y = 2x^2$.

8. Sample responses can be found in the worked solutions in the online resources; for example, $y = -0.5x^2$.

9. a. $y = 3x^2$ **b.** $y = -x^2$

10. a. -350 **b.** -350 **c.** -31.5
   d. -7.875 **e.** -16.94

11. a.

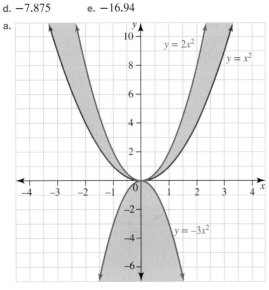

b.

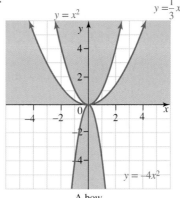

A bow

12. $y = \dfrac{2}{3}x^2$

13. a. Sample responses can be found in the worked solutions in the online resources.
   b. 6

14. a. 400 watts   **b.** 625 watts

*4. a, b.

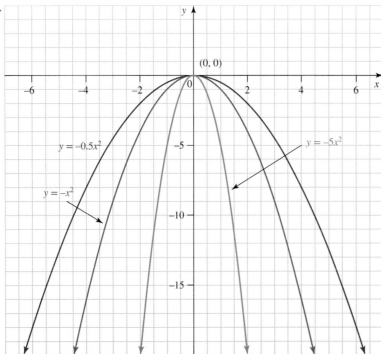

8.4 Quadratic equations of the form $y = kx^2 + c$

1. a.

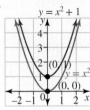

$y = x^2 + 1$

b.

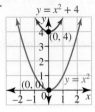

$y = x^2 + 4$

c.

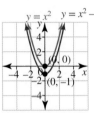

$y = x^2$ $y = x^2 - 1$

d.

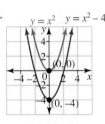

$y = x^2$ $y = x^2 - 4$

e.

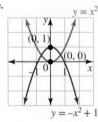

$y = x^2$
$y = -x^2 + 1$

f.

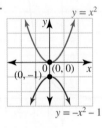

$y = x^2$
$y = -x^2 - 1$

2. A positive number moves the graph up.

3. A negative number moves the graph down.

4. a.

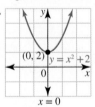

$y = x^2 + 2$
Minimum

b.

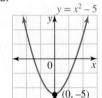

$y = x^2 - 5$
Minimum

c.

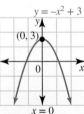

$y = -x^2 + 3$
Maximum

d.

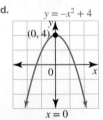

$y = -x^2 + 4$
Maximum

e.

$y = -x^2 - 3$
Maximum

f.

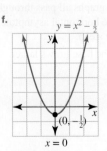

$y = x^2 - \frac{1}{2}$
Minimum

5. a. x-intercepts: $(-2, 0) (2, 0)$

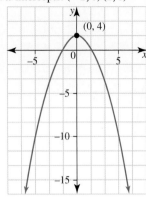

b. No x-intercepts

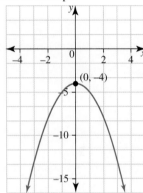

c. x-intercepts: $(-1, 0) (1, 0)$

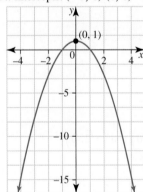

6. a. No

b. A negative sign inverts the graph.

c. $x = 0$ (the y-axis)

7. a. C   **b.** D   **c.** D   **d.** D

8. a. iii   **b.** iv   **c.** v   **d.** i

e. vi   **f.** ii

9. 5 km above sea level

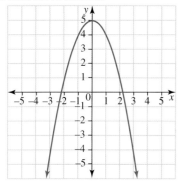

10. a. x-intercepts at $(-9, 0), (9, 0)$

  b. 243 cm

  c. 18 cm

11. a. $-2\sqrt{3} \le x \le 2\sqrt{3}$

  b.

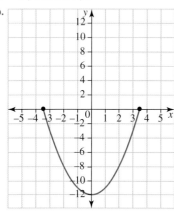

  c. 12 m

  d. 6.9 metres

12. a. Answers will vary.

  b. Answers will vary.

13. Answers will vary.

14. Intersections at $\left(\sqrt{-2}, 0\right)$ and $\left(\sqrt{2}, 0\right)$

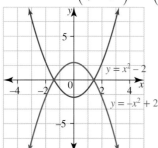

15. The cup should be placed 50 cm horizontally away from the desk.

16. $y = 2x^2 + 2$

17. The parabolas intersect symmetrically at $(-4.53, 226.11)$ and $(4.53, 226.11)$.

18. $y = 4 + 4\sqrt{5}$

19. If k becomes a larger number, the graph is narrower. Changing c changes the value of the y-intercept.

20. $y = -5x^2 + 50x$

8.5 Exponential relationships and graphs

1.

x	-3	-2	-1	0	1	2	3
y	$\frac{1}{27}$	$\frac{1}{9}$	$\frac{1}{3}$	1	3	9	27

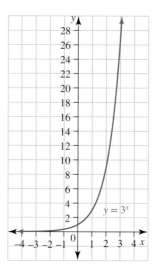

2. a. 2   **e.** 3   **f.** 4

  g. 10   **h.** a

3.

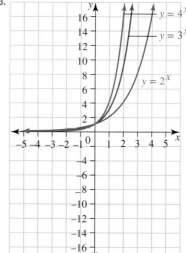

a. The graphs all pass through $(0, 1)$. The graphs have the same horizontal asymptote ($y = 0$). The graphs are all very steep.

b. As the base grows larger, the graphs become steeper.

c.

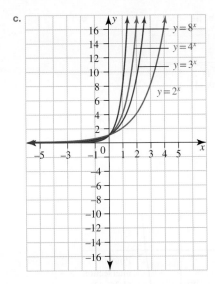

iii.

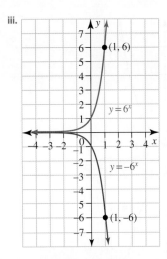

b. In each case the graphs are symmetric about the *x*-axis.

4. a. i.

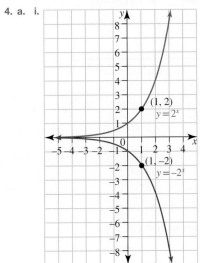

5. a. i.

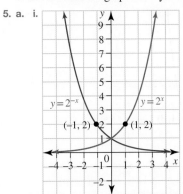

ii.

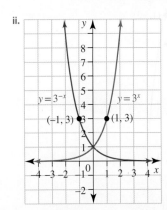

ii.

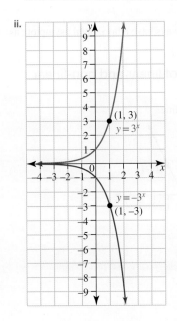

iii.

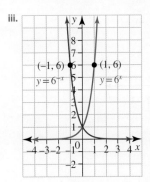

b. In each case the graphs are symmetric about the *y*-axis.

6. a, b.

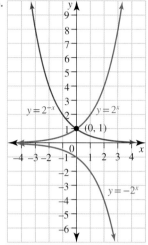

7.

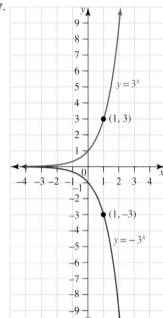

8.

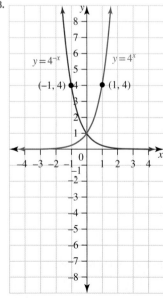

9. a. ii   **b.** iii   **c.** iv   **d.** i

10.

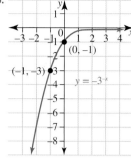

11.

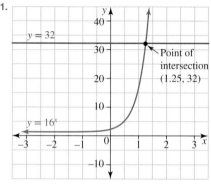

$$x = 1.25$$

12. a. 10000

    b. i. 1111

       ii. 41

       iii. 0

13. a. Yes

    b. There is a constant ratio of 1.3.

    c. 30%

    d. 3.26 million

    e. 30 million

14. Yes, $y = a^x$ will always have a horizontal asymptote as $a^x \neq 0$.

8.6 Applications of linear and non-linear relationships

1. a. $y = 2x$: straight line.

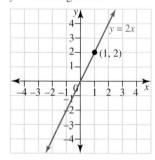

b. $y = 2x^2$: parabola

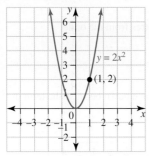

c. $y = 2^x$: exponential curve

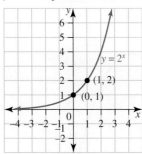

2. a. $y = 4x$: straight line

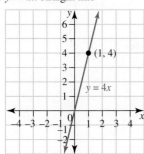

b. $y = 4x^2$: parabola

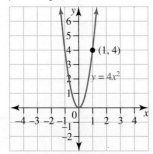

c. $y = 4^x$: exponential curve

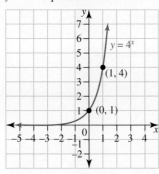

3. a. $y = -5x$: straight line

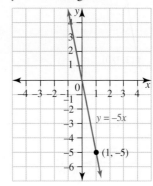

b. $y = -5x^2$: parabola

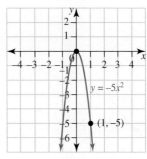

c. $y = -5^x$: exponential curve

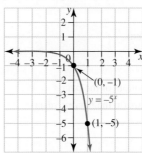

4. C

5. False

6. False

7. $(-2, 4)$ and $(3, 9)$

8. $(2, 4)$

9. No points of intersection

10. One point, $(2, 4)$

11. $(1, 0)$

12. $(0, -1)$

13. a. 20 bacteria

b. 180 bacteria

c.

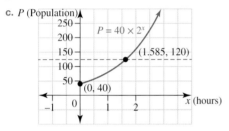

d. 1 hour

14. a. 40

  b. 1280

  c. P (Population)

  d. Approximately 1.6 hours

15. a. $20 000

  b. $23 328

  c. See the figure at the bottom of the page.*

  d. Approximately 14 years (to the nearest year)

16. a. (4, 16)

  b. 16 metres

  c. 1.6 metres or 6.5 metres from the outlet

17. a. (3, 9)

  b. 9 metres

  c. 3.5 metres

18. a. The parabola is concave down because the term in t^2 is negative.

  b. 20 m

t	0	1	2	3	4
h	0	15	20	15	0

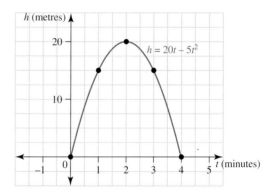

  c. Approximately 3 minutes

*15. c.

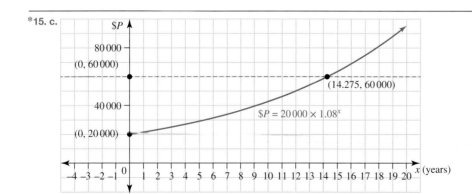

Project

1. a. See the figure at the bottom of the page.*

   b. All of the graphs have a y-intercept of $y = 1$. The graphs that have a positive coefficient of x have two positive x-intercepts. The graphs that have a negative coefficient of x have two negative x-intercepts. As the coefficient of x decreases, the x-intercepts move further left.

   c. The value of a affects the width of the graph. The greater the value, the narrower the graph. When a is negative, the graph is reflected in the x-axis. The value of c affects the position of the graph vertically. As the value of c increases, the graph moves upwards. The value of c does not affect the shape of the graph.

   d. Students may find some similarities in the effect of the coefficients on the different families of graphs. For example, the constant term always affects the vertical position of the graph, not its shape and the leading coefficient always affects the width of the graph.

   e. Student responses will vary. Students should aim to recreate the pattern as best as possible, and describe the process in mathematical terms. Sample responses can be found in the worked solutions in the online resources.

2. Student responses will vary. Students should try to use their imagination and create a pattern that they find visually beautiful.

8.7 Review

1. B
2. C
3. A
4. D
5. B
6. A

7. A
8. C
9. D
10. B
11. D
12. D
13. D
14. A
15. B
16. a. ii   b. iii   c. iv   d. i
17. a. iv   b. iii   c. i   d. ii
18. a. $x = 0$; vertex $(0, -4)$, minimum; x-intercepts are -2 and 2, y-intercept is -4.

    b. $x = 0$; vertex $(0, 9)$, maximum; x-intercepts are 3 and -3, y-intercept is 9.

    c. $x = 1$; vertex $(1, 0)$, minimum; x-intercept is 1, y-intercept is 1.

    d. $x = -3$; vertex $(-3, 0)$, maximum; x-intercept is -3, y-intercept is -9.

    e. $x = -1$; vertex $(-1, 1)$, minimum; no real x-intercepts; y-intercept is 2.

    f. $x = -1$; vertex $(-1, -4)$, minimum; x-intercepts are 1 and -3, y-intercept is -3.

*1. a.

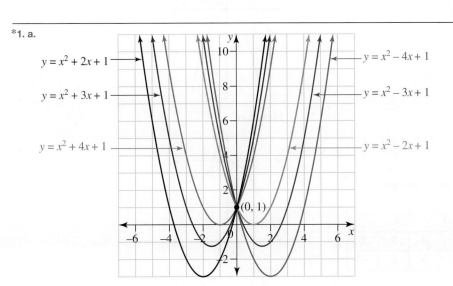

$y = x^2 + 2x + 1$ $y = x^2 + 3x + 1$ $y = x^2 + 4x + 1$ $y = x^2 - 4x + 1$ $y = x^2 - 3x + 1$ $y = x^2 - 2x + 1$

$(0, 1)$

19. a. $y = x^2 - 4x$, $-2 \le x \le 6$

x	−2	−1	0	1	2	3	4	5	6
y	12	5	0	−3	−4	−3	0	5	12

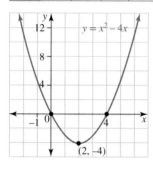

$x = 2$; vertex $(2, -4)$, minimum; x-intercepts are 0 and 4, y-intercept is 0.

b. $y = -x^2 - 2x + 8$, $-5 \le x \le 3$

x	−5	−4	−3	−2	−1	0	1	2	3
y	−7	0	5	8	9	8	5	0	−7

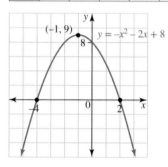

$x = -1$; vertex $(-1, 9)$, maximum; x-intercepts are −4 and 2, y-intercept is 8.

c. $y = 2x^2 - 4x + 4$, $-2 \le x \le 3$

x	−2	−1	0	1	2	3
y	20	10	4	2	4	10

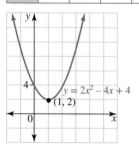

$x = 1$; vertex $(1, 2)$, minimum; no x-intercepts, y-intercept is 4.

d. $y = -x^2 + 6x - 5$, $0 \le x \le 6$

x	0	1	2	3	4	5	6
y	−5	0	3	4	3	0	−5

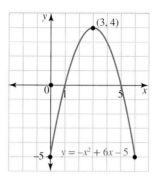

$x = 3$; vertex $(3, 4)$, maximum; x-intercepts are 1 and 5, y-intercept is −5.

20. a.

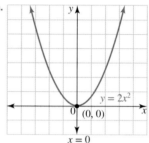

Minimum

b.

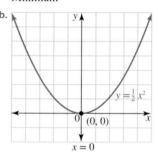

Minimum

c.

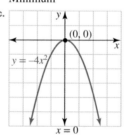

Maximum

d.

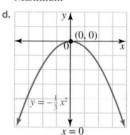

Maximum

21. a.

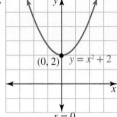

$(0, 2)$ $y = x^2 + 2$
$x = 0$

Minimum

b.

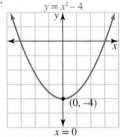

$y = x^2 - 4$
$(0, -4)$
$x = 0$

Minimum

c.

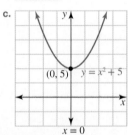

$(0, 5)$ $y = x^2 + 5$
$x = 0$

Minimum

d.

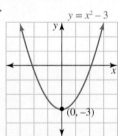

$y = x^2 - 3$
$(0, -3)$

Minimum

22. 1000 metres

23. a.

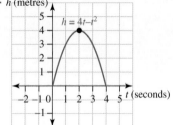

$h = 4t - t^2$

b. 4 metres
c. 2 seconds
d. 4 seconds

24. a.

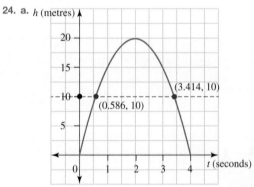

$(3.414, 10)$
$(0.586, 10)$

b. 4 seconds
c. 20 metres
d. 2.8 seconds (1 d.p.)

25. a. 20
b. 45
c.

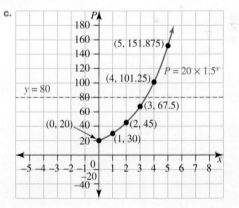

$(5, 151.875)$
$(4, 101.25)$
$P = 20 \times 1.5^x$
$y = 80$
$(3, 67.5)$
$(0, 20)$
$(2, 45)$
$(1, 30)$

d. 3.5 hours

9 Variation and rates of change (Path)

LESSON
9.1 Overview

Why learn this?

Proportion and rates are often used to compare quantities. Kilometres per hour, price per kilogram, dollars per litre and pay per hour are all examples of rates that are used in everyday life. Supermarkets give the prices of items in, for example, dollars per kilogram, so customers can compare different-sized packages of the same item and make decisions about which is the better value. Chefs use proportions when combining different ingredients from a recipe.

In geometry, the circumference of a circle is proportional to its diameter. We say, simply, that the circumference, C, is pi (π) times its diameter, d; that is, $C = \pi d$. Calculations of π date back to the time of Archimedes, but the symbol became widely used in the 1700s. The Golden Ratio, approximately 1.618, is a proportion that is found in geometry, art and architecture and has been made famous in the illustrations of Leonardo da Vinci (1452–1519).

In the construction industry, concrete is made in different proportions of gravel, sand, cement and water depending on the particular required application. Conversion rates allow businesspeople and the general population to convert Australian dollars to an equivalent amount in an overseas currency. Artists, architects and designers use the concept of proportion in many spheres of their work.

Hey students! Bring these pages to life online

Watch videos

Engage with interactivities

Answer questions and check solutions

Find all this and MORE in jacPLUS

Reading content and rich media, including interactivities and videos for every concept

Extra learning resources

Differentiated question sets

Questions with immediate feedback, and fully worked solutions to help students get unstuck

1. State whether this statement is True or False.
   The expression $y \propto x$ means y is directly proportional to x.

2. **MC** Identify which of the following relationships is directly proportional.

   A.

a	0	1	2	3
b	0	2	4	6

   B.

a	0	1	2	3
b	0	1	4	9

   C.

a	1	2	3	4
b	4	3	2	1

   D.

a	0	1	2	3
b	1	3	5	7

3. Mobile phone calls are charged at $1 per minute. State if the cost of a phone bill and the number of 1-minute time periods are directly proportional.

4. State whether this statement is True or False.
   The perimeter of a square and its side length have a relationship that is directly proportional.

5. **MC** If $a \propto b$ and $a = 3$ when $b = 9$, select the rule that links a and b.

   A. $a = \dfrac{1}{3}b$ **B.** $a = 3b$ **C.** $a = \dfrac{3}{b}$ **D.** $a = 9b$

6. The quantities x and y are related to each other by direct variation. If k is the constant of proportionality, complete the proportion statement $x = $ _____.

7. The voltage, v, that powers a mobile phone is directly proportional to the current, I amps.
   A mobile phone with consistent resistance uses a current of two amps when powered by a voltage of 5.1 volts.
   a. Determine the consistent resistance or constant of proportionality, k.
   b. Calculate the voltage required to operate a mobile phone that would use a current of 1.2 amps.

8. State whether this statement is True or False.
   The statement 'x varies directly with y' can be written as $y \propto \dfrac{1}{x}$ or $y = \dfrac{k}{x}$.

9. **MC** If y varies with the square of x and $y = 100$ when $x = 2$, then the constant of variation is:
   A. 400   **B.** 200   **C.** 100   **D.** 25

10. **MC** If b is directly proportional to c^2, select the constant of variation if $b = 72$ when $c = 12$.
    A. 0.5   **B.** 2   **C.** 6   **D.** 36

11. **MC** y is directly proportional to x.
    If x is tripled, select what y is.
    A. Doubled   **B.** Tripled
    C. Reduced to one-third   **D.** Reduced to one-ninth

12. **MC** On average, Noah kicks 2.25 goals per game of soccer. In a 16-game season, the number of goals Noah would kick is:

    A. 32   **B.** 36   **C.** 30   **D.** 38

13. Tea bags in a supermarket can be bought for \$1.45 per pack of 10 or for \$3.85 per pack of 25. Determine the cheaper way of buying tea bags.

14. Given that $y \propto x$ and $\dfrac{y}{x} = 4$, determine the rule relating y and x.

15. Determine your average speed if you drive 320 km in five hours.

LESSON
9.2 Direct linear proportion

LEARNING INTENTION

At the end of this lesson you should be able to:
- understand the concept of direct linear proportion
- identify and write direct linear proportion statements.

▶ 9.2.1 Direct linear proportion

eles-4818

- Proportion is a mathematical comparision between two numbers. It can be expressed as an equation stating that two ratios are equal.
- In **direct linear proportion**, the two variables change at the same rate.
- Suppose that ice-cream costs \$3 and that you are to buy some for your friends.
- There is a relationship between the cost of the ice-cream (C) and the number of ice-creams that you buy (n).
- The relationship between the two variables (number of ice-creams and cost of the ice-cream) can be illustrated in a table or a graph.

n	0	1	2	3	4
C (\$)	0	3	6	9	12

- This relationship is called direct linear proportion as it shows the following characteristics:
  - As n increases, so does C.
  - The graph of the relationship is a straight line passing through the origin as $n = 0$ and $C = 0$.

• We say that 'C is directly proportional to n' or 'C varies directly as n'.

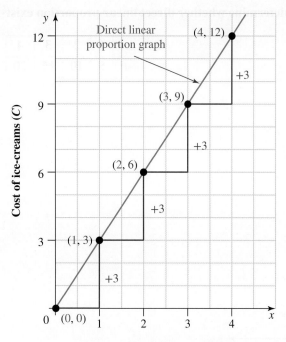

Number of ice-creams bought (n)

WORKED EXAMPLE 1 Verifying whether direct linear proportion exists

For each of the following pairs of variables, state whether direct linear proportion exists.
a. The height of a stack of photocopy paper (h) and the number of sheets (n) in the stack
b. Your Maths mark (m) and the number of hours of Maths homework you have completed (n)

THINK	WRITE
a. When n increases, so does h. When $n = 0$, $h = 0$. If graphed, the relationship would be linear.	a. $h \propto n$
b. As n increases, so does m. When $n = 0$, I may get a low mark at least, so $n \neq 0$.	b. m is not directly proportional to n.

For each of the following, determine whether direct linear proportion exists between the variables.

a.

t	0	1	2	3	4
y	0	1	3	7	15

b.

n	1	2	3	4
c	5	10	15	20

c.

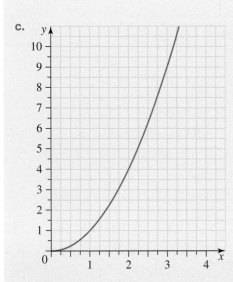

THINK

a. From the table, when t increases, so does y.
When $t = 0$, $y = 0$. The t-values increase by a constant amount but the y-values do not, so the relationship is not linear.
When t is doubled, y is not.

b. From the table, as n increases, so does c.
Extending the pattern gives $n = 0$, $c = 0$. The n-values and C-values increase by constant amounts, so the relationship is linear.

c. When x increases, so does y.
When $x = 0$, $y = 0$.
The graph is not a straight line.

WRITE

a. y is not directly proportional to t.

b. $C \alpha n$

c. y is not directly proportional to x.

DISCUSSION

How do you know when two quantities are directly proportional?

Exercise 9.2 Direct linear proportion

learn on

9.2 Quick quiz

9.2 Exercise

Individual pathways

■ PRACTISE	■ CONSOLIDATE	■ MASTER
1, 3, 6, 9, 13	2, 5, 7, 10, 11, 14	4, 8, 12, 15, 16

Fluency

1. **WE1** For each of the following pairs of variables, state whether direct linear proportion exists. If it does not exist, give a reason why.

   a. The distance (d) travelled in a car travelling at 60 km/h and the time taken (t)

   b. The speed of a swimmer (s) and the time the swimmer takes to complete one lap of the pool (t)

   c. The cost of a bus ticket (c) and the distance travelled (d)

2. For each of the following pairs of variables, state whether direct linear proportion exists. If it does not exist, give a reason why.

   a. The perimeter (p) of a square and the side length (l)

   b. The area of a square (A) and the side length (l)

   c. The total cost (C) of buying n boxes of pencils

3. For each of the following pairs of variables, state whether direct linear proportion exists. If it does not exist, give a reason why.

   a. The weight of an object in kilograms (k) and in pounds (p)

   b. The distance (d) travelled in a taxi and the cost (c)

   c. A person's height (h) and their age (a)

4. State whether the following statements are True or False.
   There is a direct linear relationship between:

   a. the total cost, C, of purchasing netballs and the number, n, purchased

   b. the circumference of a circle and its diameter

   c. the area of a semicircle and its radius.

5. **MC** For the table shown, if $y \propto x$, then the values of a and b would be:

x	1	2	4	8
y	3	6	a	b

   A. $a = 8$, $b = 16$

   B. $a = 8$, $b = 24$

   C. $a = 9$, $b = 12$

   D. $a = 12$, $b = 24$

Understanding

6. **WE2** For each of the following, determine whether direct linear proportion exists between the variables. If it does not, explain why.

a.
x	0	1	2	3	4
y	0	1	3	8	15

b.
a	0	1	2	3	6
M	0	8	16	24	48

c.
t	0	1	2	4	8
d	0	3	6	9	12

d.
n	0	1	2	3	4
C	10	20	30	40	50

7. For each of the following, determine whether direct proportion exists between the variables. If it does not, explain why.

a.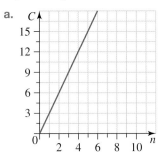

b.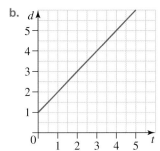

c.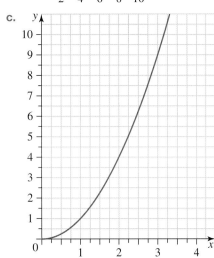

d.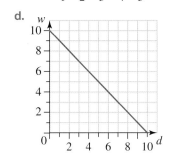

8. List five pairs of real-life variables that exhibit direct proportion.

Communicating, reasoning and problem solving

9. Explain which point must always exist in a table of values if the two variables exhibit direct linear proportionality.

10. If direct linear proportion exists between two variables m and n, fill out the table and explain your reasoning.

m	0	2	5
n	0		20

11. If the variables x and y in the following table are directly proportional, determine the values of a, b and c. Explain your reasoning.

x	2	4	9	a
y	8	b	c	128

12. If $y \propto x$, explain what happens to:

    a. y if x is doubled   **b.** y if x is halved   **c.** x if y is tripled.

13. Mobile phone calls are charged at 17 cents per 30 seconds.

    a. Does direct linear proportion exist between the cost of a phone bill and the number of 30-second time periods? Justify your answer.

    b. If a call went for 7.5 minutes, determine how much the call would cost.

14. An electrician charges $55 for every 30 minutes or part of 30 minutes for his labour on a building site.

    a. Does direct linear proportion exist between the cost of hiring the electrician and the time spent at the building site? Justify your answer.

    b. If the electrician worked for 8.5 hours, determine how much this would cost.

15. A one-litre can of paint covers five square metres of wall.

    a. Does a direct linear proportion exist between the number of litres purchased and the area of the walls to be painted? Justify your answer.

    b. Evaluate the number of cans needed to paint a wall 5 metres long and 2.9 metres high with two coats of paint.

16. Bruce is building a pergola and needs to buy treated pine timber. He wants 4.2-metre and 5.4-metre lengths of timber. If a 4.2-metre length costs $23.10 and a 5.4-metre length costs $29.16, determine if direct linear proportion exists between the cost of the timber and the length of the timber per metre.

LESSON
9.3 Direct proportion and ratio

LEARNING INTENTION

At the end of this lesson you should be able to:
- calculate the constant of proportionality
- express the constant of proportionality as a ratio of the two quantities.

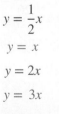 9.3.1 Constant of proportionality

eles-4819

- Proportionality relationships ($y \propto x$) are shown by the graph with four straight lines:

$$y = \frac{1}{2}x$$
$$y = x$$
$$y = 2x$$
$$y = 3x$$

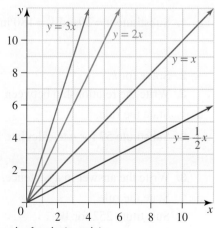

- For the equation $y = kx$:
  - k is a constant, called the **constant of proportionality**
  - y is directly proportional to x, or $y \propto x$
  - all the graphs start at $(0, 0)$
  - y is called the *dependent variable* and is normally placed on the vertical axis (y-axis)
  - x is called the *independent variable* and is normally placed on the horizontal axis (x-axis).

> **Constant of proportionality**
>
> For the equation:
>
> $$y = kx$$
>
> k is the constant of proportionality

WORKED EXAMPLE 3 Determining the constant of direct proportionality

Given that $y \propto x$ and $y = 12$ when $x = 3$, calculate the constant of proportionality and state the rule relating y and x.

THINK	WRITE
1. $y \propto x$, so write the linear rule.	$y = kx$
2. Substitute $y = 12$ and $x = 3$ into $y = kx$.	$12 = 3k$
3. Calculate the constant of proportionality by solving for k.	$k = 4$
4. Write the rule.	$y = 4x$

WORKED EXAMPLE 4 Solving problems involving direct proportionality

The weight (W) of \$1 coins in a bag varies directly as the number of coins (n). Twenty coins weigh 180 g.
a. Determine the relationship between W and n.
b. Calculate how much 57 coins weigh.
c. Determine how many coins weigh 252 g.

THINK

Summarise the information given in a table.

WRITE

$W \propto n$
$\therefore W = kn$

n	20	57	
W	180		252

a. 1. Substitute $n = 20$ and $W = 180$ into $W = kn$.

  2. Solve for k.

  3. Write the relationship between W and n.

b. 1. State the rule.

  2. Substitute 57 for n to calculate W.

  3. Write the answer in a sentence.

c. 1. State the rule.

  2. Substitute 252 for W.

a. $180 = 20k$
$$\frac{180}{20} = \frac{20k}{20}$$
$$k = 9$$

$W = 9n$

b. $W = 9n$

$W = 9 \times 57$
$\quad = 513$

Fifty-seven coins weigh 513 g.

c. $W = 9n$
$252 = 9n$

3. Solve for n.	$n = \dfrac{252}{9}$
	$= 28$
4. Write the answer in a sentence.	Twenty-eight coins weigh $252\,\text{g}$.

9.3.2 Ratio

eles-4820

- If $y \propto x$, then $y = kx$, where k is constant.
  Transposing this formula gives $\dfrac{y}{x} = k$.
- The constant of proportionality, k, is the ratio of any pair of values (x, y).
- For example, this table shows that $v \propto t$.

t	1	2	3	4
v	5	10	15	20

It is clear that $\dfrac{20}{4} = \dfrac{15}{3} = \dfrac{10}{2} = \dfrac{5}{1} = 5.$

WORKED EXAMPLE 5 Applying the constant of proportionality as a ratio

Sharon works part-time and is paid at a fixed rate per hour. If she earns \$135 for 6 hours' work, calculate how much she will earn for 11 hours.

THINK	WRITE
1. Sharon's payment is directly proportional to the number of hours worked. Write the rule.	$P \propto n$ $\therefore P = kn$
2. Summarise the information given. The value of x needs to be calculated.	<table><tr><td>n</td><td>6</td><td>11</td></tr><tr><td>P</td><td>135</td><td>x</td></tr></table>
3. Since $P = kn$, $\dfrac{P}{n}$ is constant. Solve for x.	$\dfrac{135}{6} = \dfrac{x}{11}$ $x = \dfrac{135 \times 11}{6}$ $x = 247.5$
4. Write the answer in a sentence.	Sharon earns \$247.50 for 11 hours of work.

 Resources

 Interactivities Constants of proportionality (int-6057)

Exercise 9.3 Direct proportion and ratio

9.3 Quick quiz on	9.3 Exercise

Individual pathways

■ PRACTISE	■ CONSOLIDATE	■ MASTER
1, 2, 6, 7, 12, 15	3, 4, 8, 9, 13, 16	5, 10, 11, 14, 17, 18

Fluency

1. **WE3** If a is directly proportional to b, and $a = 30$ when $b = 5$, calculate the constant of proportionality and state the rule linking a and b.

2. If $a \propto b$, and $a = 2.5$ when $b = 5$, determine the rule linking a and b.

3. If $C \propto t$, and $C = 100$ when $t = 8$, determine the rule linking C and t.

4. If $v \propto t$, and $t = 20$ when $v = 10$, determine the rule linking v and t.

5. If $F \propto a$, and $a = 40$ when $F = 100$, determine the rule linking a and F.

Understanding

6. **WE4** Springs are often used to weigh objects, because the extension of a spring (E) is directly proportional to the weight (W) of the object hanging from the spring. A 4-kg load stretches a spring by 2.5 cm.

   a. Determine the relationship between E and W.
   b. Calculate the load that will stretch the spring by 12 cm.
   c. Determine how far 7 kg will extend the spring.

7. Han finds that 40 shelled almonds weigh 52 g.

   a. Determine the relationship between the weight (W) and the number of almonds (n).
   b. Calculate how many almonds there would be in a 500g bag.
   c. Calculate how much 250 almonds would weigh.

8. Petra knows that her bicycle wheel turns 40 times when she travels 100 m.

   a. Determine the relationship between the distance travelled (d) and the number of turns of the wheel (n).
   b. Calculate how far she goes if her wheel turns 807 times.
   c. Determine how many times her wheel turns if she travels 5 km.

9. Fiona, who operates a plant nursery, uses large quantities of potting mix. Last week she used 96 kg of potting mix to place 800 seedlings in medium-sized pots.

   a. Determine the relationship between the mass of potting mix (M) and the number of seedlings (n).
   b. Calculate how many seedlings she can pot with her remaining 54 kg of potting mix.
   c. Determine how much potting mix she will need to pot 3000 more seedlings.

10. **WE5** Tamara is paid at a fixed rate per hour. If she earns $136 for five hours of work, calculate how much will she earn for eight hours of work.

11. It costs $158 to buy 40 bags of birdseed. Calculate how much 55 bags will cost.

Communicating, reasoning and problem solving

12. If 2.5 L of lawn fertiliser will cover an area of $150\,m^2$, determine how much fertiliser is needed to cover an area of $800\,m^2$ at the same rate. Justify your answer.

13. Paul paid $68.13 for 45 L of fuel. At the same rate, determine how much he would pay for 70 L. Justify your answer.

14. Rose gold is an alloy of gold and copper that is used to make high-quality musical instruments. If it takes 45 g of gold to produce 60 g of rose gold, evaluate how much gold would be needed to make 500 g of rose gold. Justify your answer.

15. If Noah takes a group of friends to the movies for his birthday and it would cost $62.50 for five tickets, determine how much it would cost if there were 12 people (including Noah) in the group.

16. Sharyn enjoys quality chocolate, so she makes a trip to her favourite chocolate shop. She is able to select her favourite chocolates for $7.50 per 150 grams. Since Sharyn loves her chocolate, she decides to purchase 675 grams. Evaluate how much she spent.

17. Anthony drives to Mildura, covering an average of 75 kilometers in 45 minutes. He has to travel 610 kilometres to get to Mildura.

   a. Determine the relationship between the distance he travelled in kilometres, D, and his driving time in hours, T.

   b. Evaluate how long it will take Anthony to complete his trip.

   c. If he stops at Bendigo, 220 kilometres from his starting position, determine how long it will take him to reach Bendigo.

   d. Determine how much longer it will take him to arrive at Mildura after leaving Bendigo.

18. The volume of a bird's egg can be determined by the formula $V = kl^3$, where V is the volume in cm^3, l is the length of the egg in cm and k is a constant. A typical ostrich egg is 15 cm long and has a volume of $7425\,cm^3$. Evaluate the volume of a chicken egg that is 5 cm long.

LESSON
9.4 Inverse proportion

LEARNING INTENTION

At the end of this lesson you should be able to:
- calculate the constant of proportionality, k, between two variables that are inversely proportional to each other
- determine the rule between two inversely proportional variables and plot the graph of their relationship.

9.4.1 Inverse proportion

eles-4875

- If 24 sweets are shared between 4 children, then each child will receive 6 sweets. If the sweets are shared by 3 children, then each will receive 8 sweets.
- The relationship between the number of children (C) and the number sweets for each child (n) can be given in a table.

C	1	2	3	4	6	8	12
n	24	12	8	6	4	3	2

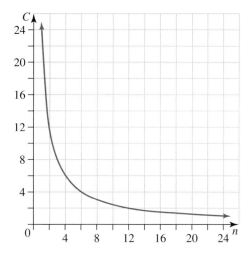

- As the number of children (C) increases, the number of sweets for each child (n) decreases. This is an example of **inverse proportion** or inverse variation.
- We say that 'n is inversely proportional to C' or 'n varies inversely as C'.
- This is written as $C \propto \dfrac{1}{n}$, or $C = \dfrac{k}{n}$, where k is a constant (the constant of proportionality). This formula can be rearranged to $Cn = k$.

  Note that multiplying any pair of values in the table (3×8, 12×2) gives the same result.
- The relationship has some important characteristics:
  - As C increases, n decreases, and vice versa.
  - The graph of the relationship is a **hyperbola**.

Expressing inverse proportion

If the x-value increases and the y-value decreases, and vice versa, then this relationship can be expressed as:

y is directly proportional to the inverse of x

or

y is inversely proportional to x

$$y \propto \dfrac{1}{x}$$

That is, $y = \dfrac{k}{x}$ where k is the constant of proportionality.

WORKED EXAMPLE 6 Solving problems involving inverse relationships

y is inversely proportional to x and $y = 10$ when $x = 2$.
a. Calculate the constant of proportionality, k, and hence the rule relating x and y.
b. Plot a graph of the relationship between x and y, for values of x from 2 to 10.

THINK	WRITE/DRAW
a. 1. Write the relationship between the variables.	a. $y \propto \dfrac{1}{x}$
2. Rewrite as an equation using k, the constant of proportionality	$y = \dfrac{k}{x}$
3. Substitute $y = 10, x = 2$, into $y = \dfrac{k}{x}$ and solve for k	$10 = \dfrac{k}{2}$ $k = 20$
4. Write the rule by substituting $k = 20$ into $y = \dfrac{k}{x}$.	$y = \dfrac{20}{x}$

b. 1. Use the rule $y = \dfrac{20}{x}$ to set up a table of values for x and y, taking values for x which are positive factors of k so that only whole number values of y are obtained. For example, $x = 4, y = \dfrac{20}{4} = 5$.

b.

x	2	4	6	8	10
y	10	5	3.3	2.5	2

2. Plot the points on a clearly labelled set of axes and join the points with a smooth curve. Label the graph.

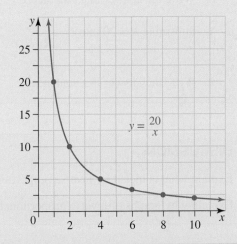

WORKED EXAMPLE 7 Determining the rule for an inversly proportional relationship

When a wire is connected to a power source, the amount of electrical current (I) passing through the wire is inversely proportional to the resistance (R) of the wire. If a current of 0.2 amperes flows through a wire of resistance 60 ohms:
a. calculate the constant of proportionality
b. determine the rule relating R and I
c. calculate the resistance if the current equals 5 amperes
d. determine the current that will flow through a wire of resistance 20 ohms.

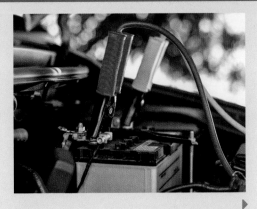

THINK			

THINK

Summarise the information in a table.

$I \propto \dfrac{1}{R}$

Write the rule.

a. 1. Substitute $R = 60$ and $I = 0.2$, into $I = \dfrac{k}{R}$.

   2. Solve for k.

b. Write the rule using $k = 12$.

c. 1. Substitute $I = 5$ into $I = \dfrac{12}{R}$.

   2. Solve for R.
   3. Write the answer in a sentence.

d. 1. Substitute $R = 20$ into $I = \dfrac{12}{R}$.

   2. Write the answer in a sentence.

WRITE

R	60		20
I	0.2	5	

$I = \dfrac{k}{R}$

a. $\quad 0.2 = \dfrac{k}{60}$

$0.2 \times 60 = k$

$k = 12$

b. $I = \dfrac{12}{R}$

c. $\quad 5 = \dfrac{12}{R}$

$5R = 12$

$R = \dfrac{12}{5}$

$= 2.4$

The resistance equals 2.4 ohms.

d. $I = \dfrac{12}{20}$

$= 0.6$

The current will be 0.6 amperes.

DISCUSSION

Explain what is meant by inverse proportion.

Explain why graphs of inverse proportionality only show the first quadrant.

 Resources

 Interactivity Inverse proportion (int-6058)

Exercise 9.4 Inverse proportion

| 9.4 Quick quiz on | 9.4 Exercise |

Individual pathways

■ PRACTISE	■ CONSOLIDATE	■ MASTER
1, 3, 6, 10, 13	2, 4, 7, 11, 14	5, 8, 9, 12, 15

Fluency

1. Decide whether inverse proportion exists between each pair of variables. If it exists, write an equation to describe the relationship.

   a. The speed of a car (s) and the time (t) it takes to complete one lap of a race circuit.
   b. The amount of money (D) that I have and the number (n) of cards that I can buy.
   c. The time (t) that it takes to make a pair of jeans and the number of pairs (p) that can be made in one day.
   d. The price (P) of petrol and the amount (L) that can be bought for $80.
   e. The price (P) of petrol and the cost (C) of buying $80\,L$.
   f. The number of questions (n) in a test and the amount of time (t) available to answer each one.

2. List three examples of inverse proportion.

3. **WE6** y varies inversely as x and $y = 100$ when $x = 10$.

   a. Calculate the constant of proportionality, k, and hence the rule relating x and y.
   b. Plot a graph of the relationship between x and y, for values of x that are positive factors of k less than 21.

4. p is inversely proportional to q and $p = 12$ when $q = 4$.

   a. Calculate the constant of proportionality, k, and hence the rule relating p and q.
   b. Plot a graph of the relationship between q and p, for values of q that are positive factors of k less than 11.

5. y varies inversely as x and $y = 42$ when $x = 1$.

   a. Calculate the constant of proportionality, k, and hence the rule relating x and y.
   b. Plot a graph of the relationship between x and y, for values of x from 1 to 10.

Understanding

6. **WE7** When a constant force is applied to an object, its acceleration is inversely proportional to its mass. When the acceleration of an object is $40\,m/s^2$, the corresponding mass is $100\,kg$.

   a. Calculate the constant of proportionality.
   b. Determine the rule relating mass and acceleration.
   c. Determine the acceleration of a $200\,kg$ object.
   d. Determine the acceleration of a $1000\,kg$ object.

7. The number of colouring pencils sold is inversely proportional to the price of each pencil.
   Two thousand pencils are sold when the price is $0.25 each.

   a. Calculate the constant of proportionality.
   b. Determine the number of pencils that could be sold for $0.20 each.
   c. Determine the number of pencils that could be sold for $0.50 each.

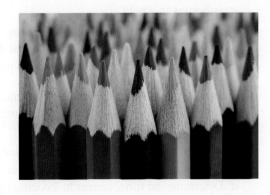

8. The time taken to complete a journey is inversely proportional to the speed travelled. A trip is completed in 4.5 hours travelling at 75 km per hour.

   a. Calculate the constant of proportionality.

   b. Determine how long, to the nearest minute, the trip would take if the speed was 85 km per hour.

   c. Determine the speed required to complete the journey in 3.5 hours, correct to 1 decimal place.

   d. Determine the distance travelled in each case.

9. The cost per person travelling in a charter plane is inversely proportional to the number of people in the charter group. It costs $350 per person when 50 people are travelling.

   a. Calculate the constant of variation.

   b. Determine the cost per person, to the nearest cent, if there are 75 people travelling.

   c. Determine how many people are required to reduce the cost to $250 per person.

   d. Determine the total cost of hiring the charter plane.

Communicating, reasoning and problem solving

10. The electrical current in a wire is inversely proportional to the resistance of the wire to that current. There is a current of 10 amperes when the resistance of the wire is 20 ohms.

   a. Calculate the constant of proportionality.

   b. Determine the current possible when the resistance is 200 ohms.

   c. Determine the resistance of the wire when the current is 15 amperes.

   d. Justify your answer to parts **b** and **c** using a graph.

11. The pressure of an ideal gas is inversely proportional to the volume taken up by the gas. A balloon is filled with air so it takes up 3 L at a pressure of 5 atmospheres.

   a. Calculate the constant of proportionality.

   b. Determine the new volume of the balloon if the pressure was dropped to 0.75 atmospheres.

   c. Determine the pressure if the same amount of air took up a volume of 6 L.

12. Two equations relating the time of a trip, T, and the speed at which they travel, S, are given. For both cases the time is inversely proportional to the speed: $T_1 = \dfrac{5}{S_1}$ and $T_2 = \dfrac{7}{S_2}$. Explain what impact the different constants of proportionality have on the time of the trip.

13. The time it takes to pick a field of strawberries is inversely proportional to the number of pickers. It takes 2 people 5 hours to pick all of the strawberries in a field.

   a. Calculate the constant of proportionality.

   b. Determine the rule relating time (T) and the number of pickers (P).

   c. Determine the time spent if there are 6 pickers.

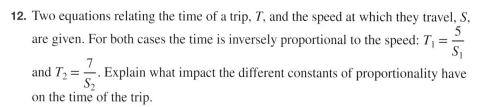

14. For a constant distance covered by a sprinter, the sprinter's speed is inversely proportional to their time. If a sprinter runs at a speed of 10.4 m/s, the corresponding time is 9.62 seconds.

   a. Calculate the constant of variation.

   b. Determine the rule relating speed (V) and time (T).

   c. Determine the time, correct to 2 decimal places, if they ran at a speed of 10.44 m/s.

   d. Determine the time, correct to 2 decimal places, if they ran at a speed of 6.67 m/s.

15. A holiday hostel is built to accommodate group
bookings of up to 45 people. It is known that it would
cost each of the individuals in a group of 20 people
$67.50 per night to rent this venue.
The cost of the venue will remain the same no matter
how many people are part of the group booking.

a. Write the rule for the cost per person (C) and the
number of people in a group booking (n).
b. If 6 people from the original group are no longer able
to attend, determine the new cost per person.
c. Calculate the cheapest possible cost per person.

LESSON
9.5 Introduction to rates

LEARNING INTENTION

At the end of this lesson you should be able to:
- calculate the rate of two related quantities
- solve word problems involving rates.

eles-4821

▶ 9.5.1 Rates

- The word **rate** occurs commonly in news reports and conversation.
  'Home ownership rates are falling.'
  'People work at different rates.'
  'What is the current rate of inflation?'
  'Do you offer a student rate?'
  'The crime rate seems to be increasing.'

- A rate compares two related quantities.
- In general, to calculate a rate (or specific ratio), one quantity is divided by another.
- More often than not, we want to express rates so that one quantity is compared to another one expressed as
  a single unit quantity. We call this a unit rate.
- Speed is common example of rate.

Average speed

- If you travel 160 km in 4 hours, your average speed is 40 km per hour.
- Speed is an example of a rate, and its unit of measurement (kilometres per hour) contains a formula.

- The word 'per' can be replaced with 'divided by', so $\text{speed} = \dfrac{\text{distance (in km)}}{\text{time (in hours)}}$.

- It is important to note the units involved.
  For example, an athlete's speed is often measured in metres per second (m/s) rather than km/h.

WORKED EXAMPLE 8 Calculating unit rates

Calculate the rates suggested by these statements.
a. A shearer shears 1110 sheep in 5 days.
b. Eight litres of fuel costs $12.56.
c. A cricket team scored 152 runs in 20 overs.

THINK	WRITE
a. 1. The rate suggested is 'sheep per day'.	a. $\text{Rate} = \dfrac{\text{number of sheep}}{\text{number of days}}$
2. Substitute the values for the number of sheep and the number of days.	$= \dfrac{1110}{5}$
3. Write the rate.	$= 222$ sheep per day
b. 1. The rate suggested is 'dollars per litre'.	b. $\text{Rate} = \dfrac{\text{number of dollars}}{\text{number of litres}}$
2. Substitute the values for the price per 8 litres and the number of litres.	$= \dfrac{12.56}{8}$
3. Write the rate.	$= \$1.57$ per litre
c. 1. The rate suggested is 'runs per over'.	c. $\text{Rate} = \dfrac{\text{number of runs}}{\text{number of overs}}$
2. Substitute the values for the number of runs and the number of overs.	$= \dfrac{152}{20}$
3. Write the rate.	$= 7.6$ runs per over

WORKED EXAMPLE 9 Solving word problems involving rates

The concentration of a solution is measured in g/L (grams per litre). Calculate the concentration of the solution when 10 g of salt is dissolved in 750 mL of water.

THINK	WRITE
1. Concentration is measured in g/L, which means that concentration = number of grams (mass) ÷ number of litres (volume).	$\text{Concentration} = \dfrac{\text{mass}}{\text{volume}}$
2. Substitute the values of mass and volume.	$= \dfrac{10}{0.75}$
3. Simplify.	$= 13.3$ g/L
4. Write the rate in a sentence.	The concentration of the solution is 13.3 g/L

 Resources

 Interactivity Speed (int-6457)

| 9.5 Quick quiz on | 9.5 Exercise |

Individual pathways

■ PRACTISE	■ CONSOLIDATE	■ MASTER
1, 3, 6, 9	2, 5, 7, 10	4, 8, 11

Fluency

1. Hayden drove from Hay to Bee. He covered a total distance of 96 km and took 1.5 hours for the trip. Calculate Hayden's average speed for the journey.

2. Calculate the average speed in km/h for each of the following.
   a. 90 km in 45 min
   b. 5500 km in 3 h 15 min (correct to 1 decimal place)

3. WE8 Calculate the rates suggested by these statements, giving your answers to 2 decimal places where necessary.

   a. It costs $736 for 8 theatre tickets.
   b. Penelope decorated 72 small cakes in 3 hours.
   c. Usain Bolt has a 100-metre world sprint record of 9.58 seconds.
   d. It takes 30 hours to fill the swimming pool to a depth of 90 cm.
   e. Peter received $260 for 15 hours' work.
   f. Yan received $300 for assembling 6 air conditioners.

Understanding

4. A metal bolt of volume 25 cm^3 has a mass of 100 g. Calculate its density (mass per unit of volume).

5. WE9 One hundred and twenty grams of sugar is dissolved in 200 mL of water. Calculate the concentration of this solution in g/L. Explain your answer.

Communicating, reasoning and problem solving

6. In a race between a tortoise and a hare, the hare ran at 72 km per hour while the giant tortoise moved at 240 cm per minute. Compare and explain the difference in the speed of the two animals.

7. A school had 300 students in 2013 and 450 students in 2015. Determine the average rate of growth in the number of students per year. Show your working.

8. The average speed of a car is determined by the distance of the journey and the time the journey takes. Explain the two ways in which the speed can be increased.

9. Mt Feathertop is Victoria's second-highest peak. To walk to the top involves an increase in height of 1500 m over a horizontal distance of 10 km (10 000 m). Determine the average gradient of the track.

10. In the 2000 Sydney Olympics, Cathy Freeman won gold in the 400-metre race. Her time was 49.11 seconds. In the 2000 Beijing Olympics, Usain Bolt set a new world record for the men's 100-metre race. His time was 9.69 seconds.
    Evaluate the average speed of the winner of each race in kilometres per hour.

11. Beaches are sometimes unfit for swimming if heavy rain has washed pollution into the water. A beach is declared unsafe for swimming if the concentration of bacteria is more than 5000 organisms per litre. A sample of 20 millilitres was tested and found to contain 55 organisms.

Evaluate the concentration in the sample (in organisms/litre) and state whether or not the beach should be closed.

LESSON
9.6 Constant and variable rates

LEARNING INTENTION

At the end of this lesson you should be able to:
- describe the rate of change of a graph as constant or variable, and increasing or decreasing
- construct a graph showing constant rates
- construct graphical representations of rates of change of quantities over time from descriptions or diagrams.

9.6.1 Constant rates

eles-4823

- A **constant rate** of change means the quantity always changes by the same amount in a given time.
- Consider a car travelling along a highway at a constant speed (rate) of 90 km/h. After one hour it will have travelled 90 km, as shown in the table.

	+1	+1	+1
Time (h)	0	1	2	3
Distance (km)	0	90	180	270
	+90	+90	+90

- The distance–time graph is also shown.

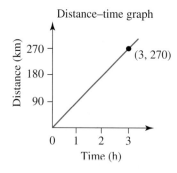

Distance–time graph

- The gradient of the graph is $\dfrac{270 - 0}{3 - 0} = 90$.

- The equation of the graph is $d = 90t$, and the gradient is equal to the speed, or rate of progress.
- A constant rate of change produces a linear (straight-line) graph.

WORKED EXAMPLE 10 Describing constant rates from a graph

Each diagram shown illustrates the distance travelled by a car over time. Describe the journey, including the speed of the car.

a.

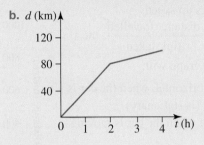

b.

THINK

a. There are three distinct sections.

1. In the first 2 hours, the car travels
   80 km: $\dfrac{80}{2} = 40$ km/h.

2. In the next hour, the car does not move.

3. In the fourth hour, the car travels 40 km.

b. There are two distinct sections.

1. In the first 2 hours, the car travels
   80 km: $\dfrac{80}{2} = 40$ km/h.

2. In the next 2 hours, the car travels
   20 km: $\dfrac{20}{2} = 10$ km/h.

WRITE

a.

The car travels at a speed of 40 km/h for 2 hours.

The car stops for 1 hour.

The car then travels for 1 hour at 40 km/h.

b.

The car travels at a speed of 40 km/h for 2 hours.

The car then travels at 10 km/h for a further 2 hours.

WORKED EXAMPLE 11 Drawing a distance–time graph to illustrate constant rates

A cyclist travels for 1 hour at a constant speed of 10 km/h, then stops for a 30-minute break before riding a further 6 km for half an hour at a constant speed.
Draw a distance–time graph to illustrate the cyclist's journey.

THINK

There are three phases to the journey.
The graph starts at $(0, 0)$.

1. In the first hour, the cyclist travels 10 km. Draw a line segment from $(0, 0)$ to $(1, 10)$.

2. For the next half-hour, the cyclist is stationary, so draw a horizontal line segment from $(1, 10)$ to $(1.5, 10)$.

3. In the next half-hour, the cyclist travels 6 km. Draw a line segment from $(1.5, 10)$ to $(2, 16)$.

DRAW

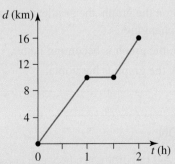

▶ 9.6.2 Variable rates

eles-4824

- A variable rate means that a quantity does not change by the same amount in a given time.
- In reality, a car tends not to travel at constant speed. It starts from rest and gradually picks up speed.
- Since speed $= \dfrac{\text{distance travelled}}{\text{time taken}}$, the distance–time graph will:
  - be flat, or horizontal, when the car is not moving (is stationary)
  - be fairly flat when the car is moving slowly
  - be steeper and steeper as the speed of the car increases.
- This is demonstrated in the distance–time graph shown.
- A variable rate of change produces a non-linear graph.

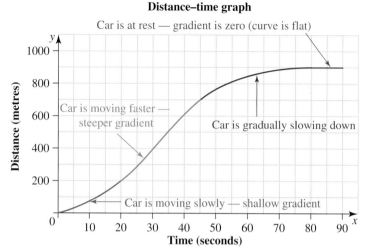

Distance–time graph

Car is at rest — gradient is zero (curve is flat)

Car is moving faster — steeper gradient

Car is gradually slowing down

Car is moving slowly — shallow gradient

Distance (metres) — 1000, 800, 600, 400, 200

Time (seconds) — 10, 20, 30, 40, 50, 60, 70, 80, 90

WORKED EXAMPLE 12 Describing graphs with variable rates

The diagram shown illustrates the distance travelled by a car over time. Describe what is happening, in terms of speed, at each of the marked points.

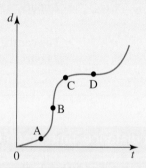

THINK	WRITE
1. At point A on the graph, the gradient is small but becoming steeper.	At point A the car is travelling slowly but accelerating.
2. At point B on the graph, the gradient is at its steepest and is not changing.	At point B the car is at its greatest speed.
3. At point C, the graph is becoming flatter.	At point C the car is slowing down.
4. At point D, the graph is horizontal.	At point D the car is stationary.

DISCUSSION

How can you tell the difference between constant and variable rates?

Explain why a vertical graph is impossible when describing a distance–time graph.

9.6.3 Rates of change

eles-6247

- Imagine that water is flowing out of a hose at a steady rate and will be used to fill several containers.

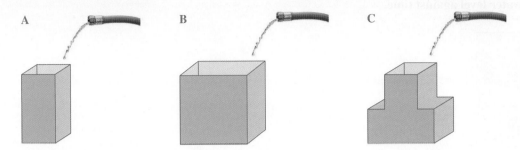

Because they are of different widths, they will fill at different rates. The narrow container will fill at a faster rate than the other two.

- Consider a graph of the water level against time for the three containers.

Container *A* is narrow, so the water level will rise quickly.

Container *B* is wide and will fill at a slower rate.

Container *C* is wide at the bottom, so the water will rise slowly at first, then quickly when it reaches the narrow part.

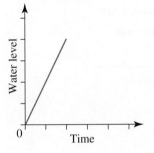

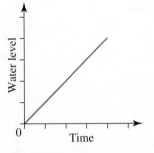

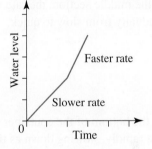

The water level changes at a constant rate.

The water level changes at a constant rate.

The water level changes at two different but constant rates.

- Here is a more complex container, with three distinct sections.

The bottom section fills slowly and the top section fills quickly.

In between, the rate changes steadily from slow to fast. In this section, the rate is increasing.

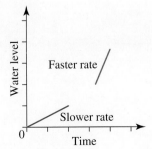

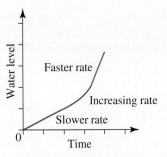

Each of these containers is being filled with water at a steady rate. For each container, sketch a graph of the water level against time.

a.

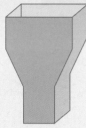

b.

THINK	DRAW

a. The container has three distinct sections.

The bottom section will fill quickly

The top section will fill slowly.

In the middle section, the rate will increase gradually from slow to quick.

a.

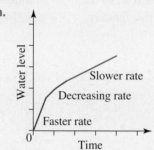

b. At the bottom of the container the water will rise rapidly, slowing down as the container becomes wider.

b.

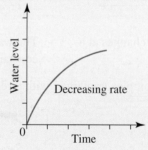

  Resources

Interactivities Constant rates (int-6060)

Variable rates (int-6061)

Rates of change (int-6062)

| 9.6 Quick quiz | on | | 9.6 Exercise |

Individual pathways

■ PRACTISE	■ CONSOLIDATE	■ MASTER
1, 3, 5, 8, 9, 14	2, 4, 10, 12, 15	6, 7, 11, 13, 16, 17, 18

Fluency

1. **WE10** Each diagram shown illustrates the distance travelled by a car over time. Describe the journey, including the speed of the car.

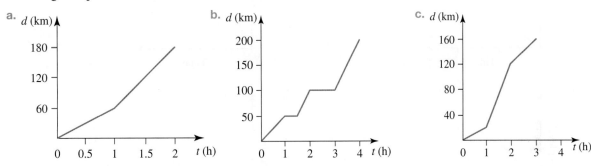

a.

b.

c.

2. Two friends take part in a 24-kilometre mini-marathon. They run at constant speed. Ali takes 2 hours and Beth takes 3 hours to complete the journey.

   a. On the same diagram, draw a distance–time graph for each runner.
   b. i. Calculate the equation for each graph.
      ii. State the difference between the two graphs.

3. **WE11** Draw a distance–time graph to illustrate each of the following journeys.

   a. A cyclist rides at 40 km/h for 30 minutes, stops for a 30-minute break, and then travels another 30 km at a speed of 15 km/h.
   b. Zelko jogs at a speed of 10 km/h for one hour, and then at half the speed for another hour.

4. **WE12** The diagrams shown illustrate the distances travelled by two cars over time.

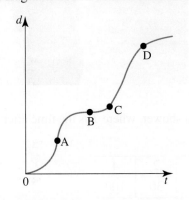
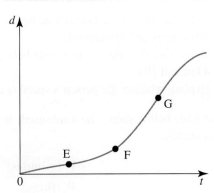

   a. Describe what is happening in terms of speed at each of the marked points.
   b. For each diagram, state at which point:
      i. the speed is the greatest
      ii. the speed is the lowest
      iii. the car is stationary.

5. These containers are filled with water at a steady rate. Match each container with the appropriate graph.

a.   b.   c.   d.

A.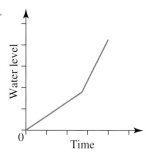

B.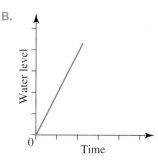

C.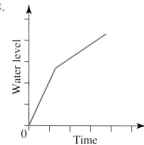

D.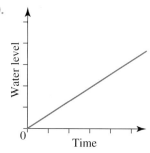

Understanding

6. The table below shows the distance travelled, D, as a person runs for R minutes.

R (minutes)	10	20	50
D (km)	2	4	10

a. Calculate the rate in km/minute between the time 10 minutes and 20 minutes.
b. Determine the rate in km/minute between the distance 4 km and 10 km.
c. Explain whether the person's speed is constant.

7. The table below shows the water used, W, after the start of a shower, where T is the time after the shower was started.

T (minutes)	0	1	2	4
W (litres)	0	20	30	100

a. Calculate the rate of water usage in L/minute for the four stages of the shower.
b. Explain whether the rate of water usage was constant.

8. Which of these graphs show:

a. a steady rate

b. an increasing rate

c. a decreasing rate?

A.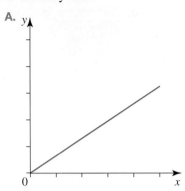

B.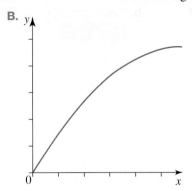

C.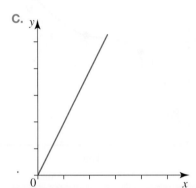

D.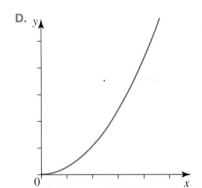

Communicating, reasoning and problem solving

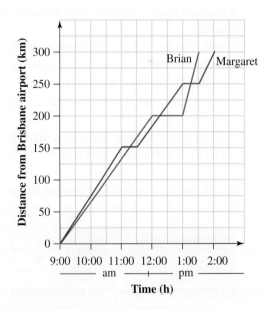

9. Margaret and Brian left Brisbane airport at 9:00 am. They travelled separately but on the same road and in the same direction.

Their journeys are represented by the travel graph below. Show your working.

a. Determine the distance from the airport at which their paths crossed.

b. Determine how far apart they were at 1:00 pm.

c. Determine for how long each person stopped on the way.

d. Determine the total time spent driving and the total distance for each person.

e. Evaluate the average speed while driving for each person.

10. Hannah rode her bike along the bay one morning. She left home at 7:30 am and covered 12 km in the first hour. She felt tired and rested for half an hour.

After resting she completed another 8 km in the next hour to reach her destination. Show your working.

a. Determine how long Hannah took for the entire journey.

b. Calculate the total distance for which she actually rode her bike.

c. Draw a travel graph for Hannah's journey.

11. **WE13** These containers are being filled with water at a steady rate. For each container, sketch a graph of water level against time and explain your reasons.

a. 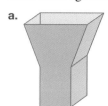 b. 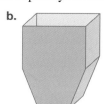 c. 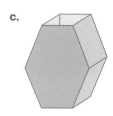 d.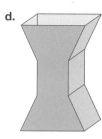

12. The distance versus time graph for a runner is shown below. Four times during the run are marked as A, B, C and D on the graph. State, with reasoning, at which of the points the runner is:

a. travelling fastest
b. travelling slowest
c. slowing down
d. speeding up
e. travelling at constant speed.

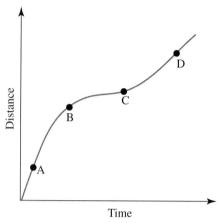

13. Rebecca and Joanne set off at the same time to jog 12 kilometres. Joanne ran the entire journey at constant speed and finished at the same time as Rebecca.

Rebecca set out at 12 km/h, stopping after 30 minutes to let Joanne catch up, then ran at a steady rate to complete the distance in 2 hours.

a. Show the progress of the two runners on a distance–time graph.
b. Determine how long Rebecca waited for Joanna to catch up. Show your working.

14. An internet service provider charges $30 per month plus $0.10 per megabyte downloaded. The table of monthly cost versus download amount is shown.

Download (MB)	0	100	200	300	400	500
Cost ($)	30	40	50	60	70	80

a. Determine how much the cost increases by when the download amount increases from:

i. 0 to 100 MB
ii. 100 to 200 MB
iii. 200 to 300 MB.

b. Explain whether the cost is increasing at a constant rate.

15. Use the graph showing the volume of water in a rainwater tank to answer the following questions.

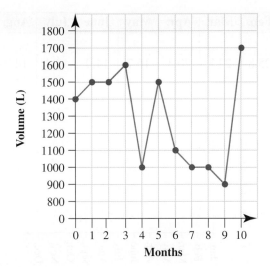

a. Determine the day(s) during which the rate of change was positive.
b. Determine the day(s) during which the rate of change was negative.
c. Determine the day(s) during which the rate of change was zero.
d. Determine the day on which the volume of water increased at the fastest rate.
e. Determine the day on which the volume of water decreased at the fastest rate.

16. The graph shows the number of soft-drink cans in a vending machine at the end of each day.

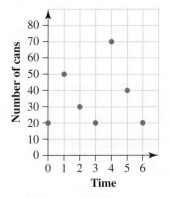

a. Determine how much the number of cans changed by in the first day.
b. Determine how much the number of cans changed by in the fifth day.
c. Explain whether the number of soft-drink cans is changing at a constant or variable rate.

17. The table and graph show Melbourne's average daily maximum temperature over the year.

	Jan.	Feb.	Mar.	Apr.	May	June	July	Aug.	Sept.	Oct.	Nov.	Dec.
Mean maximum (°C)	25.8	25.8	23.8	20.3	16.7	14.0	13.4	14.9	17.2	19.6	21.9	24.2

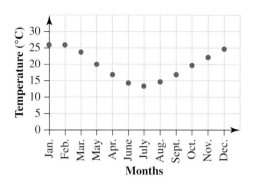

a. Determine the average maximum temperature in:
   i. February
                                           ii. June.

b. Calculate the change in temperature from:
   i. January to August
                                       ii. November to December.

c. Explain whether the temperature is changing at a constant rate.

18. Water is poured at a constant rate into the container shown.

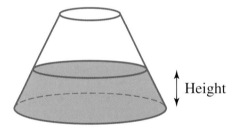

Sketch the graph of height of water against time.

LESSON
9.7 Review

9.7.1 Topic summary

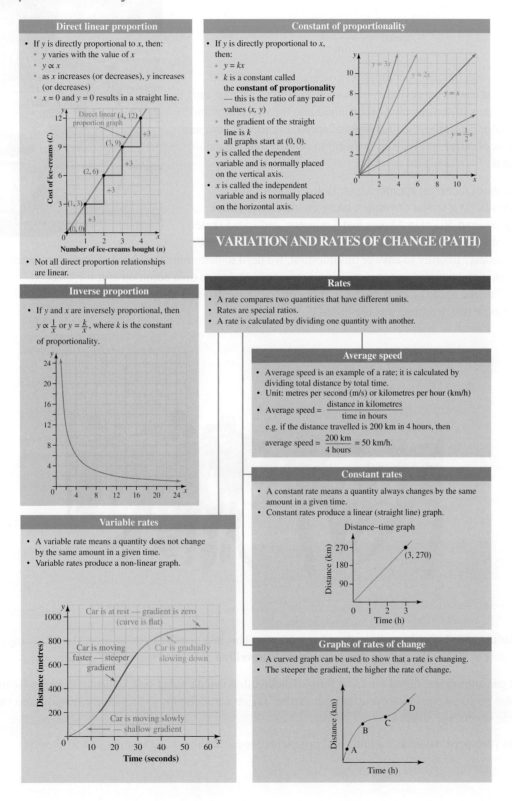

Direct linear proportion

- If y is directly proportional to x, then:
  - y varies with the value of x
  - $y \propto x$
  - as x increases (or decreases), y increases (or decreases)
  - $x = 0$ and $y = 0$ results in a straight line.

- Not all direct proportion relationships are linear.

Constant of proportionality

- If y is directly proportional to x, then:
  - $y = kx$
  - k is a constant called the **constant of proportionality** — this is the ratio of any pair of values (x, y)
  - the gradient of the straight line is k
  - all graphs start at $(0, 0)$.
- y is called the dependent variable and is normally placed on the vertical axis.
- x is called the independent variable and is normally placed on the horizontal axis.

VARIATION AND RATES OF CHANGE (PATH)

Inverse proportion

- If y and x are inversely proportional, then $y \propto \frac{1}{x}$ or $y = \frac{k}{x}$, where k is the constant of proportionality.

Rates

- A rate compares two quantities that have different units.
- Rates are special ratios.
- A rate is calculated by dividing one quantity with another.

Average speed

- Average speed is an example of a rate; it is calculated by dividing total distance by total time.
- Unit: metres per second (m/s) or kilometres per hour (km/h)
- Average speed $= \dfrac{\text{distance in kilometres}}{\text{time in hours}}$
  e.g. if the distance travelled is 200 km in 4 hours, then
  average speed $= \dfrac{200 \text{ km}}{4 \text{ hours}} = 50$ km/h.

Constant rates

- A constant rate means a quantity always changes by the same amount in a given time.
- Constant rates produce a linear (straight line) graph.

Variable rates

- A variable rate means a quantity does not change by the same amount in a given time.
- Variable rates produce a non-linear graph.

Graphs of rates of change

- A curved graph can be used to show that a rate is changing.
- The steeper the gradient, the higher the rate of change.

9.7.2 Project

To calculate the average speed of a journey, we can use the speed formula:

$$\text{speed} = \frac{\text{distance}}{\text{time}}$$

Speed is a *rate*, because it compares two quantities of different units. Speed can be expressed in units such as km/h, km/ min and m/s. The units of the quantities substituted into the numerator and denominator determine the final units of speed. In order to compare the speeds of different events, it is useful to convert them to the same units.

At a summer Olympics, sprinter Usain Bolt ran 100 m in 9.69 seconds. At a winter Olympics, cross-country skier Petter Northug covered 50 km in 2 hours and 5 minutes, and speed skater Mika Poutala covered 0.5 km in 34.86 seconds. Which competitor was the fastest? In order to answer this question, we need to determine the speed of each competitor. Since the information is quoted in a variety of units, we need to decide on a common unit for speed.

1. Calculate the speed of each athlete in m/s, correct to one decimal place.
2. Determine which athlete was the fastest.

Consider the speed of objects in the world around you. This task requires you to order the following objects from fastest to slowest, assuming that each object is travelling at its fastest possible speed:

electric car, submarine, diesel train, car ferry, skateboard, bowled cricket ball, windsurfer, served tennis ball, solar-powered car, motorcycle, aircraft carrier, Jaguar car, helicopter, airliner, rocket-powered car, the Concorde supersonic airliner

3. From your personal understanding and experience, order the above list from fastest to slowest.

We can make a more informed judgement if we have facts available about the movement of each of these objects. Consider the following facts:

- When travelling at its fastest speed, the Concorde could cover a distance of about 7000 km in 3 hours.
- The fastest airliner can cover a distance of 5174 km in 2 hours and a helicopter 600 km in 1.5 hours.
- It would take 9.5 hours for a rocket-powered car to travel 9652 km, 3.25 hours for an electric car to travel 1280 km and 1.5 hours for a Jaguar car to travel 525 km.
- On a sunny day, a solar-powered car can travel 39 km in half an hour.
- In 15 minutes, a car ferry can travel 26.75 km, an aircraft carrier 14 km and a submarine 18.5 km.
- It takes the fastest diesel train 0.42 hours to travel 100 km and a motorcyclist just 12 minutes to travel the same distance.
- In perfect weather conditions, a skateboarder can travel 30 km in 20 minutes and a windsurfer can travel 42 km in 30 minutes.
- A bowled cricket ball can travel 20 m in about 0.45 seconds, while a served tennis ball can travel 25 m in about 0.4 seconds.

4. Using the information above, decide on a common unit of speed and determine the speed of each of the objects.
5. Order the above objects from fastest to slowest.
6. Compare the order from question 5 with the list you made in question 3.
7. Conduct an experiment with your classmates to record the times that objects take to cover a certain distance. For example, record the time it takes to run 100 m or throw a ball 20 m. Compare the speeds with the speeds of the objects calculated during this investigation.

Exercise 9.7 Review questions

learn on

Fluency

1.  **MC** y is directly proportional to x and $y = 450$ when $x = 15$. The rule relating x and y is:

   A. $y = 0.033x$ **B.** $y = 30x$ **C.** $y = 60x$ **D.** $y = 6750x$

2. **MC** If $y \propto x$ and $y = 10$ when $x = 50$, the constant of proportionality is:

   A. 10   **B.** 5   **C.** 1   **D.** 0.2

3. **MC** If $y \propto x$ and $y = 10$ when $x = 50$, the value of x when $y = 12$ is:

   A. 6   **B.** 60   **C.** 40   **D.** 2.4

4. In the table below, if $y \propto x$, state the rule for the relationship between x and y.

x	0	1	2	3
y	0	4	8	12

5. **MC** If y is inversely proportional to x, then select which of the following statements is true.
   A. $x + y$ is a constant value.
   B. $y \div x$ is a constant value.
   C. $y \times x$ is a constant value.
   D. $y - x$ is a constant value.

6. In the graph below, y varies directly as x. State the rule for the relationship between x and y.

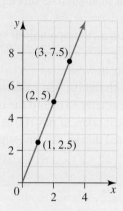

7. **MC** A speed of 60 km/h is equivalent to approximately:
   A. 17 m/ min
   B. 1 km/s
   C. 1 m/s
   D. 17 m/s

8. **MC** A metal part has a density of 37 g/mm³. If its volume is 6 mm³, it has a mass of:
   A. 6.1 g
   B. 6.1 kg
   C. 222 kg
   D. 222 g

9. **MC** Identify which of the following is *not* a rate.
   A. 50 km/h
   B. 70 beats/ min
   C. 40 kg
   D. Gradient

10. **MC** Calculate how long it takes to travel 240 km at an average speed of 60 km/h.
    A. 0.25 hours
    B. 240 minutes
    C. 180 hours
    D. 14 400 hours

11. **MC** A plane takes 2 hours to travel 1600 km. Calculate how long it takes to travel 1000 km.
    A. 150 minutes
    B. 48 minutes
    C. 2 hours
    D. 75 minutes

Understanding

12. **MC** The graph shows Sandy's bank balance. Determine the length of time (in months) that Sandy has been saving.

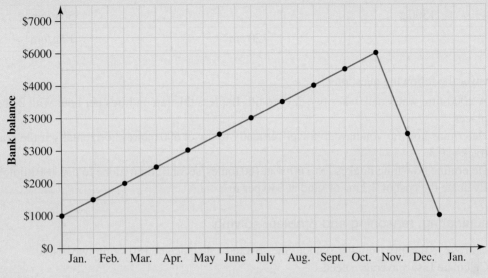

A. 2 months **B.** 5 months **C.** 10 months **D.** 12 months

13. True or False: The following graph shows that $y \propto x$. Give your reason.

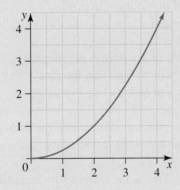

14. If $w \propto v$, and $w = 7.5$ when $v = 5$, calculate k, the constant of proportionality.

15. The wages earned by a worker for different numbers of hours are shown in the table.

Time (hours)	1	2	3	4
Wages ($)	25	50	75	100

Determine the rule relating wages, w, to the hours worked, t, if $w \propto t$. Give your answer in hours.

16. Calculate the missing quantities in the table.

Mass	Volume	Density
500 g	20 cm^3	
1500 g		50 g/cm^3
	120 cm^3	17 g/cm^3

17. Answer the following questions.

   a. A 2000-litre water tank takes two days to fill. Express this rate in litres per hour with 2 decimal places.

   b. Determine how far you have to travel vertically to travel 600 metres horizontally if the gradient of the track is 0.3.

   c. Determine how long it takes for a 60-watt (60 joules/second) light globe to use 100 kilojoules.

   d. Evaluate the cost of 2.3 m^3 of sand at $40 per m^3.

Communicating, reasoning and problem solving

18. Lisa drove to the city from her school. She covered a distance of 180 km in two hours.
   a. Calculate Lisa's average speed.
   b. Lisa travelled back at an average speed of 60 km/h. Determine the travel time.

19. Seventy grams of ammonium sulfate crystals are dissolved in 0.5 L of water.
   a. Calculate the concentration of the solution in g/mL.
   b. Another 500 mL of water is added. Determine the concentration of the solution now.

20. A skyscraper can be built at a rate of 4.5 storeys per month.
   a. Determine how many storeys will be built in 6 months.
   b. Determine how many storeys will be built in 24 months.

21. A certain kind of eucalyptus tree grows at a linear rate for its first two years of growth. If the growth rate is 5 cm per month, determine how long it will take to grow to a height of 1.07 m.

22. The pressure inside a boiler increases as the temperature increases. For each 1°C, the pressure increases by 10 units. At a temperature of 100 °C, the pressure is 600 units.
   If the boiler can withstand a pressure of 2000 units, determine the temperature at which this pressure is reached.

23. Hector has a part-time job as a waiter at a local cafe and is paid $8.50 per hour. Complete the table of values relating the amount of money Hector receives to the number of hours he works.

Number of hours	0	2	4	6	8	10
Pay						

24. Karina left home at 9:00 am. She spent some time at a friend's house, then travelled to the airport to pick up her sister. She then travelled straight back home. Her journey is shown by the travel graph.

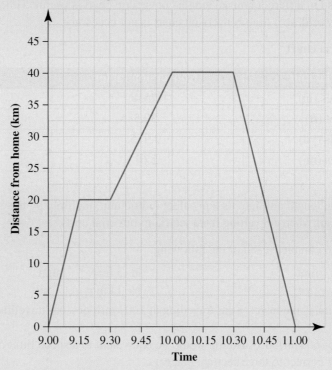

a. Determine how far Karina's house is from her friend's house.
b. Determine how much time Karina spent at her friend's place.
c. Calculate how far the airport is from Karina's house.
d. Determine how much time Karina spent at the airport.
e. Calculate how much time Karina took to drive home.
f. Calculate the average speed of Karina's journey:

   i. from her home to her friend's place
   ii. from her friend's place to the airport
   iii. from the airport to her home.

25. A commercial aircraft covers a distance of 1700 km in 2 hours 5 minutes. Evaluate the distance in metres that the aircraft travels each second.

26. A fun park charges a $10 entry fee and an additional $3 per ride. Complete the following table of values relating the total cost to the number of rides.

Rides	0	2	4	6	8	10
Cost						

Use the following information to answer questions **27–30**.

To help compare speed in different units, a length conversion chart is useful. Time is measured in the same units throughout the world, so a time conversion chart is not necessary.

Length conversion chart

Imperial unit	Conversion factor	Metric unit
Inches (in)	25.4	mm
Feet (ft)	30.5	cm
Yards (yd)	0.915	m
Miles (mi)	1.61	km

To convert an imperial length unit into its equivalent metric unit, multiply by the conversion factor. Divide by the conversion factor when converting from metric to imperial units.

27. **a.** Use the table to complete each of the following.

    i. $12\,\text{in} = \underline{\quad}$ mm   **ii.** $3\,\text{ft} = \underline{\quad}$ mm
    iii. $1\,\text{m} = = \underline{\quad}$ yd   **iv.** $1\,\text{km} = \underline{\quad}$ mi

    b. Determine which is the faster: a car travelling at 100 km/h or a car travelling at 100 miles/h.

28. The speed limit of 60 km/h in Australia would be equivalent to a speed limit of $\underline{\quad}$ miles/h in USA. Give your answer correct to three decimal places.

29. A launched rocket covers a distance of 17 miles in 10 seconds. Calculate its speed in km/h.

30. Here are some record speeds for moving objects.

Object	Speed
Motorcycle	149 m/s
Train	302 miles/h
Human skiing	244 km/h
Bullet from 38-calibre revolver	4000 ft/s

Convert these speeds to the same unit. Then place them in order from fastest to slowest.

on To test your understanding and knowledge of this topic, go to your learnON title at www.jacplus.com.au and complete the **post-test**.

Answers

Topic 9 Variation and rates of change (Path)

9.1 Pre-test

1. True
2. A
3. Yes
4. True
5. A
6. $x = \dfrac{k}{y}$
7. a. $k = 2.55$ b. $v = 3.06\,\text{amps}$
8. False
9. E
10. A
11. B
12. B
13. $1.45 for a pack of 10
14. $y = 4x$
15. 64 km/h

9.2 Direct linear proportion

1. a. Yes
   b. No; as speed increases, time decreases.
   c. No; doubling distance doesn't double the cost.
2. a. Yes
   b. No; doubling the side length doesn't double the area.
   c. Yes
3. a. Yes
   b. No; doubling distance doesn't double the cost (due to the initial fee).
   c. No; doubling age doesn't double height.
4. a. True: as the number, n, increases, so does the cost, C.
   b. True: $C = \pi D$, as C increases with D. π is a constant number.
   c. False: $A = \dfrac{1}{2}\pi r^2$, so it is not linear.
5. E
6. a. No; as the value of x increases, the value of y does not increase by a constant amount.
   b. Yes; as the value of a increases, the value of M increases by a constant amount. For every increase in a by 1, the value of M increases by 8.
   c. No; as the value of x increases, the value of y does not increase by a constant amount.
   d. No; when $n = 0$, C does not equal 0.
7. a. Yes
   b. No; when $t = 0$, d does not.
   c. No; as x doubles, y does not.
   d. No; when $d = 0$, w does not.
8. Sample responses can be found in the worked solutions in the online resources.

9. The point $(0, 0)$ must always exist in a table of values of two variables exhibiting direct proportionality.
10. The missing value is 8. The relationship between n and m can be calculated by the values 5 and 20 in the table ($n = 4m$).
11. $a = 32, b = 16, c = 36$
12. a. y is doubled.
    b. y is halved.
    c. x is tripled.
13. a. Yes, direct proportion does exist.
    b. $2.55
14. a. Yes   b. $935
15. a. Yes, direct linear proportion does exist.
    b. 6 cans of paint
16. Direct proportion does not exist. The price per metre for the 4.2-metre length is $5.50, and the price per metre for the 5.4-metre length is $5.40.

9.3 Direct proportion and ratio

1. $k = 6, a = 6b$
2. $a = 0.5b$
3. $C = 12.5t$
4. $v = 0.5t$
5. $F = 2.5a$
6. a. $E = 0.625W$ b. 19.2 kg   c. 4.375 cm
7. a. $W = 1.3n$
   b. ≈ 385 almonds
   c. 325 g
8. a. $d = 2.5n$ b. 2017.5 m   c. 2000 turns
9. a. $M = 0.12n$ b. 450 seedlings   c. 360 kg
10. $217.60
11. $217.25
12. 13.33 L
13. $105.98
14. 375 g
15. $150
16. $33.75
17. a. $D = 100T$ c. 2 hours 12 minutes
    b. 6 hours 6 minutes   d. 3 hours 54 minutes
18. 275 cm³

9.4 Inverse proportion

1. a. $s = \dfrac{k}{t}$ or $t = \dfrac{k}{s}$
   b. No
   c. $t = \dfrac{k}{p}$ or $p = \dfrac{k}{t}$
   d. $L = \dfrac{k}{p}$ or $p = \dfrac{k}{L}$
   e. No
   f. $t = \dfrac{k}{n}$ or $n = \dfrac{k}{t}$
2. Sample responses can be found in the worked solutions in the online resources.

3. a. $k = 1000, y = \dfrac{1000}{x}$

b.

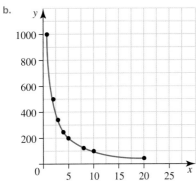

4. a. $k = 48, p = \dfrac{48}{q}$

b.

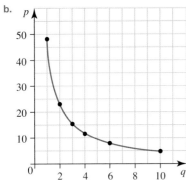

5. a. $k = 42, y = \dfrac{42}{x}$

b.

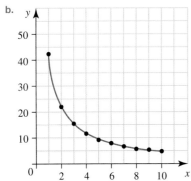

6. a. 4000   **b.** $a = \dfrac{4000}{m}$ **c.** $20\,\text{m/s}^2$

d. $4\,\text{m/s}^2$

7. a. 500   **b.** 2500 pencils   **c.** 1000 pencils

8. a. 337.5   **b.** 3.97 hours = 3 h 58 min
c. 96.4 km/h   **d.** 337.5 km

9. a. 17 500   **b.** $233.33   **c.** 70 people
d. $17 500

10. a. 200

b. 1 ampere

c. 13.3 ohms

d. Sample responses can be found in the worked solutions in the online resources.

11. a. $k = 15$

b. $V = 20\,L$

c. $P = 2.5$ atmospheres

12. The constant of proportionality represents the distance of the trip; therefore, when this value is smaller, the time taken to complete the trip at the same speed is also smaller.

13. a. 10   **b.** $T = \dfrac{10}{P}$ **c.** 1 hour 40 minutes

14. a. 100.048

b. $T = \dfrac{100.048}{V}$

c. 9.58 seconds

d. 15.00 seconds

15. a. $C = \dfrac{1350}{n}$

b. $96.43 per person

c. $30 per person

9.5 Introduction to rates

1. 64 km/h

2. a. 120 km/h

b. 1692.3 km/h

3. a. $92 per ticket

b. 24 cakes per hour

c. 10.44 m/s

d. 3 cm/h

e. $17.33/h

f. $50 per air conditioner

4. $4\,\text{g/cm}^3$

5. 600 g/L

6. The hare runs at 120 000 cm per minute; the tortoise runs at 0.144 km per hour. The hare runs 500 times faster than the tortoise.

7. 75 students/year

8. Either the distance of the journey increases and the time remains constant, or the distance of the journey remains constant and the time decreases.

9. 0.15

10. Cathy Freeman: 29.32 km/h; Usain Bolt: 37.15 km/h

11. 2750 organisms/litre. The beach should not be closed.

9.6 Constant and variable rates

1. a. 1 hour at 60 km/h, then 1 hour at 120 km/h

b. 1 hour at 50 km/h, a 30-minute stop, 30 minutes at 100 km/h, a 1-hour stop, then 1 hour at 100 km/h

c. 1 hour at 20 km/h, 1 hour at 100 km/h, 1 hour at 40 km/h

2. a.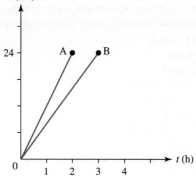

b. **i.** A: $d = 12t$; B: $d = 8t$

ii. A has a steeper gradient than B.

3. a.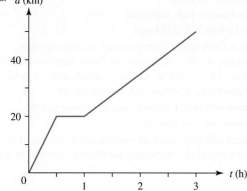

b.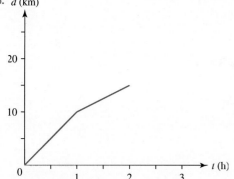

4. a. **A.** The car is moving with steady speed.

B. The car is momentarily stationary.

C. The speed is increasing.

D. The car is slowing down.

E. The car is moving at a slow steady speed.

F. The speed is increasing.

G. The car is moving at a faster steady speed.

b. **i.** A, G   **ii.** B, E   **iii.** B

5. a. B   **b.** D   **c.** C   **d.** A

6. a. 0.2 km/ min

b. 0.2 km/ min

c. Yes, both rates are the same.

7. a. 20, 10, 35, 0 L/ min

b. No

8. a. A, C   **b.** D   **c.** B

9. a. 150 km, 200 km and 250 km

b. 50 km

c. Both stop for 1 hour.

d. Brian — 300 km, 3.5 h; Margaret — 300 km, 4 h

e. Brian — 85.7 km/h; Margaret — 75 km/h

10. a. 2.5 hours

b. 20 km

c.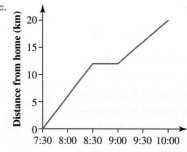

11. a.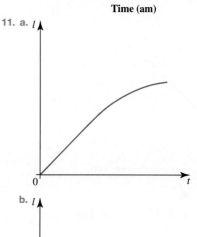

b.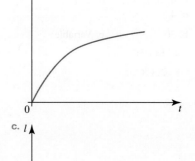

c.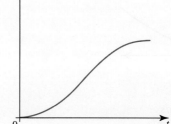

d.

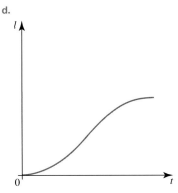

12. **a.** Point A — the gradient is greatest.
    b. Point C — the gradient is least.
    c. Point B — the gradient is decreasing.
    d. Point C — the gradient is increasing.
    e. Points A and D — the gradient is constant.

13. **a.**

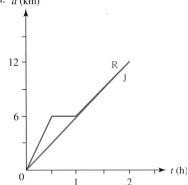

   b. 30 minutes

14. **a. i.** $10   **ii.** $10   **iii.** $10
    b. Yes

15. **a.** 1, 3, 5, 10   **b.** 4, 6, 7, 9   **c.** 2, 8
    d. 10   **e.** 4

16. **a.** 30   **b.** 30   **c.** Variable

17. **a. i.** 25.8 °C   **ii.** 14.0 °C
    b. i. 10.9 °C   **ii.** 2.3 °C
    c. No

18.

$$\text{Height of water vs Time}$$

Project

1. Usain Bolt = 10.3 m/s
   Petter Northug = 6.7 m/s
   Mika Poutala = 14.3 m/s

2. Mika Poutala

3. Answers will vary. Sample response: the Concorde supersonic airliner, rocket-powered car, Jaguar motorcar,

helicopter, airliner, motorcycle, electric car, diesel train, served tennis ball, car ferry, skateboard, bowled cricket ball, windsurfer, submarine, aircraft carrier, solar-powered car.

4. Concorde: 2333.3 km/h
   Airliner: 2587 km/h
   Helicopter: 400 km/h
   Rocket-powered car: 1016 km/h
   Electric car: 393.8 km/h
   Jaguar car: 350 km/h
   Solar-powered car: 78 km/h
   Car ferry: 107 km/h
   Aircraft carrier: 56 km/h
   Submarine: 74 km/h
   Diesel train: 238.1 km/h
   Motorcyclist: 500 km/h
   Skateboarder: 90 km/h
   Windsurfer: 84 km/h
   Bowled cricket ball: 160 km/h
   Served tennis ball: 225 km/h

5. Airliner, Concorde, rocket-powered car, motorcyclist, helicopter, electric car, Jaguar car, diesel train, served tennis ball, bowled cricket ball, oar ferry, skateboarder, windsurfer, solar-powered car, submarine, aircraft carrier

6. Answers will vary. Concorde was the fastest and the submarine was the slowest.

7. Answers will vary. Students could measure the speed of different objects by measuring the distance covered in certain time.

9.7 Review questions

1. B
2. D
3. B
4. $y = 4x$
5. C
6. $y = 2.5x$
7. D
8. D
9. C
10. B
11. D
12. C
13. False. The graph starts at $(0, 0)$ and increases, but it is not linear. Therefore, y does not vary directly with x.
14. 1.5
15. $w = 25t$
16.

Mass	Volume	Density
500 g	20 cm^3	25 g/cm^3
1500 g	30 cm^3	50 g/cm^3
2040 g	120 cm^3	17 g/cm^3

17. **a.** 41.67 L/h   **b.** 180 m
    c. 28 minutes   **d.** $92

18. **a.** 90 km/h   **b.** 3 h

19. **a.** 0.14 g/mL   **b.** 0.07 g/mL

20. a. 27   b. 108

21. 21.4 months

22. 240 °C

23.

Number of hours	0	2	4	6	8	10
Pay	$0	$17	$34	$51	$68	$85

24. a. 20 km

    b. 15 min

    c. 40 km

    d. 30 min

    e. 30 min

    f. i. 80 km/h

        ii. 40 km/h

        iii. 80 km/h

25. $226\frac{2}{3}$ m/sec

26.

Rides	0	2	4	6	8	10
Cost	$10	$16	$22	$28	$34	$40

27. a. i. 304.8 mm   ii. 915 mm

       iii. 1.09 yd   iv. 0.62 miles

    b. The one travelling at 100 miles/h

28. 37.267 miles/h

29. 9853.2 km/h

30. The conversions may vary, but the order from fastest to slowest is:

    Bullet from 38-calibre revolver

    Motorcycle

    Train

    Human skiing

10 Trigonometry

LESSON
10.1 Overview

Why learn this?

Trigonometry is the branch of geometry that is concerned with triangles. The word 'trigonometry' was created in the sixteenth century from the Greek words *trigōnon* ('triangle') and *metron* ('measure'), but the study of the geometry of triangles goes back to at least the sixth century BCE, when the ancient Greek philosopher Pythagoras of Samos developed his famous theorem.

Pythagoras was particularly interested in right-angled triangles and the relationships between their sides. Later Greek mathematicians used Pythagoras' theorem and the trigonometric ratios to calculate all kinds of distances, including Earth's circumference. Trigonometry is still the primary tool used by surveyors and geographers today when working out distances between points on Earth's surface.

None of the structures we build would be possible without our understanding of geometry and trigonometry. Engineers apply the principles of geometry and trigonometry regularly to make sure that buildings are strong, stable and capable of withstanding extreme conditions. Triangles are particularly useful to engineers and architects because they are the strongest shape. Any forces applied to a triangular frame will be distributed equally to all of its sides and joins. This fact has been known for thousands of years — triangular building frames were used as far back as the sixth century BCE.

A truss is an example of a structure that relies on the strength of triangles. Trusses are often used to hold up the roofs of houses and to keep bridges from falling down. Triangular frames can even be applied to curved shapes. Geodesic domes, like the one shown here, are rounded structures that are made up of many small triangular frames connected together. This use of triangular frames makes geodesic domes very strong, but also very light and easy to build.

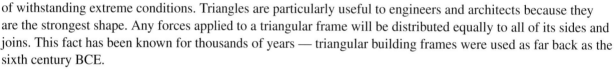

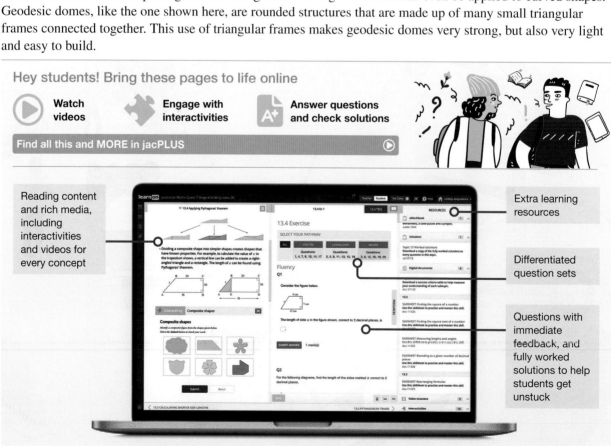

1. Calculate the value of *x* correct to 2 decimal places.

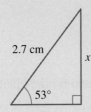

2. Calculate the value of *x*.

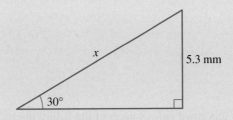

3. Calculate the value of the unknown angle, θ, correct to 2 decimal places.

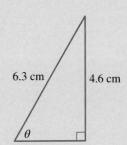

4. Calculate the value of the unknown angle, θ, correct to 2 decimal places.

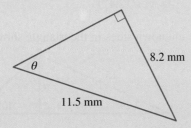

5. Evaluate, correct to 2 decimal places:
   a. $\sin^{-1}(0.75)$
   b. $\cos^{-1}(0.75)$
   c. $\tan^{-1}(0.75)$

6. **MC** Consider the right-angled triangle shown. Identify the correct option for angle θ.

A. a is the adjacent side.
B. b is the opposite side.
C. a is the hypotenuse
D. b is the adjacent side.

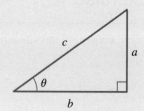

7. **MC** Identify the correct trigonometric ratio for the triangle shown.

A. $\tan \theta = \dfrac{a}{c}$

B. $\sin \theta = \dfrac{b}{a}$

C. $\cos \theta = \dfrac{c}{b}$

D. $\sin \theta = \dfrac{c}{b}$

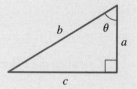

8. Calculate the length of the unknown side of each of the following triangles, correct to 2 decimal places.

a.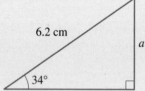

b.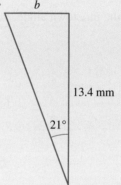

9. Evaluate, correct to 4 decimal places.

a. $\sin(62.5°)$ b. $\cos(12.1°)$ c. $\tan(74.9°)$

10. **MC** Identify the lengths of the unknown sides in the triangle shown.

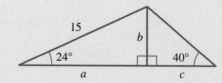

A. $a = 6.1$, $b = 13.7$ and $c = 7.3$
B. $a = 6.1$, $b = 13.7$ and $c = 5.1$
C. $a = 5.1$, $b = 6.1$ and $c = 13.7$
D. $a = 13.7$, $b = 6.1$ and $c = 7.3$

11. Calculate the value of θ in the triangle shown, correct to 1 decimal place.

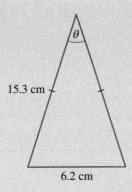

15.3 cm

6.2 cm

12. **MC** Identify the value of θ in the triangle shown, correct to 2 decimal places.
A. 25.27°
B. 28.16°
C. 28.17°
D. 61.83°

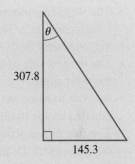

307.8

145.3

13. **MC** Determine the correct value of the side length a.
A. 7.62
B. 10.04
C. 14.34
D. 20.48

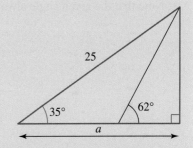

25

35°

62°

a

14. A person who is 1.54 m tall stands 10 m from the foot of a tree and records the angle of elevation (using an inclinometer) to the top of the tree as 30°.
Evaluate the height of the tree, correct to 2 decimal places.

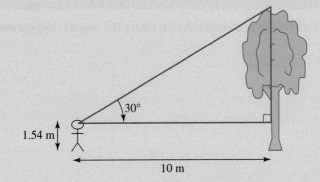

30°

1.54 m

10 m

15. The angle of depression from a scuba diver who is floating on the water's surface to a shark swimming below them on the sea floor is 35.8°. The depth of the water is 35 m.
Evaluate the horizontal distance from the scuba diver to the shark, correct to 2 decimal places.

LESSON
10.2 Trigonometric ratios and right-angled triangles

LEARNING INTENTION

At the end of this lesson you should be able to:
- identify and label the hypotenuse, opposite and adjacent sides of a right-angled triangle with respect to a given angle
- use the ratios of a triangle's sides to calculate the sine, cosine and tangent of an angle.

▶ 10.2.1 What is trigonometry?

eles-4756

- The word **trigonometry** is derived from the Greek words *trigonon* ('triangle') and *metron* ('measurement'). The literal translation of the word is 'to measure a triangle'.
- Trigonometry deals with the relationship between the sides and the angles of a triangle.
- The longest side of a right-angled triangle is called the hypotenuse. It is always located opposite the right angle.
- In order to name the remaining two sides of a triangle, another angle, called the 'given angle', θ, must be labelled on the triangle. θ is pronounced 'theta'; it is the eighth letter of the Greek alphabet.
- The side that is across from the given angle is called the **opposite side**. You can see that the opposite side does not touch the given angle at all.
- The remaining side is called the **adjacent side**.
- Note that the given angle always sits between the hypotenuse and the adjacent side.

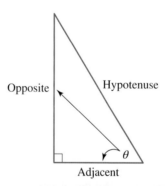

WORKED EXAMPLE 1 Labelling the sides of a right-angled triangle

Label the sides of the right-angled triangle shown using the words 'hypotenuse', 'adjacent' and 'opposite'.

THINK	WRITE

THINK

1. Identify the hypotenuse. Remember that the hypotenuse always lies opposite the right angle.

2. Look at the position of the given angle. The given angle always sits between the hypotenuse and the adjacent side; therefore, the adjacent side is the bottom side of this triangle.

3. The opposite side does not touch the given angle; therefore, it is the vertical side on the right of this triangle.

WRITE

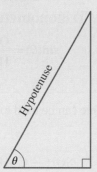

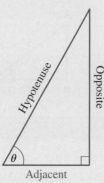

x

▶ 10.2.2 Trigonometric ratios

eles-4757

- Trigonometry is based upon the ratios between pairs of side lengths. Each of these ratios has a special name.
- In any right-angled triangle:

$$\textbf{sine}(\theta) = \frac{\text{opposite}}{\text{hypotenuse}}$$

$$\textbf{cosine}(\theta) = \frac{\text{adjacent}}{\text{hypotenuse}}$$

$$\textbf{tangent}(\theta) = \frac{\text{opposite}}{\text{adjacent}}$$

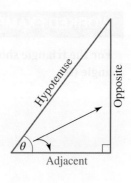

x

- These rules are often abbreviated as shown in the following.

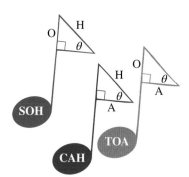

Trigonometric ratios

$$\sin(\theta) = \frac{O}{H} \qquad \cos(\theta) = \frac{A}{H} \qquad \tan(\theta) = \frac{O}{A}$$

- The following mnemonic can be used to help remember the trigonometric ratios.

$$SOH - CAH - TOA$$

- In this mnemonic:
  - **SOH** refers to $\sin(\theta) = $ **O**pposite/**H**ypotenuse
  - **CAH** refers to $\cos(\theta) = $ **A**djacent/**H**ypotenuse
  - **TOA** refers to $\tan(\theta) = $ **O**pposite/**A**djacent.

- Two similar right-angled triangles are shown below. Triangle DEF is an enlargement of triangle ABC by a factor of 2.

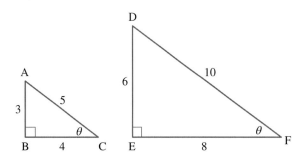

		Triangle ABC	**Triangle DEF**
$\sin\theta = $	$\dfrac{\text{Opposite}}{\text{Hypotenuse}}$	$\dfrac{3}{5}$	$\dfrac{6}{10} = \dfrac{3}{5}$
$\cos\theta = $	$\dfrac{\text{Adjacent}}{\text{Hypotenuse}}$	$\dfrac{4}{5}$	$\dfrac{8}{10} = \dfrac{4}{5}$
$\tan\theta = $	$\dfrac{\text{Opposite}}{\text{Adjacent}}$	$\dfrac{3}{4}$	$\dfrac{6}{8} = \dfrac{3}{4}$

- This table illustrates the fact that trigonometric ratios remain constant in similar right-angled triangles.

WORKED EXAMPLE 2 Identifying trigonometric ratios

For the triangle shown, write the equations for the sine, cosine and tangent ratios of the given angle (θ).

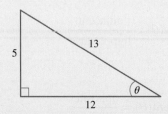

THINK

1. Label the sides of the triangle.

WRITE

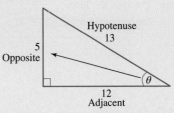

2. Write the trigonometric ratios.

$$\sin(\theta) = \frac{O}{H}, \cos(\theta) = \frac{A}{H}, \tan(\theta) = \frac{O}{A}.$$

3. Substitute the values of A, O and H into each formula.

$$\sin(\theta) = \frac{5}{13}, \cos(\theta) = \frac{12}{13}, \tan(\theta) = \frac{5}{12}$$

WORKED EXAMPLE 3 Relating sides to trigonometric ratios

For each of the following triangles, write the trigonometric ratio that relates the two given sides and the given angle.

a.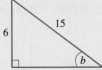

b.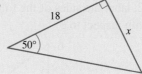

THINK

a. 1. Label the given sides.

WRITE

a.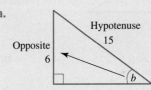

2. We are given O and H. These are used in SOH. Write the ratio.

$$\sin(\theta) = \frac{O}{H}$$

3. Substitute the values of the pronumerals into the ratio.

$$\sin(b) = \frac{6}{15}$$

4. Simplify the fraction.

$$\sin(b) = \frac{2}{5}$$

b. 1. Label the given sides.

b.

Adjacent 18, Opposite x, 50°

2. We are given A and O. These are used in TOA. Write the ratio.

$$\tan(\theta) = \frac{O}{A}$$

3. Substitute the values of the angle and the pronumerals into the ratio.

$$\tan(50°) = \frac{x}{18}$$

Exercise 10.2 Trigonometric ratios

learn on

| **10.2 Quick quiz** on | **10.2 Exercise** |

Individual pathways

■ PRACTISE	■ CONSOLIDATE	■ MASTER
1, 3, 5, 7, 9, 12, 15	2, 6, 10, 13, 16	4, 8, 11, 14, 17

Fluency

WE1 For questions 1 and 2, label the sides of the following right-angled triangles using the words 'hypotenuse', 'adjacent' and 'opposite' with respect to θ.

1. a. 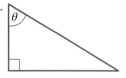   b. 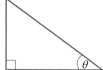   c.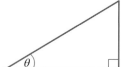

2. a.    b.    c.

3. For each of the following right-angled triangles, label the given angle θ and, where appropriate, label the hypotenuse, the adjacent side and the opposite side.

a. 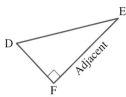   b. 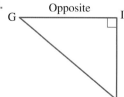   c.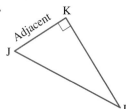

4. **MC** Identify which option correctly names the sides and angle (θ) of the triangle shown.

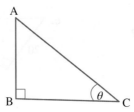

A. $\angle ABC = \theta$, AB = adjacent side, AC = hypotenuse, BC = opposite side
B. $\angle ACB = \theta$, AC = opposite side, BC = hypotenuse, AC = adjacent side
C. $\angle BAC = \theta$, AB = opposite side, BC = hypotenuse, AC = adjacent side
D. $\angle ACB = \theta$, AB = opposite side, AC = hypotenuse, BC = adjacent side

5. **WE2** Based on the given angle in each of the triangles shown, write the expressions for the ratios of sine, cosine and tangent.

a.

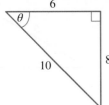

b.

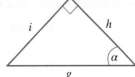

6. **WE2** Based on the given angle in each of the triangles shown, write the expressions for the ratios of sine, cosine and tangent.

a.

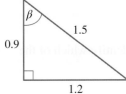

b.

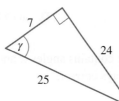

c.

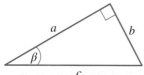

d.

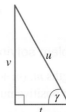

7. **WE3** Write the trigonometric ratio that relates the two given sides and the given angle in each of the following triangles.

a.

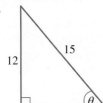

b.

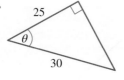

c.

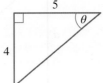

d.

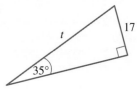

8. Write the trigonometric ratio that relates the two given sides and the given angle in each of the following triangles.

a.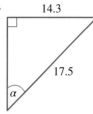
14.3

17.5

α

b.
7

x

15°

c.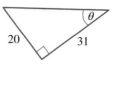
θ

20   31

d.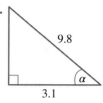
9.8

α

3.1

Understanding

9. **MC** Identify the correct trigonometric ratio for the triangle shown.

A. $\tan(\gamma) = \dfrac{a}{c}$

B. $\sin(\gamma) = \dfrac{c}{a}$

C. $\cos(\gamma) = \dfrac{c}{b}$

D. $\sin(\gamma) = \dfrac{c}{b}$

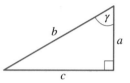

10. **MC** Identify the incorrect trigonometric ratio for the triangle shown.

A. $\sin(\alpha) = \dfrac{b}{c}$

B. $\sin(\alpha) = \dfrac{a}{c}$

C. $\cos(\alpha) = \dfrac{a}{c}$

D. $\tan(\alpha) = \dfrac{b}{a}$

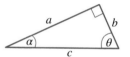

11. **MC** A right-angled triangle contains angles α and β as well as its right angle. Identify which of the following statements is always correct.

A. $\sin(\alpha) = \sin(\beta)$

B. $\tan(\alpha) = \tan(\beta)$

C. $\cos(\alpha) = 1 - \sin(\beta)$

D. $\tan(\alpha) = \dfrac{1}{\tan(\beta)}$

Communicating, reasoning and problem solving

12. If a right-angled triangle has side lengths m, $(m + n)$ and $(m - n)$, explain which of the lengths is the hypotenuse. Assume m and n are both positive numbers.

13. Given the triangle shown:

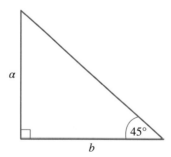
α

b

45°

a. explain why $a = b$
b. determine the value of $\tan(45°)$.

14. Using a protractor and ruler, carefully measure and draw a right-angled triangle with a base 10 cm long and an angle of 60°, as shown in the diagram.
    Measure the length of the other two sides of this triangle and mark these lengths on the diagram as well. Give the lengths of these sides to the nearest millimetre.
    Use your measurements to calculate the following three ratios, correct to 2 decimal places.

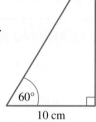

$$\frac{\text{opposite}}{\text{adjacent}} = \qquad , \quad \frac{\text{opposite}}{\text{hypotenuse}} = \qquad , \quad \frac{\text{adjacent}}{\text{hypotenuse}} = $$

Draw another triangle, similar to the first. Make the base length any length you want, but make sure all angles are equal to the angles in the first triangle.
Once this is done, measure the length of the remaining two sides.
Calculate the following three ratios correct to 2 decimal places.

$$\frac{\text{opposite}}{\text{adjacent}} = \qquad , \quad \frac{\text{opposite}}{\text{hypotenuse}} = \qquad , \quad \frac{\text{adjacent}}{\text{hypotenuse}} = $$

Comment on the conclusions you can draw from these two sets of ratios.

15. A ladder leans on a wall as shown. In relation to the angle given, identify the part of the image that represents:

    a. the adjacent side
    b. the hypotenuse
    c. the opposite side.

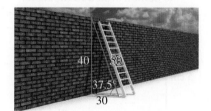

16. Consider the right-angled triangle shown.

    a. Label each of the sides using the letters O (opposite), A (adjacent) and H (hypotenuse) with respect to the 41° angle.
    b. Determine the value of each of the following trigonometric ratios, correct to 2 decimal places.
       i. $\sin(41°)$ ii. $\cos(41°)$ iii. $\tan(41°)$
    c. Calculate the value of the unknown angle, α.
    d. Determine the value of each of the following trigonometric ratios, correct to 2 decimal places.
       (*Hint:* Start by re-labelling the sides of the triangle with respect to angle α.)
       i. $\sin(\alpha)$ ii. $\cos(\alpha)$ iii. $\tan(\alpha)$
    e. Comment about the relationship between the following quantities:
       i. $\sin(41°)$ and $\cos(\alpha)$ ii. $\sin(\alpha)$ and $\cos(41°)$
    f. Make a general statement about the two angles α and 41°.

17. In relation to right-angled triangles, investigate the following.

    a. As the acute angle increases in size, determine what happens to the ratio of the length of the opposite side to the length of the hypotenuse in any right-angled triangle.
    b. As the acute angle increases in size, determine what happens to the ratio of the length of the adjacent side to the length of the hypotenuse.
       Determine what happens to the ratio of the length of the opposite side to the length of the adjacent side.
    c. Evaluate the largest possible value for the following.

       i. $\sin(\theta)$ ii. $\cos(\theta)$ iii. $\tan(\theta)$

LESSON
10.3 Calculating unknown side lengths

LEARNING INTENTION

At the end of this lesson you should be able to:
- determine approximations of the trigonometric ratios for a given angle
- determine the lengths of unknown sides in a right-angled triangle given an angle and one known side length.

▶ 10.3.1 Values of trigonometric ratios

eles-4758

- The values of the trigonometric ratios are always the same for a given angle.
  For example, if a right-angled triangle has a given angle of 30°, then the ratio of the opposite side to the hypotenuse will always be equal to sin(30°) = 0.5.
- The ratio of the adjacent side to the hypotenuse will always be equal to cos(30°) ≈ 0.87. This relationship is demonstrated in the table.

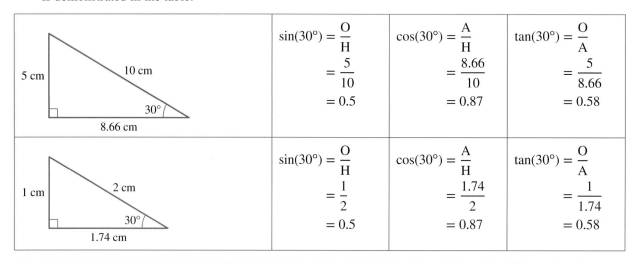

5 cm / 10 cm / 8.66 cm / 30°	$\sin(30°) = \dfrac{O}{H}$ $= \dfrac{5}{10}$ $= 0.5$	$\cos(30°) = \dfrac{A}{H}$ $= \dfrac{8.66}{10}$ $= 0.87$	$\tan(30°) = \dfrac{O}{A}$ $= \dfrac{5}{8.66}$ $= 0.58$
1 cm / 2 cm / 1.74 cm / 30°	$\sin(30°) = \dfrac{O}{H}$ $= \dfrac{1}{2}$ $= 0.5$	$\cos(30°) = \dfrac{A}{H}$ $= \dfrac{1.74}{2}$ $= 0.87$	$\tan(30°) = \dfrac{O}{A}$ $= \dfrac{1}{1.74}$ $= 0.58$

WORKED EXAMPLE 4 Evaluating trigonometric ratios using a calculator

Evaluate each of the following. Give your answers correct to 4 decimal places.
a. sin(53°)   **b. cos(31°)**   **c. tan(79°)**

THINK

a. 1. Set the calculator to degree mode. Make the calculation and then write out the first 5 decimal places.

   2. Round the answer to 4 decimal places.

b. 1. Set the calculator to degree mode. Make the calculation and then write out the first 5 decimal places.

   2. Round the answer to 4 decimal places.

c. 1. Set the calculator to degree mode. Make the calculation and then write out the first 5 decimal places.

   2. Round the answer to 4 decimal places.

WRITE

a. $\sin(53°) = 0.798\,63$

   ≈ 0.7986

b. $\cos(31°) = 0.857\,16$

   ≈ 0.8572

c. $\tan(79°) = 5.144\,55$

   ≈ 5.1446

▶ 10.3.2 Determining side lengths using trigonometric ratios

- If an angle and any side length of a right-angled triangle are known, it is possible to calculate the lengths of the other sides using trigonometry.

WORKED EXAMPLE 5 Calculating an unknown side length using the tangent ratio

Use the appropriate trigonometric ratio to calculate the length of the unknown side in the triangle shown. Give your answer correct to 2 decimal places.

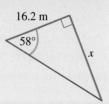

THINK	WRITE
1. Label the given sides.	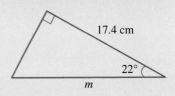
2. These sides can be used in TOA. Write the ratio.	$\tan(\theta) = \dfrac{O}{A}$
3. Substitute the values of θ, O and A into the ratio.	$\tan(58°) = \dfrac{x}{16.2}$
4. Solve the equation for x.	$16.2 \times \tan(58°) = x$
5. Calculate the value of x to 3 decimal places, then round the answer to 2 decimal places.	$x = 16.2 \tan(58°)$ $x = 25.925$ $x \approx 25.93 \text{ m}$

WORKED EXAMPLE 6 Calculating an unknown side length using the cosine ratio

Use the appropriate trigonometric ratios to calculate the length of the side marked m in the triangle shown. Give your answer correct to 2 decimal places.

THINK	WRITE
1. Label the given sides.	
2. These sides can be used in CAH. Write the ratio.	$\cos(\theta) = \dfrac{A}{H}$

▶

3. Substitute the values of θ, A and H into the ratio.	$\cos(22°) = \dfrac{17.4}{m}$
4. Solve the equation for m.	$m\cos(22°) = 17.4$ $m = \dfrac{17.4}{\cos(22°)}$
5. Calculate the value of m to 3 decimal places, then round the answer to 2 decimal places.	$m = 18.766$ $m \approx 18.77\,\text{cm}$

WORKED EXAMPLE 7 Solving worded problems using trigonometric ratios

Bachar set out on a bushwalking expedition. Using a compass, he set off on a course N 70°E (or 070°T) and travelled a distance of 5 km from his base camp.

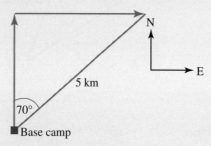

a. **Calculate how far east Bachar travelled.**
b. **Calculate how far north Bachar travelled from the base camp.**
   Give answers correct to 2 decimal places.

THINK

WRITE

a. 1. Label the eastern distance x.
   Label the northern distance y.
   Label the sides of the triangle as hypotenuse, opposite and adjacent.

a.

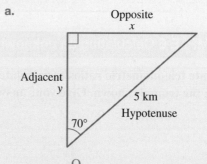

2. To calculate the value of x, use the sides of the triangle: $x = O$, $H = 5$.
   These are used in SOH. Write the ratio.

$\sin(\theta) = \dfrac{O}{H}$

3. Substitute the values of the angle and the pronumerals into the sine ratio.

$\sin(70°) = \dfrac{x}{5}$

4. Make x the subject of the equation.

$x = 5\sin(70°)$

5. Solve for x using a calculator. Write out the answer to 3 decimal places.

$= 4.698$

6. Round the answer to 2 decimal places. $\approx 4.70\,\text{km}$

7. Write the answer in sentence form.   Bachar has travelled 4.70 km east of the base
                                                camp.

b. 1. To calculate the value of y, use the   **b.** $\cos(\theta) = \dfrac{A}{H}$
     hypotenuse and the adjacent side: $y = A$,
     $H = 5$.
     These are used in CAH. Write the ratio.

   2. Substitute the values of the angle and the $\cos(70°) = \dfrac{y}{5}$
      pronumerals into the cosine ratio.

   3. Make y the subject of the equation. $y = 5\cos(70°)$

   4. Solve for y using a calculator. Write out the $= 1.710$
      answer to 3 decimal places.

   5. Round the answer to 2 decimal places. $\approx 1.71\,\text{km}$

   6. Write the answer in sentence form.   Bachar has travelled 1.71 km north of the base
                                                   camp.

Exercise 10.3 Calculating unknown side lengths   **learn** on

10.3 Quick quiz on	10.3 Exercise

Individual pathways

■ PRACTISE	■ CONSOLIDATE	■ MASTER
1, 4, 6, 9, 12, 13, 16, 19, 22	2, 5, 7, 10, 14, 17, 20, 23	3, 8, 11, 15, 18, 21, 24, 25

Fluency

1. **a.** Calculate the following. Give your answers correct to 4 decimal places.
   i. $\sin(55°)$ **ii.** $\sin(11.6°)$
   b. Complete the table shown. Use your calculator to calculate each value of $\sin(\theta)$. Give your answers
   correct to 2 decimal places.

θ	0°	15°	30°	45°	60°	75°	90°
$\sin(\theta)$							

   c. Summarise the trend in these values.

2. **a.** Calculate the following. Give your answers correct to 4 decimal places.
   i. $\cos(38°)$ **ii.** $\cos(53.71°)$
   b. Complete the table shown. Use your calculator to calculate each value of $\cos(\theta)$. Give your answers
   correct to 2 decimal places.

θ	0°	15°	30°	45°	60°	75°	90°
$\cos(\theta)$							

   c. Summarise the trend in these values.

3. a. Calculate the following. Give your answers correct to 4 decimal places.

    i. tan(18°)   **ii.** tan(51.9°)

b. Complete the table shown. Use your calculator to calculate each value of tan(θ). Give your answers correct correct to 2 decimal places.

θ	0°	15°	30°	45°	60°	75°	90°
tan(θ)							

c. Determine the value of tan(89°) and tan(89.9°).

d. Comment on these results.

For questions **4–10**, use the appropriate trigonometric ratios to calculate the length of the unknown side in each of the triangles shown. Give your answers correct to 2 decimal places.

4. `WE5`

  a.
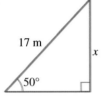

  b.

  c.
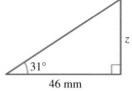

5. a.
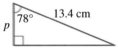

  b.
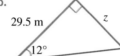

  c.

6. `WE6`

  a.

  b.

  c.
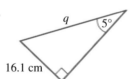

7. a.
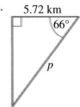

  b.
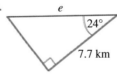

  c.

8. a.
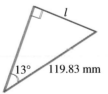

  b.
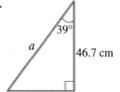

  c.
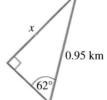

9. a.

b.

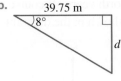

c.

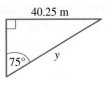

10. a.

b.

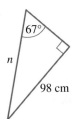

c.

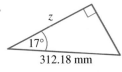

11. Calculate the lengths of the unknown sides in the triangles shown. Give your answers correct to 2 decimal places.

a.

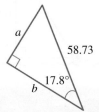

b.

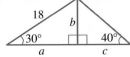

c.

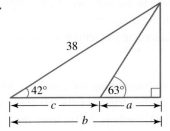

Understanding

12. **MC** Identify the value of x in the triangle shown, correct to 2 decimal places.

A. 59.65
B. 23.31
C. 64.80
D. 27.51

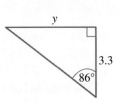

13. **MC** Identify the value of x in the triangle shown, correct to 2 decimal places.

A. 99.24 mm
B. 92.55 mm
C. 185.55 mm
D. 198.97 mm

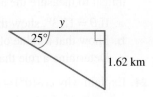

14. **MC** Identify y in the triangle shown, correct to 2 decimal places.

A. 47.19
B. 7.94
C. 1.37
D. 0.23

15. **MC** Identify the value of y in the triangle shown, correct to 2 decimal places.

A. 0.76 km
B. 1.79 km
C. 3.83 km
D. 3.47 km

16. **WE7** A ship that was supposed to travel due north veers off course and travels N 80°E (or 080°T) for a distance of 280 km instead. The ship's path is shown in the diagram.

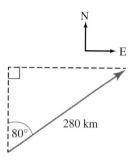

   a. Calculate the distance east that the ship has travelled.
   b. Determine the distance north that the ship has travelled.

17. A rescue helicopter spots a missing surfer drifting out to sea on their damaged board. The helicopter descends vertically to a height of 19 m above sea level and drops down an emergency rope, which the surfer grabs onto.
   Due to the high winds, the rope swings at an angle of 27° to the vertical, as shown in the diagram.
   Calculate the length of the rope.

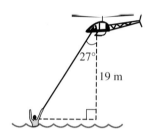

18. Walking along the coastline, Michelle (M) looks up at an angle of 55° and sees her friend Hella (H) at the lookout point on top of the cliff.
   If Michelle is 200 m from the cliff's base, determine the height of the cliff. (Assume Michelle and Hella are the same height.)

Communicating, reasoning and problem solving

19. Using a diagram, explain why $\sin(70°) = \cos(20°)$ and $\cos(70°) = \sin(20°)$. Generally speaking, explain which cosine $\sin(\theta)$ will be equal to.

20. One method for determining the distance across a body of water is illustrated in the diagram shown.

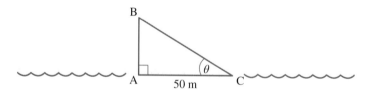

   The required distance is AB. A surveyor moves 50 m at a right angle to point C, then uses a tool called a transit to measure the angle θ ($\angle ACB$).

   a. If $\theta = 12.3°$, show that the distance from A to B is 10.90 m.
   b. Show that a value of $\theta = 63.44°$ gives a distance from A to B of 100 m.
   c. Determine a rule that can be used to calculate the distance from A to C.

21. Explain why $\cos(0°) = 1$ and $\sin(0°) = 0$, but $\cos(90°) = 0$ and $\sin(90°) = 1$

22. Calculate the value of the pronumeral in each of the triangles shown in the following photos.

a.

10 m 39° x

b.

6.2 m 29° h

c.

x 1.6 m 38°

23. A tile in the shape of a parallelogram has the measurements shown. Determine the tile's width, w, to the nearest mm.

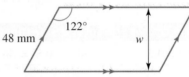

122° 48 mm w

24. A pole is supported by two wires as shown. If the length of the lower wire is 4.3 m, evaluathe following to 1 decimal place.

a. The length of the top wire
b. The height of the pole

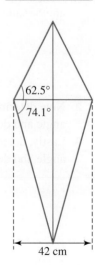

13° 48°

25. The frame of a kite is built from six wooden rods as shown. Evaluate the total length of wood used to make the frame of the kite. Give your answer to the nearest metre.

62.5° 74.1° 42 cm

LESSON
10.4 Calculating unknown angles

LEARNING INTENTION

At the end of this lesson you should be able to:
- determine the size of an angle from a given trigonometric ratio
- determine the unknown angle in a right-angled triangle when given two side lengths.

▶ 10.4.1 Inverse trigonometric ratios

eles-4760

- When we are given an angle, the sine, cosine and tangent functions let us determine the ratio of side lengths, but what if we only know the side lengths and we want to determine the angle in a right-angled triangle?
- The inverse trigonometric functions allow us to calculate angles using sine, cosine or tangent ratios.
- We have seen that $\sin(30°) = 0.5$. This means that $30°$ is the inverse sine of 0.5. This is written as $\sin^{-1}(0.5) = 30°$.

Inverse trigonometic ratios

Sine function	Inverse sine function
$\sin(30°) = 0.5$	$\sin^{-1}(0.5) = 30°$

- When reading the expression $\sin^{-1}(x)$ out loud, we say, 'the inverse sine of x'.
- When reading the expression $\cos^{-1}(x)$ out loud, we say, 'the inverse cosine of x'.
- When reading the expression $\tan^{-1}(x)$ out loud, we say, 'the inverse tangent of x'.
- You can calculate inverse trigonometric ratios using the SIN^{-1}, COS^{-1} and TAN^{-1} buttons on your calculator.

Digital technology

1. Use your calculator to determine $\sin(30°)$, then determine the inverse sine ($\sin^{-1}$) of the answer. Choose another angle and do the same thing.

sin(60)
0.8660254038

sin⁻¹(Ans)
60

2. Now determine $\cos(30°)$ and then determine the inverse cosine ($\cos^{-1}$) of the answer. Choose another angle and do the same thing.
3. Lastly, determine $\tan(45°)$ and then determine the inverse tangent ($\tan^{-1}$) of the answer. Choose another angle and do the same thing.

- The fact that sin and $\sin^{-1}$ cancel each other out is useful when solving equations like the following.

  Consider: $\sin(\theta) = 0.3$.

  Take the inverse sine of both sides
  $$\sin^{-1}(\sin(\theta)) = \sin^{-1}(0.3)$$
  $$\theta = \sin^{-1}(0.3)$$

  Consider: $\sin^{-1}(x) = 15°$.

  Take the sine of both sides.
  $$\sin^{-1}(\sin(x)) = \sin(15°)$$
  $$\theta = \sin(15°)$$
  Similarly, $\cos(\theta) = 0.522$ means that
  $$\theta = \cos^{-1}(0.522)$$
  and $\tan(\theta) = 1.25$ means that
  $$\theta = \tan^{-1}(1.25).$$

WORKED EXAMPLE 8 Evaluating inverse cosine values

Evaluate $\cos^{-1}(0.3678)$, correct to the nearest degree.

THINK	WRITE
1. Set your calculator to degree mode and make the calculation.	$\cos^{-1}(0.3678) = 68.4$
2. Round the answer to the nearest whole number and include the degree symbol in your answer.	$\approx 68°$

WORKED EXAMPLE 9 Determining angles using inverse trigonometric ratios

Determine the size of angle θ in each of the following. Give answers correct to the nearest degree.
a. $\sin(\theta) = 0.6543$
b. $\tan(\theta) = 1.745$

THINK	WRITE
a. 1. θ is the inverse sine of 0.6543.	a. $\sin(\theta) = 0.6543$ $\theta = \sin^{-1}(0.6543)$
2. Make the calculation using your calculator's inverse sine function and record the answer.	$= 40.8$
3. Round the answer to the nearest degree and include the degree symbol in your answer.	$\approx 41°$
b. 1. θ is the inverse tangent of 1.745.	b. $\tan(\theta) = 1.745$
2. Make the calculation using your calculator's inverse tangent function and record the answer.	$\theta = \tan^{-1}(1.745)$ $= 60.18$
3. Round the answer to the nearest degree and include the degree symbol in your answer.	$\approx 60°$

▶ 10.4.2 Determining the angle when two sides are known

eles-4761

- If the lengths of any two sides of a right-angled triangle are known, it is possible to determine an angle using inverse sine, inverse cosine or inverse tangent.

WORKED EXAMPLE 10 Determining an unknown angle when two sides are known

Determine the value of θ in the triangle shown. Give your answer correct to the nearest degree.

THINK	WRITE
1. Label the given sides. These are used in CAH. Write out the ratio.	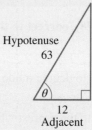 $\cos(\theta) = \dfrac{A}{H}$
2. Substitute the given values into the cosine ratio.	$\cos(\theta) = \dfrac{12}{63}$
3. θ is the inverse cosine of $\dfrac{12}{63}$.	$\theta = \cos^{-1}\left(\dfrac{12}{63}\right)$
4. Evaluate.	$= 79.0$
5. Round the answer to the nearest degree.	$\approx 79°$

WORKED EXAMPLE 11 Solving problems in real-world contexts

Roberta goes water skiing on the Hawkesbury River. She is going to try out a new ramp. The ramp rises 1.5 m above the water level and has a length of 6.4 m. Determine the magnitude (size) of the angle that the ramp makes with the water's surface. Give your answer correct to the nearest degree.

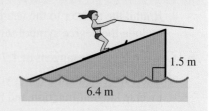

THINK	WRITE
1. Draw a simple diagram, showing the known lengths and the angle to be found.	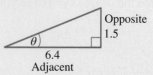

2. Label the given sides. These are used in TOA. Write out the ratio.

$$\tan(\theta) = \frac{O}{A}$$

3. Substitute the values of the pronumerals into the tangent ratio.

$$\tan(\theta) = \frac{1.5}{6.4}$$

4. θ is the inverse tangent of $\frac{1.5}{6.4}$.

$$\theta = \tan^{-1}\left(\frac{1.5}{6.4}\right)$$

5. Evaluate.

$$= 13.19$$

6. Round the answer to the nearest degree.

$$\approx 13°$$

7. Write the answer in words.

The ramp makes an angle of 13° with the water's surface.

 Resources

Interactivity Finding the angle when two sides are known (int-6046)

Exercise 10.4 Calculating unknown angles

learn on

10.4 Quick quiz on

10.4 Exercise

Individual pathways

■ PRACTISE	■ CONSOLIDATE	■ MASTER
1, 4, 7, 10, 11, 14, 18, 21	2, 5, 8, 12, 15, 17, 19, 22	3, 6, 9, 13, 16, 20, 23, 24

Fluency

1. Calculate each of the following, rounding your answers to the nearest degree.

   a. $\sin^{-1}(0.6294)$ b. $\cos^{-1}(0.3110)$ c. $\tan^{-1}(0.7409)$

2. Calculate each of the following, rounding your answers to the nearest degree.

   a. $\tan^{-1}(1.3061)$ b. $\sin^{-1}(0.9357)$ c. $\cos^{-1}(0.3275)$

3. Calculate each of the following, rounding your answers to the nearest degree.

   a. $\cos^{-1}(0.1928)$ b. $\tan^{-1}(4.1966)$ c. $\sin^{-1}(0.2554)$

4. WE9 Determine the size of the angle in each of the following, rounding your answers to the nearest degree.

   a. $\sin(\theta) = 0.3214$ b. $\sin \theta = 0.6752$ c. $\sin(\beta) = 0.8235$ d. $\cos(\beta) = 0.9351$

5. Determine the size of the angle in each of the following, rounding your answers to the nearest degree.

   a. $\cos(\alpha) = 0.6529$ b. $\cos(\alpha) = 0.1722$ c. $\tan(\theta) = 0.7065$ d. $\tan(a) = 1$

6. Determine the size of the angle in each of the following, rounding your answers to the nearest degree.

   a. $\tan(b) = 0.876$ b. $\sin(c) = 0.3936$ c. $\cos(\theta) = 0.5241$ d. $\tan(\alpha) = 5.6214$

7. **WE10** Determine the value of θ in each of the following triangles, rounding your answers to the nearest degree.

a.

b.

c.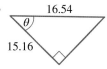

8. Determine the value of θ in each of the following triangles, rounding your answers to the nearest degree.

a.

b.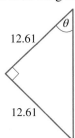

c.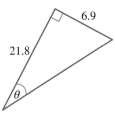

9. Determine the value of θ in each of the following triangles, rounding your answers to the nearest degree.

a.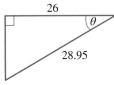

b.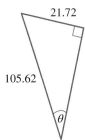

c.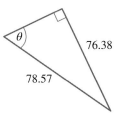

10. **MC** If $\cos(\theta) = 0.8752$, identify the value of θ, correct to 2 decimal places.

    A. $61.07°$ **B.** $41.19°$ **C.** $25.84°$ **D.** $28.93°$

11. **MC** If $\sin(\theta) = 0.5530$, identify the value of θ, correct to 2 decimal places.

    A. $56.43°$ **B.** $33.57°$ **C.** $28.94°$ **D.** $36.87°$

12. **MC** Identify the value of θ in the triangle shown, correct to 2 decimal places.

    A. $41.30°$
    B. $28.55°$
    C. $48.70°$
    D. $61.45°$

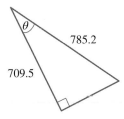

13. **MC** Identify the value of θ in the triangle shown, correct to 2 decimal places.

    A. $42.10°$
    B. $64.63°$
    C. $25.37°$
    D. $47.90°$

Understanding

14. A piece of fabric measuring 2.54 m by 1.5 m has been printed with parallel diagonal stripes. Determine the angle each diagonal makes with the length of the fabric.
    Give your answer correct to 2 decimal places.

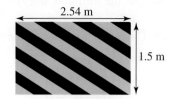

15. **WE11** Danny Dingo is perched on top of a 20-m-high cliff. They are watching Erwin Emu, who is feeding on a bush that stands 8 m away from the base of the cliff. Danny has purchased a flying contraption that they hope will help them capture Erwin.
    Calculate the angle to the cliff that Danny should follow downwards to catch their prey.
    Give your answer correct to 2 decimal places.

16. a. Complete the table shown.

x	0.0	0.1	0.2	0.3	0.4	0.5	0.6	0.7	0.8	0.9	1.0
$y = \cos^{-1}(x)$	90°					60°					0°

    b. Plot the table from part **a** on graph paper. Alternatively you can use a spreadsheet program or a suitable calculator.

17. A zipline runs from an observation platform 400 m above sea level to a landing platform on the ground 1200 m away.
    Calculate the angle that the zipline makes with the ground at the landing platform.
    Give your answer correct to 2 decimal places.

Communicating, reasoning and problem solving

18. Safety guidelines for wheelchair access ramps used to state that the gradient of a ramp had to be in the ratio 1 : 20.

    a. Using this ratio, show that the angle that any ramp needs to make with the horizontal is closest to 3°.
    b. New regulations have changed the guidelines about the ratio of the gradient of a ramp. Now the angle the ramp makes with the horizontal must be closest to 6°.
    Explain why, using this new angle size, the new ratio could be 1 : 9.5.

19. Jayani and Lee are camping with their friends Awer and Susie. Both couples have tents that are 2 m high. The top of each 2 m tent pole has to be tied with a piece of rope so that the pole stays upright.
    To make sure this rope doesn't trip anyone, Jayani and Lee decide that the angle between the rope and the ground should be 60°.

    a. Determine the length of the rope that Jayani and Lee need to run from the top of their tent pole to the ground. Give your answer correct to 2 decimal places.
    b. Awer and Susie set up their tent further into the camping ground. They want to secure their tent pole using a piece of rope that they know is somewhere between 2 m and 3 m long.

        i. Explain why Awer and Susie's rope will have to be longer than 2 m.
        ii. Show that the minimum angle Awer and Susie's rope will make with the ground will be 41.8°.

20. Use the formulas $\sin(\theta) = \dfrac{O}{H}$ and $\cos(\theta) = \dfrac{A}{H}$ to prove that $\tan(\theta) = \dfrac{\sin(\theta)}{\cos(\theta)}$.

21. Calculate the value of the pronumeral in each of the following, correct to 2 decimal places.

a.

5.4 cm

12 cm

θ

b.

1.2 m

θ

0.75 m

c.

0.75 m

x

1.8 m

22. A family is building a patio extension for their house. One section of the new patio will have a gable roof. A similar structure is shown, with the planned post heights and the span of the roof given.

To allow more light in, the family wants the peak (the highest point) of the roof to be at least 5 m above ground level.

According to building regulations, the slope of the roof (the angle that the sloping edge makes with the horizontal) must be 22°.

6 m

3.2 m

a. Use trigonometry to calculate whether the roof would be high enough if the angle was 22°.

b. Using trigonometry, determine the size of the obtuse angle formed at the peak of the roof.

23. The height of a square-based prism is twice its base length. Calculate the angle the diagonal of the prism makes with the diagonal of the base.

24. A series of seven shapes are marked out as shown inside a square with a side length of 10 cm. If the dots marked on the diagram represent the midpoints of the square's sides, determine the dimensions of each of the seven smaller shapes.

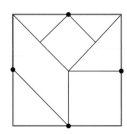

LESSON
10.5 Angles of elevation and depression

LEARNING INTENTION

At the end of this lesson you should be able to:
- use angles of depression and elevation to determine heights and distances
- use trigonometric ratios to determine angles of depression and elevation.

▶ 10.5.1 Angles of elevation and depression

eles-4762

- When looking up towards an object, the **angle of elevation** is the angle between the horizontal line and the line of vision.

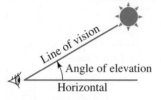

- When looking down at an object, the **angle of depression** is the angle between the horizontal line and the line of vision.

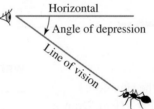

- Angles of elevation and depression are always measured from horizontal lines.
- For any two objects, A and B, the angle of elevation of B, as seen from A, is equal to the angle of depression of A, as seen from B. This can be proven with angle properties of parallel lines and alternate angles.

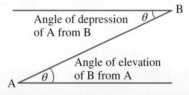

WORKED EXAMPLE 12 Determining height using angle of elevation

At a point 10 m from the base of a tree, the angle of elevation of the top of the tree is 38°. Calculate the height of the tree. Give your answer to the nearest centimetre.

THINK

1. Draw a simple diagram to represent the situation. The angle of elevation is 38° from the horizontal.

WRITE

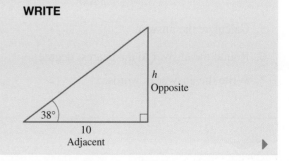

2. Label the given sides of the triangle. These sides are used in TOA. Write out the ratio.

$$\tan(38°) = \frac{h}{10}$$

3. Multiply both sides by 10.

$$10\tan(38°) = h$$

4. Calculate the answer, correct to 3 decimal places.

$$h = 7.812$$

5. Round the answer to 2 decimal places.

$$\approx 7.81$$

6. Write the answer in words.

The tree is 7.81 m tall.

WORKED EXAMPLE 13 Calculating an angle of depression

A 30-m-tall lighthouse stands on top of a cliff that is 180 m high. Determine the angle of depression (θ) of a ship from the top of the lighthouse if the ship is 3700 m from the bottom of the cliff.

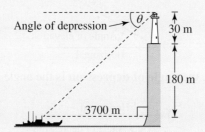

THINK

1. Draw a simple diagram to represent the situation. The height of the triangle is $180 + 30 = 210$ m. Draw a horizontal line from the top of the triangle and mark the angle of depression, θ. Mark the alternate angle as well.

WRITE

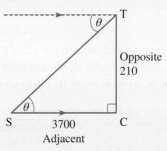

2. Label the given sides of the triangle. These sides are used in TOA. Write out the ratio.

$$\tan(\theta) = \frac{O}{A}$$

3. Substitute the given values into the ratio.

$$\tan(\theta) = \frac{210}{3700}$$

4. θ is the inverse tangent of $\frac{210}{3700}$.

$$\theta = \tan^{-1}\left(\frac{210}{3700}\right)$$

5. Calculate the answer.

$$= 3.24$$

6. Round the answer to the nearest degree.

$$\approx 3°$$

7. Write the answer in words.

The angle of depression of the ship from the top of the lighthouse is 3°.

on **Resources**

Interactivity Finding the angle of elevation and angle of depression (int-6047)

Exercise 10.5 Angles of elevation and depression   learn on

10.5 Quick quiz on	**10.5 Exercise**

Individual pathways

■ PRACTISE	■ CONSOLIDATE	■ MASTER
1, 4, 6, 9, 12	2, 5, 7, 10, 13	3, 8, 11, 14, 15, 16

Fluency

1. **WE12** A lifesaver standing on their tower 3 m above the ground spots a swimmer experiencing difficulty.

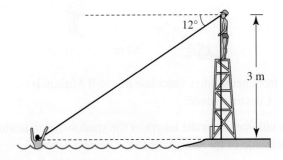

The angle of depression of the swimmer from the lifesaver is 12°.
Calculate how far the swimmer is from the lifesaver's tower.
Give your answer correct to 2 decimal places.

2. From the top of a 50-m-high lookout, the angle of depression of a camp site that is level with the base of the lookout is 37°.

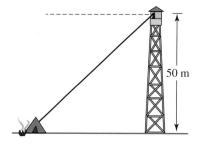

50 m

Calculate how far the camp site is from the base of the lookout.

3. Building specifications require the angle of elevation of any ramp constructed for public use to be less than 3°.

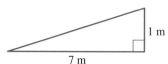

1 m

7 m

A new shopping centre is constructing its access ramps with a ratio of 7 m horizontal length to 1 m vertical height.

Calculate the angle of elevation of these new ramps. Determine whether the new ramps meet the specifications required by ramps intended for public use.

Understanding

4. From a point on the ground 60 m away from a tree, the angle of elevation of the top of the tree is 35°.

   a. Draw a labelled diagram to represent this situation.

   b. Calculate the height of the tree to the nearest metre.

5. Miriam wants to take a video of her daughter Alexandra's first attempts at crawling. When Alexandra lies on the floor and looks up at her mother, the angle of elevation is 17°.

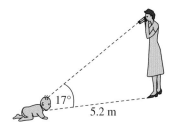

17°

5.2 m

If Alexandra is 5.2 m away from her mother, calculate how tall Miriam is.
Give your answer correct to 1 decimal place.

6. **WE13** Hien, who is 1.95 m tall, measures the length of the shadow he casts along the ground as 0.98 m. Determine the angle of depression of the sun's rays to the nearest degree.

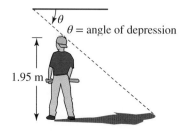

θ

θ = angle of depression

1.95 m

7. A support beam is 3.8 m tall and leaning against a wall.

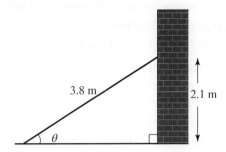

a. Determine the angle the support beam makes with the ground if it reaches 2.1 m up the wall. Give your answer to the nearest degree.

b. Determine how far the support beam is placed from the wall. Give your answer to the nearest metre.

8. **MC** A lighthouse is 78 m tall. The angle of elevation to the top of the lighthouse from point B, which is level with the base of the lighthouse, is 60°.

Select the correct diagram for this information.

A.

B.

C.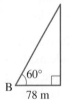

D.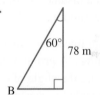

Communicating, reasoning and problem solving

9. Con and John are practising shots at goal. Con is 3.6 m away from the goal and John is 4.2 m away, as shown in the diagram.
If the height of the goal post is 2.44 m, determine the maximum angle of elevation that each player can kick the ball in order to score a goal.
Give your answer to the nearest degree.

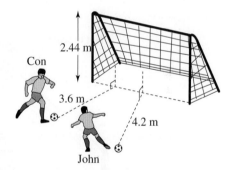

10. Lifesaver Sami is sitting in a tower 10 m from the water's edge and 4 m above sea level. They spot some dolphins playing near a marker at sea directly in front of them.
The marker the dolphins are swimming near is 20 m from the water's edge.

a. Draw a diagram to represent this information.

b. Show that the angle of depression of Sami's view of the dolphins, correct to 1 decimal place, is 7.6°.

c. As the dolphins swim towards Sami, determine whether the angle of depression would increase or decrease. Justify your answer in terms of the tangent ratios.

11. A pair of office buildings are 100 m and 75 m high. From the top of the north side of the taller building, the angle of depression to the top of the south side of the shorter building is 20°, as shown.
Show that the horizontal distance between the north side of the taller building and the south side of the shorter building is closest to 69 m.

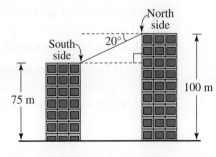

12. From a rescue helicopter 80 m above the ocean, the angles of depression of two shipwreck survivors are 40° and 60° respectively. The two survivors and the helicopter are in line with each other.

   a. Draw a labelled diagram to represent the situation.
   b. Calculate the distance between the two survivors. Give your answer to the nearest metre.

13. Rouka was hiking in the mountains when she spotted an eagle sitting up in a tree. The angle of elevation of her view of the eagle was 35°. She then walked 20 m towards the tree. From her new position, her angle of elevation was 50°. The distance between the eagle and the ground was 35.5 m.

   a. Draw a labelled diagram to represent this information.
   b. If Rouka's eyes are located 9 cm below the very top of her head, calculate how tall she is. Give your answer in metres, correct to the nearest centimetre.

14. A lookout in a lighthouse tower can see two ships approaching the coast. Their angles of depression are 25° and 30°. If the ships are 100 m apart, show that the height of the lighthouse, to the nearest metre, is 242 m.

15. As shown in the diagram, at a certain distance from an office building, the angle of elevation to the top of the building is 60°. From a distance 12 m further back, the angle of elevation to the top of the building is 45°. Show that the building is 28.4 m high.

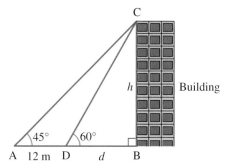

16. A gum tree stands in a courtyard in the middle of a group of office buildings. A group of three Year 9 students, Jackie, Pho and Theo, measure the angle of elevation from three different positions. They are unable to measure the distance to the base of the tree because of the steel tree guard around the base. This diagram shows the angles of elevation and the distances measured.

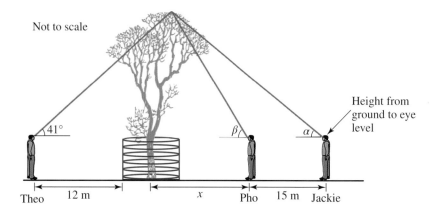

a. Show that $x = \dfrac{15 \tan(\alpha)}{\tan(\beta) - \tan(\alpha)}$, where x is the distance, in metres, from the base of the tree to Pho's position.

b. The students estimate the tree to be 15 m taller than them. Pho measured the angle of elevation to be 72°. If these measurements are correct, calculate Jackie's angle of elevation, correct to the nearest degree.

c. Theo did some calculations and determined that the tree was only about 10.4 m taller than them. Jackie claims that Theo's calculation of 10.4 m is incorrect.

   i. Decide if Jackie's claim is correct. Show how Theo calculated a height of 10.4 m.
   ii. If the height of the tree was actually 15 m above the height of the students, determine the horizontal distance Theo should have used in his calculations. Write your answer to the nearest centimetre.

LESSON
10.6 Review

10.6.1 Topic summary

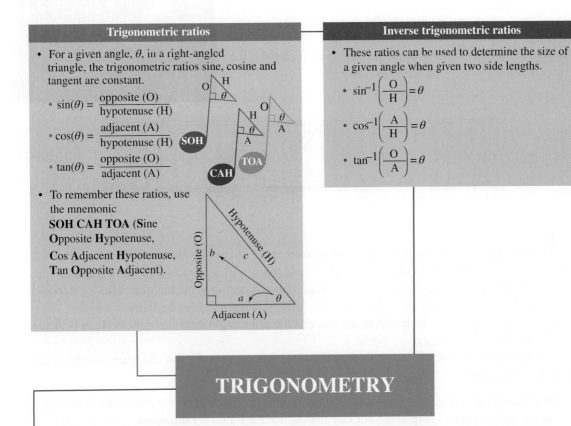

Trigonometric ratios

- For a given angle, θ, in a right-angled triangle, the trigonometric ratios sine, cosine and tangent are constant.

- $\sin(\theta) = \dfrac{\text{opposite (O)}}{\text{hypotenuse (H)}}$

- $\cos(\theta) = \dfrac{\text{adjacent (A)}}{\text{hypotenuse (H)}}$

- $\tan(\theta) = \dfrac{\text{opposite (O)}}{\text{adjacent (A)}}$

- To remember these ratios, use the mnemonic **SOH CAH TOA** (**S**ine **O**pposite **H**ypotenuse, **C**os **A**djacent **H**ypotenuse, **T**an **O**pposite **A**djacent).

Inverse trigonometric ratios

- These ratios can be used to determine the size of a given angle when given two side lengths.

- $\sin^{-1}\left(\dfrac{O}{H}\right) = \theta$

- $\cos^{-1}\left(\dfrac{A}{H}\right) = \theta$

- $\tan^{-1}\left(\dfrac{O}{A}\right) = \theta$

TRIGONOMETRY

Angles of elevation and depression

- When looking upwards at an object, the **angle of elevation** is the angle between the horizontal and the upward line of sight.

- When looking downwards at an object, the **angle of depression** is the angle between the horizontal and the downward line of sight.

- The angle of elevation when looking from A to B is equal to the angle of depression when looking from B to A.

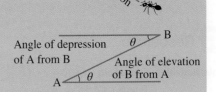

10.6.2 Project

Wall braces

Builders use braces to strengthen wall frames. These braces typically run between the top and bottom horizontal sections of the wall frame.

Building industry standards require that the acute angle that a brace makes with the horizontal sections needs to be somewhere between 37° and 53°. Sometimes more than one brace may be required if the frame is particularly long, as shown in the diagram.

37° to 53°

1. Cut out some thin strips of cardboard and arrange them in the shape of a rectangle to represent a wall frame. Place pins at the corners of the rectangle to hold the strips together. Notice that the frame can easily be moved out of shape. Attach a brace to this frame according to the building industry standards discussed. Write a brief comment to describe what effect this brace has on the frame.
2. Investigate what happens to the length of the brace as the acute angle it creates with the base is increased from 37° to 53°.
3. Use your findings from question **2** to calculate the angle that requires the shortest brace and the angle that requires the longest brace.

Most modern houses are constructed with a ceiling height (the height of the walls from floor to ceiling) of 2.4 m. Use this information to help with your calculations for the following questions.

4. Assume you are working with a section of wall that is 3.5 m long. Calculate the length of the longest possible brace. Draw a diagram and show your working to support your answer.
5. Calculate the minimum wall length for which two braces are required. Draw a diagram and show your working to support your answer.
6. Some older houses have ceilings that are over 2.4. m high. Answer the questions in **4** and **5** for a house with a ceiling that is 3 m high. Draw diagrams and show your workings to support your answers.
7. Take the measurements of a wall with no windows at your school or at home. Draw a scale drawing of the frame of this wall and show where the brace or braces for this wall might lie. Calculate the length and angle of each brace.

 Resources

Interactivities Crossword (int-0703)
Sudoku puzzle (int-3206)

Fluency

1. MC Identify which of the following correctly names the sides and angle of the triangle shown.

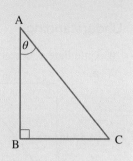

   A. $\angle C = \theta$, AB = adjacent side, AC = hypotenuse, BC = opposite side
   B. $\angle C = \theta$, AB = opposite side, BC = hypotenuse, AC = adjacent side
   C. $\angle A = \theta$, AB = opposite side, AC = hypotenuse, BC = adjacent side
   D. $\angle A = \theta$, AB = adjacent side, AC = hypotenuse, BC = opposite side

2. MC Select which of the following statements is correct.

   A. $\sin(60°) = \cos(60°)$
   B. $\cos(25°) = \cos(65°)$
   C. $\cos(60°) = \sin(30°)$
   D. $\sin(70°) = \cos(70°)$

3. MC Identify the value of x in the triangle shown, correct to 2 decimal places.

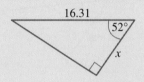

   A. 26.49
   B. 10.04
   C. 12.85
   D. 20.70

4. MC Identify which of the following could be used to calculate the value of x in the triangle shown.

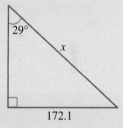

   A. $x = \dfrac{172.1}{\cos(29°)}$
   B. $x = \dfrac{172.1}{\sin(29°)}$
   C. $x = 172.1 \times \sin(29°)$
   D. $x = 172.1 \times \cos(29°)$

5. MC Identify which of the following could be used to calculate the value of x in the triangle shown.

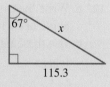

   A. $x = \dfrac{115.3}{\sin(23°)}$
   B. $x = \dfrac{115.3}{\cos(67°)}$
   C. $x = 115.3° \times \sin(67°)$
   D. $x = \dfrac{115.3}{\cos(23°)}$

6. MC Identify which of the following could be used to calculate the value of x in the triangle shown.

   A. $x = \dfrac{28.74}{\cos(17°)}$
   B. $x = 28.74 \times \sin(17°)$
   C. $x = 28.74 \times \cos(17°)$
   D. $x = 28.74 \times \cos(73°)$

7. MC Select which of the following is closest to the value of $\tan^{-1}(1.8931)$.
   A. 62°
   B. 0.0331°
   C. 1.08°
   D. 69°

8. **MC** Select the value of θ in the triangle shown, correct to 2 decimal places.

    A. $40.89°$ **B.** $60°$ **C.** $35.27°$ **D.** $30°$

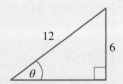

Understanding

9. Calculate x, correct to 2 decimal places.

   a. 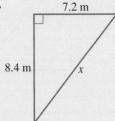   **b.** 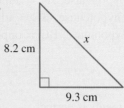   **c.**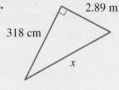

10. Calculate x, correct to 2 decimal places.

   a. 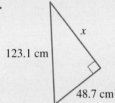   **b.** 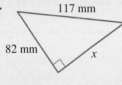   **c.**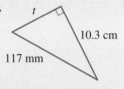

11. Label the unlabelled sides of the following right-angled triangles using the symbol θ and the words 'hypotenuse', 'adjacent' and 'opposite' where appropriate.

   a. 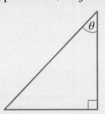   **b.**    **c.**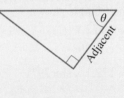

12. **a.** Write the trigonometric ratios that connect the lengths of the given sides and the size of the given angle in each of the following triangles.

     b. Use these ratios to calculate the size of the angle.

   i. 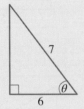   **ii.** 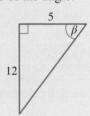   **iii.**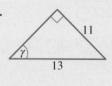

13. Use a calculator to evaluate the following trigonometric ratios, correct to 4 decimal places.

   a. $\sin(54°)$ **b.** $\cos(39°)$ **c.** $\tan(12°)$

14. Calculate the values of the pronumerals in each of the following triangles. Give your answers correct to 2 decimal places.

a.

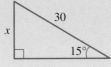

b.

c.

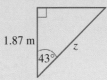

d.

e.

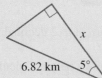

f.

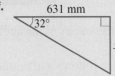

15. Evaluate each of the following, correct to the nearest degree.

    a. $\sin^{-1}(0.1572)$ **b.** $\cos^{-1}(0.8361)$ **c.** $\tan^{-1}(0.5237)$

16. Calculate the size of the angle in each of the following. Give your answers correct to the nearest degree.

    a. $\sin(\theta) = 0.5321$ **b.** $\cos(\theta) = 0.7071$ **c.** $\tan(\theta) = 0.8235$

    d. $\cos(\alpha) = 0.3729$ **e.** $\tan(\alpha) = 0.5774$ **f.** $\sin(\beta) = 0.8660$

    g. $\cos(\beta) = 0.5050$ **h.** $\tan(\beta) = 8.3791$

17. A tree is 6.7 m tall. At a certain time of the day it casts a shadow that is 1.87 m long. Determine the angle of depression of the rays of the sun at that time.
Round your answer to 2 decimal places.

Communicating, reasoning and problem solving

18. A pair of towers stand 30 m apart. From the top of tower A, the angle of depression of the base of tower B is 60° and the angle of depression of the top of tower B is 30°.
Rounding your answer to the nearest metre, calculate the height of tower B.

19. Calculate the angles of a triangle whose sides can be described using the Pythagorean triad 3, 4, 5.

20. A stack of chairs is 2 m tall. The stack needs to fit through a doorway that is 1.8 m high. The maximum angle that the stack of chairs can be safely tilted is 25° to the vertical.
Based on this information, determine if it is safe to try to move the stack of chairs through the doorway.

21. Calculate the value of the unknown angle, θ, correct to 2 decimal places.

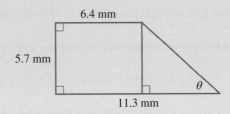

22. A kitesurfer has a kite of length 2.5 m and strings of length 7 m as shown.

 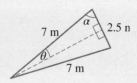

7 m   α   2.5 n
θ
7 m

Calculate the values of the angles θ and α, correct to 2 decimal places.

23. A yacht race follows a triangular course as shown below. Calculate, correct to 1 decimal place:

a.

the distance of the final leg, y

b.

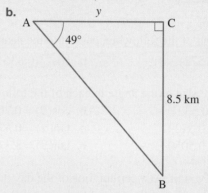

A y C
49°
8.5 km
B

the total distance of the course.

24. A truss is used to build a section of a roof.

If the vertical height of the truss is 1.5 m and the span (horizontal distance between the walls) is 8 m wide, calculate the pitch of the roof (its angle with the horizontal) correct to 1 decimal place.

25. A 2.5 m ladder is placed against a wall. The base of the ladder is 1.7 m from the wall.
   a. Calculate the angle, correct to 2 decimal places, that the ladder makes with the ground.
   b. Calculate how far the ladder reaches up the wall, correct to 2 decimal places.

 To test your understanding and knowledge of this topic go to your learnON title at www.jacplus.com.au and complete the **post-test**.

Answers

Topic 10 Trigonometry

10.1 Pre-test

1. 2.16
2. 10.6
3. 46.90°
4. 45.48°
5. a. 48.59°   b. 41.41°   c. 36.87°
6. D
7. D
8. a. 3.47 cm   b. 5.14 mm
9. a. 0.8870   b. 0.9778   c. 3.7062
10. D
11. 23.4°
12. A
13. D
14. 7.31 m
15. 48.53 m

10.2 Trigonometric ratios and right-angled triangles

1. a.   b.   c.

2. a.   b. 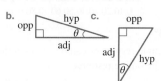   c.

3. a. DE = hypotenuse DF = opposite ∠E = θ
   b. GH = hypotenuse IH = adjacent ∠H = θ
   c. JL = hypotenuse KL = opposite ∠J = θ

4. D

5. a. $\sin(\theta) = \dfrac{4}{5}$, $\cos(\theta) = \dfrac{3}{5}$, $\tan(\theta) = \dfrac{4}{3}$

   b. $\sin(\alpha) = \dfrac{i}{g}$, $\cos(\alpha) = \dfrac{h}{g}$, $\tan(\alpha) = \dfrac{i}{h}$

6. a. $\sin(\beta) = 0.8$, $\cos(\beta) = 0.6$, $\tan(\beta) = 1.3$

   b. $\sin(\gamma) = \dfrac{24}{25}$, $\cos(\gamma) = \dfrac{7}{25}$, $\tan(\gamma) = \dfrac{24}{7}$

   c. $\sin(\beta) = \dfrac{b}{c}$, $\cos(\beta) = \dfrac{a}{c}$, $\tan(\beta) = \dfrac{b}{a}$

   d. $\sin(\gamma) = \dfrac{v}{u}$, $\cos(\gamma) = \dfrac{t}{u}$, $\tan(\gamma) = \dfrac{v}{t}$

7. a. $\sin(\theta) = \dfrac{12}{15} = \dfrac{4}{5}$ b. $\cos(\theta) = \dfrac{25}{30} = \dfrac{5}{6}$

   c. $\tan(\theta) = \dfrac{4}{5}$ d. $\sin(35°) = \dfrac{17}{t}$

8. a. $\sin(\alpha) = \dfrac{14.3}{17.5}$ b. $\sin(15°) = \dfrac{7}{x}$

   c. $\tan(\theta) = \dfrac{20}{31}$ d. $\cos(\alpha) = \dfrac{3.1}{9.8}$

9. D
10. B
11. D
12. Provided n is a positive value, $(m + n)$ would be the hypotenuse, because it has a greater value than both m and $(m - n)$.
13. a. Sample responses can be found in the worked solutions in the online resources.
    b. 1
14. The ratios remain constant when the angle is unchanged. The size of the triangle has no effect.
15. a. Ground   b. Ladder   c. Brick wall
16. a.

    b. i. $\sin(41°) = 0.65$
       ii. $\cos(41°) = 0.76$
       iii. $\tan(41°) = 0.86$
    c. $\alpha = 49°$
    d. i. $\sin(49°) = 0.76$
       ii. $\cos(49°) = 0.65$
       iii. $\tan(49°) = 1.16$
    e. i. They are equal   ii. They are equal
    f. The sine of an angle is equal to the cosine of its complement.
17. a. The ratio of the length of the opposite side to the length of the hypotenuse will increase.
    b. The ratio of the length of the adjacent side will decrease, and the ratio of the opposite side to the adjacent will increase.
    c. i. 1
       ii. 1
       iii. ∞

10.3 Calculating unknown side lengths

1. a. i. 0.8192   ii. 0.2011

   b.

θ	0°	15°	30°	45°	60°	75°	90°
$\sin(\theta)$	0	0.26	0.50	0.71	0.87	0.97	1.00

   c. As θ increases, so does $\sin(\theta)$, starting at 0 and increasing to 1.

2. a. i. 0.7880
      ii. 0.5919

   b.

θ	0°	15°	30°	45°	60°	75°	90°
$\cos(\theta)$	1.00	0.97	0.87	0.71	0.50	0.26	0

   c. As θ increases, $\cos(\theta)$ decreases, starting at 1 and decreasing to 0.

3. a. i. 0.3249

    ii. 1.2753

b.

θ	0°	15°	30°	45°	60°	75°	90°
$\tan(\theta)$	0	0.27	0.58	1.00	1.73	3.73	Undefined

c. $\tan(89°) = 57.29$, $\tan(89.9°) = 572.96$

d. As θ increases, $\tan(\theta)$ increases, starting at 0 and becoming very large. There is no value for $\tan(90°)$.

4. a. 13.02 m   b. 7.04 m   c. 27.64 mm

5. a. 2.79 cm   b. 6.27 m   c. 14.16 m

6. a. 2.95 cm   b. 25.99 cm   c. 184.73 cm

7. a. 14.06 km   b. 8.43 km   c. 31.04 m

8. a. 26.96 mm   b. 60.09 cm   c. 0.84 km

9. a. 0.94 km   b. 5.59 m   c. 41.67 m

10. a. 54.73 m   b. 106.46 cm   c. 298.54 mm

11. a. $a = 17.95$, $b = 55.92$

     b. $a = 15.59$, $b = 9.00$, $c = 10.73$

     c. $a = 12.96$, $b = 28.24$, $c = 15.28$

12. D

13. B

14. A

15. D

16. a. 275.75 km   b. 48.62 km

17. 21.32 m

18. 285.63 m

19. Sample responses can be found in the worked solutions in the online resources.

20. a. Sample responses can be found in the worked solutions in the online resources.

     b. Sample responses can be found in the worked solutions in the online resources.

     c. $AC = \dfrac{AB}{\tan(\theta)}$

21. Sample responses can be found in the worked solutions in the online resources.

22. a. $x = 12.87$ m   b. $h = 3.01$ m   c. $x = 2.60$ m

23. $w = 41$ mm

24. a. 5.9 m   b. 5.2 m

25. 4 m

10.4 Calculating unknown angles

1. a. 39°   b. 72°   c. 37°

2. a. 53°   b. 69°   c. 71°

3. a. 79°   b. 77°   c. 15°

4. a. 19°   b. 42°   c. 55°   d. 21°

5. a. 49°   b. 80°   c. 35°   d. 45°

6. a. 41°   b. 23°   c. 58°   d. 80°

7. a. 47°   b. 45°   c. 24°

8. a. 43°   b. 45°   c. 18°

9. a. 26°   b. 12°   c. 76°

10. D

11. B

12. D

13. C

14. 30.56°

15. 21.80°

16. a.

x	$y = \cos^{-1}(x)$
0.0	90°
0.1	84°
0.2	78°
0.3	73°
0.4	66°
0.5	60°
0.6	53°
0.7	46°
0.8	37°
0.9	26°
1.0	0°

b.

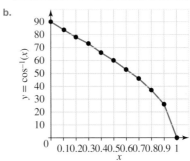

17. 18.43°

18. a. Sample responses can be found in the worked solutions in the online resources.

     b. Sample responses can be found in the worked solutions in the online resources.

19. a. 2.31 m

     b. i. Sample responses can be found in the worked solutions in the online resources.

       ii. Sample responses can be found in the worked solutions in the online resources.

20. Sample responses can be found in the worked solutions in the online resources.

21. a. $\theta = 26.01°$ b. $\theta = 32.01°$ c. $x = 22.62°$

22. a. The roof would not be high enough.

     b. 136°

23. 54.74°

24. Large square dimensions: length $= 5$ cm, width $= 5$ cm

Large triangle dimensions: base $= 5$ cm, height $= 5$ cm

Small square dimensions: length $= \dfrac{5\sqrt{2}}{2}$ cm, width $\dfrac{5\sqrt{2}}{2}$ cm

Small triangle dimensions: base $= \dfrac{5\sqrt{2}}{2}$ cm, height $\dfrac{5\sqrt{2}}{2}$ cm

Parallelogram dimensions: height $= 5$ cm, length $5\sqrt{2}$ cm

10.5 Angles of elevation and depression

1. 14.11 m

2. 66.35 m

3. The new ramps have an angle of inclination of 8.13°.
   This does not meet the required specifications.

4. a.   b. 42 m

5. 1.6 m

6. 63°

7. 34°, 3 m

8. B

9. Con: 34°, John: 30°

10. a.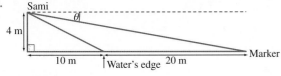

    b. $\tan(\theta) = \dfrac{3}{4}$

    $\theta = \tan^{-1}\left(\dfrac{4}{30}\right)$

    $\theta \approx 7.595$

    $\approx 7.6°$

    c.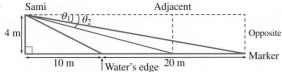

    As the dolphins swim towards Sami, the adjacent length decreases and the opposite remains unchanged.

    $\tan(\theta) = \dfrac{\text{opposite}}{\text{adjacent}}$

    Therefore, θ will increase as the adjacent length decreases.

    If the dolphins are at the water's edge,

    $\tan(\theta) = 10$

    $\theta = \tan^{-1}\left(\dfrac{4}{10}\right)$

    $\theta \approx 21.801$

    $\approx 21.8°$

11. Sample responses can be found in the worked solutions in the online resources.

12. a. 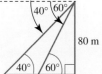   b. 49 m

13. a. Sample responses can be found in the worked solutions in the online resources.

    b. 1.64 m

14. Sample responses can be found in the worked solutions in the online resources.

15. Sample responses can be found in the worked solutions in the online resources.

16. a. Sample responses can be found in the worked solutions in the online resources.

    b. 37°

    c. i. Yes

    ii. 17.26 m

Project

1. Sample responses can be found in the worked solutions in the online resources.

2. Sample responses can be found in the worked solutions in the online resources.

3. 53° requires the shortest brace and 37° requires the longest brace.

4. 4.24 m

5. 3.62 m

6. 4.61 m, 4.52 m

7. Sample responses can be found in the worked solutions in the online resources.

10.6 Review questions

1. D

2. C

3. B

4. B

5. D

6. C

7. A

8. D

9. a. 11.06 m

   b. 12.40 cm

   c. 429.70 cm or 4.30 m

10. a. 113.06 cm

    b. 83.46 mm

    c. 55.50 mm or 5.55 cm

11. a.   b.   c.

12. a. i. $\cos(\theta) = \dfrac{6}{7}$

    ii. $\tan(\beta) = \dfrac{12}{5}$

    iii. $\sin(\gamma) = \dfrac{11}{13}$

    b. i. $\theta = 31°$

    ii. $\beta = 67°$

    iii. $\gamma = 58°$

13. a. 0.8090   b. 0.7771   c. 0.2126

14. a. 7.76   b. 36.00   c. 2.56 m
    d. 19.03   e. 6.79 km   f. 394.29 mm

15. a. 9°   b. 33°   c. 28°

16. a. 32°   b. 45°   c. 39°
    d. 68°   e. 30°   f. 60°
    g. 60°   h. 83°

17. 74.41°

18. 35 m

19. 90°, 53° and 37°

20. No. The stack of chairs must be tilted by 25.84° to fit through the doorway, which is more than the safe angle of 25°.

21. 49.32° (to 2 decimal places)

22. $\theta = 10.29°$ and $\alpha = 79.71°$ (to 2 decimal places)

23. a. 7.4 km   b. 27.2 km

24. 20.6° (to 1 decimal places)

25. a. 47.16°   b. 1.83 m

11 Measurement

LESSON SEQUENCE

LESSON
11.1 Overview

Why learn this?

We live in a world surrounded by shapes and objects. Often, we ask questions such as 'how long?', 'how far?' or 'how big?'. These questions are all answered using measurement. Measurements may be in one, two or three dimensions. Examples of measurements in one dimension would be length and perimeter, temperature and time. Measurements in two dimensions include areas of shapes including the areas of curved surfaces of objects such as drink cans. Measurements in three dimensions include volumes of objects and capacity, such as the number of millilitres in a drink can.

Since objects are all around us, measurement skills are used in many real-world situations. Carpenters, builders, concreters, landscape gardeners and construction workers are just a few of the many tradespeople who need to understand various aspects of measurement when ordering or working onsite and following plans. Designers, interior decorators and architects use measurement in their drawings and calculations. Chefs measure ingredients in their cooking. Nurses and health professionals follow instructions regarding the amount of a drug to administer to a patient. To maximise profits, manufacturers need to minimise the amount of raw materials used in production. This means knowing the measurements of various parts. Understanding the basic concepts involved in measurement is beneficial in many real-world situations.

1. Calculate the perimeter of this figure, giving your answer in centimetres.

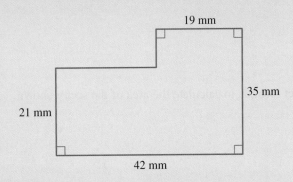

2. Calculate the area of the triangle shown, correct to 2 decimal places.

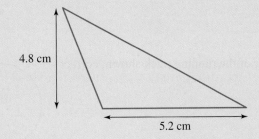

3. Calculate the perimeter of the triangle shown.

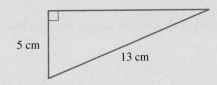

4. **MC** An area of 0.005 m^2 is equivalent to:
   A. 0.000 05 cm^2
   B. 0.5 cm^2
   C. 50 cm^2
   D. 0.005 cm^2

5. Calculate the volume of the cuboid shown in mm^3.

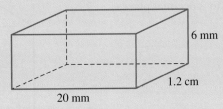

6. **MC** Select the correct formula to calculate the perimeter of the shape shown.

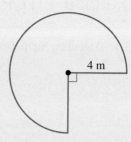

A. $P = \pi \times 4^2$
B. $P = 8 \times \pi$
C. $P = \pi \times 4^2 + 8$
D. $P = 6 \times \pi + 8$

7. **MC** Select the correct formula to calculate the area of the sector shown.

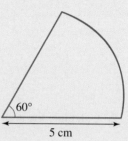

A. $A = \dfrac{60}{360} \times \pi \times 5$

B. $A = \dfrac{60}{360} \times \pi \times 25$

C. $A = \dfrac{60}{300} \times \pi \times 25$

D. $A = \dfrac{60}{360} \times 2 \times \pi \times 5$

8. **MC** Select the perimeter of the running track shown, correct to 2 decimal places.

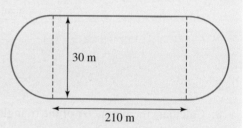

A. 94.25 m
B. 304.24 m
C. 467.12 m
D. 514.25 m

9. **MC** Select the correct formula to calculate the area of the annulus.

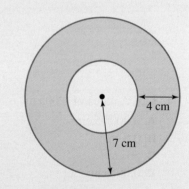

A. $A = \pi(7^2 - 4^2)$
B. $A = \pi(7^2 - 3^2)$
C. $A = \pi \times 4^2$
D. $A = 2 \times \pi \times (7 - 3)$

10. Calculate the surface area (SA) of the rectangular prism, giving your answer correct to 2 decimal places.

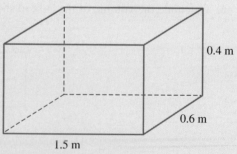

11. **MC** Select the correct formula to calculate the surface area of the triangular prism.

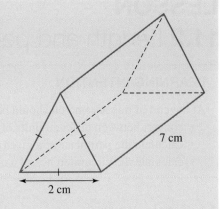

A. $SA = \dfrac{1}{2} \times 2 \times 2 \times 7 \times 3$

B. $SA = \dfrac{1}{2} \times 2 \times 2 \times 2 + 3 \times 7 \times 2$

C. $SA = \dfrac{1}{2} \times 2 \times 2 + 2 \times 7 \times 2$

D. $SA = 2 \times \sqrt{3} + 3 \times 7 \times 2$

12. Convert 3.1 L into cm^3.

13. Calculate the capacity (in mL) of two-thirds of the soup can shown, correct to 1 decimal place.

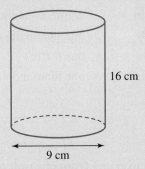

14. A garden bed is designed in the shape of a trapezium with the dimensions shown. If the area of the garden bed is $30\,m^2$, calculate its perimeter.

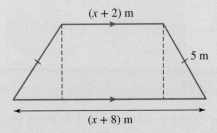

15. A label is made for a tin as shown in the diagram.

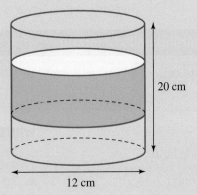

The label is symmetrically positioned so that it is 3 cm from the top and bottom of the tin. Determine the area of paper required to make the label, correct to 1 decimal place.

LESSON
11.2 Length and perimeter

LEARNING INTENTION

At the end of this lesson you should be able to:
- convert between different units of length
- calculate the perimeter of a given shape
- calculate the circumference of a circle.

▶ 11.2.1 Length

eles-4851

- In the **metric system**, units of length are based on the metre. The following units are commonly used.

millimetre	(mm)	one-thousandth of a metre
centimetre	(cm)	one-hundredth of a metre
metre	(m)	one metre
kilometre	(km)	one thousand metres

Converting units of length

The following chart is useful when converting from one unit of length to another.

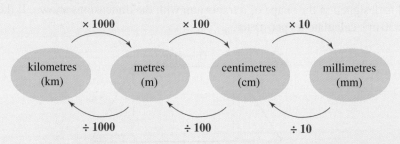

For example, $36\,000\,\text{mm} = 36\,000 \div 10 \div 100\,\text{m}$
$= 36\,\text{m}$

WORKED EXAMPLE 1 Converting units of length

Convert the following lengths into cm.
a. **37 mm**
b. **2.54 km**

THINK

a. There will be fewer cm, so divide by 10.

b. There will be more cm, so multiply.
   km → m: ×1000
   m → cm: ×100

WRITE

a. $37 \div 10 = 3.7\,\text{cm}$

b. $2.54 \times 1000 = 2540\,\text{m}$
   $2540 \times 100 = 254\,000\,\text{cm}$

⊳ 11.2.2 Perimeter

eles-4852

- The **perimeter** of a plane (flat) figure is the distance around the outside of the figure.
- If the figure has straight edges, then the perimeter can be found by simply adding all the side lengths.
- Ensure that all lengths are in the same unit.

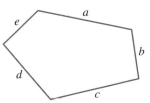

Perimeter $a + b + c + d + e$

Circumference

- **Circumference** is a special name given to the perimeter of a circle.
- Circumference is calculated using the following formulas.

Circumference of a circle

$C = \pi d$, **where** d **is the diameter of the circle**

or

$C = 2\pi r$, **where** r **is the radius of the circle.**

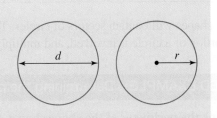

- Use the exact value of π unless otherwise directed.

WORKED EXAMPLE 2 Calculating the perimeter of a shape

Calculate the perimeter of this figure. Give your answer in centimetres.

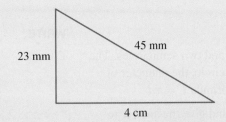

THINK	WRITE
1. State the lengths of the sides in the same unit of length. Convert mm to cm.	$23\,\text{mm} = 23 \div 10$ $\qquad = 2.3\,\text{cm}$ $45\,\text{mm} = 45 \div 10$ $\qquad = 4.5\,\text{cm}$ $2.3\,\text{cm}, 4.5\,\text{cm}, 4\,\text{cm}$
2. Add the side lengths together.	$P = 2.3 + 4.5 + 4$ $\quad = 10.8\,\text{cm}$
3. Write the answer in the correct unit.	The perimeter is $10.8\,\text{cm}$.

WORKED EXAMPLE 3 Calculating the circumference of a circle

Calculate the circumference of the circle shown. Give your answer correct to 2 decimal places.

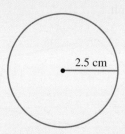

2.5 cm

THINK

1. The radius is known, so apply the formula $C = 2\pi r$.

2. Substitute $r = 2.5$ into $C = 2\pi r$.

3. Calculate the circumference to 3 decimal places and then round correct to 2 decimal places.

WRITE

$C = 2\pi r$

$C = 2 \times \pi \times 2.5$

$C \approx 15.707$
$C \approx 15.71 \, \text{cm}$

- Some shapes will contain sectors of circles. To calculate the perimeter of these shapes, determine what proportion of a circle is involved, and multiply this by the perimeter of the circle.

WORKED EXAMPLE 4 Determining the perimeter of a given shape

Determine the perimeter of the shape shown. Give your answer in cm correct to 2 decimal places.

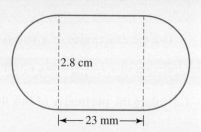

2.8 cm

|← 23 mm →|

THINK

1. There are two straight sections and two semicircles. The two semicircles make up a full circle, the diameter of which is known, so apply the formula $C = \pi d$.

2. Substitute $d = 2.8$ into $C = \pi d$ and give the answer to 3 decimal places.

3. Convert 23 mm to cm (23 mm = 2.3 cm). The perimeter is the sum of all the outside lengths.

4. Round to 2 decimal places.

WRITE

$C = \pi d$

$= \pi \times 2.8$
$\approx 8.796 \, \text{cm}$

$P = 8.796 + 2 \times 2.3$
$= 13.396$

$P \approx 13.40 \, \text{cm}$

  Resources

Interactivities Converting units of length (int-4011)
               Perimeter (int-4013)
               Circumference (int-3782)

11.2 Quick quiz on	11.2 Exercise

Individual pathways

■ PRACTISE	■ CONSOLIDATE	■ MASTER
1, 3, 6, 7, 9, 12, 14, 20, 24	2, 4, 10, 13, 15, 17, 21, 25	5, 8, 11, 16, 18, 19, 22, 23, 26

Fluency

1. **WE1** Convert the following lengths to the units shown.

   a. $5\,cm =$ _____ mm

   b. $1.52\,m =$ _____ cm

   c. $12.5\,mm =$ _____ m

   d. $0.0322\,m =$ _____ mm

2. Convert the following lengths to the units shown.

   a. $6.57\,m =$ _____ km

   b. $64\,cm =$ _____ km

   c. $0.000\,014\,35\,km =$ _____ mm

   d. $18.35\,cm =$ _____ km

3. **WE2** Calculate the perimeters of the following figures, in millimetres.

   a.
   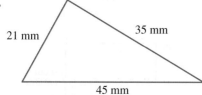

   b.
   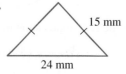

   c.
   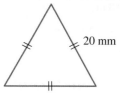

   d.
   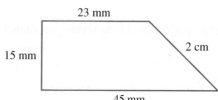

4. Calculate the perimeters of each of the following figures. Give your answers in centimetres.

   a.
   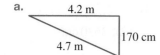

   b.
   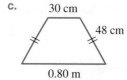

   c.

5. Calculate the perimeters of the following figures, in centimetres.

   a.
   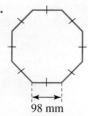

   b.
   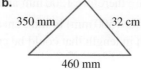

   c.
   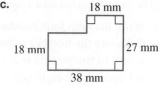

6. Calculate the perimeter of each of the squares shown.

a.
2.4 cm

b.
11.5 mm

c.
7.75 km

7. **WE3** Calculate the circumferences of the circles shown. Give your answers correct to 2 decimal places.

a.
8 cm

b.
4 m

c.
22 mm

8. Calculate the circumferences of the circles shown. Give your answers correct to 2 decimal places.

a.
7.1 cm

b.
3142 km

c.
1055 mm

9. Calculate the perimeters of the rectangles shown.

a.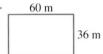
60 m
36 m

b.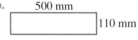
500 mm
110 mm

c.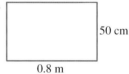
50 cm
0.8 m

10. Calculate the perimeters of the rectangles shown.

a.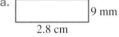
9 mm
2.8 cm

b.
3 km
1.8 km

c.
100 cm
3 m

11. **MC** A circle has a radius of 34 cm. Its circumference, to the nearest centimetre, is:

A. 102 cm   B. 214 cm   C. 107 cm   D. 3630 cm

Understanding

12. Timber is sold in standard lengths, which increase in 300-mm intervals from the smallest available length of 900 mm.
    (The next two standard lengths available are therefore 1200 mm and 1500 mm.)

    a. Write the next four standard lengths (after 1500 mm) in mm, cm and m.
    b. Calculate the number of pieces 600 mm in length that could be cut from a length of timber that is 2.4 m long.
    c. If I need to cut eight pieces of timber, each 41 cm long, determine the smallest standard length I should buy.
       Note: Ignore any timber lost due to the cuts.

13. A bridge called the Akashi-Kaikyo Bridge links the Islands of Honshu and Shikoku in Japan. Its central span covers 1.990 km.

   a. State the length of the central span, in metres.
   b. Calculate how much longer the span of the Akashi–Kaikyo Bridge is than the span of the Sydney Harbour Bridge, which is 1149 m.

14. **WE4** Determine the perimeters of the shapes shown. Give your answers correct to 2 decimal places.

   a.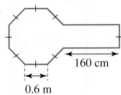

   b.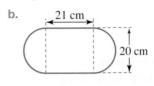

   c.

15. Determine the perimeters of the shapes shown. Give your answers correct to 2 decimal places.

   a.

   b.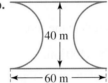

   c.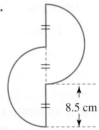

16. Determine the perimeters of the shapes shown. Give your answers correct to 2 decimal places.

   a.

   b.

   c.

17. Determine the perimeter of the racetrack shown in the plan. Give your answer correct to 1 decimal place.

18. Yacht races are often run over a triangular course, as shown. Determine the distance covered by the yachts if they completed 3 laps of the course.

19. Use Pythagoras' theorem to calculate the length of the missing side and, hence, the perimeter of the triangular frame shown.

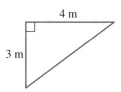

Communicating, reasoning and problem solving

20. The Hubble Space Telescope is over 13 m in length. It orbits the Earth at a height of 559 km, where it can take extremely sharp images outside the distortion of the Earth's atmosphere.

    a. If the radius of the Earth is 6371 km, show that the distance travelled by the Hubble Space Telescope in one orbit, to the nearest km, is 43 542 km.

    b. If the telescope completes one orbit in 96 minutes, show that its speed is approximately 7559 m/s.

21. A bullet can travel in air at 500 m/s.

    a. Show how the bullet travels 50 000 cm in 1 second.

    b. Calculate how long it takes for the bullet to travel 1 cm.

    c. If a super-slow-motion camera can take 100 000 pictures each second, determine how many shots would be taken by this camera to show the bullet travelling 1 cm.

22. Edward is repainting all the lines of a netball court at the local sports stadium. The dimensions of the netball court are shown.

    a. Calculate the total length of lines that need to be repainted. Give your answer to 2 decimal places.

    Edward starts painting at 8 pm when the centre is closing, and it takes him $1\frac{1}{2}$ minutes on average to paint each metre of line.

    b. Show that it will take him 233 minutes to complete the job.

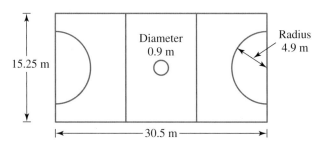

23. The radius of the Earth is accepted to be roughly 6400 km.

    a. Calculate how far, to the nearest km, you travel in one complete rotation of the Earth along the equator.

    b. As the Earth spins on its axis once every 24 hours, calculate the speed you are moving at.

    c. If the Earth is 150 000 000 km from the Sun, and it takes 365.25 days to circle around the Sun, show that the speed of the Earth's orbit around the Sun is 107 515 km/h. Give your answer to the nearest whole number.

24. One-fifth of an 80-cm length of jewellery wire is cut off. A further 22-cm length is then removed. Evaluate whether there is enough wire remaining to make a 40-cm necklace.

25. A church needs to repair one of its hexagonal stained glass windows. Use the information given in the diagram to calculate the width of the window. Give your answer to 1 decimal place.

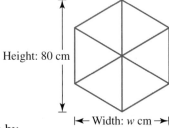

26. A spider is sitting in one top corner of a room that has dimensions 6 m by 4 m by 4 m. It needs to get to the corner of the floor that is diagonally opposite. The spider must crawl along the ceiling, then down a wall, until it reaches its destination.

    a. If the spider crawls first to the diagonally opposite corner of the ceiling, then down the wall to its destination, calculate the distance it would crawl.

    b. Determine the shortest distance from the top back corner to the lower left corner.

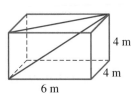

LESSON
11.3 Area

LEARNING INTENTION

At the end of this lesson you should be able to:
- convert between units of area
- calculate areas of triangles, quadrilaterals and circles using formulas
- determine areas of composite shapes.

▶ 11.3.1 Area

eles-4853

- The diagram shows a square of side length 1 cm.

1 cm

By definition it has an area of 1 cm^2 (1 square centimetre).
Note: This is a 'square centimetre', not a 'centimetre squared'.

- **Area** tells us how many squares it takes to cover a figure, so the area of the rectangle shown is 12 cm^2.
- Area is commonly measured in square millimetres $\left(\text{mm}^2\right)$, square centimetres $\left(\text{cm}^2\right)$, square metres $\left(\text{m}^2\right)$, or square kilometres $\left(\text{km}^2\right)$.

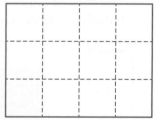

Converting units of area

The following chart is useful when converting between units of area.

$\times 1000^2$ $\times 100^2$ $\times 10^2$

| square kilometres (km^2) | square metres (m^2) | square centimetres (cm^2) | square millimetres (mm^2) |

$\div 1000^2$ $\div 100^2$ $\div 10^2$

For example, $54 \text{ km}^2 = 54 \times 1000^2 \times 100^2$
$= 540\,000\,000\,000 \text{ cm}^2$

DISCUSSION

When converting between units of area, why do we divide by the same conversions as length but squared?

- Another common unit is the **hectare** (ha), a 100 m × 100 m square equal to 10 000 m², which is used to measure small areas of land.

The hectare

1 hectare (ha) = 100 m × 100 m
= 10 000 m²

WORKED EXAMPLE 5 Converting units of area

Convert 1.3 km² into:
a. **square metres**   b. **hectares.**

THINK	WRITE
a. There will be more m², so multiply by 1000².	a. $1.3 \times 1000^2 = 1\,300\,000\,\text{m}^2$
b. Divide the result of part **a** by 10 000, as 1 ha = 10 000 m².	b. $1\,300\,000 \div 10\,000 = 130\,\text{ha}$

⏵ 11.3.2 Using formulas to calculate area

eles-4854

- There are many useful formulas that can be used to calculate the areas of simple shapes. Some common formulas are summarised here.

Square	**Rectangle**
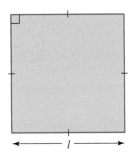 $A = l^2$	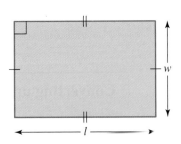 $A = lw$
Parallelogram	**Triangle**
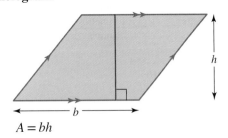 $A = bh$	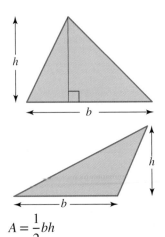 $A = \frac{1}{2}bh$

Trapezium	Kite
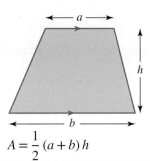 $A = \frac{1}{2}(a+b)h$	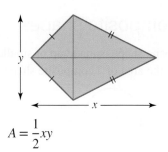 $A = \frac{1}{2}xy$
Circle	**Rhombus**
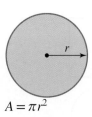 $A = \pi r^2$	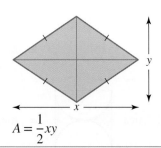 $A = \frac{1}{2}xy$

WORKED EXAMPLE 6 Calculating the areas of figures

Calculate the area of each of the following figures in cm², correct to 1 decimal place.

a.

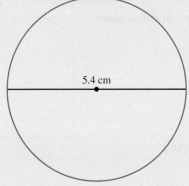

5.4 cm

b.

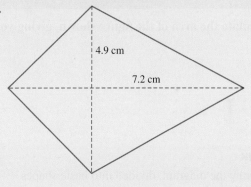

4.9 cm

7.2 cm

THINK	WRITE
a. 1. The figure is a circle. State the radius.	**a.** $r = 2.7$ cm
2. Apply the formula for area of a circle: $A = \pi r^2$.	$A = \pi r^2$
3. The radius is half the diameter. Substitute the value $r = 2.7$.	$= \pi \times 2.7^2$ ≈ 22.90
4. Round the answer to 1 decimal place.	$A \approx 22.9$ cm²
b. 1. The figure is a kite. State the lengths of the diagonals.	**b.** $x = 4.9$ cm, $y = 7.2$ cm
2. Apply the formula for area of a kite: $A = \frac{1}{2}xy$.	$A = \frac{1}{2}xy$
3. Subsitute the values $x = 4.9$ and $y = 7.2$.	$= \frac{1}{2} \times 4.9 \times 7.2$
	$= 17.64$
4. Round the answer to 1 decimal place.	$A \approx 17.6$ cm²

▶ 11.3.3 Composite shapes

eles-4855

- A composite shape is made up of smaller, simpler shapes. We can add or subtract the areas of shapes to calculate the area of a composite shape. Here are two examples.

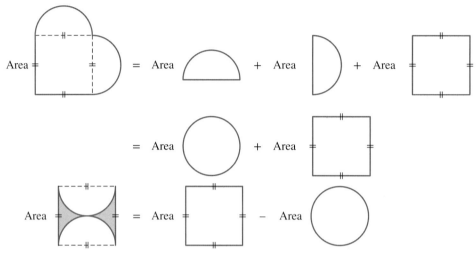

- In the first example, two semicircles are added to a square to create a composite shape.
- In the second example, the two semicircles are subtracted from the square to obtain the shaded area.

WORKED EXAMPLE 7 Calculating the area of composite shapes

Calculate the area of the figure shown, giving your answer correct to 1 decimal place.

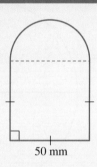

50 mm

THINK	WRITE/DRAW
1. Draw the diagram, divided into basic shapes.	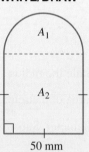 50 mm
2. A_1 is a semicircle with a radius of $\dfrac{50}{2} = 25$ mm. The area of a semicircle is half the area of a complete circle.	$A_1 = \dfrac{1}{2}\pi r^2$
3. Substitute $r = 25$ into the formula and evaluate, correct to 4 decimal places.	$A_1 = \dfrac{1}{2} \times \pi \times 25^2$ $\approx 981.7477 \text{ mm}^2$
4. A_2 is a square of side length 50 mm. Write the formula.	$A_2 = l^2$

5. Substitute $l = 50$ into the formula and evaluate A_2.

$$= 50^2$$
$$= 2500 \text{ mm}^2$$

6. Sum to calculate the total area.

$$\text{Total area} = A_1 + A_2$$
$$= 981.7477 + 2500$$
$$= 3481.7477 \text{ mm}^2$$

7. Round the final answer correct to 1 decimal place.

$$\approx 3481.7 \text{ mm}^2$$

DISCUSSION

How many ways can this composite shape be broken up into simpler shapes?

When calculating the area of this composite shape, will you get the same answer if it is broken up differently?

Support your answer with calculations.

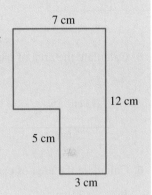

7 cm

12 cm

5 cm

3 cm

---

on Resources

▶ **Video eLesson** Composite area (eles-1886)

🧩 **Interactivities** Conversion chart for area (int-3783)
Area of rectangles (int-3784)
Area of parallelograms (int-3786)
Area of trapeziums (int-3790)
Area of circles (int-3788)
Area of rhombuses (int-3787)

Exercise 11.3 Area

learnon

11.3 Quick quiz on	**11.3 Exercise**

Individual pathways

■ PRACTISE	■ CONSOLIDATE	■ MASTER
1, 2, 4, 7, 9, 13, 17, 19, 22, 27	3, 5, 8, 10, 14, 15, 18, 20, 23, 24, 28	6, 11, 12, 16, 21, 25, 26, 29, 30

Fluency

1. **MC** To convert an area measurement from square kilometres to square metres:
   A. divide by 1000
   B. multiply by 1000
   C. divide by 1 000 000
   D. multiply by 1 000 000

2. **WE5** Convert the following to the units shown.

   a. $13\,400\,\text{m}^2 =$ _____ km^2

   b. $0.04\,\text{cm}^2 =$ _____ mm^2

   c. $3\,500\,000\,\text{cm}^2 =$ _____ m^2

   d. $0.005\,\text{m}^2 =$ _____ cm^2

3. Convert the following to the units shown.

   a. $0.043\,\text{km}^2 =$ _____ m^2

   b. $200\,\text{mm}^2 =$ _____ cm^2

   c. $1.41\,\text{km}^2 =$ _____ ha

   d. $3800\,\text{m}^2 =$ _____ ha

4. **WE6** Calculate the area of each of the following shapes.

   a.
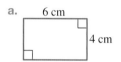
6 cm, 4 cm

   b.

4 mm

   c.
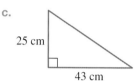
25 cm, 43 cm

5. Calculate the area of each of the following shapes.

   a.
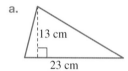
13 cm, 23 cm

   b.
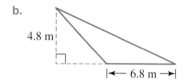
4.8 m, 6.8 m

   c.
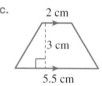
2 cm, 3 cm, 5.5 cm

6. Calculate the area of each of the following shapes.

   a.
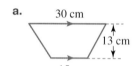
30 cm, 13 cm, 15 cm

   b.
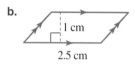
1 cm, 2.5 cm

   c.
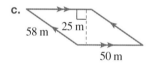
58 m, 25 m, 50 m

7. Calculate the area of each of the following shapes. Give your answers correct to 2 decimal places.

   a.
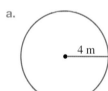
4 m

   b.

2 mm

   c.

3.4 m

8. Calculate the area of each of the following shapes. Where appropriate, give your answer correct to 2 decimal places.

   a.
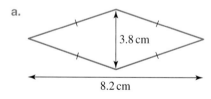
3.8 cm, 8.2 cm

   b.
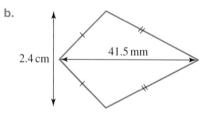
2.4 cm, 41.5 mm

   c.
10.4 m, 7.3 m

9. **WE7** Calculate the areas of the composite shapes shown. Express your answers correct to 1 decimal place.

a.
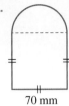
70 mm

b.
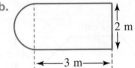
2 m
3 m

c.
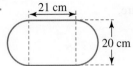
21 cm
20 cm

10. Calculate the areas of the composite shapes shown. Where appropriate, express your answers correct to 1 decimal place.

a.
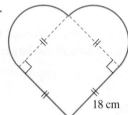
18 cm

b.

120 m
80 m

c.
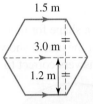
1.5 m
3.0 m
1.2 m

11. Calculate the areas of the composite shapes shown. Express your answers correct to 1 decimal place.

a.
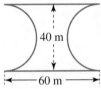
40 m
60 m

b.
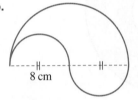
8 cm

12. Calculate the areas of the composite shapes shown. Where appropriate, express your answers correct to 1 decimal place.

a.
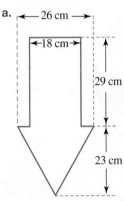
26 cm
18 cm
29 cm
23 cm

b.
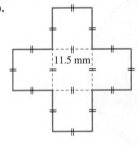
11.5 mm

Understanding

13. Calculate the area of the regular hexagon shown by dividing it into two trapeziums.

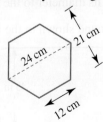
24 cm
21 cm
12 cm

14. Calculate the area of the regular octagon by dividing it into two trapeziums and a rectangle, as shown in the figure.

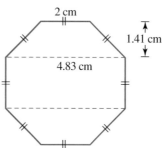

15. An annulus is a shape formed by two concentric circles (two circles with a common centre). Calculate the area of each of the annuli shown by subtracting the area of the smaller circle from the area of the larger circle. Give answers correct to 2 decimal places.

a.

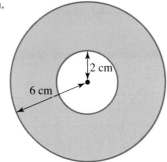

b.

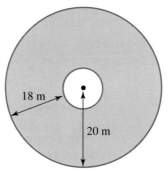

16. Calculate the area of each of the annuli shown (the shaded area). Give answers correct to 2 decimal places.

a.

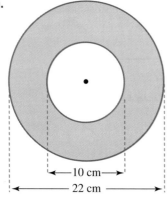

b.

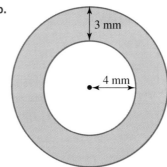

17. **MC** A pizza has a diameter of 30 cm. If your sister eats one-quarter, determine what area of pizza remains.

    A. $168.8 \, \text{cm}^2$ B. $530.1 \, \text{cm}^2$ C. $706.9 \, \text{cm}^2$ D. $176.7 \, \text{cm}^2$

18. A circle has an area of $4500 \, \text{cm}^2$. Calculate its diameter to the nearest mm.

19. Determine the cost of covering the sportsground shown in the figure with turf, if the turf costs $7.50 per square metre.

20. The Murray–Darling River Basin is Australia's largest catchment. Irrigation of farms in the Murray–Darling Basin has caused soil degradation due to rising salt levels. Studies indicate that about 500 000 hectares of the basin could be affected in the next 50 years.

   a. Convert the possible affected area to square kilometres. (1 km^2 = 100 hectares)
   b. The total area of the Murray–Darling Basin is about 1 million square kilometres, about one-seventh of the continent. Calculate what percentage of this total area may be affected by salinity.

21. The plan shows two rooms, which are to be refloored.

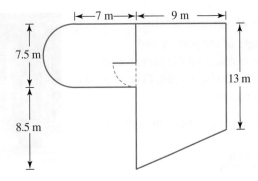

   Calculate the cost if the flooring costs $45 per square metre. Allow 10% more for wastage and round to the nearest $10.

Communicating, reasoning and problem solving

22. A sheet of paper measures 29.5 cm by 21.0 cm.
   a. Calculate the area of the sheet of paper.
   b. Determine the radius of the largest circle that could be drawn on this sheet.
   c. Calculate the area of this circle, to 2 decimal places.
   d. If the interior of the circle is shaded red, show that 56% of the paper is red.

23. A chessboard is made up of 8 rows and 8 columns of squares. Each square is 42 cm^2 in area. Show that the shortest distance from the upper right corner to the lower left corner of the chessboard is 73.32 cm.

24. Two rectangles of sides 15 cm by 10 cm and 8 cm by 5 cm overlap as shown. Show that the difference in area between the two non-overlapping sections of the rectangles is 110 cm^2.

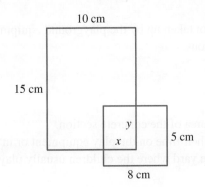

25. Answer the following questions using the information in the figure.

 a. Calculate the area of a square with side lengths 40 cm.

 b. If the midpoints of each side of the previous square are joined by straight lines to make another square, calculate the area of the smaller square.

 c. Now the midpoints of the previous square are also joined with straight lines to make another square. Calculate the area of this even smaller square.

 d. This process is repeated again to make an even smaller square. Calculate the area of this smallest square.

 e. Explain any patterns you observe.

 f. Determine the percentage of the original square's area taken up by the smallest square.

 g. Show that the area of the combined figure that is coloured pink is 1000 cm^2.

$\longmapsto$ 40 cm $\longmapsto$

26. Show that a square of perimeter $4x + 20$ has an area of $x^2 + 10x + 25$.

27. The area of a children's square playground is 50 m^2.

 a. Calculate the exact side length of the playground.

 b. Pine logs 3 m long are to be laid around the playground. Determine how many logs need to be bought. (The logs can be cut into smaller pieces if required.)

28. A sandpit is designed in the shape of a trapezium, with the dimensions shown.

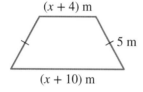

$(x + 4)$ m

5 m

$(x + 10)$ m

If the area of the sandpit is 14 m^2, determine its perimeter.

29. A rectangular classroom has a perimeter of 28 m and its length is 4 m shorter than its width. Determine the area of the classroom.

30. The playground equipment is half the length and half the width of a square kindergarten yard, as shown.

 a. Identify the fraction of the kindergarten yard that is occupied by the play equipment.

 b. During a working bee, the playground equipment area is extended 2 m in length and 1 m in width. If x represents the length of the kindergarten yard, write an expression for the area of the play equipment.

 c. Write an expression for the area of the kindergarten yard *not* taken up by the playground equipment.

Playground equipment

 d. The kindergarten yard that is not taken up by the playground equipment is divided into 3 equal-sized sections:

 • A grassed section
 • A sandpit
 • A concrete section

 i. Write an expression for the area of the concrete section.

 ii. The children usually spend their time on the play equipment or in the sandpit. Write a simplified expression for the area of the yard where the children usually play.

LESSON
11.4 Area and perimeter of a sector

LEARNING INTENTION

At the end of this lesson you should be able to:
- identify sectors, semicircles and quadrants
- calculate the area of a sector
- calculate the length of the arc and the perimeter of a sector.

▶ 11.4.1 Sectors

eles-4856

- A **sector** is the shape created when a circle is cut by two radii.
- A circle, like a pizza, can be cut into many sectors.
- Two important sectors that have special names are the **semicircle** (half-circle) and the **quadrant** (quarter-circle).

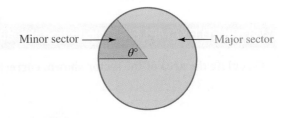

Minor sector ⟶ $\theta°$ ⟵ Major sector

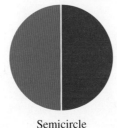

Semicircle

Quadrant

WORKED EXAMPLE 8 Calculating the area of a quadrant

Calculate the area enclosed by the figure shown, correct to 1 decimal place.

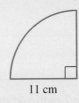

11 cm

THINK

1. The figure is a quadrant or quarter-circle. Write the formula for its area.

2. Substitute $r = 11$ into the formula.

3. Evaluate and round the answer correct to 1 decimal place. Include the units.

WRITE

$A = \dfrac{1}{4} \times \pi r^2$

$A = \dfrac{1}{4} \times \pi \times 11^2$

≈ 95.03

$\approx 95.0\,\text{cm}^2$

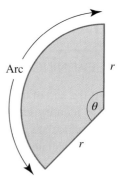

▶ 11.4.2 Area of a sector

eles-4857

- Sectors are specified by the angle (θ) between the two radii.
- For example, in a quadrant, $\theta = 90°$, so a quadrant is $\dfrac{90}{360}$ or $\dfrac{1}{4}$ of a circle.

Area of a sector

- For any value of θ, the area of the sector is given by:

$$A_{sector} = \frac{\theta}{360} \times \pi r^2$$

where r is the radius of the sector.

WORKED EXAMPLE 9 Calculating the area of a sector

Calculate the area of the sector shown, correct to 1 decimal place.

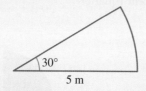

THINK	WRITE
1. This sector is $\dfrac{30}{360}$ of a circle. Write the formula for its area.	$A = \dfrac{30}{360}\pi r^2$
2. Substitute $r = 5$ into the formula.	$A = \dfrac{30}{360} \times \pi \times 5^2$
	≈ 6.54
3. Evaluate and round the answer correct to 1 decimal place. Include the units.	$\approx 6.5 \text{ m}^2$

▶ 11.4.3 Perimeter of a sector

eles-4858

- The perimeter, P, of a sector is the sum of the 2 radii and the curved section, which is called an **arc** of a circle.
- The length of the arc, l, will be $\dfrac{\theta}{360}$ of the circumference of the circle.

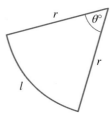

Arc length of a sector

- For any value of θ, the arc length, l, of the sector is given by:

$$l = \frac{\theta}{360} \times 2\pi r$$

where r is the radius of the sector.

- The perimeter of a sector can therefore be calculated using the following formula.

> ### Perimeter of a sector
>
> - The perimeter, P, of a sector is given by:
>
> $$P = 2r + l$$
>
> where r is the radius of the sector and l is the arc length.

WORKED EXAMPLE 10 Calculating the perimeter of a sector

Calculate the perimeter of the sector shown, correct to 1 decimal place.

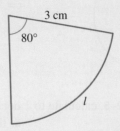

THINK

1. The sector is $\dfrac{80}{360}$ of a circle. Write the formula for the length of the curved side.

2. Substitute $r = 3$ and evaluate l. Don't round off until the end.

3. Add all the sides together to calculate the perimeter.

4. Round the answer to 1 decimal place.

WRITE

$l = \dfrac{80}{360} \times 2\pi r$

$l = \dfrac{80}{360} \times 2 \times \pi \times 3$
$\approx 4.189\,\text{cm}$

$P = 4.189 + 3 + 3$
$= 10.189\,\text{cm}$

$P \approx 10.2\,\text{cm}$

---

on Resources

Interactivities   Area of a sector (int-6076)

                 Perimeter of a sector (int-6077)

Exercise 11.4 Area and perimeter of a sector

11.4 Quick quiz on	11.4 Exercise

Individual pathways

■ PRACTISE	■ CONSOLIDATE	■ MASTER
1, 2, 7, 10, 12, 17, 20	3, 4, 8, 11, 14, 16, 18, 21	5, 6, 9, 13, 15, 19, 22

Fluency

1. Calculate the areas of the semicircles shown, correct to 2 decimal places.

a.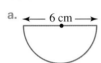
6 cm

b.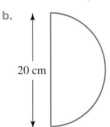
20 cm

c.
r
$r = 4.2$ cm

d.
D
$D = 24$ mm

WE8 For each of the quadrants in questions **2–5**, calculate to 1 decimal place:

    i. the perimeter   **ii.** the area enclosed.

2.
4 cm

3.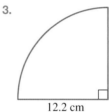
12.2 cm

4.
a
$a = 11.4$ m

5.
←1.5 m→

6. **MC** Select the correct formula for calculating the area of the sector shown.

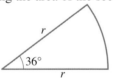
r
36°
r

  A. $A = \dfrac{1}{4} \pi r^2$ **B.** $A = \dfrac{3}{4} \pi r^2$ **C.** $A = \dfrac{1}{100} \pi r^2$ **D.** $A = \dfrac{1}{10} \pi r^2$

WE9&10 For each of the sectors shown in questions **7–9**, calculate to 1 decimal place:

    i. the perimeter   **ii.** the area.

7. a.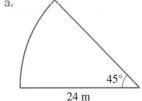
45°
24 m

b.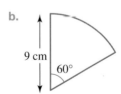
9 cm
60°

8. a.

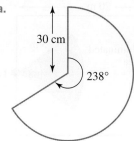

30 cm

238°

b.

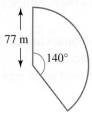

77 m

140°

9. a.

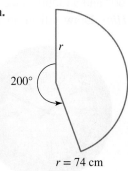

r

200°

r = 74 cm

b.

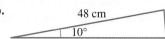

48 cm

10°

Understanding

10. A searchlight lights up the ground to a distance of 240 m. Calculate the area the searchlight illuminates if it can swing through an angle of 120°, as shown in the diagram.
Give your answer correct to 1 decimal place.

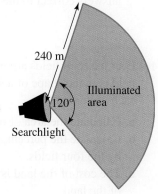

240 m

120°

Illuminated area

Searchlight

11. Calculate the perimeter, correct to 1 decimal place, of the figure shown.

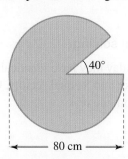

40°

80 cm

12. A goat is tethered by an 8.5-m rope to the outside of a corner post in a paddock, as shown in the diagram.

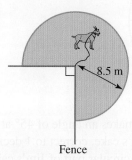

8.5 m

Fence

Calculate the area of grass (shaded) on which the goat is able to graze. Give your answer correct to 1 decimal place.

13. A beam of light is projected onto a theatre stage as shown in the diagram.

    a. Calculate the illuminated area (correct to 1 decimal place) by first evaluating the area of the sector.

    b. Calculate the percentage of the total stage area that is illuminated by the light beam.

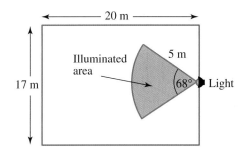

14. A sector has an angle of 80° and a radius of 8 cm; another sector has an angle of 160° and a radius of 4 cm. Determine the ratio of the area of the first sector to the area of the second sector. (*Hint:* Draw a diagram.)

15. The minute hand on a vintage clock is 20 centimetres long and the hour hand is 12 centimetres long. Round answers to the nearest centimentre.

    a. Calculate the distance travelled by the minute hand from midday until 2 pm.

    b. During this same time, calculate how far the hour hand travelled. Give your answers correct to the nearest centimetre.

16. Four baseball fields are to be constructed inside a rectangular piece of land. Each field is in the shape of a sector of a circle, as shown in light green. The radius of each sector is 80 m.

    a. Calculate the area of one baseball field, correct to the nearest whole number.

    b. Calculate the percentage, correct to 1 decimal place, of the total area occupied by the four fields.

    c. The cost of the land is $24 000 per hectare. Calculate the total purchase price of the land.

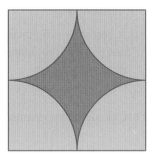

Communicating, reasoning and problem solving

17. John and Jim are twins, and on their birthday they have identical birthday cakes, each with a diameter of 30 cm. Grandma Maureen cuts John's cake into 8 equal sectors. Grandma Mary cuts Jim's cake with a circle in the centre and then 6 equal portions from the rest.

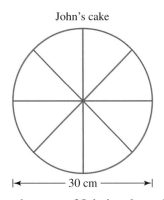

John's cake

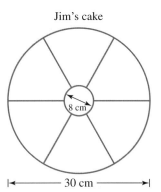

Jim's cake

    a. Show that each sector of John's cake makes an angle of 45° at the centre of the cake.

    b. Calculate the area of one slice of John's cake, correct to 1 decimal place.

    c. Calculate the area of the small central circular part of Jim's cake, correct to 1 decimal place.

    d. Calculate the area of one of the larger portions of Jim's cake, correct to 1 decimal place.

    e. If each boy eats one slice of the largest part of his own cake, does John eat more cake? Explain your answer.

18. A lighthouse has a light beam in the shape of a sector of a circle that rotates at 10 revolutions per minute and covers an angle of 40°. A person stands 200 m from the lighthouse and observes the beam. Show that the time between the end of oneflash and the start of the next is approximately 5.33 seconds.

19. Answer the following questions. Where appropriate, give all answers to the nearest whole number.

    a. A donkey inside a square enclosure is tethered to a post at one of the corners. Show that a rope of length 120 m is required to prevent the donkey from eating more than half the grass in the enclosure.
    b. Suppose two donkeys are tethered at opposite corners of the square region shown. Calculate how long each rope should be so that the donkeys together can graze half of the area.
    c. This time four donkeys are tethered, one at each corner of the square region. Calculate how long each rope should be so that all the donkeys can graze only half of the area.
    d. Another donkey is tethered to a post inside an enclosure in the shape of an equilateral triangle. The post is at one of the vertices. Show that a rope of length 64 m is required so that the donkey eats only half of the grass in the enclosure.
    e. This time the donkey is tethered halfway along one side of the equilateral triangular region shown in the diagram. Calculate how long the rope should be so that the donkey can graze half of the area.

150 m

100 m   100 m

100 m

20. A new logo, divided into 6 equal sectors, has been designed based on two circles with the same centre, as shown in the diagram.
The radius of the larger circle is 60 cm and the radius of the smaller circle is 20 cm. The shaded sections are to be coated with a special paint costing $145 a square metre. The edge of the shaded sections is made from a special material costing $36 per metre.
Calculate the cost to make the shaded sections. Give your answer to the nearest dollar.

21. The metal washer shown has an inner radius of r cm and an outer radius of $(r+1)$ cm.

    a. State, in terms of r, the area of the circular piece of metal that was cut out of the washer.
    b. State, in terms of r, the area of the larger circle.
    c. Show that the area of the metal washer in terms of r is $\pi(2r+1)$ cm^2.
    d. If r is 2 m, what is the exact area of the washer?
    e. If the area of the washer is 15π cm^2, show that the radius would be 7 cm.

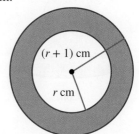

$(r+1)$ cm

r cm

22. The area of a sector of a circle is π cm^2, and the length of its arc is 2 cm. Determine the radius of the circle (in terms of π).

LESSON
11.5 Surface area of rectangular and triangular prisms

LEARNING INTENTION

At the end of this lesson you should be able to:
• calculate the surface area of rectangular and triangular prisms.

▶ 11.5.1 Prisms

eles-4859

• A **prism** is a three-dimensional figure with a uniform (unchanging) **cross-section** and all faces flat.
• A right prism has two opposite faces identical and the remaining faces are rectangles.
• An oblique prism has two identical opposite faces, and the remaining faces are parallelograms.

   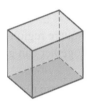   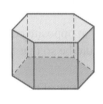   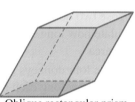

Right triangular prism   Right rectangular prism (cuboid)   Right hexagonal prism   Oblique rectangular prism

• A prism can be sliced (cross-sectioned) in such a way that each 'slice' has an identical base.

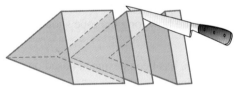

'Slicing' a prism into pieces produces congruent same shape, same size cross-sections.

• The following objects are not prisms because they do not have uniform cross-sections.

         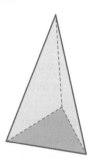

Cone   Sphere   Square pyramid   Triangular pyramid

▶ 11.5.2 Nets of right prisms

eles-6245

• A right prism can be drawn as a two-dimensional figure. This shape is called a **net**. In other words, a net is a two-dimensional figure that can be folded to form a three-dimensional object.

Net of a cube

- The diagram displays the net of a cube. Notice that when folded, the faces do not overlap.

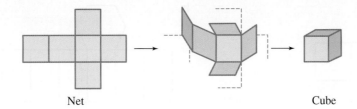

Net   Cube

The following nets are also nets of a cube.

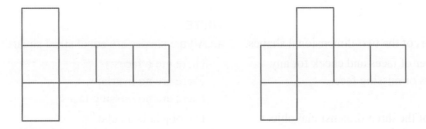

Net of a rectangular prism

- The following nets can be folded to form a rectangular prism. Notice that the top and the bottom faces of the rectangular prism are on either side of the net. These two faces cannot be drawn on the same side of the net. The left and right faces alternate with the back and front faces.

Nets of other prisms

- A net can be drawn for any type of prism.

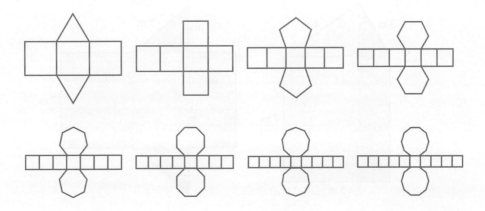

State the names of the three-dimensional objects that can be formed by folding the following nets.

a.

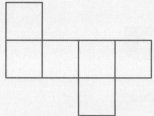

b.

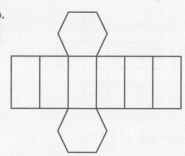

THINK	WRITE
a. 1. State any features of the two-dimensional shapes.	**a.** All the shapes are equal sized squares.
2. Count the number of faces and check for any possible overlaps or missing faces.	There are 6 faces. There are no overlaps. There are no missing faces.
3. State the name of the three-dimensional object.	The object is a cube.
b. 1. State any features of the two-dimensional shapes.	**b.** There are six identical rectangles and two identical regular hexagons.
2. Count the number of faces and check for any possible overlaps or missing faces.	There are 8 faces. There are no overlaps. There are no missing faces.
3. State the name of the base and the name of the three-dimensional object.	The base is a regular hexagon. The object is a hexagonal prism.

⏵ 11.5.3 Surface area of a prism

eles-6164

- Consider the **triangular prism** shown below.
- It has 5 faces: 2 bases, which are right-angled triangles, and 3 rectangular sides. The net of the prism is drawn next to it.

  Note: A net can be drawn in several different ways to represent the same prism.

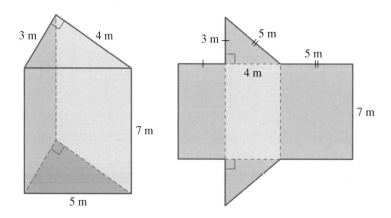

- Drawing the net of a prism creates a composite shape made up of several simple shapes.
- The area of the net is the same as the total surface area (SA) of the prism.

Calculate the surface area of the rectangular prism (cuboid) shown.

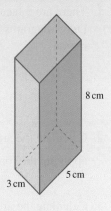

THINK

1. There are 6 faces: 2 rectangular bases and 4 rectangular sides. Draw diagrams for each pair of faces and label each region.

2. Calculate the area of each rectangle by applying the formula $A = lw$, where l is the length and w is the width of the rectangle.

3. The total surface area (SA) is the sum of the area of 2 of each shape. Write the answer.

WRITE/DRAW

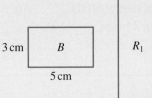
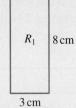
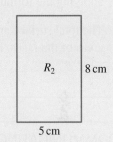

$B = 3 \times 5$ $R_1 = 3 \times 8$ $R_2 = 5 \times 8$
$= 15$ $= 24$ $= 40$

$SA = 2 \times B + 2 \times R_1 + 2 \times R_2$
$= 30 + 48 + 80$
$= 158 \, \text{cm}^2$

Calculate the surface area of the right-angled triangular prism shown.

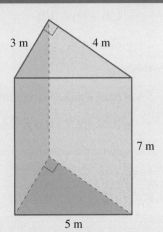

THINK	WRITE/DRAW

1. There are 5 faces: 2 triangles and 3 rectangles. Draw diagrams for each face and label each region.

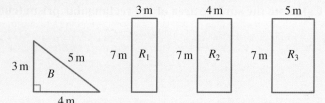

2. Calculate the area of the triangular base by applying the formula $A = \dfrac{1}{2}bh$.

$$B = \dfrac{1}{2} \times 3 \times 4$$
$$= 6\,\text{m}^2$$

3. Calculate the area of each rectangular face by applying the formula $A = lw$.

$$R_1 = 3 \times 7 \quad R_2 = 4 \times 7 \quad R_3 = 5 \times 7$$
$$= 21 \qquad\quad = 28 \qquad\quad = 35$$

4. The total surface area is the sum of all the areas of the faces, including 2 bases.

$$SA = 2 \times B + R_1 + R_2 + R_3$$
$$= 12 + 21 + 28 + 35$$
$$= 96\,\text{m}^2$$

COMMUNICATING — COLLABORATIVE TASK: Who do you agree with?

Three students, Menali, Ryan and Olivia, were asked to calculate the surface area of the following triangular prism, showing all working out.

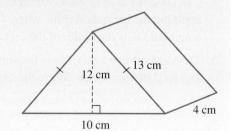

Here are the three student's working out.

Menali:

SA = triangles + base + sides

$= 10 \times 12 + 10 \times 4 + 13 \times 4 \times 2$

$= 120 + 40 + 104$

$= 264\,\text{cm}^2$

Ryan:

SA = base + sides + front and back

$= 10 \times 4 + 13 \times 4 \times 2 + 2(1/2 \times 10 \times 12)$

$= 264\,\text{cm}^2$

Olivia:

SA = 3 × rectangles + 2 triangles

$= 10 \times 4 \times 3 + 2(1/2 \times 10 \times 13)$

$= 250\,\text{cm}^2$

Which students, working out do you agree with and why?

What advice could you give the student(s) who answered the question incorrectly?

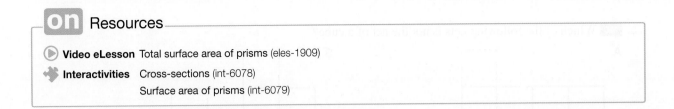

Exercise 11.5 Surface area of rectangular and triangular prisms

learn on

11.5 Quick quiz **on**	11.5 Exercise

Individual pathways

■ PRACTISE	■ CONSOLIDATE	■ MASTER
1, 3, 5, 7, 11, 13, 14, 15, 18, 20, 22, 24	2, 4, 6, 8, 10, 12, 16, 21, 25	9, 17, 19, 23, 26

Fluency

1. **WE11** Determine the shapes that can be constructed using the following nets.

a.

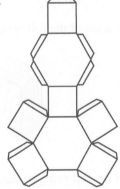

b.

2. State the names of the three-dimensional objects that can be formed by folding the following nets.

a.

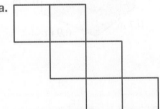

b.

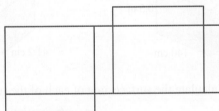

3. **MC** Which of the following nets is not the net of a cube?

A.

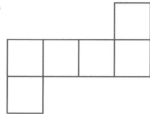

B.

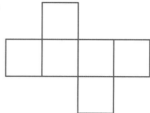

C.

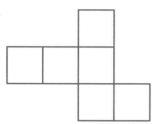

D.

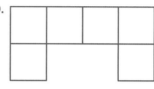

4. What three-dimensional objects can be formed by folding the following nets?

a.

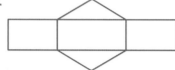

b.

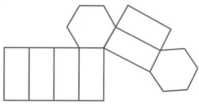

5. **WE12** Calculate the surface areas of the following rectangular prisms (cuboids).

a.

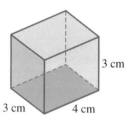

3 cm
3 cm   4 cm

b.

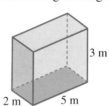

3 m
2 m   5 m

c.

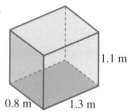

1.1 m
0.8 m   1.3 m

6. Calculate the surface area of the following rectangular prisms (cuboids).

a.

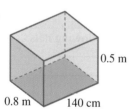

0.5 m
0.8 m   140 cm

b.

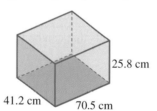

25.8 cm
41.2 cm   70.5 cm

c.

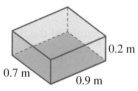

0.2 m
0.7 m   0.9 m

7. **WE13** Calculate the surface area of each of the following triangular prisms.

a.

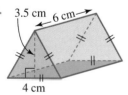

3.5 cm 6 cm
4 cm

b.

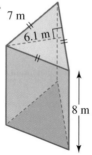

7 m
6.1 m
8 m

c.

2.5 cm
1 cm
h
$h = 0.87$ cm

8. Calculate the surface area of each of the triangular prisms shown.

a.

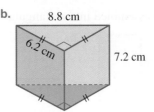

17 cm   15 cm
8 cm   18 cm

b.

8.8 cm
6.2 cm   7.2 cm

c.

44 mm
36 mm   25 mm   14 mm

9. A shipping company is planning to buy and paint the outside surface of one of these shipping containers. Determine how many cans of paint the company should buy if the base of the container is not painted and each can of paint covers about 40 m².

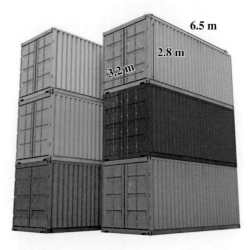

6.5 m
2.8 m
3.2 m

10. The aim of the Rubik's cube puzzle is to make each face of the cube one colour. Calculate the surface area of the Rubik's cube if each small coloured square is 1.2 cm in length. Assume that there are no gaps between the squares.

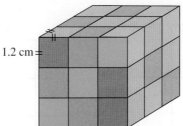

1.2 cm

11. Determine how many square metres of iron sheet are needed to construct the water tank shown.

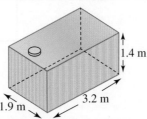

1.4 m
3.2 m
1.9 m

12. Calculate the surface area of the tank in question 7 if no top is made.

Understanding

13. State whether the following nets can be assembled into rectangular prisms.

a.

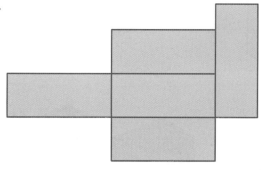

b.

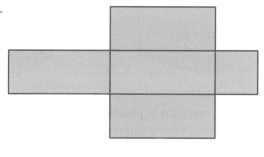

c.

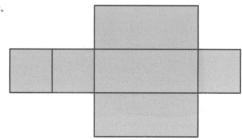

d.

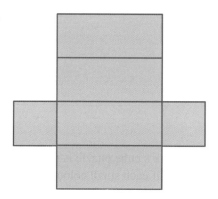

14. Calculate the surface areas of these nets.

a.

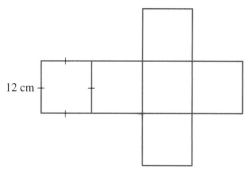

12 cm

b.

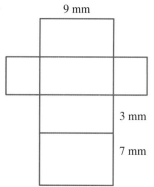

9 mm

3 mm

7 mm

c.

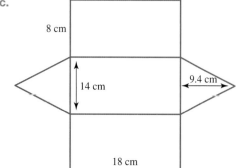

8 cm

14 cm

9.4 cm

18 cm

15. Calculate the area of cardboard that would be needed to construct a box to pack this prism, assuming that no overlap occurs.

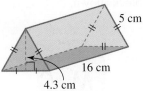

16. An aquarium is a triangular prism with the dimensions shown. The top of the tank is open. Calculate the area of glass that was required to construct the tank. Give your answer correct to 2 decimal places.

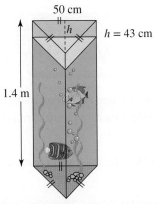

17. A tent is constructed as shown. Calculate the area of canvas needed to make the tent, including the floor.

18. a. Determine how many square centimetres of cardboard are needed to construct the shoebox shown, ignoring the overlap on the top.
    b. Draw a sketch of a net that could be used to make the box.

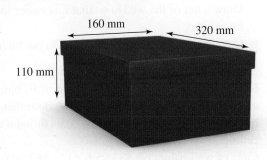

19. Calculate the surface area of a square-based prism of height 4 cm, given that the side length of its base is 6 cm.

Communicating, reasoning and problem solving

20. **a.** Calculate the surface area of the toy block shown.

5 cm

b. If two of the blocks are placed together as shown, calculate the surface area of the prism that is formed.

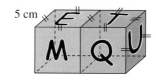

5 cm

c. Calculate the surface area of the prism formed by three blocks.

d. Use the pattern to determine the surface area of a prism formed by eight blocks arranged in a line. Explain your reasoning.

21. A cube has a side length of 2 cm. Show that the least surface area of a solid formed by joining eight such cubes is 96 cm².

22. Calculate the surface area of these triangular prisms, rounding your answer to 2 decimal places.

a.

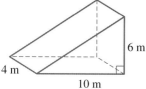

6 m
4 m
10 m

b.

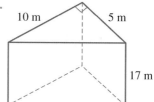

10 m   5 m
17 m

23. A prism has an equilateral triangular base with a perimeter of 12 cm. If the total surface area of the prism is 302 cm², show that the length of the prism is approximately 24 cm.

Problem solving

24. A wedge in the shape of a triangular prism, as shown in the diagram, is to be painted.

a. Draw a net of the wedge so that it is easier to calculate the area to be painted.

b. Calculate the area to be painted. (Do not include the base.)

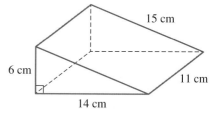

15 cm
6 cm
11 cm
14 cm

25. Ken wants to paint his son's bedroom walls blue and the ceiling white. The room measures 3 m by 4 m, with a ceiling height of 2.6 m. There is one door that measures 1 m by 2 m and one window that measures 1.8 m by 0.9 m. Each surface takes two coats of paint and 1 L of paint covers 16 m² on the walls and 12 m² on the ceiling.

Cans of wall paint cost $33.95 for 1 L, $63.90 for 4 L, $147 for 10 L and $174 for 15 L. Ceiling paint costs $24 for 1 L and $60 for 4 L. Determine the cheapest options for Ken to paint the room.

26. A swimming pool has a length of 50 m and a width of 28 m. The shallow end of the pool has a depth of 0.80 m, which increases steadily to 3.8 m at the deep end.

  a. Calculate how much paint would be needed to paint the floor of the pool.

  b. If the pool is to be filled to the top, calculate how much water will be needed.

LESSON
11.6 Surface area of a cylinder

LEARNING INTENTION

At the end of this lesson you should be able to:
- draw the net of a cylinder.
- calculate the curved surface area of a cylinder
- calculate the total surface area of a cylinder.

▶ 11.6.1 Surface area of a cylinder

eles-4861

- A **cylinder** is a solid object with two identical flat circular ends and one curved side. It has a uniform cross-section.
- The net of a cylinder has two circular bases and one rectangular face. The rectangular face is the curved surface of the cylinder.
- Because the rectangle wraps around the circular base, the width of the rectangle is the same as the circumference of the circle. Therefore, the width is equal to $2\pi r$.
- The area of each base is πr^2, and the area of the rectangle is $2\pi rh$.

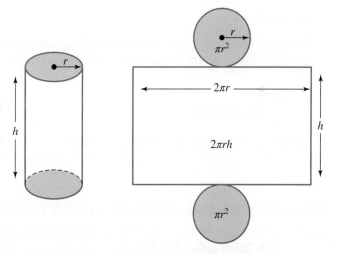

Surface area of a closed cylinder

$$SA = 2\pi r^2 + 2\pi rh$$

- *Note:* For an open cylinder, we would change the formula to exclude either one or both of the bases, area πr^2.

a. Use the formula $A = 2\pi rh$ to calculate the area of the curved surface of the cylinder, correct to 1 decimal place.
b. Use the formula $SA = 2\pi rh + 2\pi r^2$ to calculate the surface area of the closed cylinder, correct to 1 decimal place.

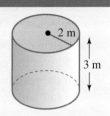

THINK

WRITE

a. 1. Write the formula for the curved surface area.

a. $A = 2\pi rh$

2. Identify the values of the pronumerals.

$r = 2, h = 3$

3. Substitute $r = 2$ and $h = 3$.

$A = 2 \times \pi \times 2 \times 3$

4. Evaluate, round to 1 decimal place and include units.

≈ 37.69

$\approx 37.7 \text{ m}^2$

b. 1. Write the formula for the surface area of a cylinder.

b. $SA = 2\pi rh + 2\pi r^2$

2. Identify the values of the pronumerals.

$r = 2, h = 3$

3. Substitute $r = 2$ and $h = 3$.

$SA = (2 \times \pi \times 2 \times 3) + (2 \times \pi \times 2^2)$

4. Evaluate, round to 1 decimal place and include units.

≈ 62.83

$\approx 62.8 \text{ m}^2$

DISCUSSION

A yidaki (didgeridoo) is a cylinder-shaped musical instrument. Yidakis can come in various sizes. Discuss the effect that different surface areas would have on the tone the instrument makes.

 Resources

 Interactivity Surface area of a cylinder (int-6080)

Exercise 11.6 Surface area of a cylinder

learn

11.6 Quick quiz on	11.6 Exercise

Individual pathways

■ PRACTISE	■ CONSOLIDATE	■ MASTER
1, 2, 7, 9, 13, 15, 18	3, 4, 8, 10, 14, 16, 19	5, 6, 11, 12, 17, 20, 21

Fluency

WE14 For each of the cylinders in questions **1–6**, answer the following questions, rounding answers to 1 decimal place.

a. Use the formula $A = 2\pi rh$ to calculate the area of the curved surface of the cylinder.

b. Use the formula $SA = 2\pi rh + 2\pi r^2$ to calculate the total surface area of the cylinder.

1.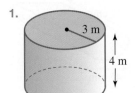

3 m

4 m

2.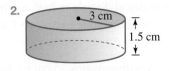

3 cm

1.5 cm

3.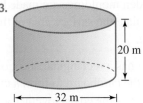

20 m

32 m

4.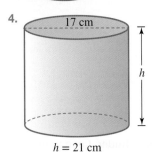

17 cm

h

h = 21 cm

5.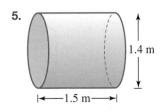

1.4 m

1.5 m

6.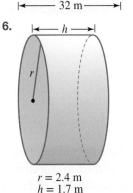

h

r

r = 2.4 m
h = 1.7 m

7. A can of energy drink has a height of 130 mm and a radius of 24 mm.

   a. Draw a net of the can.

   b. Calculate the surface area of the can, to the nearest cm².

8. A cylinder has a radius of 15 cm and a height of 45 mm.
   Determine its surface area correct to 1 decimal place.

Understanding

9. A cylinder has a surface area of 2000 cm² and a radius of 8 cm.
   Determine the cylinder's height correct to 2 decimal places.

10. A can of asparagus is 137 mm tall and has a diameter of 66 mm;
    a can of tomatoes is 102 mm tall and has a diameter of 71 mm;
    and a can of beetroot is 47 mm tall with a diameter of 84 mm.

    a. Determine which can has the largest surface area and
       which has the smallest surface area.
    b. Calculate the difference between the largest and smallest
       surface areas, correct to the nearest cm².

11. A 13-m-high storage tank was constructed from stainless steel
    (including the lid and the base). The diameter is 3 metres, as
    shown. Give your answer correct to 2 decimal places.

    a. Calculate the surface area of the tank.
    b. Determine the cost of the steel for the side of the tank if it
       comes in sheets 1 m wide that cost $60 a metre.

13 m

3 m

12. The concrete pipe shown in the diagram has the following measurements:
    $t = 30$ mm, $D = 18$ cm, $l = 27$ cm.
    Answer the following, giving your answers correct to 2 decimal places.

    a. Calculate the outer curved surface area.
    b. Calculate the inner curved surface area.
    c. Calculate the total surface area of both ends.
    d. Hence calculate the surface area for the entire shape.

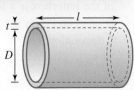

t

D

l

13. Wooden mouldings are made by cutting cylindrical dowels in half, as shown. Calculate the surface area of the moulding, rounding your answer to 1 decimal place.

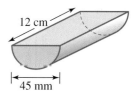

14. Kiara has a rectangular sheet of cardboard with dimensions 25 cm by 14 cm. She rolls the cardboard to form a cylinder so that the shorter side, 14 cm, is its height, and glues the edges together with a 1-cm overlap.

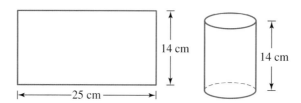

a. Determine the radius of the circle Kiara needs to construct to put at the top of her cylinder. Give your answer correct to 2 decimal places.
b. Determine the total surface area of her cylinder if she also makes the top and bottom of her cylinder out of cardboard. Give your answer correct to 1 decimal place.

Communicating, reasoning and problem solving

15. If the radius of a cylinder is twice its height, write a formula for the surface area in terms of its height only.

16. Cylinder A has a 10% greater radius and a 10% greater height than Cylinder B. Show that the ratio of their surface areas is 121 : 100.

17. Two identical cylinders (with height h cm and radius r cm) are modified slightly.
The first cylinder's radius is increased by 10% and the second cylinder's height is increased by 20%. After these modifications have been made, the surface area of the second cylinder is greater than that of the first cylinder. Show that $h > 2.1r$.

18. An above-ground swimming pool has the following shape, with semicircular ends.
Calculate how much plastic would be needed to line the base and sides of the pool. Give your answer to 1 decimal place.

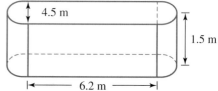

19. A over-sized solid wooden die is constructed for a children's playground. The side dimensions of the die are 50 cm. The number on each side of the die will be represented by cylindrical holes that will be drilled out of each side. Each hole will have a diameter of 10 cm and depth of 2 cm. All surfaces on the die will be painted (including the die holes). Show that the total area required to be painted is 1.63 m^2.

20. The following letterbox is to be spray-painted on the outside. Calculate the total area to be spray-painted. Assume that the end of the letterbox is a semicircle above a rectangle. The letter slot is open and does not require painting. Give your answer to 1 decimal place.

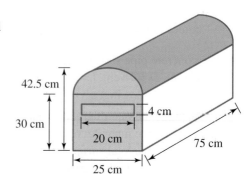

21. A timber fence is designed as shown. Evaluate how many square metres of paint are required to completely paint the fence front, back, sides and top with 2 coats of paint. Assume each paling is 2 cm in thickness and that the top of each paling is a semicircle.

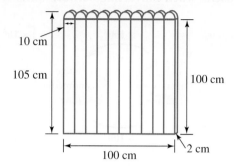

LESSON
11.7 Volume of prisms and cylinders

LEARNING INTENTION

At the end of this lesson you should be able to:
- calculate the volume of prisms and cylinders
- convert between units of volume
- convert between units of capacity and units of volume.

▶ 11.7.1 Volume and capacity

eles-4862

Volume

- The volume of a solid is the amount of space it fills or occupies.
- The diagram at far right shows a single cube of side length 1 cm. By definition, the cube has a **volume** of 1 cm³ (1 cubic centimetre).
  Note: This is a 'cubic centimetre', not a 'centimetre cubed'.
- The volume of some solids can be found by dividing them into cubes with 1 cm sides.
- Volume is commonly measured in cubic millimetres (mm³), cubic centimetres (cm³), cubic metres (m³) or cubic kilometres (km³).

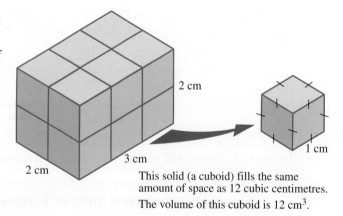

This solid (a cuboid) fills the same amount of space as 12 cubic centimetres. The volume of this cuboid is 12 cm³.

Converting units of volume

The following chart is useful when converting between units of volume.

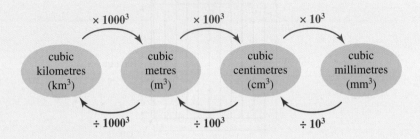

For example, $3\,m^3 = 3 \times 100^3 \times 10^3\,mm^3$

$= 3\,000\,000\,000\,mm^3$

Capacity

- **Capacity** is another term for volume, which is usually applied to the measurement of liquids and containers.
- The capacity of a container is the volume of liquid that it can hold.
- The standard measurement for capacity is the litre (L).
- Other common units are the millilitre (mL), kilolitre (kL), and megalitre (ML), where:

$$1\,L = 1000\,mL$$
$$1\,kL = 1000\,L$$
$$1\,ML = 1\,000\,000\,L$$

Converting units of capacity

The following chart is useful when converting between units of capacity.

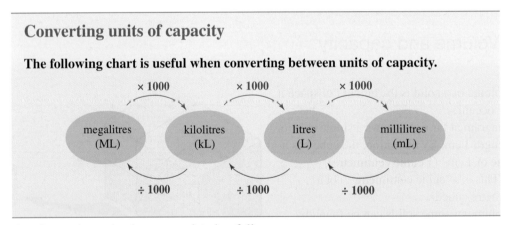

- The units of capacity and volume are related as follows.

Converting between units of volume and capacity

$$1\,cm^3 = 1\,mL$$
$$1000\,cm^3 = 1\,L$$
$$1\,m^3 = 1000\,L = 1\,kL$$

Convert:
a. **13.2 L into cm³**
b. **3.1 m³ into litres**
c. **0.13 cm³ into mm³**
d. **3.8 kL into m³.**

THINK	WRITE
a. 1. $1\,L = 1000\,mL$, so multiply by 1000.	a. $13.2 \times 1000 = 13\,200\,mL$
2. $1\,mL = 1\,cm^3$. Convert to cm^3.	$= 13\,200\,cm^3$
b. $1\,m^3 = 1000\,L$, so multiply by 1000.	b. $3.1 \times 1000 = 3100\,L$
c. There will be more mm^3, so multiply by 10^3.	c. $0.13 \times 1000 = 130\,mm^3$
d. $1\,kL = 1\,m^3$. Convert to m^3.	d. $3.8\,kL = 3.8\,m^3$

▶ 11.7.2 Volume of a prism

eles-4863

- The volume of a prism can be found by multiplying its cross-sectional area (A) by its height (h).

Volume of a prism

$$V_{prism} = A \times h$$

where:

A is the area of the base or cross-section of the prism

h is the perpendicular height of the prism.

- The cross-section (A) of a prism is often referred to as the base, even if it is not at the bottom of the prism.
- The height (h) is always measured perpendicular to the base, as shown in the following diagrams.

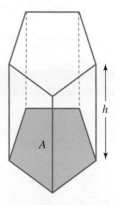
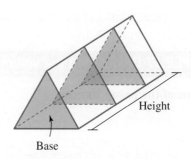

Volume of a cube or a cuboid

- A specific formula can be developed for the volumes of cubes and cuboids.

Cube	Cuboid (rectangular prism)
	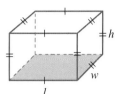
Volume = base area × height $= l^2 \times l$ $= l^3$	Volume = base area × height $= lw \times h$ $= lwh$

WORKED EXAMPLE 16 Calculating the volume of a prism

Calculate the volume of the hexagonal prism shown.

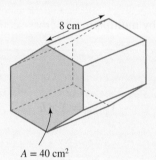

8 cm

$A = 40$ cm²

THINK	WRITE
1. Write the formula for the volume of a prism.	$V = A \times h$
2. Identify the values of the pronumerals.	$A = 40, \ h = 8$
3. Substitute $A = 40$ and $h = 8$ into the formula and evaluate. Include the units.	$V = 40 \times 8$ $= 320$ cm³

WORKED EXAMPLE 17 Calculating the volume of a prism

Calculate the volume of the prism shown.

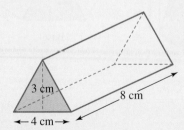

3 cm

8 cm

←4 cm→

THINK	WRITE
1. The base of the prism is a triangle. Write the formula for the area of the triangle.	$A = \dfrac{1}{2}bh$
2. Substitute $b = 4$, $h = 3$ into the formula and evaluate.	$= \dfrac{1}{2} \times 4 \times 3$
	$= 6 \text{ cm}^3$
3. Write the formula for the volume of a prism.	$V = A \times h$
4. State the values of A and h.	$A = 6$, $h = 8$
5. Substitute $A = 6$ and $h = 8$ into the formula and evaluate. Include the units.	$V = 6 \times 8$
	$= 48 \text{ cm}^3$

⏵ 11.7.3 Volume of a cylinder

eles-4860

- A cylinder is a special kind of prism that has a circular base and uniform cross-section.
- A formula for the volume of a cylinder is shown.
- Volume = base area × height

$$= \text{area of circle} \times \text{height}$$
$$= \pi r^2 \times h$$
$$= \pi r^2 h$$

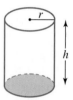

Volume of a cylinder

$$V = \pi r^2 h$$

where:

r **is the radius**

h **is the height.**

WORKED EXAMPLE 18 Calculating the capacity of a cylinder

Calculate the capacity, in litres, of a cylindrical water tank that has a diameter of 5.4 m and a height of 3 m. Give your answer correct to the nearest litre.

THINK	WRITE/DRAW
1. Draw a labelled diagram of the tank.	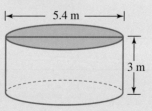
2. The base is a circle, so $A = \pi r^2$. Write the formula for the volume of a cylinder.	$V = \pi r^2 h$

3. Recognise the values of r and h. Recall that the radius, r, is half the diameter. $r = \dfrac{5.4}{2} = 2.7 \text{ m}$

$r = 2.7,\ h = 3$

4. Substitute $r = 2.7$, $h = 3$ into the formula and calculate the volume.

$V = \pi \times (2.7)^2 \times 3$

$\approx 68.706\,631 \text{ m}^3$

5. Convert this volume to litres (multiply by 1000).

$V = 68.706\,631 \times 1000$

$= 68\,706.631$

$\approx 68\,707 \text{ L}$

WORKED EXAMPLE 19 Calculating the volume of a composite solid

Calculate the total volume of this solid, to the nearest cubic centimetre.

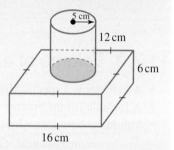

THINK

WRITE/DRAW

1. The solid is made from two objects. Draw and label each object.
Let V_1 = volume of the square prism.
Let V_2 = volume of the cylinder.

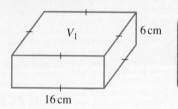

2. Calculate V_1 (square prism).

$V_1 = Ah$

$= 16^2 \times 6$

$= 1536 \text{ cm}^3$

3. Calculate V_2 (cylinder).

$V_2 = \pi r^2 h$

$= \pi \times 5^2 \times 12$

$\approx 942.478 \text{ cm}^3$

4. To calculate the total volume, add the volumes found above.

$V = V_1 + V_2$

$= 1536 + 942.478$

$= 2478.478$

$\approx 2478 \text{ cm}^3$

DISCUSSION

Consider two cylinders where one cylinder is twice as wide but only half as high as the other cylinder.

Do these cylinders have the same volume?

Exercise 11.7 Volume of prisms and cylinders

learn

11.7 Quick quiz on	**11.7 Exercise**

Individual pathways

■ **PRACTISE**	■ **CONSOLIDATE**	■ **MASTER**
1, 4, 5, 8, 9, 12, 22, 25	2, 6, 10, 13, 14, 16, 17, 20, 23, 26	3, 7, 11, 15, 18, 19, 21, 24, 27, 28

Fluency

1. **WE15** Convert the following units into mL.

   a. $325 \, cm^3$　　　 b. $2.6 \, m^3$　　　 c. $5.1 \, L$　　　 d. $0.63 \, kL$

2. Convert the following units into cm^3.

   a. $5.8 \, mL$　　　 b. $6.1 \, L$　　　 c. $3.2 \, m^3$　　　 d. $59.3 \, mm^3$

3. Convert the following units into kL.

   a. $358 \, L$　　　 b. $55.8 \, m^3$　　　 c. $8752 \, L$　　　 d. $5.3 \, ML$

4. Calculate the volumes of the cuboids shown. Assume that each small cube has sides of 1 cm.

   a. 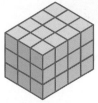　　　 b. 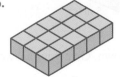　　　 c.

5. **WE16** Calculate the volumes of the following objects.

   a.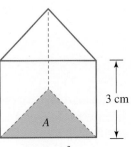

   $A = 4 \, cm^2$

   3 cm

   b.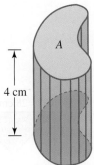

   4 cm

   $A = 17 \, cm^2$

   c.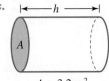

   $A = 3.2 \, m^2$

   $h = 3.0 \, m$

6. Calculate the volumes of the following objects. When appropriate, round answers to 1 decimal place.

a.
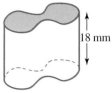
18 mm

Base area = 35 mm²

b.
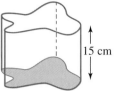
15 cm

Base area = 28 cm²

c.
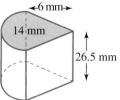
←6 mm→
14 mm
26.5 mm

7. Calculate the volumes of the following objects. Round your answers to 1 decimal place.

a.
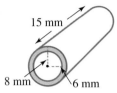
15 mm
8 mm   6 mm

b.
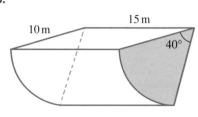
10 m   15 m
40°

c.
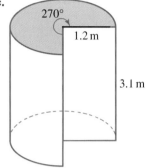
270°
1.2 m
3.1 m

8. Calculate the volumes of the following rectangular prisms.

a.
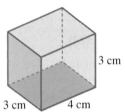
3 cm
3 cm   4 cm

b.
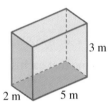
3 m
2 m   5 m

c.
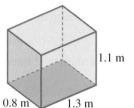
1.1 m
0.8 m   1.3 m

9. **WE17** Calculate the volumes of the prisms shown. Give your answers correct to 2 decimal places.

a.
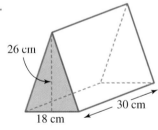
26 cm
30 cm
18 cm

b.
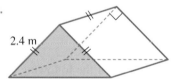
2.4 m

c.
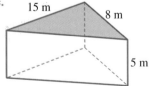
15 m   8 m
5 m

d.
←—— 28 cm ——→
17 cm
37 cm

10. Calculate the volumes of the following cylinders. Give your answers correct to 1 decimal place.

a.
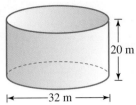
20 m

32 m

b.
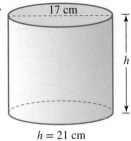
17 cm

h

$h = 21$ cm

c.
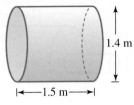
1.4 m

1.5 m

11. Calculate the volumes of the following cylinders. Give your answers correct to 1 decimal place.

a.
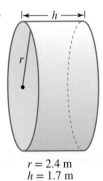
h

r

$r = 2.4$ m
$h = 1.7$ m

b.
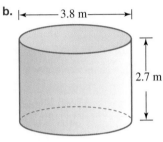
3.8 m

2.7 m

c.
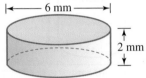
6 mm

2 mm

12. **WE18** Calculate the volume of water, in litres (L), that can fill a cylindrical water tank that has a diameter of 3.2 m and a height of 1.8 m.

13. Calculate the capacity of the Esky shown, in litres.

0.5 m
0.42 m
0.84 m

14. **WE19** Calculate the volume of each solid to the nearest cm³.

a.
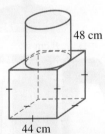
48 cm

44 cm

b.
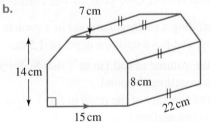
7 cm

14 cm

8 cm

15 cm

22 cm

c.
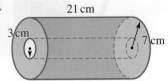
21 cm
3 cm
7 cm

15. Calculate the volume of each solid to the nearest cm³.

a.

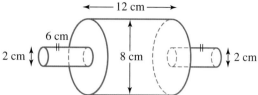

b.

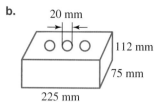

c.

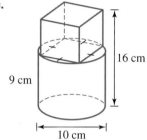

Understanding

16. Calculate the capacity, in litres, of the cylindrical storage tank shown. Give your answer correct to 1 decimal place.

17. Calculate the capacity (in mL) of this cylindrical coffee plunger when it is filled to the level shown. Give your answer correct to 1 decimal place.

18. Sudhira is installing a rectangular pond in a garden. The pond is 1.5 m wide and 2.2 m long, and has a uniform depth of 1.5 m.

a. Calculate the volume of soil $\left(\text{in m}^3\right)$ that Sudhira must remove to make the hole before installing the pond.

b. Calculate the capacity of the pond in litres. (Ignore the thickness of the walls in this calculation.)

Communicating, reasoning and problem solving

19. Until its closure in 2001, Fresh Kills on Staten Island outside New York City was one of the world's biggest landfill garbage dumps. (New Yorkers throw out about $100\,000$ tonnes of refuse weekly.)
Calculate the approximate volume (in m^3) of the Fresh Kills landfill if it covers an area of 1215 hectares and is about 240 m high.
Express your answer in scientific notation. (*Note:* 1 hectare $= 10\,000\,m^2$)

20. Calculate the internal volume (in cm^3) of the wooden chest shown, correct to 2 decimal places. (Ignore the thickness of the walls.)

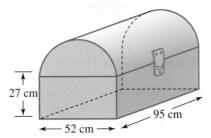

21. Consider the refrigerator shown.

    a. Calculate the volume of the refrigerator in m^3.
    b. Determine the capacity of the refrigerator if the walls are all 5 cm thick.

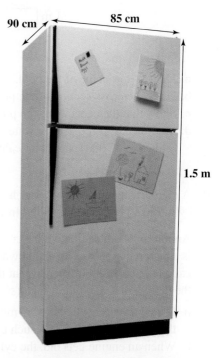

22. a. Calculate the volume of plastic needed to make the door wedge shown. Give your answer correct to 2 decimal places.

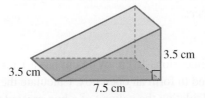

    b. The wedges can be packed snugly into cartons with dimensions $45\,cm \times 70\,cm \times 35\,cm$. Determine how many wedges fit into each carton.

23. A cylindrical glass is designed to hold 1.25 L.

Show that its height is 132 mm if it has a diameter of 110 mm.

24. Mark is responsible for the maintenance of the Olympic (50 m) pool at an aquatic centre. The figure shows the dimensions of an Olympic pool.

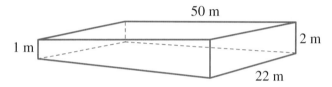

a. Identify the shape of the pool.
b. Draw the cross-section of the prism and calculate its area.
c. Show that the capacity of the pool is 1 650 000 L.
d. Mark needs to replace the water in the pool every 6 months. If the pool is drained at 45 000 L per hour and refilled at 35 000 L per hour, calculate how long it will take to:
   i. drain
   ii. refill (in hours and minutes).

25. A cylindrical container of water has a diameter of 16 cm and is 40 cm tall. Determine how many full cylindrical glasses can be filled from the container if the glasses have a diameter of 6 cm and are 12 cm high.

26. An internal combustion engine consists of 4 cylinders. Each cylinder is a piston that moves up and down. The diameter of each cylinder is called the bore and the height that the piston moves up and down within the cylinder is called the stroke (stroke = height).

a. If the bore of a cylinder is 84 mm and the stroke is 72 mm, calculate the volume (in litres) of 4 such cylinders, correct to 2 decimal places.
b. When an engine gets old, the cylinders have to be 're-bored'; that is, the bore is increased by a small amount (and new pistons put in them). If the re-boring increases the diameter by 1.1 mm, determine the increase in volume of the four cylinders (in litres). Give your answer correct to 3 decimal places.

27. Answer the following questions.

a. A square sheet of metal with dimensions 15 cm by 15 cm has a 1-cm square cut out of each corner. The remainder of the square is folded to form an open box. Calculate the volume of the box.
b. Write a general formula for calculating the volume of a box created from a metal sheet of any size with any size square cut out of the corners.

28. A water cooler in an office holds 15 L of water. A cylindrical glass measures 8 cm in diameter and is 15 cm high. Determine how many glasses of water can be completely filled from a full water cooler.

LESSON
11.8 Review

11.8.1 Topic summary

MEASUREMENT

Units of measurement

- Length:

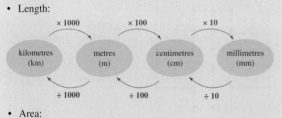

- Area:

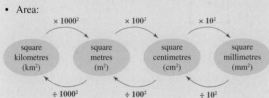

- Volume:

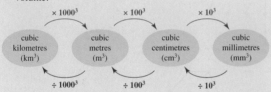

- Capacity

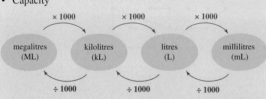

- When converting between units of capacity and volume:
  - $1 \text{ cm}^3 = 1 \text{ mL}$
  - $1000 \text{ cm}^3 = 1 \text{ L}$
  - $1 \text{ m}^3 = 1000 \text{ L} = 1 \text{ kL}$

Area and perimeter of a sector

- The curved part of a sector is often labeled l. Its length can be calculated by the formula $l = \dfrac{\theta}{360} \times 2\pi r$, where r is the radius and θ is the angle between the two straight sides.
- The perimeter of a sector is: $P = 2r + l$
- The area of a sector is: $A = \dfrac{\theta}{360} \times \pi r^2$

Perimeter and circumference

- The **perimeter** is the distance around a shape.
- The perimeter of a circle is called the **circumference**.
- The circumference of a circle can be calculated using either of the following formulas, where r is the radius of the circle and d is the diameter of the circle:
  - $C = 2\pi r$
  - $C = \pi d$

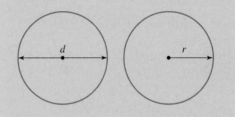

Area

- The area of a shape is the amount of space it takes up.
- The areas of common shapes can be calculated using the following formulas:
  - Square: $A = l^2$
  - Rectangle: $A = lw$
  - Parallelogram: $A = bh$
  - Triangle: $A = \dfrac{1}{2} bh$
  - Trapezium: $A = \dfrac{1}{2}(a + b)h$
  - Kite: $A = \dfrac{1}{2} xy$
  - Circle: $A = \pi r^2$
  - Rhombus: $A = \dfrac{1}{2} xy$

Surface area

- The surface area of an object can be calculated by adding the areas of all its faces.
- The formula for the surface area of a closed cylinder is:
  $SA = 2\pi r^2 + 2\pi rh = 2\pi r(r + h)$

Volume

- The volume of a prism is equal to its cross-sectional area multiplied by its height: $V_{\text{prism}} = A \times h$
- The volume of a cube is: $V_{\text{cube}} = l^3$
- The volume of a cuboid is: $V_{\text{cuboid}} = lwh$
- The volume of a cylinder is: $V_{\text{cylinder}} = \pi r^2 h$
- Capacity is another term for volume, and is usually applied to the measurement of liquids and containers.

11.8.2 Project

Areas of polygons

The area of plane figures can be found using a formula. For example, if the figure is a rectangle, its area is found by multiplying its length by its width. If the figure is a triangle, the area is found by multiplying half the base length by its perpendicular height. If the figures are complex, break them down into simple shapes and determine the area of each of these shapes. The area of a complex shape is the sum of the area of its simple shapes. Drawing polygons onto grid paper is one method that can be used to determine their area.

Consider the following seven polygons drawn on 1-cm grid paper.

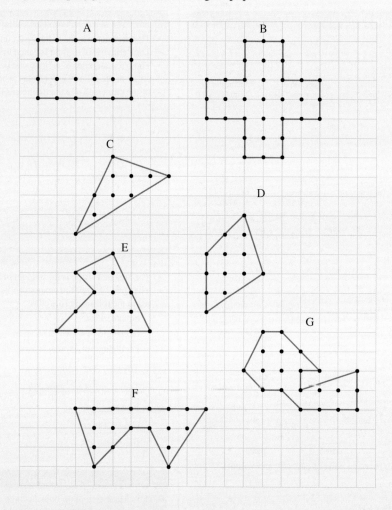

1. By counting the squares or half-squares, determine the area of each polygon in cm². In some cases, it may be necessary to divide some sections into half-rectangles in order to determine the exact area.

Formulas define a relationship between dimensions of figures. In order to search for a formula to calculate the area of polygons drawn on the grid paper, consider the next question.

2. For each of the polygons provided above, complete the following table. Count the number of dots on the perimeter of each polygon and count the number of dots that are within the perimeter of each polygon.

Polygon	Dots on perimeter (b)	Dots within perimeter (i)	Area of polygon (A)
A			
B			
C			
D			
E			
F			
G			

3. Choose a pronumeral to represent the headings in the table. Investigate and determine a relationship between the area of each polygon and the dots on and within the perimeter. Test that the relationship determined works for each polygon. Write the relationship as a formula.

4. Draw some polygons on grid paper. Use your formula to determine the area of each shape. Confirm that your formula works by counting the squares in each polygon. The results of both methods (formula and method) should be the same.

 Resources

 Interactivities Crossword (int-2709)

Sudoku puzzle (int-3211)

Exercise 11.8 Review questions

learnon

Fluency

1. **MC** Identify which of the following is true.
   A. 5 cm is 100 times as big as 5 mm.
   B. 5 metres is 100 times as big as 5 cm.
   C. 5 km is 1000 times as small as 5 metres.
   D. 5 mm is 100 times as small as 5 metres.

2. **MC** The circumference of a circle with a diameter of 12.25 cm is:
   A. 471.44 cm   B. 384.85 mm   C. 76.97 cm   D. 117.86 cm

3. **MC** The area of the given shape is:

   A. 216 m²   B. 140 m²   C. 150 m²   D. 90 m²

   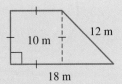

4. **MC** The area of a circle with diameter 7.5 cm, to 1 decimal place, is:
   A. 176.7 cm²   B. 47.1 cm²   C. 23.6 cm²   D. 44.2 cm²

5. **MC** The perimeter of the shape shown, to 1 decimal place, is:
   A. 1075.2 cm
   B. 55.5 cm
   C. 153.2 cm
   D. 66.1 cm

6. **MC** The surface area and volume of a cube with side length 7 m are respectively:

   A. $294\,m^2$, $343\,m^3$ B. $49\,m^2$, $343\,m^3$ C. $147\,m^2$, $49\,m^3$ D. $28\,m^2$, $84\,m^3$

7. **MC** The surface area of a rectangular box with dimensions 7 m, 3 m, 2 m is:

   A. $42\,m^2$ B. $72\,m^2$ C. $82\,cm^2$ D. $82\,m^2$

8. **MC** The surface area and volume of a cylinder with radius 35 cm and height 40 cm are:

   A. $16\,493.36\,cm^2$ and $153\,938\,cm^3$ B. $8796.5\,cm^2$ and $11\,246.5\,cm^3$

   C. $153\,938\,cm^2$ and $11\,246.5\,cm^3$ D. $8796.5\,cm^2$ and $153\,938\,cm^3$

9. Convert the following lengths into the required units.

   a. $26\,mm =$ _____ cm   b. $1385\,mm =$ _____ cm

   c. $1.63\,cm =$ _____ mm   d. $1.5\,km =$ _____ m

   e. $0.077\,km =$ _____ m   f. $2850\,m =$ _____ km

Understanding

10. Calculate the circumferences of circles with the following dimensions (correct to 1 decimal place).

    a. Radius 4 cm   b. Radius 5.6 m   c. Diameter 12 cm

11. Calculate the perimeters of the following shapes (correct to 1 decimal place).

    a.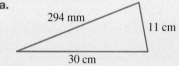

    294 mm   11 cm   30 cm

    b.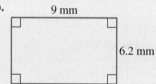

    9 mm   6.2 mm

    c.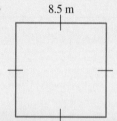

    8.5 m

    d.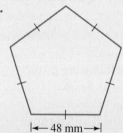

    ⟵ 48 mm ⟶

12. Calculate the perimeters of the following shapes (correct to 1 decimal place).

    a.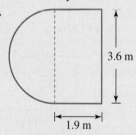

    3.6 m   1.9 m

    b.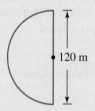

    120 m

    c.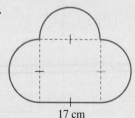

    17 cm

    d.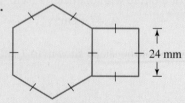

    24 mm

13. Calculate the areas of the following shapes.

a.
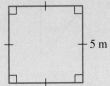
5 m

b.
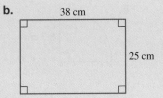
38 cm
25 cm

c.
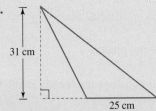
31 cm
25 cm

d.
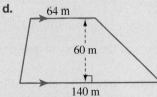
64 m
60 m
140 m

14. Calculate the areas of the following shapes (correct to 1 decimal place).

a.
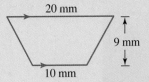
20 mm
9 mm
10 mm

b.
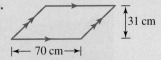
31 cm
70 cm

c.
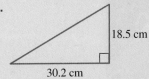
18.5 cm
30.2 cm

d.
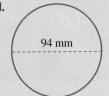
94 mm

15. Calculate the areas of the following 2-dimensional shapes by dividing them into simpler shapes. (Where necessary, express your answer correct to 1 decimal place.)

a.
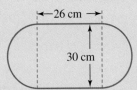
26 cm
30 cm

b.
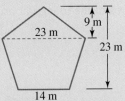
9 m
23 m
23 m
14 m

c.
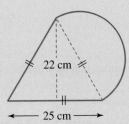
22 cm
25 cm

d.
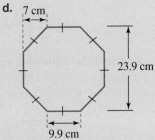
7 cm
23.9 cm
9.9 cm

16. Calculate the areas of the sectors shown (correct to 2 decimal places).

a.
←20 m→

b.
8.7 cm

c.
70°
←18 cm→

d.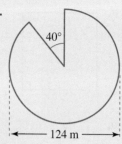
40°
← 124 m →

17. Calculate the perimeters of the sectors in question **16**, correct to 2 decimal places.

18. Calculate the inner surface area of the grape-collecting vat using the dimensions for length, width and depth shown.

70 cm
82 cm
1.6 m

19. Calculate the surface area of the triangular prism shown.

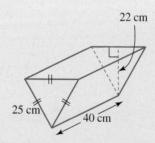
22 cm
25 cm
40 cm

20. Calculate the volume of each of the following, giving your answers to 2 decimal places where necessary.

a.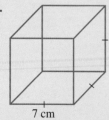
7 cm

b.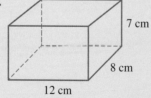
7 cm
8 cm
12 cm

c.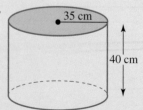
35 cm
40 cm

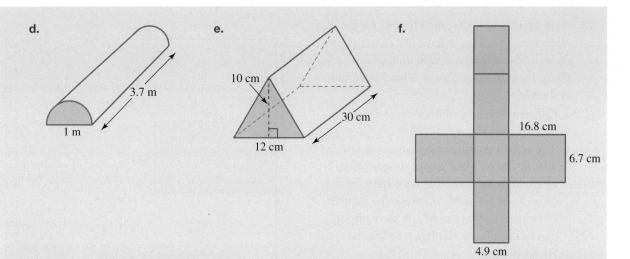

d. 3.7 m, 1 m

e. 10 cm, 30 cm, 12 cm

f. 16.8 cm, 6.7 cm, 4.9 cm

21. Calculate the volume of water that can completely fill the inside of this washing machine. Give your answer in litres, correct to 2 decimal places.

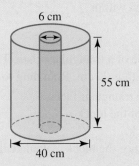

6 cm, 55 cm, 40 cm

22. Calculate the capacity of the object shown (to the nearest litre).

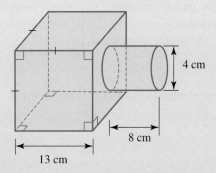

4 cm, 8 cm, 13 cm

Communicating, reasoning and problem solving

23. Calculate the cost of painting an outer surface (including the lid) of a cylindrical water tank that has a radius of 2.2 m and a height of 1.6 m if the paint costs $1.90 per square metre.

24. A harvester travels at 11 km/h. The comb (harvesting section) is 8.7 m wide. Determine how many hectares per hour can be harvested with this comb.

25. Nina decides to invite some friends for a sleepover. She plans to have her guests sleep on inflatable mattresses with dimensions 60 cm by 170 cm. A plan of Nina's bedroom is shown.

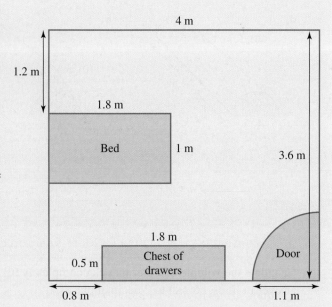

a. Calculate the area of floor space available, to 2 decimal places.

b. The area of an inflatable mattress is 1.02 m². Nina's best friend suggests that, by dividing the available floor space by the area of one inflatable mattress, the number of mattresses that can fit into the bedroom can be calculated. Explain whether the friend is correct.

c. Determine how many self-inflatable mattresses can fit in the room as it is.

d. Explain how your answer to c would change if the bed could be moved within the bedroom.

26. A block of chocolate is in the shape of a triangular prism. The face of the prism is an equilateral triangle of 5 cm sides, and the length is 22 cm. Rounding to 2 decimal places where necessary.

a. Determine the area of one of the triangular faces.

b. Calculate the total surface area of the block.

c. Calculate the volume of the block.

d. If it cost $0.025 per cm³ to produce the block, calculate the total cost of production.

27. An A4 sheet of paper has dimensions 210 mm × 297 mm and can be rolled two different ways (by rotating the paper) to make baseless cylinders, as illustrated.
Compare the volumes to decide which shape has the greatest volume.

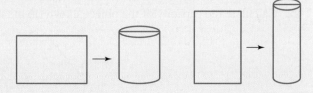

28. These congruent squares have shaded parts of circles inside them.

a.

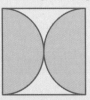

b.

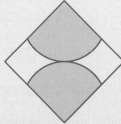

Compare the shaded area in figure **a** with that in figure **b**. Determine which figure has a larger shaded area.

29. The widespread manufacture of soft drinks in cans only really started in the 1960s. Today cans of soft drink are very common. They can be purchased in a variety of sizes. You can also purchase them as a single can or bulk-buy them in larger quantities. Does the arrangement of the cans in a multi-pack affect the amount of packaging used to wrap them?

For this investigation we will look at the packaging of 375-mL cans. This size of can has a radius of approximately 3.2 cm and a height of approximately 12 cm. We will consider packaging these cans in cardboard in the shape of a prism, ignoring any overlap for glueing. Round answers to 2 decimal places.

   a. A 6-can pack could be packed in a single row of cans.
      i. Draw a diagram to show what this pack would look like.
      ii. Calculate the amount of cardboard required for the package.
   b. The cans could also be packaged in a 3×2 arrangement.
      i. Show the shape of this pack.
      ii. Calculate the amount of cardboard required for this pack.
   c. Evaluate which of the two 6-can pack arrangements is the more economical in terms of packaging.
   d. These cans can be packaged as a single layer of 4×3. Calculate how much cardboard this would require.
   e. They could also have a two-level arrangement of 3×2 each layer. Determine what packaging this would require.
   f. It is also possible to have a three-level arrangement of 2×2 each layer. Determine how much packaging would be needed in this case.
   g. Comment on the best way to package a 12-can pack so that it uses the minimum amount of wrapping. Provide diagrams and mathematical evidence to support your conclusion.

on To test your understanding and knowledge of this topic, go to your learnON title at www.jacplus.com.au and complete the **post-test**.

Answers

Topic 11 Measurement

Exercise 11.1 Pre-test

1. $15.4 \, \text{cm}$
2. $12.48 \, \text{cm}^2$
3. $30 \, \text{cm}$
4. C
5. $1440 \, \text{mm}^3$
6. D
7. B
8. D
9. B
10. $3.48 \, \text{m}^2$
11. D
12. $3100 \, \text{cm}^3$
13. $678.6 \, \text{mL}$
14. $25 \, \text{m}$
15. $527.8 \, \text{cm}^2$

Exercise 11.2 Length and perimeter

1. a. $50 \, \text{mm}$ b. $152 \, \text{cm}$ c. $0.0125 \, \text{m}$ d. $32.2 \, \text{mm}$
2. a. $0.006 \, 57 \, \text{km}$ b. $0.000 \, 64 \, \text{km}$
   c. $14.35 \, \text{mm}$ d. $0.000 \, 183 \, 5 \, \text{km}$
3. a. $101 \, \text{mm}$ b. $54 \, \text{mm}$ c. $60 \, \text{mm}$ d. $103 \, \text{mm}$
4. a. $1060 \, \text{cm}$ b. $85.4 \, \text{cm}$ c. $206 \, \text{cm}$
5. a. $78.4 \, \text{cm}$ b. $113 \, \text{cm}$ c. $13 \, \text{cm}$
6. a. $9.6 \, \text{cm}$ b. $46 \, \text{mm}$ c. $31 \, \text{km}$
7. a. $25.13 \, \text{cm}$ b. $25.13 \, \text{m}$ c. $69.12 \, \text{mm}$
8. a. $44.61 \, \text{cm}$ b. $19 \, 741.77 \, \text{km}$ c. $3314.38 \, \text{mm}$
9. a. $192 \, \text{m}$ b. $1220 \, \text{mm}$ c. $260 \, \text{cm}$ or $2.6 \, \text{m}$
10. a. $74 \, \text{mm}$ or $7.4 \, \text{cm}$
    b. $9.6 \, \text{km}$
    c. $8 \, \text{m}$ or $800 \, \text{cm}$
11. B
12. a. $1800 \, \text{mm}$, $2100 \, \text{mm}$, $2400 \, \text{mm}$, $2700 \, \text{mm}$, $180 \, \text{cm}$, $210 \, \text{cm}$, $240 \, \text{cm}$, $270 \, \text{cm}$, $1.8 \, \text{m}$, $2.1 \, \text{m}$, $2.4 \, \text{m}$, $2.7 \, \text{m}$
    b. 4
    c. $3300 \, \text{mm}$
13. a. $1990 \, \text{m}$ b. $841 \, \text{m}$
14. a. $127.12 \, \text{cm}$ b. $104.83 \, \text{cm}$ c. $61.70 \, \text{cm}$
15. a. $8 \, \text{m}$ or $800 \, \text{mm}$ b. $480 \, \text{mm}$ c. $405.35 \, \text{cm}$
16. a. $125.66 \, \text{cm}$ b. $245.66 \, \text{m}$ c. $70.41 \, \text{cm}$
17. $222.5 \, \text{m}$
18. $37.5 \, \text{km}$
19. $12 \, \text{m}$
20. a. $r = 6371 + 559$
       $= 6930 \, \text{km}$

$C = 2\pi r$
$\quad = 2 \times \pi \times 6930$
$\quad = 13 \, 860\pi$
$\quad = 43 \, 542.474$
$\quad \approx 43 \, 542 \, \text{km}$

b. 96 minutes $= 96 \times 60$
   $\qquad\qquad\quad = 5760$ seconds
   $43 \, 542 \, \text{km} = 43 \, 542 \, 000 \, \text{m}$
   $\quad \text{Speed} = 43 \, 542 \, 000 \div 5760$
   $\qquad\qquad = 7559.375 \ldots$
   $\qquad\qquad \approx 7559 \, \text{m/s}$

21. a. $500 \, \text{m/s} = (500 \times 100) \, \text{cm/s}$
       $\qquad\qquad = 50 \, 000 \, \text{cm/s}$
    b. $0.000 \, 02$ seconds
    c. 2
22. a. $155.62 \, \text{m}$
    b. $155.62 \, \text{m} \times 1.5$ minutes/metre $= 233.43$ minutes
23. a. $40 \, 212 \, \text{km}$
    b. $1676 \, \text{km/h}$
    c. Sample responses can be found in the worked solutions in the online resources.
24. Yes, $42 \, \text{cm}$ of wire remains.
25. $69.3 \, \text{cm}$
26. a. $\left(2\sqrt{13} + 4\right) \, \text{m}$ b. $2\sqrt{17} \, \text{m}$

Exercise 11.3 Area

1. D
2. a. $13 \, 400 \, \text{m}^2 = 0.0134 \, \text{km}^2$
   b. $0.04 \, \text{cm}^2 = 4 \, \text{mm}^2$
   c. $3 \, 500 \, 000 \, \text{cm}^2 = 350 \, \text{m}^2$
   d. $0.005 \, \text{m}^2 = 50 \, \text{cm}^2$
3. a. $0.043 \, \text{km}^2 = 43 \, 000 \, \text{m}^2$
   b. $200 \, \text{mm}^2 = 2 \, \text{cm}^2$
   c. $1.41 \, \text{km}^2 = 141 \, \text{ha}$
   d. $3800 \, \text{m}^2 = 0.38 \, \text{ha}$
4. a. $24 \, \text{cm}^2$ b. $16 \, \text{mm}^2$ c. $537.5 \, \text{cm}^2$
5. a. $149.5 \, \text{cm}^2$ b. $16.32 \, \text{m}^2$ c. $11.25 \, \text{cm}^2$
6. a. $292.5 \, \text{cm}^2$ b. $2.5 \, \text{cm}^2$ c. $1250 \, \text{m}^2$
7. a. $50.27 \, \text{m}^2$ b. $3.14 \, \text{mm}^2$ c. $36.32 \, \text{m}^2$
8. a. $15.58 \, \text{cm}^2$ b. $4.98 \, \text{cm}^2$ or $498 \, \text{mm}^2$ c. $65 \, \text{m}^2$
9. a. $6824.2 \, \text{mm}^2$ b. $7.6 \, \text{m}^2$ c. $734.2 \, \text{cm}^2$
10. a. $578.5 \, \text{cm}^2$ b. $7086.7 \, \text{m}^2$ c. $5.4 \, \text{m}^2$
11. a. $1143.4 \, \text{m}^2$ b. $100.5 \, \text{cm}^2$
12. a. $821 \, \text{cm}^2$ b. $661.3 \, \text{mm}^2$
13. $378 \, \text{cm}^2$
14. $\approx 19.3 \, \text{cm}^2$
15. a. $100.53 \, \text{cm}^2$ b. $1244.07 \, \text{m}^2$
16. a. $301.59 \, \text{cm}^2$ b. $103.67 \, \text{mm}^2$
17. B
18. $75.7 \, \text{cm}$
19. $\$29 \, 596.51$

20. a. 5000 km^2 b. 0.5%

21. $10 152

22. a. 619.5 cm^2

   b. 10.5 cm

   c. 346.36 cm^2

   d. Sample responses can be found in the worked solutions in the online resources.

23. Sample responses can be found in the worked solutions in the online resources.

24. Sample responses can be found in the worked solutions in the online resources.

25. a. 1600 cm^2

   b. 800 cm^2

   c. 400 cm^2

   d. 200 cm^2

   e. The area halves each time.

   f. 12.5%

   g. Sample responses can be found in the worked solutions in the online resources.

26. Sample responses can be found in the worked solutions in the online resources.

27. a. $\sqrt{50} = 5\sqrt{2} \text{ m}$ b. 10 logs

28. 17 m

29. 45 m^2

30. a. $\dfrac{1}{4}$ b. $\dfrac{1}{4}\left(x^2 + 6x + 8\right)$

   c. $\dfrac{1}{4}\left(3x^2 - 6x - 8\right)$

   d.   i. $\dfrac{1}{12}\left(3x^2 - 6x - 8\right)$

      ii. $\dfrac{1}{6}\left(3x^2 + 6x + 8\right)$

Exercise 11.4 Area and perimeter of a sector

1. a. 14.14 cm^2 b. 157.08 cm^2
   c. 27.71 cm^2 d. 226.19 mm^2

2. i. 14.3 cm ii. 12.6 cm^2

3. i. 43.6 cm ii. 116.9 cm^2

4. i. 40.7 m ii. 102.1 m^2

5. i. 5.4 m ii. 1.8 m^2

6. D

7. a. i. 66.8 m ii. 226.2 m^2

   b. i. 27.4 cm ii. 42.4 cm^2

8. a. i. 184.6 cm ii. 1869.2 cm^2

   b. i. 342.1 m ii. 7243.6 m^2

9. a. i. 354.6 cm ii. 7645.9 cm^2

   b. i. 104.4 cm ii. 201.1 cm^2

10. $60 318.6 \text{ m}^2$

11. 303.4 cm

12. 170.2 m^2

13. a. 14.8 m^2 b. 4.4%

14. $2 : 1$

15. a. 251 cm b. 13 cm

16. a. 5027 m^2 b. 78.5% c. $61 440

17. a. Sample responses can be found in the worked solutions in the online resources.

   b. 88.4 cm^2

   c. 50.3 cm^2

   d. 109.4 cm^2

   e. No, Jim eats more cake

18. Sample responses can be found in the worked solutions in the online resources.

19. a. Sample responses can be found in the worked solutions in the online resources.

   b. 85 m each

   c. 60 m each

   d. Sample responses can be found in the worked solutions in the online resources.

   e. 37 m

20. $250

21. a. $\pi r^2 \text{ cm}^2$

   b. $\pi(r + 1)^2 \text{ cm}^2$

   c, e. Sample responses can be found in the worked solutions in the online resources.

   d. $5\pi \text{ cm}^2$

22. Radius $= \pi \text{ cm}$

Exercise 11.5 Surface area of rectangular and triangular prisms

1. a. Triangular prism
   b. Hexagonal prism

2. a. Cube
   b. Rectangular prism

3. D

4. a. Triangular prism
   b. Hexagonal prism

5. a. 66 cm^2 b. 62 m^2 c. 6.7 m^2

6. a. 4.44 m^2 b. $11 572.92 \text{ cm}^2$ c. 1.9 m^2

7. a. 86 cm^2 b. 210.7 m^2 c. 8.37 cm^2

8. a. 840 cm^2 b. 191.08 cm^2 c. 2370 mm^2

9. 2 cans of paint

10. 77.76 cm^2

11. 26.44 m^2

12. 20.36 m^2

13. a. No   b. No   c. No   d. Yes

14. a. 864 cm^2 b. 222 mm^2 c. 923.6 cm^2

15. 261.5 cm^2

16. 2.21 m^2

17. 15.2 m^2

18. a. 2080 cm^2

   b. Sample responses can be found in the worked solutions in the online resources.

19. 168 cm^2

20. a. 150 cm^2 b. 250 cm^2 c. 350 cm^2 d. 850 cm^2

21. The solid formed is a cube with side length 4 cm.

22. a. $170.65\,\text{m}^2$ b. $495.07\,\text{cm}^2$

23. Area of base $= \dfrac{1}{2}bh$

$$= \dfrac{1}{2} \times 4 \times \sqrt{12}$$
$$= 2\sqrt{12}$$

Total surface area $= 3 \times l \times h + 2 \times \text{area of base}$
$$302 = 3 \times 4 \times h + 2 \times 2\sqrt{12}$$
$$302 = 12h + 4\sqrt{12}$$
$$302 = 12h + 13.86$$
$$12h = 288.14$$
$$h = \dfrac{288.14}{12}$$
$$h = 24.01$$
$$h \approx 24\,\text{cm}$$

24. a.

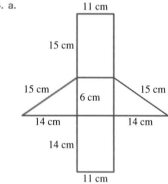

b. $315\,\text{cm}^2$

25. $145.85 (2 L of ceiling paint, 1 L + 4 L for walls)

26. a. $1402.52\,\text{m}^2$ b. $3220\,\text{m}^3$

Exercise 11.6 Surface area of a cylinder

1. a. $75.4\,\text{m}^2$ b. $131.9\,\text{m}^2$
2. a. $28.3\,\text{cm}^2$ b. $84.8\,\text{cm}^2$
3. a. $2010.6\,\text{m}^2$ b. $3619.1\,\text{m}^2$
4. a. $1121.5\,\text{cm}^2$ b. $1575.5\,\text{cm}^2$
5. a. $6.6\,\text{m}^2$ b. $9.7\,\text{m}^2$
6. a. $25.6\,\text{m}^2$ b. $61.8\,\text{m}^2$
7. a.

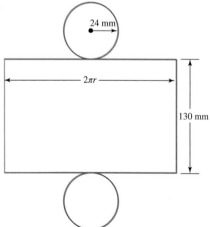

b. $232\,\text{cm}^2$

8. $\approx 1837.8\,\text{cm}^2$

9. $\approx 31.79\,\text{cm}$

10. a. Asparagus is largest; beetroot is smallest.
    b. $118\,\text{cm}^2$

11. a. $\approx 136.66\,\text{m}^2$ b. $7351.33

12. a. $\approx 2035.75\,\text{cm}^2$ b. $\approx 1526.81\,\text{cm}^2$
    c. $\approx 395.84\,\text{cm}^2$ d. $\approx 3958.41\,\text{cm}^2$

13. $154.73\,\text{cm}^2$

14. a. $3.82\,\text{cm}$ b. $427.7\,\text{cm}^2$

15. $12\pi h^2$

16. Sample responses can be found in the worked solutions in the online resources.

17. Sample responses can be found in the worked solutions in the online resources.

18. $83.6\,\text{m}^2$

19. The area to be painted is $1.63\,\text{m}^2$.

20. $11\,231.12\,\text{cm}^2$

21. $4.34\,\text{m}^2$

Exercise 11.7 Volume of prisms and cylinders

1. a. $325\,\text{mL}$ b. $2\,600\,000\,\text{mL}$
   c. $5100\,\text{mL}$ d. $630\,000\,\text{mL}$

2. a. $5.8\,\text{cm}^3$ b. $6100\,\text{cm}^3$
   c. $3\,200\,000\,\text{cm}^3$ d. $0.0593\,\text{cm}^3$

3. a. $0.358\,\text{kL}$ b. $55.8\,\text{kL}$ c. $8.752\,\text{kL}$ d. $5300\,\text{kL}$

4. a. $36\,\text{cm}^3$ b. $15\,\text{cm}^3$ c. $72\,\text{cm}^3$

5. a. $12\,\text{cm}^3$ b. $68\,\text{cm}^3$ c. $9.6\,\text{m}^3$

6. a. $630\,\text{mm}^3$ b. $420\,\text{cm}^3$ c. $\approx 3152.7\,\text{mm}^3$

7. a. $\approx 1319.5\,\text{mm}^3$ b. $\approx 523.6\,\text{m}^3$ c. $\approx 10.5\,\text{m}^3$

8. a. $36\,\text{cm}^3$ b. $30\,\text{m}^3$ c. $1.144\,\text{m}^3$

9. a. $7020\,\text{cm}^3$ b. $6.91\,\text{m}^3$ c. $300\,\text{m}^3$ d. $8806\,\text{cm}^3$

10. a. $16\,085.0\,\text{m}^3$ b. $4766.6\,\text{cm}^3$ c. $2.3\,\text{m}^3$

11. a. $30.8\,\text{m}^3$ b. $30.6\,\text{m}^3$ c. $56.5\,\text{mm}^3$

12. $14\,476.5\,\text{L}$

13. $176.4\,\text{L}$

14. a. $158\,169\,\text{cm}^3$ b. $4092\,\text{cm}^3$ c. $2639\,\text{cm}^3$

15. a. $641\,\text{cm}^3$ b. $1784\,\text{cm}^3$ c. $1057\,\text{cm}^3$

16. $75\,322.8\,\text{L}$

17. $1145.1\,\text{mL}$

18. a. $4.95\,\text{m}^3$ b. $4950\,\text{L}$

19. $2.916 \times 10^9\,\text{m}^3$

20. $234\,256.54\,\text{cm}^3$

21. a. $1.1475\,\text{m}^3$ b. $840\,\text{L}$

22. a. $45.94\,\text{cm}^3$
    b. 2400 (to the nearest whole number)

23. Sample responses can be found in the worked solutions in the online resources.

24. a. Prism

 b. $75\,\text{m}^2$

 c. Sample responses can be found in the worked solutions in the online resources.

 d. i. 36 h 40 min   ii. 47 h 9 min

25. 23

26. a. 1.60 L   b. 0.042 L

27. a. $169\,\text{cm}^3$

 b. $V = x(l - 2x)(w - 2x)$

28. 19 full glasses

Project

1. A. $15\,\text{cm}^2$ B. $20\,\text{cm}^2$ C. $6\,\text{cm}^2$ D. $8.5\,\text{cm}^2$
   E. $10.5\,\text{cm}^2$ F. $11.5\,\text{cm}^2$ G. $12.5\,\text{cm}^2$

2.

Polygon	Dots on perimeter (b)	Dots within perimeter (i)	Area of polygon (A)
A	16	8	15
B	24	9	20
C	4	6	6
D	7	6	8.5
E	11	6	10.5
F	13	6	11.5
G	15	6	12.5

3. Sample responses can be found in the worked solutions in the online resources.

4. Sample responses can be found in the worked solutions in the online resources.

Exercise 11.8 Review questions

1. B
2. B
3. B
4. D
5. D
6. A
7. D
8. A
9. a. 26 mm = 2.6 cm   b. 1385 mm = 138.5 cm
   c. 1.63 cm = 16.3 mm   d. 1.5 km = 1500 m
   e. 0.077 km = 77 m   f. 2850 m = 2.85 km

10. a. 25.1 cm   b. 35.2 m   c. 37.7 cm

11. a. 70.4 cm or 704 mm   b. 30.4 mm
    c. 34 m   d. 240 mm

12. a. 13.1 m   b. 308.5 m
    c. 97.1 cm   d. 192 mm

13. a. $25\,\text{m}^2$ b. $950\,\text{cm}^2$
    c. $387.5\,\text{cm}^2$ d. $6120\,\text{m}^2$

14. a. $135\,\text{mm}^2$ b. $2170\,\text{cm}^2$
    c. $279.4\,\text{cm}^2$ d. $6939.8\,\text{mm}^2$

15. a. $1486.9\,\text{cm}^2$ b. $362.5\,\text{m}^2$
    c. $520.4\,\text{cm}^2$ d. $473.2\,\text{cm}^2$

16. a. $628.32\,\text{m}^2$ b. $59.45\,\text{cm}^2$
    c. $197.92\,\text{cm}^2$ d. $10\,734.47\,\text{m}^2$

17. a. 102.83 m   b. 31.07 cm
    c. 57.99 cm   d. 470.27 m

18. $60\,120\,\text{cm}^2$ or $6.012\,\text{m}^2$

19. $3550\,\text{cm}^2$

20. a. $343\,\text{cm}^3$ b. $672\,\text{cm}^3$
    c. $153\,938.04\,\text{cm}^3$ d. $1.45\,\text{m}^3$
    e. $1800\,\text{cm}^3$ f. $551.54\,\text{cm}^2$

21. 67.56 L

22. 2 L

23. $99.80

24. 9.57 hectares

25. a. $10.75\,\text{m}^2$
    b. No
    c. 8
    d. 9 mattresses could fit in the room.

26. a. $10.83\,\text{cm}^2$ b. $351.65\,\text{cm}^2$
    c. $238.26\,\text{cm}^3$ d. $5.95

27. The cylinder of height 210 mm has the greater volume.

28. The shaded parts have the same area.

29. a. i.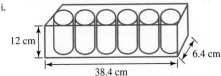

 ii. $1566.72\,\text{cm}^2$

 b. i.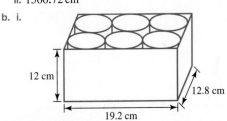

 ii. $1259.52\,\text{cm}^2$

 c. 3×2 arrangement

 d. $2058.24\,\text{cm}^2$

 e. $2027.52\,\text{cm}^2$

 f. Yes, $2170.88\,\text{cm}^2$

 g. The smallest amount of packaging required is 3×2, with two levels of packaging.

12 Properties of geometrical figures

LESSON
12.1 Overview

Why learn this?

Geometry allows us to explore our world in very precise ways. It is also one of the oldest areas of mathematics. It involves the study of points, lines and angles and how they can be combined to make different shapes. Similarity and congruence are two important concepts in geometry. When trying to determine whether two shapes are exactly the same, or if they are enlargements of each other, the answers can be found by considering the sides and angles of those shapes.

When you take a photo on your phone it is very small. If you save it to your computer you can make it larger. The larger photo is an enlargement of the original photo — this is an example of how similar figures work in the everyday world.

The principle of similar triangles can be used to work out the height of tall objects by calculating the length of their shadows. This technique was extremely important in early engineering and architecture. Today, architects and designers still prepare scale diagrams before starting the building process.

Hey students! Bring these pages to life online

▶ Watch videos

🧩 Engage with interactivities

A+ Answer questions and check solutions

Find all this and MORE in jacPLUS ▶

Reading content and rich media, including interactivities and videos for every concept

Extra learning resources

Differentiated question sets

Questions with immediate feedback, and fully worked solutions to help students get unstuck

1. Express the ratio $1\frac{2}{3} : 2\frac{3}{5}$ in simplest form.

$$1\frac{2}{3} : 2\frac{3}{5} = \boxed{} : \boxed{}$$

2. Determine the value of x in the proportion $2 : 3 = 9 : x$

3. **MC** Select the correct symbol for similarity.
   A. $=$ B. $\approx$ C. $\sim$ D. $\simeq$

4. All equilateral triangles are similar. State whether this statement is True or False.

5. Determine the simplest ratio of $y : z$ if $3x = 2y$ and $4x = 3z$.

6. The ratio $1.4 : 0.2$ in its simplest form is $14 : 2$. State whether this statement is True or False.

7. Determine the value of x if $x : 8 = 4 : 3$

8. The scale on a floor plan is $1 : 100$. If a room is 3.5 metres in length, what is its length on the plan?

9. **MC** Select which similarity test can be used to prove $\triangle ABC \sim \triangle ADE$.

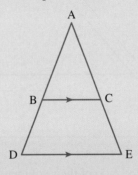

   A. AAA   B. RHS   C. SSS   D. ASA

10. **MC** $\triangle ABC$ and $\triangle CDE$ are similar.

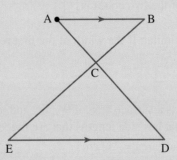

   Choose which of the following statements is true.
   A. $\dfrac{AB}{ED} = \dfrac{BC}{CD}$ B. $\dfrac{AB}{ED} = \dfrac{AC}{CD}$ C. $\dfrac{AC}{EC} = \dfrac{BC}{CD}$ D. $\dfrac{AB}{ED} = \dfrac{BC}{CE}$

11. ΔABC and ΔCDE are similar.

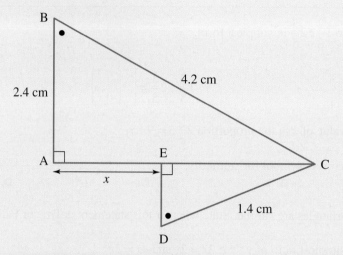

Determine the value of the length x correct to 1 decimal place.

12. ΔABC and ΔCDE are similar. Determine the value of b.

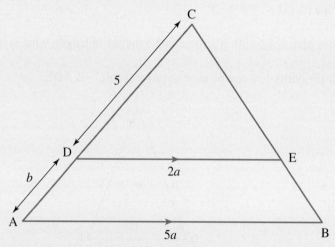

13. **PATH** A pair of cones are similar. The ratio of their volumes is 27 : 125.
   a. The perpendicular height ratio of the two cones is □ : □.

   b. The areas of the base ratio of the two cones is □ : □.

14. **PATH** A pair of rectangles are similar. If the width of the first rectangle is 3 times the width of the other, the ratio of their areas is:
   larger area : smaller area = □ : □.

15. **PATH** **MC** A rectangular box has a surface area of 94 cm² and volume of 60 cm³. Select the volume and surface area of a similar box that has side lengths that are twice the size of the original.
   A. Volume = 480 cm³ and surface area = 752 cm²
   B. Volume = 480 cm³ and surface area = 376 cm²
   C. Volume = 120 cm³ and surface area = 188 cm²
   D. Volume = 240 cm³ and surface area = 376 cm²

LESSON
12.2 Ratio and scale

LEARNING INTENTION

At the end of this lesson you should be able to:
- compare two quantities of the same type using ratios
- simplify ratios
- enlarge or reduce a figure by applying a scale factor
- calculate and use a scale factor, including plans.

▶ 12.2.1 Ratio

eles-4727

- **Ratios** are used to compare quantities of the same kind, measured in the same unit.
- The ratio '1 is to 4' can be written in two ways: as 1 : 4 or as $\frac{1}{4}$.
- The order of the numbers in a ratio is important.
- In simplest form, a ratio is written using the smallest whole numbers possible.

WORKED EXAMPLE 1 Expressing ratios in simplest form

A lighthouse is positioned on a cliff that is 80 m high. A ship at sea is 3600 m from the base of the cliff.
a. Write the following ratios in simplest form.
  i. The height of the cliff to the distance of the ship from shore
  ii. The distance of the ship from shore to the height of the cliff
b. Compare the distance of the ship from shore with the height of the cliff.

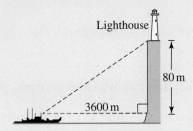

THINK	WRITE
a. i. 1. The height and distance are in the same units (m). Write the height first and the distance second.	**a. i.** Height of cliff : distance of ship from shore = 80 : 3600
2. Simplify the ratio by dividing both terms by the highest common factor (80).	= 1 : 45
ii. 1. Write the distance first and the height second.	**ii.** Distance of ship from shore : height of cliff $= \dfrac{3600}{80}$

2. Simplify by dividing both terms by the highest common factor (80).

$$= \frac{45}{1}$$

Note: Do not write $\frac{45}{1}$ as 45, because a ratio is a comparison between two numbers.

b. 1. Write the ratio 'distance of the ship from shore to height of the cliff'.

b. 45 : 1

2. Write the answer.

The distance of the ship from shore is 45 times the height of the cliff.

WORKED EXAMPLE 2 Simplifying ratios into simplest form

Express each of the following ratios in simplest form.

a. **24 : 8**   b. **3.6 : 8.4**   c. $1\frac{4}{9} : 1\frac{2}{3}$

THINK

a. Divide both terms by the highest common factor (8).

b. 1. Multiply both terms by 10 to obtain whole numbers.

2. Divide both terms by the highest common factor (12).

c. 1. Change both mixed numbers into improper fractions.

2. Multiply both terms by the lowest common denominator (3) to obtain whole numbers.

WRITE

a. 24 : 8
$= 3 : 1$

b. 3.6 : 8.4
$= 36 : 84$

$= 3 : 7$

c. $1\frac{4}{9} : 1\frac{2}{3}$

$= \frac{13}{9} : \frac{5}{3}$

$= 13 : 15$

- A **proportion** is a statement that indicates that two ratios are equal. A proportion can be written in two ways, for example in the format used for $4 : 7 = x : 15$ or in the format $\frac{4}{7} = \frac{x}{15}$.

WORKED EXAMPLE 3 Calculating a value in a proportion

Find the value of x in the proportion $4 : 9 = 7 : x$.

THINK

1. Write the ratios as equal fractions.

2. Multiply both sides by x.

3. Solve the equation to obtain the value of x.

4. Write the answer.

WRITE

$\frac{4}{9} = \frac{7}{x}$

$\frac{4x}{9} = 7$

$4x = 63$

$x = 15.75$

12.2.2 Scale

eles-4728

- Ratios are used when creating scale drawings or maps.
- Consider the situation in which we want to enlarge a triangle ABC (the **object**) by a **scale factor** of 2 (this means we want to make it twice its size). The following is one method that we can use.

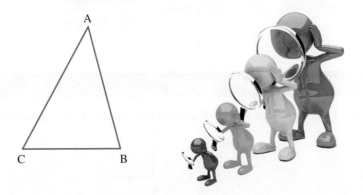

1. Mark a point O somewhere outside the triangle and draw the lines OA, OB and OC as shown.

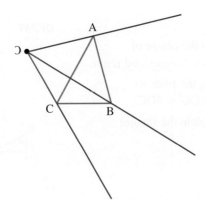

2. Measure the length of OA and mark in the point A′ (this is called the **image** of A) so that the distance OA′ is twice the distance of OA.
3. In the same way, mark in points B′ and C′. $(OB' = 2 \times OB,$ and $OC' = 2 \times OC.)$

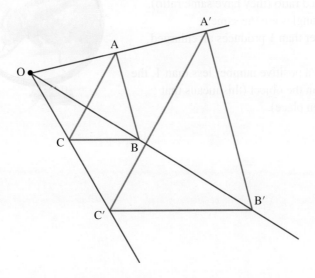

4. Joining A′B′C′ gives a triangle that has side lengths double those of △ABC. △A′B′C′ is called the image of △ABC.

Scale factor

$$\text{Scale factor} = \frac{\text{image length}}{\text{object length}}$$

WORKED EXAMPLE 4 Enlarging a figure

Enlarge triangle ABC by a scale factor of 3, with the centre of enlargement at point O.

THINK

1. Join each vertex of the triangle to the centre of enlargement (O) with straight lines, then extend them.
2. Locate points A′, B′ and C′ along the lines so that OA′ = 3OA, OB′ = 3OB and OC′ = 3OC.
3. Join points A′, B′ and C′ to complete the image.

DRAW

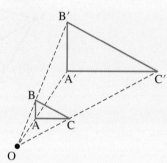

- Enlargements have the following properties.
  - The corresponding side lengths of the enlarged figure are changed in a fixed ratio (they have same ratio).
  - The corresponding angles are the same.
  - A scale factor greater than 1 produces an enlarged figure.
  - If the scale factor is a positive number less than 1, the image is smaller than the object (this means that a reduction has taken place).

A triangle PQR has been enlarged to create triangle P′ Q′ R′. PQ = 4 cm, PR = 6 cm, P′ Q′ = 10 cm and Q′ R′ = 20 cm. Calculate:

a. the scale factor for the enlargement
b. the length of P′R′
c. the length of QR.

THINK **WRITE/DRAW**

a. 1. Draw a diagram. a.

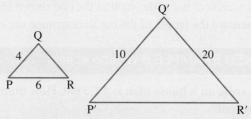

2. Find two corresponding sides. P′Q′ corresponds to PQ.

$$\text{Scale factor} = \frac{\text{image length}}{\text{object length}}$$

$$= \frac{P'Q'}{PQ}$$

$$= \frac{10}{4}$$

$$= 2.5$$

b. 1. Apply the scale factor. P′R′ = 2.5 × PR

b. $P'R' = 2.5 \times PR$
$$= 2.5 \times 6$$
$$= 15$$

2. Write the answer. P′R′ is 15 cm long.

c. 1. Apply the scale factor. Q′R′ = 2.5 × QR

c. $\dfrac{Q'R'}{QR} = \dfrac{20\,\text{cm}}{x\,\text{cm}} = 2.5$

$$Q'R' = 2.5 \times QR$$

$$20 = 2.5 \times QR$$

$$QR = \frac{20}{2.5}$$

$$= 8$$

2. Write the answer. QR is 8 cm long.

⏵ 12.2.3 Scales in photographs, plans and drawings

eles-6165

- A scale on a photograph, plan or drawing describes the ratio between the distance or size on the photograph, plan or drawing and the actual distance or size.
- A scale may be written in several ways.
  - A ratio such as 1 : 100 000 means that 1 cm on the map or plan represents a distance of 100 000 cm on the Earth, 1 mm on the map or plan represents 100 000 mm on the Earth, etc.
  - A statement such as 1 cm ⇔ 500 km means that 1 cm on the map or plan represents a distance of 500 km on the Earth.
  - A graphical bar scale, such as the one shown in Worked example 6, can be used. In this case, we need to measure the length of the bar to determine the scale ratio.

WORKED EXAMPLE 6 Writing scales

a. **The scale on a house plan is 1 : 2000. How many metres does a length of 1 cm on the map represent?**
b. **Write the scale 1 cm ⇔ 1500 m as a ratio scale.**
c. **A map shows a scale of 1 cm ⇔ 1500 m. What distance is represented by 18 mm on the map?**
d. **Use the following graphical scale to determine the distance that 1 cm on the map represents.**

$$0 \qquad 5 \qquad 10 \qquad 15 \text{ kilometres}$$

THINK	WRITE
a. 1. Use the ratio to add the same units to both sides. Since the answer asks for cm, use this unit.	a. 1 : 2000 is the same as: 1 cm ⇔ 2000 cm
2. Bring the length to an appropriate unit. In this case, the answer is required in metres.	2000 cm = (2000 ÷ 100) m = 20 m So, 1 cm ⇔ 20 m.
3. Write the answer.	1 cm on the map represents a length of 20 m in reality.
b. 1. Convert the ground units to the same as those of the map.	b. 5000 m = (5000 × 100) cm = 500 000 cm
2. Write the scale as a ratio.	So, 1 cm ⇔ 5000 m is the same as 1 cm ⇔ 500 000 cm. Expressed as a ratio, this is 1 : 500 000.
c. 1. Write a statement explaining this scale.	c. 1 cm ⇔ 1500 m means that 1 cm on the map represents a distance of 1500 m.
2. Change the question distance to the same units as those for the map.	18 mm = (18 ÷ 10) cm = 1.8 cm
3. Multiply by this scale factor.	1.8 cm represents (1.8 × 1500) m, which is 2700 m or 2.7 km.
d. 1. Measure the length of the bars with a ruler.	d. The bars measure 3.5 cm.
2. Relate this measurement to the length shown in the scale.	This represents a length of 15 km.
3. Write as a scale.	3.5 cm ⇔ 15 km
4. Calculate a 1 cm map distance and write the answer.	So, 1 cm ⇔ $\frac{15}{35}$ km. 1 cm on the map represents a distance of 4.3 km.

- Sometimes it is necessary to do the reverse, and convert from a real length to a plan measurement.

WORKED EXAMPLE 7 Calculating length from a scale

The length of a wall for a particular house is 12 m. If the scale of a house plan is 1 : 200, what length, on the plan, would represent the actual length of the wall?

THINK	WRITE
1. Write the scale as a ratio using units of cm.	1 : 200 means that 1 cm on the plan represents an actual length of 200 cm. $1\,\text{cm} \Leftrightarrow 200\,\text{cm}$
2. Convert the actual length from cm to m units.	$200\,\text{cm} = (200 \div 100)\,\text{m}$ $\phantom{200\,\text{cm}} = 2\,\text{m}$ So 1 cm on the plan represents an actual length of 2 m.
3. Divide the actual length by 2 to find how many lots of 2 m there are in 12 m.	$\dfrac{12\,\text{m}}{2\,\text{m}} = 6$
4. Multiply both sides of the scale by this scale factor, then simplify.	$1\,\text{cm} \times 6 \Leftrightarrow 2\,\text{m} \times 6$ $\phantom{1\,\text{cm}}6\,\text{cm} \Leftrightarrow 12\,\text{m}$
5. Answer the question.	The length of the wall on the house plan would be 6 cm.

DISCUSSION

The image below is a photo of the largest dinosaur footprint ever found. It was discovered on the north-western coast of Western Australia and is a footprint of a sauropod. In the photograph, the footprint is 6 cm. If the scale factor of the photograph to the real footprint is 28.33, what is the actual length of the footprint to the nearest cm?

▶ 12.2.4 House plans

eles-6166

- Whenever a house is being constructed, floor plans are supplied to the builder so that the footings, walls, roof and other parts of the house are all placed in the correct positions and are the correct length.
- The scale of the plan is generally given as a ratio. This enables us to calculate all dimensions within the house.

WORKED EXAMPLE 8 Using a house plan to calculate dimensions within a house

A plan for a house is shown.
a. **Calculate the dimensions of the house.**
b. **Calculate the area of the lounge room.**

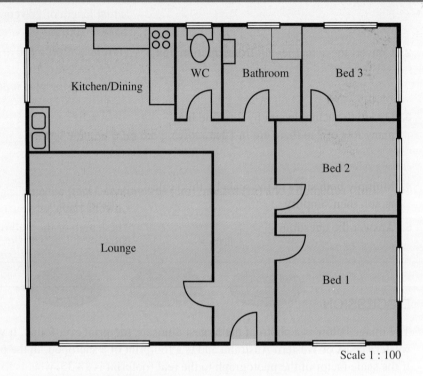

Scale 1 : 100

THINK	WRITE
a. 1. Measure the length and width of the house on the plan.	a. Length of house on plan = 12 cm Width of house on plan = 10 cm
2. Multiply each of these measurements by 100.	Actual length of house = 12 cm × 100 = 1200 cm = 12 m Actual width = 10 cm × 100 = 1000 cm = 10 m
3. Write your answer.	The dimensions of the house are 12 m by 10 m.
b. 1. Measure the length and width of the lounge room on the plan.	b. Length of lounge room on plan = 6 cm Width of lounge room on plan = 6 m
2. Multiply each of these measurements by 100.	Actual length of lounge room = 6 cm × 100 = 600 cm = 6 m The actual width of the lounge room is also 6 m.

3. Calculate the area of the lounge room.
$$\text{Area} = 6^2$$
$$= 36\,\text{m}^2$$

4. Write your answer.

The area of the lounge room is $36\,\text{m}^2$.

COMMUNICATING — COLLABORATIVE TASK: Using technology to construct scale drawings

With an understanding of scale drawings, you should now be able to draw plans of objects familiar to you, such as your home, bedroom or classroom.

Using software such as Google SketchUp, in pairs create a 2D or 3D model of your classroom or another familiar building of your choice.

 Resources

Interactivities Proportion (int-3735)
Introduction to ratios (int-3733)
Scale factors (int-6041)
Scales (int-4146)
House plans (int-4147)

Exercise 12.2 Ratio and scale

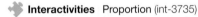 **learn** on

12.2 Quick quiz on	12.2 Exercise

Individual pathways

■ PRACTISE	■ CONSOLIDATE	■ MASTER
1, 3, 5, 10, 13, 16, 17, 18, 22, 24, 27, 29	2, 4, 6, 8, 11, 15, 19, 21, 23, 26, 28, 30	7, 9, 12, 14, 20, 25, 31, 32

Fluency

1. The horse track shown is 1200 m long and 35 m wide.

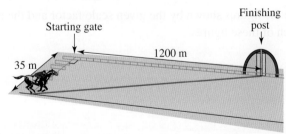

a. Write the following ratios in simplest form.
   i. Track length to track width
   ii. Track width to track length

b. Compare the distance of the length of the track with the width of the track.

2. A dingo perched on top of a cliff spots an emu on the ground below.

   a. Write the following ratios in simplest form.

      i. Cliff height to distance between cliff base and emu
      ii. Distance between emu and cliff base to cliff height

   b. Compare the height of the cliff with the horizontal distance between the base of the cliff and the emu.

3. **WE2a** Express each of the following ratios in simplest form.

   a. $12 : 18$
   b. $8 : 56$
   c. $9 : 27$
   d. $14 : 35$

4. Express each of the following ratios in simplest form.

   a. $16 : 60$
   b. $200 : 155$
   c. $32 : 100$
   d. $800 : 264$

5. **WE2b** Express each of the following ratios in simplest form.

   a. $1.2 : 0.2$
   b. $3.9 : 4.5$
   c. $9.6 : 2.4$
   d. $18 : 3.6$

6. Express each of the following ratios in simplest form.

   a. $1.8 : 3.6$
   b. $4.4 : 0.66$
   c. $0.9 : 5.4$
   d. $0.35 : 0.21$

7. Express each of the following ratios in simplest form.

   a. $6 : 1.2$
   b. $12.1 : 5.5$
   c. $8.6 : 4$
   d. $0.07 : 14$

8. **WE2c** Write each of the following ratios in the simplest form.

   a. $1\frac{1}{2} : 2$
   b. $2 : 1\frac{3}{4}$
   c. $1\frac{1}{3} : 2$
   d. $1\frac{2}{5} : 1\frac{1}{4}$

9. Write each of the following ratios in the simplest form.

   a. $\frac{4}{7} : 2$
   b. $5 : 1\frac{1}{2}$
   c. $2\frac{3}{4} : 1\frac{1}{3}$
   d. $3\frac{5}{6} : 2\frac{1}{2}$

10. **WE3** Determine the value of the pronumeral in each of the following proportions.

    a. $a : 15 = 3 : 5$
    b. $b : 18 = 4 : 3$
    c. $24 : c = 3 : 4$

11. Determine the value of the pronumeral in each of the following proportions.

    a. $e : 33 = 5 : 44$
    b. $6 : f = 5 : 12$
    c. $3 : 4 = g : 5$

12. Determine the value of the pronumeral in each of the following proportions.

    a. $11 : 3 = i : 8$
    b. $7 : 20 = 3 : j$
    c. $15 : 13 = 12 : k$

For questions **13** to **15**, enlarge the figures shown by the given scale factor and the centre of enlargement marked with O. Show the image of each of these figures.

13. **WE4** Scale factor = 2

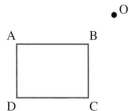

14. Scale factor = 1.5

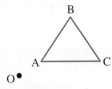

15. Scale factor = $\frac{1}{2}$

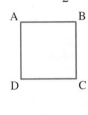

Understanding

16. **WE5** A quadrilateral ABCD is enlarged to A′B′C′D′. AB = 7 cm, AD = 4 cm, A′B′ = 21 cm, B′C′ = 10.5 cm. Determine:

   a. the scale factor for enlargement
   b. A′D′
   c. BC.

17. **WE6&7** If a scale of 1 : 50 is used to make each of the following drawings:

   a. what would be the width of the drawing of a house that is 25 m wide
   b. what would be the height of the drawing of a man who is 2 m tall
   c. what would be the dimensions of the drawing of a rectangular pool 50 m by 20 m
   d. what would be the dimensions of the drawing of a table 4 m by 1.5 m?

18. If a scale of 1 : 50 is used to make each of the following drawings:

   a. what is the actual height of a tree whose height in the drawing is 10 cm
   b. what is the actual length of a truck whose length in the drawing is 25 cm
   c. what are the dimensions of a television set whose dimensions on the drawing are 10 mm by 8 mm
   d. what are the dimensions of a bed whose dimensions on the drawing are 4.5 cm long by 3 cm wide by 1.2 cm high?

19. A rectangular room has a width of 3 m and a length of 5 m.

   a. What are the dimensions of the diagram on a scale of 1 : 100?
   b. What is the area of the rectangle on the diagram?
   c. The diagram has been enlarged so that the dimensions of the new diagram are double the dimensions of the first diagram. What is the area of the rectangle on the new diagram?
   d. Calculate the area of a rectangle with dimensions three times bigger than the original diagram.
   e. How does enlargement affect the area of a shape?

20. Athena the artist paints portraits. For her present subject, Seth, she uses the following measurement relationships.

   • Seth's arm span is the same length as his height.
   • The length of Seth's hand is one-tenth of his height.
   • The distance across Seth's shoulders is one-quarter his height.

What is the relationship between Seth's arm span, the length of his hand and the distance across his shoulders?

21. Mikaela wants to buy a new fridge that is 800 mm deep and 1 m wide. However, she is restricted by the kitchen space. The place for the fridge on her drawing of her flat has a depth of 1.5 cm and a width of 2 cm. If the scale of the drawing is 1 : 50, will this fridge fit in the space given? State the actual dimensions of the fridge space.

22. The estimated volume of Earth's salt water is 1 285 600 000 cubic kilometres. The estimated volume of fresh water is about 35 000 000 cubic kilometres.

　　a. Determine the ratio of fresh water to salt water (in simplest form).
　　b. Determine the value of x, to the nearest whole number, when the ratio found in part a is expressed in the form 1 : x.

23. Super strength glue comes in two tubes that contain 'Part A' and 'Part B' pastes. These pastes have to be mixed in the ratio 1 : 4 for maximum strength. Determine how many mL of Part A should be mixed with 10 mL of Part B.

24. A recipe for a tasty cake says the butter and flour need to be combined in the ratio 2 : 7. Determine the amount of butter that should be mixed with 3.5 kg of flour.

25. **WE8** The diagram shown lays out the floor plan of a house. The actual size of bedroom 1 is 8 m × 4 m.

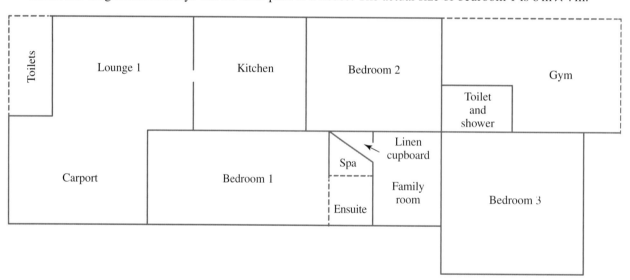

　　a. If the dimensions of bedroom 1 as it appears on the ground plan are 4 cm by 2 cm, calculate the scale factor when the actual house (object) is built from the plan (image).
　　b. Determine the real-life dimensions of bedroom 3 if the dimensions as shown on the ground plan are 3 cm × 3 cm.
　　c. Determine the real-life dimensions of the kitchen if the dimensions as shown on the ground plan are 2.5 cm × 2.5 cm.

26. Pure gold is classed as 24-carat gold. This kind of gold is too soft to use for making jewellery, so it gets combined with other metals to form an alloy.
    The ratio of gold to other metals in 18-carat gold is 18 : 6.
    The composition of 18-carat rose gold is 75% gold, 22.25% copper and 2.75% silver.

    a. Show that the mass of silver in a 2.5-gram rose gold bracelet is 0.07 g.
    b. Determine the composition of metals in a rose gold bracelet that contains 0.5 g of copper.

27. The angles of a triangle have the ratio 3 : 4 : 5. Show that the sizes of the three angles are 45°, 60° and 75°.

28. The dimensions of a rectangular box have the ratio 2 : 3 : 5. The box's volume is 21 870 cm³. Show that the dimensions of the box are 18, 27 and 45 cm.

29. Sharnee, a tourist at Kakadu National Park, takes a picture of a 2-metre-long crocodile beside a cliff. When they develop the pictures, they can see that on the photo the crocodile is 2.5 cm long and the cliff is 8.5 cm high.
    Determine the actual height of the cliff in cm.

30. The quantities P and Q are in the ratio 2 : 3. If P is reduced by 1, the ratio becomes $\frac{1}{2}$. Determine the values of P and Q.

31. The ratio of boys to girls among the students who signed up for a basketball competition is 4 : 3.
    If 3 boys drop out of the competition and 4 girls join, there will be the same number of boys and girls.
    Evaluate the number of students who have signed up for the basketball competition.

32. In a group of students who voted in a Year 9 class president election, the ratio of girls to boys is 2 : 3.
    If 10 more girls and 5 more boys had voted, the ratio would have been 3 : 4.
    Evaluate the number of students who voted altogether.

LESSON
12.3 Similar figures

LEARNING INTENTION

At the end of this lesson you should be able to:
- identify and describe the properties of similar figures
- determine the scale factor or proportion between similar figures
- show that two triangles are similar using the appropriate similarity test.

12.3.1 Similar figures and similarity condition

eles-4732

- **Similar figures** are identical in shape, but different in size.
- The corresponding angles in similar figures are equal in size and the corresponding sides are in the same ratio.
- The symbol used to denote similarity is ~. When reading this symbol out loud we say, 'is similar to'.
- In the triangles shown △ABC is similar to △UVW. That is, △ABC~△UVW.
- The ratio of side lengths is known as the scale factor.

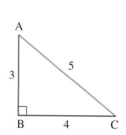
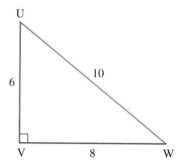

Scale factor

$$\text{scale factor} = \frac{\text{length of image}}{\text{length of object}}$$

- **If the scale factor is > 1, an enlargement has occurred.**
- **If the scale factor is < 1, a reduction has occurred.**

- The scale factor for the triangles shown is 2. Each side in UVW is twice the length of the corresponding side in ABC.
- Enlargements and reductions are transformations that create similar figures.
- The method for creating enlarged figures that is explained in lesson 12.2 can also be used to create similar figures.

WORKED EXAMPLE 9 Enlarging by a scale factor

Enlarge the shape shown by a factor of 2.

THINK	DRAW
1. Select a point, O, somewhere inside the given shape and join it with straight-line segments to each vertex. Extend the lines beyond the shape.	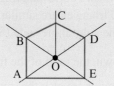
2. Measure the distance OA and mark in the point A′ so that $OA' = 2 \times OA$. Repeat this for the other vertices.	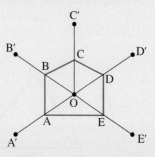
3. Join the image vertices A′B′C′D′E′ with straight lines.	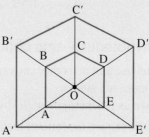

▶ 12.3.2 Testing triangles for similarity (Path)

eles-6167

- To determine whether two triangles are similar, it is not necessary to know that all pairs of corresponding sides are in the same ratio and that all corresponding angles are equal.
- Certain minimum conditions can guarantee that two triangles are similar.

Angle–angle–angle condition of similarity (AAA)

- If two angles of a triangle are equal to two angles of another triangle, then they are similar as the third angles must also be equal.
- This is known as the angle–angle–angle (AAA) condition for similarity.
- In the diagram shown, $\triangle ABC \sim \triangle RST$ (AAA).

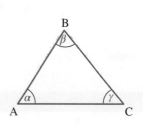
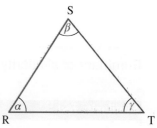

Side–side–side condition for similarity (SSS)

- If two triangles have a constant ratio for all corresponding side lengths, then the two triangles are similar.
- This is known as the side–side–side (SSS) condition for similarity.
- In the diagram shown, the ratios of all corresponding side lengths are equal $\left(\dfrac{9}{6} = \dfrac{15}{10} = \dfrac{10.5}{7} = 1.5\right)$; therefore, $\triangle ABC \sim \triangle RST$ (SSS).

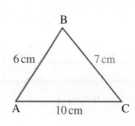
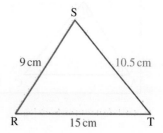

Side–angle–side condition for similarity (SAS)

- If two triangles have two corresponding sides in the same ratio and the included angles of those sides are equal, then the two triangles are similar.
- This is known as the side–angle–side (SAS) condition for similarity.
- In the diagram shown, the ratio of the triangles' two corresponding side lengths are equal $\left(\dfrac{9}{6} = \dfrac{15}{10} = 1.5\right)$ and the included angles are also the same; therefore, $\triangle ABC \sim \triangle RST$ (SAS).

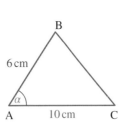 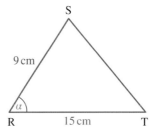

Right angle–hypotenuse–side condition for similarity (RHS)

- If the hypotenuse and one other corresponding side of two right-angled triangles are in the same ratio, then the two triangles are similar.
- This is known as the the right angle–hypotenuse–side (RHS) condition for similarity.
- In the diagram shown, the ratio of the hypotenuses and one other pair of corresponding sides are equal $\left(\dfrac{12}{6} = \dfrac{10}{5} = 2\right)$; therefore, $\triangle ABC \sim \triangle RST$ (RHS).

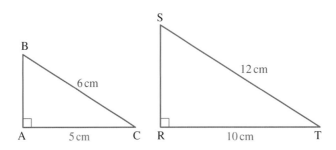

Summary of similarity tests

Similarity test	Description	Abbreviation
	• Two corresponding angles are equal in size	AAA (angle-angle-angle)

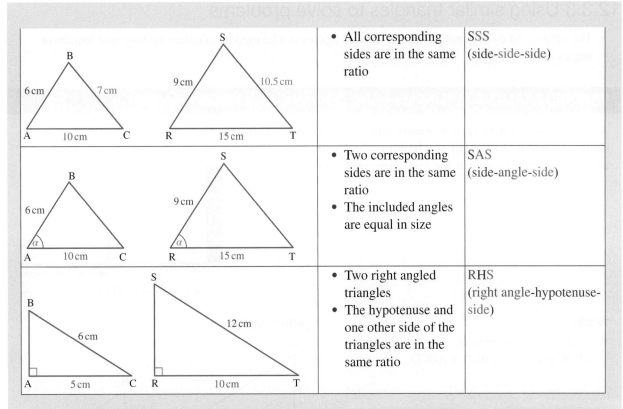

	• All corresponding sides are in the same ratio	SSS (side-side-side)
	• Two corresponding sides are in the same ratio • The included angles are equal in size	SAS (side-angle-side)
	• Two right angled triangles • The hypotenuse and one other side of the triangles are in the same ratio	RHS (right angle-hypotenuse-side)

- *Note:* When using the AAA test, it is sufficient to show that two corresponding angles are equal. Since the sum of the interior angles in any triangle is 180°, the third corresponding angle will automatically be equal.

PATH **WORKED EXAMPLE 10 Identifying similar triangles**

Identify a pair of similar triangles from the triangles shown. Give a reason for your answer.

a.

b.

c.
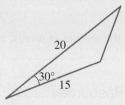

THINK

1. In each triangle we know the size of two sides and the included angle, so the SAS test can be applied. Since all included angles are equal (30°), we need to determine the ratios of the corresponding sides, looking at two triangles at a time.

2. Write the answer.

WRITE

For triangles **a** and **b**:
$$\frac{15}{10} = \frac{9}{6} = 1.5$$

For triangles **a** and **c**:
$$\frac{20}{10} = 2, \frac{15}{6} = 2.5$$

Triangle **a** ~ triangle **b** (SAS)

12.3.3 Using similar triangles to solve problems

eles-6246

- The ratio of the corresponding sides in similar figures can be used to calculate missing side lengths or angles in those figures.

WORKED EXAMPLE 11 Solving worded problems using similar triangles

A pole 1.5 metres high casts a shadow 3 metres long, as shown. Calculate the height of a building that casts a shadow 15 metres long at the same time of the day.

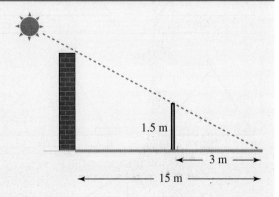

THINK	WRITE/DRAW
1. Represent the given information on a diagram. $\angle BAC = \angle EDC$; $\angle BCA = \angle ECD$	
2. Triangles ABC and DEC are similar. This means the ratios of corresponding sides are the same. Write the ratios.	$\triangle ABC \sim \triangle DEC$ (AAA) $$\frac{h}{1.5} = \frac{15}{3}$$
3. Solve the equation for h.	$$h = \frac{15 \times 1.5}{3}$$ $$= 7.5$$
4. Write the answer in words, including units.	The building is 7.5 metres high.

---

on Resources

▶ **Video eLesson** Similar triangles (eles-1925)

12.3 Quick quiz on	12.3 Exercise

Individual pathways

■ PRACTISE	■ CONSOLIDATE	■ MASTER
1, 2, 3, 7, 8, 12, 15, 16, 19	4, 5, 9, 13, 17, 20, 21	6, 10, 11, 14, 18, 22, 23, 24

Fluency

1. **WE9** Enlarge (or reduce) the following shapes by the scale factor given.

a.

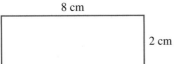

8 cm

2 cm

Scale factor = 3

b.

5 cm

Scale factor = 2

c.

2.5 m

4 m

Scale factor = 0.5

WE10 **PATH** For questions **2** to **6**, identify the pair of similar triangles among those shown. Give reasons for your answers.

2. a.

40° 60°

b.

50° 60°

c.

40° 60°

3. a.

4

3

2

b.

8

6

4

c.

7

5

4

4. a.

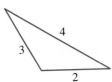

4

20° 5

b.

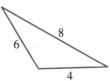

2

20° 2.5

c.

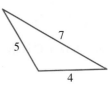

8

20° 12

5. a.

b.

c.

6. a.

b.

c.

Understanding

7. Name two similar triangles in each of the following figures, ensuring that vertices are listed in the correct order.

a.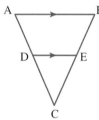

b.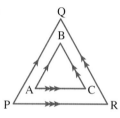

c.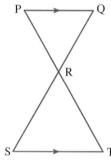

d.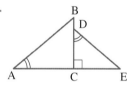

e.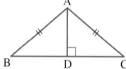

8. In the diagram shown, C is the centre of the circle. Complete this statement: △ABC is similar to …

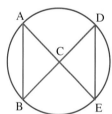

9. For the diagram shown:

a. complete the statement: $\dfrac{AB}{AD} = \dfrac{BC}{\square} = \dfrac{\square}{AE}$

b. determine the values of the pronumerals.

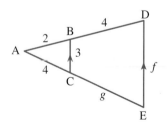

10. Using the diagram shown:

    a. determine the values of h and i

    b. determine the values of j and k.

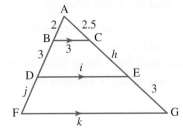

11. Determine the value of the pronumeral in the diagram shown.

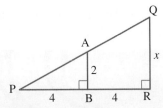

12. If the two triangles shown are similar, determine the values of the pronumerals x and y.

13. Determine the values of the pronumerals x and y in the diagram shown.

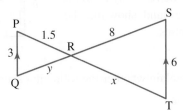

14. Determine the value of each pronumeral in the following triangles. Show how you arrived at your answers.

    a.

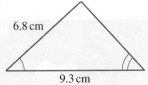

    b.

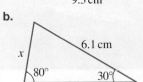

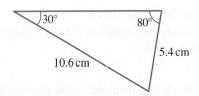

    c.

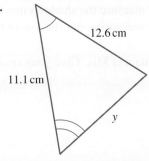

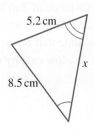

Communicating, reasoning and problem solving

15. **WE11** A ladder just touches a bench and also leans on a wall that is 4 metres high, as shown in the diagram.
    If the bench is 50 centimetres high and 1 metre from the base of the ladder, show that the base of the ladder is 8 metres from the wall.

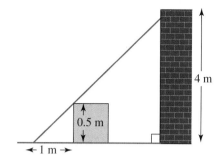

16. Natalie is 1.5 metres tall. They cast a shadow 2 metres long at a certain time of the day. If Alex is 1.8 metres tall, show that their shadow would be 2.4 metres long at the same time of day.

17. A string 50 metres long is pegged to the ground and tied to the top of a flagpole. It just touches Maz on the top of her head. If Maz is 1.5 metres tall and 5 metres away from the point where the string is held to the ground, show that the height, h, of the flagpole is 14.37 metres.

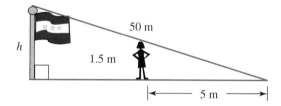

18. Using diagrams or other methods, explain whether the following statements are True or False.
    a. All equilateral triangles are similar.
    b. All isosceles triangles are similar.
    c. All right-angled triangles are similar.
    d. All right-angled isosceles triangles are similar.

19. Paw and Thuy play tennis at night under floodlights. When Paw stands 2.5 m from the base of the floodlight, her shadow is 60 cm long.

    a. If Paw is 1.3 m tall, evaluate the height of the floodlight in metres, correct to 2 decimal places.
    b. If Thuy, who is 1.6 m tall, stands in the same place, calculate her shadow length in cm.

20. To determine the height of a flagpole, Jenna and Mia decide to measure the shadow cast by the flagpole. They place a 1-m ruler at a distance of 3 m from the base of the flagpole and measure the shadows that both the ruler and flagpole cast. Both shadows finish at the same point. After measuring the shadow of the flagpole, Jenna and Mia calculate that the height of the flagpole is 5 m.
    Determine the length of the shadow cast by the flagpole, as measured by Jenna and Mia. Give your answer in metres.

21. Use the diagram shown to determine the value of a if $XZ = 8$ cm, $X'Z' = 12$ cm, $X'X = a$ cm and $XY = (a + 1)$ cm.

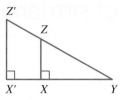

22. AB and CD are parallel lines in the figure shown.

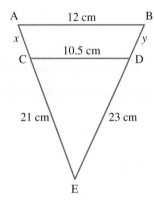

a. State the similar triangles.
b. Determine the values of x and y.

23. PQ is the diameter of the circle shown. The circle's centre is located at S. R is any point on the circumference. T is the midpoint of PR.

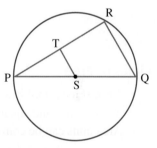

a. Write down everything you know about this figure.
b. Explain why ΔPTS is similar to ΔPRQ.
c. Determine the length of TS if RQ is 8 cm.
d. Determine the length of every other side, given that PT is 3 cm and the angle PRQ is a right angle.

24. For the diagram shown, show that, if the base of the triangle is raised to half of the height of the triangle, the length of the base of the newly formed triangle will be half of its original length.

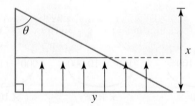

LESSON
12.4 Area and volume of similar figures (Path)

LEARNING INTENTION

At the end of this lesson you should be able to:
- convert between different units of length, area and volume
- determine the area of similar figures
- determine the volume of similar figures
- solve problems involving areas and volumes of similar shapes and solids.

▶ 12.4.1 Converting between units of length, area and volume

eles-4733
- Recall the conversions of length, area and volume from the previous topic.

Units of length
- Metric units of length include millimetres (mm), centimetres (cm), metres (m) and kilometres (km).
- Length units can be converted using the chart shown.

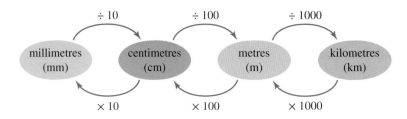

Units of area
- **Area** is measured in square units, such as square millimetres (mm^2), square centimetres (cm^2), square metres (m^2) and square kilometres (km^2).
- Area units can be converted using the chart shown.

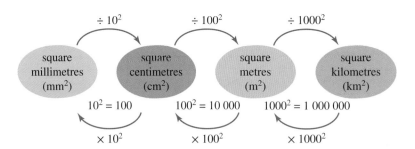

- Area units are the squares of the corresponding length units.

Units of volume

- **Volume** is measured in cubic units, such as cubic millimetres (mm^3), cubic centimetres (cm^3) and cubic metres (m^3).
- Volume units can be converted using the chart shown.

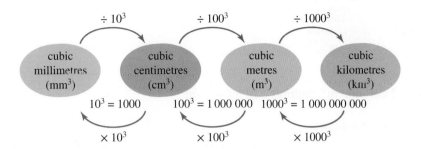

- Volume units are the cubes of the corresponding length units.

WORKED EXAMPLE 12 Converting between units of measurements

a. Convert 9 m into mm.
b. Convert 150 cm² into mm².
c. A cube has a side length of 8 cm. Calculate the volume of the cube in m³.

THINK	WRITE
a. 1. To convert m to mm:	**a.** $1\,m = 1000\,mm$

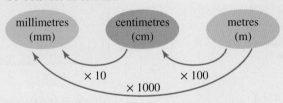

2. To convert 9 m to mm, multiply 9 with 1000.	$9\,m = 9 \times 1000\,mm$
3. Simplify and write the answer.	$9000\,mm$
b. 1. To convert cm² into mm²:	**b.** $1\,cm^2 = 100\,mm^2$

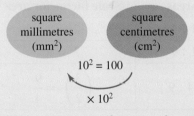

2. To convert 150 cm² into mm², multiply 150 by 10^2 (100).	$150\,cm^2 = 150 \times 100\,mm^2$
3. Simplify and write the answer.	$15\,000\,mm^2$
c. 1. Write the formula for the volume of a cube.	**c.** Volume of a cube $(V) = l \times l \times l$, where l is the side length.
2. Substitute the value of the side length (l) in the volume formula.	$V = 8 \times 8 \times 8$ $= 512\,cm^3$

3. The question states that the answer should be given in m³. To convert cm³ into m³:

$$1\,cm^3 = \frac{1}{100^3}\,m^3$$

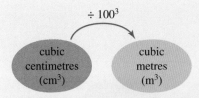

4. To convert 512 cm³ into m³, multiply 512 by $\frac{1}{100^3}$ m³.

$$512\,cm^3 = \frac{512}{100^3}\,m^3$$

5. Simplify and write the answer.

$0.000\,512\,m^3$

eles-4734

▶ 12.4.2 Area and volume of similar figures

Area and surface area of similar figures

- If the side lengths of similar figures are in the ratio of $a : b$, then the area of the similar figures is in the ratio $a^2 : b^2$.

  For example, consider the squares shown.

A

2 cm

B
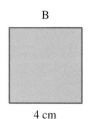
4 cm

C

6 cm

Area A $= 2 \times 2$
$ = 4\,cm^2$

Area B $= 4 \times 4$
$ = 16\,cm^2$

Area C $= 6 \times 6$
$ = 36\,cm^2$

- The scale factors for the side lengths and the scale factors for the areas are calculated in the table shown.

Squares	Ratio of lengths	Ratio of areas	Scale factor for area
A : B	$2 : 4 = 1 : 2$	$4 : 16 = 1 : 4$	$2^2 = 4$
A : C	$2 : 6 = 1 : 3$	$4 : 36 = 1 : 9$	$3^2 = 9$
B : C	$4 : 6 = 2 : 3$	$16 : 36 = 4 : 9$	$\left(\dfrac{3}{2}\right)^2 = \dfrac{9}{4}$

- The surface area of a 3D object also increases by the square of the length scale factor. For example, consider the cubes shown.

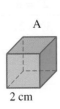

A

2 cm

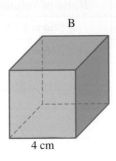

B

4 cm

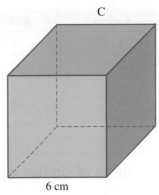

C

6 cm

Surface area A $= 6 \times 4$
$= 24 \, \text{cm}^2$

Surface area B $= 6 \times 16$
$= 96 \, \text{cm}^2$

Surface area C $= 6 \times 36$
$= 216 \, \text{cm}^2$

- The scale factors for the side lengths and the scale factors for the surface areas are calculated in the table shown.

Cubes	Ratio of lengths	Ratio of surface areas	Scale factor for surface area
A : B	$2 : 4 = 1 : 2$	$24 : 96 = 1 : 4$	$2^2 = 4$
A : C	$2 : 6 = 1 : 3$	$24 : 216 = 1 : 9$	$3^2 = 9$
B : C	$4 : 6 = 2 : 3$	$96 : 216 = 4 : 9$	$\left(\dfrac{3}{2}\right)^2 = \dfrac{9}{4}$

Areas of similar figures

When side lengths are increased by a factor of n, the area increases by a factor of n^2.

Volume of similar figures

- If the side lengths of any solid are in the ratio $a : b$, then the volume of similar solids is in the ratio $a^3 : b^3$. For example, consider the cubes shown.

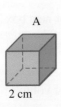

A

2 cm

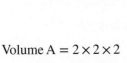

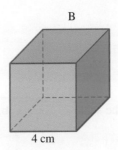

B

4 cm

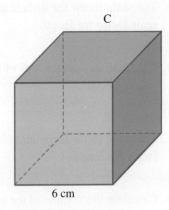

C

6 cm

Volume A $= 2 \times 2 \times 2$
$= 8 \, \text{cm}^3$

Volume B $= 4 \times 4 \times 4$
$= 64 \, \text{cm}^3$

Volume C $= 6 \times 6 \times 6$
$= 216 \, \text{cm}^3$

- The scale factors for the side lengths and the scale factors for the volumes are calculated in the table shown.

Cubes	Ratio of lengths	Ratio of volumes	Scale factor for volume
A : B	$2 : 4 = 1 : 2$	$8 : 64 = 1 : 8$	$2^3 = 8$
A : C	$2 : 6 = 1 : 3$	$8 : 216 = 1 : 27$	$3^3 = 27$
B : C	$4 : 6 = 2 : 3$	$64 : 216 = 8 : 27$	$\left(\dfrac{3}{2}\right)^3 = \dfrac{27}{8}$

Volumes of similar figures

When side lengths are increased by a factor of n, the volume increases by a factor of n^3. If the scale factor of similar figures' sides is $a : b$ then the scale factor of their volumes is $a^3 : b^3$.

WORKED EXAMPLE 13 Calculating area and volume of similar figures

The side lengths of a box have been increased by a factor of 3.

a. **Calculate the surface area of the new box if the original surface areawas 94 cm².**
b. **Determine the volume of the new box if the original volume was 60 cm³.**

THINK	WRITE
a. 1. State the scale factor for side length used to increase the size of the original box and express as a ratio.	a. Scale factor for side length $= 3$ Ratio $= 1 : 3$
2. The scale factor for surface area is the square of the scale factor for length.	Scale factor for surface area ratio $= 1^2 : 3^2$ $\qquad = 1 : 9$
3. Calculate the surface area of the new box.	Surface area of new box $= 94 \times 9$ $= 846 \text{ cm}^2$
b. 1. The scale factor for volume is the cube of the scale factor for length.	b. Scale factor for volume ratio $= 1^3 : 3^3$ $\qquad = 1 : 27$
2. Calculate the volume of the new box.	Volume of new box $= 60 \times 27$ $= 1620 \text{ cm}^3$

A small cone has a radius of 5 cm and a height of 15 cm.
a. If the scale factor is 4, determine the dimensions of a larger similar cone.
b. Giving your answers both in exact form and correct to 3 decimal places, calculate the volume of:
   i. the smaller cone   ii. the larger cone.
c. Show that the volumes of the two cones are in the ratio 1 : 64.

THINK	WRITE
a. The scale factor is 4, so multiply the radius and height by 4. Write the answers.	a. For the larger cone: radius $= 5\,\text{cm} \times 4 = 20\,\text{cm}$ height $= 15\,\text{cm} \times 4 = 60\,\text{cm}$
b. i. Use your calculator to work out the volume of the smaller cone by substituting the following values into the formula: $r = 5$ $h = 15$.	b. i. $V_s = \dfrac{1}{3}\pi r^2 h$ $= \dfrac{1}{3}\pi \times 5^2 \times 15$ $= 125\pi\,\text{cm}^3$ $\approx 392.699\,\text{cm}^3$ (to 3 d.p)
ii. Use your calculator to work out the volume of the larger cone (V_L) by substituting the following values into the formula: $r = 20$ $h = 60$.	ii. $V_L = \dfrac{1}{3}\pi r^2 h$ $= \dfrac{1}{3}\pi \times 20^2 \times 60$ $= 8000\pi$ $\approx 25\,132.741\,\text{cm}^3$ (to 3 d.p)
c. Use your calculator to evaluate the ratio of the volumes of the smaller cone and the larger cone. $\dfrac{V_s}{V_L} = \dfrac{125\pi}{8000\pi}$	c. $\dfrac{V_s}{V_L} = \dfrac{1}{64}$ The volumes are in the ratio 1 : 64.

COMMUNICATING — COLLABORATIVE TASK: Paper sizes and magnification factors (shown on printers)

1. When enlarging an A4 sheet of paper to an A3 sheet of paper, the scale factor is 141%. Investigate and prove this scale factor.
2. When reducing an A3 sheet of paper to an A4 sheet of paper, what is the scale factor?
3. What is the scale factor when reducing an A3 sheet of paper to an A5 sheet of paper?

Discuss your findings as a class.

on Resources

  Interactivities   Units of length (int-3779)
                           Area of similar figures (int-6043)
                           Volume and surface area of similar figures (int-6044)

12.4 Quick quiz on	12.4 Exercise

Individual pathways

■ PRACTISE	■ CONSOLIDATE	■ MASTER
1, 3, 4, 8, 9, 13	2, 5, 7, 10, 14, 15	6, 11, 12, 16, 17, 18

Fluency

1. **WE13a** The side lengths of the following shapes have all been increased by a factor of 3. Copy and complete the table shown.

Original surface area	Enlarged surface area
$100 \, cm^2$	a.
$7.5 \, cm^2$	b.
$95 \, mm^2$	c.
d.	$918 \, cm^2$
e.	$45 \, m^2$
f.	$225 \, mm^2$

2. A rectangular box has a surface area of $96 \, cm^2$ and volume of $36 \, cm^3$. Calculate the volume and surface area of a similar box that has side lengths that are double the size of the original box.

3. **WE13b** The side lengths of the following shapes have all been increased by a factor of 3. Copy and complete the table shown.

Original volume	Enlarged volume
$200 \, cm^3$	a.
$12.5 \, cm^3$	b.
$67 \, mm^3$	c.
d.	$2700 \, cm^3$
e.	$67.5 \, m^3$
f.	$27 \, mm^3$

Understanding

4. The area of a bathroom as drawn on a house plan is $5 \, cm^2$. Calculate the area of the actual bathroom if the map has a scale of 1 : 100.

5. The area of a kitchen is $25 \, m^2$.

   a. **WE12** Convert $25 \, m^2$ to cm^2.
   b. Calculate the area of the kitchen as drawn on a plan if the scale of the plan is 1 : 120. (Give your answer correct to 1 decimal place.)

6. The volume of a swimming pool as it appears on its construction plan is $20 \, cm^3$. Determine the actual volume of the pool if the plan has a scale of 1 : 75.

7. The total surface area of an aeroplane's wings is $120 \, m^2$.
   a. Convert $120 \, m^2$ to cm^2.
   b. Calculate the total surface area of the wings of a scale model of the aeroplane if the model is built using the scale 1 : 80.

Communicating, reasoning and problem solving

8. A triangle ABC maps to triangle $A'B'C'$ under an enlargement.
   $AB = 7 \, cm$, $AC = 5 \, cm$, $A'B' = 21 \, cm$, $B'C' = 30 \, cm$.
   a. Show that the scale factor for enlargement is 3.
   b. Determine BC.
   c. Determine $A'C'$.
   d. If the area of $\triangle ABC$ is $9 \, cm^2$, show that the area of $\triangle A'B'C'$ is $81 \, cm^2$.

9. A pentagon has an area of $20 \, cm^2$. If all the side lengths are doubled, show that the area of the enlarged pentagon is $80 \, cm^2$.

10. Two rectangles are similar. If the width of the first rectangle is twice of width of the other, prove that the ratio of their areas is 4 : 1.

11. A cube has a surface area of $253.5 \, cm^2$. (Give answers correct to 1 decimal place where appropriate.)
    a. Show that the side length of the cube is $6.5 \, cm$.
    b. Show that the volume of the cube is $274.625 \, cm^3$.
    c. Determine the volume of a similar cube that has side lengths twice as long.
    d. Determine the volume of a similar cube that has side lengths half as long.
    e. Determine the surface area of a similar cube that has side lengths one third as long.

12. In the diagram shown, a light is shining through a hole, resulting in a circular bright spot with a radius of $5 \, cm$ on the screen. The hole is $10 \, mm$ wide.

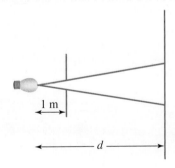

If the light is 1 m behind the hole, show that the light is $10 \, m$ from the screen.

13. The areas of two similar trapeziums are 9 and 25. Determine the ratio of one pair of these trapeziums' corresponding side lengths.

14. Answer the following questions.
    a. Calculate the areas of squares with sides 2 cm, 5 cm, 10 cm and 20 cm.
    b. State in words how the ratio of the areas of these squares is related to the ratio of their side lengths.

15. Two cones are similar. The ratio of these cones' volumes is 27 : 64. Determine the ratio of:
    a. the perpendicular heights of the cones
    b. the areas of the bases of the cones.

16. Rectangle A has dimensions 5 by 4 units, rectangle B has the dimensions 4 by 3 units, and rectangle C has the dimensions 3 by 2.4 units.
    a. Determine which of these rectangles are similar. Explain your answer.
    b. Evaluate the area scale factor for the similar rectangles that you have identified.

17. A balloon in the shape of a sphere has an initial volume of $840\,\text{cm}^3$. Its volume is then increased to $430\,080\,\text{cm}^3$. Determine the increase in the radius of the balloon.

18. **WE14** The bottom half of an egg timer, which is shaped like two cones connected at their apexes, has sand poured into it as shown by the blue section of this diagram.

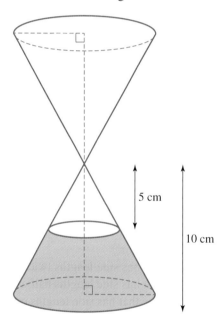

Using the measurements given, evaluate the ratio of the volume of sand in the bottom half of the egg timer to the volume of empty space that is left in the bottom half of the egg timer. You could also use technology to evaluate the answer.

LESSON
12.5 Review

12.5.1 Topic summary

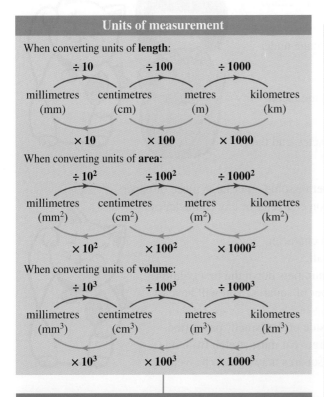

Units of measurement

When converting units of **length**:

$\div 10$ $\div 100$ $\div 1000$

millimetres   centimetres   metres   kilometres
(mm)   (cm)   (m)   (km)

$\times 10$ $\times 100$ $\times 1000$

When converting units of **area**:

$\div 10^2$ $\div 100^2$ $\div 1000^2$

millimetres   centimetres   metres   kilometres
(mm^2)   (cm^2)   (m^2)   (km^2)

$\times 10^2$ $\times 100^2$ $\times 1000^2$

When converting units of **volume**:

$\div 10^3$ $\div 100^3$ $\div 1000^3$

millimetres   centimetres   metres   kilometres
(mm^3)   (cm^3)   (m^3)   (km^3)

$\times 10^3$ $\times 100^3$ $\times 1000^3$

Ratio and scale factors

- Ratios compare quantities of the same type.
- Always simplify ratios. For example: $8 : 24 = 1 : 3$
- Scale factor $= \dfrac{\text{image length}}{\text{object length}}$.
- For a scale factor of n:
  - if $n > 1$, the image is larger than the object
  - if $n > 1$, the image is smaller than the object.
- House plans are drawn with a ratio as the scale factor.
- Using the scale factor, we can calculate the dimensions of the house from the plans.

PROPERTIES OF GEOMETRICAL FIGURES

Similar figures

- Similar figures have the same shape but different sizes.
- The symbol for similarity is ~.

Area and volume of similar figures (Path)

If the scale factor of similar figures' sides is $a : b$ then:
- the scale factor of their areas is $a^2 : b^2$
- the scale factor of their volumes is $a^3 : b^3$

Tests for similar triangles (Path)

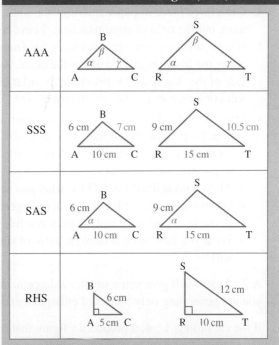

AAA	(triangles with angles α, β, γ)
SSS	6 cm, 7 cm, 10 cm and 9 cm, 10.5 cm, 15 cm
SAS	6 cm, 10 cm, α and 9 cm, 15 cm, α
RHS	6 cm, 5 cm and 12 cm, 10 cm

- *Note:* Sides of similar triangles are not equal. They are proportional, or have the same scale factor.

12.5.2 Project

Enlargement activity

Enlargement is the construction of a bigger picture from a small one.
The picture is identical to the other except that it is bigger.
The new picture is often called the image. This can also be called creating a similar figure.

The geometrical properties shared by a shape and its image under enlargement can be listed as:

- lines are enlarged as lines
- sides are enlarged to corresponding sides by the same factor
- matching angles on the two shapes are equal.

In this activity, we will start with a small cartoon character, and then 'blow it up' to almost life-size.

Equipment: ruler, pencil, cartoon print, butcher's paper or some other large piece of paper.

1. Do some research on the internet and select a cartoon character or any character of your choice.
2. Draw a grid of 2-cm squares over the small cartoon character.
   Example: The cat is 9 squares wide and 7 squares tall.
3. Label the grids with letters across the top row and numbers down the first column.
4. Get a large piece of paper and draw the same number of squares. You will have to work out the ratio of similitude (e.g. 2 cm : 8 cm).
5. If your small cartoon character stretches from one side of the 'small' paper (the paper the image is printed on) to the other, your 'large' cat must stretch from one side of the 'big' paper to the other. Your large grid squares may have to be 8 cm by 8 cm or larger, depending on the paper size.
6. Draw this enlarged grid on your large paper. Use a metre ruler or some other long straight-edged tool. Be sure to keep all of your squares the same size.
   - At this point, you are ready to draw. Remember, you do NOT have to be an artist to produce an impressive enlargement.
   - All you do is draw EXACTLY what you see in each small cell into its corresponding large cell.
   - For example, in cell B3 of the cat enlargement, you see the tip of his ear, so draw this in the big grid.
   - If you take your time and are very careful, you will produce an extremely impressive enlargement.
   - What you have used is called a 'ratio of similitude'. This ratio controls how large the new picture will be.

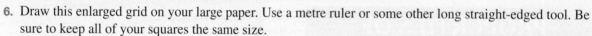

A 2 : 5 ratio will give you a smaller enlargement than a 2 : 7 ratio, because for every 2 units on the original you are generating only 5 units of enlargement instead of 7.

If the cat ratio is 1 : 4, it produces a figure that has a linear measure that is four times bigger.

The big cat's overall **area**, however, will be **16 times larger** than the small cat's. This is because area is found by taking length times width.

The length is 4 times longer and the width is 4 times longer. Thus, the **area** is $4 \times 4 = 16$ times **larger** than the original cat.

The overall **volume** will be $4 \times 4 \times 4$ or **64 times larger!** This means that the big cat will weigh 64 times more than the small cat.

Exercise 12.5 Review questions

Fluency

1. Express each of the following ratios in simplest form.
   a. 8 : 16   b. 24 : 16   c. 27 : 18

2. Express each of the following ratios in simplest form.
   a. 56 : 80   b. 8 : 20   c. 49 : 35

3. **MC** There are 9 girls and 17 boys in a Year 9 maths class. Select the ratio of boys to girls.
   A. 9 : 17   B. 17 : 9   C. 17 : 26   D. 9 : 26

4. Jan raised $15 for a charity fundraising event, while her friend Lara raised $25. Calculate the ratio of:
   a. the amount Jan raised to the amount Lara raised
   b. the amount Jan raised to the total amount raised by the pair
   c. the amount the pair raised to the amount Lara raised.

5. A pigeon breeder has 45 pigeons. These include 15 white, 21 speckled pigeons, and the rest grey.
   Calculate the following ratios.
   a. White pigeons to grey pigeons
   b. Grey pigeons to speckled pigeons
   c. White pigeons to the total number of pigeons

6. Express each of the following ratios in simplest form.
   a. $4\frac{1}{2} : 1$ b. $6\frac{1}{8} : 9$ c. $7\frac{1}{4} : 3\frac{1}{2}$

7. Express each of the following ratios in simplest form.
   a. 8.4 : 7.2   b. 0.2 : 2.48   c. 6.6 : 0.22

8. **MC** Jack has completed 2.5 km of a 4.5 km race. Choose the ratio of the distance Jack has completed to the remaining distance he has left to run.
   A. 9 : 5   B. 4 : 5   C. 4 : 9   D. 5 : 4

9. **MC** If $x : 16 = 5 : 4$, select the value of the pronumeral x.
   A. 1   B. 4   C. 16   D. 20

10. **MC** If $40 : 9 = 800 : y$, select the value of the pronumeral y.
    A. 20   B. 40   C. 180   D. 450

11. Determine the value of the pronumeral in each of the following.
    a. $a : 15 = 2 : 5$ b. $b : 20 = 5 : 8$ c. $9 : 10 = 12 : c$

12. Determine the value of the pronumeral in each of the following.
    a. $11 : 9 = d : 5$ b. $7 : e = 4 : 5$ c. $3 : 4 = 8 : f$

Understanding

13. $\Delta PQR \sim \Delta DEF$. Determine the length of the missing side in each of the following combinations.
    a. Calculate DF if PQ = 10 cm, DE = 5 cm and PR = 6 cm
    b. Calculate EF if PQ = 4 cm, DE = 12 cm and QR = 5 cm
    c. Calculate QR if DE = 4 cm, PQ = 6 cm and EF = 8 cm
    d. Calculate PQ if DF = 5 cm, PR = 8 cm and DE = 6 cm
    e. Calculate DE if QR = 16 cm, EF = 6 cm and PQ = 12 cm

14. Copy each of the following shapes and enlarge (or reduce) them by the given factor.

    a.
    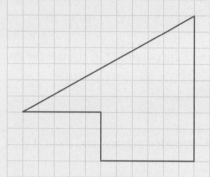
    Enlarge by a factor of 2.

    b.
    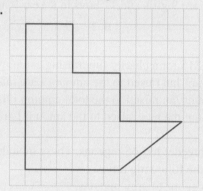
    Reduce by a factor of 3.

    c.
    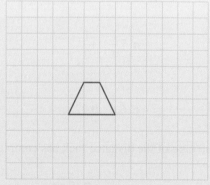
    Enlarge by a factor of 4.

15. Determine the enlargement factors that have been used on the following shapes.

a.

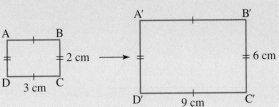

b.

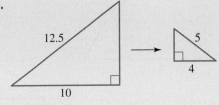

c.

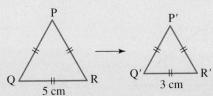

d.

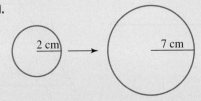

16. The scale on a map is given as 1 : 100 000. Find the actual distances between:
    a. two schools that are 5 cm apart on the map
    b. two parks that are 2.5 cm apart on the map
    c. two farms that are 4 cm apart on the map.

17. A scale of 1 : 200 is used to make models of a tree and a car.
    a. What would be the length of the model car if the car is actually 2.5 metres long?
    b. What would be the height of the model tree if the tree is actually 6.5 metres tall?

18. The area of a family room is 16 m^2 and the length of the room is 6.4 m. Calculate the area of the room if it is drawn on a plan that uses a scale of 1 : 20.

19. Each of the diagrams shown shows a pair of similar triangles. Calculate the value of x in each case.

a.

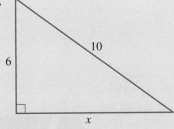

b.

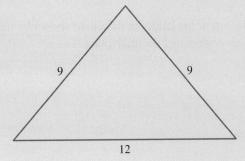

c.

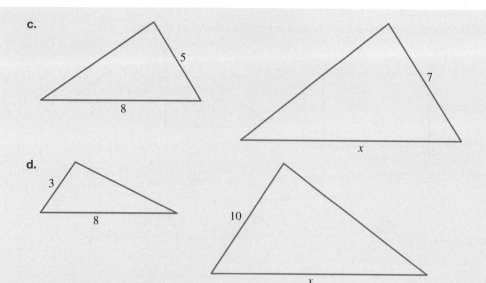

d.

Communicating, reasoning and problem solving

20. The diagram shows a ramp made by Jinghua for her automotive class. The first post has a height of 0.25 m and is placed 2 m from the end of the ramp.
    If the second post is 1.5 m high, determine the distance it should be placed from the first post.

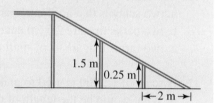

21. Consider the figure shown.

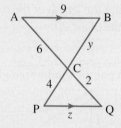

    Determine the values of the pronumerals.

22. *Slocum* is a yacht with a length of 12 m and a beam (width) of 2.5 m. A model of *Slocum* is constructed for a museum. If the length and width of the model are one fifth of the yacht's actual length and width, determine how the volume of the yacht and the volume of the model compare.

23. Calculate the height of the top of the ladder in the photo shown by using similar triangles. Give your answer correct to 1 decimal place.

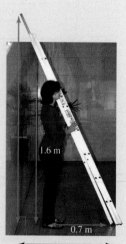

24. Poh is given a 1-m ruler and asked to estimate the height of a palm tree. She places the ruler vertically so that its shadow ends at exactly the same point as the shadow of the palm tree.
The ruler's shadow is 2.5 m long and the palm tree's shadow is 12.5 m long.

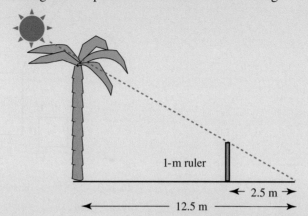

Poh performed some calculations using similar triangles and calculated the height of the palm tree to be 4 m. Her friend Mikalya said that she thought Poh's calculations were incorrect, and that the answer should be 5 m.

a. State the correct answer.

b. Explain the error that was made by the person with the incorrect answer.

25. A flagpole casts a shadow 2 m long. If a 50-cm ruler is placed upright at the base of the flagpole, it casts a shadow 20 cm long. Evaluate the height of the flagpole.

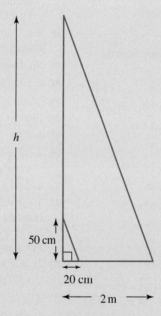

on To test your understanding and knowledge of this topic go to your learnON title at www.jacplus.com.au and complete the **post-test**.

Answers

Topic 12 Properties of geometrical figures

12.1 Pre-test

1. 25 : 39
2. $x = 13.5$
3. C
4. True
5. $y : z = 9 : 8$
6. False
7. $\dfrac{32}{3}$
8. 35 mm
9. A
10. B, D
11. $x = 2.3$ cm
12. $b = 7.5$
13. a. 3 : 5   b. 9 : 25
14. 9 : 1
15. B

12.2 Ratio and scale

1. a. i. 240 : 7   ii. 7 : 240

   b. The track is $34\dfrac{2}{7}$ times as long as it is wide.

2. a. i. 5 : 2   ii. 2 : 5

   b. The cliff is 2.5 times as high as the distance from the base of the cliff to the emu.

3. a. 2 : 3   b. 1 : 7   c. 1 : 3   d. 2 : 5
4. a. 4 : 15   b. 40 : 31   c. 8 : 25   d. 100 : 33
5. a. 6 : 1   b. 13 : 15   c. 4 : 1   d. 5 : 1
6. a. 1 : 2   b. 20 : 3   c. 1 : 6   d. 5 : 3
7. a. 5 : 1   b. 11 : 5   c. 43 : 20   d. 1 : 200
8. a. 3 : 4   b. 8 : 7   c. 2 : 3   d. 28 : 25
9. a. 2 : 7   b. 10 : 3   c. 33 : 16   d. 23 : 15
10. a. $a = 9$ b. $b = 24$ c. $c = 32$
11. a. $e = 3\dfrac{3}{4}$ b. $f = 14\dfrac{2}{5}$ c. $g = 3\dfrac{3}{4}$
12. a. $i = 29\dfrac{1}{3}$ b. $j = 8\dfrac{4}{7}$ c. $k = 10\dfrac{2}{5}$

13.

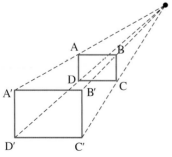

14.

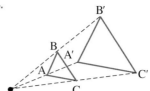

15.

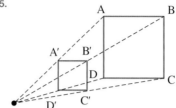

16. a. 3   b. 12 cm   c. 3.5 cm
17. a. 50 cm   b. 4 cm

    c. 100 cm × 40 cm   d. 8 cm × 3 cm
18. a. 5 m   b. 12.5 m

    c. 50 cm × 40 cm   d. 2.25 m × 1.5 m × 0.6 m
19. a. 3 cm and 5 cm

    b. 15 m^2

    c. 60 m^2

    d. 135 m^2

    e. If the dimensions double, the area becomes four times larger. If the dimensions are three times bigger, the area becomes nine times larger. If the dimensions are enlarged by a factor of n, the area is enlarged by a factor of n^2.
20. Seth's arm span is 10 times the length of his hand and 4 times the distance across his shoulders.
21. No. Depth is 750 cm, width is 1 m.
22. a. 175 : 6428   b. 37
23. 2.5 mL
24. 1000 g
25. a. 200   b. 6 m × 6 m   c. 5 m × 5 m
26. a. Sample responses can be found in the worked solutions in the online resources.

    b. 1.69 g gold, 0.5 g copper, 0.06 g silver
27. $3 + 4 + 5 = 12$

    $180 \div 12 = 15$

    $3 \times 15 = 45; 4 \times 15 = 60; 5 \times 15 = 75$

    The 3 angles are 45°, 60° and 75°.
28. $2k \times 3k \times 5k = 30k^3$

    $$30k^3 = 21870$$

    $$k^3 = 729$$

    $$k = 9$$

    Substituting k into the ratio $2k : 3k : 5k$, the dimensions are 18 cm, 27 cm and 45 cm.
29. 680 cm
30. $P = 4, Q = 6$
31. 49 students
32. 125 students

12.3 Similar figures

1. a.

24 cm

6 cm

b. 10 cm

c. 1.25 m

2 m

2. a and c, AAA

3. a and b, SSS

4. a and b, SAS

5. a and c, RHS

6. a and c, SSS

7. a. $\triangle ABC$ and $\triangle DEC$ **b.** $\triangle PQR$ and $\triangle ABC$
c. $\triangle PQR$ and $\triangle TSR$ **d.** $\triangle ABC$ and $\triangle DEC$
e. $\triangle ADB$ and $\triangle ADC$

8. a. $\triangle EDC$

9. a. $\dfrac{AB}{AD} = \dfrac{BC}{DE} = \dfrac{AC}{AE}$ **b.** $f = 9, g = 8$

10. a. $h = 3.75, i = 7.5$ **b.** $j = 2.4, k = 11.1$

11. $x = 4$

12. $x = 20°, y = 2\dfrac{1}{4}$

13. $x = 3, y = 4$

14. a. $x = 7.1$ **b.** $x = 3.1$ **c.** $x = 7.5, y = 7.7$

15. Sample responses can be found in the worked solutions in the online resources.

16. Sample responses can be found in the worked solutions in the online resources.

17. Sample responses can be found in the worked solutions in the online resources.

18. a. True **b.** False **c.** False **d.** True

19. a. 6.72 m **b.** 78 cm

20. 3.75 m

21. $a = 1$ cm

22. a. $\triangle EDC$ and $\triangle EBA$
b. $x = 3$ cm, $y \approx 3.29$ cm

23. a. Sample response can be found in the worked solutions in the online resources.
b. SAS
c. 4 cm
d. PR 6 cm, PS 5 cm and PQ 10 cm

24. The triangles are similar (AAA). $l = \dfrac{y}{2}$.

12.4 Area and volume of similar figures (Path)

1. a. 900 cm² **b.** 67.5 cm² **c.** 855 mm²
d. 102 cm² **e.** 5 m² **f.** 25 mm²

2. SA = 384 cm², V = 288 cm³

3. a. 5400 cm³ **b.** 337.5 cm³ **c.** 1809 mm³
d. 100 cm³ **e.** 2.5 m³ **f.** 1 mm³

4. 50 000 cm²

5. a. 250 000 cm² **b.** 17.4 cm²

6. 8 437 500 cm³

7. a. 1 200 000 cm² **b.** 187.5 cm²

8. a. Sample responses can be found in the worked solutions in the online resources.
b. 10 cm
c. 15 cm
d. Sample responses can be found in the worked solutions in the online resources.

9. Sample responses can be found in the worked solutions in the online resources.

10. Sample responses can be found in the worked solutions in the online resources.

11. a. Sample responses can be found in the worked solutions in the online resources.
b. Sample responses can be found in the worked solutions in the online resources.
c. 2197 cm³
d. 34.3 cm³
e. 28.2 cm²

12. Sample responses can be found in the worked solutions in the online resources.

13. 3 : 5

14. a. 4 cm², 25 cm², 100 cm², 400 cm²
b. The ratio of the areas is equal to the square of the ratio of the side lengths.

15. a. 3 : 4 **b.** 9 : 16

16. a. A and C are similar rectangles by the ratio 5 : 3 or scale factor $\dfrac{5}{3}$.
b. $\dfrac{25}{9}$

17. The new radius is 8 times the old radius.

18. The ratio is 7 : 1.

Project

Students will apply the knowledge of deductive geometry to enlarge a cartoon character to almost life-size.

12.5 Review

1. a. 1 : 2 **b.** 3 : 2 **c.** 3 : 2

2. a. 7 : 10 **b.** 2 : 5 **c.** 7 : 5

3. B

4. a. 3 : 5 **b.** 3 : 8 **c.** 8 : 5

5. a. 5 : 3 **b.** 3 : 7 **c.** 1 : 3

6. a. 9 : 2 **b.** 49 : 72 **c.** 29 : 14

7. a. 7 : 6 **b.** 5 : 62 **c.** 30 : 1

8. D

9. D

10. C

11. a. $a = 6$ **b.** $b = 12.5$ **c.** $c = 13\dfrac{1}{3}$

12. a. $d = 6\frac{1}{9}$ **b.** $e = 8\frac{3}{4}$ **c.** $f = 10\frac{2}{3}$

13. a. 3 cm   **b.** 15 cm   **c.** 12 cm
    d. 9.6 cm   **e.** 4.5 cm

14. a.

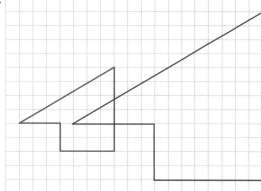

b.

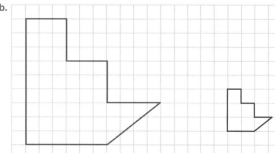

c.

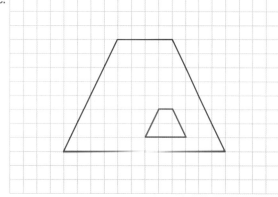

15. a. 3   **b.** 0.4   **c.** $\frac{3}{5}$ **d.** 3.5

16. a. 5 km   **b.** 2.5 km   **c.** 4 km

17. a. 1.25 cm   **b.** 3.25 cm

18. 400 cm^2

19. a. $x = 8$ **b.** $x = 4$ **c.** $x = 11\frac{1}{5}$ **d.** $x = 26\frac{2}{3}$

20. 10 m

21. $y = 12, z = 3$

22. The model is $\dfrac{1}{125}$ the volume of the yacht.

23. 2.1 m

24. a. 5 m

    b. Mikalya had the correct answer. Poh used the distance of 10 m (from the ruler to the tree) in her calculations instead of 12.5 m (the whole length of the tree's shadow).

25. 5 m

13 Data analysis

13.1 Overview

Why learn this?

According to the novelist Mark Twain, 'There are three kinds of lies: lies, damned lies and statistics.' Statistics can easily be used to manipulate people unless they have an understanding of the basic concepts involved.

Statistics, when used properly, can be an invaluable aid to good decision-making. However, deliberate distortion of the data or meaningless pictures can be used to support almost any claim or point of view. Whenever you read an advertisement, hear a news report or are given some data by a friend, you need to have a healthy degree of scepticism about the reliability of the source and nature of the data presented. A solid understanding of statistics is crucially important, as it is very easy to fall prey to statistics that are designed to confuse and mislead.

In 2020 when the COVID-19 pandemic hit, news and all forms of media were flooded with statistics. These statistics were used to inform governments worldwide about infection rates, recovery rates and all sorts of other important information. These statistics guided the decision-making process in determining the restrictions that were imposed or relaxed to maintain a safe community.

Statistics are also used to provide more information about a population in order to inform government policies. For example, the results of a census might indicate that the people in a particular city are fed up with traffic congestion. With this information now known, the government might prioritise works on public roads, or increase funding for public transport to try to create a more viable alternative to driving.

Hey students! Bring these pages to life online

- Watch videos
- Engage with interactivities
- Answer questions and check solutions

Find all this and MORE in jacPLUS

Reading content and rich media, including interactivities and videos for every concept

Extra learning resources

Differentiated question sets

Questions with immediate feedback, and fully worked solutions to help students get unstuck

1. The following data show the number of cars in each of the 12 houses along a street.

$$2, 3, 3, 2, 2, 3, 2, 4, 3, 1, 1, 0$$

   Calculate the median number of cars.

2. Calculate the range of the following data set: $5, 15, 23, 6, 31, 24, 26, 14, 12, 34, 18, 9, 17, 32$.

3. The frequency table shows the scores obtained by 100 professional golfers in the final round of a tournament.

Score	Frequency
67	2
68	6
69	7
70	11
71	16
72	23
73	17
74	11
75	9

   Identify the modal score.

4. A sample of 15 people was selected at random from those attending a local swimming pool. Their ages (in years) were recorded as follows:

$$19, 7, 83, 41, 17, 23, 62, 55, 15, 25, 32, 29, 11, 18, 10$$

   Calculate the mean age of people attending the swimming pool, correct to 1 decimal place.

5. Prepare a five-number summary for the following data:

$$7, 12, 14, 15, 16, 16, 17, 20$$

6. At Einstein Secondary School a Year 9 mathematics class has 22 students. The following were the test scores for the class.

$$34, 47, 54, 59, 60, 63, 66, 69, 73, 77, 78, 78, 79, 80, 82, 83, 85, 86, 88, 89, 90, 91$$

   Calculate the interquartile range (IQR).

7. The mean of a set of five scores is 11.8. If four of the scores are $17, 9, 14$ and 6, calculate the fifth score.

8. The box plot below shows the price of a meal for one person from ten fast-food shops.

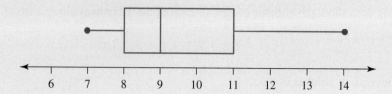

State whether the data is negatively skewed, positively skewed or symmetrical.

9. A frequency table for the time taken by 20 people to put together an item of flat-pack furniture is shown.

Time taken (min)	Frequency
0–4	1
5–95	3
10–14	5
15–19	2
20–24	4
25–29	2
30–34	2
35–39	1

Calculate the cumulative frequency to put together an item of flat-pack furniture in less than 20 minutes.

10. MC The frequency table below shows the scores obtained by 100 professional golfers in the final round of a tournament.

Score	Frequency
67	2
68	5
69	8
70	11
71	16
72	22
73	14
74	13
75	9

Select the median score.

A. 71   B. 71.75   C. 72   D. 72.5

11. The heights of six basketball players (in cm) are:

    178.1 185.6 173.3 193.4 183.1 193.0

    Calculate the mean and standard deviation, correct to 2 decimal places.

12. **MC** A group of 22 people recorded how many cans of soft drink they drank in a day. The table shows the number of cans drunk by each person.

0	2	2	2	1	1	3	4	4	2	1
2	4	1	6	3	3	5	4	1	2	5

    Select the statement that is not true.
    A. The maximum number of soft drinks cans drank is 6.
    B. The minimum number of soft drink cans drank is 0.
    C. The interquartile range is 3.
    D. The median number of soft drink cans is 2.5.

13. **MC** Select the approximate median in the cumulative frequency percentage graph shown.

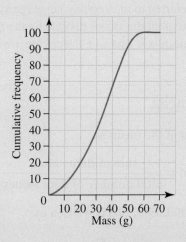

    A. 30   B. 34   C. 40   D. 50

14. The following table shows the typing speed in words per minute (wpm) of 30 Year 9 students.

14	22	30	41	50	60	42	19	23	33
53	60	26	34	45	55	61	28	29	35
48	57	37	49	38	49	38	58	35	48

    Calculate the mean typing speed and interquartile range for Year 9.

15. A netball team has played 10 games of netball so far this season. The scores are shown below.

    64, 68, 65, 71, 58, 66, 65, 77, 59, 57

    Calculate the standard deviation, correct to 2 decimal places.

LESSON
13.2 Measures of central tendency

LEARNING INTENTION

At the end of this lesson you should be able to:
- calculate the mean, median and mode of data presented as ungrouped data (in a single list), frequency distribution tables and grouped data.

▶ 13.2.1 Mean, median and mode

eles-4949

- **Measures of central tendency** are summary statistics that measure the middle (or centre) of the data. These are known as the mean, median and mode.
  - The **mean** is the average of all observations in a set of data.
  - The **median** is the middle observation in an ordered set of data.
  - The **mode** is the most frequent observation in a data set.

The mean

- The *mean* of a set of data is what is referred to in everyday language as the *average*.
- The mean of a set of values is the sum of all the values divided by the number of values.
- The symbol we use to represent the mean is $\bar{x}$; that is, a lower-case x with a bar on top (pronounced "x bar").

Calculating the mean

The formal definition of the mean is:

$$\text{Mean} = \frac{\text{sum of data values}}{\text{total number of data values}}$$

Using mathematical notation, this is written as:

$$\bar{x} = \frac{\sum x}{n}$$

The median

- The median represents the *middle* score when the data values are placed in ascending order.
- When there are an odd number of data values, the median is the middle value.

$$1 \qquad 1 \qquad 3 \qquad \boxed{4} \qquad 6 \qquad 7 \qquad 8$$

median = 4

- When there are an even number of data values, the median is the average of the two middle values.

$$2 \qquad 3 \qquad 3 \qquad \boxed{5} \qquad \boxed{6} \qquad 6 \qquad 7 \qquad 9$$

$$\text{median} = \frac{5+6}{2} = 5.5$$

Calculating the median

When calculating the median:
1. **Arrange the data values in order (usually in ascending order).**
2. **The *position* of the median is the $\left(\dfrac{n+1}{2}\right)$th data value, where n is the total number of data values.**

Note: **If there are an even number of data values then there will be two middle values. In this case, the median is the average of those data values.**

The mode

- The mode is the score that occurs most often.
- The data set can have no modes, one mode, two modes (**bimodal**) or more than two modes (**multimodal**).

Calculating the mode

When determining the mode:
1. **Arrange the data values in ascending order (smallest to largest). This step is optional but does help.**
2. **Look for the number that occurs most often (has the highest frequency).**

- If no value in a data set appears more than once then there is no mode.
- If a data set has multiple values that appear the most then it has multiple modes.

For example, the set 1, 2, 2, 4, 5, 5, 7 has two modes, 2 and 5.

WORKED EXAMPLE 1 Calculating mean, median and mode

For the data 6, 2, 4, 3, 4, 5, 4, 5, calculate the:

a. mean b. median c. mode.

THINK	WRITE
a. 1. Calculate the sum of the scores; that is, $\sum x$.	a. $\sum x = 6 + 2 + 4 + 3 + 4 + 5 + 4 + 5$ $= 33$
2. Count the number of scores; that is, n.	$n = 8$
3. Write the rule for the mean.	$\bar{x} = \dfrac{\sum x}{n}$
4. Substitute the known values into the rule.	$= \dfrac{33}{8}$
5. Evaluate.	$= 4.125$
6. Write the answer.	The mean is 4.125.

b. 1. Write the scores in ascending numerical order.

b. 2 3 4 4 4 5 5 6

2. Locate the position of the median using the rule $\dfrac{n+1}{2}$, where $n = 8$. This places the median as the 4.5th score; that is, between the 4th and 5th score.

$$\text{Median} = \dfrac{n+1}{2}\text{th score}$$

$$= \dfrac{8+1}{2}\text{th score}$$

$$= 4.5\text{th score}$$

2 3 **4 4** 4 5 6

3. Obtain the average of the two middle scores.

$$\text{Median} = \dfrac{4+4}{2}$$

$$= \dfrac{8}{2}$$

$$= 4$$

4. Write the answer

The median is 4.

c. 1. Systematically work through the set and make note of any repeated values (scores). 4 is repeated 3 times and 5 is repeated twice. As 4 is repeated the most, this will be our mode.

c.

$$\begin{array}{ccccccccc} & & & & & \downarrow & \downarrow & \\ 2 & 3 & 4 & 4 & 4 & 5 & 5 & 6 \\ & & \uparrow & \uparrow & \uparrow & & & \end{array}$$

2. Write the answer.

The mode is 4.

DIGITAL TECHNOLOGY

A scientific calculator is able to determine the mean of a data set.

Using the Statistics function:

Press Mode → 2 STAT → 1: 1-VAR

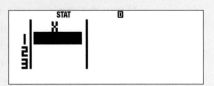

Enter each data value and press = Once complete, press AC.

Note: You will be able to see "STAT" at the top of the screen to know your data saved.

To access the statistics menu, press SHIFT → 1.

4: Var will bring up the variance menu.

2: $\bar{x}$ to calculate the mean.

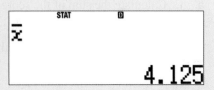

▶ 13.2.2 Calculating mean, median and mode from a frequency distribution table

- If data is provided in the form of a frequency distribution table we can determine the mean, median and mode using slightly different methods.
- The mode is the score with the highest **frequency**.
- To calculate the median, add a cumulative frequency column to the table and use it to determine the score that is the $\left(\dfrac{n+1}{2}\right)$ th data value.

- To calculate the mean, add a column that is the score multiplied by its frequency $f \times x$. The following formula can then be used to calculate the mean, where $\sum (f \times x)$ is the sum of the $(f \times x)$ column. $\sum$ is the uppercase Greek letter sigma.

> **Calculating the mean from a frequency table**
>
> $$\bar{x} = \frac{\sum (f \times x)}{n}$$

WORKED EXAMPLE 2 Calculations from a frequency distribution table

Using the frequency distribution table, calculate the:

a. **mean**   b. **median**   c. **mode.**

Score (x)	Frequency (f)
4	1
5	2
6	5
7	4
8	3
Total	15

THINK

1. Rule up a table with four columns titled Score (x), Frequency (f), Frequency × score ($f \times x$) and Cumulative frequency (cf).

2. Enter the data and complete both the $f \times x$ and cumulative frequency columns.

WRITE

Score (x)	Frequency (f)	Frequency × score ($f \times x$)	Cumulative frequency (cf)
4	1	4	1
5	2	10	$1 + 2 = 3$
6	5	30	$3 + 5 = 8$
7	4	28	$8 + 4 = 12$
8	3	24	$12 + 3 = 15$
	$n = 15$	$\sum (f \times x) = 96$	

a. **1.** Write the rule for the mean.

a. $\bar{x} = \dfrac{\sum (f \times x)}{n}$

2. Substitute the known values into the rule and evaluate.

$\bar{x} = \dfrac{96}{15}$

$= 6.4$

3. Write the answer.

The mean of the data set is 6.4.

b. **1.** Locate the position of the median using the rule $\dfrac{n+1}{2}$, where $n = 15$.

This places the median as the 8th score.

b. The median is the $\left(\dfrac{15+1}{2}\right)$th or 8th score.

2. Use the cumulative frequency column to find the 8th score and write the answer.

The median of the data set is 6.

c. **1.** The mode is the score with the highest frequency.

c. The score with the highest frequency is 6.

2. Write the answer.

The mode of the data set is 6.

DIGITAL TECHNOLOGY

A scientific calculator is able to determine the mean and median of a data set.

Using the Statistics function:

Press Mode → 2 STAT → 1: 1-VAR

To turn on the frequency column, press SHIFT → MODE → Down arrow → 3: STAT → 1: ON

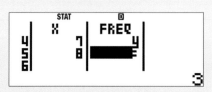

Enter each data value and press = Use the arrows to move around the table. Once complete, press AC.

```
1:Type    2:Data
3:Sum     4:Var
5:MinMax
```

To access the statistics menu, press SHIFT → 1.

Mean:

4: Var will bring up the variance menu.

2: $\bar{x}$ to calculate the mean.

```
1:n       2:x̄
3:σx      4:sx
```

Median:

5: MinMax

4: med

```
1:minX    2:maxX
3:Q1      4:med
5:Q3
```

13.2.3 Mean, median and mode of grouped data

eles-6168

- When the data are grouped into class intervals, the actual values (or data) are lost. In such cases, we have to approximate the real values with the midpoints of the intervals into which these values fall.
  For example, if in a grouped frequency table showing the heights of different students, four students had a height between 180 and 185 cm, we have to assume that each of those four students is 182.5 cm tall.

Mean

- The formula for calculating the mean is the same as the formula used when the data is displayed in a frequency distribution table:

$$\bar{x} = \frac{\sum(f \times x)}{n}$$

  Here, x represents the midpoint (or class centre) of each class interval, f is the corresponding frequency and n is the total number of observations in a set.

Median

- The median is found by drawing a **cumulative frequency** curve (ogive) of the data and estimating the median from the 50th percentile (see section 13.2.3).

Modal class

- The **modal class** is the class interval that has the highest frequency.

13.2.4 Cumulative frequency graphs (ogives)

eles-4951

- Data from a cumulative frequency table can be plotted to form a **cumulative frequency graph** (sometimes referred to as cumulative frequency polgons), which is also called an **ogive** (pronounced '*oh-jive*').
- To plot an ogive for data that is in class intervals, the maximum value for the class interval is used as the value against which the cumulative frequency is plotted.

For example, the following table and graph show the mass of cartons of eggs ranging from 55 g to 65 g.

Mass (g)	Frequency (f)	Cumulative frequency (cf)	Percentage cumulative frequency ($\%cf$)
55–<57	2	2	6%
57–<59	6	2 + 6 = 8	22%
59–<61	12	8 + 12 = 20	56%
61–<63	11	20 + 11 = 31	86%
63–<65	5	31 + 5 = 36	100%

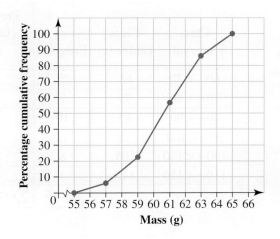

Quantiles

- An ogive can be used to divide the data into any given number of equal parts called **quantiles**.
- Quantiles are named after the number of parts that the data are divided into.
  - **Percentiles** divide the data into 100 equal-sized parts.
  - **Quartiles** divide the data into 4 equal-sized parts. For example, 25% of the data values lie at or below the first quartile.

Percentile	Quartile and symbol	Common name
25th percentile	First quartile, Q_1	Lower quartile
50th percentile	Second quartile, Q_2	Median
75th percentile	Third quartile, Q_3	Upper quartile
100th percentile	Fourth quartile, Q_4	Maximum

- A percentile is named after the percentage of data that lies at or below that value. For example, 60% of the data values lie at or below the 60th percentile.
- Percentiles can be read off a percentage cumulative frequency curve.
- A percentage cumulative frequency curve is created by:
  - writing the cumulative frequencies as a percentage of the total number of data values
  - plotting the percentage cumulative frequencies against the maximum value for each interval.

WORKED EXAMPLE 3 Estimating mean, median and modal class in grouped data

For the given data:
a. **estimate the mean** b. **estimate the median** c. **determine the modal class.**

Class interval	Frequency
$60 - <70$	5
$70 - <80$	7
$80 - <90$	10
$90 - <100$	12
$100 - <110$	8
$110 - <120$	3
Total	45

THINK

1. Draw up a table with 5 columns headed Class interval, Class centre (x), Frequency (f), Frequency × class centre ($f \times x$) and Cumulative frequency (cf).

2. Complete the x, $f \times x$ and cf columns.

WRITE

Class interval	Class centre (x)	Freq. (f)	Frequency × class centre ($f \times x$)	Cumulative frequency (cf)
$60 - <70$	65	5	325	5
$70 - <80$	75	7	525	12
$80 - <90$	85	10	850	22
$90 - <100$	95	12	1140	34
$100 - <110$	105	8	840	42
$110 - <120$	115	3	345	45
		$n = 45$	$\sum(f \times x) = 4025$	

a. 1. Write the rule for the mean.

a. $\bar{x} = \dfrac{\sum(f \times x)}{n}$

2. Substitute the known values into the rule and evaluate.

$\bar{x} = \dfrac{4025}{45}$

$\simeq 89.4$

3. Write the answer.

The mean for the given data is approximately 89.4.

b. 1. Draw a combined cumulative frequency histogram and ogive, labelling class centres on the horizontal axis and cumulative frequency on the vertical axis. Join the end-points of each class interval with a straight line to form the ogive.

b.

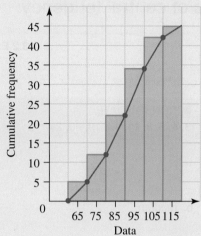

2. Locate the middle of the cumulative frequency axis, which is 22.5.

3. Draw a horizontal line from this point to the ogive and a vertical line to the horizontal axis.

4. Read off the value of the median from the *x*-axis and write the answer.

The median for the given data is approximately 90.

c. 1. The modal class is the class interval with the highest frequency.

c. The class internal 90–100 occurs twelve times, which is the highest frequency.

2. Write the answer.

The modal class is the 90–100 class interval.

Exercise 13.2 Measures of central tendency

learn on

13.2 Quick quiz on | **13.2 Exercise**

Individual pathways

■ PRACTISE	■ CONSOLIDATE	■ MASTER
1, 2, 7, 10, 14, 17, 18, 23	3, 4, 6, 8, 11, 15, 19, 20, 24	5, 9, 12, 13, 16, 21, 22, 25

Fluency

WE1 For questions **1** to **5**, calculate the:

a. mean **b.** median **c.** mode.

1. 3, 5, 6, 8, 8, 9, 10

2. 4, 6, 7, 4, 8, 9, 7, 10

3. 17, 15, 48, 23, 41, 56, 61, 52

4. 4.5, 4.7, 4.8, 4.8, 4.9, 5.0, 5.3

5. $7\frac{1}{2}$, $10\frac{1}{4}$, 12, $12\frac{1}{4}$, 13, $13\frac{1}{2}$, $13\frac{1}{2}$, 14

6. The stem-and-leaf plot below shows the test results of 25 Year nine students in Mathematics. Calculate the mean, median and mode.

Key: 3 | 2 = 32

Stem	Leaf:
3	2 9
4	0 6 8
5	1 3 5
6	2 6 7 9
7	3 6 7 8
8	0 4 4 6 8 9
9	2 5 8

WE2 Using the frequency distribution tables shown in questions **7** and **8**, calculate the:

a. mean   **b.** median   **c.** mode.

7.

Score (x)	Frequency (f)
4	3
5	6
6	9
7	4
8	2
Total	24

8.

Score (x)	Frequency (f)
12	4
13	5
14	10
15	12
16	9
Total	40

9. The following data show the number of bedrooms in each of the 10 houses in a particular neighbourhood:

$$2, 1, 3, 4, 2, 3, 2, 2, 3, 3.$$

a. Calculate the mean and median number of bedrooms.

b. A local motel contains 20 rooms. Add this observation to the set of data and recalculate the values of the mean and median.

c. Compare the answers obtained in parts **a** and **b** and complete the following statement:

When the set of data contains an unusually large value(s), called an outlier, the _____ (mean/median) is the better measure of central tendency, as it is less affected by this extreme value.

10. **WE3** For the given data:

a. estimate the mean   **b.** estimate the median   **c.** determine the modal class.

Class interval	Frequency
$40 - <50$	2
$50 - <60$	4
$60 - <70$	6
$70 - <80$	9
$80 - <90$	5
$90 - <100$	4
Total	30

11. Calculate the mean of the grouped data shown in the table below, correct to 2 decimal places.

Class interval	Frequency
100–109	3
110–119	7
120–129	10
130–139	6
140–149	4
Total	30

12. Determine the modal class of the data shown in the table below.

Class interval	Frequency
$50 - <55$	1
$55 - <60$	3
$60 - <65$	4
$65 - <70$	5
$70 - <75$	3
$75 - <80$	2
Total	18

13. The number of textbooks sold by various bookshops during the second week of December was recorded. The results are summarised in the table below.

Number of books sold	Frequency
220–229	2
230–239	2
240–249	3
250–259	5
260–269	4
270–279	4
Total	20

a. **MC** The modal class of the data is given by the class interval(s):

   A. 220–229 and 230–239   **B.** 250–259

   C. 260–269 and 270–279   **D.** of both A and

b. **MC** The class centre of the first class interval is:

   A. 224   **B.** 224.5   **C.** 224.75   **D.** 225

c. **MC** The median of the data is in the interval:

   A. 230–239   **B.** 240–249   **C.** 250–259   **D.** 260–269

d. **MC** The estimated mean of the data is:

   A. 251   **B.** 252   **C.** 253   **D.** 254

Understanding

14. A random sample was taken, composed of 30 people shopping at a supermarket on a Tuesday night. The amount of money (to the nearest dollar) spent by each person was recorded as follows:

    6, 32, 66, 17, 45,   1, 19, 52, 36, 23, 28, 20,   7, 47, 39,
    6, 68, 28, 54,   9, 10, 58, 40, 12, 25, 49, 74, 63, 41, 13

    a. Calculate the mean and median amount of money spent at the checkout by the people in this sample.
    b. Group the data into class intervals of 10 and complete the frequency distribution table. Use this table to estimate the mean amount of money spent.
    c. Add the cumulative frequency column to your table and fill it in. Hence, construct the ogive. Use the ogive to estimate the median.
    d. Compare the mean and the median of the original data from part a with the mean and the median obtained for grouped data in parts b and c. Explain if the estimates obtained in parts b and c were good enough.

15. Answer the following question and show your working.

    a. Add one more number to the set of data 3, 4, 4, 6 so that the mean of a new set is equal to its median.
    b. Design a set of five numbers so that mean = median = mode = 5.
    c. In the set of numbers 2, 5, 8, 10, 15, change one number so that the median remains unchanged while the mean increases by 1.

16. Thirty men were asked to reveal the number of hours they spent doing housework each week. The results are detailed below.

    1,   5,   2, 12,   2,   6,   2,   8, 14, 18,
    0,   1,   1,   8, 20, 25,   3,   0,   1,   2,
    7, 10, 12,   1,   5,   1, 18,   0,   2,   2

    a. Present the data in a frequency distribution table. (Use class intervals of 0–4, 5–9 etc.)
    b. Use your table to estimate the mean number of hours that the men spent doing housework.
    c. Determine the median class for hours spent by the men at housework.
    d. Identify the modal class for hours spent by the men at housework.

Communicating, reasoning and problem solving

17. **MC** In a set of data there is one score that is extremely small when compared to all the others. This outlying value is most likely to:

    A. have greatest effect upon the mean of the data
    B. have greatest effect upon the median of the data
    C. have greatest effect upon the mode of the data
    D. have very little effect on any of the statistics as we are told that the number is extremely small

18. The data shown give the age of 25 patients admitted to the emergency ward of a hospital.

18,	16,	6,	75,	24,
23,	82,	75,	25,	21,
43,	19,	84,	76,	31,
78,	24,	20,	63,	79,
80,	20,	23,	17,	19

a. Present the data in a frequency distribution table. (Use class intervals of $0 - <15$, $15 - <30$ and so on.)
b. Draw a histogram of the data.
c. Suggest a word to describe the pattern of the data in this distribution.
d. Use your table to estimate the mean age of patients admitted.
e. Determine the median class for age of patients admitted.
f. Identify the modal class for age of patients admitted.
g. Draw an ogive of the data.
h. Use the ogive to determine the median age.
i. Explain if any of your statistics (mean, median or mode) give a clear representation of the typical age of an emergency ward patient.
j. Give some reasons that could explain the pattern of the distribution of data in this question.

19. The batting scores for two cricket players over 6 innings are as follows:

Player A   31, 34, 42, 28, 30, 41
Player B   0, 0, 1, 0, 250, 0

a. Calculate the mean score for each player.
b. State which player appears to be better, based upon mean result. Justify your answer.
c. Determine the median score for each player.
d. State which player appears to be better when the decision is based on the median result. Justify your answer.
e. State which player do you think would be the most useful to have in a cricket team. Justify your answer. Explain how can the mean result sometimes lead to a misleading conclusion.

20. The following frequency table gives the number of employees in different salary brackets for a small manufacturing plant.

Position	Salary ($)	Number of employees
Machine operator	18 000	50
Machine mechanic	20 000	15
Floor steward	24 000	10
Manager	62 000	4
Chief executive officer	80 000	1

a. Workers are arguing for a pay rise but the management of the factory claims that workers are well paid because the mean salary of the factory is $22 100.
Explain whether the management is being honest.
b. Suppose that you were representing the factory workers and had to write a short submission in support of the pay rise.
How could you explain the management's claim? Quote some other statistics in favour of your case.

21. The resting pulse rate of 20 female athletes was measured. The results are detailed below.

| 50, | 52, | 48, | 52, | 71, | 61, | 30, | 45, | 42, | 48, |
| 43, | 47, | 51, | 62, | 34, | 61, | 44, | 54, | 38, | 40 |

   a. Construct a frequency distribution table. (Use class sizes of $1-<10$, $10-<20$ and so on.)
   b. Use your table to estimate the mean of the data.
   c. Determine the median class of the data.
   d. Identify the modal class of the data.
   e. Draw an ogive of the data. (You may like to use a graphics calculator for this.)
   f. Use the ogive to determine the median pulse rate.

22. Design a set of five numbers with:
   a. mean = median = mode
   b. mean > median > mode
   c. mean < median = mode.

23. The numbers 15, a, 17, b, 22, c, 10 and d have a mean of 14. Calculate the mean of a, b, c and d.

24. The numbers m, n, p, q, r, and s have a mean of a while x, y and z have a mean of b. Calculate the mean of all nine numbers.

25. The mean and median of six two-digit prime numbers is 39 and the mode is 31. The smallest number is 13. Determine the six numbers.

LESSON
13.3 Measures of spread: range and interquartile range

LEARNING INTENTION

At the end of this lesson you should be able to:
   • calculate the range and interquartile range of a data set.

▶ 13.3.1 Measures of spread

eles-4952
   • **Measures of spread** describe how far data values are spread from the centre or from each other.
   • A shoe store proprietor has stores in Newcastle and Wollongong. The number of pairs of shoes sold each day over one week is recorded below.

| Newcastle: | 45, | 60, | 50, | 55, | 48, | 40, | 52 |
| Wollongong: | 20, | 85, | 50, | 15, | 30, | 60, | 90 |

In each of these data sets consider the measures of central tendency.

Newcastle:	Mean = 50	Wollongong:	Mean = 50
	Median = 50		Median = 50
	No mode		No mode

- With these measures being the same for both data sets we could come to the conclusion that both data sets are very similar; however, if we look at the data sets, they are very different.
  We can see that the data for Newcastle are very clustered around the mean, whereas the Wollongong data are more spread out.
- The data from Newcastle are between 40 and 60, whereas the Wollongong data are between 15 and 90.
- **Range** and **interquartile range (IQR)** are both measures of spread.

Range

- The range is defined as the difference between the highest and the lowest values in the set of data.

> **Calculating the range of a data set**
>
> **Range = highest score − lowest score**
>
> $$= X_{max} - X_{min}$$

WORKED EXAMPLE 4 Calculating the range of a data set

Calculate the range of the given data set: 2.1, 3.5, 3.9, 4.0, 4.7, 4.8, 5.2.

THINK	WRITE
1. Identify the lowest score (X_{min}) of the data set.	Lowest score $= 2.1$
2. Identify the highest score (X_{max}) of the data set.	Highest score $= 5.2$
3. Write the rule for the range.	Range $= X_{max} - X_{min}$
4. Substitute the known values into the rule.	$= 5.2 - 2.1$
5. Evaluate and write the answer.	$= 3.1$

Interquartile range

- The interquartile range (IQR) is the range of the middle 50% of all the scores in an ordered set. When calculating the interquartile range, the data are first organised into quartiles, each containing 25% of the data.
- The word 'quartile' comes from the word 'quarter'.

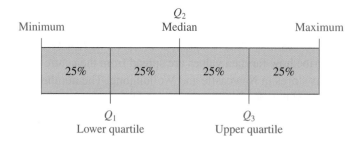

- The lower quartile (Q_1) is the median of the lower half of the data set.
- The upper quartile (Q_3) is the median of the upper half of the data set.

Calculating the IQR

Interquartile range (IQR) = upper quartile − lower quartile

$$= Q_{\text{upper}} - Q_{\text{lower}}$$

$$= Q_3 - Q_1$$

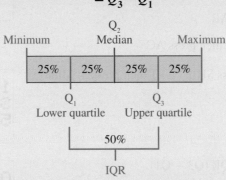

- The IQR is not affected by extremely large or extremely small data values (**outliers**), so in some circumstances the IQR is a better indicator of the spread of data than the range.

WORKED EXAMPLE 5 Calculating the IQR of a data set

Calculate the interquartile range (IQR) of the following set of data:
$$3, \quad 2, \quad 8, \quad 6, \quad 1, \quad 5, \quad 3, \quad 7, \quad 6.$$

THINK	WRITE
1. Arrange the scores in order.	1 2 3 3 5 6 6 7 8
2. Locate the median and use it to divide the lower 50% in half and the upper 50% in half. *Note:* The median is the 5th score in this data set and should not be included in the lower or upper ends of the data. This is the case for any data set with an odd amount of scores.	1 2 3 3 5 6 6 7 8 min $\quad Q_1$ Median Q_3 max
3. Calculate Q_1, the median of the lower half of the data.	$Q_1 = \dfrac{2+3}{2}$ $= \dfrac{5}{2}$ $= 2.5$
4. Calculate Q_3, the median of the upper half of the data.	$Q_3 = \dfrac{6+7}{2}$ $= \dfrac{13}{2}$ $= 6.5$
5. Calculate the interquartile range.	IQR $= Q_3 - Q_1$ $= 6.5 - 2.5$
6. Write the answer.	$= 4$

DIGITAL TECHNOLOGY

A scientific calculator is able to determine the interquartile range (IQR).

Using the Statistics function:

Press Mode → 2 STAT → 1: 1-VAR

Input each data point and press = then press AC.

To access the statistics menu, press SHIFT → 1 then press 5: MinMax for the five-number summary.

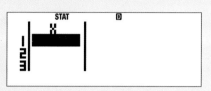

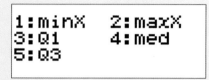

Use Q1 and Q3 to calculate the IQR (Q3 − Q1).

Determining the IQR from a graph

- When data are presented in a frequency distribution table, either ungrouped or grouped, the interquartile range is found by drawing an ogive.

WORKED EXAMPLE 6 Calculating the IQR from a graph

The following frequency distribution table gives the number of customers who order different volumes of concrete from a readymix concrete company during the course of a day.

Calculate the interquartile range of the data.

Volume (m³)	Frequency
0.0 − < 0.5	15
0.5 − < 1.0	12
1.0 − < 1.5	10
1.5 − < 2.0	8
2.0 − < 2.5	2
2.5 − < 3.0	4

THINK

1. To calculate the 25th and 75th percentiles from the ogive, first add a class centre column and a cumulative frequency column to the frequency distribution table and fill them in.

WRITE/DRAW

Volume	Class centre	f	cf
0.0 − < 0.5	0.25	15	15
0.5 − < 1.0	0.75	12	27
1.0 − < 1.5	1.25	10	37
1.5 − < 2.0	1.75	8	45
2.0 − < 2.5	2.25	2	47
2.5 − < 3.0	2.75	4	51

2. Draw the cumulative frequency graph. A percentage axis will be useful.

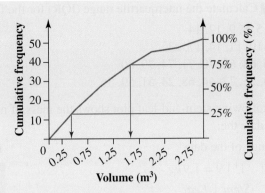

3. Identify the upper quartile (75th percentile) and lower quartile (25th percentile) from the ogive.

$Q_3 = 1.6 \, \text{m}^3$

$Q_1 = 0.4 \, \text{m}^3$

4. The interquartile range is the difference between the upper and lower quartiles.

$\text{IQR} = Q_3 - Q_1$
$= 1.6 - 0.4$
$= 1.2 \, \text{m}^3$

COMMUNICATING — COLLABORATIVE TASK: Measures of centre and spread

In pairs or small groups, design an A3 poster that includes how to calculate the mean, mode, median, range and interquartile range for a set of data. Use the same data set across your poster. Work collaboratively to choose an appropriate data set. Don't forget to ensure your poster is engaging and coherent.

 Resources

 Interactivities Range (int-3822)

The interquartile range (int-4813)

Exercise 13.3 Measures of spread: range and interquartile range

learn

| 13.3 Quick quiz on | | 13.3 Exercise |

Individual pathways

■ PRACTISE	■ CONSOLIDATE	■ MASTER
1, 4, 7, 10, 13	2, 6, 8, 11, 14	3, 5, 9, 12, 15

Fluency

1. **WE4** Calculate the range for each of the following sets of data.

   a. 4, 3, 9, 12, 8, 17, 2, 16

   b. 49.5, 13.7, 12.3, 36.5, 89.4, 27.8, 53.4, 66.8

   c. $7\frac{1}{2}$, $12\frac{3}{4}$, $5\frac{1}{4}$, $8\frac{2}{3}$, $9\frac{1}{6}$, $3\frac{3}{4}$

2. **WE5** Calculate the interquartile range (IQR) for the following sets of data.
   a. 3, 5, 8, 9, 12, 14
   b. 7, 10, 11, 14, 17, 23
   c. 66, 68, 68, 70, 71, 74, 79, 80
   d. 19, 25, 72, 44, 68, 24, 51, 59, 36

3. The following stem-and-leaf plot shows the mass of newborn babies (rounded to the nearest 100g). Calculate the:

   a. range of the data

   b. IQR of the data.

   Key: $1^* | 9 = 1.9\,kg$

Stem	Leaf
1*	9
2	2 4
2*	6 7 8 9
3	0 0 1 2 3 4
3*	5 5 6 7 8 8 8 9
4	0 1 3 4 4
4*	5 6 6 8 9
5	0 1 2 2

4. Use the cumulative frequency graph shown to calculate the interquartile range of the data.

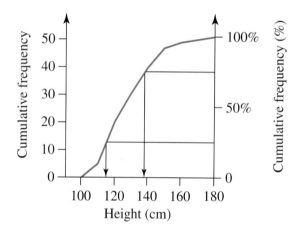

5. **WE6** The following frequency distribution table gives the amount of time spent by 50 people shopping for Christmas presents.

Time (h)	$0-<0.5$	$0.5-<1$	$1-<1.5$	$1.5-<2$	$2-<2.5$	$2.5-<3$	$3-<3.5$	$3.5-<4$
Frequency	1	2	7	15	13	8	2	2

   Estimate the IQR of the data.

6. **MC** Calculate the interquartile range of the following data:

   17, 18, 18, 19, 20, 21, 21, 23, 25

   A. 8   B. 18   C. 4   D. 20

Understanding

7. The following frequency distribution table shows the life expectancy in hours of 40 household batteries.

Life (h)	50 – < 55	55 – < 60	60 – < 65	65 – < 70	70 – < 75	75 – < 80
Frequency	4	10	12	8	5	1

a. Draw a cumulative frequency graph that represents the data in the table above.

b. Use the cumulative frequency graph to answer the following questions.

   i. Calculate the median score.

   ii. Determine the upper and lower quartiles.

   iii. Calculate the interquartile range.

   iv. Identify the number of batteries that lasted less than 60 hours.

   v. Identify the number of batteries that lasted 70 hours or more.

8. Calculate the IQR for the following data.

Class interval	Frequency
120 – < 130	2
130 – < 140	3
140 – < 150	9
150 – < 160	14
160 – < 170	10
170 – < 180	8
180 – < 190	6
190 – < 200	3

9. For each of the following sets of data, state:

  i. the range and

  ii. the IQR of each set.

  a. 6, 9, 12, 13, 20, 22, 26, 29

  b. 7, 15, 2, 26, 47, 19, 9, 33, 38

  c. 120, 99, 101, 136, 119, 87, 123, 115, 107, 100

Communicating, reasoning and problem solving

10. Explain what the measures of spread tell us about a set of data.

11. As newly appointed coach of Terrorolo's Meteors netball team, Kate decided to record each player's statistics for the previous season. The number of goals scored by the leading goal shooter was:

$$1, \quad 3, \quad 8, \quad 18, \quad 19, \quad 23, \quad 25, \quad 25, \quad 25, \quad 26, \quad 27, \quad 28,$$
$$28, \quad 28, \quad 28, \quad 29, \quad 29, \quad 30, \quad 30, \quad 33, \quad 35, \quad 36, \quad 37, \quad 40$$

a. Calculate the mean of the data, correct to one decimal place.

b. Calculate the median of the data.

c. Calculate the range of the data.

d. Determine the interquartile range of the data.

e. There are three scores that are much lower than most. Explain the effect these scores have on the summary statistics.

12. The following stem-and-leaf plot shows the ages of 30 people going to a gym. Rounding answers to one decimal place when necessary.

Key: 1|6 = 16 years old

Stem	Leaf:
1	67789
2	001234567789
3	01223479
4	1248
5	2

Determine the mean, median, range and interquartile range of the data set.

13. Calculate the mean, median, mode, range and IQR of the following data collected when the temperature of the soil around 25 germinating seedlings was recorded:
28.9, 27.4, 23.6, 25.6, 21.1, 22.9, 29.6, 25.7, 27.4, 23.6, 22.4, 24.6, 21.8, 26.4, 24.9, 25.0, 23.5, 26.1, 23.6, 25.3, 29.5, 23.5, 22.0, 27.9, 23.6.

14. Four positive numbers a, b, c and d have a mean of 12, a median and mode of 9 and a range of 14. Determine the values of a, b, c and d.

15. A set of five positive integer scores have the following summary statistics:
    - range = 9
    - median = 6
    - $Q_1 = 3$ and $Q_3 = 9$.

    a. Explain whether the five scores could be 1, 4, 6, 8 and 10.
    b. A sixth score is added to the set. Determine whether there is a score that will maintain the summary statistics given above. Justify your answer.

LESSON
13.4 Measures of spread: Standard deviation

LEARNING INTENTION

At the end of this lesson you should be able to:
- calculate the standard deviation using technology
- interpret the mean and standard deviation of data
- identify the effect of outliers on the standard deviation.

▶ 13.4.1 Standard deviation

eles-4958

- The **standard deviation** for a set of data is a measure of spread of how far the data values are spread out (deviate) from the mean. The value of the standard deviation tells you the average deviation of the data from the mean.
- **Deviation** is the difference between each data value and the mean $(x - \bar{x})$. The standard deviation is calculated from the square of the deviations.

Standard deviation formula

- **Standard deviation is denoted by the lowercase Greek letter sigma, σ, and can be calculated by using the following formula.**

$$\sigma = \sqrt{\frac{\sum (x - \bar{x})^2}{n}}$$

where:
Σ = sum of
x = each value in the data set
$\bar{x}$ = mean
n = number of data values.

- A low standard deviation indicates that the data values tend to be close to the mean.
- A high standard deviation indicates that the data values tend to be spread out over a large range, away from the mean.
- Standard deviation can be calculated using a scientific or graphics calculator.

DIGITAL TECHNOLOGY

A scientific calculator is able to calculate the standard deviation

Using the Statistics function:

Press Mode → 2 STAT → 1: 1-VAR

Input each data point and press = Then press AC.

To access the calculate the standard deviation, press SHIFT → 1 → 4: VAR → 3: σx

Note: 3: σx calculates the population standard deviation, 4: sx calculates the sample standard deviation.

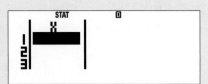

WORKED EXAMPLE 7 Calculating the standard deviation by hand

The number of lollies in each of eight packets is 11, 12, 13, 14, 16, 17, 18, 19.
Calculate the mean and standard deviation correct to 2 decimal places.

THINK

1. Calculate the mean.

WRITE

$$\bar{x} = \frac{11 + 12 + 13 + 14 + 16 + 17 + 18 + 19}{8}$$

$$= \frac{120}{8}$$

$$= 15$$

2. To calculate the deviations $(x - \bar{x})$, set up a frequency table as shown and complete.

No. of lollies (x)	$(x - \bar{x})$
11	$11 - 15 = -4$
12	-3
13	-2
14	-1
16	1
17	2
18	3
19	4
Total	

3. Add another column to the table to calculate the square of the deviations, $(x - \bar{x})^2$. Then sum the results: $\Sigma(x - \bar{x})^2$.

No. of lollies (x)	$(x - \bar{x})$	$(x - \bar{x})^2$
11	$11 - 15 = -4$	16
12	-3	9
13	-2	4
14	-1	1
16	1	1
17	2	4
18	3	9
19	4	16
Total		$\displaystyle\sum(x - \bar{x})^2 = 60$

4. To calculate the standard deviation, divide the sum of the squares by the number of data values, then take the square root of the result.

$$\sigma = \sqrt{\dfrac{\sum(x - \bar{x})^2}{n}}$$

$$= \sqrt{\dfrac{60}{8}}$$

$$\approx 2.74 \text{ (correct to 2 decimal places)}$$

5. Check the result using a calculator.

The calculator returns an answer of $\sigma_n = 2.73861$. Answer confirmed.

6. Interpret the result.

The average (mean) number of lollies in each pack is 15 with a standard deviation of 2.74, which means that the number of lollies in each pack differs from the mean by an average of 2.74.

Standard deviations of populations and samples

- So far we have calculated the standard deviation for a population of data, that is, for complete sets of data. There is another formula for calculating standard deviation for samples of data, that is, data that have been randomly selected from a larger population.
- The sample standard deviation is more commonly used in day-to-day life, as it is usually impossible to collect data from an entire population.
- For example, if you wanted to know how much time Year 9 students across the country spend on social media, you would not be able to collect data from every student in the country. You would have to take a sample instead.

- The sample standard deviation is denoted by the letter s, and can be calculated using the following formula.

> **Sample standard deviation formula**
>
> $$s = \sqrt{\frac{\sum (x - \bar{x})^2}{n - 1}}$$

- Calculators display both values for the standard deviation, so it is important to understand the difference between them.

13.4.2 Effects on standard deviation

eles-4959

- The standard deviation is affected by extreme values, those values that are much smaller or larger than most of the data set (outliers).

WORKED EXAMPLE 8 Interpreting the effects on standard deviation

On a particular day, Lucy played golf brilliantly and scored 60.
The scores in her previous 12 games of golf were 87, 88, 88, 89, 90, 90, 90, 92, 93, 93, 95 and 97.
Comment on the effect this latest score has on the standard deviation.

THINK	WRITE
1. Use a calculator to calculate the mean and the standard deviation without the latest score.	$\bar{x} = 91$ $\sigma = 2.1955$ ≈ 2.92
2. Use a calculator to calculate the mean and standard deviation with the latest score.	$\bar{x} = 88.6154$ $\sigma = 8.7225$ ≈ 88.62 ≈ 8.72
3. Interpret the result.	In the first 12 games Lucy's mean score was 91 with a standard deviation of 2.92. This implied that Lucy's scores on average were 2.92 either side of her average of 91. Lucy's latest performance resulted in a mean score of 88.62 with a standard deviation of 8.72. This indicates a slightly lower mean score, but the much higher standard deviation indicates that the data are now much more spread out and that the extremely good score of 60 is an anomaly.

13.4.3 Properties of standard deviation

eles-6169

- If a constant c is added to all data values in a set, the deviations $(x - \bar{x})$ will remain unchanged and consequently the standard deviation remains unchanged.
- If all data values in a set are multiplied by a constant k, the deviations $(x - \bar{x})$ will be multiplied by k, that is $k(x - \bar{x})$; consequently the standard deviation is increased by a factor of k.

- The standard deviation can be used to measure consistency.
- When the standard deviation is low we are able to say that the scores in the data set are more consistent with each other.

WORKED EXAMPLE 9 Calculating numerical changes to the standard deviation

For the data 5, 9, 6, 11, 10, 7:
a. calculate the standard deviation, correct to 2 decimal places
b. calculate the standard deviation if 4 is added to each data value. Comment on the effect.
c. calculate the standard deviation if each data value is doubled. Comment on the effect.

THINK	WRITE
a. Use a calculator to calculate the standard deviation.	a. $\sigma \approx 2.16$ (correct to 2 decimal places)
b. 1. Add 4 to each data value in the set.	b. 9, 13, 10, 15, 14, 11
2. Use a calculator to calculate the standard deviation.	$\sigma \approx 2.16$ (correct to 2 decimal places)
3. Comment on the effect of adding 4 to each data value.	Adding 4 to each data value had no effect on the standard deviation, which remained at 2.16.
c. 1. Double each data value in the set.	c. 10, 18, 12, 22, 20, 14
2. Use a calculator to calculate the standard deviation.	$\sigma \approx 4.32$ (correct to 2 decimal places)
3. Comment on the effect of doubling each data value.	Doubling each data value doubled the standard deviation to 4.32.

 Resources

 Interactivities The standard deviation for a sample (int-4814)

Exercise 13.4 Measures of spread: Standard deviation   learn

<table>
<tr><td>13.4 Quick quiz on</td><td>13.4 Exercise</td></tr>
</table>

Individual pathways

■ PRACTISE	■ CONSOLIDATE	■ MASTER
1, 2, 4, 8, 10, 12	3, 5, 7, 11, 13	6, 9, 14, 15

Fluency

1. **WE7** Calculate the standard deviation of each of the following data sets, correct to 2 decimal places.
   a. 3, 5, 8, 2, 7, 1, 6, 5
   b. 11, 8, 7, 12, 10, 11, 14
   c. 25, 15, 78, 35, 56, 41, 17, 24
   d. 5.2, 4.7, 5.1, 12.6, 4.8

2. Calculate the standard deviation of each of the following data sets, correct to 2 decimal places.

a.

Score (x)	Frequency (f)
1	1
2	5
3	9
4	7
5	3

b.

Score (x)	Frequency (f)
16	15
17	24
18	26
19	28
20	27

c.

Score (x)	Frequency (f)
8	15
10	19
12	18
14	7
16	6
18	2

d.

Score (x)	Frequency (f)
65	15
66	15
67	16
68	17
69	16
70	15
71	15
72	12

3. Complete the following frequency distribution table and use it to calculate the standard deviation of the data set.

Class	Class centre (x)	Frequency (f)
1 – 10		6
11 – 20		15
21 – 30		25
31 – 40		8
41 – 50		6

4. First-quarter profit increases for 8 leading companies are given below as percentages.

$$2.3, \ 0.8, \ 1.6, \ 2.1, \ 1.7, \ 1.3, \ 1.4, \ 1.9$$

Calculate the mean score and the standard deviation for this set of data.

5. The heights in metres of a group of army recruits are given.

$$1.8, \ 1.95, \ 1.87, \ 1.77, \ 1.75, \ 1.79, \ 1.81, \ 1.83, \ 1.76, \ 1.80, \ 1.92, \ 1.87, \ 1.85, \ 1.83$$

Calculate the mean score and the standard deviation for this set of data.

6. Times (to the nearest tenth of a second) for the heats in the open 100 m sprint at the school sports are given in the stem-and-leaf plot shown.
Calculate the standard deviation for this set of data.

Key: 11|0 = 11.0 s

Stem	Leaf
11	0
11	2 3
11	4 4 5
11	6 6
11	8 8 9
12	0 1
12	2 2 3
12	4 4
12	6
12	9

7. The number of outgoing phone calls from an office each day over a 4-week period is shown in the stem-and-leaf plot.
Calculate the standard deviation for this set of data.

Key: 1|3 = 13 calls

Stem	Leaf
0	8 9
1	3 4 7 9
2	0 1 3 7 7
3	3 4
4	1 5 6 7 8
5	3 8

8. **MC** A new legal aid service has been operational for only 5 weeks. The number of people who have made use of the service each day during this period is set out in the stem-and-leaf plot shown.
The standard deviation of these data is:

A. 6.00
B. 6.34
C. 6.47
D. 15.44

Key: 1|6 = 16 people

Stem	Leaf
0	2 4
0	7 7 9
1	0 1 4 4 4 4
1	5 6 6 7 8 8 9
2	1 2 2 3 3 3
2	7

Understanding

9. **WE8** The speeds, in km/h, of the first 25 cars caught by a roadside speed camera on a particular day were:

82, 82, 84, 84, 84, 84, 85, 85, 85, 86, 86, 87, 89, 89, 89, 90, 91, 91, 92, 94, 95, 96, 99, 100, 102

The next car that passed the speed camera was travelling at 140 km/h.
Comment on the effect of the speed of this last car on the standard deviation for the data.

10. Explain what the standard deviation tells us about a set of data.

11. **WE9** For the data 1, 4, 5, 9, 11:

   a. calculate the standard deviation
   b. calculate the standard deviation if 7 is added to each data value. Comment on the effect.
   c. calculate the standard deviation if all data values are multiplied by 3. Comment on the effect.

Communicating, reasoning and problem solving

12. If the mean for a set of data is 45 and the standard deviation is 6, determine how many standard deviations above the mean is a data value of 57.

13. Five numbers a, b, c, d and e have a mean of 12 and a standard deviation of 4.

   a. If each number is increased by 3, calculate the new mean and standard deviation.
   b. If each number is multiplied by 3, calculate the new mean and standard deviation.

14. Show using an example the effect, if any, on the standard deviation of adding a data value to a set of data that is equivalent to the mean.

15. Twenty-five students sat a test and the results for 24 of the students are given in the following stem-and-leaf plot.

   a. If the average mark for the test was 27.84, determine the mark obtained by the 25th student.
   b. Determine how many students scored higher than the median score.
   c. Calculate the standard deviation of the marks, giving your answer correct to 2 decimal places.

   Key: $1|2 = 12$ marks

Stem	Leaf
0	8 9
1	1 2 3 7 8 9
2	2 3 5 6 8
3	0 1 2 4 6 8
4	0 2 5 6 8

LESSON
13.5 Box plots

LEARNING INTENTION

At the end of this lesson you should be able to:
- calculate the five-number summary for a set of data
- draw a box plot showing the five-number summary of a data set
- calculate outliers in a data set
- describe skewness of distributions
- compare box plots to dot plots or histograms
- draw parallel box plots and compare sets of data.

⏵ 13.5.1 Five-number summary

eles-4953

- A five-number summary is a list consisting of the lowest score (X_{min}), lower quartile (Q_1), median (Q_2), upper quartile (Q_3) and greatest score (X_{max}) of a set of data.

WORKED EXAMPLE 10 Calculations using the five-number summary

From the following five-number summary, calculate:

a. the interquartile range   **b.** the range.

X_{min}	Q_1	Median (Q_2)	Q_3	X_{max}
29	37	39	44	48

THINK

a. The interquartile range is the difference between the upper and lower quartiles.

b. The range is the difference between the greatest score and the lowest score.

WRITE

a. $IQR = Q_3 - Q_1$
$= 44 - 37$
$= 7$

b. $Range = X_{max} - X_{min}$
$= 48 - 29$
$= 19$

▶ 13.5.2 Box plots

eles-4954

- A **box plot** is a graph of the five-number summary.
- Box plots consist of a central divided box with attached whiskers.
- The box spans the interquartile range.
- The median is marked by a vertical line drawn inside the box.
- The whiskers indicate the range of scores:

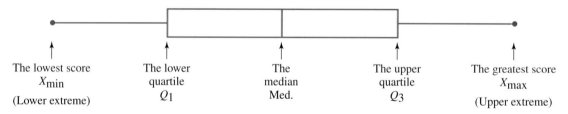

- Box plots are *always drawn to scale*.
- They are presented either with the five-number summary figures attached as labels (diagram at right) or with a scale presented alongside the box plot like the diagram below. They can also be drawn vertically.

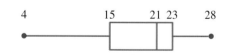

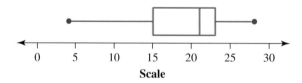

Identification of extreme values or outliers

- If an extreme value or outlier occurs in a set of data, it can be denoted by a small cross on the box plot. The whisker is then shortened to the next largest (or smallest) figure.

- The box plot below shows that the lowest score was 5. This was an extreme value as the rest of the scores were located within the range 15 to 42.

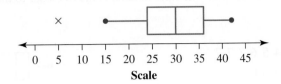

Scale

- Outliers are still included when calculating the range of the data.
- Outliers sit $1.5 \times IQR$ or greater away from Q_1 or Q_3.

Identifying outliers

Lower limit $= Q_1 - 1.5 \times IQR$

Upper limit $= Q_3 + 1.5 \times IQR$

Any scores that sit outside these limits are considered outliers.

Symmetry and skewness in distributions

- A **symmetrical** plot has data that are evenly spaced around a central point. The median and mean are in the centre of the plot.
  Examples of a stem-and-leaf plot and a symmetrical box plot are shown below.

Stem	Leaf
26*	6
27	0 1 3
27*	5 6 8 9
28	0 1 1 1 2 4
28*	5 7 8 8
29	2 2 2
29*	5

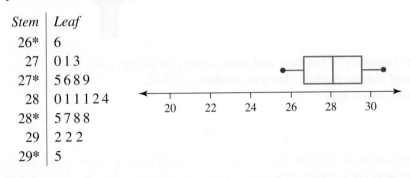

- A **negatively skewed** plot has larger amounts of data at the higher end. The mean is smaller than the median.
- This is illustrated by the stem-and-leaf plot below where the leaves increase in length as the data increase in value. It is illustrated on the box plot when the median is much closer to the maximum value than the minimum value.

Stem	Leaf
5	1
6	2 9
7	1 1 2 2
8	1 4 4 5 6 6
9	5 3 4 4 5 6 7 7 7

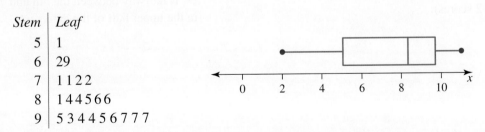

- A **positively skewed** plot has larger amounts of data at the lower end. The mean is larger than the median.
- This is illustrated on the stem-and-leaf plot below where the leaves increase in length as the data decrease in value. It is illustrated on the box plot when the median is much closer to the minimum value than the maximum value.

Stem	Leaf
5	1 3 4 4 5 6 7 7 7
6	2 4 4 5 6 6
7	1 1 2 2
8	1 6
9	5

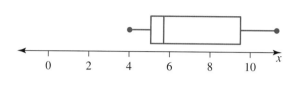

WORKED EXAMPLE 11 Drawing a box plot

The following stem-and-leaf plot gives the speed of 25 cars caught by a roadside speed camera.

Key: $8\,|\,2 = 82\,\text{km/h}, 8^*\,|\,6 = 86\,\text{km/h}$

Stem	Leaf
8	2 2 4 4 4 4
8*	5 5 6 6 7 9 9 9
9	0 1 1 2 4
9*	5 6 9
10	0 2
10*	
11	4

a. **Prepare a five-number summary of the data and draw a box plot to represent it.**
b. **Identify any outliers and redraw the box plot with outliers marked.**
c. **Describe the distribution of the data.**

THINK

a. 1. First identify the positions of the median and upper and lower quartiles. There are 25 data values. The median is the $\left(\dfrac{n+1}{2}\right)$ th score. The lower quartile is the median of the lower half of the data. The upper quartile is the median of the upper half of the data (each half contains 12 scores).

WRITE

a. The median is the $\left(\dfrac{25+1}{2}\right)$ th score — that is, the 13th score.

Q_1 is the $\left(\dfrac{12+1}{2}\right)$ th score in the lower half — that is, the 6.5th score. That is, halfway between the 6th and 7th scores.

Q_3 is halfway between the 6th and 7th scores in the upper half of the data.

2. Mark the positions of the median and upper and lower quartiles on the stem-and-leaf plot.

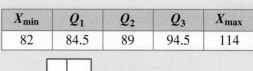

Key: $8 | 2 = 82$ km/h
$8* | 6 = 86$ km/h

Stem	Leaf
8	2 2 4 4 4 4
8*	5 5 6 6 7 9 ⑨ 9
9	0 1 1 2 4
9*	5 6 9
10	0 2
10*	
11	4

Q_1, Median, Q_3

3. Write the five-number summary:
The lowest score is 82.
The lower quartile is between 84 and 85; that is, 84.5.
The median is 89.
The upper quartile is between 94 and 95; that is, 94.5.
The greatest score is 114.
Draw the box plot for this summary.

Five-number summary:

X_{min}	Q_1	Q_2	Q_3	X_{max}
82	84.5	89	94.5	114

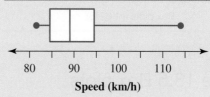

Speed (km/h)

b. 1. Calculate the IQR.

b. $IQR = Q_3 - Q_1$
$= 94.5 - 84.5$
$= 10$

2. Calculate the lower and upper limits.

Lower limit $= 84.5 - 1.5 \times 10$
$= 69.5$
Upper limit $= 94.5 + 1.5 \times 10$
$= 109.5$

3. Identify the outliers.

114 is above the upper limit of 109.5, so it is an outlier.

4. Redraw the box plot, including the outlier marked as a cross. Draw the whisker to the next largest figure, 102.

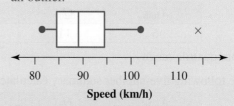

Speed (km/h)

c. Describe the distribution.

c. The data is positively skewed.

A scientific calculator is able to calculate the five-number summary.

Using the Statistics function, press:

Press Mode → 2 STAT → 1: 1-VAR

Input each data point and press = then press AC.

To access the statistics menu, press SHIFT → 1 → 5: MinMax

This will bring up the five-number summary.

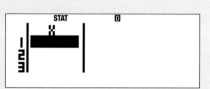

Exercise 13.5 Box plots

learn

13.5 Quick quiz on | **13.5 Exercise**

Individual pathways

■ **PRACTISE**	■ **CONSOLIDATE**	■ **MASTER**
1, 4, 6, 9, 12, 14, 17	2, 5, 7, 10, 13, 15	3, 8, 11, 16, 18

Fluency

1. **WE10** From the following five-number summary calculate:

X_{min}	Q_1	Median	Q_3	X_{max}
6	11	13	16	32

    a. the interquartile range   **b.** the range.

2. From the following five-number summary calculate:

X_{min}	Q_1	Median	Q_3	X_{max}
101	119	122	125	128

    a. the interquartile range   **b.** the range.

3. From the following five-number summary calculate:

X_{min}	Q_1	Median	Q_3	X_{max}
39.2	46.5	49.0	52.3	57.8

    a. the interquartile range   **b.** the range.

4. The box plot shows the distribution of final points scored by a football team over a season's roster.

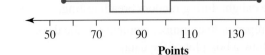

a. Identify the team's greatest points score.
b. Identify the team's least points score.
c. Calculate the team's median points score.
d. Calculate the range of points scored.
e. Calculate the interquartile range of points scored.

5. The box plot shows the distribution of data formed by counting the number of gummy bears in each of a large sample of packs.

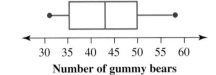

a. Identify the largest number of gummy bears in any pack.
b. Identify the smallest number of gummy bears in any pack.
c. Identify the median number of gummy bears in any pack.
d. Calculate the range of numbers of gummy bears per pack.
e. Calculate the interquartile range of gummy bears per pack.

Questions 6 to 8 refer to the following box plot:

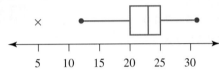

6. **MC** The median of the data is:

   A. 20   **B.** 23   **C.** 25   **D.** 31

7. **MC** The interquartile range of the data is:

   A. 23   **B.** 26   **C.** 5   **D.** 20 to 25

8. **MC** Select the statement that is not true of the data represented by the box plot.

   A. One-quarter of the scores are between 5 and 20.
   B. Half of the scores are between 20 and 25.
   C. Most of the data are contained between the scores of 5 and 20.
   D. One-third of the scores are between 5 and 20.

Understanding

9. The number of sales made each day by a salesperson is recorded over a 2-week period:

$$25, 31, 28, 43, 37, 43, 22, 45, 48, 33$$

a. Prepare a five-number summary of the data. (There is no need to draw a stem-and-leaf plot of the data. Just arrange them in order of size.)
b. Draw a box plot of the data.
c. Identify any outliers.

10. The data below show monthly rainfall in millimetres.

J	F	M	A	M	J	J	A	S	O	N	D
10	12	21	23	39	22	15	11	22	37	45	30

a. Prepare a five-number summary of the data.
b. Draw a box plot of the data.

11. **WE11** The stem-and-leaf plot shown details the age of 25 offenders who were caught during random breath testing.

a. Prepare a five-number summary of the data.
b. Draw a box plot of the data.
c. Identify any outliers.
d. Describe the distribution of the data.

Key: 1|8 = 18 years

Stem	Leaf
1	8 8 9 9 9
2	0 0 0 1 1 3 4 6 9
3	0 1 2 7
4	2 5
5	3 6 8
6	6
7	4

12. The following stem-and-leaf plot details the price at which 30 blocks of land in a particular suburb sold for.

Key: 12|4 = $124 000

Stem	Leaf
12	4 7 9
13	0 0 2 5 5
14	0 0 2 3 5 5 7 9 9
15	0 0 2 3 7 7 8
16	0 2 2 5 8
17	5

a. Prepare a five-number summary of the data.
b. Draw a box plot of the data.

13. An investigation into the transport needs of an outer suburb community recorded the number of passengers boarding a bus during each of its journeys, as follows.

12, 43, 76, 24, 46, 24, 21, 46, 54, 109, 87, 23, 78, 37, 22, 139, 65, 78, 89, 52, 23, 30, 54, 56, 32, 66, 49

Display the data by constructing a histogram using class intervals of 20 and a comparative box plot on the same axis.

Communicating, reasoning and problem solving

14. Explain the advantages and disadvantages of box plots as a visual form of representing data.

15. The following data detail the number of hamburgers sold by a fast food outlet every day over a 4-week period.

M	T	W	T	F	S	S
125	144	132	148	187	172	181
134	157	152	126	155	183	188
131	121	165	129	143	182	181
152	163	150	148	152	179	181

a. Prepare a stem-and-leaf plot of the data. (Use a class size of 10.)
b. Draw a box plot of the data.
c. Comment on what these graphs tell you about hamburger sales.

16. The following data show the ages of 30 mothers upon the birth of their first baby.

$$
\begin{array}{cccccccccc}
22, & 21, & 18, & 33, & 17, & 23, & 22, & 24, & 24, & 20, \\
25, & 29, & 32, & 18, & 19, & 22, & 23, & 24, & 28, & 20, \\
31, & 22, & 19, & 17, & 23, & 48, & 25, & 18, & 23, & 20
\end{array}
$$

   a. Prepare a stem-and-leaf plot of the data. (Use a class size of 5.)
   b. Draw a box plot of the data. Indicate any extreme values appropriately.
   c. Describe the distribution in words. Comment on what the distribution says about the age that mothers have their first baby.

17. Sketch a histogram for the box plot shown.

18. Fifteen French restaurants were visited by three newspaper restaurant reviewers. The average price of a meal for a single person was investigated. The following box plot shows the results.

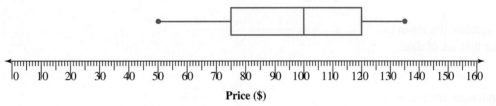

Price ($)

   a. Identify the price of the cheapest meal.
   b. Identify the price of the most expensive meal.
   c. Identify the median cost of a meal.
   d. Calculate the interquartile range for the price of a meal.
   e. Determine the percentage of the prices that were below the median.

LESSON
13.6 Describing and comparing data sets

13.6.1 Describing and comparing data sets

eles-4962

• Besides locating the centre of the data (the mean, median or mode), any analysis of data must measure the extent of the spread of the data (range, interquartile range and standard deviation). Two data sets may have centres that are very similar but be quite differently distributed.
• Decisions need to be made about which measure of centre and which measure of spread to use when describing, analysing and comparing data.
• The mean is calculated using every data value in the set. The median is the middle score of an ordered set of data, so it is a more useful measure of centre when a set of data contains outliers.

- The range is determined by calculating the difference between the maximum and minimum data values, so it includes outliers. It provides only a rough idea about the spread of the data and is inadequate in providing sufficient detail for analysis. It is useful, however, when we are interested in extreme values such as high and low tides or maximum and minimum temperatures.
- The interquartile range is the difference between the upper and lower quartiles, so it does not include every data value in its calculation, but it will overcome the problem of outliers skewing data.
- The standard deviation is calculated using every data value in the set.

WORKED EXAMPLE 12 Interpreting mean and standard deviation

For the two sets of data 6, 7, 8, 9, 10 and 12, 4, 10, 11, 3:
a. calculate the mean
b. calculate the standard deviation
c. comment on the similarities and differences.

THINK	WRITE
a. 1. Calculate the mean of the first set of data.	a. $\bar{x}_1 = \dfrac{6+7+8+9+10}{5}$ $= 8$
2. Calculate the mean of the second set of data.	$\bar{x}_2 = \dfrac{12+4+10+11+3}{5}$ $= 8$
b. 1. Calculate the standard deviation of the first set of data by hand or using a calculator.	b. $\sigma_1 = \sqrt{\dfrac{(6-8)^2+(7-8)^2+(8-8)^2+(9-8)^2+(10-8)^2}{5}}$ ≈ 1.41
2. Calculate the standard deviation of the second set of data by hand or using a calculator.	$\sigma_2 = \sqrt{\dfrac{(12-8)^2+(4-8)^2+(10-8)^2+(11-8)^2+(3-8)^2}{5}}$ ≈ 3.74
c. Comment on the findings.	c. For both sets of data the mean was the same, 8. However, the standard deviation for the second set (3.74) was much higher than the standard deviation of the first set (1.41), implying that the second set is more widely distributed than the first. This is confirmed by the range, which is $10-6=4$ for the first set and $12-3=9$ for the second.

WORKED EXAMPLE 13 Comparing data sets

Below are the scores achieved by two students in eight Mathematics tests throughout the year.
John: 45, 62, 64, 55, 58, 51, 59, 62
Penny: 84, 37, 45, 80, 74, 44, 46, 50
a. Determine the most appropriate measure of centre and measure of spread to compare the performance of the students.
b. Identify the student who performed better over the eight tests. Justify your answer.
c. Identify the student who was more consistent over the eight tests. Justify your answer.

▶ 13.6.2 Comparing different graphical representations

eles-4956

Box plots and dot plots

- Box plots are a concise summary of data. A box plot can be directly related to a dot plot.
- **Dot plots** display each data value represented by a dot placed on a number line.
- The following data are the amount of money (in $) that a group of 27 five-year-olds had with them on a day visiting the zoo with their parents.

0	0.85	0	1.8	1.65	8.45	3.75	0.55	4.1	2.4	2.15
1.2	1.35	0.9	3.45	1	0	0	1.45	1.25	1.7	2.65
1.85	4.75	3.9	1.15							

- The dot plot below and its comparative box plot show the distribution of these data.

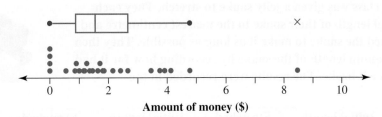

- Both graphs indicate that the data is positively skewed and both graphs indicate the presence of the outlier. However, the box plot provides an excellent summary of the centre and spread of the distribution.

Box plots and histograms

- **Histograms** are graphs that display continuous numerical variables and do not retain all original data.
- The following data are the number of minutes, rounded to the nearest minute, that forty Year 9 students take to travel to their school on a particular day.

15	22	14	12	21	34	19	11	13	0	16
4	23	8	12	18	24	17	14	3	10	12
9	15	20	5	19	13	17	11	16	19	24
12	7	14	17	10	14	23				

- The data are displayed in the histogram and box plot shown.
- Both graphs indicate that the data is slightly positively skewed. The histogram clearly shows the frequencies of each class interval. Neither graph displays the original values. The histogram does not give precise information about the centre, but the distribution of the data is visible. However, the box plot shows the presence of an outlier and provides an excellent summary of the centre and spread of the distribution.

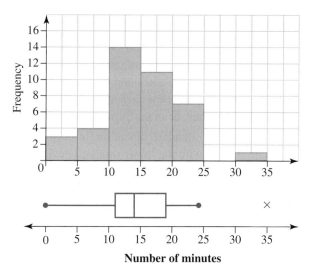

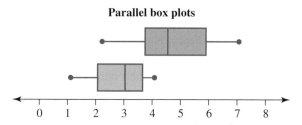

Number of minutes

Parallel box plots

- When multiple data displays are used to display similar sets of data, comparisons and conclusions can then be drawn about the data.
- We can use **parallel box plots** and multiple or parallel box plots to help compare statistics such as the median, range and interquartile range.

Parallel box plots

WORKED EXAMPLE 14 Comparing two sets of data

Each member of a class was given a jelly snake to stretch. They each measured the initial length of their snake to the nearest centimetre and then slowly stretched the snake to make it as long as possible. They then measured the maximum length of the snake by recording how far it had stretched at the time it broke. The results were recorded in the following table.

Initial length (cm)	Stretched length (cm)	Initial length (cm)	Stretched length (cm)
13	29	14	27
14	28	13	27
17	36	15	36
10	24	16	36
14	35	15	36
16	36	16	34
15	37	17	35
16	37	12	27
14	30	9	17
16	33	16	41
17	36	17	38
16	38	16	36
17	38	17	41
14	31	16	33
17	40	11	21

The above data was drawn on parallel box plots as shown below.

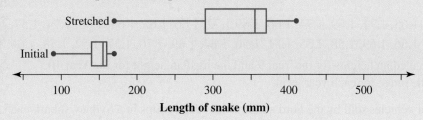

Length of snake (mm)

Compare the data sets and draw your conclusion for the stretched snake.

THINK	WRITE
1. Determine the median in the case of the initial and stretched length of the snake.	The change in the length of the snake when stretched is evidenced by the increased median and spread shown on the box plots. The median snake length before being stretched was 15.5 cm, but the median snake length after being stretched was 35 cm.
2. Draw your conclusion.	The range increased after stretching, as did the IQR.

Exercise 13.6 Describing and comparing data sets

learnon

13.6 Quick quiz on	13.6 Exercise

Individual pathways

■ PRACTISE	■ CONSOLIDATE	■ MASTER
1, 5, 6, 10, 12, 13, 19	2, 4, 8, 11, 14, 15, 20, 23	3, 7, 9, 16, 17, 18, 21, 22, 24

Where necessary, round your answers correct to 2 decimal places.

Fluency

1. **WE12** For the two sets of data, 65, 67, 61, 63, 62, 60 and 56, 70, 65, 72, 60, 55:
   a. calculate the mean
   b. calculate the standard deviation
   c. comment on the similarities and differences.

2. A bank surveys the average morning and afternoon waiting times for customers. The figures were taken each Monday to Friday in the morning and afternoon for one month. The waiting times are measured in minutes and are listed below.

   Morning waiting times: 0.7, 1.8, 1.6, 1.3, 1.1, 1.1, 2.9, 2.6, 2.6, 2.6, 2.5, 2.5, 2.4, 2.4, 2.3, 2.1, 3.9, 3.5, 3.2, 4.5

   Afternoon waiting times: 0.7, 0.8, 0.8, 1.1, 1.1, 1.2, 1.4, 1.4, 1.5, 1.6, 1.6, 1.6, 1.7, 2.2, 2.5, 2.5, 2.8, 3.1, 3.6, 5.7

   a. Identify the median morning waiting time and the median afternoon waiting time.
   b. Calculate the range for morning waiting times and the range for afternoon waiting times.
   c. Use the information to comment about the average waiting time at the bank in the morning compared with the afternoon.

3. In a class of 30 students there are 15 Year 9 and 15 Year 10. Their heights are measured in metres and are listed below.

   Year 9:   1.65, 1.71, 1.59, 1.74, 1.66, 1.69 1.72, 1.66, 1.65, 1.64, 1.68, 1.74, 1.57, 1.59, 1.60
   Year 10:   1.66, 1.69, 1.58, 1.55, 1.51, 1.56, 1.64, 1.69, 1.70, 1.57, 1.52, 1.58, 1.64, 1.68, 1.67

   a. Identify the median heights for the Year 9 and the median height for the Year 10.
   b. Calculate the range for each year level.

4. The number of vehicles sold by the Ford and Hyundai dealerships in a Sydney suburb each week for a three-month period are listed below.

   a. State the median of both distributions.
   b. Calculate the range of both distributions.
   c. Calculate the interquartile range of both distributions.
   d. Show both distributions on a box plot.

   Ford: 7, 4, 19, 15, 12, 12 , 11, 10, 28, 25, 24, 24, 30
   Hyundai: 3, 9, 11, 11, 11, 16, 16, 18, 22, 22, 27, 29, 35

5. The box plot drawn below displays statistical data of two AFL teams over a season.

   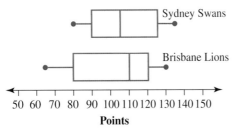
   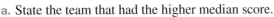

   a. State the team that had the higher median score.
   b. Determine the range of scores for each team.
   c. For each team calculate the interquartile range.
   d. Comment on the skewness of both box plots.

Understanding

6. Prepare comparative box plots for the following dot plots (using the same axis) and describe what each plot reveals about the data.

   Number of sick days taken by workers last year at factory A

   Number of sick days taken by workers last year at factory B

7. **WE14** At a weight-loss clinic, the following weights (in kilograms) were recorded before and after treatment.

Before	75	80	75	140	77	89	97	123	128	95	152	92
After	69	66	72	118	74	83	89	117	105	81	134	85

| Before | 85 | 90 | 95 | 132 | 87 | 109 | 87 | 129 | 135 | 85 | 137 | 102 |
| After | 79 | 84 | 90 | 124 | 83 | 102 | 84 | 115 | 125 | 81 | 123 | 94 |

a. Prepare a five-number summary for weight before and after treatment.
b. Draw parallel box plots for weight before and after treatment.
c. Comment on the comparison of weights before and after treatment.

8. Tanya measures the heights (in m) of a group of Year 9 boys and girls and produces the following five-point summaries for each data set.
   Boys: 1.45, 1.56, 1.62, 1.70, 1.81
   Girls: 1.50, 1.55, 1.62, 1.66, 1.73

a. Draw a box plot for both sets of data and display them on the same scale.
b. Calculate the median of each distribution.
c. Calculate the range of each distribution.
d. Calculate the interquartile range for each distribution.
e. Comment on the spread of the heights among the boys and the girls.

9. The box plots show the average daily sales of cold drinks at the school canteen in summer and winter.

a. Calculate the range of sales in both summer and winter.
b. Calculate the interquartile range of the sales in both summer and winter.
c. Comment on the relationship between the two data sets, both in terms of measures of centre and measures of spread.

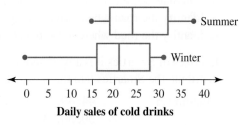

Daily sales of cold drinks

10. **MC** Andrea surveys the age of people at two movies being shown at a local cinema. The box plot shows the results. Select the conclusion that could be drawn based on the information shown in the box plot.

A. Movie A attracts an older audience than Movie B
B. Movie B attracts an older audience than Movie A.
C. Movie A appeals to a wider age group than Movie B.
D. Movie B appeals to a wider age group than Movie A.

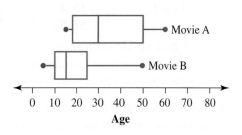

Age

11. **MC** The figures below show the age of the first 10 men and women to finish a marathon.
   Men: 28, 34, 25, 36, 25, 35, 22, 23, 40, 24
   Women: 19, 27, 20, 26, 30, 18, 28, 25, 28, 22
Choose the correct statement from the following.

A. The mean age of the men is greater than the mean age of the women.
B. The range is greater among the men than among the women.
C. The interquartile range is greater among the men than among the women.
D. The standard deviation is less among the men than among the women.

Communicating, reasoning and problem solving

12. **WE13** Cory recorded his marks for each test that he did in English and Science throughout the year.

    English: 55, 64, 59, 56, 62, 54, 65, 50

    Science: 35, 75, 81 32, 37, 62, 77, 75

    a. Determine the most appropriate measure of centre and measure of spread to compare Cory's performance in the two subjects.
    b. Identify the subject in which Cory received a better average. Justify your answer.
    c. Identify the subject in which Cory performed more consistently. Justify your answer.

13. Draw an example of a graph that is:

    a. symmetrical
    b. positively skewed with one mode
    c. negatively skewed with two modes.

14. The police set up two radar speed checks on a back street of Sydney and on a main road. In both places the speed limit is 60 km/h. The results of the first 10 cars that have their speed checked are given below.

    Back street: 60, 62, 58, 55, 59, 56, 65, 70, 61, 64

    Main road: 55, 58, 59, 50, 40, 90, 54, 62, 60, 60

    a. Calculate the mean and standard deviation of the readings taken at each point.
    b. Identify the road where drivers are generally driving faster. Justify your answer.
    c. Identify the road where the spread of readings is greater. Justify your answer.

15. In boxes of Smarties it is advertised that there are 50 Smarties in each box. Two machines are used to distribute the Smarties into the boxes. The results from a sample taken from each machine are listed below.

    Machine A: 44, 49, 49, 48, 47, 47, 46, 46, 45, 54, 53, 52, 52, 52, 51, 51, 51, 50, 50, 50,
                        50, 50, 50

    Machine B: 45, 47, 48, 49, 49, 49, 49, 49, 49, 49, 49, 50, 50, 50, 50, 50, 51, 51, 51, 51,
                        51, 52, 52, 53, 59

    a. Display the data from both machines on parallel box plots.
    b. Calculate the mean and standard deviation of the number of Smarties distributed from both machines.
    c. State which machine is the more dependable. Justify your answer.

16. Nathan and Timana are wingers in their local rugby league team. The number of tries they have scored in each season are listed below.

    Nathan: 25, 23, 13, 36, 1, 8, 0, 9, 16, 20
    Timana: 5, 10, 12, 14, 18, 11, 8, 14, 12, 19

    a. Calculate the mean number of tries scored by each player.
    b. Calculate the range of tries scored by each player. Justify your answer.
    c. Calculate the interquartile range of tries scored by each player. Justify your answer.
    d. State which player would you consider to be the more consistent. Justify your answer.

17. Year 9 students at Merrigong High School sit exams in
Science and Maths. The results are shown in the table below.

Mark	Number of students in Science	Number of students in Maths
51 – 60	7	6
61 – 70	10	7
71 – 80	8	12
81 – 91	8	9
91 – 100	2	6

a. Determine if either distribution is symmetrical.
b. If either distribution is not symmetrical, state whether it is positively or negatively skewed.
c. Discuss the possible reasons for any skewness.
d. State the modal class of each distribution.
e. Determine which subject has the greater standard deviation greater. Explain your answer.

18. A new drug for the relief of cold symptoms has been developed.
To test the drug, 40 people were exposed to a cold virus. Twenty
patients were then given a dose of the drug while another 20
patients were given a placebo. (In medical tests a control group
is often given a *placebo* drug. The subjects in this group believe
that they have been given the real drug but in fact their dose
contains no drug at all.)
All participants were then asked to indicate the time when they
first felt relief of symptoms. The number of hours from the time
the dose was administered to the time when the patients first felt
relief of symptoms are detailed below.

Group A (drug)

| 25, | 29, | 32, | 45, | 18, | 21, | 37, | 42, | 62, | 13, |
| 42, | 38, | 44, | 42, | 35, | 47, | 62, | 17, | 34, | 32 |

Group B (placebo)

| 25, | 17, | 35, | 42, | 35, | 28, | 20, | 32, | 38, | 35, |
| 34, | 32, | 25, | 18, | 22, | 28, | 21, | 24, | 32, | 36 |

a. Display the data for both groups on a parallel box plot.
b. Make comparisons of the data. Use statistics in your answer.
c. Explain if the drug works. Justify your answer.
d. Determine other considerations that should be taken into account when trying to draw conclusions from an experiment of this type.

19. The heights of Year 9 and Year 12 students (to the nearest centimetre) are being investigated. The results of some sample data are shown below.

Year 9	160	154	157	170	167	164	172	158	177	180	175	168	159	155	163	163	169	173	172	170
Year 12	160	172	185	163	177	190	183	181	176	188	168	167	166	177	173	172	179	175	174	108

a. Draw a parallel box plot.
b. Comment on what the parallel box plot tell you about the heights of Year 9 and Year 12 students.

20. Kloe compares her English and Maths marks. The results of eight tests in each subject are shown below.

English: 76, 64, 90, 67, 83, 60, 85, 37
Maths: 80, 56, 92, 84, 65, 58, 55, 62

    a. Calculate Kloe's mean mark in each subject.
    b. Calculate the range of marks in each subject.
    c. Calculate the standard deviation of marks in each subject.
    d. Based on the above data, determine the subject that Kloe has performed more consistently in.

21. A sample of 50 students was surveyed on whether they owned an iPad or a mobile phone. The results showed that 38 per cent of the students owned both. Sixty per cent of the students owned a mobile phone and there were four students who had an iPad only.
Evaluate the percentage of students that did not own a mobile phone or an iPad.

22. The life expectancy of non-Aboriginal and non–Torres Strait Islander people in Australian states and territories is shown on the box plot below.

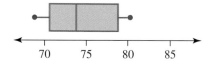

70   75   80   85

**Life expectancy of non-Aboriginal and non–Torres Strait
Islander people in Australian states and territories**

The life expectancies of Aboriginal and Torres Strait Islander people in each of the Australian states and territories are 56, 58.4, 51.3, 57.8, 53.9, 55.4 and 61.0.

    a. Draw parallel box plots on the same axes. Compare and comment on your results.
    b. Comment on the advantage and disadvantage of using a box plot.

23. Consider the box plots below that show the number of weekly sales of houses by two real estate agencies.

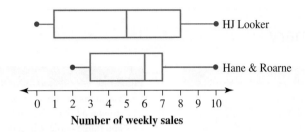

Number of weekly sales

a. Determine the median number of weekly sales for each real estate agency.
b. State which agency had the greater range of sales. Justify your answer.
c. State which agency had the greater interquartile range of sales. Justify your answer.
d. State which agency performed better. Explain your answer.

24. The following data give the box plots for three different age groups in a triathlon for under thirities.

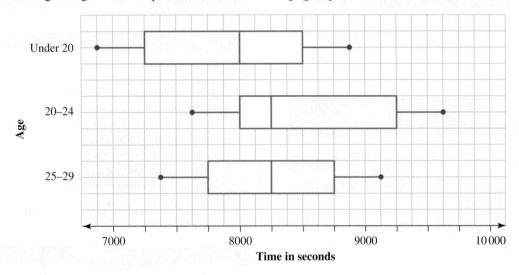

Time in seconds

a. Identify the slowest time for the 20–24 year olds.
b. Estimate the difference in time between the fastest triathlete in:

   i. the under 20s and the 20–24 group
  ii. the under 20s and the 25–29 group
  iii. the under 20–24 group and the 25–29 group.

c. Comment on the overall performance of the three groups.

13.7 Review

13.7.1 Topic summary

Symmetry and skewness in distributions

- A symmetrical plot has data that are evenly spaced around a central point. The median and mean are in the centre of the plot

- A negatively skewed plot has larger amounts of data at the higher end. The mean is smaller than the median.

- A positively skewed plot has larger amounts of data at the lower end. The mean is larger than the median.

DATA ANALYSIS

Measures of central tendency

- The three measures of central tendency are the mean, median and mode.
- The mean is the average of all values in a set of data. It is therefore affected by extreme values.

$$\bar{x} = \frac{\sum x}{n}$$

- The median is the middle value in an ordered set of data. It is located at the $\left(\frac{n+1}{2}\right)$th score.
- The mode is the most frequent value in a set of data.
- For the data set 2, 4, 5, 7, 7:

  The mean is $\dfrac{2+4+5+7+7}{5} = 5$.

  The median is the middle value, 5.
  The mode is 7.

Comparing data sets

- Measures of centre and measures of spread are used to compare data sets.
- It is important to consider which measure of centre or spread is the most relevant when comparing data sets.
- For example, if there are outliers in a data set, the median will likely be a better measure of centre than the mean, and the IQR would be a better measure of spread than the range.
- Parallel box plots can also be used to compare the spread of two or more data sets.

Measures of spread

- Measures of spread describe how far the data values are spread from the centre or from each other.
- The range is the difference between the maximum and minimum data values.

  Range = maxiumum value – minimum value

- The interquartile range (IQR) is the range of the middle 50% of the scores in an ordered set:

  IQR $= Q_3 - Q_1$

  where Q_1 and Q_3 are the first and third quartiles respectively.

Standard deviation

- The standard deviation is a more sophisticated measure of spread.
- The standard deviation measures how far, on average, each data value is away from the mean.
- The deviation of a data value is the difference between it and the mean $(x - \bar{x})$.
- The formula for the standard deviation is:

  $$\sigma = \sqrt{\frac{\sum (x - \bar{x})^2}{n}}$$

- The standard deviation is more easily calculated using a scientific calculator.

Box plots

- The five-number summary of a data set is a list containing:
  - the minimum value
  - the lower quartile, Q_1
  - the median
  - the upper quartile, Q_3
  - the maximum value.
- Boxplots are graphs of the five-number summary.

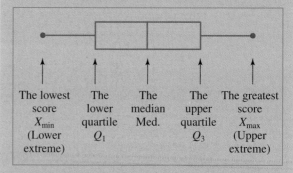

| The lowest score $X_{\min}$ (Lower extreme) | The lower quartile Q_1 | The median Med. | The upper quartile Q_3 | The greatest score $X_{\max}$ (Upper extreme) |

- Outliers are calculated and marked on the box plot. A score is considered an outlier if it falls outside the upper or lower boundary.

  Lower boundary $= Q_1 - 1.5 \times \text{IQR}$

  Upper boundary $= Q_3 + 1.5 \times \text{IQR}$

13.7.2 Project

Cricket scores

Data are used to predict, analyse, compare and measure many aspects of the game of cricket. Attendance is tallied at every match. Players' scores are analysed to see if they should be kept on the team. Comparisons of bowling and batting averages are used to select winners for awards. Runs made, wickets taken, no-balls bowled, the number of ducks scored in a game as well as the number of 4s and 6s are all counted and analysed after the game. Data of all sorts are gathered and recorded, and measures of central tendency and spread are then calculated and interpreted.

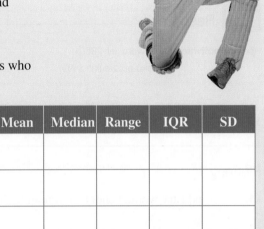

Sets of data have been made available for you to analyse, and decisions based on the resultant measures can be made.

Batting averages

The following table shows the runs scored by four cricketers who are vying for selection to the state team.

Player	Runs in the last 25 matches	Mean	Median	Range	IQR	SD
Will	13, 18, 23, 21, 9, 12, 31, 21, 20, 18, 14, 16, 28, 17, 10, 14, 9, 23, 12, 24, 0, 18, 14, 14, 20					
Rohit	2, 0, 112, 11, 0, 0, 8, 0, 10, 0, 56, 4, 8, 164, 6, 12, 2, 0, 5, 0, 0, 0, 8, 18, 0					
Marnus	12, 0, 45, 23, 0, 8, 21, 32, 6, 0, 8, 14, 1, 27, 23, 43, 7, 45, 2, 32, 0, 6, 11, 21, 32					
Ben	2, 0, 3, 12, 0, 2, 5, 8, 42, 0, 12, 8, 9, 17, 31, 28, 21, 42, 31, 24, 30, 22, 18, 20, 31					

1. Calculate the mean, median, range, IQR and standard deviation scored for each cricketer.
2. You need to recommend the selection of two of the four cricketers. For each player, write two points as to why you would or would not select them. Use statistics in your comments.

Bowling averages

The bowling average is the number of runs per wicket taken:

$$\textbf{Bowling average} = \frac{\textbf{number of runs scored}}{\textbf{number of wicket taken}}$$

The smaller the average, the better the bowler has performed.

Josh and Ravi were competing for three bowling awards:
- Best in semifinal
- Best in final
- Best overall

The following table gives their scores.

	Semifinal		Final	
	Runs scored	Wickets taken	Runs scored	Wickets taken
Josh	12	5	28	6
Ravi	10	4	15	3

2. Calculate the bowling averages for the following and fill in the table below.
   - Semifinal
   - Final
   - Overall

	Semifinal average	Final average	Overall average
Josh			
Ravi			

3. Explain how Ravi can have the better overall average when Josh has the better average in both the semifinal and final.

 Resources

 Interactivities Crossword (int-2860)
   Sudoku puzzle (int-3599)

Exercise 13.7 Review questions

learnon

Fluency

1. What industry would benefit from finding the mode of a data set?

2. Can the exact interquartile range be calculated from a grouped frequency distribution table?

3. Calculate the mean, median and mode for each of the following sets of data:
   a. 7, 15, 8, 8, 20, 14, 8, 10, 12, 6, 19

   b. Key: $1|2 = 12$

Stem	Leaf
1	2 6
2	1 7 8
3	0 3 3 4 6 8
4	0 1 1 5 9
5	1 3 6

   c.

Score (x)	Frequency (f)
70	2
71	6
72	9
73	7
74	4

4. For each of the following data sets, calculate the range.

a. 4, 3, 6, 7, 2, 5, 8, 4, 3

b.

x	13	14	15	16	17	18	19
f	3	6	7	12	6	7	8

c. Key: $1|8 = 18$

Stem	Leaf
1	7 8 8 9
2	1 2 4 4 5 7 7 7 8 9 9
3	0 0 0 1 3 4 7

5. For each of the following data sets, calculate the interquartile range.

a. 18, 14, 15, 19, 20, 11 16, 19, 18, 19

b. Key: $9|8 = 9.8$

Stem	Leaf
8	7 8 8 9
9	0 2 4 4 5 7 7 7 8 9 9
10	0 1 1 1 3

6. Consider the box plot shown.

a. Calculate the median.

b. Calculate the range.

c. Determine the interquartile range.

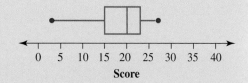

Understanding

7. The typing speed in words per minute (wpm) of 30 Year 8 and Year 9 students are listed below.

Year 8: 9, 9, 19, 18, 16, 15, 14, 12, 10, 29, 28, 28, 26, 24, 22,
21, 20, 20, 39, 37, 37,36, 34, 31, 30, 48, 46, 45, 42, 40

Year 9: 17, 19, 22, 23, 26, 28 29, 30, 32, 35, 35, 37, 38, 38, 41,
42, 45, 48, 48, 49, 49,50, 53, 55, 57, 58, 60, 60, 63

a. Using a calculator or otherwise, construct a pair of parallel box plots to represent the two sets of data.

b. Calculate the mean, median, range, interquartile range and standard deviation of each set.

c. Compare the two distributions, using your answers to parts **a** and **b**.

8. The following data give the amount of cut meat (in kg) obtained from each of 20 lambs.

4.5	6.2	5.8	4.7	4.0	3.9	6.2	6.8	5.5	6.1
5.9	5.8	5.0	4.3	4.0	4.6	4.8	5.3	4.2	4.8

a. Create a stem-and-leaf plot of the data. (Use a class size of 0.5 kg.)

b. Prepare a five-point summary of the data.

c. Draw a box plot of the data.

9. Calculate the standard deviation of each of the following data sets
   a. 58, 12, 98, 45, 60, 34, 42, 71, 90, 66
   b.

x	1	2	3	4	5
f	2	6	12	8	5

   c. Key: $1|4 = 14$

Stem	Leaf
0	1 3 4 4 5 7 8
1	0 0 0 1 2 2 4 5 7 8 9
2	0 2 2 3 5 7

10. **MC** The Millers obtained a number of quotes on the price of having their home painted. The quotes, to the nearest hundred dollars, were:

    4200, 5100, 4700, 4600, 4800, 5000, 4700, 4900

    The standard deviation for this set of data, to the nearest whole dollar, is:

    A. 260   **B.** 278   **C.** 324   **D.** 325

11. **MC** The number of Year 12 students who, during semester 2, spent all their spare periods studying in the resource centre is shown on the stem-and-leaf plot below.

    Key: $2|5 = 25$ students

Stem	Leaf
0	8
1	
2	5 6 6 7
3	0 2 3 6 9
4	7 9
5	6
6	1

    The standard deviation for this set of data, to the nearest whole number is:

    A. 12   **B.** 14   **C.** 17   **D.** 35

12. Each week, varying amounts of a chemical are added to a filtering system. The amounts required (in mL) over the past 19 weeks are shown in the stem-and-leaf plot.

    Key: $3|8$ represents 0.38 ml

Stem	Leaf
2	1
2	2 2
2	4 4 4 5
2	6 6
2	8 8 99
3	0
3	2 2
3	4
3	6
3	8

Calculate the standard deviation of the amounts used.

13. Calculate the mean, median and mode of this data set: 2, 5, 6, 2, 5, 7, 8. Comment on the shape of the distribution.

14. The box plot shows the heights (in cm) of Year 12 students in a Maths class.

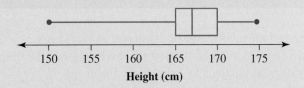

Height (cm)

   a. State the median class height.
   b. Calculate the range of heights.
   c. Calculate the interquartile range of the heights.

Communicating, reasoning and problem solving

15. **MC** A data set has a mean of 75 and a standard deviation of 5. Another score of 50 is added to the data set. Choose the correct statement from the following.
   A. The mean will increase and the standard deviation will increase.
   B. The mean will increase and the standard deviation will decrease.
   C. The mean will decrease and the standard deviation will increase.
   D. The mean will decrease and the standard deviation will decrease.

16. **MC** A data set has a mean of 60 and a standard deviation of 10. A score of 100 is added to the data set. This score becomes the highest score in the data set. Identify which of the following will increase.
   Note: There may be more than one correct answer.
   A. Mean
   B. Standard deviation
   C. Range
   D. Interquartile range

17. A sample of 30 people was selected at random from those attending a local swimming pool. Their ages (in years) were recorded as follows:

19, 7, 58, 41, 17, 23, 62, 55, 40, 37, 32, 29, 21, 18, 16
10, 40, 36, 33, 59, 65, 68, 15, 9, 20, 29, 38, 24, 10, 30

a. Calculate the mean and the median age of the people in this sample.
b. Group the data into class intervals of 10 (0–9 etc.) and complete the frequency distribution table.
c. Use the frequency distribution table to estimate the mean age.
d. Calculate the cumulative frequency and hence plot the ogive.
e. Estimate the median age from the ogive.
f. Compare the mean and median of the original data in part **a** with the estimates of the mean and the median obtained for the grouped data in parts **c** and **e**.
g. Determine if the estimates were good enough. Explain your answer.

18. The table below shows the number of cars that are garaged at each house in a certain street each night.

Number of cars	Frequency
1	9
2	6
3	2
4	1
5	1

a. Show these data in a frequency histogram.
b. State if the data is positively or negatively skewed. Justify your answer.

19. Consider the data set represented by the frequency histogram shown.
    a. Explain if the data is symmetrical.
    b. State if the mean and median of the data can be seen. If so, determine their values.
    c. Evaluate the mode of the data.

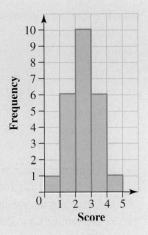

20. There are $3m$ values in a data set for which $\bar{x} = m$ and $\sigma = \dfrac{m}{2}$.

    a. Comment on the changes to the mean and standard deviation if each value of the data set is multiplied by m.
    b. An additional value is added to the original data set, giving a new mean of $m + 2$. Evaluate the additional value.

21. The following data show the number of pets in each of the 12 houses in Coral Avenue, Rosebud.
    $$2, 3, 3, 2, 2, 3, 2, 4, 3, 1, 1, 0$$

    a. Calculate the mean and median number of pets.
    b. The empty block of land at the end of the street was bought by a Cattery and now houses 20 cats. Recalculate the mean and median.
    c. Explain why the answers are so different, and which measure of central tendency is best used for certain data.

22. The number of Year 9 students in all the 40 schools in the Northern District of the Education Department was recorded as follows:

    56, 134, 93, 67, 123, 107, 167, 124, 108, 78, 89, 99, 103, 107, 110, 45, 112, 127, 106, 111, 127, 145, 87, 75, 90, 123, 100, 87, 116, 128, 131, 106, 123, 87, 105, 112, 145, 115, 126, 92

    a. Using an interval of 10, produce a table showing the frequency for each interval.
    b. Use the table to estimate the mean.
    c. Calculate the mean of the ungrouped data.
    d. Compare the results from parts b and c and explain any differences.

23. The ages of a group of 30 golf players and 30 tennis players as they enter hospital for the first time are listed below.

Golf players: 9, 8, 19, 19, 18, 189, 8, 18, 16, 13, 12, 11, 28, 27, 27, 26, 24, 23, 22, 20, 38, 36, 31, 30, 47, 45, 42, 55, 53, 78

Tennis players: 5, 17, 17, 18, 19, 19 20, 20, 21, 22, 24, 25, 25, 26, 27, 29, 30, 31, 33, 33, 35, 38, 42, 43, 46, 48, 51, 53, 54, 62

a. Construct a pair of parallel box plots to represent the two sets of data, showing working out for the median and 1st and 3rd quartiles.
b. Calculate the mean, range and IQR for both sets of data.
c. Determine any outliers if they exist.
d. Write a short paragraph comparing the data.

24. The times, in seconds, of the duration of 20 TV advertisements shown in the 6–8 pm time slot are recorded below.

16, 60, 35, 23, 45, 15 25, 55, 33, 20, 22, 30, 28, 38, 40, 18, 29, 19, 35, 75

a. From the data, determine the:
  i. mode
  ii. median
  iii. mean
  iv. range
  v. lower quartile
  vi. upper quartile
  vii. interquartile range.
b. Using your results from part a, construct a box plot for the time, in seconds, for the 20 TV advertisements in the 6–8 pm time slot.
c. From your box plot, determine:
  i. the percentage of advertisements that are more than 39 seconds in length
  ii. the percentage of advertisements that last between 21 and 39 seconds
  iii. the percentage of advertisements that are more than 21 seconds in length

The types of TV advertisements during the 6–8 pm time slot were categorised as Fast food, Supermarkets, Program information and Retail (clothing, sporting goods, furniture).
A frequency table for the frequency of these advertisements is shown below.

Type	Frequency
Fast food	7
Supermarkets	5
Program information	3
Retail	5

d. State the type of data that has been collected in the table.
e. Determine the percentage of advertisements that are advertisements for fast food outlets.
f. Suggest a good option for a graphical representation of this type of data.

25. The speeds, in km/h, of 55 cars travelling along a major road are recorded below.

Speed	Frequency
60–64	1
65–69	1
70–74	10
75–79	13
80–84	9
85–89	8
90–94	6
95–99	3
100–104	2
105–109	1
110–114	1
Total	55

a. By calculating the midpoint for each class interval, determine the mean speed, in km/h, of the cars travelling along the road.

Write your answer correct to 2 decimal places.

b. The speed limit along the road is 75 km/h. A speed camera is set to photograph the license plates of cars travelling 7% more than the speed limit. A speeding fine is automatically sent to the owners of the cars photographed.

Based on the 55 cars recorded, determine the number of speeding fines that were issued.

c. Drivers of cars travelling 5 km/h up to 15 km/h over the speed limit are fined $135. Drivers of cars travelling more than 15 km/h and up to 25 km/h over the speed limit are fined $165 and drivers of cars recorded travelling more than 25 km/h and up to 35 km/h are fined $250. Drivers travelling more than 35 km/h pay a $250 fine in addition to having their driver's license suspended.

If it is assumed that this data is representative of the speeding habits of drivers along a major road and there are 30 000 cars travelling along this road on any given month.

i. Determine the amount, in dollars, collected in fines throughout the month. Write your answer correct to the nearest cent.

ii. Evaluate the number of drivers that would expect to have their licenses suspended throughout the month.

To test your understanding and knowledge of this topic, go to your learnON title at www.jacplus.com.au and complete the **post-test**.

Answers

Topic 13 Data analysis

13.1 Pre-test

1. 2
2. 29
3. 72
4. 29.8
5. (7, 13, 15.5, 16.5, 20)
6. 22
7. 13
8. Positively skewed
9. 11
10. C
11. $\bar{x} = 184.42$ and $\sigma = 7.31$
13. D
13. B
14. The mean typing speed is 40.57 and IQR is 20 for Year 9.
15. 5.83

13.2 Measures of central tendency

1. a. 7   b. 8   c. 8
2. a. 6.875   b. 7   c. 4, 7
3. a. 39.125   b. 44.5   c. No mode
4. a. 4.857   b. 4.8   c. 4.8
5. a. 12   b. 12.625   c. 13.5
6. Maths: mean = 69.12, median = 73, mode = 84
7. a. 5.83   b. 6   c. 6
8. a. 14.425   b. 15   c. 15
9. a. Mean = 2.5, median = 2.5
   b. Mean = 4.09, median = 3
   c. Median
10. a. $72\frac{2}{3}$ b. 75   c. 70 − < 80
11. 124.83
12. 65 − < 70
13. a. B   b. B   c. C   d. D
14. a. Mean = $32.93, median = $30
    b.

Class interval	Frequency	Cumulative frequency
0 − 9	5	5
10 − 19	5	10
20 − 29	5	15
30 − 39	3	18
40 − 49	5	23
50 − 59	3	26
60 − 69	3	29
70 − 79	1	30
Total	30	

Mean = $32.50, median = $30

c.

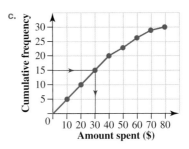

d. The mean is slightly underestimated; the median is exact. The estimate is good enough as it provides a guide only to the amount that may be spent by future customers.

15. a. 3
    b. 4, 5, 5, 5, 6 (one of the possible solutions)
    c. One possible solution is to exchange 15 with 20.
16. a. Frequency column: 16, 6, 4, 2, 1, 1
    b. 6.8
    c. 0 − 4 hours
    d. 0 − 4 hours
17. A
18. a. Frequency column: 1, 13, 2, 0, 1, 8
    b.

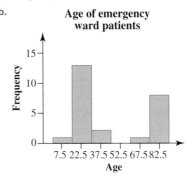

c. Asymmetrical or bimodal (as if the data come from two separate graphs).
d. 44.1
e. 15 − <30
f. 15 − <30
g.

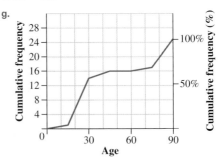

h. 28
i. No
j. Sample responses can be found in the worked solutions in the online resources.

19. a. Player A mean $= 34.33$, Player B mean $= 41.83$

b. Player B

c. Player A median $= 32.5$, Player B median $= 0$

d. Player A

e. Player A is more consistent. One large score can distort the mean.

20. a. Sample responses can be found in the worked solutions in the online resources.

b. Sample responses can be found in the worked solutions in the online resources.

21. a. Frequency column: 3, 8, 5, 3, 1

b. 50.5

c. $40 - < 50$

d. $40 - < 50$

e.

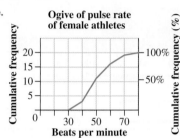

f. Approximately 48 beats/ min

22. Answers will vary. Sample responses include:

a. 3, 4, 5, 5, 8

b. 4, 4, 5, 10, 16

c. 2, 3, 6, 6, 12

23. 12

24. $\dfrac{2a + b}{3}$

25. 13, 31, 31, 47, 53, 59

13.3 Measures of spread: Range and interquartile range

1. a. 15   **b.** 77.1   **c.** 9

2. a. 7   **b.** 7   **c.** 8.5   **d.** 39

3. a. 3.3 kg   **b.** 1.5 kg

4. 22 cm

5. 0.8

6. C

7. a.

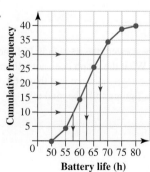

b. i. 62.5

ii. $Q_1 = 58$, $Q_3 = 67$

iii. 9

iv. 14

v. 6

8. $IQR = 23$

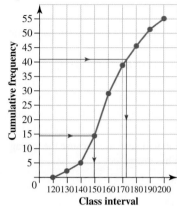

9. a. i. Range $= 23$

ii. $IQR = 13.5$

b. i. Range $= 45$

ii. $IQR = 27.5$

c. i. Range $= 49$

ii. $IQR = 20$

10. Measures of spread tell us how far apart the values (scores) are from one another.

11. a. 25.5

b. 28

c. 39

d. 6

e. The three lower scores affect the mean and range but not the median and the IQR.

12. mean $= 29.1$; median $= 27.5$; range $= 36$; IQR $= 13$

13.

| Mean $= 25.036$, | Median $= 24.9$, | Mode $= 23.6$ |
| Range $= 8.5$, | IQR $= 3.4$ | |

14. One possible solution is: $a = 22$, $b = 9$, $c = 9$ and $d = 8$

15. a. Yes

Range $= 10 - 1 = 9$

Median (middle score) $= 6$

b. No there is not. If 6 was added the range and the median would stay the same but Q1 and Q3 would change.

13.4 Measures of spread: Standard deviation

1. a. 2.29   **b.** 2.19   **c.** 20.17   **d.** 3.07

2. a. 1.03   **b.** 1.33   **c.** 2.67   **d.** 2.22

3. 10.82

4. Mean $= 1.64\%$ Std dev. $= 0.45\%$

5. Mean $= 1.83$ Std dev. $= 0.06$ m

6. 0.49 s

7. 15.10 calls

8. B

9. The mean of the first 25 cars is 89.24 km/h with a standard deviation of 5.60. The mean of the first 26 cars is 91.19 with a standard deviation of 11.20, indicating that the extreme speed of 140 km/h is an anomaly. By including the 26th car it has more than doubled the standard deviation.

10. The standard deviation tells us how spread out the data is from the mean

11. a. $\sigma \approx 3.58$

   b. The mean is increased by 7 but the standard deviation remains at $\sigma \approx 3.58$.

12. 57 is two standard deviations above the mean.

13. a. New mean is the old mean increased by 3 (15) but no change to the standard deviation.

   b. New mean is 3 times the old mean (36) and new standard deviation is 3 times the old standard deviation (12).

14. The standard deviation will decrease because the average distance to the mean has decreased.
   Set 1: 1, 3, 7, 12, 12
   Mean: 7; SD: 4.52
   Set 2: 1, 3, 7, 7, 12, 12
   Mean; 7; SD: 4.12

15. a. 43   b. 12   c. 12.19

13.5 Box plots

1. a. 5   b. 26

2. a. 6   b. 27

3. a. 5.8   b. 18.6

4. a. 140   b. 56   c. 90   d. 84   e. 26

5. a. 58   b. 31   c. 43   d. 27   e. 14

6. B

7. C

8. C, D

9. a. (22, 28, 35, 43, 48)

   b.
   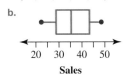
   Sales

   c. No outliers

10. a. (10, 13.5, 22, 33.5, 45)

   b.
   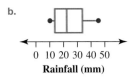
   Rainfall (mm)

11. a. (18, 20, 26, 43.5, 74)

   b.
   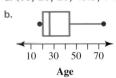
   Age

   c. No outliers

   d. The distribution is positively skewed, with most of the offenders being young drivers.

12. a. (124 000, 135 000, 148 000, 157 000, 175 000)

   b.
   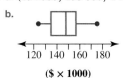
   ($ × 1000)

13.
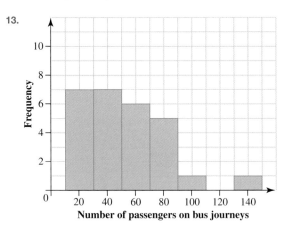
Number of passengers on bus journeys

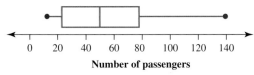
Number of passengers

14. The advantages of box plots is that they are clear visual representations of 5-number summary, display outliers and can handle a large volume of data. The disadvantage is that individual scores are lost.

15. a. Key: 12|1 = 121.

Stem	Leaf
12	1569
13	124
14	3488
15	022257
16	35
17	29
18	1112378

   b.
   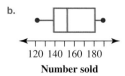
   Number sold

   c. The data is negatively skewed, the median is 152 and there are no outliers. What the graph does not tell you is on most days the hamburger sales are less than 160. Over the weekend the sales figures spike beyond this.

16. a. Key: 1*|7 = 17 years

Stem	Leaf
1*	7788899
2	000122223333444
2*	5589
3	123
3*	
4	
4*	8

b.

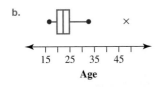

Age

c. The distribution is positively skewed, with most of the first-time mothers being under the age of 30, half are between the age of 20-25. There is one outlier (48) in this group.

17.

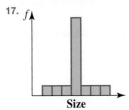

Size

18. a. $50   **b.** $135   **c.** $100   **d.** $45   **e.** 50%

13.6 Describing and comparing data sets

1. a. The mean of the first set is 63. The mean of the second set is 63.

b. The standard deviation of the first set is. 2.38 The standard deviation of the second set is 6.53.

c. For both sets of data the mean is the same, 63. However, the standard deviation for the second set (6.53) is much higher than the standard deviation of the first set (2.38), implying that the second set is more widely distributed than the first. This is confirmed by the range, which is $67 - 60 = 7$ for the first set and $72 - 55 = 17$ for the second.

2. a. Morning: median = 2.45; afternoon: median = 1.6

b. Morning: range = 3.8; afternoon: range = 5

c. The waiting time is generally shorter in the afternoon. One outlier in the afternoon data causes the range to be larger. Otherwise, the afternoon data are far less spread out.

3. a. Year 9: Median = 1.66
Year 10: Median = 1.64

b. Year 9: Range = 0.17
Year 10: Range = 0.19

c. The shape of the Year 10 distribution is positively skewed. Without 1.70 m height, it would be bimodal. The shape of the Year 9 distribution is negatively skewed.

4. a. Ford: median = 15; Hyundai: median = 16

b. Ford: range = 26; Hyundai: range = 32

c. Ford: IQR = 14; Hyundai: IQR = 13.5

d.

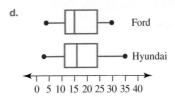

5. a. Brisbane Lions

b. Brisbane Lions: range = 65;
Sydney Swans: range = 55

c. Brisbane Lions: IQR = 40;
Sydney Swans: IQR = 35

d. Brisbane lions: negatively skewed
Sydney Swans: slightly positively skewed/almost symmetrical

6.

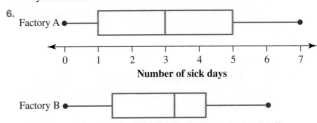

Both graphs indicate that the data is slightly negatively skewed. However, the box plot provides an excellent summary of the centre and spread of the distribution.

7.

	X_{min}	Q_1	Median	Q_3	X_{max}
Before	75	86	95	128.5	152
After	66	81	87	116	134

b.

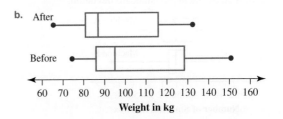

Weight in kg

c. As a whole, the program was effective. The median weight dropped from 95 kg to 87 kg, a loss of 8 kg. A noticeable shift in the graph shows that after the program 50% of participants weighed between 66 and 87 kg, compared to 25% of participants weighing between 75 and 86 kg before they started. Before the program the range of weights was 77 kg (from 75 kg to 152 kg); after the program the range had decreased to 68 kg. The IQR also diminished from 42.5 kg to 35 kg.

8. a.

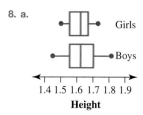

Height

b. Boys: median = 1.62; girls: median = 1.62

c. Boys: range = 0.36; girls: range = 0.23

d. Boys: IQR = 0.25; girls: IQR = 0.17

e. Although boys and girls have the same median height, the spread of heights is greater among boys as shown by the greater range and interquartile range.

9. a. Summer: range = 23; winter: range = 31

b. Summer: IQR = 14; winter: IQR = 11

c. There are generally more cold drinks sold in summer as shown by the higher median. The spread of data is similar as shown by the IQR although the range in winter is greater.

10. A

11. A, B, C

12. a. In order to include all data values in the calculation of measures of centre and spread, calculate the mean and standard deviation.

b. Cory achieved a better average mark in Science (59.25) than he did in English (58.13).

c. Cory was more consistent in English ($\sigma = 4.94$) than he was in Science ($\sigma = 19.74$)

13. Sample responses can be found in the worked solutions in the online resources.

14. a. Back street: $\bar{x} = 61$, $\sigma = 4.27$; main road: $\bar{x} = 58.8$, $\sigma = 12.06$

b. The drivers are generally driving faster on the back street.

c. The spread of speeds is greater on the main road as indicated by the higher standard deviation.

15. a.

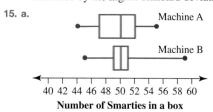

Number of Smarties in a box

b. Machine A: mean = 49.88, standard deviation = 2.87; Machine B: mean = 50.12, standard deviation = 2.44

c. Machine B is more reliable, as shown by the lower standard deviation and IQR. The range is greater on machine B only because of a single outlier.

16. a. Nathan: mean = 15.1; Timana: mean = 12.3

b. Nathan: range = 36; Timana: range = 14

c. Nathan: IQR = 15; Timana: IQR = 4

d. Timana's lower range and IQR shows that he is the more consistent player.

17. a. Yes — Maths

b. Science: positively skewed

c. The Science test may have been more difficult.

d. Science: 61 − 70, Maths: 71 − 80

e. Maths has a greater standard deviation (12.64) compared to Science (11.94). Maths data is more spread out from the mean, Sciences data is more consistent (not by much).

18. a. Five-point summary

Group A:	13	27	36	43	62
Group B:	17	23	30	35	42

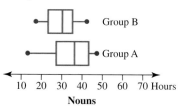

Nouns

b. Sample responses can be found in the worked solutions in the online resources.

c. Sample responses can be found in the worked solutions in the online resources.

d. Sample responses can be found in the worked solutions in the online resources.

19. a.

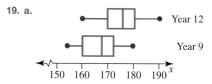

b. On average, the Year 12 students are about 6 − 10 cm taller than the Year 9 students. The heights of the majority of Year 12 students are between 170 cm and 180 cm, whereas the majority of the Year 9 students are between 160 and 172 cm in height.

20. a. English: mean = 70.25; Maths: mean = 69

b. English: range = 53; Maths: range = 37

c. English: $\sigma = 16.06$; Maths: $\sigma = 13.35$

d. Kloe has performed more consistently in Maths as the range and standard deviation are both lower.

21. 32%

22. a.

The parallel box plots show a significant gap between the life expectancy of Aboriginal and Torres Strait Islander people and that of non-Aboriginal and non–Torres Strait Islander people. Even the maximum median age of Aboriginal and Torres Strait Islander people is much lower than the minimum of non-Aboriginal and non–Torres Strait Islander people.

b. The advantage of box plots is that it gives a clear graphical representation of the results and in this case shows a significant difference between the median life expectancy of Aboriginal and Torres Strait Islander people and non-Aboriginal and non–Torres Strait Islander people. The disadvantage is that we lose the data for individual states and territories.

23. a. HJ Looker: median = 5; Hane and Roarne: median = 6

b. HJ Looker

c. HJ Looker

d. Hane and Roarne had a higher median and a lower spread and so they appear to have performed better.

24. a. 9625 seconds

b. i. Under 20 − (20–24): 750 seconds difference

ii. Under 20 − (25–29): 500 seconds difference

iii. (20–24) − (25–29): 250 seconds difference

c. The under-20s performed best of the three groups, with the fastest time for each metric (minimum time, first quartile, median, third quartile and maximum time). The next best performing group was the 25–29-year-olds. They had the same median as the 20–24-year-olds, but outperformed them in all of the other metrics. The 24–29-year-olds were the most consistent group, with a range of 1750 seconds compared to the range of 2000 seconds of the other groups.

Project

1. See the table at the bottom of the page*

2. a. Will has a similar mean and median, which shows he was fairly consistent. The range and IQR values are low indicating that his scores remain at the lower end with not much deviation for the middle 50%.

b. Rohit has the best average but a very low median indicating his scores are not consistent. The range is extremely high and the IQR very low in comparison showing he can score very well at times but is not a consistent scorer.

c. Marnus has a similar mean to Will and Ben but a lower median, indicating his scores are sometimes high but generally are lower than the average. The range and IQR show a consistent batting average and spread with only a few higher scores and some lower ones.

d. Ben has a similar mean and median, which shows he was a consistent player. The range and IQR show a consistent batting average and spread.

Players to be selected:
Would recommend **Will** if the team needs someone with very consistent batting scores every game but no outstanding runs.

Would recommend **Rohit** if the team needs someone who might score very high occasionally but in general fails to score many runs.

Would recommend **Marnus** if the team needs someone who is fairly consistent but can score quite well at times and the rest of the time does OK.

Would recommend **Ben** if the team needs someone who is fairly consistent but can score quite well at times and the rest of the time has a better median than Glenn.

3.

	Semifinal average	Final average	Overall average
Josh	2.4	4.67	3.64
Ravi	2.5	5	3.57

4. In the final, wickets were more costly than in the semifinal. That is, Josh conceded many runs in getting his six wickets. This affected the overall mean. In reality, Josh was the most valuable player overall, but this method of combining the data of the two matches led to this unexpected result.

13.7 Review questions

1. Sales industry. For example, a shop that sells shoes. The mode would be the most helpful as it tells us what shoe size is the most popular and therefore which size to order more stock of.

2. No, as the raw data is not listed. It has been placed into groups and to calculate the interquartile range the group centre would be used, giving us an aproximation.

3. a. Mean = 11.55; median = 10; mode = 8

b. Mean = 36; median = 36; mode = 33, 41

c. Mean = 72.18; median = 72; mode = 72

4. a. 6   **b.** 6   **c.** 20

5. a. 4   **b.** 0.85

6. a. 20   **b.** 24   **c.** 8

7. a.

*1.

Player	Runs in the last 25 matches	Mean	Median	Range	IQR	SD
Will	13, 18, 23, 21, 9, 12, 31, 21, 20, 18, 14, 16, 28, 17, 10, 14, 9, 23, 12, 24, 0, 18, 14, 14, 20	16.76	17	31	8.5	6.51
Rohit	2, 0, 112, 11, 0, 0, 8, 0, 10, 0, 56, 4, 8, 164, 6, 12, 2, 0, 5, 0, 0, 0, 8, 18, 0	17.04	4	164	10.5	38.10
Marnus	12, 0, 45, 23, 0, 8, 21, 32, 6, 0, 8, 14, 1, 27, 23, 43, 7, 45, 2, 32, 0, 6, 11, 21, 32	16.76	12	45	25.5	14.60
Ben	2, 0, 3, 12, 0, 2, 5, 8, 42, 0, 12, 8, 9, 17, 31, 28, 21, 42, 31, 24, 30, 22, 18, 20, 31	16.72	17	42	25	12.90

b. Year 8: mean = 26.83, median = 27, range = 39,
IQR = 19, sd = 11.45
Year 9: mean = 40.9, median = 41, range = 46, IQR = 22, sd = 12.98

c. The typing speed of Year 9 students is about 13 to 14 wpm faster than that of Year 8 students. The spread of data in Year 8 is slightly less than in Year 9.

8. a. Key: 3*|9 = 3.9 kg

Stem	Leaf
3*	9
4	0 0 0 2 3
4*	5 6 7 8
5	0 3
5*	5 8 8 9
6	1 2 2
6*	8

b. (3.9, 4.4, 4.9, 5.85, 6.8)

c.

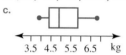

3.5 4.5 5.5 6.5   kg

9. a. 24.45   b. 1.10   c. 7.35

10. A

11. B

12. 0.05 ml

13. Mean = 5, median = 5, mode = 2 and 5.
The distribution is positively skewed and bimodal.

14. a. Median height = 167 cm
b. Range = 25 cm
c. IQR = 5 cm

15. C

16. A, B and C

17. a. Mean = 32.03; median = 29.5

b.

Class interval	Frequency
0 − 9	2
10 − 19	7
20 − 29	6
30 − 39	6
40 − 49	3
50 − 59	3
60 − 69	3
Total	30

c. Mean = 31.83

d.

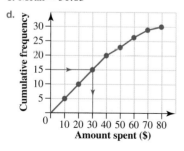

e. Median = 30

f. Estimates from parts c and e were fairly accurate.

g. Yes, they were fairly close to the mean and median of the raw data.

18. a.

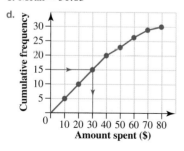

b. Positively skewed — a greater number of scores is distributed at the lower end of the distribution.

19. a. Yes, as the shape of the histogram is a bell curve. The mean, median and mode are located at the same point (in the middle).

b. Yes. Both are 3

c. 3

20. a. $\bar{x} = m^2$ $\sigma = \dfrac{m^2}{2}$ b. $7m + 2$

21. a. Mean = 2.17, median = 2

b. Mean = 3.54, median = 2

c. The median relies on the middle value of the data and won't change much if an extra value is added. The mean however has increased because this large value will change the average of the numbers. The mean is used as a measure of central tendency if there are no outliers or if the data are symmetrical. The median is used as a measure of central tendency if there are outliers or the data are skewed.

22. a.

Interval	Frequency (f)	Midpoint × (f)
40 − 49	1	44.5 × 1 = 44.5
50 − 59	1	54.5 × 1 = 54.5
60 − 69	1	64.5 × 1 = 64.5
70 − 79	2	74.5 × 2 = 149
80 − 89	4	84.5 × 4 = 338
90 − 99	4	94.5 × 4 = 378
100 − 109	8	104.5 × 8 = 836
110 − 119	6	114.5 × 6 = 687
120 − 129	8	124.5 × 8 = 996
130 − 139	2	134.5 × 2 = 269
140 − 149	2	144.5 × 2 = 289
150 − 159	0	154.5 × 0 = 0
160 − 169	1	164.5 × 1 = 164.5
Total	40	4270

b. 106.75

c. 107.15

d. The differences in this case were minimal; however, the grouped data mean is not based on the actual data but on the frequency in each interval and the interval midpoint. It is unlikely to yield an identical value to the actual mean. The spread of the scores within the class interval has a great effect on the grouped data mean.

23. a.

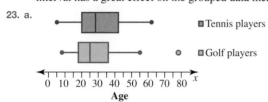

Tennis players

Golf players

b.

	Golf players	Tennis players
Mean	28.2	31.1
Range	70	57
IQR	18	22

c. There is one outlier — a golf player aged 78.

d. Typically golf players seem to enter hospital for the first time at a younger age than tennis players. Tennis players enter hospitals over a larger age range than golf players.

20–25-year-olds had the most first-time visits for golf players, tennis players were older.

24. a. i. 35 s

   ii. 29.5 s

   iii. 33.05 s

   iv. 60 s

   v. 21 s

   vi. 39 s

   vii. 18 s

b.

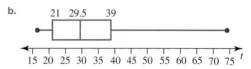

c. i. 25%

   ii. 50%

   iii. 75%

d. Categorical

e. 35%

f. Pictogram, pie chart or bar chart.

25. a. 82.73 km/h

b. 30 cars

c. i. $2 607 272.73

   ii. About 545

14 Probability

LESSON
14.1 Overview

Why learn this?

New technologies that predict human behaviour are all based on probability. On 11 May 1997, IBM's supercomputer Deep Blue made history by defeating chess grandmaster Gary Kasparov in a six-game match under standard time controls. This was the first time that a computer had defeated the highest ranked chess player in the world. Deep Blue won by evaluating millions of possible positions each second and determining the probability of victory from each possible choice. This application of probability allowed artificial intelligence to defeat a human who had spent his life mastering the game.

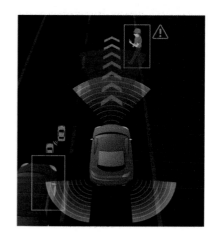

Combining advanced computing with probability is no longer used only to play games. From the daily convenience of predictive text, through to assisting doctors with cancer diagnosis, companies spend vast amounts of money on developing probability-based software. Self-driving cars use probability functions to predict the behaviour of other cars on the road and pedestrians. The more accurate the predictive functions, the safer self-driving cars can become. Even politicians use campaign data analysis to develop models that produce predictions about individual citizens' likelihood of supporting specific candidates and issues, and the likelihood of these citizens changing their support if they're targeted with various campaign interventions.

As technology improves, so will its predictive power in determining the likelihood of certain outcomes occurring. It is important that we study and understand probability so we know how technology is being used and the impact it will have on our day-to-day life.

Hey students! Bring these pages to life online

▶ Watch videos

🧩 Engage with interactivities

A+ Answer questions and check solutions

Find all this and MORE in jacPLUS ▶

Reading content and rich media, including interactivities and videos for every concept

Extra learning resources

Differentiated question sets

Questions with immediate feedback, and fully worked solutions to help students get unstuck

1. **MC** A six-sided die is rolled, and the number uppermost is noted. The event of rolling an even number is:

   A. $\{1, 2, 3, 4, 5, 6\}$ **B.** $\{0, 1, 2, 3, 4, 5, 6\}$ **C.** $\{2, 4, 6\}$ **D.** $\{1, 3, 5\}$

2. The coloured spinner shown is spun once and the colour is noted.

   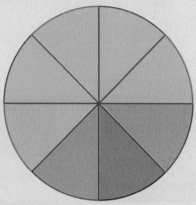

   Written in its simplest form, state the probability of spinning:
   a. an orange and a blue   **b.** an orange or a pink.

3. A coin is tossed in an experiment and the outcomes are recorded.

Outcome	Heads	Tails
Frequency	72	28

   a. Identify how many trials there were.
   b. Calculate the observed probability for tossing a Tail, giving your answer in simplest form.
   c. Calculate the theoretical probability for tossing a Tail with a fair coin, giving your answer in simplest form.
   d. In simplest form, calculate the difference in this experiment between the theoretical and observed probabilities for tossing a Tail.

4. A random number is selected from:
   $S = \{1, 2, 3, 4, 5, 6, 7, 8, 9, 10, 11, 12, 13, 14, 15, 16, 17, 18, 19, 20, 21, 22, 23, 24, 25, 26, 27, 28, 29, 30\}$
   Calculate the exact probability of selecting a number that is a multiple of 3, giving your answer in simplest form.

5. Identify whether the following statement is True or False.
   If a large number of trials is conducted in an experiment, the relative (observed) frequency of each outcome will be very close to its theoretical probability.

6. **MC** The spinner is used to simulate success of an advertising company. The probability of obtaining a number less than 5 on the spinner represents the chance of the advertising company being successful. This probability is:

   A. 1 **B.** $\dfrac{5}{4}$ **C.** $\dfrac{1}{5}$ **D.** $\dfrac{4}{5}$

   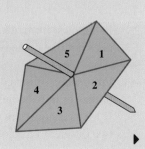

▶

7. **PATH** **MC** The information presented in the Venn diagram can be shown on a 2-way frequency table as:

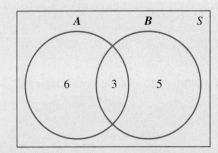

A.

	A	A′	Total
B	5	3	8
B′	0	6	6
Total	5	9	14

B.

	A	A′	Total
B	3	6	9
B′	5	0	5
Total	8	6	14

C.

	A	A′	Total
B	3	5	8
B′	6	0	6
Total	9	5	14

D.

	A	A′	Total
B	3	2	5
B′	3	3	6
Total	6	5	11

8. **MC** If $S = \{$numbers between 1 and 20 inclusive$\}$, identify the complement of the event $A = \{$multiples of 5 and prime numbers$\}$.
   A. $\{1, 4, 6, 8, 9, 12, 14, 16, 18\}$
   B. $\{5, 10, 15, 20\}$
   C. $\{2, 3, 5, 7, 11, 13, 17, 19\}$
   D. $\{2, 3, 5, 7, 10, 11, 13, 15, 17, 19, 20\}$

9. The jacks and aces from a deck of cards are shuffled, then two cards are drawn. Calculate the exact probability, in simplest form, that two aces are chosen:
   a. if the first card is replaced
   b. if the first card is not replaced.

10. **PATH** The Venn diagram shows the results of a survey where students were asked to indicate whether they prefer drama (D) or comedy (C) movies.

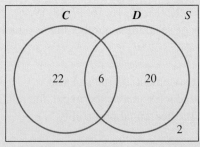

In simplest form, determine the probability that a student selected at random prefers drama movies but does not like comedy.

11. From a bag of mixed lollies, students can choose from three different types of lollies: red frogs, milk bottles or jelly babies. The bag contains five of each type of lolly.
   If Mahsa chooses two lollies without looking into the bag, calculate the probability that she will choose two different types of lollies.

12. **PATH** The students in a class were asked about their sport preferences — whether they played basketball or tennis or neither. The information was recorded in a 2-way frequency table.

	Basketball (*B*)	No basketball (*B'*)
Tennis (*T*)	25	20
No tennis (*T'*)	10	5

a. **MC** The student sport preferences recorded in the 2-way frequency table are represented on a Venn diagram as:

A.
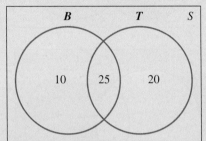

B.
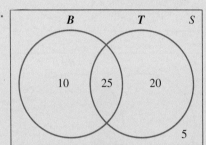

C.
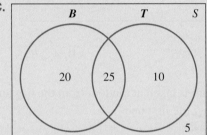

D.
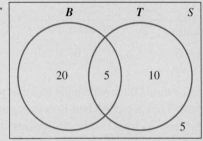

b. If one student is selected at random, calculate the probability that the student plays tennis only, correct to 2 decimal places.

13. A survey of Year 8, 9 and 10 students asked the students to choose dinner options for their school camp.

Year level	Lasagna	Stir-fry
8	82	75
9	67	90
10	89	45
Total	240	210

Calculate the following probabilities.
a. The probability that a randomly selected student chose lasagne.
b. The probability that a randomly selected student was in Year 10.
c. The probability that a randomly selected student was in Year 8 and chose stir-fry.
d. The probability that a randomly selected student who chose lasagna was in Year 9.

14. **PATH** Match the following Venn diagrams to the correct set notation for the shaded regions.

Venn diagrams

a.

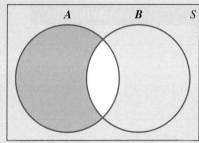

b.

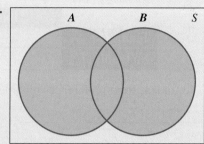

c.

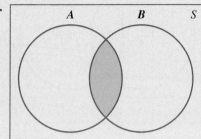

d.

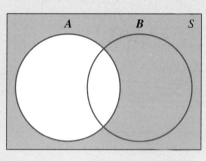

Set notation

   A. $A \cap B$ **B.** A' **C.** $A \cap B'$ **D.** $A \cup B$

15. **MC** A bag contains three blue balls and two red balls. A ball is taken at random from the bag and its colour noted. Then a second ball is drawn, without replacing the first one.
Identify the tree diagram that best represents this sample space.

A.

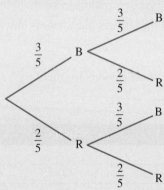

B.

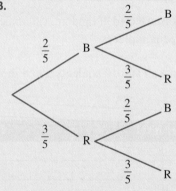

C.

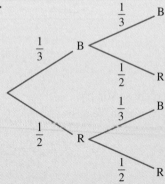

D.

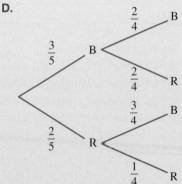

LESSON
14.2 Theoretical probability

▶ 14.2.1 The language of probability

eles-4877

- The **probability** of an event is a measure of the likelihood that the event will take place.
- If an event is certain to occur, then it has a probability of 1.
- If an event is impossible, then it has a probability of 0.
- The probability of any other event taking place is given by a number between 0 and 1.
- An event is likely to occur if it has a probability between 0.5 and 1.
- An event is unlikely to occur if it has a probability between 0 and 0.5.

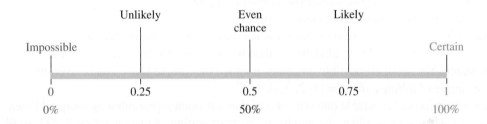

WORKED EXAMPLE 1 Placing events on a probability scale

On the probability scale given, insert each of the following events at appropriate points.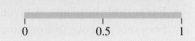
a. **The Sun will rise tomorrow morning.**
b. **You will come to school the next Monday during a school term.**
c. **It will snow in Victoria this winter.**

THINK	WRITE/DRAW
a. 1. Carefully read the given statement and label its position on the probability scale.	a.
2. Write the answer and provide reasoning.	The Sun rises every morning, so it is a certainty it will rise tomorrow morning.
b. 1. Carefully read the given statement and label its position on the probability scale.	b. 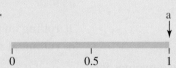

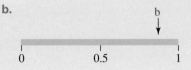

2. Write the answer and provide reasoning.	It is very likely but not certain that I will come to school on a Monday during term. Circumstances such as illness or public holidays may prevent me from coming to school on a specific Monday during a school term.
c. 1. Carefully read the given statement and label its position on the probability scale.	**c.** 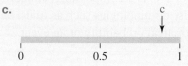
2. Write the answer and provide reasoning.	It is highly likely but not certain that it will snow in Victoria during winter.

14.2.2 Key terms of probability

eles-4878

- The study of probability uses many special terms that must be clearly understood.
- **Chance experiment**: A chance experiment is a process, such as rolling a die, that can be repeated many times.
- **Trial**: A trial is one performance of an experiment to get a result.
  For example, each roll of the die is called a trial.
- **Outcome**: The outcome is the result obtained when the experiment is conducted.
  For example, when a normal six-sided die is rolled the outcome can be 1, 2, 3, 4, 5 or 6.
- **Sample space**: The set of all possible outcomes is called the sample space and is given the symbol S.
  For the example of rolling a die, $S = \{1, 2, 3, 4, 5, 6\}$.
- **Event**: An event is the favourable outcome of a trial and is often represented by a capital letter.
  For example, when a die is rolled, A could be the event of getting an even number; $A = \{2, 4, 6\}$.
- **Complement of an event**: The complement of an event is every outcome not in the event.
  For example, when a die is rolled, if A is the event of getting a 1 or a 2, $A = \{1, 2\}$, the complement of A (A') is the event of getting a 3, 4, 5, 6: $A' = \{3, 4, 5, 6\}$.
- **Favourable outcome**: A favourable outcome for an event is any outcome that belongs to the event.
  For event A above (rolling an even number), the favourable outcomes are 2, 4 and 6.

WORKED EXAMPLE 2 Identifying sample space, events and outcomes

For the chance experiment of rolling a die:
a. list the sample space
b. list the events for:

  i. rolling a 4   **ii. rolling an even number**
  iii. rolling at least 5   **iv. rolling at most 2**

c. list the favourable outcomes for:

  i. {4, 5, 6}   **ii. not rolling 5**
  iii. rolling 3 or 4   **iv. rolling 3 and 4.**

THINK	WRITE
a. The outcomes are the numbers 1 to 6.	**a.** $S = \{1, 2, 3, 4, 5, 6\}$
b. **i.** This describes only one outcome.	**b.** **i.** $\{4\}$
ii. The possible even numbers are 2, 4 and 6.	**ii.** $\{2, 4, 6\}$
iii. 'At least 5' means 5 is the smallest.	**iii.** $\{5, 6\}$
iv. 'At most 2' means 2 is the largest.	**iv.** $\{1, 2\}$
c. **i.** The outcomes are shown inside the brackets.	**c.** **i.** $4, 5, 6$
ii. 'Not 5' means everything except 5.	**ii.** $1, 2, 3, 4, 6$
iii. The event is $\{3, 4\}$.	**iii.** $3, 4$
iv. There is no number that is both 3 and 4.	**iv.** There are no favourable outcomes.

14.2.3 Theoretical probability

eles-4879

- When a coin is tossed, there are two possible outcomes, Heads or Tails. That is, $S = \{H, T\}$.
- We assume the coin is unbiased, meaning each outcome is **equally likely** to occur.
- Since each outcome is equally likely, they must have the same probability.
- The total of all probabilities is 1, so the probability of an outcome is given by:

  - $P(\text{an outcome}) = \dfrac{1}{\text{total number of outcomes}}$

  For example:

  - $P(\text{Heads}) = \dfrac{1}{2}$ and $P(\text{Tails}) = \dfrac{1}{2}$

- The probability of an event A is found by adding up all the probabilities of the favourable outcomes in event A.

Theoretical probability

When determining probabilities of equally likely outcomes, use the following:

$$P(\text{event A}) = \frac{\text{number of favourable outcomes}}{\text{total number of outcomes}}$$

WORKED EXAMPLE 3 Calculating theoretical probability

A die is rolled and the number uppermost is noted. Determine the probability of each of the following events.

a. $A = \{1\}$ **b.** $B = \{\text{odd numbers}\}$ **c.** $C = \{4 \text{ or } 6\}$

THINK	WRITE
There are six possible outcomes.	
a. A has one favourable outcome.	**a.** $P(A) = \dfrac{1}{6}$

b. *B* has three favourable outcomes: 1, 3 and 5.

b. $P(B) = \dfrac{3}{6}$

$\quad = \dfrac{1}{2}$

c. *C* has two favourable outcomes.

c. $P(C) = \dfrac{2}{6}$

$\quad = \dfrac{1}{3}$

 Resources

 Interactivities Probability scale (int-3824)

Theoretical probability (int-6081)

Exercise 14.2 Theoretical probability

learn

14.2 Quick quiz on	14.2 Exercise

Individual pathways

■ PRACTISE	■ CONSOLIDATE	■ MASTER
1, 3, 5, 8, 10, 14, 15, 19, 22	2, 4, 6, 9, 11, 16, 20, 23	7, 12, 13, 17, 18, 21, 24, 25

Fluency

1. **WE1** On the given probability scale, insert each of the following events at appropriate points. Indicate the chance of each event using one of the following terms: certain, likely, unlikely, impossible.

   a. The school will have a lunch break on Friday.
   b. Australia will host two consecutive Olympic Games.

2. On the given probability scale, insert each of the following events at appropriate points. Indicate the chance of each event using one of the following terms: certain, likely, unlikely, impossible.

   a. At least one student in a particular class will obtain an A for Mathematics.
   b. Australia will have a swimming team in the Commonwealth Games.

3. On the given probability scale, insert each of the following events at appropriate points. Indicate the chance of each event using one of the following terms: certain, likely, unlikely, impossible.

   a. Mathematics will be taught in secondary schools.
   b. In the future most cars will run without LPG or petrol.

4. On the given probability scale, insert each of the following events at appropriate points. Indicate the chance of each event using one of the following terms: certain, likely, unlikely, impossible.

   a. Winter will be cold.
   b. Bean seeds, when sown, will germinate.

5. **WE2a** For each chance experiment below, list the sample space.
   a. Rolling a die.
   b. Tossing a coin.
   c. Testing a light bulb to see whether it is defective or not.
   d. Choosing a card from a normal deck and noting its colour.
   e. Choosing a card from a normal deck and noting its suit.

6. **WE2b** A normal six-sided die is rolled. List each of the following events.
   a. Rolling a number less than or equal to 3.
   b. Rolling an odd number.
   c. Rolling an even number or 1.
   d. Not rolling a 1 or 2.
   e. Rolling at most a 4.
   f. Rolling at least a 5.

7. **WE2c** A normal six-sided die is rolled. List the favourable outcomes for each of the following events.
   a. $A = \{3, 5\}$
   b. $B = \{1, 2\}$
   c. $C =$ 'rolling a number greater than 5'
   d. $D =$ 'not rolling a 3 or a 4'
   e. $E =$ 'rolling an odd number or a 2'
   f. $F =$ 'rolling an odd number and a 2'
   g. $G =$ 'rolling an odd number and a 3'

8. A card is selected from a normal deck of 52 cards and its suit is noted.
   a. List the sample space.
   b. List each of the following events.
      i. Drawing a black card.
      ii. Drawing a red card.
      iii. Not drawing a heart.
      iv. Drawing a black or a red card.

9. Determine the number of outcomes there are for:
   a. rolling a die
   b. tossing a coin
   c. drawing a card from a standard deck
   d. drawing a card and noting its suit
   e. noting the remainder when a number is divided by 5.

10. A card is drawn at random from a standard deck of 52 cards.

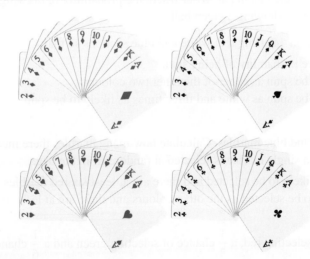

Note: 'At random' means that every card has the same chance of being selected.
Calculate the probability of selecting:
   a. an ace
   b. a king
   c. the 2 of spades
   d. a diamond.

11. **WE3** A card is drawn at random from a deck of 52. Determine the probability of each event below.
    a. $A = \{5 \text{ of clubs}\}$
    b. $B = \{\text{black card}\}$
    c. $C = \{5 \text{ of clubs or queen of diamonds}\}$
    d. $D = \{\text{hearts}\}$
    e. $E = \{\text{hearts or clubs}\}$

12. A card is drawn at random from a deck of 52. Determine the probability of each event below.
    a. $F = \{\text{hearts and 5}\}$
    b. $G = \{\text{hearts or 5}\}$
    c. $H = \{\text{aces or kings}\}$
    d. $I = \{\text{aces and kings}\}$
    e. $J = \{\text{not a 7}\}$

13. A letter is chosen at random from the letters in the word PROBABILITY. Determine the probability that the letter is:
    a. B
    b. not B
    c. a vowel
    d. not a vowel.

14. The following coloured spinner is spun and the colour is noted. Determine the probability of each of the events given below.
    a. $A = \{\text{blue}\}$
    b. $B = \{\text{orange}\}$
    c. $C = \{\text{orange or pink}\}$
    d. $D = \{\text{orange and pink}\}$
    e. $E = \{\text{not blue}\}$

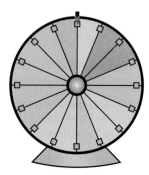

Understanding

15. A bag contains four purple balls and two green balls.
    a. If a ball is drawn at random, then calculate the probability that it will be:
        i. purple
        ii. green.
    b. Design an experiment like the one in part **a** but where the probability of drawing a purple ball is 3 times that of drawing a green ball.

16. Design spinners (see question **14**) using red, white and blue sections so that:
    a. each colour has the same probability of being spun
    b. red is twice as likely to be spun as either of the other two colours
    c. red is twice as likely to be spun as white and three times as likely to be spun as blue.

17. A bag contains red, green and blue marbles. Calculate how many marbles there must be in the bag for the following to be true when a single marble is selected at random from the bag.
    a. Each colour is equally likely to be selected and there are at least six red marbles in the bag.
    b. Blue is twice as likely to be selected as the other colours and there are at least five green marbles in the bag.
    c. There is a $\frac{1}{2}$ chance of selecting red, a $\frac{1}{3}$ chance of selecting green and a $\frac{1}{6}$ chance of selecting blue from the bag when there are between 30 and 40 marbles in the bag.

18. **a.** A bag contains seven gold and three silver coins. If a coin is drawn at random from the bag, calculate the probability that it will be:

  i. gold   **ii.** silver.

  b. After a gold coin is taken out of the bag, a second coin is then selected at random. Assuming the first coin was not returned to the bag, calculate the probability that the second coin will be:

  i. gold   **ii.** silver.

Communicating, reasoning and problem solving

19. Do you think that the probability of tossing Heads is the same as the probability of tossing Tails if your friend tosses the coin? Suggest some reasons that it might not be.

20. If the following four probabilities were given to you, explain which two you would say were not correct.

$$0.725, -0.5, 0.005, 1.05$$

21. A coin is going to be tossed five times in a row. During the first four flips the coin comes up Heads each time. What is the probability that the coin will come up Heads again on the fifth flip? Justify your answer.

22. Consider the spinner shown. Discuss whether the spinner has an equal chance of falling on each of the colours.

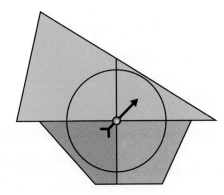

23. A box contains two coins. One is a double-headed coin, and the other is a normal coin with Heads on one side and Tails on the other. You draw one of the coins from a box and look at one of the sides. It is Heads. Determine the probability that the other side also shows Heads.

24. 'Unders and overs' is a game played with two normal six-sided dice. The two dice are rolled, and the numbers uppermost added to give a total. Players bet on the outcome being 'under 7', 'equal to 7' or 'over 7'.
    If you had to choose one of these outcomes, which would you choose?
    Explain why.

25. Justine and Mary have designed a new darts game for their Year 9 Fete Day. Instead of a circular dart board, their dart board is in the shape of two equilateral triangles.
    The inner triangle (bullseye) has a side length of 3 cm, while the outer triangle has side length 10 cm.
    Given that a player's dart falls in one of the triangles, determine the probability that it lands in the bullseye.
    Write your answer correct to 2 decimal places.

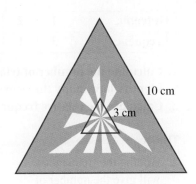

LESSON
14.3 Observed probability and simulations

LEARNING INTENTION

At the end of this lesson you should be able to:
- explore the relative frequency of an outcome
- compare and understand the difference between theoretical probability and observed probability
- use digital tools to simulate situations involving probabilities.

14.3.1 Relative frequency

eles-4880

- A die is rolled 12 times and the outcomes are recorded in the table shown.

Outcome	1	2	3	4	5	6
Frequency	3	1	1	2	2	3

In this chance experiment there were 12 trials.
The table shows that the number 1 was rolled 3 times out of 12.

- So the **relative frequency** of 1 is 3 out of 12, or $\dfrac{3}{12} = \dfrac{1}{4}$.

As a decimal, the relative frequency of 1 is equal to 0.25.

> **Relative frequency**
>
> The relative frequency of an outcome is given by:
>
> $$\text{Relative frequency} = \frac{\text{frequency of an outcome}}{\text{total number of trials}}$$

- As the number of trials becomes larger, the relative frequency of each outcome will become very close to the theoretical probability.

WORKED EXAMPLE 4 Calculating relative frequency

For the chance experiment of rolling a die, the following outcomes were noted.

Outcome	1	2	3	4	5	6
Frequency	3	1	4	6	3	3

a. **Calculate the number of trials.**
b. **Identify how many threes were rolled.**
c. **Calculate the relative frequency for each number written as a decimal.**

THINK	WRITE
a. Adding the frequencies will give the number of trials.	a. $1 + 3 + 4 + 6 + 3 + 3 = 20$ trials
b. The frequency of 3 is 4.	b. 4 threes were rolled.

c. Add a relative frequency row to the table and complete it. The relative frequency is calculated by dividing the frequency of the outcome by the total number of trials.

c.

Outcome	1	2	3	4	5	6
Frequency	3	1	4	6	3	3
Relative frequency	$\dfrac{3}{20} =$ 0.15	$\dfrac{1}{20} =$ 0.05	$\dfrac{4}{20} =$ 0.2	$\dfrac{6}{20} =$ 0.3	$\dfrac{3}{20} =$ 0.15	$\dfrac{3}{20} =$ 0.15

14.3.2 Observed probability

cle3-4881

- When it is not possible to calculate the theoretical probability of an outcome, carrying out simulations involving repeated trials can be used to determine the **observed probability**, also called experimental probability.
- The relative frequency of an outcome is the observed probability.

> **Observed probability**
>
> The observed probability of an outcome is given by:
>
> $$\text{Observed probability} = \frac{\text{frequency of an outcome}}{\text{total number of trials}}$$

- Consider the spinner shown. The outcomes are not distributed symetrically.
- The observed probability of each outcome can be found by using the spinner many times and recording the outcomes. As more trials are conducted, the observed probability will become more accurate and closer to the true probability of each section.

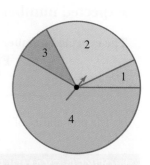

WORKED EXAMPLE 5 Calculating observed probabilities

The spinner shown was spun 100 times and the following results were achieved.

Outcome	1	2	3	4
Frequency	7	26	9	58

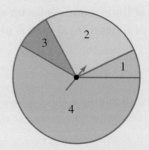

a. **Calculate the number of trials.**
b. **Calculate the observed probability of each outcome.**
c. **Calculate or recognise the sum of the four probabilities.**

THINK

a. Adding the frequencies will determine the number of trials.

WRITE

a. $7 + 26 + 9 + 58 = 100$ trials

b. The observed probability equals the relative frequency. This is calculated by dividing the frequency of the outcome by the total number of trials.

b. $P(1) = \frac{7}{100}$
$= 0.07$
$P(2) = \frac{26}{100}$
$= 0.26$
$P(3) = \frac{9}{100}$
$= 0.09$
$P(4) = \frac{58}{100}$
$= 0.58$

c. Add the probabilities (they should equal 1).

c. $0.07 + 0.26 + 0.09 + 0.58 = 1$

14.3.3 Expected number of results

eles-4882

- If we tossed a coin 100 times, we would expect there to be 50 Heads, since $P(\text{Heads}) = \frac{1}{2}$.

Expected number of results

The expected number of favourable outcomes from a series of trials is found from:

Expected number = probability of outcome × number of trials

- The probability of an outcome can be the theoretical probability or an observed probability.

WORKED EXAMPLE 6 Expected number of outcomes

Calculate the expected number of results in the following situations.
a. The number of Tails after flipping a coin 250 times
b. The number of times a one comes up after rolling a dice 120 times
c. The number of times a royal card is picked from a deck that is reshuffled with the card replaced 650 times

THINK

a. The probability of getting a Tail is $P(\text{Tails}) = \frac{1}{2}$.

WRITE

a. Number of Tails $= P(\text{Tails}) \times 250$
$= \frac{1}{2} \times 250$
$= 125$ times

b. The probability of getting a one is $P(\text{one}) = \frac{1}{6}$.

b. Number of ones $= P(\text{one}) \times 120$
$= \frac{1}{6} \times 120$
$= 20$ times

c. A deck of cards has 52 cards and 12 of them are royal cards. This means the probability of getting a royal is $P(\text{royal}) = \dfrac{12}{52} = \dfrac{3}{13}$.

c. Number of royals $= P(\text{royal}) \times 650$

$$= \dfrac{3}{13} \times 650$$

$$= 150 \text{ times}$$

14.3.4 Simulations

- Simulating experiments using a manual device such as dice and spinners can take a lot of time. A more efficient method of collecting results is to use a list of randomly generated numbers.
- Random number generators can generate a series of numbers between two given values, for example integers from 1 to 6.
- The digital technology box below shows how to create a random set of data on a scientific calculator.

Digital technology

Scientific calculators have a random function that allows the user to generate integers at random. This can be used to create a random set of data for probability or statistics questions.

The RanInt function (press the ALPHA button and the decimal point to access it) can be used to generate random whole numbers.

```
RanInt#(1,10)        Math ▲
                          4
RanInt#(1,10)        Math ▲
                          1
```

ranInt#(1, 10) will generate a random integer between 1 and 10.

The comma button is found by pressing the SHIFT button and the right bracket button .

COMMUNICATING — COLLABORATIVE TASK: Construct a spinner

Equipment: Paper, cardboard, pens, toothpick

1. Construct an irregular spinner using cardboard and a toothpick. By carrying out a number of trials, estimate the probability of each outcome.
2. Swap your spinner with one of your classmates' spinners and estimate the probability of each outcome for their spinner.
3. Compare your results.
4. As a class, discuss the influence of the number of trials on the accuracy of the observed probabilities.

 Resources

 Interactivity Experimental probability (int-3825)

14.3 Quick quiz **on**	14.3 Exercise

Individual pathways

■ PRACTISE	■ CONSOLIDATE	■ MASTER
1, 3, 8, 10, 11, 15, 17, 19, 22	2, 4, 6, 9, 12, 18, 20, 23, 24	5, 7, 13, 14, 16, 21, 25, 26, 27

Fluency

1. **WE4** Each of the two tables shown contains the results of a chance experiment (rolling a die). For each table, calculate:

   i. the number of trials held
   ii. the number of fives rolled
   iii. the relative frequency for each outcome, correct to 2 decimal places
   iv. the sum of the relative frequencies.

   a.

Number	1	2	3	4	5	6
Frequency	3	1	5	2	4	1

   b.

Number	1	2	3	4	5	6
Frequency	52	38	45	49	40	46

2. A coin is tossed in two chance experiments. The outcomes are recorded in the tables shown. For each experiment, calculate:

   i. the relative frequency of both outcomes
   ii. the sum of the relative frequencies.

   a.

Outcome	H	T
Frequency	22	28

   b.

Outcome	H	T
Frequency	31	19

3. **WE5** An unbalanced die was rolled 200 times and the following outcomes were recorded.

Number	1	2	3	4	5	6
Frequency	18	32	25	29	23	73

   Using these results, calculate:

   a. P(6)
   b. P(odd number)
   c. P(at most 2)
   d. P(not 3).

4. A box of matches claims on its cover to contain 100 matches. A survey of 200 boxes established the following results.

Number of matches	95	96	97	98	99	100	101	102	103	104
Frequency	1	13	14	17	27	55	30	16	13	14

   If you were to purchase a box of these matches, calculate the probability that:

   a. the box would contain 100 matches
   b. the box would contain at least 100 matches
   c. the box would contain more than 100 matches
   d. the box would contain no more than 100 matches.

5. A packet of chips is labelled as weighing 170 grams. This is not always the case and there will be some variation in the weight of each packet. A packet of chips is considered underweight if its weight is below 168 grams. Chips are made in batches of 1000 at a time.
A sample of the weights from a particular batch are shown below.

Weight (grams)	166	167	168	169	170	171	172	173	174
Frequency	2	3	13	17	27	18	9	5	1

Calculate the probability that:

a. a packet of chips is its advertised weight
b. a packet of chips is above its advertised weight
c. a packet of chips is underweight.
A batch of chips is rejected if more than 50 packets in a batch are classed as underweight.
d. Explain if the batch that this sample of chips is taken from should be rejected.

Understanding

6. Here is a series of statements based on observed probability. If a statement is not reasonable, give a reason why.

a. I tossed a coin five times and there were four Heads, so P(Heads) = 0.8.
b. Sydney Roosters have won 1064 matches out of the 2045 that they have played, so P(Sydney will win their next game) = 0.52.
c. P(the sun will rise tomorrow) = 1
d. At a factory, a test of 10 000 light globes showed that 7 were faulty. Therefore, P(faulty light globe) = 0.0007.
e. In Sydney it rains an average of 143.7 days each year, so P(it will rain in Sydney on the 17th of next month) = 0.39.

7. At a birthday party, some cans of soft drink were put in a container of ice. There were 16 cans of Coke, 20 cans of Sprite, 13 cans of Fanta, 8 cans of Sunkist and 15 cans of Pepsi.
If a can was picked at random, calculate the probability that it was:

a. a can of Pepsi
b. not a can of Fanta.

8. **WE6** Calculate the expected number of Tails if a fair coin is tossed 400 times.

9. Calculate the expected number of threes if a fair die is rolled 120 times.

10. **MC** In Tattslotto, six numbers are drawn from the numbers 1, 2, 3, ... 45. The number of different combinations of six numbers is 8 145 060.
If you buy one ticket, what is the probability that you will win the draw?

A. $\dfrac{1}{8\,145\,060}$ B. $\dfrac{1}{45}$

C. $\dfrac{45}{8\,145\,060}$ D. $\dfrac{1}{6}$

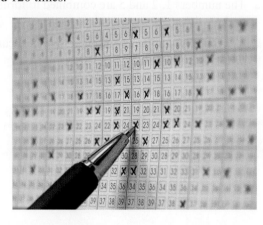

11. **MC** A survey of high school students asked 'Should Saturday be a normal school day?' 350 students voted yes, and 450 voted no.
From the following, select the probability that a student chosen at random said no.

A. $\dfrac{7}{16}$ B. $\dfrac{9}{16}$ C. $\dfrac{7}{9}$ D. $\dfrac{9}{14}$

12. In a poll of 200 people, 110 supported party M, 60 supported party N and 30 were undecided. If a person is chosen at random from this group of people, calculate the probability that he or she:

a. supports party M
b. supports party N
c. supports a party
d. is not sure what party to support.

13. A random number is picked from $N = \{1, 2, 3, \ldots 100\}$. Calculate the probability of picking a number that is:

a. a multiple of 3
b. a multiple of 4 or 5
c. a multiple of 5 and 6.

14. The numbers 3, 5 and 6 are combined to form a three-digit number such that no digit may be repeated.

a. i. Determine how many numbers can be formed.
   ii. List them.
b. Determine P(the number is odd).
c. Determine P(the number is even).
d. Determine P(the number is a multiple of 5).

15. **MC** In a batch of batteries, 2 out of every 10 in a large sample were faulty. At this rate, calculate how many batteries are expected to be faulty in a batch of 1500.

A. 2   B. 150   C. 200   D. 300

16. Svetlana, Sarah, Leonie and Trang are volleyball players. The probabilities that they will score a point on serve are 0.6, 0.4, 0.3 and 0.2 respectively.
Calculate how many points on serve are expected from each player if they serve 10 times each.

17. **MC** A survey of the favourite leisure activity of 200 Year 9 students produced the following results.

Activity	Playing sport	Fishing	Watching TV	Video games	Surfing
Number of students	58	26	28	38	50

The probability (given as a percentage) that a student selected at random from this group will have surfing as their favourite leisure activity is:

A. 50%   B. 100%   C. 25%   D. 0%

18. The numbers 1, 2 and 5 are combined to form a three-digit number, allowing for any digit to be repeated up to three times.

a. Recognise how many different numbers can be formed.
b. List the numbers.
c. Determine P(the number is even).
d. Determine P(the number is odd).
e. Determine P(the number is a multiple of 3).

Communicating, reasoning and problem solving

19. John has a 12-sided die numbered 1 to 12 and Lisa has a 20-sided die numbered 1 to 20. They are playing a game where the first person to get the number 10 wins.
They are rolling their dice individually.

    a. Calculate P(John gets a 10).
    b. Calculate P(Lisa gets a 10).
    c. Explain whether the game is fair.

20. At a supermarket checkout, the scanners have temporarily broken down and the cashiers must enter in the bar codes manually. One particular cashier overcharged 7 of the last 10 customers she served by entering the incorrect bar code.

    a. Based on the cashier's record, determine the probability of making a mistake with the next customer.
    b. Explain if another customer should have any objections with being served by this cashier.

21. If you flip a coin six times, determine how many of the possible outcomes could include a Tail on the second toss.

22. In a jar, there are 600 red balls, 400 green balls, and an unknown number of yellow balls. If the probability of selecting a green ball is $\frac{1}{5}$, determine how many yellow balls are in the jar.

23. In a jar there are an unknown number of balls, N, with 20 of them green. The other colours contained in the jar are red, yellow and blue, with P(red or yellow) $= \frac{1}{2}$, P(red or green) $= \frac{1}{4}$ and P(blue) $= \frac{1}{3}$.
Determine the number of red, yellow and blue balls in the jar.

24. The biological sex of babies in a set of triplets is simulated by flipping three coins. If a coin lands Tails up, the baby is male. If a coin lands Heads up, the baby is female.
In the simulation, the trial is repeated 40 times.
The following results show the number of Heads obtained in each trial:

    0, 3, 2, 1, 1, 0, 1, 2, 1, 0, 1, 0, 2, 0, 1, 0, 1, 2, 3, 2, 1, 3, 0, 2, 1, 2, 0, 3, 1, 3, 0, 1, 0, 1, 3, 2, 2, 1, 2, 1

    a. Calculate the probability that exactly one of the babies in a set of triplets is female.
    b. Calculate the probability that more than one of the babies in the set of triplets is female.

25. Use your calculator to generate three random sets of numbers between 1 and 6 to simulate rolling a dice. The number of trials in each set will be:

    a. 10
    b. 25
    c. 50

    Calculate the relative frequency of each outcome (1 to 6) and comment on what you notice about observed and theoretical probability.

26. A survey of the favourite foods of Year 9 students is recorded, with the following results.

Meal	Tally
Hamburger	45
Fish and chips	31
Macaroni and cheese	30
Lamb souvlaki	25
BBQ pork ribs	21
Cornflakes	17
T-bone steak	14
Banana split	12
Corn-on-the-cob	9
Hot dogs	8
Garden salad	8
Veggie burger	7
Smoked salmon	6
Muesli	5
Fruit salad	3

a. Estimate the probability that macaroni and cheese is the favourite food of a randomly selected Year 9 student.
b. Estimate the probability that a vegetarian dish is a randomly selected student's favourite food.
c. Estimate the probability that a beef dish is a randomly selected student's favourite food.

27. A spinner has six sections of different sizes. Steven conducts an experiment and records the following results.

Section	1	2	3	4	5	6
Frequency	6	24	15	30	48	12

Determine the angle size of each section of the spinner using the results above.

LESSON
14.4 Tree diagrams and two-step experiments

LEARNING INTENTION

At the end of this lesson you should be able to:
- construct the sample space for a two-step experiment using arrays and tree diagrams
- solve problems using tree diagrams for experiments with and without replacement.

▶ 14.4.1 The sample space of two-step experiments

eles-4888

- A two-step experiment involves two separate actions. It may be the same action repeated (tossing a coin twice) or two separate actions (tossing a coin and rolling a dice).
- Imagine two bags (that are not transparent) that contain coloured counters. The first bag has a mixture of black and white counters, and the second bag holds blue, magenta (pink) and purple counters. In a probability experiment, one counter is to be selected at random from each bag and its colour noted.

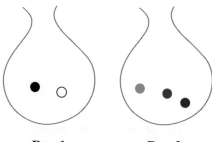

Bag 1   **Bag 2**

- The sample space for this experiment can be found using a table called an **array** that systematically displays all the outcomes.

			Second action		
			Bag 2		
			B	M	P
First action	Bag 1	B	B B	B M	B P
		W	W B	W M	W P

The sample space, $S = \{BB, BM, BP, WB, WM, WP\}$.
- The sample space can also be found using a **tree diagram**, a branching diagram that lists all the possible outcomes.

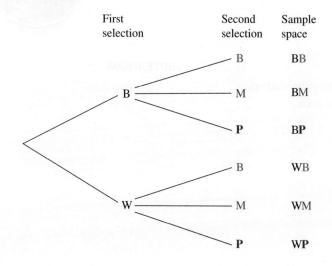

Two dice are rolled and the numbers uppermost are noted.
List the sample space in an array.
a. Recognise how many outcomes there are.
b. Calculate how many outcomes contain at least one 5.
c. Calculate P(at least one 5).

THINK

Draw an array (table) showing all the possible outcomes.

WRITE/DRAW

		Second die					
		1	2	3	4	5	6
First die	1	1,1	1,2	1,3	1,4	1,5	1,6
	2	2,1	2,2	2,3	2,4	2,5	2,6
	3	3,1	3,2	3,3	3,4	3,5	3,6
	4	4,1	4,2	4,3	4,4	4,5	4,6
	5	5,1	5,2	5,3	5,4	5,5	5,6
	6	6,1	6,2	6,3	6,4	6,5	6,6

a. The table shows 36 outcomes.

b. Count the outcomes that contain 5. The cells are shaded in the table.

c. There are 11 favourable outcomes and 36 in total.

a. There are 36 outcomes.

b. Eleven outcomes include 5.

c. $P(\text{at least one } 5) = \dfrac{11}{36}$

Two coins are tossed and the outcomes are noted.
Show the sample space on a tree diagram.
a. Recognise how many outcomes there are.
b. Calculate the probability of tossing at least one Head.

THINK

1. Draw a tree representing the outcomes for the toss of the first coin.

WRITE/DRAW

First coin

H

T

2. For the second coin the tree looks like this:

Second coin

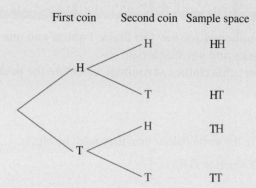

Add this tree to both ends of the first tree.

3. List the outcomes.

  a. Count the outcomes in the sample space.

  a. There are four outcomes (HH, HT, TH, TT).

  b. Three outcomes have at least one Head.

  b. P(at least one Head) = $\dfrac{3}{4}$

14.4.2 Two-step probabilities

eles-4889

- When a coin is tossed, P(H) = $\dfrac{1}{2}$, and when a die is rolled, P(3) = $\dfrac{1}{6}$.

- If a coin is tossed and a die is rolled, what is the probability of getting a Head *and* a 3?
- Consider the sample space.

		Second action					
		1	**2**	**3**	**4**	**5**	**6**
First action	**H**	H, 1	H, 2	**H, 3**	H, 4	H, 5	H, 6
	T	T, 1	T, 2	T, 3	T, 4	T, 5	T, 6

There are 12 outcomes, and P(Head and 3) = $\dfrac{1}{12}$.

- In this case, P(Head and 3) = P(H) × P(3); that is, $\dfrac{1}{12} = \dfrac{1}{2} \times \dfrac{1}{6}$.

Two-step probabilities

If A is the outcome of the first action and B is the outcome of the second action in a two-step experiment, then:
- **$A \cap B$ is the outcome of A followed by B**
- **$P(A \cap B) = P(A) \times P(B)$.**

In one cupboard Joe has two black T-shirts and one yellow one. In his drawer there are three pairs of white socks and one black pair.

If he selects his clothes at random, calculate the probability that his socks and T-shirt will be the same colour.

THINK

If they are the same colour then they must be black.

P(black T-shirt) $= P(B_t) = \dfrac{2}{3}$

P(black socks) $= P(B_s) = \dfrac{1}{4}$

WRITE

$$P(B_t \cap B_s) = P(B_t) \times P(B_s)$$
$$= \dfrac{2}{3} \times \dfrac{1}{4}$$
$$= \dfrac{1}{6}$$

▶ 14.4.3 Experiments with replacement

eles-4890

- If a two-step experiment requires an object to be selected, say from a bag, the person doing the selecting has two options after the first selection.
  - Place the object bag back in the bag, meaning the experiment is being carried out **with replacement**. In this case, the number of objects in the bag remains constant.
  - Permanently remove the object from the bag, meaning the experiment is being carried out **without replacement**. In this situation, the number of the objects in the bag is reduced by 1 after every selection.

A bag contains three pink and two blue counters. A counter is taken at random from the bag, its colour is noted, then it is returned to the bag and a second counter is chosen.

a. Show the outcomes on a tree diagram.
b. Calculate the probability of each outcome.
c. Calculate the sum of the probabilities.

THINK

a. 1. Draw a tree for the first trial. Write the probability on the branch.
   Note: The probabilities should sum to 1.

2. For the second trial the tree is the same. Add this tree to both ends of the first tree, then list the outcomes.

WRITE/DRAW

a.

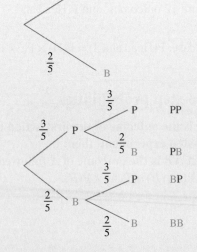

b. For both draws $P(P) = \dfrac{3}{5}$ and $P(B) = \dfrac{2}{5}$.

Use the rule $P(A \cap B) = P(A) \times P(B)$ to determine the probabilities.

b. $P(P \cap P) = P(P) \times P(P)$

$$= \dfrac{3}{5} \times \dfrac{3}{5}$$

$$= \dfrac{9}{25}$$

$P(P \cap B) = P(P) \times P(B)$

$$= \dfrac{3}{5} \times \dfrac{2}{5}$$

$$= \dfrac{6}{25}$$

$P(B \cap P) = P(B) \times P(P)$

$$= \dfrac{2}{5} \times \dfrac{3}{5}$$

$$= \dfrac{6}{25}$$

$P(B \cap B) = P(B) \times P(B)$

$$= \dfrac{2}{5} \times \dfrac{2}{5}$$

$$= \dfrac{4}{25}$$

c. Add the probabilities.

c. $\dfrac{9}{25} + \dfrac{6}{25} + \dfrac{6}{25} + \dfrac{4}{25} = 1$

- In Worked example 10, $P(P) = \dfrac{3}{5}$ and $P(B) = \dfrac{2}{5}$ for both trials.

  This would not be true if a counter was selected *but not replaced.*

▶ 14.4.4 Experiments without replacement

eles-4891

- Let us consider again the situation described in Worked example 10, and consider what happens if the first marble is not replaced.
- Initially the bag contains three pink and two blue counters, and either a pink counter or a blue counter will be chosen.

- $P(P) = \dfrac{3}{5}$ and $P(B) = \dfrac{2}{5}$.

- If the counter is not replaced, then the sample space is affected as follows:

If the first counter randomly selected is pink, then the sample space for the second draw looks like this:

So $P(P) = \dfrac{2}{4}$ and $P(B) = \dfrac{2}{4}$.

If the first counter randomly selected is blue, then the sample space for the second draw looks like this:

So $P(P) = \dfrac{3}{4}$ and $P(B) = \dfrac{1}{4}$.

WORKED EXAMPLE 11 Calculating two-step experiments without replacement

A bag contains three pink and two blue counters. A counter is taken at random from the bag and its colour is noted, then a second counter is drawn, without replacing the first one.
a. Show the outcomes on a tree diagram.
b. Calculate the probability of each outcome.
c. Calculate the sum of the probabilities.

THINK

a. Draw a tree diagram, listing the probabilities.

WRITE/DRAW

a.

$$
\begin{array}{ccc}
 & \frac{2}{4} & P \quad PP \\
\frac{3}{5}\; P & & \\
 & \frac{2}{4} & B \quad PB \\
 & & \\
 & \frac{3}{4} & P \quad BP \\
\frac{2}{5}\; B & & \\
 & \frac{1}{4} & B \quad BB
\end{array}
$$

b. Use the rule $P(A \cap B) = P(A) \times P(B)$ to determine the probabilities.

b. $P(P \cap P) = P(P) \times P(P)$

$$= \frac{3}{5} \times \frac{2}{4}$$

$$= \frac{6}{20}$$

$$= \frac{3}{10}$$

$P(P \cap B) = P(P) \times P(B)$

$$= \frac{3}{5} \times \frac{2}{4}$$

$$= \frac{6}{20}$$

$$= \frac{3}{10}$$

$$P(B \cap P) = P(B) \times P(P)$$

$$= \frac{2}{5} \times \frac{3}{4}$$

$$= \frac{6}{20}$$

$$= \frac{3}{10}$$

$$P(B \cap B) = P(B) \times P(B)$$

$$= \frac{2}{5} \times \frac{1}{4}$$

$$= \frac{2}{20}$$

$$= \frac{1}{10}$$

c. Add the probabilities.

c. $\dfrac{3}{10} + \dfrac{3}{10} + \dfrac{3}{10} + \dfrac{1}{10} = 1$

WORKED EXAMPLE 12 Language of two-step experiments

Consider the situation presented in Worked example 11. Use the tree diagram to calculate the following probabilities.
a. P(a pink counter and a blue counter)
b. P(a pink counter then a blue counter)
c. P(a matching pair)
d. P(different colours)

THINK

WRITE

a. Think of the outcomes that have a pink and a blue counter. There are two: PB and BP. Add their probabilities together to calculate the answer.

a. P(a pink counter and a blue counter)
$= P(PB) + P(BP)$

$= \dfrac{3}{10} + \dfrac{3}{10}$

$= \dfrac{3}{5}$

b. In this case, we have to take the order the counters are selected into account. The only outcome that has a pink then a blue is PB.

b. P(a pink counter then a blue counter)
$= P(PB)$
$= \dfrac{3}{10}$

c. A matching pair implies both counters are the same colour. These outcomes are BB and PP.

c. P(a matching pair)
$= P(BB) + P(PP)$

$= \dfrac{1}{10} + \dfrac{3}{10}$

$= \dfrac{2}{5}$

d. One way to approach this is to think that if the colours are different, this is all outcomes not counted in part **c**. Thus, we can calculate the answer by subtracting the answer for part **c** from 1.

d. P(different colours)
$$= 1 - \text{P(a matching pair)}$$
$$= 1 - \frac{2}{5}$$
$$= \frac{3}{5}$$

DISCUSSION

How does replacement affect the probability of an event occurring?

---

on Resources

▶ **Video eLesson** Tree diagrams (eles-1894)

✦ **Interactivity** Two-step experiments (int-6083)

Exercise 14.4 Tree diagrams and two-step experiments   learn on

14.4 Quick quiz **on**	14.4 Exercise

Individual pathways

■ PRACTISE	■ CONSOLIDATE	■ MASTER
1, 2, 4, 5, 8, 9, 12, 15, 18, 21, 22	3, 7, 10, 13, 16, 19, 23, 24	6, 11, 14, 17, 20, 25, 26, 27

Fluency

1. **WE7** If two dice are rolled and their *sum* is noted, complete the array below to show the sample space.

		Die 1					
		1	**2**	**3**	**4**	**5**	**6**
Die 2	**1**	2					
	2					7	
	3						
	4						
	5						
	6			9			

a. Calculate P(rolling a total of 5).
b. Calculate P(rolling a total of 1).
c. Determine the most probable outcome.

2. In her cupboard Rosa has three scarves (red, blue and pink) and two beanies (brown and purple). If she randomly chooses one scarf and one beanie, show the sample space in an array.

3. A ten-sided die is rolled and then a coin is flipped.
   a. Use an array to determine all the outcomes.
   b. Calculate:
      i. P(even and a Head)
      ii. P(even or a Head)
      iii. P(a Tail and a number greater than 7)
      iv. P(a Tail or a number greater than 7).

4. One box contains red and blue pencils, and a second box contains red, blue and green pencils. If one pencil is chosen at random from each box and the colours are noted, draw a tree diagram to show the sample space.

5. **WE8** A bag contains three discs labelled 1, 3 and 5, and another bag contains two discs, labelled 2 and 4, as shown below. A disc is taken from each bag and the larger number is recorded.

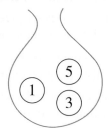 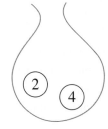

   a. Complete the tree diagram below to list the sample space.

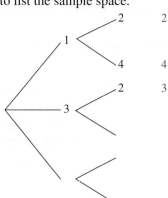

   b. Calculate:
      i. P(2)   ii. P(1)   iii. P(odd number)

6. Two dice are rolled and the difference between the two numbers is found.
   a. Use an array to determine all the outcomes.
   b. Calculate:
      i. P(odd number)   ii. P(0)
      iii. P(a number more than 2)   iv. P(a number no more than 2).

7. **WE9** A die is rolled twice. Calculate the probability of rolling:

   a. a 6 on the first roll
   b. a double 6
   c. an even number on both dice
   d. a total of 12.

8. A coin is tossed twice.

   a. Show the outcomes on a tree diagram.
   b. Calculate:
      i. P(2 Tails)
      ii. P(at least 1 Tail).

9. **WE10** A bag contains three red counters and one blue counter. A counter is chosen at random. A second counter is drawn with replacement.

   a. Show the outcomes and probabilities on a tree diagram.
   b. Calculate the probability of choosing:
      i. a red counter then a blue counter
      ii. two blue counters.

10. A bag contains five red, six green and four blue counters. A counter is chosen at random. A second counter is drawn with replacement. Use a tree diagram to calculate the probability of choosing:

    a. a red and a green counter
    b. a red counter then a green counter
    c. a red or green counter
    d. a matching pair of the same colour.

11. A bag contains eight black, nine white and three red counters. A counter is chosen at random. A second counter is drawn with replacement. Use a tree diagram to calculate the probability of choosing:

    a. at least one black counter
    b. a black counter and a white counter
    c. a black counter or a white counter
    d. a matching pair of the same colour
    e. different colours.

12. **WE11** A bag contains three black balls and two red balls. If two balls are selected, randomly, without replacement:

    a. show the outcomes and their probabilities on a tree diagram
    b. calculate P(2 red balls).

13. A bag contains five red, three blue and two green balls. If two balls are randomly selected without replacement, calculate:

    a. P(two red balls)
    b. P(a red ball then a green ball)
    c. P(a red ball and a green ball)
    d. P(the same colour selected twice).

14. A bag contains five black, one white and six red balls. If two balls are randomly selected without replacement, calculate:

    a. P(two red balls)
    b. P(a red ball then a white ball)
    c. P(a red ball and a black ball)
    d. P(the same colour selected twice).

Understanding

15. The kings and queens from a deck of cards are shuffled, then two cards are chosen. Calculate the probability that two kings are chosen:

    a. if the first card is replaced
    b. if the first card is not replaced.

16. The 12 royal cards from a deck of cards are shuffled together, then two cards are chosen. Determine the probability that:

    a. two queens are chosen if the first card is replaced

    b. a matching pair (same value, not suit) is chosen if the first card is not replaced

    c. two cards from different suits are chosen if the first card is not replaced.

17. **WE12** Each week John and Paul play two sets of tennis against each other. They each have an equal chance of winning the first set.
    When John wins the first set, his probability of winning the second set rises to 0.6, but if he loses the first set, he has only a 0.3 chance of winning the second set.

    a. Show the possible outcomes on a tree diagram.
    b. Calculate:
       i. P(John wins both sets)
       ii. P(Paul wins both sets)
       iii. P(they win one set each).

Communicating, reasoning and problem solving

18. A bag contains four red and six yellow balls. If the first ball drawn is yellow, explain the difference in the probability of drawing the second ball if the first ball was replaced compared to not being replaced.

19. James has six different ties, five different shirts and three different suits that he can choose from when getting ready for work.

    a. Determine how many days he can go without repeating an item of clothing.
    b. Determine how many possible combinations of clothing there are.
    c. James receives a new shirt and tie for his birthday. Determine how many more combinations of clothing are now possible.

20. Three dice are tossed and the total is recorded.

    a. State the smallest and largest possible totals.
    b. Calculate the probabilities for all possible totals.

21. You draw two cards, one after the other without replacement, from a deck of 52 cards. Calculate the probability of:

    a. drawing two aces
    b. drawing two face cards (J, Q, K)
    c. getting a 'pair' $(22, 33, 44 \ldots QQ, KK, AA)$.

22. A chance experiment involves flipping a coin and rolling two dice. Determine the probability of obtaining Tails and two numbers whose sum is greater than 4.

23. In a jar there are 10 red balls and 6 green balls. Jacob takes out two balls, one at a time, without replacing them. Calculate the probability that both balls are the same colour.

24. A coin is being tossed repeatedly. Determine how many possible outcomes there will be if it is tossed:

    a. 3 times
    b. 5 times
    c. 4 times
    d. x times.

25. A box of chocolates contains milk chocolates and dark chocolates. The probability of selecting a dark chocolate from a full box is $\frac{1}{6}$. Once a dark chocolate has been taken from the box, the chance of selecting a second dark chocolate drops to $\frac{1}{7}$.

   a. Calculate how many chocolates are in the box altogether.
   b. If two chocolates were randomly selected from the box, calculate the probability of getting two of the same type of chocolate.

26. Claire's maths teacher decided to surprise the class with a four-question multiple-choice quiz. If each question has four possible options, calculate the probability the Claire passes the test given she guesses every question.

27. In the game of 'Texas Hold'Em' poker, five cards are progressively placed face up in the centre of the table for all players to use.
   At one point in the game there are three face-up cards (two hearts and one diamond). You have two diamonds in your hand for a total of three diamonds. Five diamonds make a flush.
   Given that there are 47 cards left, determine the probability that the next two face-up cards are both diamonds.

LESSON
14.5 Exploring Venn diagrams and 2-way tables (Path)

LEARNING INTENTION

At the end of this lesson you should be able to:
- construct and interpret Venn diagrams
- construct and interpret 2-way tables
- convert between Venn diagrams and 2-way tables.

⊙ 14.5.1 Exploring Venn diagrams

eles-4478
- A **Venn diagram** is made up of a rectangle and one or more circles.
- A Venn diagram contains all possible outcomes in the sample space, S.
- A Venn diagram is used to illustrate the relationship between sets of objects or numbers.
- All outcomes for a given event will be contained within a specific circle.
- Outcomes that belong to multiple events will be found in the overlapping region of two or more circles.
- The overlapping region of two circles is called the intersection of the two events and is represented using the ∩ symbol.
- If the circles do not overlap, then these events are called **mutually exclusive** events.

The following Venn diagram represents the languages studied by Year 9 students at a school.

Languages studied by Year 8 students

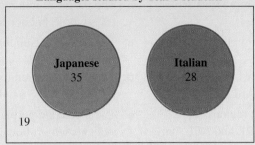

a. Identify whether there are any students who study both Japanese and Italian.
b. Calculate how many students there were overall.
c. Identify how many students studied Japanese.
d. Identify how many students studied Italian.
e. Identify how many students studied neither Japanese nor Italian.

THINK

a. The circles representing Japanese and Italian do not overlap, so these attributes are mutually exclusive.

b. Add all of the numbers in the diagram together. This represents all of the students.

c. Look at the region represented by Japanese.

d. Look at the region represented by Italian.

e. Look at the region outside of the circle.

WRITE/DRAW

a. No, there are no students who study both Japanese and Italian.

b. $19 + 35 + 28 = 82$
There are 82 students overall.

c. There are 35 students who study Japanese.

d. There are 28 students who study Italian.

e. There are 19 students who study neither Japanese nor Italian.

- When two regions in a Venn diagram overlap, the events are not mutually exclusive, and the overlapping region represents outcomes that are common to both events.

Describe the following regions in this Venn diagram.

Sports followed by Year 9 students

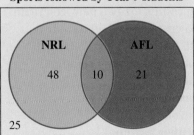

a. The green region
b. The purple region
c. The orange region
d. The region outside the circles
e. The green and purple regions combined
f. The green, purple and orange regions combined

THINK	WRITE/DRAW
a. Look at the green region. It represents NRL only.	**a.** The green region represents the students who follow NRL but not AFL.
b. Look at the purple region. It represents AFL only.	**b.** The purple region represents the students who follow AFL but not NRL.
c. Look at the orange region. It represents both AFL and NRL.	**c.** The orange region represents the students who follow both AFL and NRL.
d. Look at the region outside the circles. It represents neither NRL nor AFL.	**d.** The outside region represents the students who follow neither NRL nor AFL.
e. Look at the green and purple regions. Combined, they represent NRL and AFL, but not both.	**e.** The green and purple regions represent the students who follow NRL or AFL, but not both.
f. Look at the green, purple and orange regions. Combined, they represent NRL, AFL and both.	**f.** The green, purple and orange regions represent the students who follow NRL, AFL and both.

- Venn diagrams can also cover 3 attributes by including 3 circles.
- The diagram shows the lunch preferences of Year 9 students.
- The light blue section represents those students who like pizza, sandwiches and salad, while the dark pink section represents those students who like pizza and salad, but not sandwiches.
- Try to work out what the other colours represent!

Lunch preferences of Year 9 students

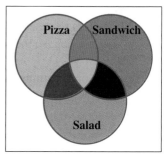

▶ 14.5.2 Constructing Venn diagrams

eles-4479

- Venn diagrams can be constructed from given data to represent all possible combinations of two attributes.
- The language used in the description is very important. It indicates whether an outcome is included or not in the set, or attribute.

WORKED EXAMPLE 15 Constructing a Venn diagram

An ice-creamery conducted a survey of 60 customers on a Monday and obtained the following results for two new ice-cream flavours. The results showed that 35 customers liked flavour A, 40 liked flavour B, and 24 liked both flavours.

a. Draw a Venn diagram to illustrate the above information.

b. Use the Venn diagram to determine:
   i. how many customers liked flavour A only
   ii. how many customers liked flavour B only
   iii. how many customers liked neither flavour.

c. If a customer was selected at random on that Monday morning, calculate:
   i. the probability that they liked both flavours
   ii. the probability that they liked neither flavour
   iii. the probability that they liked flavour A
   iv. the probability that they liked only flavour A.

THINK	WRITE
a. 1. Draw and label two overlapping circles within a rectangle to represent flavour A and flavour B. *Note:* The circles for flavours A and B overlap because some customers liked both flavours.	**a.** 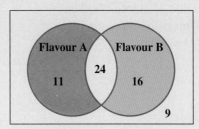

a. 2. Working from the overlapping area outwards, determine the number of customers in each region.
Note: The total must equal the number of customers surveyed, which is 60.

There are 24 customers who liked both flavours. The flavour A circle contains 11 customers (35−24) who liked flavour A but not flavour B.
The flavour B circle contains 16 customers (40−24) who liked flavour B but not flavour A.
The remaining 9 customers $60 - (11 + 24 + 16)$ liked neither flavour.

b. i. The non-overlapping part of flavour A's circle in the Venn diagram refers to the customers who liked flavour A only.

b. i. Eleven customers liked flavour A only.

ii. The non-overlapping part of flavour B's circle in the Venn diagram refers to the customers who liked flavour B only.

ii. Sixteen customers liked flavour B only.

iii. The region outside the circles in the Venn diagram refers to the customers who liked neither flavour.

iii. Nine customers liked neither flavour.

c. i. We are told that 24 customers like both flavours. This is the overlapping section in the Venn diagram.

c. i. $P(\text{liked both flavours}) = \dfrac{24}{60}$
$$= \dfrac{2}{5}$$
$$= 0.4 \text{ or } 40\%$$
There is a 40% chance that a customer liked both new flavours.

ii. Nine customers do not like either flavour. This is the section inside the rectangle but not inside any of the circles.

ii. $P(\text{liked neither flavour}) = \dfrac{9}{60}$
$$= \dfrac{3}{20}$$
$$= 0.15 \text{ or } 15\%$$
There is a 15% chance that a customer liked neither flavour.

iii. We are told that 35 customers liked flavour A. This is the entire circle for flavour A (11 + 24).

iii. $P(\text{liked flavour A}) = \dfrac{35}{60}$

$= \dfrac{7}{12}$

$= 0.58\dot{3} \text{ or } 58\frac{1}{3}\%$

There is a $58\frac{1}{3}\%$ chance that a customer liked flavour A.

iv. There are 11 customers who liked flavour A only. This is the pink section in the Venn diagram.

iv. $P(\text{liked flavour A only}) = \dfrac{11}{60}$

$= 0.18\dot{3} \text{ or } 18\frac{1}{3}\%$

There is an $18\frac{1}{3}\%$ chance that a customer liked flavour A only.

COMMUNICATING — COLLABORATIVE TASK: Eye and hair colour probability

Equipment: Post-it notes, marker pens

1. On the board, create a large Venn diagram with three overlapping circles. Label the circles 'blue eyes', 'brown hair' and 'black hair'.
2. Students write their initials on a Post-it note using a marker pen and stick the Post-it note in the appropriate place in the Venn diagram according to their relevant physical attributes.
3. Count the number of Post-it notes in each section. Remove the Post-it notes and write the number of Post-it notes for that section in the section on the board.
4. As a class, discuss the following. If a student is chosen randomly, determine the probability that they have:
   a. blue eyes
   b. neither blue eyes nor brown hair
   c. brown hair
   d. black hair
   e. not blue eyes and brown hair
   f. blue eyes and black hair
   g. not blue eyes
   h. not black hair.

14.5.3 Exploring 2-way tables

eles-4480

- **2-way tables** can also be used to represent the relationship between non-mutually exclusive events.
- In a 2-way table, the rows indicate one of the events and the columns indicate the other event.
- Consider the following simple example of a 2-way table that shows 100 students split by their biological sex and whether they are right-handed or left-handed.

	Right-handed	Left-handed
Male	20	19
Female	30	31

The number 20 tells us that there are 20 right-handed males.

The number 31 tells us that there are 31 left-handed females.

The following 2-way table shows the relationship between age and height of Year 9 students.

Age compared to height of Year 9 students

		Height		
		Below 160 cm	Above 160 cm	
Age	Younger than 14	25	11	36
	14 years and older	9	24	33
		34	35	69

a. Identify how many students are younger than 14 overall.
b. Identify how many students 14 and older are taller than 160 cm.
c. Identify how many students younger than 14 are taller than 160 cm.
d. Are there more students younger than 14 below 160 cm than students 14 and older above 160 cm?
e. Determine how many students are either 14 and older or taller than 160 cm.
f. Calculate the probability that a randomly selected student is taller than 160 cm. Give your answer correct to 4 decimal places.

THINK

a. Look at the row represented by the younger-than-14 attribute. The number at the end of this row contains the sum of this row's data.

b. Look at the region represented by the intersection between the 14-and-older attribute and the above–160 cm attribute.

c. Look at the region represented by the intersection between the younger-than-14 attribute and the above–160 cm attribute.

d. Compare the two required regions.

e. Identify all of the required regions (the row represented by the 14-and-older attribute and the intersection of the younger-than-14 attribute and the above–160 cm attribute).

f. Looking at the 2-way table, there are 35 students whose height is above 160 cm. There are 69 students in total.

WRITE/DRAW

a. There are 36 students younger than 14 overall.

b. There are 24 students 14 and older who are taller than 160 cm.

c. There are 11 students younger than 14 and taller than 160 cm.

d. Yes, there are more students younger than 14 below 160 cm than students 14 and older above 160 cm (25 compared to 24).

e. $33 + 11 = 44$
There are 44 students who are either 14 and older or above 160 cm.

f. $P(\text{above } 160\,\text{cm}) = \dfrac{35}{69}$
$= 0.5072$

Note: In part **e** of Worked example 16 we only need to add the number of students younger than 14 and above 160 cm, as the 14-and-older students above 160 cm have already been included in the 14-and-older attribute.

There are 1400 penguins in an Australian penguin colony. Of these, 675 are male and 370 are up to 5 years old. Of the female penguins, 410 are over 5 years old.

a. Draw a 2-way table to illustrate the above information.
b. Use the 2-way table to calculate how many:
   i. penguins are up to 5 years old
   ii. penguins are over 5 years old
   iii. male penguins are over 5 years old.

THINK

a. 1. Draw the 2-way table, listing one attribute on the left-hand side and one on the top.

2. Enter the information given in the question.

3. Calculate the missing information and complete the table.

WRITE/DRAW

a.

	Up to 5	Over 5	
Male			
Female			

There are 1400 penguins overall.
There are 675 male penguins, of which 370 are up to 5 years old.
There are 410 female penguins over 5 years old.

	Up to 5	Over 5	
Male	370		675
Female		410	
			1400

There are 305 male penguins who are over 5 years old.
$(675 - 370 = 305)$
There are 725 female penguins.
$(1400 - 675 = 725)$
There are 315 female penguins who are up to 5 years old.
$(725 - 410 = 315)$
There are 685 penguins who are up to 5 years old.
$(370 + 315 = 685)$
There are 715 penguins who are over 5 years old.
$(305 + 410 = 715)$

	Up to 5	Over 5	
Male	370	305	675
Female	315	410	725
	685	715	1400

b. **i.** Look at the column represented by the up-to-5 attribute. The number at the end of this column contains the sum of this column's data.

ii. Look at the column represented by the over-5 attribute. The number at the end of this column contains the sum of this column's data.

iii. Look at the cell represented by the intersection between the male and over-5 attributes.

b. **i.** 685 penguins are up to 5 years old.

ii. 715 penguins are over 5 years old.

iii. 305 of the male penguins are over 5 years old.

14.5.4 Converting between Venn diagrams and 2-way tables

eles-4481

- When representing the relationships between two events, we can convert from Venn diagrams to 2-way tables and vice versa.

WORKED EXAMPLE 18 Converting Venn diagrams to a 2-way table

The following Venn diagram shows the relationship between students who wear glasses or not and preference for cats or dogs.

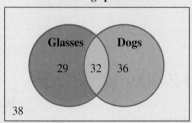

The relationship between glasses and cats/dogs preference

Glasses 29 | 32 | 36 Dogs
38

Convert this information into a 2-way table.

THINK

1. Draw the 2-way table, listing one attribute on the left-hand side and one on the top.

2. Calculate the total number of people.

3. Interpret the information in the Venn diagram.

WRITE

	Cats	Dogs	
Glasses			
Not glasses			

$29 + 32 + 36 + 38 = 135$

There are 32 students who wear glasses and who prefer dogs.

There are 36 students who don't wear glasses and who prefer dogs.

There are 29 students who wear glasses and who prefer non-dogs (cats).

There are 38 students who don't wear glasses and who prefer non-dogs (cats).

4. Enter the information into the 2-way table.

	Cats	Dogs	
Glasses	29	32	
Not glasses	38	36	
			135

5. Calculate the missing values by summing the rows and columns, then enter this information into the 2-way table.

$29 + 32 = 61$
$38 + 36 = 74$
$29 + 38 = 67$
$32 + 36 = 68$

	Cats	Dogs	
Glasses	29	32	61
Not glasses	38	36	74
	67	68	135

DISCUSSION

Are there certain situations that are better suited to Venn diagrams than 2-way tables or vice versa?

---

on Resources

▶ **Video eLesson** Venn diagrams (eles-1934)

✦ **Interactivities** Venn diagrams (int-3828)

Intersection and union of sets (int-3829)

Exercise 14.5 Exploring Venn diagrams and 2-way tables (Path)

learn on

14.5 Quick quiz on	**14.5 Exercise**

Individual pathways

■ PRACTISE	■ CONSOLIDATE	■ MASTER
1, 3, 6, 8, 10, 13, 16, 20	2, 4, 7, 11, 14, 17, 21	5, 9, 12, 15, 18, 19, 22

Fluency

1. **WE13** The following Venn diagram represents the favourite colours of primary school children.

   a. Identify whether there are any children who chose both red and blue.
   b. Calculate how many children there were overall.
   c. Identify how many children preferred red.
   d. Identify how many children preferred blue.
   e. Identify how many children preferred neither red nor blue.

Favourite colours of primary school children

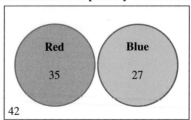

2. The following Venn diagram represents a study of the most popular exercise classes at a gym.

   a. Determine how many people were surveyed overall.
   b. Identify how many people went to spin classes.
   c. Identify how many people went to weights classes.
   d. Identify how many people went to neither spin nor weights classes.
   e. Determine how many people went to either the spin or the weights class, but not both.

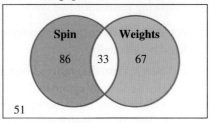
Most popular exercise classes

3. **WE14** Describe the following regions in this Venn diagram.

   a. The combined pink and orange regions
   b. The white region
   c. The purple region
   d. The combined pink, purple and orange regions
   e. The combined pink and purple regions

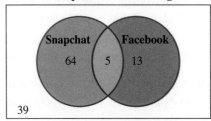
Social media platform that teenagers like

4. A survey of a Year 9 class found the numbers of classmates who play basketball, cricket and soccer. Use the Venn diagram shown to calculate the number of students who:

   a. were in the class
   b. play basketball
   c. play cricket and basketball
   d. play cricket and basketball but not soccer
   e. play soccer but not cricket.

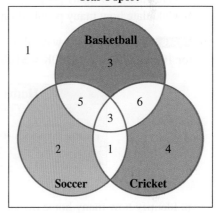
Year 8 sport

5. A survey of a Year 9 class found the numbers of class members who play basketball, cricket and soccer. Use the Venn diagram from question 4 to calculate the number of students who:

   a. play all three sports
   b. do not play cricket, basketball or soccer
   c. do not play cricket
   d. play cricket or basketball or both
   e. play at least one of basketball or cricket or soccer.

Understanding

6. **WE15** A cosmetics shop surveyed 100 customers about two new bath bomb products. The results showed that 63 customers liked product A, 28 liked product B, and 11 liked both.

   a. Draw a Venn diagram to illustrate the above information.
   b. Use the Venn diagram to determine:
      i. how many customers liked product A only
      ii. how many customers liked product B only
      iii. how many customers liked neither product.
   c. If a customer was selected at random, determine the probability that they liked both new products.

7. A tyre manufacturer conducting a survey of 2200 customers obtained the following results for two tyres:

   A total of 1390 customers preferred tyre A, 1084 preferred tyre B, and 496 liked both equally.

   a. Draw a Venn diagram to illustrate the above information.
   b. Use the Venn diagram to determine:
      i. how many customers preferred tyre A only
      ii. how many customers preferred tyre B only
      iii. how many customers preferred neither tyre.

8. The favourite cruise destinations of 120 Australian tourists were as follows: 55 people chose Fiji, 37 chose Tahiti and 42 chose another destination.

   a. Draw a Venn diagram to illustrate the above information.
   b. Calculate how many of the tourists chose both Fiji and Tahiti.

9. A survey asked 300 people which music streaming service they subscribe to. A total of 91 people subscribed to Apple Music and 143 people subscribed to Spotify.

   Of the 143 people who subscribed to Spotify, 130 didn't subscribe to Apple Music.

   a. Calculate how many people subscribed to Apple Music only.
   b. Calculate how many people didn't subscribe to either Apple Music or Spotify.

10. **WE16** The following 2-way table shows the relationship between home ownership and household income for 100 individuals. ($100K = $100 000)

Home ownership compared to household income			
	Under $100K	**Over $100K**	
Home ownership	11	33	44
Renting	38	18	56
	49	51	100

   a. Identify how many home owners there are overall.
   b. Identify how many home owners earn under $100K.
   c. Identify how many renters earn over $100K.
   d. Determine how many people are renters or earning above $100K.
   e. Calculate the probability that a randomly selected individual earns under $100K. Give your answer correct to 4 decimal places.

11. **WE17** There are 150 pop and rap songs in rotation on a popular radio channel. This includes 92 songs from this year, of which 63 are pop songs. Of the older songs, 21 can be classified as rap.

   a. Draw a 2-way table to illustrate the above information.
   b. Use the 2-way table to calculate how many:
      i. old songs are pop songs
      ii. songs from this year are rap songs
      iii. of all the songs are pop songs.

12. There are 180 pigs and cattle on Farmer Smith's farm. 85 of the animals are male, of which 59 are pigs. Of the cattle, 31 are female.

   a. Draw a 2-way table to illustrate the above information.
   b. Use the 2-way table to calculate how many:
      i. cattle are male
      ii. females are pigs.
   c. If you randomly selected an animal from Farmer Smith's farm, calculate the probability that the animal is:
      i. a pig
      ii. male.

13. The following 2-way table shows the relationship between sport participation of gamers and whether they prefer PlayStation or Xbox.

Sport participation compared to console preference			
	PlayStation	**Xbox**	
Play sport	24	19	**43**
Don't play sport	21	17	**38**
	45	**36**	**81**

   a. Convert this information into a Venn diagram.
   b. Determine how many people this table represents.
   c. Determine how many gamers don't play sport.
   d. If you randomly selected one of the gamers, what is the probability that they do not play sport?

14. **WE18** The following Venn diagram shows the relationship between students who live in metropolitan or regional areas and whether they are meat-eaters or vegetarians.

Relationship between location and diet

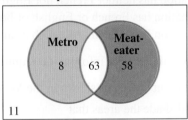

Convert this information into a 2-way table.

15. The following Venn diagram shows the relationship between hair colour and eye colour.

Relationship between hair and eye colour

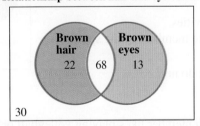

Convert this information into a 2-way table.

Communicating, reasoning and problem solving

16. Year 9 students were asked about their fruit preferences and the data was then recorded in the Venn diagram shown.

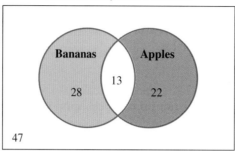

Fruits liked by Year 8 students

a. Convert this information into a 2-way table.
b. Calculate the probability that a student prefers apples.
c. Determine the probability that a student likes neither apples nor bananas.
d. Determine the probability that a student likes apples or bananas, but not both.

17. The following 2-way table shows the relationship between nations where French is an official language and the nations where English is an official language.

	English	Not English	
French	5	24	**29**
Not French	49	117	**166**
	54	**141**	**195**

a. Calculate the probability that a nation has neither French nor English as an official language.
b. Determine the probability that a nation has French or English or both as an official language.
c. Determine the probability that a nation has French or English as an official language, but not both.

18. Explain what is the minimum number of pieces of information required to complete a 2-way table.

19. Members of a sporting club play different sports, as shown in the Venn diagram.

a. Copy the given Venn diagram and shade the areas that represent:
   i. members who play tennis only
   ii. members who walk only
   iii. members who both play tennis and go walking.
b. Calculate how many members:
   i. play volleyball
   ii. are involved in all three activities.
c. Determine how many people are members of the sporting club.
d. Determine how many members do not:
   i. play tennis
   ii. go walking.
e. Determine the probability that a member likes playing volleyball or tennis but does not like walking.
f. Evaluate the probability that a member likes playing volleyball and tennis but does not like walking.

Sports played by club members

20. The following -way table represents the relationship between whether a household is located in a metropolitan or regional area and if it has any pets.

Location compared to pet ownership			
	Have pets	**Don't have pets**	
Metro	*c*	18	*a*
Regional	45	*d*	55
	72	*b*	**100**

   a. Identify the missing number labelled *a*.
   b. Identify the missing number labelled *b*.
   c. Identify the missing number labelled *c*.
   d. Identify the missing number labelled *d*.
   e. According to this 2-way table, determine what percentage of households have pets.

21. Year 9 students recorded what 100 people were wearing. The data was supposed to be recorded in a Venn diagram, but the students didn't finish putting in their results. The incomplete Venn diagram is shown.

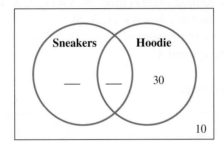

   a. Determine how many people in total were wearing sneakers.

   b. If $\dfrac{1}{3}$ of the people wearing sneakers were also wearing a hoodie, evaluate how many people were wearing both sneakers and a hoodie.

   c. Use your answers to a and b to fill in the gaps and complete the Venn diagram.

22. A survey of 140 fifteen-year-olds investigated how many read magazines (*M*), crime novels (*C*) and science fiction (*S*). It found:
   • 23 read both magazines and science fiction
   • 21 read both magazines and crime novels
   • 25 read both crime novels and science fiction
   • 15 read all three
   • 40 read magazines only
   • 38 read crime novels only
   • 10 read science fiction only.

   a. Show this information on a fully labelled Venn diagram.
   b. Determine how many fifteen-year-olds read magazines.
   c. Evaluate how many fifteen-year-olds read only crime.
   d. Determine how many fifteen-year-olds read science fiction.
   e. Evaluate how many fifteen-year-olds read none of these three.

LESSON
14.6 Venn diagrams and 2-way tables using set notation (Path)

LEARNING INTENTION

At the end of this lesson you should be able to:
- identify the complement of an event A
- construct and interpret Venn diagrams and 2-way tables
- use a Venn diagram or 2-way table to determine $A \cap B$ or $A \cup B$
- use a Venn diagram or 2-way table to solve probability problems.

▶ 14.6.1 The complement of an event

eles-4883

- Suppose that a die is rolled. The sample space is given by $S = \{1, 2, 3, 4, 5, 6\}$.
- If A is the event 'rolling an odd number', then $A = \{1, 3, 5\}$.
- There is another event called 'the **complement** of A', or not A. This event contains all the outcomes that do not belong to A. It is given by the symbol A'.
- In this case $A' = \{2, 4, 6\}$.
- A and A' can be shown on a Venn diagram.

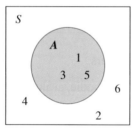
A is shaded.

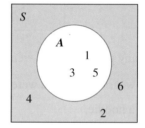
A' (not A) is shaded.

Complementary events

For the event A, the complement is denoted A', and the two are related by the following:

$$P(A) + P(A') = 1$$

WORKED EXAMPLE 19 Determining the complement

For the sample space $S = \{1, 2, 3, 4, 5\}$, list the complement of each of the following events.
a. $A = \{$multiples of 3$\}$
b. $B = \{$square numbers$\}$
c. $C = \{1, 2, 3, 5\}$

THINK	WRITE
a. The only multiple of 3 in the set is 3. Therefore, $A = \{3\}$. A' is every other element of the set.	a. $A' = \{1, 2, 4, 5\}$

---

b. The only square numbers are 1 and 4. Therefore, $B = \{1, 4\}$. B' is every other element of the set.

b. $B' = \{2, 3, 5\}$

c. $C = \{1, 2, 3, 5\}$. C' is every other element of the set.

c. $C' = \{4\}$

▶ 14.6.2 Venn diagrams: the regions

eles-4884

- A Venn diagram consists of a rectangle and one or more circles.
- A Venn diagram contains all possible outcomes in the sample space and will have the S symbol in the top left corner.
- A Venn diagram is used to illustrate the relationship between sets of objects or numbers.
- All outcomes for a given event will be contained within a specific circle.
- Outcomes that belong to multiple events will be found in the overlapping region of two or more circles.
- The overlapping region of two circles is called the intersection of the two events and is represented using the ∩ symbol.
- In describing a region, care needs to be taken with words such as 'only' or 'not in'.
- The four regions for two events are shown below.

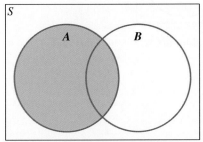

The circle on the left contains all outcomes in event A.

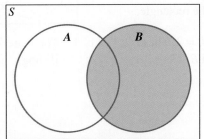

The circle on the right contains all outcomes in event B.

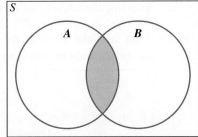

The overlap or intersection of the two circles contains the outcomes that are in event A 'and' in event B. This is denoted by $A \cap B$.

A Venn diagram for two events A and B has four distinct regions.

- $A \cap B'$ contains the outcomes in event A and not in event B.
- $A \cap B$ contains the outcomes in event A and in event B.
- $A' \cap B$ contains the outcomes not in event A and in event B.
- $A' \cap B'$ contains the outcomes not in event A and not in event B.

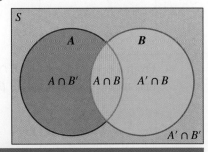

WORKED EXAMPLE 20 Setting up a Venn diagram

In a class of 20 students, 5 study Art, 9 study Biology and 2 students study both.

Let $A = \{$students who study Art$\}$ and $B = \{$students who study Biology$\}$.

a. Create a Venn diagram to represent this information.

b. Identify the number of students represented by the following and state what these regions represent:

   i. $A \cap B$ ii. $A \cap B'$ iii. $A' \cap B$ iv. $A' \cap B'$

▶

THINK	WRITE/DRAW
a. Draw a sample space with events *A* and *B*.	**a.** 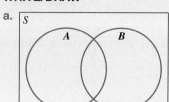
Place a 2 in the intersection of both circles since we know 2 students take both subjects.	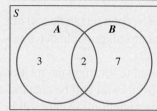
Since 5 study Art and there are 2 already in the middle, place a 3 in the remaining section of circle A.	
Since 9 study Biology and there are 2 already in the middle, place a seven in the remaining section of circle *B*.	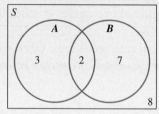
The total number inside the three circles is $3 + 2 + 7 = 12$. This means there must be $20 - 12 = 8$ outside of the two circles. Place an 8 outside the circles, within the rectangle.	

b. i. From the Venn diagram, $A \cap B = 2$. These are the students in both *A* and *B*.

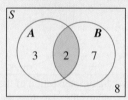

b. i. There are 2 students who study Art and Biology.

ii. From the Venn diagram, $A \cap B' = 3$. These are the students in *A* and not in *B*.

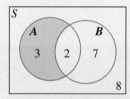

ii. There are 3 students who study Art and not Biology, i.e. 3 students study Art only.

iii. From the Venn diagram, $A' \cap B = 7$. These are the students not in *A* and in *B*.

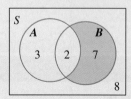

iii. There are 7 students who do not study Art and do study Biology, i.e. 7 students study Biology only.

iv. From the Venn diagram, $A' \cap B' = 8$. These are the students not in *A* and not in *B*.

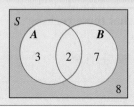

iv. There are 8 students who do not study Art and do not study Biology.

14.6.3 2-way tables

eles-4885

- The information in a Venn diagram can also be represented using a 2-way table. The relationship between the two is shown below.

	Event B	Event B'	Total
Event A	$A \cap B$	$A \cap B'$	A
Event A'	$A' \cap B$	$A' \cap B'$	A'
Total	B	B'	

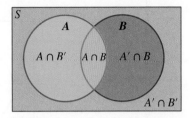

WORKED EXAMPLE 21 Creating a 2-way table

In a class of 20 students, 5 study Art, 9 study Biology and 2 students study both. Create a 2-way table to represent this information.

THINK	WRITE
1. Create an empty 2-way table.	

	Biology	Not Biology	Total
Art			
Not Art			
Total			

2. Fill the table in with the information provided in the question.
   - 2 students study both subjects.
   - 5 in total take Art.
   - 9 in total take Biology.
   - 20 students in the class.

	Biology	Not Biology	Total
Art	2		5
Not Art			
Total	9		20

3. Use the totals of the rows and columns to fill in the gaps in the table.
   - $9 - 2 = 7$ Biology and not Art
   - $5 - 2 = 3$ Art and not Biology
   - $20 - 9 = 11$ not Biology total
   - $20 - 5 = 15$ not Art total

	Biology	Not Biology	Total
Art	2	3	5
Not Art	7	8	15
Total	9	11	20

4. The last value, not Biology and not Art, can be found from either of the following:
   - $11 - 3 = 8$
   - $15 - 7 = 8$

14.6.4 Number of outcomes

eles-4886

- If event A contains seven outcomes or members, this is written as $n(A) = 7$.
- So $n(A \cap B') = 3$ means that there are three outcomes that are in event A and not in event B.

For the Venn diagram shown, write down the number of outcomes in each of the following.

a. M b. M' c. $M \cap N$ d. $M \cap N'$ e. $M' \cap N'$

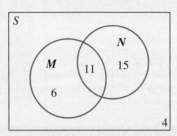

THINK

a. Identify the regions showing M and add the outcomes.

b. Identify the regions showing M' and add the outcomes.

c. $M \cap N$ means 'M and N'. Identify the region.

d. $M \cap N'$ means 'M and not N'. Identify the region.

WRITE/DRAW

a.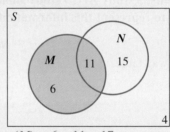

$n(M) = 6 + 11 = 17$

b.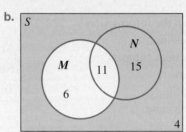

$n(M') = 4 + 15 = 19$

c.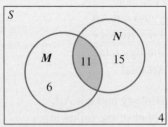

$n(M \cap N) = 11$

d.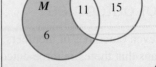

$n(M \cap N') = 6$

e. $M' \cap N'$ means 'not M and not N'. Identify the region.

e.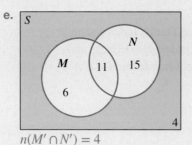

$n(M' \cap N') = 4$

WORKED EXAMPLE 23 Using a Venn diagram to create a 2-way table

Show the information from the Venn diagram on a 2-way table.

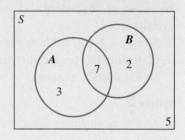

THINK

1. Draw a 2×2 table and add the labels A, A', B and B'.

2. There are 7 elements in A and B.
   There are 3 elements in A and 'not B'.
   There are 2 elements in 'not A' and B.
   There are 5 elements in 'not A' and 'not B'.

3. Add in a column and a row to show the totals.

WRITE

	A	A'
B		
B'		

	A	A'
B	7	2
B'	3	5

	A	A'	Total
B	7	2	9
B'	3	5	8
Total	10	7	17

WORKED EXAMPLE 24 Using a 2-way table to create a Venn diagram

Show the information from the 2-way table on a Venn diagram.

	Left-handed	Right-handed
Blue eyes	7	20
Not blue eyes	17	48

THINK		**DRAW**

THINK

Draw a Venn diagram that includes a sample space and events L for left-handedness and B for blue eyes. (Right-handedness $= L'$.)

$n(L \cap B) = 7$

$n(L \cap B') = 17$

$n(L' \cap B) = 20$

$n(L' \cap B') = 48$

DRAW

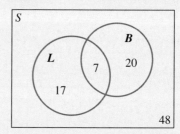

WORKED EXAMPLE 25 Probability from a 2-way table

In a class of 30 students, 15 swim for exercise and 20 run for exercise and 5 participate in neither activity.

a. Create a 2-way table to represent this information.

b. Calculate the probability that a randomly chosen student from this class does running and swimming for exercise.

THINK

a. 1. Create an empty 2-way table.

WRITE

a.

	Swim	Not swim	Total
Run			
Not run			
Total			

2. Fill the table in with the information provided in the question.
- 5 students do neither activity.
- 15 in total swim.
- 20 in total run.
- 30 students are in the class.

	Swim	Not swim	Total
Run			20
Not run		5	
Total	15		30

3. Use the totals of the rows and columns to fill in the gaps in the table.

	Swim	Not swim	Total
Run	10	10	20
Not run	5	5	10
Total	15	15	30

b. The probability a student runs and swims is given by:

$$P(\text{run} \cap \text{swim}) = \frac{\text{number in run} \cap \text{swim}}{\text{total in the class}}$$

b. $P(\text{run} \cap \text{swim}) = \dfrac{\text{number in run} \cap \text{swim}}{\text{total in the class}}$

$$= \frac{10}{30}$$

$$= \frac{1}{3}$$

14.6.5 Venn diagrams: the intersection and union of events

eles-4887

- The **intersection** of two events ($A \cap B$) is all outcomes in event A 'and' in event B.
- The **union** of two events ($A \cup B$) is all outcomes in events A 'or' in event B.

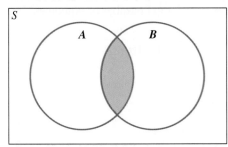

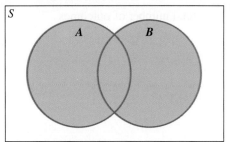

The intersection of the two circles contains the outcomes that are in event A 'and' in event B. This is denoted by $A \cap B$.

Everything contained within the two circles is an outcome that is in event A 'or' in event B. This is denoted by $A \cup B$.

WORKED EXAMPLE 26 Calculating the union of two events

Use the Venn diagram shown to calculate the value of the following.
a. $n(A)$
b. $P(B)$
c. $n(A \cup B)$
d. $P(A \cap B)$

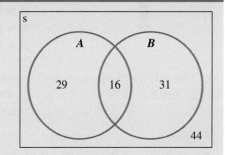

THINK

a. Identify the number of outcomes in the A circle.

b. Identify the number of outcomes in the B circle.

$P(B) = \dfrac{\text{number of favourable outcomes in } B}{\text{total number of outcomes}}$

c. Identify the number of outcomes in either of the two circles.

WRITE

a. $n(A) = 29 + 16 = 45$

b. $P(B) = \dfrac{\text{favourable outcomes in } B}{\text{total number of outcomes}}$

$= \dfrac{n(B)}{\text{total number of outcomes}}$

$= \dfrac{16 + 31}{29 + 16 + 31 + 44}$

$= \dfrac{47}{120}$

c. $n(A \cup B) = 29 + 16 + 31$
$= 76$

d. Identify the number of outcomes in the intersection of the two circles.

$$P(A \cap B) = \frac{\text{favourable outcomes in } A \cap B}{\text{total number of outcomes}}$$

d. $P(A \cap B) = \dfrac{\text{number of favourable outcomes in } A \cap B}{\text{total number of outcomes}}$

$$= \frac{n(A \cap B)}{\text{total number of outcomes}}$$

$$= \frac{16}{29 + 16 + 31 + 44}$$

$$= \frac{16}{120}$$

$$= \frac{2}{15}$$

WORKED EXAMPLE 27 Completing a Venn diagram

a. **Place the elements of the following sets of numbers in their correct position in a single Venn diagram.**

$$S = \{\text{Number 1 to 20 inclusive}\}$$
$$A = \{\text{Multiples of 3 from 1 to 20 inclusive}\}$$
$$B = \{\text{Multiples of 2 from 1 to 20 inclusive}\}$$

b. **Use this Venn diagram to determine the following.**

   i. $A \cap B$ **ii.** $A \cup B$ **iii.** $A \cap B'$ **iv.** $A' \cup B'$

THINK

a.   Write out the numbers in each event A and B:

$$A = \{3, 6, 9, 12, 15, 18\}$$

$$A = \{2, 4, 6, 8, 10, 12, 14, 16, 18, 20\}$$

Identify the numbers that appear in both sets. In this case it is 6, 12 and 18. These numbers will be placed in the overlap of the two circles for A and B.

All numbers not in A or B are placed outside the two circles.

After placing the numbers in the Venn diagram, check that all numbers from 1 to 20 are written down.

b.   **i.** $A \cap B$ are the numbers in A 'and' in B.

    ii. $A \cup B$ are the numbers in A 'or' in B.

    iii. $A \cap B'$ are the number in A and not in B. Refer to the four sections of the Venn diagram to locate this region.

    iv. $A' \cup B'$ is any number that is not in A 'or' not in B. This ends up being any number not in $A \cap B$.

WRITE/DRAW

a.

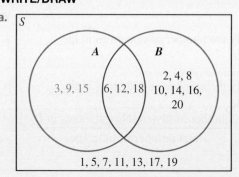

b.   **i.** $A \cap B = \{6, 12, 18\}$

    ii. $A \cup B = \{2, 3, 4, 6, 8, 9, 10, 12, 14, 15, 16, 18, 20\}$

    iii. $A \cap B' = \{3, 9, 15\}$

    iv. $A' \cup B' = \left\{ \begin{matrix} 1, 2, 3, 4, 5, 7, 8, 9, 10, 11, 13, \\ 14, 15, 16, 17, 19, 20 \end{matrix} \right\}$

In a class of 24 students, 11 students play basketball, 7 play tennis, and 4 play both sports.
a. Show the information on a Venn diagram.
b. If one student is selected at random, then calculate the probability that:
   i. the student plays basketball
   ii. the student plays tennis or basketball
   iii. the student plays tennis or basketball but not both.

THINK	WRITE/DRAW
a. 1. Draw a sample space with events B and T.	a. 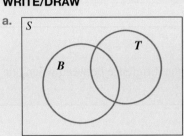
2. $n(B \cap T) = 4$ $n(B \cap T') = 11 - 4 = 7$ $n(T \cap B') = 7 - 4 = 3$ So far, 14 students out of 24 have been placed. $n(B' \cap T') = 24 - 14 = 10$	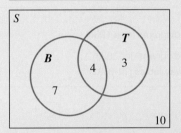
b. i. Identify the number of students who play basketball. 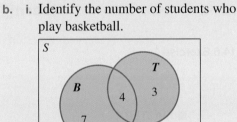 $P(B) = \dfrac{\text{number of favourable outcomes}}{\text{total number of outcomes}}$	b. i. $P(B) = \dfrac{\text{number of students who play basketball}}{\text{total number of students}}$ $= \dfrac{n(B)}{24}$ $= \dfrac{11}{24}$
ii. Identify the number of students who play tennis or basketball.	ii. $P(T \cup B) = \dfrac{n(T \cup B)}{24}$ $= \dfrac{14}{24}$ $= \dfrac{7}{12}$

iii. Identify the number of students who play tennis or basketball but not both.

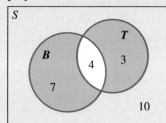

iii. $n(B \cap T') + n(B' \cap T) = 3 + 7$
$= 10$

P(tennis or basketball but not both)
$= \dfrac{10}{24}$
$= \dfrac{5}{12}$

DISCUSSION

How will you remember the difference between when one event and another occurs and when one event or another occurs?

 Resources

 Video eLesson Venn diagrams (eles-1934)

Interactivities Venn diagrams (int-3828)
Two-way tables (int-6082)

Exercise 14.6 Venn diagrams and 2-way tables using set notation (Path)

learn**on**

14.6 Quick quiz **on**	14.6 Exercise

Individual pathways

■ PRACTISE	■ CONSOLIDATE	■ MASTER
1, 4, 5, 10, 13, 16, 17, 20, 24	2, 6, 7, 9, 11, 14, 18, 21, 25	3, 8, 12, 15, 19, 22, 23, 26, 27

Fluency

1. **WE19** For the sample space $S = \{1, 2, 3, 4, 5, 6, 7, 8, 9, 10\}$, list the complement of each of the following events.

   a. $A = \{$evens$\}$
   c. $C = \{$squares$\}$
   b. $B = \{$multiples of 5$\}$
   d. $D = \{$numbers less than 8$\}$

2. If $S = \{11, 12, 13, 14, 15, 16, 17, 18, 19, 20\}$, list the complement of each of the following events.

   a. $A = \{$multiples of 3$\}$
   c. $C = \{$prime numbers$\}$
   b. $B = \{$numbers less than 20$\}$
   d. $D = \{$odd numbers or numbers greater than 16$\}$

3. If $S = \{1, 2, 3, 4, 5, 6, 7, 8, 9, 10, 11, 12, 13, 14, 15, 16, 17, 18, 19, 20\}$, list the complement of each of the following events.

   a. $A = \{$multiples of 4$\}$
   c. $C = \{$even and less than 13$\}$
   b. $B = \{$primes$\}$
   d. $D = \{$even or greater than 13$\}$

4. **WE22** For the Venn diagram shown, write down the number of outcomes in:
   a. S
   b. Q
   c. T
   d. $T \cap Q$
   e. $T \cap Q'$
   f. $Q' \cap T'$

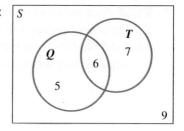

5. **WE23** Show the information from question 4 on a 2-way table.

6. **WE24** Show the information from this 2-way table on a Venn diagram.

	S	S'
V	21	7
V'	2	10

7. For each of the following Venn diagrams, use set notation to write the name of the region coloured in:
   i. blue
   ii. pink.

   a.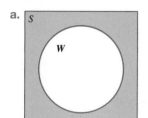
   b.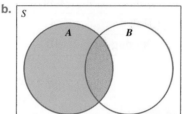
   c.

8. **WE21&25** The membership of a tennis club consists of 55 men and 45 women. There are 27 left-handed people, including 15 men.

   a. Show the information on a 2-way table.
   b. Show the information on a Venn diagram.
   c. If one member is chosen at random, calculate the probability that the person is:
      i. right-handed
      ii. a right-handed man
      iii. a left-handed woman.

9. **WE26** Using the information given in the Venn diagram, if one outcome is chosen at random, determine:

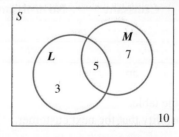

   a. $P(L)$
   b. $P(L')$
   c. $P(L \cap M)$
   d. $P(L \cap M')$

10. **WE27** Place the elements of the following sets of numbers in their correct position in a single Venn diagram.
    $S = \{$numbers between 1 to 10 inclusive$\}$
    $A = \{$odd numbers from 1 to 10$\}$
    $B = \{$squared numbers between 1 to 10 inclusive$\}$

11. Place the elements of the following sets of numbers in their correct position in a single Venn diagram.
    $S = \{$numbers between 1 to 25 inclusive$\}$
    $A = \{$multiples of 3 from 1 to 25$\}$
    $B = \{$numbers that are odd or over 17 from 1 to 25 inclusive$\}$

12. Place the elements of the following sets of numbers in their correct position in a single Venn diagram.
    $A = \{$prime numbers from 1 to 20$\}$
    $B = \{$even numbers from 1 to 20$\}$
    $C = \{$multiples of 5 from 1 to 20$\}$
    $S = \{$numbers between 1 and 20 inclusive$\}$

13. Using the information given in the table, if one family is chosen at random, calculate the probability that they own:

Pets owned by families		
	Cat	**No cat**
Dog	4	11
No dog	16	9

    a. a cat
    b. a cat and a dog
    c. a cat or a dog or both
    d. a cat or a dog but not both
    e. neither a cat nor a dog.

14. Using the information given in the table, if a person is chosen at random, calculate the probability that for exercise, this person:

Type of exercise		
	Cycling	**No cycling**
Running	12	19
No running	13	6

    a. cycles
    b. cycles and runs
    c. cycles or runs
    d. cycles or runs but not both.

15. A barista decides to record what the first 45 customers order on a particular morning. The information is partially filled out in the table shown.

	Croissant	**No croissant**	**Total**
Coffee	27		
No coffee		7	
Total	36		

    a. Fill in the missing information in the table.
    b. Using the table, calculate the probability that the next customer:
        i. orders a coffee and a croissant
        ii. order a coffee or a croissant
        iii. orders a coffee or a croissant but not both.
    c. The barista serves an average of 315 customers a day.
        If the café is open six days a week, calculate how many coffees the barista will make each week.

Understanding

16. A group of athletes was surveyed and the results were shown on a
Venn diagram.
$S = \{\text{sprinters}\}$ and $L = \{\text{long jumpers}\}$.

    a. Write down how many athletes were included in the survey.
    b. If one of the athletes is chosen at random, calculate the probability that
    the athlete competes in:
       i. long jump
       ii. long jump and sprints
       iii. long jump or sprints
       iv. long jump or sprints but not both.

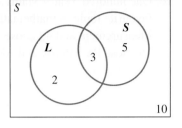

17. **WE20&28** In a class of 40 students, 26 take a train to school, 19 take a bus and 8 take neither of these.
Let $T = \{\text{takes the train}\}$ and $B = \{\text{takes the bus}\}$.

    a. Show the information on a Venn diagram.
    b. If one student is selected at random, calculate the probability that:
       i. the student takes the bus
       ii. the student takes the train or the bus
       iii. the student takes the train or the bus but not both.

18. **WE20** If $S = \{\text{children}\}$, $W = \{\text{swimmers}\}$ and $R = \{\text{runners}\}$, describe in words each of the following.

    a. W' b. $W \cap R$ c. $R' \cap W'$ d. $R \cup W$

19. A group of 12 students was asked whether they liked hip hop (H) and whether they liked classical music (C).
The results are shown in the table.

	C	H
Ali	✓	✓
Anu		
Chris		✓
George		✓
Imogen		✓
Jen	✓	✓
Luke	✓	✓
Pam	✓	
Petra		
Roger	✓	
Seedevi		✓
Tomas		

    a. Show the results on:
       i. a Venn diagram
       ii. a 2-way table.

    b. If one student is selected at random, calculate:
       i. P(H)
       ii. P($H \cup C$)
       iii. P($H \cap C$)
       iv. P(student likes classical or hip hop but not both).

Communicating, reasoning and problem solving

20. One hundred Year 9 Maths students were asked to indicate their favourite topic in mathematics. Sixty chose Probability, 50 chose Measurement and 43 chose Algebra. Some students chose two topics: 15 chose Probability and Algebra, 18 chose Measurement and Algebra, and 25 chose Probability and Measurement. Five students chose all three topics.

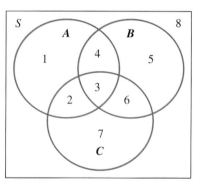

    a. Copy and complete the Venn diagram shown.
    b. Calculate how many students chose Probability only.
    c. Calculate how many students chose Algebra only.
    d. Calculate how many students chose Measurement only.
    e. Calculate how many students chose any two of the three topics.

A student is selected at random from this group. Calculate the probability that this student has chosen:

    f. Probability
    g. Algebra
    h. Algebra and Measurement
    i. Algebra and Measurement but not Probability
    j. all of the topics.

21. Create a Venn diagram using two circles to accurately describe the relationships between the following quadrilaterals: rectangle, square and rhombus.

22. Use the Venn diagram shown to write the numbers of the correct regions for each of the following problems.

    a. $A' \cup (B' \cap C)$
    b. $A \cap (B \cap C')$
    c. $A' \cap (B' \cap C')$
    d. $A \cup (B \cap C)'$

23. A recent survey taken at a cinema asked 90 teenagers what they thought about three different movies. In total, 47 liked 'Hairy Potter', 25 liked 'Stuporman' and 52 liked 'There's Something About Fred'.
16 liked 'Hairy Potter' only.
4 liked 'Stuporman' only.
27 liked 'There's Something About Fred' only.
There were 11 who liked all three films and 10 who liked none of them.

    a. Construct a Venn diagram showing the results of the survey.
    b. Calculate the probability that a teenager chosen at random liked 'Hairy Potter' and 'Stuporman' but not 'There's Something About Fred'.

24. 120 children attended a school holiday program during September. They were asked to select their favourite board game from Cluedo, Monopoly and Scrabble. They all selected at least one game, and four children chose all three games.
In total, 70 chose Monopoly and 55 chose Scrabble.
Some children selected exactly two games — 12 chose Cluedo and Scrabble, 15 chose Monopoly and Scrabble, and 20 chose Cluedo and Monopoly.

    a. Draw a Venn diagram to represent the children's selections.
    b. Calculate the probability that a child selected at random did not choose Cluedo as a favourite game.

25. Valleyview High School offers three sports at Year 9: baseball, volleyball and soccer. There are 65 students in Year 9.

2 have been given permission not to play sport due to injuries and medical conditions.

30 students play soccer.

9 students play both soccer and volleyball but not baseball.

9 students play both baseball and soccer (including those who do and don't play volleyball).

4 students play all three sports.

12 students play both baseball and volleyball (including those who do and don't play soccer).

The total number of players who play baseball is one more than the total of students who play volleyball.

a. Determine the number of students who play volleyball.

b. If a student was selected at random, calculate the probability that this student plays soccer and baseball only.

26. A Venn diagram consists of overlapping ovals that are used to show the relationships between sets. Consider the numbers 156 and 520.

Show how a Venn diagram could be used to determine their:

a. HCF   b. LCM.

27. A group of 200 shoppers was asked which type of fruit they had bought in the last week. The results are shown in the table.

Fruit	Number of shoppers
Apples (A) only	45
Bananas (B) only	34
Cherries (C) only	12
A and B	32
A and C	15
B and C	26
A and B and C	11

a. Display this information in a Venn diagram.

b. Calculate $n(A \cap B' \cap C)$.

c. Determine how many shoppers purchased apples and bananas but not cherries.

d. Calculate the relative frequency of shoppers who purchased:
   i. apples
   ii. bananas or cherries.

e. Estimate the probability that a shopper purchased cherries only.

LESSON
14.7 Mutually and non-mutually exclusive events (Path)

LEARNING INTENTION

At the end of this lesson you should be able to:
- identify and describe mutually exclusive and non-mutually exclusive events
- recognise the $P(A \cap B) = 0$ for mutually exclusive events
- identify and describe independent and dependent events.

⊙ 14.7.1 Mutually exclusive events

eles-6171

- If two events cannot both occur at the same time, then it is said the two events are **mutually exclusive**. For example, when rolling a die, the events 'getting a 1' and 'getting a 5' are mutually exclusive.
- If two sets are disjoint (have no elements in common), then the sets are mutually exclusive. For example, if $A = \{\text{prime numbers} > 10\}$ and $B = \{\text{even numbers}\}$, then A and B are mutually exclusive.
- If A and B are two mutually exclusive events (or sets), then $P(A \cap B) = 0$.
- Consider the Venn diagram shown.

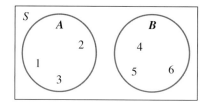

Since A and B are disjoint (do not overlap), then A and B are mutually exclusive sets.
- If two events A and B are mutually exclusive, then $P(A \text{ or } B) = P(A \cup B) = P(A) + P(B)$.

Mutually exclusive events

For two mutually exclusive events, A and B:

$P(A \text{ or } B) = P(A \cup B) = P(A) + P(B)$

Examples of mutually exclusive events

- Draw a card from a standard deck: the drawn card is a heart or a club.
  - Reason: it is impossible to get both a heart and a club at the same time.
- Record the time of arrival of overseas flights: a flight is late, on time or it is early.
  - Reason: it is impossible for the flight to arrive late, on time or early all at the same time.

Examples of non-mutually exclusive events

- Draw a card from a standard deck: the drawn card is a heart or a king.
  - Reason: it is possible to draw the king of hearts.
- Record the mode of transport of school students: count students walking or going by bus.
  - Reason: a student can walk (to the bus stop) and take a bus.

DISCUSSION

Can you think of more instances of mutually and non-mutually exclusive events?

WORKED EXAMPLE 29 Recognise mutually exclusive events

A card is drawn from a pack of 52 cards. What is the probability that the card is a diamond or a spade?

THINK	WRITE
1. The events are mutually exclusive because diamonds and spades cannot be drawn at the same time.	The two events are mutually exclusive as $P(A \cap B) = 0$.
2. Determine the probability of drawing a diamond and the probability of drawing a spade.	Number of diamonds $= 13$ Number of spades $= 13$ Number of cards in sample $= 52$ $P(\text{diamond}) = \dfrac{13}{52} \qquad P(\text{spade}) = \dfrac{13}{52}$ $\qquad\qquad = \dfrac{1}{4} \qquad\qquad\qquad = \dfrac{1}{4}$
3. Write the probability.	$P(A \cap B) = P(A) + P(B)$ $P(\text{diamond or spade}) = P(\text{diamond}) + P(\text{spade})$
4. Evaluate and simplify.	$= \dfrac{1}{4} + \dfrac{1}{4}$ $= \dfrac{1}{2}$

▶ 14.7.2 Independent events

eles-6172

- Two events are considered independent if the outcome of one event is not dependent on the outcome of the other event.
  For example, if A = {rolling a 4 on a first die} and B = {rolling a 2 on a second die}, the outcome of event A is not influenced by the outcome of event B, so the events are independent.

> **Independent events**
>
> For two independent events, *A* and *B*
>
> $P(A \text{ and } B) = P(A \cap B) = P(A) \times P(B)$

WORKED EXAMPLE 30 Independent events

Three coins are flipped simultaneously. Draw a tree diagram for the experiment. Calculate the following probabilities.

a. P(3 Heads) **b.** P(2 Heads) **c.** P(at least 1 Head)

▶

THINK	WRITE/DRAW

1. Use branches to show the individual outcomes for the first part of the experiment (flipping the first coin).

2. Link each outcome of the first flip with the outcomes of the second part of the experiment (flipping the second coin).

3. Link each outcome from the second flip with the outcomes of the third part of the experiment (flipping the third coin).

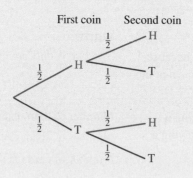

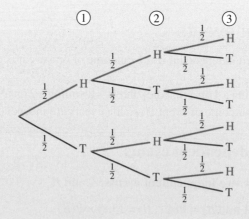

4. Determine the probability of each outcome. *Note*: The probability of each result is found by multiplying along the branches and in each case this will be $\frac{1}{2} \times \frac{1}{2} \times \frac{1}{2} = \frac{1}{8}$.

Outcomes	Probability
HHH	$\frac{1}{2} \times \frac{1}{2} \times \frac{1}{2} = \frac{1}{8}$
HHT	$\frac{1}{2} \times \frac{1}{2} \times \frac{1}{2} = \frac{1}{8}$
HTH	$\frac{1}{2} \times \frac{1}{2} \times \frac{1}{2} = \frac{1}{8}$
HTT	$\frac{1}{2} \times \frac{1}{2} \times \frac{1}{2} = \frac{1}{8}$
THH	$\frac{1}{2} \times \frac{1}{2} \times \frac{1}{2} = \frac{1}{8}$
THT	$\frac{1}{2} \times \frac{1}{2} \times \frac{1}{2} = \frac{1}{8}$
TTH	$\frac{1}{2} \times \frac{1}{2} \times \frac{1}{2} = \frac{1}{8}$
TTT	$\frac{1}{2} \times \frac{1}{2} \times \frac{1}{2} = \frac{1}{8}$
	1

a. The probability of three heads is P(H, H, H)

a. $P(3 \text{ Heads}) = \frac{1}{8}$

b. 1. {2 Heads} has 3
satisfactory outcomes:
(H, H, T), (H, T, H)
and (T, H, H), which are
mutually exclusive.

b. P(2 Heads)
$= P(H, H, T) + P(H, T, H) + P(T, H, H)$
$= \dfrac{1}{8} + \dfrac{1}{8} + \dfrac{1}{8}$
$= \dfrac{3}{8}$

2. Write your answer.

The probability of obtaining exactly 2 Heads is $\dfrac{3}{8}$.

c. 1. At least 1 Head means
any outcome that
contains one or more
Head. This is every
outcome except three
Tails. That is, it is the
complementary event to
obtaining 3 Tails.

c. P(at least 1 Head)
$= 1 - P(T, T, T)$
$= 1 - \dfrac{1}{8}$
$= \dfrac{7}{8}$

2. Write your answer.

The probability of obtaining at least 1 Head is $\dfrac{7}{8}$.

Note: The probabilities of all outcomes add to 1.

⏵ 14.7.3 Dependent events

eles-6173

- Many real-life events have some dependence upon each other, and their probabilities are likewise affected.
  Examples include:
  - the chance of rain today and the chance of a person taking an umbrella to work
  - the chance of growing healthy vegetables and the availability of good soil
  - the chance of Victory Soccer Club winning this week and winning next week
  - drawing a card at random, not replacing it, and drawing another card.
- It is important to be able to recognise the difference between *dependent* events and *independent* events.

A jar contains three black marbles, five red marbles, and two white marbles. Determine the probability of choosing a black marble (with replacement), then choosing another black marble.

THINK

1. The events are independent because the result of the second draw is not dependent on the result of the first draw.
   Demonstrate using a tree diagram.

WRITE/DRAW

The events are independent

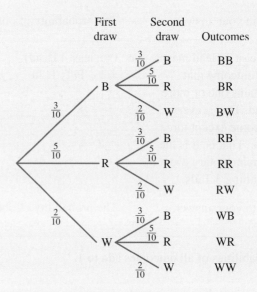

2. Determine the probability.
   Evaluate and simplify.

$$P(\text{black and black}) = P(\text{black}) \times P(\text{black})$$

$$P(\text{black and black}) = \frac{3}{10} \times \frac{3}{10}$$

$$= \frac{9}{100}$$

• If the first marble had not been replaced in the previous worked example, the second draw would be dependent on the outcome of the first draw, and so it follows that the sample space for the second draw is different from that for the first draw.

Repeat worked example 31 without replacing the first marble before the second one is drawn.

THINK

1. The words 'without replacing' indicate that the two events are dependent.
   Write the sample space and state the probability of choosing a black marble on the first selection.

WRITE/DRAW

There are 10 marbles and 3 of these are black.
The sample space is
$\{B, B, B, R, R, R, R, R, W, W\}$.
$P(\text{black}) = \dfrac{3}{10}$

2. Assume that a black marble was chosen in the first selection. Determine how many black ones remain, and the total number of remaining marbles.
   Write the sample space and state the probability of choosing a black marble on the second selection.

A black one was chosen, leaving 2 black ones and a total of 9 marbles.
The sample space is
$\{B, B, R, R, R, R, R, W, W\}$.
$P(\text{black}) = \dfrac{2}{9}$

3. Demonstrate using a tree diagram.

First draw   Second draw   Outcomes

$\dfrac{3}{10}$ B
- $\dfrac{2}{9}$ B   BB
- $\dfrac{5}{9}$ R   BR
- $\dfrac{2}{9}$ W   BW

$\dfrac{5}{10}$ R
- $\dfrac{3}{9}$ B   RB
- $\dfrac{4}{9}$ R   RR
- $\dfrac{2}{9}$ W   RW

$\dfrac{2}{10}$ W
- $\dfrac{3}{9}$ B   WB
- $\dfrac{5}{9}$ R   WR
- $\dfrac{1}{9}$ W   WW

4. Multiply the probabilities.

$P(\text{black and black}) = P(B_1) \times P(B_2)$
$$= \dfrac{3}{10} \times \dfrac{2}{9}$$
$$= \dfrac{1}{15}$$

5. Answer the question.

The probability of choosing two black marbles without replacing the first marble is $\dfrac{1}{15}$.

DISCUSSION

Your friend has just tossed a coin 4 times in a row and heads has been uppermost those 4 times. What is the probability of your friend tossing a head with her fifth toss?

 Resources

Interactivity Mutually exclusive events (int-6084)

| 14.7 Quick quiz on | 14.7 Exercise |

Individual pathways

■ PRACTISE	■ CONSOLIDATE	■ MASTER
1, 2, 3, 7, 8, 11, 12, 13, 20, 25	4, 5, 6, 9, 10, 14, 18, 19, 21, 26	15, 16, 17, 22, 23, 24, 27, 28, 29, 30, 31

Fluency

1. **MC** If a card is drawn from a pack of 52 cards, what is the probability that the card is not a queen?

   A. $\dfrac{4}{52}$ **B.** $\dfrac{4}{48}$ **C.** $\dfrac{13}{12}$ **D.** $\dfrac{48}{52}$

2. **MC** Which events are not mutually exclusive?

   A. Drawing a queen and drawing a jack from 52 playing cards
   B. Drawing a red card and drawing a black card from 52 playing cards
   C. Drawing a vowel and drawing a consonant from cards representing the 26 letters of the alphabet
   D. Obtaining a total of 8 and rolling doubles (when rolling two dice)

3. When a six-sided die is rolled 3 times, the probability of getting 3 sixes is $\dfrac{1}{216}$. What is the probability of not getting 3 sixes?

4. **MC** Eight athletes compete in a 100-m race.

   A. $\dfrac{1}{5}$ **B.** $\dfrac{5}{8}$ **C.** $\dfrac{8}{5}$ **D.** $\dfrac{4}{5}$

   The probability that the athlete in lane 1 will win is $\dfrac{1}{5}$. What is the probability that one of the other athletes wins? (Assume that there are no dead heats.)

5. A pencil case has 4 red pens, 3 blue pens and 5 black pens. If a pen is randomly drawn from the pencil case, calculate:

   a. P(drawing a blue pen)
   b. P(not drawing a blue pen).

6. Seventy Year 9 students were surveyed. Their ages ranged from 13 years to 15 years, as shown in the table below.

Age	13	14	15	Total
Boys	10	20	9	39
Girls	7	15	9	31
Total	17	35	18	70

A student from the group is selected at random. Calculate:

a. P(selecting a student of the age of 13 years)
b. P(not selecting a student of the age of 13 years)
c. P(selecting a 15-year-old boy)
d. P(not selecting a 15-year-old boy).

7. **WE29** A card is drawn from a pack of 52 cards. What is the probability that the card is a king or an ace?

8. **MC** A die is rolled. Calculate the probability of getting an even number or a 3.

A. $\dfrac{3}{6}$ B. $\dfrac{4}{6}$ C. $\dfrac{1}{6}$ D. $\dfrac{5}{6}$

9. If you spin the following spinner, what is the probability of obtaining:

a. a 1 or a 3
b. an even number or an odd number?

10. The probabilities of Dale placing 1st, 2nd, 3rd or 4th in the local surf competition are:

$$1\text{st} = \dfrac{1}{6} \qquad 2\text{nd} = \dfrac{1}{5} \qquad 3\text{rd} = \dfrac{2}{5} \qquad 4\text{th} = \dfrac{7}{30}$$

Determine the probability that Dale places:

a. 1st or 2nd   b. 3rd or 4th   c. 1st, 2nd or 3rd   d. not 1st.

Understanding

11. **WE30** A circular spinner that is divided into two equal halves, coloured red and blue, is spun 3 times.

a. Draw a tree diagram for the experiment.
b. Calculate the following probabilities.

   i. P(3 red sectors)   ii. P(2 red sectors)   iii. P(1 red sector)
   iv. P(0 red sectors)   v. P(at least 1 red sector)

12. **WE31** There are two yellow tickets, three green tickets, and four black tickets in a jar. Choose one ticket, replace it, then choose another ticket. Determine the probability that a yellow ticket is drawn first, then a black ticket.

13. **WE32** Repeat question 12 with the first ticket not being replaced before the second ticket is drawn.

14. A coin is tossed two times. Determine P(a Head and a Tail in any order).

15. A coin is tossed three times. Determine $P\,(H, H, \ T)$ (in that order).

16. A coin and a die are tossed. What is the probability of a Heads–2 outcome?

17. Holty is tossing two coins. He claims that flipping two Heads and flipping zero Heads are complementary events. Is he correct? Explain your answer.

18. Each of the numbers 1, 2, 3, ... 20 is written on a card and placed in a bag. If a card is drawn from the bag, calculate:
    a. P(drawing a multiple of 3 or a multiple of 10)
    b. P(drawing an odd number or a multiple of 4)
    c. P(drawing a card with a 5 or a 7)
    d. P(drawing a card with a number less than 5 or more than 16).

19. From a shuffled pack of 52 cards, a card is drawn. calculate:
    a. P(hearts or the jack of spades)
    b. P(a queen or a jack)
    c. P(a 7, a queen or an ace)
    d. P(neither a club nor the king of spades).

20. **MC** Which of the following are not mutually exclusive?
    A. Obtaining an odd number on a die and obtaining a 4 on a die
    B. Obtaining a Head on a coin and obtaining a Tail on a coin
    C. Obtaining a red card and obtaining a black card from a pack of 52 playing cards
    D. Obtaining a diamond and obtaining a king from a pack of 52 playing cards

21. Greg has a 30% chance of scoring an A on an examination, Carly has 70% chance of scoring an A on the examination, and Chilee has a 90% chance of scoring an A on the examination. What is the probability that all three can score an A on the examination?

22. From a deck of playing cards, a card is drawn at random, noted and replaced, then another card is drawn at random. Determine the probability that:
    a. both cards are spades
    b. neither card is a spade
    c. both cards are aces
    d. both cards are the ace of spades
    e. neither card is the ace of spades.

23. Repeat question **22** with the first drawn card not being replaced before the second card is drawn.

24. Assuming that it is equally likely that a boy or a girl will be born, answer the following.
    a. Show the gender possibilities of a 3-child family on a tree diagram.
    b. In how many ways is it possible to have exactly 2 boys in the family?
    c. What is the probability of getting exactly 2 boys in the family?
    d. Which is more likely, 3 boys or 3 girls in the family?
    e. What is the probability of having at least 1 girl in the family?

Communicating, reasoning and problem solving

25. Give an example of mutually exclusive events that are not complementary events using:
    a. sets
    b. a Venn diagram.

26. Explain why all complementary events are mutually exclusive but not all mutually exclusive events are complementary.

27. A married couple plans to have four children.

   a. List the possible outcomes in terms of boys and girls.

   b. What is the probability of them having exactly two boys?

   c. Another couple plans to have two children. What is the probability that they have exactly one boy?

28. A bag contains 6 marbles, 2 of which are red, 1 is green and 3 are blue. A marble is drawn, the colour is noted, the marble is replaced and another marble is drawn.

   a. Show the possible outcomes on a tree diagram.

   b. List the outcomes of the event 'the first marble is red'.

   c. Calculate P(the first marble is red).

   d. Calculate P(2 marbles of the same colour are drawn).

29. A tetrahedron (prism with 4 identical triangular faces) is numbered 1, 1, 2, 3 on its 4 faces. It is rolled twice. The outcome is the number facing downwards.

   a. Show the results on a tree diagram.

   b. Are the outcomes 1, 2 and 3 equally likely?

   c. Determine the following probabilities.

     i. $P(1, 1)$

     ii. P(1 is first number)

     iii. P(both numbers the same)

     iv. P(both numbers are odd)

30. Robyn is planning to watch 3 league games on one weekend. She has a choice of two games on Friday night: (A) Broncos vs Eels and (B) Bulldogs vs Cowboys. On Saturday, she can watch one of three games: (C) Knights vs Storm, (D) Titans vs Sharks and (E) Rabbitohs vs Raiders.

On Sunday, she also has a choice of three games: (F) Roosters vs Tigers, (G) Dragons vs Panthers and (H) Warriors vs Sea Eagles. She plans to watch one game each day and will choose a game at random.

   a. To determine the different combinations of games she can watch, Robyn draws a tree diagram using the codes A, B, ... H. List the sample space for Robyn's selections.

   b. Robyn's favourite team is the Broncos. What is the probability that one of the games Robyn watches involves the Broncos?

   c. Robyn has a good friend who plays for the Roosters. What is the probability that Robyn watches both the matches involving the Broncos and the Roosters?

31. What is the difference between independent events and mutually exclusive events?

LESSON
14.8 Review

14.8.1 Topic summary

Probability of an outcome

- The probability of any outcome always falls between 0 and 1.
- An outcome that is **certain** has a probability of 1.
- An outcome that is **impossible** has a probability of 0.
- An outcome that is **likely** has a probability between 0.5 and 1.
- An outcome that is **unlikely** has a probability between 0 and 0.5.

PROBABILITY

Observed probability

- **Observed probability** is used when it is difficult or impossible to determine the theoretical probability.
- **Observed probability** = relative frequency
- **Relative frequency** = $\dfrac{\text{frequency of outcome}}{\text{total number of trials}}$
- The more trials conducted, the more accurate the observed probability will be.
- **Expected number** = probability of outcome × number of trials

Key terms

- **Trial:** A single performance of an experiment to produce a result, such as rolling a die
- **Sample space:** The set of all possible outcomes, represented by the symbol S. When rolling a 6-sided die, $S = \{1, 2, 3, 4, 5, 6\}$.
- **Event:** A set of favourable outcomes
  For example, A can represent the event of rolling an even number on a 6-sided die: $A = \{2, 4, 6\}$.
- **Complement:** All outcomes that are not part of an event. The complement of event A above is denoted by A': $A' = \{1, 3, 5\}$.

Venn diagrams and 2-way tables

- These are two different methods that can be used to represent a sample space and visualise the interaction of different events.
- For example, in a class of 20 students, 5 study Art (event A), 9 study Biology (event B) and 2 study both subjects.
  - Venn diagram:

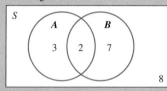

  - 2-way table:

	B	B'	Total
A	2	3	5
A'	7	8	15
Total	9	11	20

- Using these we can see:
  - $n(A' \cap B) = 7$
  - $P(A \cap B') = \dfrac{3}{20}$

Two-step experiments

- Two-step experiments involve two separate actions performed one after the other.
- The sample space of a two-step experiment can be represented with an array or a tree diagram.
- An experiment can be conducted with replacement or without replacement.
- If an experiment is performed without replacement, the probabilities will change for the second action.

		Bag 2		
		B	M	P
Bag 1	**B**	BB	BM	BP
	W	WB	WM	WP

First selection	Second selection	Sample space
B	B	BB
	M	BM
	P	BP
W	B	WB
	M	WM
	P	WP

Sections of a Venn diagram

- A Venn diagram can be split into four distinct sections, as shown.

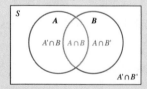

Intersection and union

- The intersection of two events A and B is written $A \cap B$. These are the outcomes that are both in A and in B.

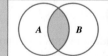

- The union of two events A and B is written $A \cup B$. These are the outcomes that are either in A or in B.

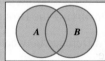

Mutually exclusive events

- Mutually exclusive events cannot occur together.

$P(A \text{ or } B) = P(A \cup B) = P(A) + P(B)$

Independent events

- When the outcome of one event does not influence the outcome of another event, the events are independent.

$P(A \text{ and } B) = P(A \cap B) = P(A) \times P(B)$

14.8.2 Project

Sand-rings

A class of students and their teacher spent a day at the beach as part of a school excursion.

Part of the day was devoted to activities involving puzzles in the sand. One of the popular — and most challenging — puzzles was 'sand-rings'. Sand-rings involves drawing three rings in the sand, as shown in the diagram.

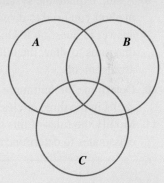

The first sand-rings puzzle requires eight shells to be arranged inside the circles, so that four shells appear inside circle A, five shells appear inside circle B and six shells appear inside circle C. The overlapping of the circles shows that the shells can be counted in two or three circles. One possible arrangement is shown below. Use this diagram to answer questions 1 to 4.

1. How many shells appear inside circle A, but not circle B?
2. How many shells appear in circles B and C, but not circle A?

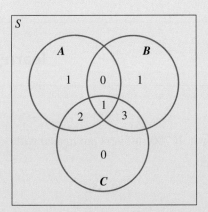

A shell is selected at random from the sand.

3. Calculate the probability it came from circle A.
4. Calculate the probability it was not in circle C.
5. The class was challenged to determine the rest of the arrangements of the eight shells. (*Remember:* Four shells need to appear in circle A, five in circle B and six in circle C.)

After completing the first puzzle, the students are given new rules. The number of shells to be arranged in the circles is reduced from eight to six. However, the number of shells to be in each circle remains the same; that is, four shells in circle *A*, five shells in circle *B* and six shells in circle *C*.

6. Using six shells, calculate how many ways the shells can be arranged so that there are four, five and six shells in the three circles.
7. Explain the system or method you used to determine your answer to question 6 above. Draw diagrams to show the different arrangements.
8. Using seven shells, calculate how many ways the shells can be arranged so that there are four, five and six shells in the three circles.
9. Again, explain the system or method you used to determine your answer to question 8. Draw diagrams to show the different arrangements.
10. Determine the minimum number of shells required to play sand-rings, so that there are four, five and six shells in the three circles.
11. Modify the rules of this game so that different totals are required for the three circles. Challenge your classmates to determine all possible solutions to your modified game.

 Resources

 Interactivities   Crossword (int-2712)
Sudoku puzzle (int-3212)

Exercise 14.8 Review questions

learn on

Fluency

1. **MC** In a trial, it was found that a drug cures $\frac{2}{5}$ of those treated by it. If 700 sufferers are treated with the drug, calculate how many of them are not expected to be cured.

A. 280   B. 420   C. 140   D. 350

2. **MC** If a die is rolled and a coin tossed, determine the probability of a 6–Heads result.
   A. $\frac{1}{6}$ B. $\frac{1}{2}$ C. $\frac{1}{8}$ D. $\frac{1}{12}$

3. **MC** Twelve nuts are taken from a jar containing macadamias and cashews. If three macadamias are obtained, the observed probability of obtaining a cashew is:

A. $\dfrac{1}{12}$
B. $\dfrac{1}{4}$
C. $\dfrac{1}{3}$
D. $\dfrac{3}{4}$

4. **MC** From a normal pack of 52 playing cards, one card is randomly drawn and replaced. If this is done 208 times, the number of red or picture cards expected to turn up is:

A. 150
B. 130
C. 128
D. 120

5. **MC** A cubic die with faces numbered 2, 3, 4, 5, 6 and 6 is rolled. The probability of rolling an even number is:

A. $\dfrac{1}{3}$
B. $\dfrac{2}{3}$
C. $\dfrac{1}{6}$
D. $\dfrac{1}{2}$

6. **MC** The probability of rolling an odd number or a multiple of 2 using the die in question **5** is:

A. 1
B. $\dfrac{1}{3}$
C. $\dfrac{1}{4}$
D. $\dfrac{3}{4}$

Questions **7** and **8** refer to the following information.

Students in a Year 9 class chose the following activities for a recreation day.

Activity	Tennis	Fishing	Golf	Bushwalking
Number of students	8	15	5	7

7. **MC** If a student is selected at random from the class, the probability that the student chose fishing is:

A. $\dfrac{1}{7}$
B. $\dfrac{2}{7}$
C. $\dfrac{3}{7}$
D. $\dfrac{4}{7}$

8. **MC** If a student is selected at random, the probability that the student did not choose bushwalking is:

A. $\dfrac{1}{35}$
B. $\dfrac{2}{5}$
C. $\dfrac{3}{5}$
D. $\dfrac{4}{5}$

9. **MC** Which one of the following does not represent independent events?
A. Flipping a coin, then rolling a die
B. The colour of your hair, and your marks in school
C. Choosing a card from a standard deck of cards without replacing it, then choosing another card from the same deck
D. Flipping a coin ten times

Understanding

10. The mass of 40 students in a Year 9 Maths class was recorded in a table.

Mass (kg)	Less than 50	50–<55	55–<60	60–<65	65 and over
Number of students	4	6	10	15	5

Calculate the observed probability of selecting a student who has:
a. a mass of 55 kg or more, but less than 60 kg
b. a mass less than 50 kg
c. a mass of 65 kg or greater.

11. Calculate the following expected values.
   a. The number of Heads in 80 tosses of a coin.
   b. The number of sixes in 200 rolls of a die.
   c. The number of hearts if a card is picked from a reshuffled pack and replaced 100 times.

12. A normal six-sided die is rolled. Calculate the probability of getting an odd number or a multiple of 4.

13. A card is drawn from a pack of 52 cards. Calculate the probability that the card is a heart or a club.

14. Insert each of the letters **a** to **d** to represent the following events at appropriate places on the probability scale shown.

   a. You will go to school on Christmas Day.
   b. All Year 9 students can go to university without doing Year 10.
   c. Year 9 students will study Maths.
   d. An Australian TV channel will telecast the news at 6:00 pm.

15. **PATH** Indicate the set that each of the shaded regions represents.

   a. Subject preference

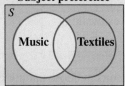

   b. Leisure activity

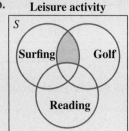

   c. Favourite drinks

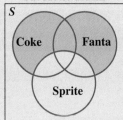

16. An equal number of red (R), black (B) and yellow (Y) counters makes up a total of 30 in a bag.
   a. List the sample space for picking a counter from the bag.
   b. Event A is 'draw a yellow counter, then randomly draw another counter from the bag'. List the sample space of event A.
   c. Explain whether 'choosing a green counter' is an outcome.

Communicating, reasoning and problem solving

17. **PATH** Teachers at a school opted for the choice of morning recess refreshments shown in the Venn diagram.
   a. State how many teachers are in the set 'cake ∩ coffee'.
   b. State the total number of teachers surveyed.
   c. If a teacher is selected at random, calculate the probability that the teacher:
      i. chose tea
      ii. chose coffee only
      iii. chose milk
      iv. did not choose tea, coffee, cake or milk
      v. did not choose coffee.

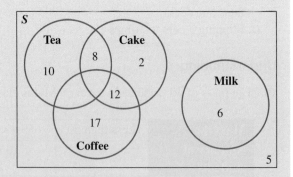

18. **PATH** Thirty-two students ordered fried rice, chicken wings and dim sims for lunch. Four students ordered all three, two ordered fried rice and chicken wings only, three ordered fried rice and dim sims only, and three ordered chicken wings and dim sims only.

    When the waiter organised the orders, he found that 16 students ordered fried rice and 18 students ordered dim sims.

    a. Show this information on a Venn diagram.

    b. Calculate how many students ordered chicken wings only.

    c. If a student is picked at random, calculate the probability that the student has:
       i. ordered chicken wings and dim sims
       ii. ordered fried rice
       iii. not ordered dim sims.

19. The following are options for dorm rooms at a university. You are required to choose one option from each of the four categories.
    • Upstairs or downstairs rooms
    • Single or double rooms
    • Male or female roommates
    • A choice of 10 different locations on campus

    Calculate how many different combinations of rooms there are to choose from.

20. A future king is the oldest male child. The future king of Mainland has two siblings. Determine the probability that he has an older sister.

21. A witness described a getaway car as having a NSW registration plate starting with TLK. The witness could not remember the three digits that followed, but recalled that all three digits were different. Calculate how many cars in NSW could have a registration plate with these letters and numbers.

22. There are 12 people trying out for a badminton team. Five of them are girls. Calculate the probability that a team chosen at random to play is a mixed doubles team.

23. An ace is chosen from a deck of standard cards and not replaced. A king is then chosen from the deck. Calculate the probability of choosing an ace and a king in this order.

24. If you randomly select one number from 1 to 500 (inclusive), calculate the probability that the selected number will have at least one 4 in the digits.

25. If you flip a coin eight times, calculate how many of the possible outcomes would you expect to have a Head on the second toss.

on To test your understanding and knowledge of this topic go to your learnON title at www.jacplus.com.au and complete the **post-test**.

Answers

Topic 14 Probability

14.1 Pre-test

1. C
2. a. Zero   b. $\dfrac{3}{8}$ or 0.375
3. a. 100   b. $\dfrac{7}{25}$ c. $\dfrac{1}{2}$ d. $\dfrac{11}{50}$
4. $\dfrac{1}{3}$
5. True
6. D
7. C
8. A
9. a. $\dfrac{1}{4}$ b. $\dfrac{3}{14}$
10. P(drama not comedy) = $\dfrac{2}{5}$
11. $\dfrac{5}{7}$
12. a. B   b. 0.33
13. a. $\dfrac{8}{15}$ b. $\dfrac{67}{225}$ c. $\dfrac{1}{6}$ d. $\dfrac{67}{240}$
14. a. C   b. D   c. A   d. B
15. D

14.2 Theoretical probability

1.
   a. Certain   b. Unlikely

2.
   a. Likely   b. Certain

3.
   a. Certain   b. Likely

4.
   a. Likely   b. Likely

5. a. $\{1, 2, 3, 4, 5, 6\}$
   b. $\{H, T\}$
   c. $\{\text{defective, not defective}\}$
   d. $\{\text{red, black}\}$
   e. $\{\text{hearts, clubs, diamonds, spades}\}$

6. a. $\{1, 2, 3\}$ b. $\{1, 3, 5\}$
   c. $\{1, 2, 4, 6\}$ d. $\{3, 4, 5, 6\}$
   e. $\{1, 2, 3, 4\}$ f. $\{5, 6\}$

7. a. $3, 5$
   b. $1, 2$
   c. 6
   d. $1, 2, 5, 6$
   e. $1, 2, 3, 5$
   f. No favourable outcomes
   g. 3

8. a. $\{\text{hearts, clubs, diamonds, spades}\}$
   b.   i. $\{\text{clubs, spades}\}$
       ii. $\{\text{hearts, diamonds}\}$
       iii. $\{\text{clubs, diamonds, spades}\}$
       iv. $\{\text{hearts, clubs, diamonds, spades}\}$

9. a. 6   b. 2   c. 52
   d. 4   e. 5

10. a. $\dfrac{1}{13}$ b. $\dfrac{1}{13}$ c. $\dfrac{1}{52}$ d. $\dfrac{1}{4}$

11. a. $\dfrac{1}{52}$ b. $\dfrac{1}{2}$ c. $\dfrac{1}{26}$ d. $\dfrac{1}{4}$ e. $\dfrac{1}{2}$

12. a. $\dfrac{1}{52}$ b. $\dfrac{4}{13}$ c. $\dfrac{2}{13}$ d. 0   e. $\dfrac{12}{13}$

13. a. $\dfrac{2}{11}$ b. $\dfrac{9}{11}$ c. $\dfrac{4}{11}$ d. $\dfrac{7}{11}$

14. a. $\dfrac{1}{4}$ b. $\dfrac{1}{8}$ c. $\dfrac{5}{16}$ d. 0   e. $\dfrac{3}{4}$

15. a. i. $\dfrac{2}{3}$ ii. $\dfrac{1}{3}$
   b. Sample responses can be found in the worked solutions in the online resources.

16. Sample responses can be found in the worked solutions in the online resources.

17. a. 18   b. 20   c. 36

18. a. i. $\dfrac{7}{10}$ ii. $\dfrac{3}{10}$
   b. i. $\dfrac{2}{3}$ ii. $\dfrac{1}{3}$

19. Sample responses can be found in the worked solutions in the online resources.

20. Probabilities must be between 0 and 1, so -0.5 and 1.05 can't be probabilities.

21. $\dfrac{1}{2}$

22. The coloured portions outside the arc of the spinner shown are of no consequence. The four colours within the arc of the spinner are of equal area (each $\dfrac{1}{4}$ circle), so there is equal chance of falling on each of the colours.

23. $\dfrac{2}{3}$

24. There are 36 outcomes, 15 under 7, 6 equal to 7 and 15 over 7. So, you would have a greater chance of winning if you chose 'under 7' or 'over 7' rather than 'equal to 7'.

25. 0.09

14.3 Observed probability and simulations

1. **a.** **i.** 16

   ii. 4

   iii.

Outcome	1	2	3	4	5	6
Relative frequency	0.19	0.06	0.31	0.13	0.25	0.06

   iv. 1

   b. **i.** 270

   ii. 40

   iii.

Outcome	1	2	3	4	5	6
Relative frequency	0.19	0.14	0.17	0.18	0.15	0.17

   iv. 1

2. **a.** **i.** r.f. (H) = 0.44, r.f. (T) = 0.56

   ii. 1

   b. **i.** r.f. (H) = 0.62, r.f. (T) = 0.38

   ii. 1

3. **a.** 0.365 **b.** 0.33 **c.** 0.25 **d.** 0.875

4. **a.** 0.275 **b.** 0.64 **c.** 0.365 **d.** 0.635

5. **a.** $\dfrac{27}{95}$ **b.** $\dfrac{33}{95}$

   c. $\dfrac{1}{19}$ **d.** Yes, reject this batch.

6. **a.** Not reasonable; not enough trials were held.

   b. Not reasonable; the conditions are different under each trial.

   c. Reasonable; the Sun rises every morning, regardless of the weather or season.

   d. Reasonable; enough trials were performed under the same conditions.

   e. Not reasonable; monthly rainfall in Sydney is not consistent throughout the year.

7. **a.** $\dfrac{5}{24}$ **b.** $\dfrac{59}{72}$

8. 200

9. 20

10. A

11. B

12. **a.** $\dfrac{11}{20}$ **b.** $\dfrac{3}{10}$ **c.** $\dfrac{17}{20}$ **d.** $\dfrac{3}{20}$

13. **a.** $\dfrac{33}{100}$ **b.** $\dfrac{40}{100} = \dfrac{2}{5}$ **c.** $\dfrac{3}{100}$

14. **a.** **i.** 6

    ii. {356, 365, 536, 563, 635, 653}

    b. $\dfrac{2}{3}$ **c.** $\dfrac{1}{3}$ **d.** $\dfrac{1}{3}$

15. D

16. Svetlana 6, Sarah 4, Leonie 3, Trang 2

17. C

18. **a.** 27

    b. $\left\{ \begin{array}{l} 111, 112, 115, 121, 122, 125, 151, 152, 155, 211, 212, \\ 215, 221, 222, 225, 251, 252, 255, 511, 512, 515, 521, \\ 522, 525, 551, 552, 555 \end{array} \right\}$

    c. $\dfrac{1}{3}$ **d.** $\dfrac{2}{3}$ **e.** $\dfrac{1}{3}$

19. **a.** $\dfrac{1}{12}$

    b. $\dfrac{1}{20}$

    c. No, because John has a higher probability of winning.

20. **a.** $\dfrac{7}{10}$

    b. Yes, far too many mistakes

21. 32

22. 1000 balls

23. Red = 10, yellow = 50, blue = 40

24. **a.** $\dfrac{7}{20}$ **b.** $\dfrac{2}{5}$

25. Sample responses can be found in the worked solutions in the online resources.

26. **a.** $\dfrac{30}{241}$ **b.** $\dfrac{91}{241}$ **c.** $\dfrac{59}{241}$

27. 16, 64, 40, 80, 128 and 32 degrees.

14.4 Tree diagrams and two-step experiments

1. See the table at the foot of the page.*

   a. $\dfrac{1}{9}$ **b.** 0 **c.** A total of 7

---

*1.

		Die 1					
		1	**2**	**3**	**4**	**5**	**6**
Die 2	**1**	2	3	4	5	6	7
	2	3	4	5	6	7	8
	3	4	5	6	7	8	9
	4	5	6	7	8	9	10
	5	6	7	8	9	10	11
	6	7	8	9	10	11	12

2.

		Scarves		
		R	Bl	Pi
Beanies	Br	Br, R	Br, Bl	Br, Pi
	Pu	Pu, R	Pu, Bl	Pu, Pi

3. a. See the table at the foot of the page.*

  b. i. $\frac{1}{4}$ ii. $\frac{3}{4}$ iii. $\frac{3}{20}$ iv. $\frac{13}{20}$

4.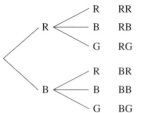

5. a.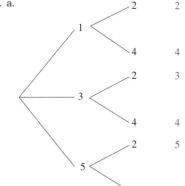

  b. i. $\frac{1}{6}$ ii. 0   iii. $\frac{1}{2}$

6. a. See the table at the foot of the page.*

  b. i. $\frac{1}{2}$ ii. $\frac{1}{6}$ iii. $\frac{1}{3}$ iv. $\frac{2}{3}$

7. a. $\frac{1}{6}$ b. $\frac{1}{36}$ c. $\frac{1}{4}$ d. $\frac{1}{36}$

8. a.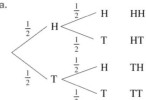

  b. i. $\frac{1}{4}$ ii. $\frac{3}{4}$

9. a.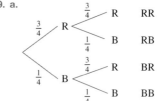

  b. i. $\frac{3}{16}$ ii. $\frac{1}{16}$

10. a. $\frac{4}{15}$ b. $\frac{2}{15}$ c. $\frac{209}{225}$ d. $\frac{77}{225}$

11. a. $\frac{16}{25}$ b. $\frac{9}{25}$ c. $\frac{391}{400}$ d. $\frac{77}{200}$ e. $\frac{123}{200}$

12. a.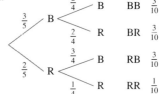

  b. $\frac{1}{10}$

*3. a.

		Dice roll									
		1	2	3	4	5	6	7	8	9	10
Coin toss	H	1, H	2, H	3, H	4, H	5, H	6, H	7, H	8, H	9, H	10, H
	T	1, T	2, T	3, T	4, T	5, T	6, T	7, T	8, T	9, T	10, T

*6. a.

		Die 1					
		1	2	3	4	5	6
Die 2	1	0	1	2	3	4	5
	2	1	0	1	2	3	4
	3	2	1	0	1	2	3
	4	3	2	1	0	1	2
	5	4	3	2	1	0	1
	6	5	4	3	2	1	0

13. a. $\dfrac{2}{9}$ b. $\dfrac{1}{9}$ c. $\dfrac{2}{9}$ d. $\dfrac{14}{45}$

14. a. $\dfrac{5}{22}$ b. $\dfrac{1}{22}$ c. $\dfrac{5}{11}$ d. $\dfrac{25}{66}$

15. a. $\dfrac{1}{4}$ b. $\dfrac{3}{14}$

16. a. $\dfrac{1}{9}$ b. $\dfrac{3}{11}$ c. $\dfrac{8}{11}$

17. a.

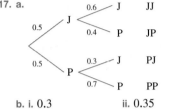

b. i. 0.3 ii. 0.35 iii. 0.35

18. If the first ball is replaced, the probability of drawing a yellow ball stays the same on the second draw, i.e. $\left(\dfrac{3}{5}\right)$.

If the first ball isn't replaced, the probability of drawing a yellow ball on the second draw decreases, i.e. $\left(\dfrac{5}{9}\right)$.

19. a. 3 days b. 90 combinations
    c. 36 new combinations

20. a. Smallest total: 3, largest total: 18

b.

Total	Probability
3	$\dfrac{1}{216}$
4	$\dfrac{3}{216}$
5	$\dfrac{6}{216}$
6	$\dfrac{10}{216}$
7	$\dfrac{15}{216}$
8	$\dfrac{21}{216}$
9	$\dfrac{25}{216}$
10	$\dfrac{27}{216}$
11	$\dfrac{27}{216}$
12	$\dfrac{25}{216}$
13	$\dfrac{21}{216}$
14	$\dfrac{15}{216}$
15	$\dfrac{10}{216}$
16	$\dfrac{6}{216}$
17	$\dfrac{3}{216}$
18	$\dfrac{1}{216}$

21. a. $\dfrac{1}{221}$ b. $\dfrac{11}{221}$ c. $\dfrac{1}{17}$

22. $\dfrac{5}{12}$

23. $\dfrac{1}{2}$

24. a. 8 b. 32 c. 16 d. 2^x

25. a. 36 b. $\dfrac{5}{7}$

26. $\dfrac{67}{256}$

27. $\dfrac{45}{1081}$

14.5 Exploring Venn diagrams and 2-way tables (Path)

1. a. No b. 104 c. 35 d. 27 e. 42

2. a. 237 b. 119 c. 100 d. 51 e. 153

3. a. Teenagers who like Snapchat
   b. Teenagers who like neither Facebook nor Snapchat
   c. Teenagers who like Facebook but not Snapchat
   d. Teenagers who like Facebook or Snapchat, or both
   e. Teenagers who like Facebook or Snapchat, but not both

4. a. 25 b. 17 c. 9 d. 6 e. 7

5. a. 3 b. 1 c. 11 d. 22 e. 24

6. a.

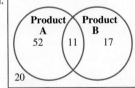

b. i. 52 ii. 17 iii. 20

c. $\dfrac{11}{100}$

7. a.

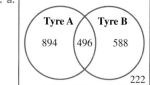

Tyre A 894 | 496 | Tyre B 588

222

b. i. 894   **ii.** 588   **iii.** 222

8. a.

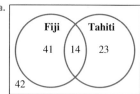

Fiji 41 | 14 | Tahiti 23

42

b. 14

9. a. 78   **b.** 79

10. a. 44   **b.** 11   **c.** 18   **d.** 89   **e.** 0.4900

11. a.

	This year	Older	
Rap	29	21	**50**
Pop	63	37	**100**
	92	**58**	**150**

b. 37

c. 29

d. 100

12. a.

	Male	Female	
Pigs	59	64	**123**
Cattle	26	31	**57**
	85	**95**	**180**

b. i. 26   **ii.** 64

c. i. $\dfrac{123}{180}$ **ii.** $\dfrac{85}{180} = \dfrac{17}{36}$

13. a.

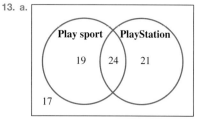

Play sport 19 | 24 | PlayStation 21

17

b. 81

c. 38

d. $\dfrac{38}{81}$

14.

	Meat-eater	Vegetarian	
Metro	63	8	**71**
Regional	58	11	**69**
	121	**19**	**140**

15.

	Brown eyes	Not brown eyes	
Brown hair	68	22	**90**
Not brown hair	13	30	**43**
	81	**52**	**133**

16. a.

	Apples	Not apples	
Bananas	13	28	**41**
Not bananas	22	47	**69**
	35	**75**	**110**

b. $\dfrac{35}{110} = \dfrac{7}{22}$

c. $\dfrac{47}{110}$

d. $\dfrac{50}{110} = \dfrac{5}{11}$

17. a. $\dfrac{117}{195} = \dfrac{3}{5}$ **b.** $\dfrac{78}{195} = \dfrac{2}{5}$ **c.** $\dfrac{73}{195}$

18. At least 4 pieces of information are required to complete a 2-way table. Without 4 pieces of information, some of the cells cannot be completed.

19. a. i.

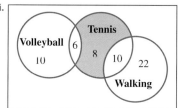

Volleyball 10 | 6 | Tennis 8 | 10 | Walking 22

ii.

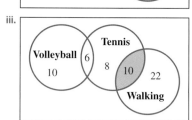

Volleyball 10 | 6 | Tennis 8 | 10 | Walking 22

iii.

Volleyball 10 | 6 | Tennis 8 | 10 | Walking 22

b. i. 16   **ii.** 0

c. 56

d. i. 32   **ii.** 24

e. $\dfrac{24}{56} = \dfrac{3}{7}$

f. $\dfrac{6}{56} = \dfrac{3}{28}$

20. a. 45   **b.** 28   **c.** 27   **d.** 10   **e.** 72%

21. a. 60
b. 20
c.

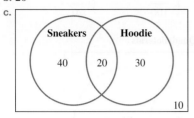

22. a.

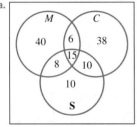

b. 69
c. 38
d. 43
e. 21

14.6 Venn diagrams and 2-way tables using set notation (Path)

1. a. $A' = \{1, 3, 5, 7, 9\}$
b. $B' = \{1, 2, 3, 4, 6, 7, 8, 9\}$
c. $C' = \{2, 3, 5, 6, 7, 8, 10\}$
d. $D' = \{8, 9, 10\}$

2. a. $A' = \{11, 13, 14, 16, 17, 19, 20\}$
b. $B' = \{20\}$
c. $C' = \{12, 14, 15, 16, 18, 20\}$
d. $D' = \{12, 14, 16\}$

3. a. $A' = \{1, 2, 3, 5, 6, 7, 9, 10, 11, 13, 14, 15, 17, 18, 19\}$
b. $B' = \{1, 4, 6, 8, 9, 10, 12, 14, 15, 16, 18, 20\}$
c. $C' = \{1, 3, 5, 7, 9, 11, 13, 14, 15, 16, 17, 18, 19, 20\}$
d. $D' = \{1, 3, 5, 7, 9, 11, 13\}$

4. a. 27   **b.** 11   **c.** 13   **d.** 6
e. 7   **f.** 9

5.

	T	T'
Q	6	5
Q'	7	9

6.

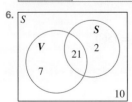

7. a. i. W' **ii.** None
b. i. $A \cap B'$ **ii.** $A \cap B$
c. i. $A' \cap B'$ **ii.** $B \cap A'$

8. a.

	Left-handed	Right-handed
Male	15	40
Female	12	33

b.

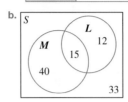

c. i. $\dfrac{73}{100}$ **ii.** $\dfrac{2}{5}$ **iii.** $\dfrac{3}{25}$

9. a. $\dfrac{8}{25}$ **b.** $\dfrac{17}{25}$ **c.** $\dfrac{1}{5}$ **d.** $\dfrac{3}{25}$

10.

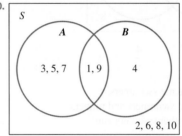

11.

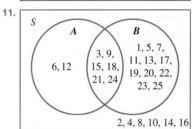

12.

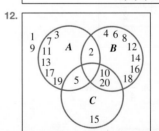

13. a. $\dfrac{1}{2}$ **b.** $\dfrac{1}{10}$ **c.** $\dfrac{31}{40}$ **d.** $\dfrac{27}{40}$ **e.** $\dfrac{9}{40}$

14. a. $\dfrac{1}{2}$ **b.** $\dfrac{6}{25}$ **c.** $\dfrac{22}{25}$ **d.** $\dfrac{16}{25}$

15. a.

	Croissant	No croissant	Total
Coffee	27	2	**29**
No coffee	9	7	**16**
Total	**36**	**9**	**45**

b. i. $\dfrac{3}{5}$ **ii.** $\dfrac{38}{45}$ **iii.** $\dfrac{11}{45}$

c. 1218

16. a. 16

b. i. $\dfrac{5}{16}$ **ii.** $\dfrac{3}{16}$ **iii.** $\dfrac{5}{8}$ **iv.** $\dfrac{7}{16}$

17. a.

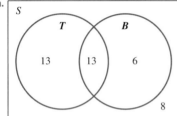

b. i. $\dfrac{29}{40}$ **ii.** $\dfrac{4}{5}$ **iii.** $\dfrac{19}{40}$

18. a. Children who are not swimmers
  b. Children who are swimmers and runners
  c. Children who neither swim nor run
  d. Children who swim or run or both

19. a. i.

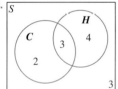

ii.

	H	**H′**
C	3	2
C′	4	3

b. i. $\dfrac{7}{12}$ **ii.** $\dfrac{3}{4}$ **iii.** $\dfrac{1}{4}$ **iv.** $\dfrac{1}{2}$

20. a.

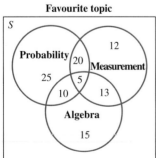

b. 25 **c.** 15 **d.** 12

e. 43 **f.** $\dfrac{3}{5}$ **g.** $\dfrac{43}{100}$

h. $\dfrac{9}{50}$ **i.** $\dfrac{13}{100}$ **j.** $\dfrac{1}{20}$

21.

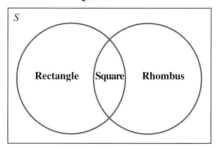

22. a. 2, 5, 6, 7, 8 **b.** 4
  c. 8 **d.** 1, 2, 3, 4, 5, 7, 8

23. a.

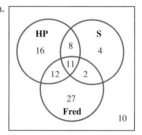

b. $\dfrac{8}{90} = \dfrac{4}{45}$

24. a.

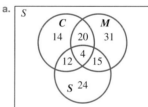

b. $\dfrac{7}{12}$

25. a. 31 students **b.** $\dfrac{1}{13}$

26.

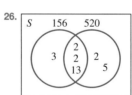

a. HCF $= 2 \times 2 \times 13 = 52$
b. LCM $= 3 \times 2 \times 2 \times 13 \times 2 \times 5 = 1560$

27. a.

b. 4

c. 21

d. i. $\dfrac{81}{200}$ ii. $\dfrac{97}{200}$

e. $\dfrac{3}{50}$

14.7 Mutually and non-mutually exclusive events (Path)

1. D

2. A

3. $\dfrac{215}{216}$

4. D

5. a. $\dfrac{1}{4}$ b. $\dfrac{3}{4}$

6. a. $\dfrac{17}{40}$ b. $\dfrac{53}{70}$ c. $\dfrac{9}{70}$ d. $\dfrac{61}{70}$

7. $\dfrac{2}{13}$

8. B

9. a. $\dfrac{1}{2}$ b. 1

10. a. $\dfrac{11}{30}$ b. $\dfrac{19}{30}$ c. $\dfrac{23}{30}$ d. $\dfrac{5}{6}$

11. a.

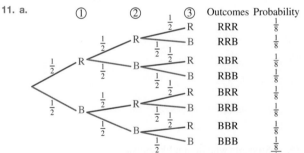

b. i. $\dfrac{1}{8}$ ii. $\dfrac{3}{8}$ iii. $\dfrac{3}{8}$
   iv. $\dfrac{1}{8}$ v. $\dfrac{7}{8}$ vi. $\dfrac{1}{2}$

12. $\dfrac{8}{81}$

13. $\dfrac{1}{9}$

14. $\dfrac{1}{2}$

15. $\dfrac{1}{8}$

16. $\dfrac{1}{12}$

17. No, because there is also the possibility of 1 Head (HT or TH).

18. a. $\dfrac{8}{20}$ b. $\dfrac{15}{20}$ c. $\dfrac{4}{20}$ d. $\dfrac{8}{20}$

19. a. $\dfrac{14}{52}$ b. $\dfrac{2}{13}$ c. $\dfrac{3}{13}$ d. $\dfrac{38}{52}$

20. D

21. $\dfrac{189}{1000}$ or 0.189

22. a. $\dfrac{1}{16}$ b. $\dfrac{9}{16}$ c. $\dfrac{1}{169}$ d. $\dfrac{1}{2704}$ e. $\dfrac{2601}{2704}$

23. a. $\dfrac{1}{17}$ b. $\dfrac{19}{34}$ c. $\dfrac{1}{221}$ d. 0 e. $\dfrac{25}{26}$

24. a.

	①	②	③	Outcomes	Probability
			B	BBB	$\frac{1}{8}$
		B	G	BBG	$\frac{1}{8}$
	B		B	BGB	$\frac{1}{8}$
		G	G	BGG	$\frac{1}{8}$
			B	GBB	$\frac{1}{8}$
	G	B	G	GBG	$\frac{1}{8}$
			B	GGB	$\frac{1}{8}$
		G	G	GGG	$\frac{1}{8}$

b. 3

c. $\dfrac{3}{8}$

d. They are equally likely.

e. $\dfrac{7}{8}$

25. a. $S = \{1, 2, 3, 4, 5, 6, 7, 8\}, A = \{1, 2, 3\}, B = \{4, 5, 6\}$

b.

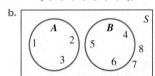

26. If two events are complementary, they cannot occur at the same time; thus, their intersection is ϕ, the same as mutually exclusive sets. However, if events are mutually exclusive, they do not need to have a sum equal to 1.

27. a. {BBBB, BBBG, BBGB, BBGG, BGBB, BGBG GBBB, GBBG, GBGB, GBGG, GGGG}

b. $\dfrac{3}{8}$

c. $\dfrac{1}{2}$

28. a.

	①	②	Outcomes	Probability
		R	RR	$\frac{1}{9}$
	R	G	RG	$\frac{1}{18}$
		B	RB	$\frac{1}{6}$
		R	GR	$\frac{1}{18}$
	G	G	GG	$\frac{1}{36}$
		B	GB	$\frac{1}{12}$
		R	BR	$\frac{1}{6}$
	B	G	BG	$\frac{1}{12}$
		B	BB	$\frac{1}{4}$

b. {(R, R),(R, G),(R, B)}

c. $\frac{1}{3}$

d. $\frac{7}{18}$

29. a.

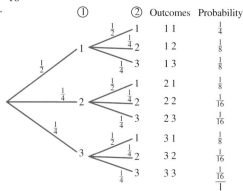

	①	②	Outcomes	Probability
		1	1 1	$\frac{1}{4}$
	1	2	1 2	$\frac{1}{8}$
		3	1 3	$\frac{1}{8}$
		1	2 1	$\frac{1}{8}$
	2	2	2 2	$\frac{1}{16}$
		3	2 3	$\frac{1}{16}$
		1	3 1	$\frac{1}{8}$
	3	2	3 2	$\frac{1}{16}$
		3	3 3	$\frac{1}{16}$
				$\overline{1}$

b. No

c. i. $\frac{1}{4}$ ii. $\frac{1}{2}$ iii. $\frac{3}{8}$ iv. $\frac{9}{16}$

30. a.

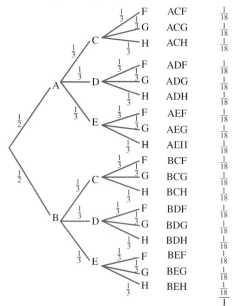

b. Sample space = {ACF, ACG, ... , BEG, BEH}

c. $\frac{1}{2}$

d. $\frac{1}{6}$

31. Events are independent if the outcome of one event is not dependent on the outcome of the other event. Mutually exclusive events cannot occur at the same time or have no elements in common.

Project

1. 3

2. 4

3. $\frac{1}{2}$

4. $\frac{1}{4}$

5. 18

6. 2

7. Sample responses can be found in the worked solutions in the online resources.

8. 8

9. Sample responses can be found in the worked solutions in the online resources.

10. Sample responses can be found in the worked solutions in the online resources.

11. 6

14.8 Review questions

1. B

2. D

3. D

4. C

5. B

6. A

7. C

8. D

9. C

10. a. $\frac{1}{4}$ b. $\frac{1}{10}$ c. $\frac{1}{8}$

11. a. 40   b. 33   c. 25

12. $\frac{2}{3}$

13. $\frac{1}{2}$

14. b

a   d   c

0   0.5   1

15. a. Students who do not like music

b. People who like surfing and golf as leisure activities, but not reading

c. People who like Coke or Fanta or both but not Sprite

16. a. {R, B, Y}
    b. {YR, YY, YB}
    c. No, there is no green counter.

17. a. 12
    b. 60
    c. i. $\frac{3}{10}$ ii. $\frac{17}{60}$ iii. $\frac{1}{10}$
    iv. $\frac{1}{12}$ v. $\frac{31}{60}$

18. a.

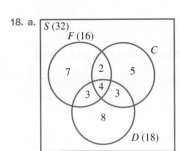

b. 5

c. i. $\dfrac{7}{32}$ ii. $\dfrac{1}{2}$ iii. $\dfrac{7}{16}$

19. 80

20. $\dfrac{3}{7}$

21. 720

22. $\dfrac{35}{66}$

23. $\dfrac{4}{663}$

24. $\dfrac{176}{500}$

25. 128

Semester review 2

The learnON platform is a powerful tool that enables students to complete revision independently and allows teachers to set mixed and spaced practice with ease.

Student self-study

Review the **Course Content** to determine which topics and lessons you studied throughout the year. Notice the green bubbles showing which elements were covered.

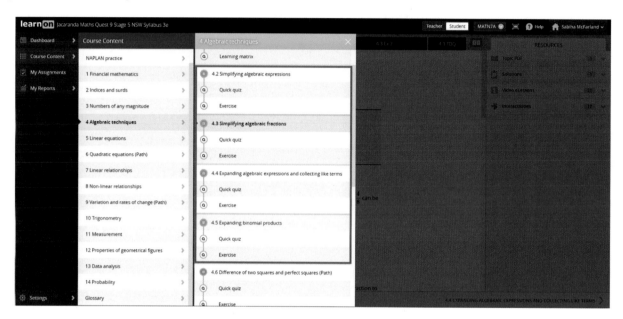

Review your results in **My Reports** and highlight the areas where you may need additional practice.

Use these and other tools to help identify areas of strengths and weakness and target those areas for improvement.

Teachers

It is possible to set questions that span multiple topics. These assignments can be given to individual students, to groups or to the whole class in a few easy steps.

Go to **Menu** and select **Assignments** and then **Create Assignment**. You can select questions from one or many topics simply by ticking the boxes as shown below.

Once your selections are made, you can assign to your whole class or subsets of your class, with individualised start and finish times. You can also share with other teachers.

More instructions and helpful hints are available at www.jacplus.com.au.

GLOSSARY

2-way table a table listing all the possible outcomes of a probability experiment in a logical manner

absolute error half of the precision (the smallest unit marked on a measuring instrument)

adjacent side the side next to the reference angle in a right-angled triangle

angle of depression the angle measured down from a horizontal line (through the observation point) to the line of vision

angle of elevation the angle measured up from the horizontal line (through the observation point) to the line of vision

approximation a value that is close to an exact value

arc a section of the circumference of a circle

area amount of surface enclosed by a shape; measured in square units, such as square metres (m^2) and square kilometres (km^2)

asymptote a line that a graph approaches but never touches

axis of symmetry the straight line that sits midway between two halves of a symmetrical graph, or between an object and its image. An object is reflected along an axis of symmetry (mirror).

balanced describes a valid equation (the expressions on both sides of the equals sign are equal)

bimodal having two modes

bimodal distribution a distribution of data that has two modes

binomial an expression containing two terms, for example $x + 3$ or $2y - z^2$

box plot a graphical representation of the 5-number summary; that is, the lowest score, lower quartile, median, upper quartile and highest score, for a particular set of data

capacity the maximum amount of fluid contained in an object, usually applied to the measurement of liquids and measured in units such as millilitres (mL), litres (L) and kilolitres (kL)

Cartesian plane an area that consists of two number lines at right angles to each other and includes all the space between the two number lines

chance experiment a process, such as rolling a die, that can be repeated many times

circumference the distance around the outside of a circle, given by the rule $2\pi r$ or πD, where r is the radius and D is the diameter of the circle

coefficient the number part of a term

conversion factor the value used to convert between units (by multiplying or dividing)

commission a payment to a salesperson calculated as a percentage of the value of goods sold

complement of an event every outcome not in the event

complement (of a set) the complement of a set, A, written A'; the set of elements that are in ξ but not in A

compound interest interest that is paid on the sum of the principal plus preceding interest over time

congruent figures identical figures with exactly the same shape and size

constant a term with a fixed value

constant of proportionality the number multiplying x in the equation $y = kx$, which is equal to the gradient of the corresponding graph; also called the constant of variation

constant rate rate of change that changes a quantity by the same amount for a given unit of time

convert to change a measurement or value to different units

critical digit the digit to the right of the last digit that is to be kept when rounding a number

cross-section the 'slice' produced when a cut is made across a prism parallel to its ends

cumulative frequency the total of all frequencies up to and including the frequency for a particular score in a frequency distribution

cumulative frequency graph a line graph that is formed when the cumulative frequencies of a set of data are plotted against the end points of their respective class intervals and then joined up by straight-line segments. It is also called an ogive.

cylinder a solid object with two identical flat circular ends and one curved side

dependent variable a variable with a value that changes in response to changes in the value of other variables

deviation the difference between each value in a data set and the mean

diameter a straight line passing through the centre of a circle from one side of the circumference to the other

difference of two squares an expression in which one perfect square is subtracted from another, for example $x^2 - 100$; this can be factorised by $(a + b)(a - b) = a^2 - b^2$

dilation occurs when a graph is made thinner or wider

direct linear proportion describes a particular relationship between two variables or quantities; as one variable increases, so does the other variable

Distributive Law the product of one number with the sum of two others equals the sum of the products of the first number with each of the others; for example, $4(6 + 2) = 4 \times 6 + 4 \times 2$. It is also applicable to algebra; for example, $3x(x + 4) = 3x^2 + 12x$.

dot plot this graphical representation uses one dot to represent a single observation. Dots are placed in columns or rows, so that each column or row corresponds to a single category or observation.

equally likely describes outcomes that have the same probability of occurring in an experiment

equation a mathematical statement that shows two equal expressions

event set of favourable outcomes in each trial of a probability experiment

exact describes a mathematical value that is known with certainty and is not rounded or approximated

expanded using the Distributive Law to remove the brackets from an expression

experimental probability probability determined by observing an experiment and gathering data, expressed by the formula: $\text{Pr(event)} = \dfrac{\text{number of ways the event can occur}}{\text{total number of equally likely outcomes}}$

expression group of terms separated by $+$ or $-$ signs, but which do not contain equals signs

factorising writing a number or term as the product of a pair of its factors

favourable outcome any outcome of a trial that belongs to the event; for example, if A is the event of getting an even number when rolling a die, the favourable outcomes are 2, 4 and 6

First Index Law when terms with the same base are multiplied, the indices are added; therefore, $a^m \times a^n = a^{m+n}$

Fourth Index Law to remove brackets, multiply the indices inside the brackets by the index outside the brackets. Where no index is shown, assume that it is 1; therefore, $(a^m)^n = a^{mn}$

frequency the number of times a particular score appears in a data set

Fundamental Theorem of Algebra theorem that states that quadratic equations can have a maximum of two solutions

gradient (slope) (m) the measure of how steep something is; the gradient of a straight line is given by $m = \dfrac{\text{rise}}{\text{run}}$ or $m = \dfrac{y_2 - y_1}{x_2 - x_1}$ and is constant anywhere along that line

gross salary a person's income before tax has been taken out

hectare (ha) a unit of area equal to the space enclosed by a square with side lengths of 100 m (1 ha $= 10\,000$ m^2); often used to measure land area. There are 100 hectares in 1 square kilometre.

highest common factor (HCF) the largest of the set of factors common to two or more numbers; for example, the HCF of 16 and 24 is 8

histogram graph that displays continuous numerical variables and does not retain all original data

hyperbola the shape of the graph of an inverse proportion relationship

image an enlarged or reduced copy of a shape (the original shape is called the object)

income tax a tax levied on a person's financial income and based on an income tax table

independent variable variable that does not change value in response to changes in the value of other variables

index notation a short way of writing a number or variable when it is multiplied by itself repeatedly (for example, $6 \times 6 \times 6 \times 6 = 6^4$)

interest a fee charged for the use of someone else's money, normally as a percentage of the amount borrowed

interquartile range (IQR) the difference between the upper (or third) quartile, Q_{upper} (or Q_3), and the lower (or first) quartile, Q_{lower} (or Q_1); that is, $\text{IQR} = Q_{\text{upper}} - Q_{\text{lower}} = Q_3 - Q_1$. It is the range of approximately the middle half of the data.

intersection in probability, an intersection of two events (written as $A \cap B$) describes all outcomes in event A and in event B

inverse operation an operation that reverses the action of another; for example, $+$ is the inverse of $-$, and $\times$ is the inverse of $\div$

inverse proportion describes a particular relationship between two variables or quantities; as one variable increases, the other variable decreases

irrational number (I) a number that cannot be written as a fraction. Examples of irrational numbers include surds, π and non-terminating, non-recurring decimals.

like terms terms containing exactly the same pronumeral part, including the power; for example, $3ab$ and $7ab$ are like terms, but $5a$ is not

linear equations equations in which the variable has an index (power) of 1

linear graph a relationship between two variables where a straight line is formed when the points are plotted on a graph

loading a payment made to a worker in addition to their usual wage or salary for working in difficult or hazardous conditions

magnitude the first digit of a number expressed in scientific notation

mean one measure of the centre of a set of data. It is given by $\text{mean} = \dfrac{\text{sum of all scores}}{\text{number of scores}}$ or $\bar{x} = \dfrac{\sum x}{n}$.

When data are presented in a frequency distribution table, $\bar{x} = \dfrac{\sum (f \times x)}{n}$.

measures of central tendency summary statistics that measure the middle (or centre) of a set of data; the mean, median and mode

measures of spread summary statistics that measure how far data values are spread either from the centre or from each other

median one measure of the centre of a set of data. It is the middle score for an odd number of scores arranged in numerical order. If there is an even number of scores, the median is the mean of the two middle scores when they are ordered. Its location is determined by the rule $\dfrac{n+1}{2}$.

For example, the median value of the set 1 3 3 4 5 6 8 9 9 is 5, while the median value for the set 1 3 3 4 5 6 8 9 9 10 is the mean of 5 and 6 (5.5).

metric system the system of measurement based on the metre

midpoint halfway point between the end points of a line segment

modal class term used when analysing grouped data, given by the class interval with the highest frequency

mode one measure of the centre of a set of data. It is the score that occurs most often. There may be no mode, one mode or more than one mode (two or more scores occur equally frequently).

monic describes a polynomial in which the coefficient of the term with the highest degree is 1; for example, $1x^2$ is a monic quadratic polynomial

multimodal having more than two modes

multi-step equation an equation that performs more than two operations on a pronumeral

mutually exclusive describes two events that cannot both occur at the same time

negatively skewed showing larger amounts of data as the values of the data increase

net a two-dimensional figure that can be folded to form a three-dimensional object

net salary a person's income after tax has been taken out

Null Factor Law (NFL) law used to solve quadratic equations, which states that if $a \times b = 0$, then either $a = 0$ or $b = 0$ or both $a = 0$ and $b = 0$

object the original shape before it is copied and the copy is enlarged or reduced (the changed shape is called the image)

observed probability the probability of an event based on the outcomes of experiments, simulations or surveys; also called experimental probability

ogive see cumulative frequency graph

opposite side the side opposite the reference angle in a right-angled triangle

outcome result obtained from a probability experiment

outliers values in a data set that are extremely large or extremely small compared to other values

overtime the time in which a worker is working hours in addition to their regular hours of employment

parabola graph of a quadratic function, for example $y = x^2$ or $y = x^2 + 8$

parallel box plots box plots drawn on the same scale to compare two or more similar data sets

parallel lines lines with the same gradient

percentile the value below which a given percentage of all scores lie. For example, the 20th percentile is the value below which 20% of the scores in the set of data lie.

perfect square a number or expression that is the result of the square of a whole number

perimeter distance around the outside of a shape

perpendicular lines lines that intersect at right angles to each other

piecework a system of payment by which a worker is paid a fixed amount for each job or task they complete

positively skewed showing smaller amounts of data as the values of the data decrease

principal the amount of money you start with when you put money into a financial institution such as a bank or credit union

precision (of a measuring instrument) the smallest unit or division displayed on the instrument; (of data) how close multiple measurements of the same investigation are to each other

prism solid shape with identical opposite ends joined by straight edges; a 3-dimensional figure with uniform cross-section

prefixes word parts added at the start of words to modify the meaning. For example, 'kilo-' means 'one thousand of'; hence, a kilometre is one thousand metres.

probability likelihood or chance of a particular event (result) occurring, given by the formula:

$$\text{Pr(event)} = \frac{\text{number of favourable outcomes}}{\text{number of possible outcomes}}.$$ The probability of an event occurring ranges from 0 (impossible) to 1 (certain) inclusive.

proportion equality of two or more ratios

quadrant a quarter of a circle; one of the four regions of a Cartesian plane produced by the intersection of the x- and y-axes

quadratic equation an equation in which the term with the highest power is a squared term; for example, $x^2 + 2x - 7 = 0$

quadratic trinomial an expression of the form $ax^2 + bx + c$, where a, b and c are constants

quantiles equal parts into which a data set is divided, for example percentiles or quartiles

quartile values that divide an ordered set into four (approximately) equal parts. There are three quartiles — the first (or lower) quartile Q_1, the second quartile (or median) Q_2 and the third (or upper) quartile Q_3.

radius (plural: radii) the distance from the centre of a circle to its circumference

range the difference between the maximum and minimum values in a data set

range (functions and relations) the set of y-values produced by the function; (statistics) the difference between the highest and lowest scores in a set of data; that is, range = highest score − lowest score

rate a ratio that compares the size of one quantity with that of another quantity; a method of measuring the change in one quantity in response to another

ratio comparison of two or more quantities of the same kind

rational number (Q) a number that can be expressed as a fraction with whole numbers both above and below the dividing sign

real number (R) any number from the set of all rational and irrational numbers

recurring decimal a decimal number with one or more digits repeated continuously; for example, 0.999 ... A rational number can be expressed exactly by placing a dot or horizontal line over the repeating digits; for example, $8.343434 \ldots = 8.\dot{3}\dot{4}$ or $8.\overline{34}$.

relative frequency frequency of a particular score divided by the total sum of the frequencies; given by the rule:

$$\text{Relative frequency of a score} = \frac{\text{frequency of the score}}{\text{total sum of frequencies}}$$

royalty a payment made to a person who owns a copyright

salary a fixed annual amount of payment to an employee, usually paid in fortnightly or monthly instalments

sample space the set of all possible outcomes from a probability experiment, written as ξ or S. A list of every possible outcome is written within a pair of curled brackets {}.

scale factor the factor by which an object is enlarged or reduced compared to the image

scientific notation (standard form) the format used to express very large or very small numbers. To express a number in standard form, write it as a number between 1 and 10 multiplied by a power of 10; for example, $64\,350\,000$ can be written as 6.435×10^7 in standard form.

Second Index Law when terms with the same base are divided, the indices are subtracted; therefore, $a^m \div a^n = a^{m-n}$

sector the shape created when a circle is cut by two radii

semicircle half of a circle

significant figures the number of digits that are important in a measurement

similar figures identical shapes with different sizes. Their corresponding angles are equal in size and their corresponding sides are in the same ratio, called a scale factor.

simple interest the interest accumulated when the interest payment in each period is a fixed fraction of the principal. The formula used to calculate simple interest is $I = \dfrac{Pin}{100}$, where I is the interest earned (in $) when a principal of P is invested at an interest rate of i% p.a. for a period of n years.

standard deviation a measure of how far the values in a data set are spread out from the mean

standard form *see* scientific notation

surd any nth root of a number that results in an irrational number

symmetrical the identical size, shape and arrangement of parts of an object on opposite sides of a line or plane

term part of an expression. Terms may contain one or more pronumerals, such as $6x$ or $3xy$, or they may consist of a number only.

terminating decimal a decimal that has a fixed number of places, for example 0.6 or 2.54

Third Index Law any term (excluding 0) with an index of 0 is equal to 1; therefore, $a^0 = 1$

translate (quadratics) to move a parabola horizontally (left/right) or vertically (up/down)

tree diagram diagram used to list all possible outcomes of two or more events; the branches show the possible links between one outcome and the next outcome

trial one performance of an experiment to get a result; for example, each roll of a die is a trial

triangular prism a prism with a triangular base

trigonometry the branch of mathematics that deals with the relationship between the sides and the angles of a triangle

turning point point at which the graph of a quadratic function (parabola) changes direction (either up or down)

two-step equation an equation in which two operations have been performed on the pronumeral; for example, $2y + 4 = 12, 6 - x = 8, \dfrac{x}{3} - 4 = 2$

undefined a numerical value that cannot be calculated

union in probability, the union of two events (written as $A \cup B$) is all possible outcomes in events A or in event B

variable (algebra) a letter or symbol in an equation or expression that may take many different values; (programming) a named container or memory location that holds a value

Venn diagram a series of circles, representing sets, within a rectangle, representing the universal set. They show the relationships between the sets.

vertex (geometry) point where two rays or arms of an angle meet; (quadratics) *see* turning point

volume amount of space a 3-dimensional object occupies; measured in cubic units, such as cubic centimetres (cm^3) and cubic metres (m^3)

wage a payment to an employee based on a fixed rate per hour

without replacement describes an experiment carried out with each event reducing the number of possible events; for example, taking an object permanently out of a bag before taking another object out of the bag. In this case, the number of the objects in the bag is reduced by one after every selection.

with replacement describes an experiment carried out without each event reducing the number of possible events; for example, taking an object out of a bag and then placing it back in the bag before taking another object out of the bag. In this case, the number of objects in the bag remains constant.

x-intercept point where a graph cuts the *x*-axis

y-intercept point where a graph cuts the *y*-axis. In the equation of a straight line, $y = mx + c$, the constant term, *c*, represents the *y*-intercept of that line

INDEX

ratios 571–2
  constant of proportionality as 419
  definition 571
  direct proportion and 418–23
  and right-angled triangles 460–8
  and scale 573–86
  in simplest form 573
  trigonometric *see* trigonometric
    ratio
real numbers 126
  to significant figures 129
rectangular prism, surface area
    of 531
relative frequency
  calculating 698
  definition 698
  of outcome 699
replacement 710–11
RHS condition 586
right-angled triangles
  ratios and 460–8
  sides of 460
right angle–hypotenuse–side condition
    for similarity 586
rounding numbers 126–32
  of decimal places 127–8
  of significant figures 128–30

S

salaries 5–9
  definition 5
  employees 5–7
sample space 692
SAS condition 586
scale factors 573–86, 598–600
  area 596
  length from 579
  in photographs 578–9
  in plans and drawings 578–9
  ratio and 573–86
  surface area 597
  volume 598
  writing 578
scientific notation
  comparing numbers in 134
  to decimal notation 135
  definition 132
Second Index Law 62
  with non-cancelling
    coefficients 63
  simplifying using 62
sectors 521
  arc length of 522
  area of 521–2, 528
  definition 521
  perimeter of 521–4, 528

semicircle 521
Seventh Index Law 74
shape, perimeter 505
side–angle–side condition for
    similarity 586
side–side–side condition for similarity
    585
significant figures 128–30
  counting 129
  definition 128
  real numbers to 129
  rounding to number of 128–30
similar figures
  area and volume of 596–8
  and similarity condition 585–7
similarity condition 585–7
  angle–angle–angle 585
  side–side–side 585
  side–angle–side 586
  right angle–hypotenuse–side 586
  testing triangles for 587–9
simple interest 31–5, 39
  calculating 32
  formula for 32
  interest 31–2
  on item 34
  principal 31–2
  technology and 34–5
simplifying algebraic expressions
    154–60
simplifying, expanding and 168–9
simulations 698–702, 707
simultaneous linear equations 382–3
sine 461
Sixth Index Law 69–70
sketching linear graphs 303–11
  horizontal and vertical lines
    307–8
skewness, in distributions 649–52
specific ratio 428
squares, concentric 96
SSS condition 585
standard deviation 640–7, 677–8
  of populations and samples
    642–3
  properties of 643–4
standard form, definition 132
straight line
  and coordinates of one point
    295–6
  general equation of 295
  in graph 296–8
subtraction
  algebraic fractions 160–2
  like terms 154–5
  of surds 84–6

surds
  addition and subtraction
    of 84–6
  comparing 84
  definition 80
  entire 84
  expressions 85
  identifying 80–1
  multiplying and dividing 81–3
  simplifying 83–4
surface area (SA)
  of cylinder 539–40, 543
  of prisms 530–3
  of rectangular prism 531
  of similar figures 598–9
  of triangular prism 531
symmetry, in
    distributions 649–52

T

tangent 461
technology
  linear graphs using 311
  and simple interest 34–5
terms
  factorising 184
theoretical probability 693
Third Index Law 63, 64
  First Index Law and 64
time of an investment 33
time sheets 10–12
transcendental number 126
translates 369
tree diagrams 707–18
  definition 707
trial 692
triangles 460–8, 585–9
triangular frames 456
triangular prism 530
  surface area of 531
trigonometric ratios 460–78
  cosine 461
  identifying 461–2
  inverse 476–8
  relating sides to 463
  and right-angled triangles
    460–8
  sine 461
  tangent 461
  side lengths using 469–71
  trigonometric values of 468–9
  using calculator 468
trigonometry 460–85
  definition 460
turning point 355
two-step equations 213